ADVANCED TOPICS IN MASS TRANSFER

Edited by **Mohamed El-Amin**

INTECHOPEN.COM

Advanced Topics in Mass Transfer
Edited by Mohamed El-Amin

CBS, Edition 2016

Published by InTech
Janeza Trdine 9, 51000 Rijeka, Croatia

Publishing Process Manager Iva Lipovic

Technical Editor Teodora Smiljanic

Cover Designer InTech Design Team

Image Copyright clarusvisus, 2010. Used under license from Shutterstock.com

Additional hard copies can be obtained from orders@intechopen.com

Advanced Topics in Mass Transfer, Edited by Mohamed El-Amin
p. cm.
ISBN 978-953-307-333-0

INTECH

open science | open minds

Contents

Preface IX

Part 1 Mass Transfer in Microchannels, Turbulence, Waves and Plasma 1

Chapter 1 **Mass-Transfer Characteristics of a Double-Y-Type Microchannel Device 3**
Susumu NII

Chapter 2 **Turbulent Buoyant Jet of a Low-Density Gas Leaks Into a High-Density Ambient: Hydrogen Leakage in Air 19**
M.F. El-Amin and Shuyu Sun

Chapter 3 **Controlled Mixing and Transport in Comb-Like and Random Jet Array Stirring Systems 43**
S. Delbos, E. Chassaing, P. P. Grand, V. Weitbrecht and T. Bleninger

Chapter 4 **Application of Airborne Sound Waves for Mass Transfer Enhancement 61**
Sergey V. Komarov

Chapter 5 **Mass-transfer in the Dusty Plasma as a Strongly Coupled Dissipative System: Simulations and Experiments 87**
Xeniya Koss, Olga Vaulina, Oleg Petrov and Vladimir Fortov

Part 2 Mass Transfer and Hydro-, Magnetohydro- and Electro- Dynamics 111

Chapter 6 **Forced Convection Mass-Transfer Enhancement in Mixing Systems 113**
Rafał Rakoczy and Stanisław Masiuk

Chapter 7 **Unsteady Magnetohydrodynamic Convective Heat and Mass Transfer Past an Infinite Vertical Plate in a Porous Medium with Thermal Radiation, Heat Generation/Absorption and Chemical Reaction 145**
Stanford Shateyi and Sandile Motsa

Chapter 8 **Mass Transfer, and Effects of Magnetic Fields on the Mass Transfer in Close Binary System 163**
Davood Manzoori

Chapter 9 **A Mass Transfer Study with Electrolytic Gas Production 175**
Eudésio O. Vilar, Eliane B.Cavalcanti and Izabelle L.T. Albuquerque

Chapter 10 **Mass Transfer Equation and Hydrodynamic Effects in Erosion-Corrosion 191**
A. Yabuki

Chapter 11 **Hydrodynamics and Mass Transfer in Heterogeneous Systems 211**
Radmila Garić-Grulovic, Nevenka Bošković-Vragolović, Željko Grbavčić and Rada Pjanović

Chapter 12 **Flow and Mass Transfer inside Networks of Minichannels 229**
Florian Huchet

Chapter 13 **Simulation of Hydrodynamics and Mass Transfer in a Valve Tray Distillation Column Using Computational Fluid Dynamics Approach 265**
A. Jafari, S.M. Mousavi, H. Moteshaffi, H. Roohian and H. Hamedi Sangari

Part 3 Mass Transfer in Food 281

Chapter 14 **Modeling Moisture Movement in Rice 283**
Bhagwati Prakash and Zhongli Pan

Chapter 15 **Mass Transfer Mechanisms during Dehydration of Vegetable Food: Traditional and Innovative Approach 305**
Derossi A., Severini C. and Cassi D.

Chapter 16 **Mass Transfer in Cheese 355**
Jorge F. Vélez-Ruiz

Part 4 Mass Transfer in Large-Scale Applications 371

Chapter 17 **Comprehensive Survey of Multi-Elements in Coastal Sea and Stream Sediments in the Island Arc Region of Japan: Mass Transfer from Terrestrial to Marine Environments 373**
Atsuyuki Ohta and Noboru Imai

Chapter 18 **Mass Transfers and Sedimentary Budgets in Geomorphologic Drainage Basin Studies 399**
Achim A. Beylich

Part 5 Combined Heat and Mass Transfer 423

Chapter 19 **Successive Linearisation Solution of Free Convection Non-Darcy Flow with Heat and Mass Transfer 425**
Sandile Motsa and Stanford Shateyi

Chapter 20 **Explicit and Approximated Solutions for Heat and Mass Transfer Problems with a Moving Interface 439**
Domingo Alberto Tarzia

Chapter 21 **Boundary Control Problems for Oberbeck–Boussinesq Model of Heat and Mass Transfer 485**
Gennady Alekseev, Dmitry Tereshko and Vladislav Pukhnachev

Chapter 22 **Heat and Mass Transfer in Desiccant Wheels 513**
Celestino Ruivo, José Costa and António Rui Figueiredo

Chapter 23 **Oscillatory Regimes of Solutocapillary Marangoni Convection 535**
Konstantin Kostarev, Andrew Zuev and Antonio Viviani

Chapter 24 **Aerodynamics of Ceramic Regular Packing for Heat-Massexchenge Processes 557**
Alexandr Pushnov

Chapter 25 **Topics in Heat and Mass Transfer in Porous Media: Cross-Diffusion, Thermophoresis and Reactive Surfaces 577**
Adrian Postelnicu

Chapter 26 **Mass and Heat Transfer During Thin-Film Evaporation of Liquid Solutions 611**
Janusz Dziak

Preface

This book introduces a number of selected advanced topics in mass transfer phenomenon and covers its theoretical, numerical, modeling and experimental aspects. The 26 chapters of this book are divided into five parts. The first is devoted to the study of some problems of mass transfer in microchannels, turbulence, waves and plasma, while the chapters regarding mass transfer with hydro-, magnetohydro- and electrodynamics are collected in the second part. The third part deals with mass transfer in food, such as rice, cheese, fruits and vegetables, while the fourth focuses on mass transfer in some large-scale applications such as geomorphologic studies. The last part introduces several issues of combined heat and mass transfer phenomena. The book is considered as a rich reference for researchers and engineers working in the field of mass transfer and related topics.

Chapter 1 focuses on mass transfer characteristics of a microchannel device of double-Y type. The second chapter studies the problem of a low-density gas jet injected into a high-density ambient with an example of hydrogen leakage in air, which is considered an important issue facing hydrogen-energy developing. Controlled mixing and transport in comb-like and random jet array stirring systems are studied in the third chapter. The goal of the fourth chapter is to give a deeper insight into the possibilities and limitations of airborne sound waves as a tool to enhance the rates of gas-phase mass transfer and its related phenomena. In Chapter 5 simulations and experiments are performed to investigate mass-transfer in dusty plasma as a strongly coupled dissipative system.

Chapter 6 studies forced convection mass transfer processes under various types of augmentation technique, i.e. rotational and reciprocating mixers and rotating magnetic field. The seventh chapter studies unsteady magnetohydrodynamic convective heat and mass transfer past an infinite vertical plate in a porous medium with thermal radiation, heat generation/absorption and chemical reaction. Effects of magnetic field on mass transfer in close binary systems are investigated in Chapter 8, while Chapter 9 studies the effect of hydrodynamic condition over mass transfer gas-electrodes for two cathodes geometries during the hydrogen production at chlor-alkali cell by diaphragm process in laboratory scale. In Chapter 10 mass transfer equation and hydrodynamic effects are used to predict the Erosion-Corrosion damage. Hydrodynamics and mass transfer in heterogeneous systems are presented in the eleventh chapter. The objective of the experimental work presented in the twelfth chapter is using several methods in order to characterize flow and mass transfer inside networks composed of crossing minichannels. Chapter 13 provides a CFD simulation of hydrodynamics and mass transfer in a valve tray distillation column.

The purpose of Chapter 14 is to illustrate empirical and theoretical approaches based upon principles of mass and heat transfer pursued in modeling of drying processes in rice. The fifteenth chapter provides the conventional and emerging theories on mass transfer during traditional and innovative drying technologies of fruits and vegetables. The mass transfer in cheese is introduced in Chapter 16.

Chapter 17 provides a comprehensive survey of multi-elements in coastal sea and stream sediments in the island arc region of Japan and mass transfer from terrestrial to marine environments. Mass transfer and sedimentary budgets in geomorphologic drainage basin studies are covered in Chapter 18.

In the nineteenth chapter a numerical perturbation scheme was developed for solving complex nonlinear boundary value problems arising in problems of heat and mass transfer, while the goal of Chapter 20 is to give a survey of some explicit and approximated solutions for heat and mass transfer in free or moving interface. The boundary control problems for Oberbeck–Boussinesq model of heat and mass transfer are discussed in the twenty-first chapter. In Chapter 22 the results of a detailed numerical model are used to determine the effectiveness parameters for the coupled heat and mass transfer processes in desiccant wheels. Chapter 23 describes experiments revealed by the self-oscillatory regimes of solutal convection in an aqueous solution of surfactant with vertically stratified concentration. Aerodynamics of ceramic regular packing for heat and mass exchange processes are studied in the twenty-fourth chapter. Chapter 25 covers cross-diffusion, chemical reaction, thermophoresis and reactive surfaces in the area of heat and mass transfer convective processes in fluid-saturated porous media. The last chapter of this book, Chapter 26, introduces mass and heat transfer during thin-film evaporation of liquid solutions applied in heat exchangers.

Mohamed Fathy El-Amin
Physical Sciences and Engineering Division
King Abdullah University of Science and Technology (KAUST)

Part 1

Mass Transfer in Microchannels, Turbulence, Waves and Plasma

1

Mass-Transfer Characteristics of a Double-Y-Type Microchannel Device

Susumu NII
Nagoya University
Japan

1. Introduction

Microfluidic devices have been extensively applied to high-throughput chemical analyses or reactions using hazardous substances (van den Berg et al., 2000). These systems are characterized by a large area per volume and a well-defined laminar flow. As a result of their high potential for the precise manipulation of chemical reactions with respect to controlled temperature, contacting time, and molecular orientation at the interface, researches on microfluidic devices for the production of highly valuable substances are increasing in importance (Ajmera et al., 2001, Maruyama et al., 2003, Löb et al., 2004). The production system is often referred to as a desktop chemical plant and the keys to realizing the plant are understanding and successfully manipulating the flow in microchannels (Hetsroni et al., 2005).

A unique feature of the microchannels is the formation of liquid layers when multiple liquids are brought into one channel. Reactions and/or mass-transfers occur in the vicinity of the interface. Liquid-liquid extractions (Brody and Yager, 1997, Tokeshi et al., 2000) are reported for systems of aqueous and organic solutions. Even mutually miscible liquids form the layers because of the laminar-flow nature (Yamaguchi et al., 2004) and a mass-transfer occurs between the liquids under well-controlled conditions. For the purpose of chemical processing, the stacked liquid layers should be separated in each layer for further treatment. However, flow splitting in microchannel devices is rarely found in published papers. In reported works for the analytical application of microchannels, the solute's concentration profile was directly measured within the channel using a laser-based refractive index detector (Costin and Synovec, 2002) or a microscope for measuring fluorescence emission (Costin et al., 2004). A strategy for flow splitting has not yet been developed for the application of microchannel devices to analytical chemistry.

In the present study, two aqueous layers were brought into contact and separated into two streams after a mass-transfer between the layers. The amount of solute transferred was obtained with a simple analysis of liquid samples taken at the channel outlet. The mass-transfer characteristics were examined through the measurement of the diffusion coefficients for solutes. Even under carefully controlled liquid feed conditions, equal flow splitting was difficult to obtain. Proposed is a method to choose solute concentration data for equal flow splitting using the ratio of the flow rate of one liquid to the other. The mass-transfer of benzoic acid was successfully analyzed with a conventional penetration model with an additional parameter. Furthermore, the diffusion coefficients of sucrose, glycine, tryptophan, and urea were measured using a basic equation obtained with benzoic acid. The values are in accordance with the literature or the correlated values. Since the same parameter is applied for obtaining the diffusion coefficients of the four different solutes, the additional parameter should characterize the flow behavior in the microchannel. The proposed method is also useful for determining the unknown diffusivities of solutes. Furthermore, to investigate the flow characteristics, the value of an additional parameter is determined in various channels having different lengths, depths and knife-edge angles. A Comparison of this parameter helps us to understand the effect of flow on mass-transfer.

2. Determination of diffusivity

2.1 Microchannel module and materials

Benzoic acid, sucrose, tryptophan, urea, and glycine were selected as transferring solutes. The aqueous solutions were prepared as feedstock, and the concentration of each of these was set at 1.19×10^{-3} kmol/m^3, 1.46×10^{-2} kmol/m^3, 1.19×10^{-3} kmol/m^3, 8.7×10^{-2} kmol/m^3, and 6.7×10^{-2} kmol/m^3 respectively. The solutions were filtered prior to use to remove the suspended solid to prevent the clogging of the channel.

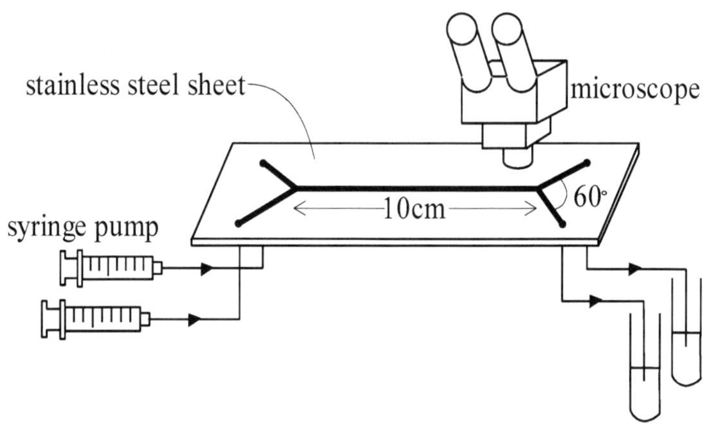

Fig. 1. Schematic representation of microchannel

The microchannel used was fabricated on a stainless steel sheet as shown in Fig. 1. The thickness of the sheet was 200μm. The channel width was 200μm for both the branches and the main portion. As showen in Fig. 2, the stainless steel sheet was set in between gaskets of a polyethylene film (100μm thick) to form the flow channels of rectangular cross section. An aqueous solution of the solute and deionized water were supplied with syringe pumps at exactly the same flow rate. These two liquids were introduced to both inlets and brought into contact at the main portion of the channel. The two liquids split at a knife-edge, and each liquid flowed into a branch. The liquids were taken out from the end of the branch. The side where the feed solution flowed is called the feed-side, and the other side where the deionized water flowed is the recovery-side. Liquid flow at the main portion of the channel was monitored with a microscope to confirm that there was no flow disturbance. Figure 3 shows a photograph of the flow splitting point, which was taken at the examination of the flow profile with the supply of a Cibacron blue solution and water in each inlet. The blue solution formed a layer and a stable interface is observed at the center of the channel. The two solutions were then split with a knife-edge at the end of the main portion of the channel.

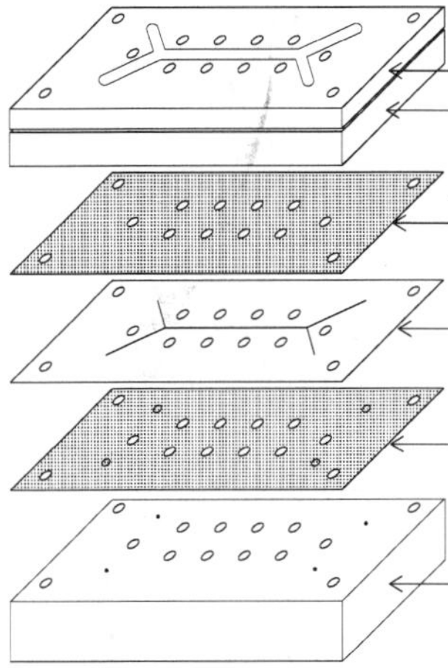

Fig. 2. Expanded view of microchannel module

Fig. 3. Flow profile within microchannel (Cibacron blue solution and water)

The solute in the feed solution began to diffuse into the deionized water right after the contact. A predetermined amount of samples was taken at the channel outlet, and the samples were analyzed with a UV spectrophotometer (UV-1600, Shimadzu Co., Ltd.) or a refractive index meter (830-RI, Japan Spectroscopic Co., Ltd.). The liquid flow rates on both sides were determined by weighing the liquid samples collected for a defined time. Although the inlet liquids were fed at the same rate, unequal flow rates were observed at the outlets. In order to adjust the flow rate of each side, capillaries were installed at the effluent tubing. All experiments were carried out at a temperature of 298±1K.

2.2 Mass-transfer model

Within the range of liquid velocities examined, the Reynolds number varies from 17 to 172. The liquid flow examined is laminar. Even when two kinds of aqueous solutions are brought into contact at the same velocity in one microchannel, a contacting plane is formed. If a concentration difference exists between the two liquids, the solute will diffuse from the concentrated to the less concentrated side across the contacting plane. Under conditions of short contact time for a mass-transfer and a considerably larger solute concentration in the feed-side than the recovery-side, the penetration model (Higbie, 1935) can be applied for solute diffusion.

$$N_{\mathrm{A}} = 2\sqrt{\frac{D}{\pi\,\theta}}\left(C_{\mathrm{i}} - C_{\mathrm{b}}\right) \tag{1}$$

where N_A is the mass-transfer rate at the interface; D, the diffusion coefficient; θ, the contact time; C_i the concentration at the contacting plane; C_b, the concentration of bulk liquid. The contact time; θ, is calculated from the channel length of main portion of the channel, l divided by the average velocity of the contacting plane, $v_{int,av}$. Eq. (1) then reduces to

$$N_{\mathrm{A}} = 2\sqrt{\frac{D\,v_{\mathrm{int,av}}}{\pi\,l}}\left(C_{\mathrm{i}} - C_{\mathrm{b}}\right) \tag{2}$$

The value of $v_{int,av}$ is determined from the simulated velocity profile within the microchannel. Details will be presented later. At the contacting plane, a parabolic profile is obtained by solving the Navier-Stokes equation with a finite element method.

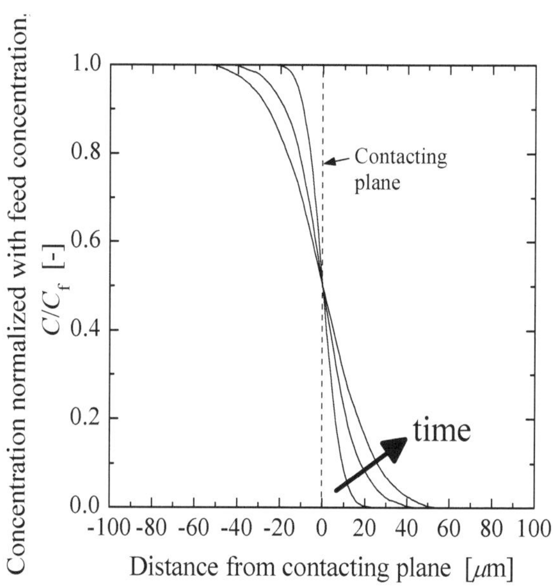

Fig. 4. Schematic representation of concentration profile within channel

The mass-transfer across the contacting plane is modeled on the basis of the diffusion in a semifinite slab. For unsteady diffusion in a semifinite slab, the concentration profile at the cross section of the main portion is schematically represented in Fig. 4. Since the two liquids are aqueous solutions, a symmetrical profile is formed in the vicinity of the contacting plane. The profile decays with the progress of contact time with maintaining a constant concentration at contacting plane, C_i. When deionized water is used as the recovery liquid, C_i equals to $0.5C_f$, and the bulk concentration, C_b, can be assumed to be zero. Then Eq. (2) is simplified to

$$N_A = \sqrt{\frac{D\, v_{int,av}}{\pi\, l}}\, C_f \tag{3}$$

The mass balance around the recovery-side of the channel can be expressed as

$$N_A\, A = v_{av}\, S\,(C_{out} - C_{in}) \tag{4}$$

where A is the area of the contacting plane; v_{av}, the average liquid velocity in the main portion of the channel, which is defined as the total volumetric flow rate divided by the cross-sectional area ($200\mu m \times 200\mu m$); S, the cross-sectional area of the recovery-side, which is half the cross-sectional area of the main portion; C_{out}, the outlet concentration of the recovery liquid; and C_{in}, the concentration at the inlet of the recovery liquid. The area, A, is expressed as the channel height, h, multiplied by the length of the main portion of the channel, l, and the area S is written as hL, where L is half the channel width. Hence, Eq. (4) becomes,

$$N_A = \frac{L\, v_{av}(C_{out} - C_{in})}{l} \tag{5}$$

For the supply of deionized water to the recovery-side, C_{in}, is set to zero. Therefore, the mass balance around the recovery-side is written as

$$N_A = \frac{L\, v_{av} C_{out}}{l} \tag{6}$$

Attention should be paid to the difference between v_{av} and $v_{int,av}$. The former is a three-dimensional average, while the latter is two dimensional. Velocities calculated at the contacting plane, v_{int}, are normalized with v_{av}, and the values of v_{int}/v_{av} are plotted in Fig. 5. A parabolic profile is obtained and the velocity of the contacting plane is averaged to determine the value of $v_{int,\,av}/v_{av}$. As is presented in the figure, the value of $v_{int,av}/v_{av}$ is 1.4. The result suggests that the two-dimensional average of $v_{int,av}$ is 1.4 times bigger than the three-dimensional average, v_{av}. Combining Eqs. (3) and (6) gives the outlet concentration of the recovery-side,

$$C_{out} = \frac{\sqrt{D}}{L}\sqrt{\frac{1.4l}{\pi\, v_{av}}}\, C_f \tag{7}$$

This is the fundamental equation for determing the solute diffusion coefficient from the measurement of the solute concentration at the outlet of the recovery-side.

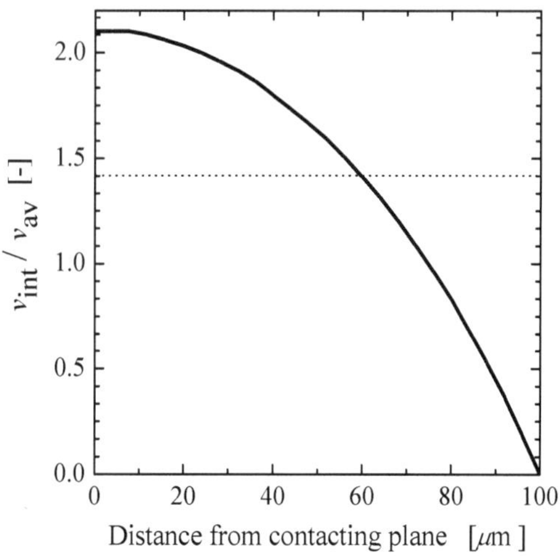

Fig. 5. Simulated velocity profile at contacting plane

2.3 Effect of contacting and splitting flow on mass-transfer

A typical experimental result for the benzoic acid concentration at the outlet of the recovery-side is presented in Fig. 6. The horizontal axis is taken as $v_{av}^{-0.5}$ in order to examine the validity of the penetration model. The observed concentration scatters at a given liquid velocity. Although the two liquids were supplied at the same velocity, the outlet liquid velocities on each side were unequal. The results clearly show the difficulty in flow splitting.

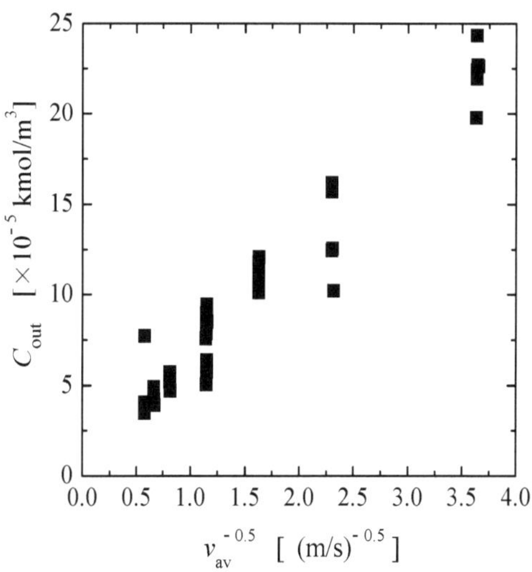

Fig. 6. Typical experimental results of benzoic acid concentration at recovery-side outlet

The concentration profile shown in Fig. 4 suggests a very steep curve around the contacting plane. When the liquid split equally, the separation was right at the contacting plane. Becouse of the big concentration change around the contacting plane, a small disturbance results in a large variation in the concentration. One of the possible reasons for the disturbance is the lack of precision in the machining of knife-edge fabrication or the attachment of fine air bubbles to the walls of the branch channels.

Splitting flows in microchannel devices seems to be a key technique for the practical application to chemical processing. Our results suggest that special attention should be paid to the fabrication of the knife-edge and the channel wall. This implies that the success of microchannel devices depends strongly on the accuracy of fabrication. This limitation act as a barrier to the development of microchannel devices. We seek a method that can use rather simple devices for exploiting the characteristics of microchannel devices.

The measurement of diffusion coefficients was carried out to understand the mass-transfer characteristics of microchannels. To acquire the solute concentration for equal flow splitting, an intentional change was given to the ratio of the flow rate of the feed-side to the recovery-side, and the solute-concentration data in the recovery-side was recorded. The flow rate were changed by setting capillaries on tubing to obtain a ratio of around unity. The solute concentration corresponding to the ratio of unity was taken to be as the value for equal splitting. Hence, this value is referred to as the relevant concentration.

Figure 7 shows a typical example of determining the relevant concentration. The ratio, R_{out}, is defined by Q_r / Q_f, where Q_r and Q_f are the flow rates of the recovery-side and the feed-side, respectively. The ratio varied within the range of $\pm 10\,\%$ around unity. The concentration value for the ratio unity is found from the curve fitting of the observed data. The measurements were repeated for each liquid velocity in order to determine the relevant concentration. The values are plotted in Fig. 8. It can be seen that C_{out} increases linearly with $v_{av}^{-0.5}$, and that this dependence demonstrates the validity of the penetration model.

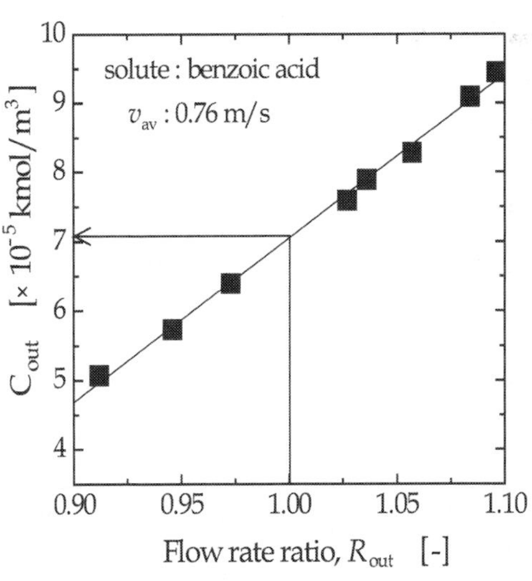

Fig. 7. Determination of relevant concentration at recovery-side outlet for equal flow splitting

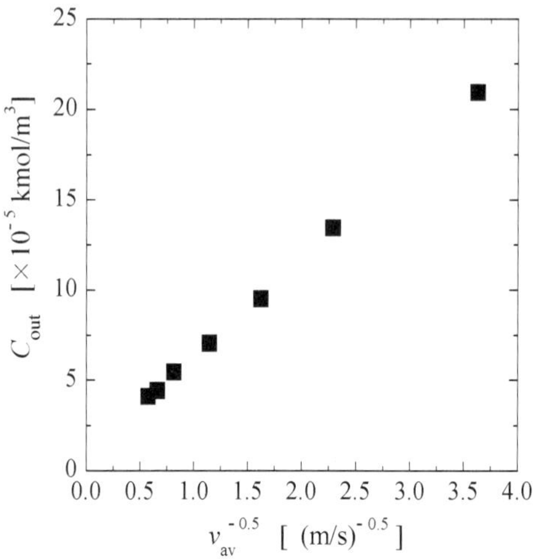

Fig. 8. Relevant concentration of benzoic acid at recovery-side outlet

According to Eq. (7), the diffusion coefficient of the solute can be determined from the slope of C_{out} plotted against $v_{av}^{-0.5}$.

$$\text{The slope} = \frac{\sqrt{D}}{L}\sqrt{\frac{1.4l}{\pi}}\,C_f \tag{8}$$

Since the diffusion coefficient of benzoic acid has been well investigated, it is used as a standard material for determining the diffusion coefficients. The slope calculated from the reported diffusion coefficient, $9.0\text{x}10^{-10}$ m^2/s at 298K (Yang and Matthews, 2000), is $7.54\text{x}10^{-5}$.

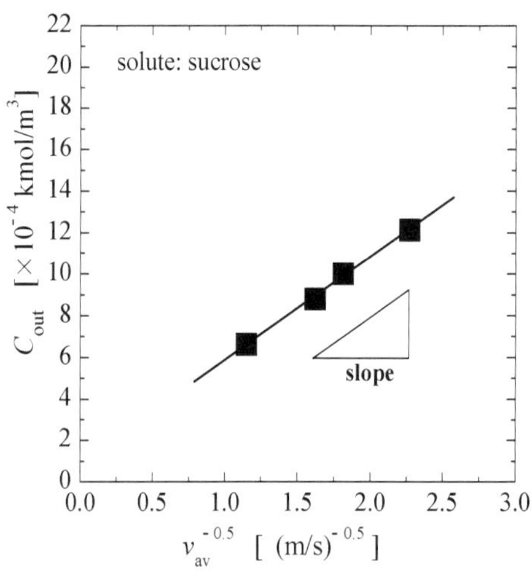

Fig. 9. Determination of diffusion coefficient of sucrose

For experiments with benzoic acid, the observed slope determined from the least mean squares approximation is 5.35x10^{-5}. The agreement of this value with this predicted one is poor. Hence, a correction factor, *P*, should be multiplied with the observed slope value to fit the predicted slope value. The value of *P* is found to be 1.4. To verify the applicability of the basic equation (8), diffusion coefficients of other solutes are determined with this method and the values are compared with the reported ones. Four solutes - sucrose, glycine, tryptophan, and urea - are selected. The results of C_{out} plotted against $v_{av}^{-0.5}$ are shown in Figs. 9 to 12.

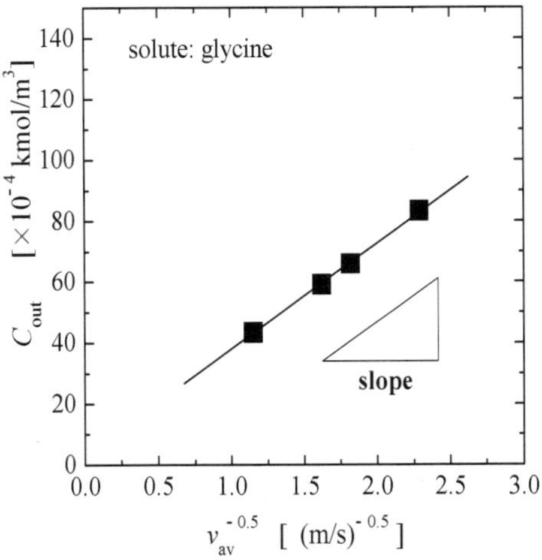

Fig. 10. Determination of diffusion coefficient of glycine

Solute	Diffusion coefficient (observed)[m²/s] at 298K	Diffusion coefficient (literature)[m²/s] at 298K
Benzoic acid	9.0 x 10^{-10}	* 9.0 x 10^{-10}
Sucrose	5.2 ± 0.2 x 10^{-10}	** 5.2 x 10^{-10}
Glycine	1.2 ± 0.1 x 10^{-9}	** 1.1 x 10^{-9}
Urea	1.4 x 10^{-9}	**1.4 x 10^{-9}
Tryptophan	5.6 x 10^{-10}	***5.7 x 10^{-10}

*(Yang and Matthews, 2000),
**(Jost, 1952),
***Calculated value from the correlation (Wilke and Chang, 1955)

Table 1. Comparison of the diffusion coefficient.

The diffusion coefficients are calculated from the corrected slope of the plots, and the results are compared with the reported diffusion coefficients in Table 1. The observed diffusion coefficients agree well with the reported values or the correlated values. The validity of the basic equation is thus confirmed for the determination of the unknown diffusion coefficient of solutes. The correction factor, *P*, to be applied to Eq. (8) expresses the characteristics of the

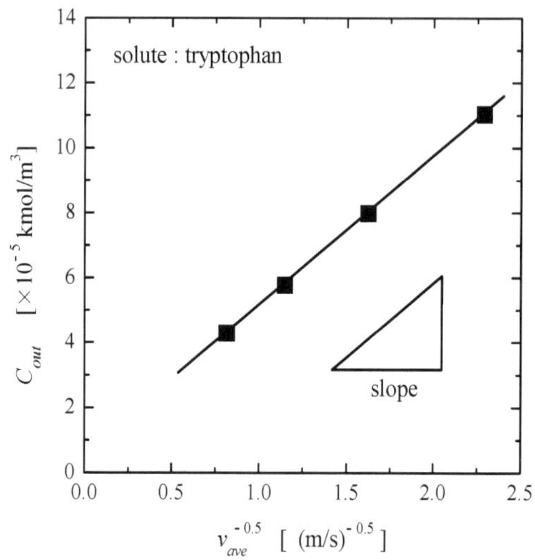

Fig. 11. Determination of diffusion coefficient of tryptophan

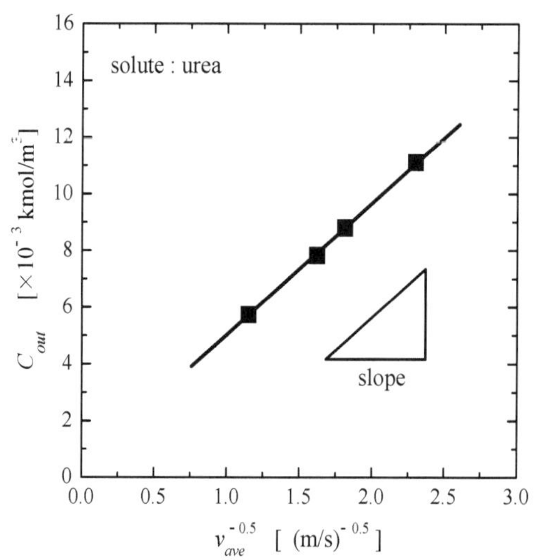

Fig. 12. Determination of diffusion coefficient of urea

mass-transfer in this microchannel. Since the *P* value is independent of the diffusing species, it can be influenced by the flow behavior in the microchannel. The value of the correction factor is bigger than one, which implies that the observed amount of the solute transferred is smaller than the theoretical amount. In the following chapter, the correction factor is determined by using microchannels of various configurations. The values are compared to discuss the flow behavior in a double-Y-type microchannel.

3. Flow characteristics in double-Y-type microchannel

3.1 Channel configurations and materials

Four types of microchannels were fabricated on the stainless sheet as shown in Fig. 11. The type(a) channel is used as a benchmark; it was used in the previous chapter. For type(b), (c) and(d) channels, the depth of the channels, which is determined from the thickness of the stainless sheet is set at 200μm. There are three channels of different depths for type (a); depths are 100, 200, and 300μm. Type (b) and (c) have different angles of knife-edge. Type (d) has a short contacting length for two liquid flows. For all types, the channel width is 200μm. As was described in the section 2.1, the channel was set in the module, and mass-transfer experiments were carried out using benzoic acid and sucrose as transferring material. On the feed side, an aqueous solution of each material was supplied, and deionized water flowed on the recovery side.

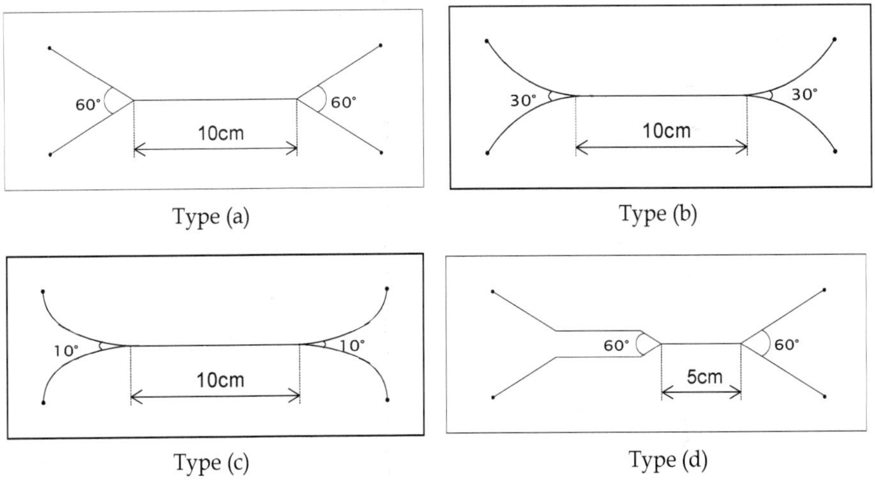

Fig. 11. Configuration of microchannels

3.2 Effect of channel thickness

Sucrose was transferred in three type(a) microchannels of different depths (100, 200, and 300μm). The concentration values at the outlet of the recovery side were compared to examine the effect of the channel depth on the mass-transfer. Since the channel width, 200μm, is constant, mass-transfer area increases linearly with the depth. The volumetric liquid flow rate is also proportional to the depth. Therefore, the amount of solute transferred is doubled when the depth doubled. It is expected that there will be no influence of channel depth on the outlet concentration. This trend is also predicted by the simulation. The experimental result is shown in Fig. 12. All the data located on a single line and the mass-transfer behavior was the same as the one expected.

The correction factor, *P*, was determined for the three channels by using the literature value of the diffusion coefficient for sucrose. For all the channels examined, the *P* value of 1.4 was to obtain the appropriate diffusion coefficient of sucrose. This fact suggests that the flow behaviour is similar for all microchannels used.

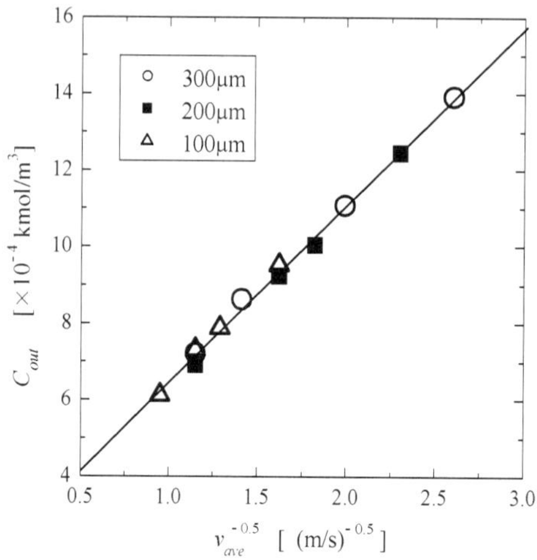

Fig. 12. Effect of channel thickness on outlet concentration of sucrose

3.3 Effect of channel length

The residence time of the liquid in contact changes with a change in the channel length. Therefore, without any disturbance in flow, the amount of solute transferred is unaffected by the channel length but by the residence time of the liquid. Sucrose was transferred

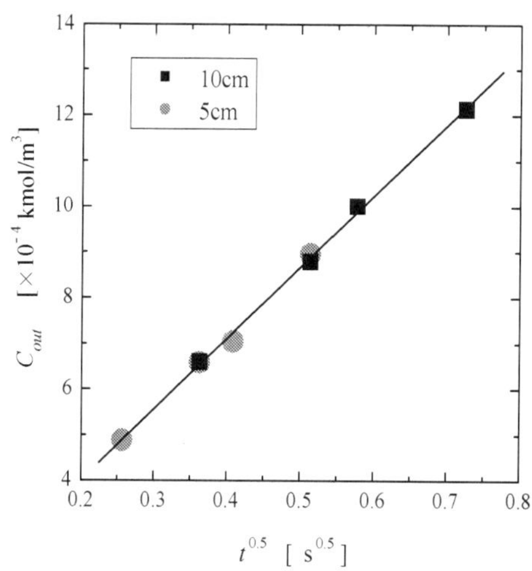

Fig. 13. Effect of channel length on outlet concentration of sucrose

in the type (d) channel having a channel length of 5cm, and the result was compared with that for the type (a). The outlet concentration was plotted against the square root of average residence time of the liquid, $t^{0.5}$ in Fig. 13. Equation (7) is rewritten as

$$C_{out} = \frac{\sqrt{D}}{L}\sqrt{\frac{1.4l}{\pi\, v_{av}}} \quad C_f = \left(\frac{C_f}{L}\sqrt{\frac{1.4D}{\pi}}\right) t^{0.5} \tag{9}$$

The factor, *P* for the type (d) channel was 1.4, which was the same for the type (a) and also as channels of different thickness of type (a). The accordance demonstrates that the factor is not an intrinsic value of a specific channel, and characterizes a general flow behavior found in the double-Y-type channels.

3.4 Effect of knife-edge angle

Using type (b) and (c) channels, we carried out sucrose transfer experiments were carried out to determine the each *P* value. The simulation predicts no effect of the knife-edge angle on the mass-transfer. In contrast with the simulation, the experimental result suggested that the mass-transfer was affected by the angle. The sucrose concentration at the outlet was plotted against $v_{av}^{-0.5}$ in Fig. 14. Since each curve had a different slope, the correction factor, *P*, had to be modified to fit the diffusion coefficient of sucrose.

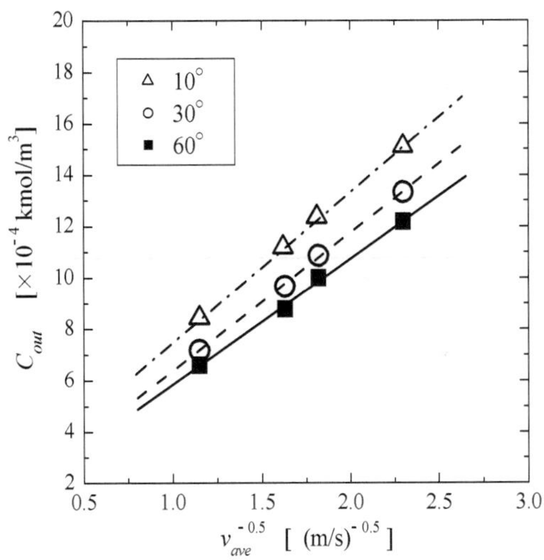

Fig. 14. Effect of knife-edge angle on outlet concentration of sucrose

Figure 15 shows a comparison of the *P* values for the three channels, type (a), (b), and (c). The *P* value changed with the knife-edge angle, and the result for benzoic acid showed a tendency similar to that in the case of the sucrose. With decreasing the knife-edge angle, the *P* value approaches unity, which corresponds with the pure penetration model. Now, it becomes clear that the factor, *P*, illustrates the characteristics of the contacting and the splitting flow in the double-Y-type channels.

How does the flow affect mass-transfer? This is a very important question; however, the answer to it is still difficult to find. One possible explanation is the effect of a secondary flow occurring in curved channels. At the knife-edge of the double-Y-type channels, two bent flows come together, or one flow is bended to split into two streams. It seems that the phenomena the occur at the contacting point have relatively strong impact on the mass-transfer. If the secondary flow induced a rotating motion whose direction was normal to the flow axis, the mass-transfer across the interface could be influenced. In addition to the flow in the bulk region, there can be some wall effects on the flow such as the slip of the liquid or surface roughness, which were pointed out by Tretheway (2002) and Mala (1999).

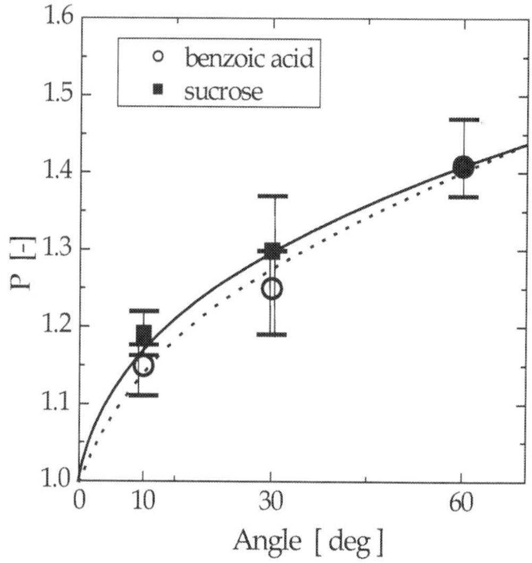

Fig. 15. Effect of knife-edge angle on the correction parameter, P

4. Conclusion

A double-Y-type microchannel device has been used for the mass-transfer from aqueous solutions to deionized water. Although the two liquids were carefully supplied at the same rate, the effluent flow rate for each side was sometimes unequal. Because of the unequal flow splitting, the observed values of the solute concentration of the recovery side were scattered. A method using the ratio of thte flow rates has been proposed for determine a relevant concentration for equal flow splitting. On the basis of the relevant concentration, the diffusion of benzoic acid in water was successfully analyzed with the penetration model with a correction factor. Furthermore, the basic equation obtained with benzoic acid was successfully applied to measure the diffusion coefficients for sucrose, glycine, tryptophan, and urea. The observed values agreed fairly well with the reported or correlated values. This fact demonstrates that the presented method is also suitable for determine the unknown diffusivities of solutes.

To pursue the physical meaning of the correction factor, various channels with different knife-edge angles, channel lengths and depths were used for the mass-transfer experiment. The value of the correction factor is affected only by the knife-edge angle, and is independent of the channel length and depth. This factor probably characterizes the flow behavior at the knife-edge.

5. Nomenclature

A	area of contacting plane	[m^2]
C_b	concentration of bulk liquid on the recovery-side	[kmol/m^3]
C_f	concentration of liquid on the feed-side	[kmol/m^3]
C_i	concentration at the contacting plane	[kmol/m^3]
C_{in}	concentration of recovery liquid at the inlet	[kmol/m^3]
C_{out}	concentration of recovery liquid at the outlet	[kmol/m^3]
D	diffusion coefficient	[m^2/s]
D_{obs}	observed value of diffusion coefficient	[m^2/s]
D_{lit}	literature value of diffusion coefficient	[m^2/s]
L	half width of the channel	[m]
l	length of the main portion of the channel	[m]
N_A	mass-transfer rate	[mol/(m^{2}s)]
S	cross sectional area of the recovery-side	[m^2]
t	average contact time defined by l/v_{av}	[s]
P	correction parameter for Eq/(8)	[-]
R_{out}	flow rate ratio defined as Q_r/Q_f	[-]
Q_f	flow rate for the feed-side	[m^3/s]
Q_r	flow rate for the recovery-side	[m^3/s]
v_{av}	average liquid velocity	[m/s]
$v_{int, av}$	average liquid velocity of the contacting plane	[m/s]

Greek symbol

θ	contact time defined by $l/v_{int, av}$	[s]

6. References

Ajmera, S. K.; Losey, M. W.; Jensen, K. F.; Schmidt, M. (2001). Microfabricated packed-bed reactor for phosgene synthesis. *AIChE J.*, 47(7), 1639-1647, ISSN 1547-5905

Brody, J. & P.; Yager, P. (1997). Diffusion-based extraction in a microfabricated device. *Sensors and Actuators A: Physical*, 58(1), 13-18 ISSN 0924-4247

Costin, C. D.; & Synovec, R. E. (2002). Measuring the transverse concentration gradient between adjacent laminar flows in a microfluidic device by a laser-based refractive index gradient detector. *Talanta*, 58(3), 551-560 ISSN 0039-9140

Costin, C. D.; McBrady, A. D.; McDonnell, M. E.; Synovec, R. E. (2004). Theoretical modeling and experimental evaluation of a microscale molecular mass sensor. *Anal. Chem.*, 76(10), 2725-2733 ISSN 0003-2700

Hetsroni, G.; Mosyak, A.; Pogrebnyak, E.; Yarin, L. P. (2005). Fluid flow in micro-channels. *Int. J. Heat and Mass Transfer*, *48*(10), 1982-1998, ISSN 0017-9310

Higbie, R. (1935). The rate of absorption of a pure gas into a still liquid during short periods of exposure *Trans. AIChE*, 31, 365-389

Jost, W.(1952) *Diffusion in solids, liquids, gases*, Academic Press Inc., New York.

Löb, P.; Löwe, H.; Hessel, V. (2004). Fluorinations, chlorinations and brominations of organic compounds in micro reactors. *J. Fluorine Chem.*, 125(11), 1677-1694, ISSN 0022-1139

Maruyama, T.; Uchida, J.; Ohkawa, T.; Futami, T.; Katayama, K.; Nishizawa, K.; Sotowa K.: Kubota, F.; Kamiya, N.; Goto, M. (2003). Enzymatic degradation of *p*-chlorophenol in a two-phase flow microchannel system. *Lab on a Chip* 3(4), 308-312, ISSN 1473-0197

Mala, G. M. & Li, D. (1999). Flow characteristics of water in microtubes, *Int. J. Heat and Fluid Flow,* 20(2) 142-148, ISSN 0142-727X

Tokeshi, M.; Minagawa, T.; Kitamori, T. (2000). Integration of a microextraction system on a glass chip: ion-pair solvent extraction of Fe(II) with 4,7-diphenyl-1,10-phenanthrolinedisulfonic acid and tri-*n*-octylmethylammonium chloride. *Anal. Chem., 72*(7), 1711-1714 ISSN 0003-2700

Tretheway, D. C. & Meinhart, C. D. (2002). Apparent fluid slip at hydrophobic microchannel walls, *Physics of fluids,* 14(3), L9-L12, ISSN 1070-6631

Yamaguchi, Y.; Takagi, F.; Watari, T.; Yamashita, K.; Nakamura, H.; Shimizu, H.; Maeda, H. (2004). Interface configuration of the two layered laminar flow in a curved microchannel. *Chem. Eng. J.,* 101(1-3), 367-372, ISSN 1385-8947

Yang, X. & Matthews, M. A. (2000). Diffusion coefficients of three organic solutes in aqueous sodium dodecyl sulfate solutions. *J. Col. and Int. Sci.,* 229(1), 53-61 ISSN 0021-9797

van den Berg, A.; Olthuis, W.; Bergveld, P. (2000). *Micro total analysis systems 2000,* Springer, ISBN 0792363876, Dordrecht

Wilke, C. R. & Chang P. (1955). Correlation of diffusion coefficients in dilute solutions, *AIChE J.,* 1(2), 133 – 272, ISSN 1547-5905

2

Turbulent Buoyant Jet of a Low-Density Gas Leaks Into a High-Density Ambient: Hydrogen Leakage in Air

M.F. El-Amin and Shuyu Sun
King Abdullah University of Science and Technology (KAUST), Thuwal
Kingdom of Saudi Arabia

1. Introduction

The low-density gas jet injected into a high-density ambient has particular interest in several industrial applications such as fuel leaks, engine exhaust, diffusion flames, materials processing, as well as natural phenomena such fires and volcanic eruptions. The most interesting application of this problem nowaday is the hydrogen leaks in air; since when it mixes with air, fire or explosion can result. The expected extensive usage of hydrogen increases the probability of its accidental release from hydrogen vessel infrastructure. Hydrogen energy has much promise as a new clean energy and is expected to replace fossil fuels; however, hydrogen leakage is considered to be an important safety issue and is a serious problem that hydrogen researchers must address. Hydrogen leaks may occur from loose fittings, o-ring seals, pinholes, or vents on hydrogen-containing vehicles, buildings, storage facilities, or other hydrogen-based systems. Hydrogen leakage may be divided into two classes, the first is a rapid leak causing combustion, while the other is an unignited slow-leak. However, hydrogen is ignited in air by some source of ignition such as static electricity (autoignition) or any external source. Classic turbulent jet flame models can be used to model the first class of hydrogen leakage; cf. (Schefer et al., 2006; Houf & Schefer, 2007; Swain et al., 2007; Takeno et al., 2007). This work is focused on the second class of unignited slow-leaks. Previous work introduced a boundary layer theory approach to model the concentration layer adjacent to a ceiling wall at the impinging and far regions in both planar and axisymmetric cases for small-scale hydrogen leakage El-Amin et al. (2008); El-Amin & Kanayama (2009a; 2008).

This kind of buoyant jets 'plume' is classed as non-Boussinesq; since the initial fractional density difference is high, which is defined as, $\Delta\rho_0 = (\rho_\infty - \rho_0)$, where ρ_0 is the initial centerline density (density at the source) and ρ_∞ is the ambient density. Generally, for binary selected low-densities gases at temperature $15^\circ C$, the initial fractional density differences are 0.93 for $H_2 - Air$, 0.86 for $He - Air$, 0.43 for $CH_4 - Air$ and 0.06 for $C_2H_2 - N_2$. (Crapper & Baines, 1977) suggested that the upper bound of applicability of the Boussinesq approximation is that the initial fractional density difference $\Delta\rho_0/\rho_\infty$ is 0.05. In general, one can say that the Boussinesq approximation is valid for small initial fractional density difference, $\Delta\rho_0/\rho_\infty << 1$. However, in these cases of invalid Boussinesq approximation a density equation must be used. Moreover, a discussion of this classification is given by (Spiegel & Veronis, 1960); and the non-Boussinesq plume was studied, for example, in (Woods, 1997) and (Carlotti & Hunt, 2005).

The integral method was used by (Agrawal & Prasad, 2003) to derive similarity solutions for several quantities of interest including the cross-stream velocity, Reynolds stress, the dominant turbulent kinetic energy production term, and eddy diffusivities of momentum and heat for axisymmetric and planar turbulent jets, plumes, and wakes. (El-Amin & Kanayama, 2009b) used the experimental results by (Schefer et al., 2008b) and others such as the streamwise and centerline velocities, and the concentration at any downstream location with their empirical constants for the momentum-dominated regime of a buoyant jet resulting from hydrogen leakage. In (El-Amin & Kanayama, 2009c), the similarity form of a non-Boussinesq hydrogen-air buoyant jet resulting from a small-scale hydrogen leakage in the air was developed. In (El-Amin, 2009), non-Boussinesq hydrogen-air buoyant jet resulting from a small-scale hydrogen leakage in the air was investigated. Recently, (El-Amin et al., 2010) introduced analysis of a turbulent buoyant confined jet using realizable $k - \epsilon$ model and several empirical models were developed.

In this chapter, we study the problem of low-density gas jet injected into high-density ambient numerically with an example of hydrogen leakage in air which is important issue faces hydrogen energy developing. The resulting buoyant jet can be divided into two regimes, momentum-dominated and buoyant-dominated. We study each regime indvidualy under certain assumptions and governing laws. Firstly, we used the integral method to derive a complete set of results and expressions for selected physical turbulent properties of the momentum-dominated buoyant jet regime of small-scale hydrogen leakage. In the second part, the buoyancy-dominated regime is investigated.

A jet/plume represents an example of free shear flows, which is produced when the fluid exits a nozzle with initial momentum. In the ideal case of jet/plume, the initial volume flux is assumed to be zero; however, in the real cases, the jet/plume has a finite source size and initial volume flux. In the pure jet "non-buoyant jet" the initial momentum flux as a high velocity injection causes the turbulent mixing. In a pure plume the buoyancy flux causes a local acceleration which lead to turbulent mixing. In the general case of a buoyant jet "forced plume" a combination of initial momentum flux and buoyancy flux is responsible for turbulent mixing. In general, jets are considered the most popular flows on which new theories of turbulence are tested. The popularity can be gauged by the extensive literature that exits about them. Comprehensive review of jets was presented by (Rajaratnam, 1976; List, 1982; Fischer et al., 1979). The behavior of the turbulent vertical round jet and plume were considered by (Turner, 1973; Shabbir & Georg, 1994). Observations show that, after some distance about $10d$, where d is the inlet diameter, from the source, the flow becomes turbulent and expands conically; where the conical head (origin) is virtually located close to the source. Upper that 10d distance, the flow tends to be self-similar (i.e., the shapes of mean quantities become independent of the axial location when normalized by scales of velocity and length; (Hinze, 1975)). The instantaneous flow field is unsteady, complex and three-dimensional; however, when flow field is averaged over a period that is much larger than the timescale (the ration of the characteristic length scale to the velocity scale) in the flow, the resulting mean flow field is steady and two-dimensional (Papantoniou & List, 1989; Venkatakrishnan et al., 1999; Bhat & Narasimha, 1996). The vertical axis passing through the center of the inlet is known as centerline and the mean flow as axisymmetric (symmetric around the centerline). It is worth mentioning that the axisymmetric jets/plumes are the best way for testing both experimental and numerical work. The distance along the centerline is denoted by z and that

	b	$U_{cl}(z)$	$C_{cl}(z)$	c_m	A	α
Jet	$c_m(z-z_0)$	$\frac{AdU_0}{(z-z_0)}$	$\frac{5.6\sqrt{\pi}dC_0}{2(z-z_0)}$	0.1 - 0.13	5-7	$\frac{c_m}{2}$
Plume	$c_m(z-z_0)$	$\frac{AF_0^{1/2}}{(z-z_0)^{1/3}}$	$\frac{9.4Y}{F_0^{1/3}(z-z_0)^{5/3}}$	0.1 - 0.13	3.9 - 4.7	$\frac{5c_m}{6}$

Table 1. Forms of jet/plume mean flow variables and their related coefficients.

in the horizontal by r. Mostly, the mean centerline velocity U_{cl} is taken as the velocity scale, and the jet/plume width, denoted by b, is taken as the length scale. The following forms in Table 1, show the mean flow variables, which are known as similarity solutions and suggested by both experimental observations and dimensional analyses (e.g. (Fischer et al., 1979; Turner, 1973; Hussein et al., 1994; Dai et al., 1995)). Note that, $F_0 = Kd^2U_0g\Delta\rho_0/\rho_\infty$, $K = \pi/(1+\lambda^2)$, $\lambda = c_m/c_c$ and, $Y = QC_0 = (\pi/4)d^2U_0C_0$, $Q = (\pi/4)d^2U_0$ In the above equations, d is the nozzle diameter, c_m is spread rate of the momentum. U_0 and C_0 are, respectively, the vertical velocity and concentration at the inlet. Q, F_0 and Y are respectively, volume flux, buoyancy flux and mass flux at the inlet. z_0 is the virtual origin, that is, the distance above/below the orifice where the flow appears to originate. In this study we will assume that both the virtual origin and the jet source have the same location, i.e. $z_0 = 0$, therefore, $z - z_0$ in the above expressions will replaced by z.

2. Momentum-dominated regime

The turbulent round jet is the most suitable model to describe unintended hydrogen releases with circular leak geometries, such as pinholes, through which the flow of hydrogen is an axisymmetric jet. Measurements of the hydrogen distribution in a laboratory-scale hydrogen leak under flow conditions and neglecting the buoyancy effects are described by (Schefer et al., 2008b). They reported a Froude number of 268, which is in the momentum-dominated regime where the effects of buoyancy-generated momentum are small and the Reynolds number was sufficient for fully developed turbulent flow. Their results showed that hydrogen jets behave similar to jets of helium and conventional hydrocarbon fuels like methane and propane in the momentum-dominated regime. As with any jet flow, the hydrogen mass fraction centerline decay rate shows z^{-1} dependence, where z is the axial distance from the jet exit. In this section, we used the integral method to derive a complete set of results and expressions for selected physical turbulent properties of momentum-dominated buoyant jet regime of small-scale hydrogen leakage El-Amin & Kanayama (2009b). Several quantities of interest, including the cross-stream velocity, Reynolds stress, velocity-concentration correlation (radial flux), dominant turbulent kinetic energy production term, turbulent eddy viscosity and turbulent eddy diffusivity, are obtained. In addition, the turbulent Schmidt number is estimated and the normalized jet-feed material density and the normalized momentum flux density are correlated. Experimental results from (?) and other works for the momentum-dominated jet resulting from small-scale hydrogen leakage are used in the integral method. For a non-buoyant jet or momentum-dominated regime of a buoyant jet, both the centerline velocity and centerline concentration are proportional with z^{-1}. The effects of buoyancy-generated momentum are assumed to be small, and the Reynolds number is sufficient for fully developed turbulent flow. The hydrogen-air momentum-dominated regime or non-buoyant jet is compared with the air-air jet as an example of non-buoyant jets. Good agreement was found between the current results and experimental results from the literature. In addition, the turbulent Schmidt number was shown to depend solely on the ratio of the momentum

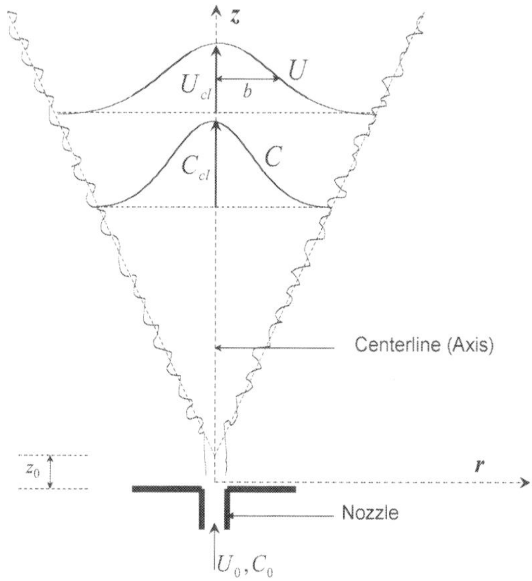

Fig. 1. Schematic of turbulent jet geometry.

spread rate to the material spread rate. Standard empirical Gaussian expressions for the mean streamwise velocity U and concentration C are substituted into the governing equations of the momentum-dominated regime of a buoyant hydrogen jet, in order to derive a number of turbulent quantities using integral methods. In the first step, the expression for U is substituted into the continuity equation and a centerline velocity variation is assumed, and the equation is integrated to determine the mean cross-stream velocity profile V. In the second step, the expressions for U and V are substituted into the simplified momentum equation and integrated to determine the Reynolds stress. In the third step, the expressions for U and V are substituted into the simplified concentration equation and integrated to determine the velocity-concentration correlation.

2.1 Governing equations

Viscous stresses are assumed to be much smaller than the corresponding turbulent shear stresses provided that the nozzle Reynolds number is about a few thousand (Rajaratnam, 1976). In self-similar region, the axisymmetric streamwise governing equations in cylindrical coordinates (see Fig.1) obtained using simplifications of (Refs. (Agrawal & Prasad, 2003; Bhat & Narasimha, 1996; Chen & Rodi, 1980; Tennekes & Lumley, 1972; Gebhart et al., 1988; Rajaratnam, 1976)) the continuity, momentum, and concentration equations in cylindrical coordinates for a vertical axisymmetric momentum-dominated regime of buoyant hydrogen free jet can be written as,

$$\frac{\partial (rV)}{\partial r} + \frac{\partial U}{\partial z} = 0 \qquad (1)$$

$$V\frac{\partial U}{\partial r}+U\frac{\partial U}{\partial z}+\frac{1}{r}\frac{\partial(r\overline{uv})}{\partial r}=0 \tag{2}$$

$$V\frac{\partial C}{\partial r}+U\frac{\partial C}{\partial z}+\frac{1}{r}\frac{\partial(r\overline{vc})}{\partial r}=0 \tag{3}$$

The overbar denotes that the time-averaged quantities u and v are the components of velocity fluctuations on the axisymmetric coordinates r and z, respectively. c is the concentration fluctuation, ν is the kinematic viscosity, and D is the mass molecular diffusivity of hydrogen. The streamwise velocity and concentration at any downstream location for a self-similar axisymmetric jet, can be approximated by a Gaussian distribution (Agrawal & Prasad, 2003; Bhat & Narasimha, 1996; Turner, 1986; List, 1982; Chen & Rodi, 1980; Agrawal & Prasad, 2002),

$$U(r,z)=U_{cl}(z)\exp\left(-\frac{r^2}{b^2(z)}\right) \tag{4}$$

$$C(r,z)=C_{cl}(z)\exp\left(-\lambda^2\frac{r^2}{b^2(z)}\right) \tag{5}$$

The centerline velocity U_{cl} and the centerline concentration C_{cl} vary with z^{-1}, while the jet width b increases linearly with z (Agrawal & Prasad, 2003; Tennekes & Lumley, 1972; Bhat & Krothapalli, 2000). Dimensional arguments together with experimental observations suggest that the mean flow variables, which are known as similarity solutions, are as follows ((Fischer et al., 1979; Hussein et al., 1994; Shabbir & Georg, 1994; Schefer et al., 2008b)),

$$U_{cl}(z)=\frac{AdU_0}{z-z_0} \tag{6}$$

$$C_{cl}(z)=\frac{d^*}{K_C(z-z_0)}=\frac{d(\rho_{H_2}/\rho_{air})^{1/2}}{K_C(z-z_0)} \tag{7}$$

$$b(z)=c_m(z-z_0) \tag{8}$$

In the above equations, d and U_0 are the nozzle diameter and the vertical velocity at the inlet, respectively (Fig.1).z_0 is the virtual origin, which is the distance above/below the orifice where the flow appears to originate. The experimentally measured spread rate c_m varies in the range 0.1-0.13 and the constant A varies between 5-7 Fischer et al. (1979); Hussein et al. (1994). For this investigation, the value of the spread rate for the hydrogen jet of $c_m = 0.103$ was used, where this was experimentally determined by Schefer et al. [1]. The spread rate for the concentration c_c is given in the formula $C = C_{cl}\exp(-r^2/[c_c^2(z-z_0)])$. It is well known that $c_c \neq c_m$, $\lambda = c_m/c_c$, meaning that the velocity and concentration spread have different rates. The corresponding streamwise concentration for the hydrogen free jet given by Schefer et al. (2008b) is $C = C_{cl}\exp(-59r^2/(z-z_0))$, which gives $59 = 1/c_c^2$ and $c_c = 0.13$, $\lambda = 0.79$. The expression for the streamwise concentration may then be rewritten as:

$$C(r,z)=C_{cl}(z)\exp(-\lambda^2\eta^2)=C_{cl}(z)\exp(-0.63\eta^2) \tag{9}$$

where $\eta = r/b$.

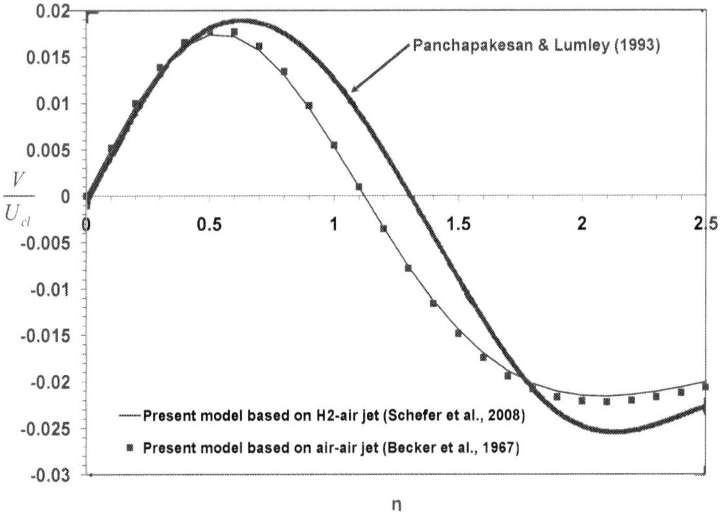

Fig. 2. Cross-stream velocity profile for the momentum-dominated jet.

2.2 Mean and turbulent quantities

The continuity equation for the time-averaged velocities can be solved by substituting from (4) and (6) into (1) to obtain the cross-stream velocity in the form:

$$\frac{V}{U_{cl}} = \frac{c_m}{2\eta}\left(-1+\exp(-\eta^2)+2\eta^2\exp(-\eta^2)\right) \tag{10}$$

Fig.2 shows the normalized cross-stream velocity profiles V/U_{cl} plotted against the normalized coordinate η. Since the case under study is a momentum-dominated (non-buoyant) jet, it is comparable with the air-air jet. The normalized cross-stream velocity profiles V/U_{cl} (Eq. (12)) are based on measurements of a hydrogen-air jet by (Schefer et al., 2008b) and an air-air jet by (Becker et al., 1967), which are compared with measurements of an air-air jet by (Panchapakesan & Lumley, 1993), as illustrated in Fig. 2. From this figure, it can be seen that the cross-stream velocity vanishes at the centerline and becomes outward in the neighborhood of the centerline ($0 \leq \eta \leq 1.12$ for a hydrogen-air jet and an air-air jet according to Eq. (12), and $0 \leq \eta \leq 1.31$ for measurements of an air-air jet from (Panchapakesan & Lumley, 1993)). A maximum is then reached (0.0173 for a hydrogen-air jet and 0.0178 for an air-air jet according to Eq. (12) at $\eta = 0.5$, and 0.0188 at $\eta = 0.62$ for measurements of an air-air jet (Panchapakesan & Lumley, 1993)), and then declines back to zero ($\eta = 1.12$ for a hydrogen-air jet and an air-air jet according to Eq. (10), and $\eta = 1.31$ for measurements of an air-air jet Panchapakesan & Lumley (1993)). In addition, the cross-stream profile's flow becomes inward ($V/U_{cl} < 0$), reaches a minimum (-0.0216 for a hydrogen-air jet and -0.0222 for an air-air jet at $\eta = 2.1$ according to Eq. (10), and -0.026 at $\eta = 2.15$ for measurements of an air-air jet Panchapakesan & Lumley (1993)) and reaches an asymptote of 0 as $\eta \to \infty$. The differences between maximum and minimum values of the cross-stream velocities noted above are attributed to the high Reynolds number ($Re = 11000$) used in Ref. (Panchapakesan & Lumley, 1993). In addition, it is noteworthy that the cross-stream velocity asymptotes to zero ($V \to 0$) much more slowly than the streamwise velocity U does. Thus, although the central

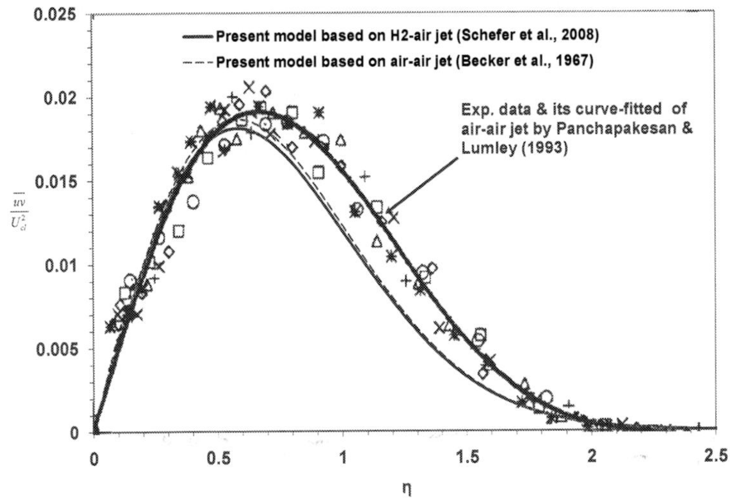

Fig. 3. Reynolds stress profile for the momentum-dominated jet.

region is dominated by the axial component of velocity, the cross-stream flow predominates far away from it (Agrawal & Prasad, 2003). From Eq. (10), $\lim_{\eta\to\infty}(V/U_{cl}) = -c_m/2\eta$. The inward extension of the curve $c_m/2\eta$ intersects the edge of the jet ($\eta = 1$) with a value of $c_m/2$ (0.0515 for the hydrogen-air jet and 0.053 for the air-air jet). The incremental volume flux can be used to define the entrainment coefficient α (see Turner (1986)):

$$\frac{d\mu}{dz} = 2\pi b U_{cl}\alpha \tag{11}$$

where $/mu$ is the volume flux for axisymmetric jet given by:

$$\mu = \int_0^\infty 2\pi r U(r) dr \tag{12}$$

From (Agrawal & Prasad, 2003), $d\mu/dz$ is the incremental volume flux entering the jet through a circular control surface at large r,

$$\frac{d\mu}{dz} = \lim_{\eta\to\infty}(-2\pi r V) = \pi b c_m U_{cl} \tag{13}$$

Eqs. (11) and (13) give the entrainment coefficient $\alpha = c_m/2 = 0.0515$ for the momentum-dominated hydrogen-air jet and $\alpha = c_m/2 = 0.053$ for the air-air jet based on measurements by (Becker et al., 1967).

By inserting the time-averaged profiles of U and V, Eq. (4) and Eq. (10), respectively, into the momentum equation, Eq. (4), we obtain the time-averaged profile for the Reynolds stress $\overline{uv}$ in the form:

$$\frac{\overline{uv}}{U_{cl}^2} = \frac{c_m}{2\eta}\exp(-\eta^2)\left(1 - \exp(-\eta^2)\right) \tag{14}$$

The normalized Reynolds stress $\overline{uv}/U_{cl}^2$ (Eq. (14)), based on measurements of a hydrogen-air jet by (Schefer et al., 2008b) and an air-air jet by (Becker et al., 1967), are compared with measurements of an air-air jet by (Panchapakesan & Lumley, 1993), as shown in Fig. 3. This figure shows that the maximum Reynolds stress $\overline{uv}/U_{cl}^2$ has a value of (0.0181 for a hydrogen-air jet and 0.0186 for an air-air jet according to Eq. (14) and at $\eta = 0.6$, and 0.019 at $\eta \approx 0.68$ for measurements of an air-air jet (Panchapakesan & Lumley, 1993)). It is notable that the fitted-curve for the measurements of (Panchapakesan & Lumley, 1993) is used for this comparison.

In the same manner used above, the time-averaged profiles of U and C are inserted into Eq. (3), and the time-averaged profile for the velocity-concentration correlation (radial flux) $\overline{vc}$ takes the form:

$$\frac{\overline{vc}}{U_{cl}^2} = \frac{c_m}{2\eta}\exp(-\lambda\eta^2)\left(1-\exp(-\eta^2)\right) \tag{15}$$

The normalized velocity-concentration correlation (radial flux) $\overline{vc}/U_{cl}^2$ (Eq. (15)) based on measurements of hydrogen-air jet by (Schefer et al., 2008b) and air-air jet by (Becker et al., 1967) are compared with measurements of a helium-air jet by (Panchapakesan & Lu, 1993), as shown in Fig. 4. This figure shows that the maximum radial flux has a value of 0.021 for the hydrogen-air jet and 0.023 for the air-air jet, according to Eq. (14) at $\eta = 0.7$, and 0.02 at $\eta \approx 0.88$ for measurements (fitted-curve) of a helium-air jet (Panchapakesan & Lu, 1993).

The dominant kinetic energy production term is defined by $\overline{uv}\partial U/\partial\eta$. Using Eqs. (4), (6) and (14), gives:

$$\frac{\overline{uv}}{U_{cl}^3}\frac{\partial U}{\partial\eta} = c_m\exp(-2\eta^2)\left(1-\exp(-\eta^2)\right) \tag{16}$$

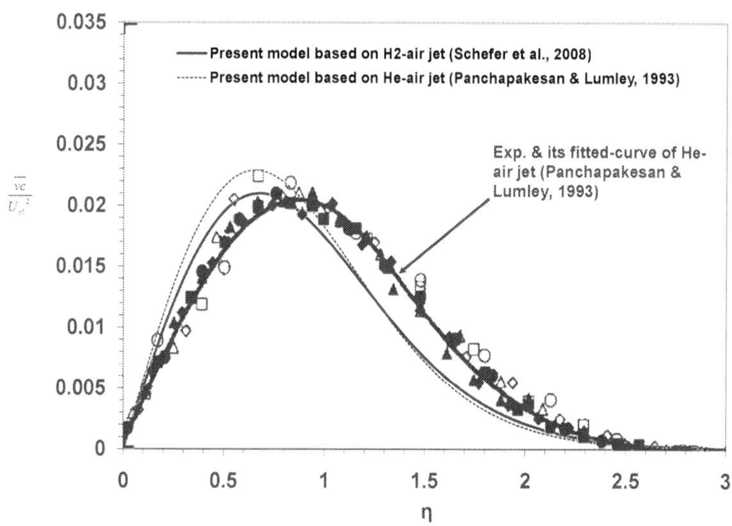

Fig. 4. Velocity-concentration correlation profile for the momentum-dominated jet.

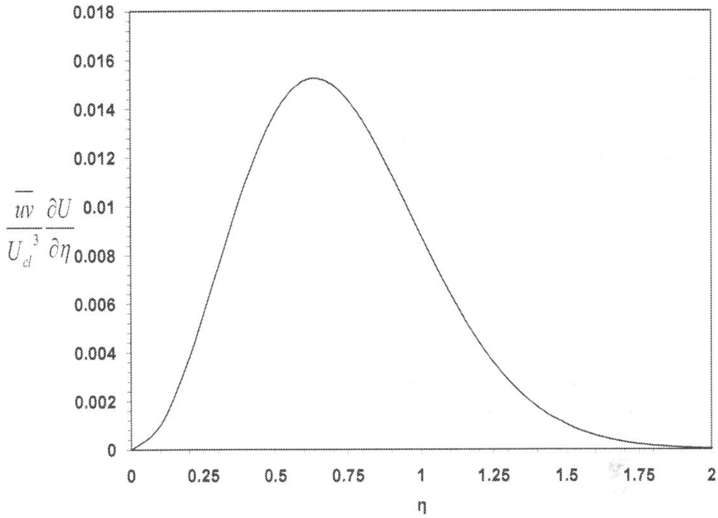

Fig. 5. Dominant kinetic energy production term profile for the momentum-dominated jet.

The complete turbulent kinetic energy equation can be found in the literature (e.g., see Wygnanski). Fig. 5 indicates that the dominant kinetic energy production term reduces to zero for large η, and has a maximum value of 0.015 at $\eta = 0.6$ for the hydrogen-air jet based on the measurements by (Schefer et al., 2008b). The following definitions for the turbulent eddy viscosity ν_t and turbulent eddy diffusivity D_t may also be used,

$$\overline{uv} = -\nu_t \frac{\partial U}{\partial r} \tag{17}$$

and

$$\overline{vc} = -D_t \frac{\partial C}{\partial r} \tag{18}$$

to derive corresponding expressions as,

$$\frac{\nu_t}{U_{cl} b} = \frac{c_m}{4\eta^2}\left(1 - \exp(-\eta^2)\right) \tag{19}$$

and

$$\frac{D_t}{U_{cl} b} = \frac{c_m}{4\lambda^2\eta^2}\left(1 - \exp(-\eta^2)\right) \tag{20}$$

respectively. Eqs. (19) and (20) indicate that the turbulent eddy viscosity and turbulent eddy diffusivity are not independent of η, as assumed in works such as (Tennekes & Lumley, 1972), but are variable as a function of η, as suggested by others (Lessen, 1978). Both the normalized turbulent eddy viscosity $\nu_t/(U_{cl}b)$, Eq. (19), and the normalized turbulent eddy diffusivity $D_t/(U_{cl}b)$, Eq. (20) for the hydrogen-air jet based on the measurements by (Schefer et al., 2008b) are plotted in Fig.15. This figure shows that both $\nu_t/(U_{cl}b)$ and $D_t/(U_{cl}b)$ have

maximum values of 0.026 and 0.041, respectively, at the centerline ($\eta = 0$) of the jet and decay to 0 for large η. It is also interesting that the turbulent eddy diffusivity is greater than turbulent eddy viscosity. If the effects of molecular transport are neglected, the turbulent Schmidt number in the free turbulent flow may be defined as the quantitative difference between the turbulent transport rate of material and momentum. In other words, the turbulent Schmidt number, $Sc_t = \nu_t / D_t$, is a ratio of the eddy viscosity of the turbulent flow ν_t to the eddy material diffusivity D_t. It is very well known from literature that the spreading rate of material in free jets is greater than that of momentum; thus, the turbulent Schmidt number is less than unity and has a significant effect on the predictions of species spreading rate in jet flows.
The turbulent eddy viscosity ν_t and the turbulent eddy diffusivity D_t, given by Eqs. (19) and (20), respectively, give the turbulent Schmidt number as:

$$Sc_t = \frac{\nu_t}{D_t} = \lambda^2 \tag{21}$$

It is interesting to note that the turbulent Schmidt number depends only on the coefficient λ, which, in turn, depends on the material. Using Eq. (21), the turbulent Schmidt number Sc_t for hydrogen momentum-dominated jet is 0.63, based on the experimental results given by (Schefer et al., 2008b). However, the same formula (Eq. (21)) used with experimental results of the helium-air jet, $c_m = 0.116, c_c = 0.138$ (Panchapakesan & Lu, 1993), the turbulent Schmidt number Sc_t becomes 0.7. Irrespective of the turbulence structure, the constant value of the turbulent Schmidt number across the whole flow field implies that the momentum process is similar to the material-transport process (He et al., 1999; Yimer et al., 2002).
According to Eq. (21), if λ^2 is replaced by Sc_t in Eq. (9), the streamwise concentration for the momentum-dominated jet may be rewritten as:

$$C(r,z) = C_{cl}(z)\exp(-Sc_t\eta^2) \tag{22}$$

Eqs. (4) and (5) give,

$$\frac{U}{U_{cl}} = \exp(-\eta^2) \tag{23}$$

Therefore, Eq. (22) becomes:

$$\frac{C}{C_{cl}} = \left(\frac{U}{U_{cl}}\right)^{Sc_t} \tag{24}$$

The normalized streamwise momentum flux density across an isothermal jet is given by:

$$G = \left(\frac{U}{U_{cl}}\right)^2 \tag{25}$$

and the normalized jet-feed material density is given by:

$$W = \frac{UC}{U_{cl}C_{cl}} \tag{26}$$

Therefore, the mean profiles of jet-feed concentration and streamwise velocity are related by:

$$W = G^{(1+Sc_t)/2} \tag{27}$$

Fig. 6 shows a plot of the correlation of the normalized jet-feed material density against the normalized momentum flux density for the hydrogen-air jet, $Sc_t = 0.63$, helium-air jet, $Sc_t = 0.7$, and air-air jet, $Sc_t = 0.76$.

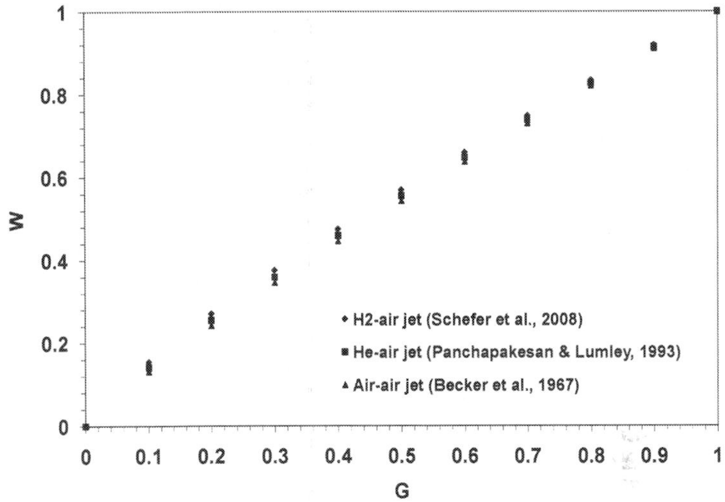

Fig. 6. Correlation of the normalized jet-feed material density against the normalized momentum flux density for hydrogen, helium and air jets.

3. Buoyancy-dominated regime

In this part, we study the buoyancy-dominated regime. It is assumed that the local rate of entrainment is consisted from two components; one is the component of entrainment due to jet momentum while the other is the component of entrainment due to buoyancy. The integral models of the mass, momentum and concentration fluxes are obtained and transformed to a set of ordinary differential equations using some similarity transformations El-Amin (2009); El-Amin & Kanayama (2009c). The resulting system is solved to determine the centerline quantities which are used to get the mean axial velocity, mean concentration and mean density of the jet. Therefore, the centerline and mean quantities are used together with the governing equation to determine some important turbulent quantities such as, cross-stream velocity, Reynolds stress, velocity-concentration correlation, turbulent eddy viscosity and turbulent eddy diffusivity.

3.1 Governing equations

Consider a vertical axisymmetric non-Boussinesq buoyant jet resulting from injecting a low-density gas jet into high-density ambient. Using cylindrical polar coordinates (z,r) with the z-axis vertical and the source at $z = 0$ (see Fig. 1, the continuity, momentum and concentration equations of the steady vertical axisymmetric buoyant free jet can be written as (Agrawal & Prasad, 2003; Bhat & Narasimha, 1996; Chen & Rodi, 1980; Gebhart et al., 1988; Rajaratnam, 1976; Tennekes & Lumley, 1972; El-Amin, 2009),

$$\frac{\partial (r\rho V)}{\partial r} + \frac{\partial (r\rho U)}{\partial z} = 0 \tag{28}$$

$$\frac{\partial (r\rho UV)}{\partial r} + \frac{\partial (r\rho U^2)}{\partial z} + \frac{\partial (r\rho \overline{uv})}{\partial r} = gr(\rho - \rho_\infty) \tag{29}$$

$$\frac{\partial(r\rho VC)}{\partial r}+\frac{\partial(r\rho UC)}{\partial z}+\frac{\partial(r\rho\overline{vc})}{\partial r}=0 \tag{30}$$

where U is the mean streamwise velocity, V is the mean cross-stream velocity profile, C is the hydrogen concentration (mass fraction). The overbar denotes the time-averaged quantities, u, v are the components of velocity fluctuations on the axisymmetric coordinates r, z, respectively. c is the concentration fluctuation, ρ is the mixture density and D is the mass molecular diffusivity of the hydrogen in air. On the other hand, from the experimental observations, the equations for the vertical velocity, density deficiency and mass fraction profiles with assuming that the hydrogen-air mixture behaves as an ideal gas, are as follows,

$$U(r,z)=U_{cl}(z)\exp\left(-\frac{r^2}{b^2(z)}\right) \tag{31}$$

$$\rho_\infty-\rho(r,z)=(\rho_\infty-\rho_{cl}(z))\exp\left(-\lambda^2\frac{r^2}{b^2(z)}\right) \tag{32}$$

$$\rho(r,z)C(r,z)=\rho_{cl}(z)C_{cl}(z)\exp\left(-\lambda^2\frac{r^2}{b^2(z)}\right) \tag{33}$$

where $U(r,z)$ and $\rho(r,z)$ are the mean velocity and mean density at any point of the jet body, $U_{cl}(z)$ and $\rho_{cl}(z)$ are the centerline velocity and density. The buoyancy spreading factor $\lambda = c_m/c_c$ expresses the difference in spreading rates between the velocity and density deficiency profiles. It is well known that $c_m \neq c_c$, i.e., velocity and density spread in different rates. The mass fraction concentration C is related to the mixture density by,

$$\rho=\frac{1}{[(1/\rho_0)-1/\rho_\infty)]C+(1/\rho_\infty)} \tag{34}$$

while, the mole fraction concentration X is related to the mixture density by,

$$\rho=\rho_\infty(1-X)+\rho_0 X \tag{35}$$

Therefore, the mass fraction and mole fraction can be related as follows,

$$C=\frac{\rho_0 X}{\rho_\infty(1-X)+\rho_0 X}=\frac{R_\rho X}{1+(R_\rho-1)X} \tag{36}$$

where $R_\rho=\frac{\rho_0}{\rho_\infty}$ is the density ratio. Also, one can note that $\rho C=\rho_0 X$, and therefore the equation of concentration, Eq. (30), can be rewritten in terms of mole fraction X as follows,

$$\frac{\partial(r\rho VX)}{\partial r}+\frac{\partial(r\rho UX)}{\partial z}+\frac{1}{\rho_0}\frac{\partial(r\rho\overline{vc})}{\partial r}=0 \tag{37}$$

or,

$$\frac{\partial(r\rho VX)}{\partial r}+\frac{\partial(r\rho UX)}{\partial z}+\frac{\partial(r\rho\overline{vx})}{\partial r}=0 \tag{38}$$

where x is the corresponding turbulent fluctuation of the mole fraction concentration.

3.2 Entrainment coefficient

The turbulent entrainment of the turbulent buoyant jet is defined as; the ambient fluid is mixed across the jet edge and becomes incorporated into the body of the jet. This process has the effect of increasing the jet volume flux and increasing/decreasing the jet density. The turbulent entrainment is usually parameterized by relating the inflow velocity to the mean flow in the jet body; instead of using details depend on transfers of mass and momentum at small scales which are very difficult to compute. (Houf & Schefer, 2008) have reported from that the local rate of entrainment increases as the jet leaves the momentum-dominated region and enters a region where the effects of buoyancy become more pronounced. The entrainment coefficient is given by,

$$\alpha = \frac{E}{2\pi U_{cl} b} \tag{39}$$

According to this expression, the value of α increases gradually as the height increases from jet region to plume region passing through the intermediate region. However, it has been experimentally observed that the value of the entrainment constant α approaches the limiting value of $\alpha = 0.82$ for pure plume. Therefore the expression, Eq. (39), is used until reach the value 0.082 then it will be enforced to be equal to this value from this point and after. The local rate of entrainment may be given in the form,

$$E = E_{mom} + E_{buoy} \tag{40}$$

$$E_{mom} = 0.282 \left(\frac{\pi d^2}{4} \frac{\rho_0 U_0^2}{\rho_\infty} \right)^{\frac{1}{2}} \tag{41}$$

Here, E_{mom} is the component of entrainment due to jet momentum (Houf & Schefer, 2008; Ricou & Spalding, 1961), while, E_{buoy} is the component of entrainment due to buoyancy (Hirst, 1971).

$$E_{buoy} = \frac{2\pi U_{cl} b a_2}{Fr_1} \tag{42}$$

where Fr_1 is the local Froude number which is defined as,

$$Fr_1 = \frac{U_{cl}^2 \rho_0}{gd(\rho_\infty - \rho_{cl})} \tag{43}$$

The local Froude number Fr_1 is high in the momentum-dominate region therefore E_{buoy} term can be neglected compared to E_{mom} in the total entrainment expression, Eq. (40). However, as the buoyancy effect become more important, the local Froude number Fr_1 decreases and E_{buoy} term begins to contribute. Determination of the constant a_2 was developed by (Houf & Schefer, 2008), based on the experiment of the vertical hydrogen jet, to be dependent on the densimetric Froude number Fr.

$$a_2 = \begin{cases} 17.313 - 0.11665 Fr + 2.0771 \times 10^{-4} Fr & \forall Fr < 268 \\ 0.97 & \forall Fr \geq 268 \end{cases} \tag{44}$$

The densimetric Froude number Fr is given by,

$$Fr = \left(\frac{U_0^2\rho_0}{gd(\rho_\infty - \rho_0)}\right)^{\frac{1}{2}} \quad (45)$$

3.3 Similarity solutions

Integrating the continuity equation (28) radially, gives,

$$\frac{d}{dz}\int_0^\infty rU(r,z)\rho(r,z)\,dr = -[r\rho V(r,z)]_{r=\infty} = -\rho_\infty[rV(r,z)]_{r=\infty} \quad (46)$$

Since $U(r,z)$ is negligible for $r > b$,then integrating Eq. (28) for $b < r < \infty$ gives,

$$\int_b^\infty \frac{\partial}{\partial r}(rV(r,z)\rho(r,z)\,dr = 0 \quad (47)$$

Therefore, this implies to,

$$-\,[rV(r,z)]_{r=\infty} = bV_e \quad (48)$$

where V_e denotes the inflow velocity at the plume edge which is known as entrainment velocity. Thus we have,

$$\frac{d}{dz}\int_0^\infty rU(r,z)\rho(r,z)\,dr = bV_e\rho_\infty \quad (49)$$

This equation indicates that the increase in plume volume flux is supplied by a radial influx from the far field which in turn implies a flow across the plume boundary b. (Batchelor, 1954) concluded that a vigorous entrainment of the ambient will be gotten as the density ratio tends to unity, $\rho_{cl}/\rho_\infty \to 1$, while as the density ratio tends to zero, $\rho_{cl}/\rho_\infty \to 0$, the entrainment falls to zero and as the density ratio varies between the two limits there will be a smooth transition. The experiments by (Ricou & Spalding, 1961) suggests that the entrainment assumption for the entrainment coefficient α, and arbitrary density ratio should be given in the form,

$$V_e = \alpha\left(\frac{\rho_{cl}}{\rho_\infty}\right)^{\frac{1}{2}} U_{cl}, \quad V_0(\rho_{cl}/\rho_\infty \to 1) = \alpha U_{cl}, \quad V_0(\rho_{cl}/\rho_\infty \to 0) = 0 \quad (50)$$

Also, (Morton, 1965) assumed that the rate of entrainment into a strongly buoyant plume is a function of both density ratio ρ_{cl}/ρ_∞ and depends on Reynolds stresses which have local magnitude proportional to $\rho_{cl}U_{cl}^2$ and hence on dimensional grounds, it seems reasonable to assume a local entrainment velocity of $\alpha(\rho_{cl}/\rho_\infty)^{1/2}U_{cl}$. A similar assumption has been made by (Thomas, 1963) and (Steward, 1970); and results were suggested theoretically by (Townsend, 1966). Therefore, the Eq. (49) can be written in the form,

$$\frac{d}{dz}\int_0^\infty rU(r,z)\rho(r,z)\,dr = b\rho_\infty\alpha\left(\frac{\rho_{cl}(z)}{\rho_\infty}\right)^{1/2} U_{cl}(z) \quad (51)$$

Finally, the mass flux equation becomes,

$$\frac{d}{dz}\int_0^\infty 2\pi rU(r,z)\rho(r,z)\,dr = 2\pi b\rho_\infty\alpha\left(\frac{\rho_{cl}(z)}{\rho_\infty}\right)^{1/2} U_{cl}(z) \quad (52)$$

Now, let us integrate Eq. (29) with respect to r from $r = 0$ to $r = \infty$, with noting that $|r\rho UV|_0^\infty = 0$ and $|r\rho\overline{uv}|_0^\infty = 0$, we get,

$$\frac{d}{dz}\int_0^\infty rU^2(r,z)\rho(r,z)\,dr = \int_0^\infty rg\left(\rho_\infty - \rho(r,z)\right)U_{cl}(r,z)\,dr \tag{53}$$

or,

$$\frac{d}{dz}\int_0^\infty 2\pi rU^2(r,z)\rho(r,z)\,dr = \int_0^\infty 2\pi rg\left(\rho_\infty - \rho(r,z)\right)U_{cl}(r,z)\,dr \tag{54}$$

Similarly, integrating Eq. (30) with respect to r from $r=0$ to $r=\infty$, with noting that $|r\rho CV|_0^\infty = 0$ and $|r\rho\overline{uc}|_0^\infty = 0$, we get,

$$\frac{d}{dz}\int_0^\infty rU(r,z)\rho(r,z)C(r,z)\,dr = 0 \tag{55}$$

or,

$$\frac{d}{dz}\int_0^\infty 2\pi rU(r,z)\rho(r,z)C(r,z)\,dr = 0 \tag{56}$$

This equation may be equivalent to the buoyancy flux equation which can be given the following form (Crapper & Baines, 1977),

$$\frac{d}{dz}\int_0^\infty 2\pi rU(r,z)\left(\rho_\infty - \rho(r,z)\right)dr = 0 \tag{57}$$

Substituting Eqs. (31)- (33) into Eq. (57) it reduces to,

$$\frac{d}{dz}\left(b^2U_{cl}(z)\left(1-\frac{\rho_{cl}(z)}{\rho_\infty}\right)\right) = 0 \tag{58}$$

Now, let us define the buoyancy flux F as,

$$F = Kb^2(z)U_{cl}(z)g\left(1-\frac{\rho_{cl}(z)}{\rho_\infty}\right) \tag{59}$$

where $K = \pi/(1+\lambda^2)$
Since the buoyancy flux is conserved, we have

$$F = F_0 = Kd^2U_{0}g\left(1-\frac{\rho_0}{\rho_\infty}\right) = Kd^2U_{0}g\frac{\Delta\rho_0}{\rho_\infty} \tag{60}$$

Substituting Eqs. (32), (33), (58) and (60) into Eqs. (52), (54) and (56) gives,

$$\frac{d}{dz}\left(b^2(z)U_{cl}(z)\right) = 2\alpha bU_{cl}\left(1-\frac{F_0}{Kgb^2(z)U_{cl}(z)}\right)^{1/2} \tag{61}$$

$$\frac{d}{dz}\left(b^2(z)U_{cl}^2(z)\right) = \frac{2F_0}{\lambda^2KU_{cl}(z)} + \frac{2F_0}{(2+\lambda^2)gK}\frac{dU_{cl}}{dz} \tag{62}$$

Using the following non-dimensional transformations (Crapper & Baines, 1977), to eliminate F_0 from Eqs. (55) and (56),

$$\hat{b} = b\left(\frac{g^3}{F_0^2}\right)^{1/5}, \hat{u} = \frac{U_{cl}}{(gF_0)^{1/5}}, \hat{z} = z\left(\frac{g^3}{F_0^2}\right)^{1/5}, N = \hat{b}^2\hat{u}, M = \hat{b}^2\hat{u}^2 \tag{63}$$

where $\hat{b}$, $\hat{u}$, $\hat{z}$ are known as top-hat variable (Thomas, 1963; Rooney & Linden, 1996) and are defined respectively as dimensionless radius, centerline velocity and vertical coordinate of the plume. N and M are the dimensionless mass and momentum respectively. Using the transformations (63), Eqs. (61) and (62) can be given in the form,

$$\frac{dN}{d\hat{z}} = 2\alpha M^{1/2}\left(1 - \frac{1}{KN}\right)^{1/2} \tag{64}$$

$$\frac{dM}{d\hat{z}} = \left(\frac{1}{KN - A}\right)\left(\frac{2N}{\lambda^2 M} - 2\alpha A \frac{M^{3/2}}{N}\left(1 - \frac{1}{KN}\right)^{1/2}\right) \tag{65}$$

where $A = 2/(2 + \lambda^2)$.
The corresponding initial conditions are,

$$N_0 = \frac{1}{(K(\rho_\infty - \rho_0)/\rho_\infty)}, \quad M_0 = \frac{(U_0^2/gb_0)^{2/5}}{(K(\rho_\infty - \rho_0)/\rho_\infty)^{6/5}} \tag{66}$$

It is common in literature to use the dimensionless height z/d to express the jet traveling distance from the source; so it will be more convenience to do the same in the current study. Using formulas (63), we can obtain the following relation between $\hat{z}$ and z,

$$\frac{z}{d} = \frac{\hat{z}}{\hat{b}_0} = \frac{\hat{z}\sqrt{M_0}}{N_0} \tag{67}$$

Also, the dimensionless centerline velocity and the jet/plume width can be related to their top-hat variables by the following relations,

$$\frac{b}{d} = \frac{\hat{b}}{\hat{b}_0} = \frac{N}{N_0}\sqrt{\frac{M_0}{M}} \tag{68}$$

$$\frac{U_{cl}}{U_0} = \frac{\hat{u}}{\hat{u}_0} = \frac{M}{M_0}\frac{N_0}{N} \tag{69}$$

3.4 Centerline quantities

In order to determine of the centerline quantities, the above non-linear first order ordinary differential system (64)- (65) with the associated initial conditions (66) are integrated numerically. With the aid of the above definition of the buoyancy flux F as well as the relation between density and mass fraction the centerline velocity and jet width and therefore centerline density and centerline concentration can be determined. The solutions are obtained for hydrogen-air buoyant jet, for various values of Froude number $Fr = 99$ ($U_0 = 49.7$), $Fr = 152$ ($Fr = 76$) and $Fr = 268$ ($U_0 = 134$); as used for experiments by (Houf & Schefer, 2008; Schefer et al., 2008a), at $d = 0.00191$, $\lambda = 0.83$, and $\Delta\rho_0/\rho_\infty = 0.93$. The entrainment coefficient α is considered as a variable given by Eq. (41). The centerline hydrogen mass fraction C_{cl} for H_2-air jet at $Fr = 99$, $Fr = 152$ and $Fr = 268$ are plotted in Fig.7 and compared with the theoretical asymptotic limits for momentum-dominated jets ($1/C_{cl} \propto z$) and the buoyancy-dominated plume limit ($1/C_{cl} \propto z^{5/3}$).

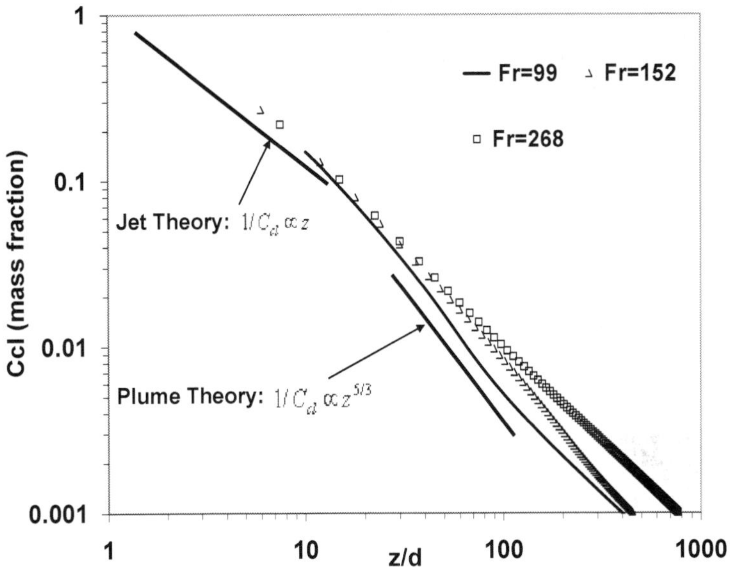

Fig. 7. Comparison of the current model and jet theory of the centerline mass fraction (log-log plot) for $H_2 - Air$ jet, with different values of Froude number.

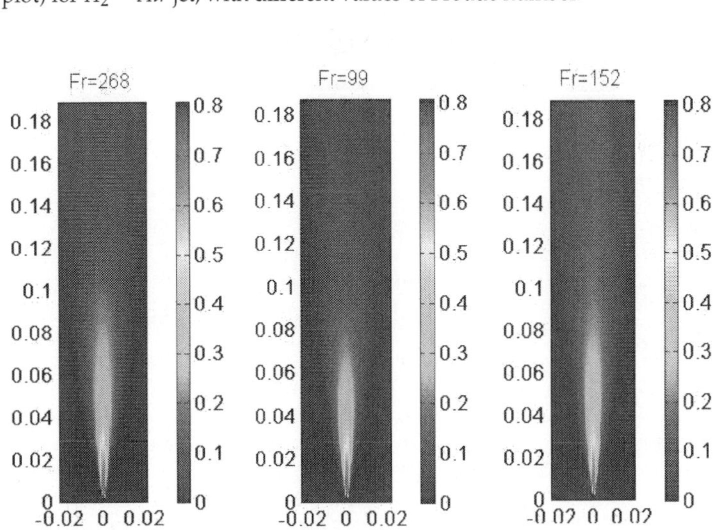

Fig. 8. Contours of the mean mole fraction for $H_2 - Air$ jet with different values of Froude number, with different values of Froude number.

3.5 Mean quantities

Using the mean centerline quantities obtained above with the Gaussian profiles assumptions of axial velocity, density deficiency and mass fraction, Eqs. (31)- (33), we can create a map of the jet concentration and velocity. Using the relation between the mass fraction, mole fraction and density, contours of mole fraction concentration are plotted in Fig.8. Fig. 9 shows a comparison between the present calculations and experiments results (Schefer et al., 2008b) for radial profiles of the mean mass fraction concentration at several axial locations for momentum-dominated ($Fr = 268$) $H_2 - Air$ jet. The maximum centerline mass fraction at each axial location occurs and decreases outward toward the surrounding air. Also, the same figure indicates that the centerline concentration decreases in the downstream direction due to mixing with the ambient air.

Variations of the normalized mean density across the He-Air jet for axial locations from $z/d =$ 50 to 120 and compared with the experimental results by (Panchapakesan & Lu, 1993) at $d =$ 0.00612, $\lambda = 0.84$, and $\Delta\rho_0/\rho_\infty = 0.86$ are shown in Fig.10. The maximum deviation of the present model from the experimental results in (Panchapakesan & Lu, 1993) is around 4% for the normalized density. It is worth mentioning that for $He - Air$ jet the entrainment coefficient is taken as a constant while it is considered as a variable as explained above for calculations of $H_2 - Air$ jet. The mean cross-stream velocity can be obtained by solving the continuity equation for the time-averaged velocities by substituting from (33) and (32) into (28). Fig.12 shows the normalized cross-stream velocity profiles V/U_{cl} plotted against the normalized coordinate r/cz for $H_2 - Air$ jet with various values of Froude number $Fr = 268$, 152 and 99. It can be seen from this figure that the cross-stream velocity vanishes at the centerline and becomes outward in the neighborhood of the centerline. A maximum is then reached, then declines back to zero and then becomes inward ($V/U_{cl} < 0$), reaches a minimum and reaches an asymptote of zero as $r/cz \to \infty$. The cross-stream velocity asymptotes to zero much more slowly than the streamwise velocity does, thus, the central region is dominated by the axial component of velocity, while the cross-stream flow predominates far away from it. Also,

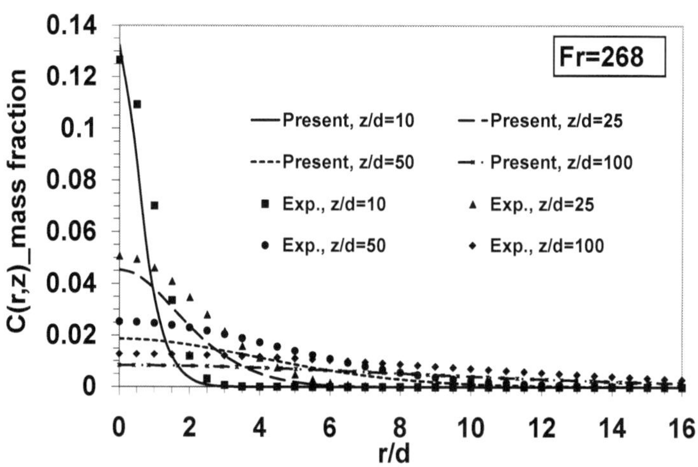

Fig. 9. Radial profiles of the mean mass fraction for momentum-dominated (Fr=268) $H_2 - Air$ jet at several axial locations, compared with experimental results (Schefer et al., 2008b).

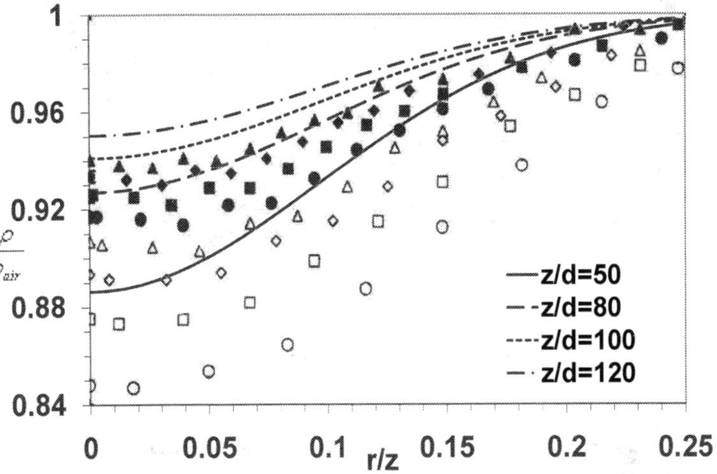

Fig. 10. Radial profiles of the normalized mean density deficiency for $He - Air$ jet at several axial locations. Dots refer to experimental results (Panchapakesan & Lu, 1993) at axial locations from z/d=50 to 120.

Fig.11 indicates that as Froude number decreases and the distance from source increases the spreading rate increases and therefore the outward velocity increases.

3.6 Selected turbulent quantities

In this section, we determine some of the most important turbulent quantities such as Reynolds stress, velocity-concentration correlation, etc. By substituting the time-averaged

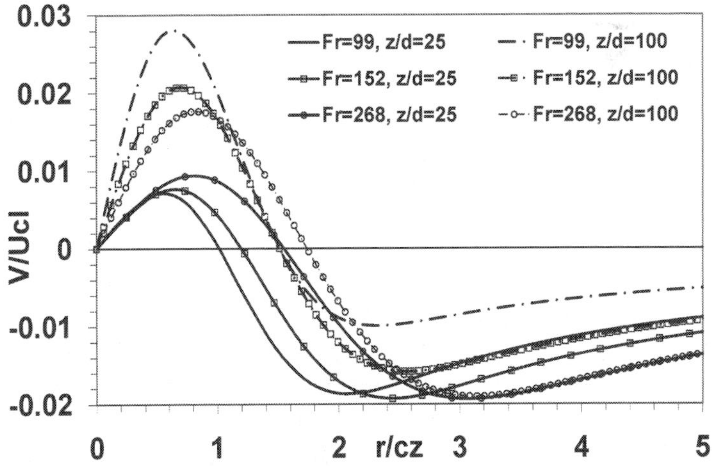

Fig. 11. Radial profiles of the mean cross-stream velocity of $H_2 - Air$ jet with different values of Froude number at several axial locations.

profiles of U, ρ and V obtained above into the momentum equation (29), we obtain the time-averaged profile for the Reynolds stress $\overline{uv}$. The radial profiles of normalized Reynolds stress are plotted against the normalized coordinate r/cz at several axial locations for $H_2 - Air$ jet with various values of Froude number $Fr = 268$, 152 and 99, as shown in Fig.12. It is observed from this figure that the Reynolds stress $\overline{uv}/U_{cl}^2$ increases as the Froude number decreases due to increasing of jet spreading rate.

Again, the time-averaged profile for the velocity-concentration correlation (radial flux) $\overline{vc}$ is determined by inserting the time-averaged profiles of U, ρ and C into the concentration equation (30). The radial profiles of normalized velocity-concentration correlation are plotted in Fig.13 against the normalized coordinate r/cz at several axial locations for $H_2 - Air$ jet with various values of Froude number $Fr = 268$, 152 and 99. This figure indicates that the velocity-concentration correlation increases as the Froude number decreases. The turbulent eddy viscosity ν_t is related to Reynolds stress and the mean axial velocity by the relation,

$$\overline{uv} = -\nu_t \frac{\partial U}{\partial r} \tag{70}$$

while, the turbulent eddy diffusivity D_t are related to the radial flux and the mean concentration by the relation,

$$\overline{vc} = -D_t \frac{\partial C}{\partial r} \tag{71}$$

Using current calculation we can determine the turbulent eddy viscosity ν_t and turbulent eddy diffusivity D_t. Also, both the normalized turbulent eddy viscosity $\nu_t/U_{cl}b$, and the normalized turbulent eddy diffusivity $D_t/U_{cl}b$, are plotted in Fig.14 against the normalized coordinate r/cz at several axial locations for momentum-dominated $H_2 - Air$ jet ($Fr = 268$). Fig.14 illustrates that both $\nu_t/U_{cl}b$ and $D_t/U_{cl}b$ have maximum values at the centerline of the jet and decay to zero for large r. If the effects of molecular transport are neglected, the turbulent Schmidt number in the free turbulent flow may be defined as the quantitative difference between the turbulent transport rate of material and momentum. In other words, the turbulent Schmidt number, $Sc_t = \nu_t D_t$, is a ratio of the eddy viscosity of the turbulent flow

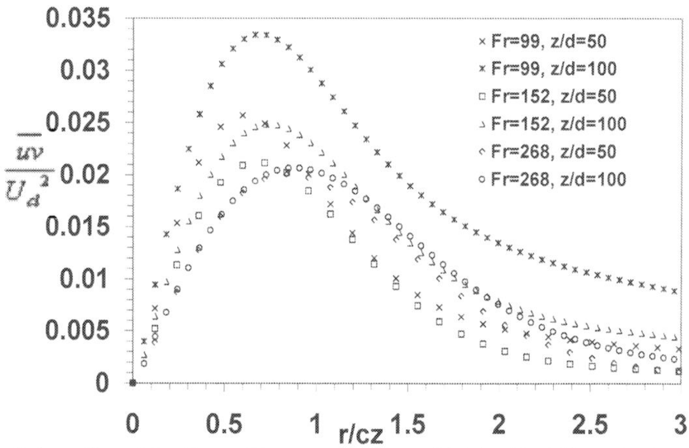

Fig. 12. Radial profiles of the Reynolds stress for $H_2 - Air$ jet with different values of Froude number at several axial locations.

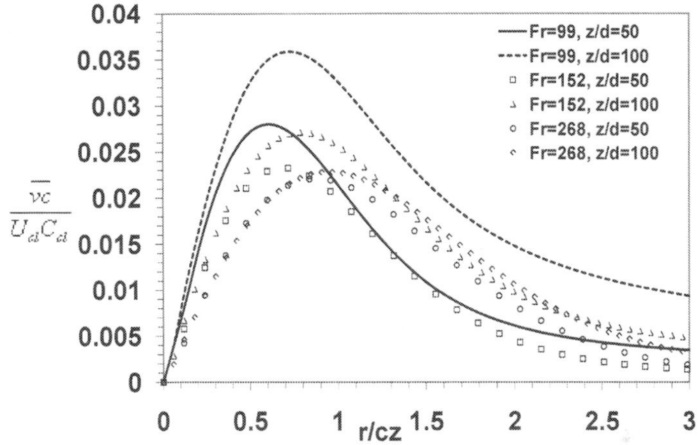

Fig. 13. Radial profiles of the velocity-concentration correlation for $H_2 - Air$ jet with different values of Froude number at several axial locations.

ν_t to the eddy material diffusivity D_t. It is very well known from literature that the spreading rate of material in free jets is greater than that of momentum; thus, the turbulent Schmidt number is less than unity and has a significant effect on the predictions of species spreading rate in jet flows. The turbulent eddy viscosity ν_t and the turbulent eddy diffusivity D_t, give the turbulent Schmidt number as,

$$Sc_t = \frac{\nu_t}{D_t} \tag{72}$$

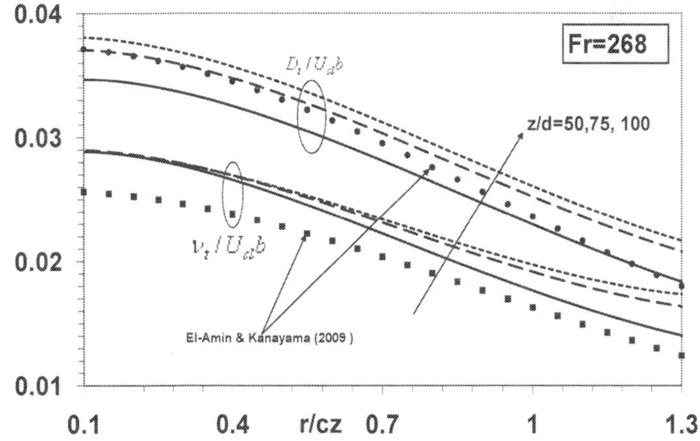

Fig. 14. Radial profiles of turbulent eddy viscosity and turbulent eddy diffusivity for $H_2 - Air$ with $Fr = 268$ (momentum-dominated) at several axial locations, compared with (El-Amin, 2009).

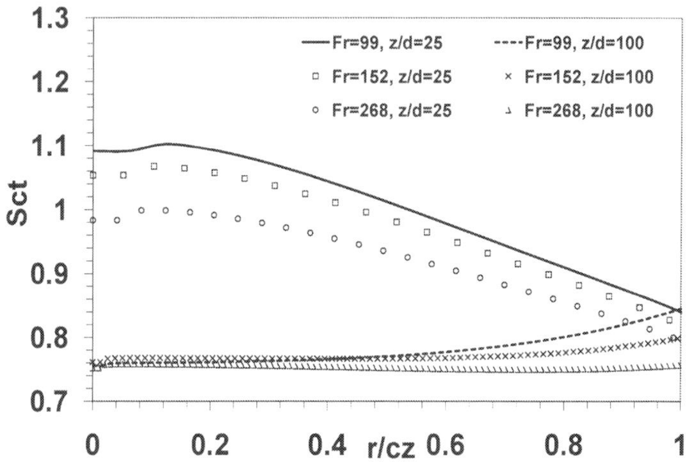

Fig. 15. Radial profiles of turbulent Schmidt number for $H_2 - Air$ jet with different values of Froude number at several axial locations.

Turbulent Schmidt number is plotted in Fig.15 against the normalized coordinate r/cz at several axial locations for $H_2 - Air$ jet with various values of Froude number $Fr = 268$, 152 and 99. In the momentum-dominated region the turbulent Schmidt number seems to be constant and the opposite is true when the buoyancy effects take place for small Froude number or far distance from the jet source. Irrespective of the turbulence structure, the constant value of the turbulent Schmidt number across the whole flow field implies that the momentum process is similar to the material-transport process. The opposite is true for the variable values of the turbulent Schmidt number indicate that the momentum process is not similar to the material-transport process. Now, let us define the normalized streamwise momentum flux density across an isothermal jet as, $G = (U/U_{cl})^2$, and the normalized jet-feed material density as, $W = UC/U_{cl}C_{cl}$. Also, by considering the analytical correlation of jet-feed concentration and streamwise velocity derived by (El-Amin & Kanayama, 2009b) as, $W = G^{(1 + Sc_t)/2}$ (see Fig.6).

4. References

Agrawal, A. & Prasad, A. K. (2002). Properties of vortices in the self-similar turbulent jet, *Exp. Fluids* 33: 565–577.

Agrawal, A. & Prasad, A. K. (2003). Integral solution for the mean flow profiles of turbulent jets, plumes, and wakes, *ASME J Fluids Engineering* 125: 813–822.

Batchelor, G. K. (1954). Heat convection and buoyancy effects in fluids, *Q. J. R. Met. Soc.* 80: 339–358.

Becker, H. A., Hottel, H. C. & Williams, G. C. (1967). The nozzle-fluid concentration of the round turbulent free jet, *J. Fluid Mech* 30: 285–303.

Bhat, G. S. & Krothapalli, A. (2000). Simulation of a round jet and a plume in a regional atmospheric model, *Monthly Weather Review* 128: 4108–4117.

Bhat, G. S. & Narasimha, R. A. (1996). A volumetrically heated jet: Large-eddy structure and entrainment characteristics, *J. Fluid Mechanics* 325: 303–330.

Carlotti, P. & Hunt, G. R. (2005). Analytical solutions for turbulent non-boussinesq plumes, *J. Fluid Mechanics* 538: 343–359.

Chen, C. J. & Rodi, W. (1980). *Vertical turbulent buoyant jets – a review of experimental data*, Pergamon Press, Oxford, UK.

Crapper, P. F. & Baines, W. D. (1977). Non-boussinesq forced plumes, *Atmospheric Environment* 11: 415–420.

Dai, Z., Tseng, L. K. & Faeth, G. M. (1995). Velocity statistics of round, fully developed, buoyant turbulent plumes, *Trans. ASME, J. Heat Transfer* 117: 138–145.

El-Amin, M. F. (2009). Non-boussinesq turbulent buoyant jet resulting from hydrogen leakage in air, *Int. J. Hydrogen Energy* 34: 7873–7882.

El-Amin, M. F., Inoue, M. & H., K. (2008). Boundary layer theory approach to the concentration layer adjacent to a ceiling wall of a hydrogen leakage: far region, *Int. J. Hydrogen Energy* 33: 7642–7647.

El-Amin, M. F. & Kanayama, H. (2008). Boundary layer theory approach to the concentration layer adjacent to a ceiling wall at impinging region of a hydrogen leakage, *Int. J. Hydrogen Energy* 33: 6393–6400.

El-Amin, M. F. & Kanayama, H. (2009a). Boundary layer theory approach to the concentration layer adjacent to the ceiling wall of a hydrogen leakage: axisymmetric impinging and far regions, *Int. J. Hydrogen Energy* 34: 1620–1626.

El-Amin, M. F. & Kanayama, H. (2009b). Integral solutions for selected turbulent quantities of small-scale hydrogen leakage: a non-buoyant jet or momentum-dominated buoyant jet regime, *Int. J. Hydrogen Energy* 34: 1607–1612.

El-Amin, M. F. & Kanayama, H. (2009c). Similarity consideration of the buoyant jet resulting from hydrogen leakage, *Int. J. Hydrogen Energy* 34: 5803–5809.

El-Amin, M. F., Sun, S., Heidemann, W. & Mller-Steinhagen, H. (2010). Analysis of a turbulent buoyant confined jet using realizable k-¡ep¿ model, *Heat and Mass Transfer* To appear: 1–20.

Fischer, H. B., List, E. J., Koh, R. C. Y., Imberger, J. & Brooks, N. H. (1979). *Mixing in Inland and Coastal Waters*, Academic Press.

Gebhart, B., Jaluria, Y., Mahajan, R. L. & Sammakia, B. (1988). *Buoyancy-induced flows and transport*, Hemisphere. Washington, DC.

He, G., Yanhu, G. & A., H. (1999). The effect of schmidt number on turbulent scalar mixing in a jet-in-cross flow, *Int. J. Heat Mass Transfer* 42: 3727–3738.

Hinze, J. O. (1975). *Turbulence, 2nd ed.*, McGraw-Hill.

Hirst, E. A. (1971). Analysis of buoyant jets within the zone of flow establishment, *Technical report*, Oak Ridge National Laboratory, Report ORNL-TM-3470.

Houf, W. G. & Schefer, R. W. (2007). Predicting radiative heat fluxes and flammability envelopes from unintended releases of hydrogen, *Int. J. Hydrogen Energy* 32: 136–151.

Houf, W. G. & Schefer, R. W. (2008). Analytical and experimental investigation of small-scale unintended releases of hydrogen, *Int. J. Hydrogen Energy* 33: 1435–1444.

Hussein, J. H., Capp, S. P. & George, W. K. (1994). Velocity measurements in a high-reynolds-number, momentum-conserving, axisymmetric, turbulent jet, *J. Fluid Mechanics* 258: 31–75.

Lessen, M. (1978). On the power laws for turbulent jets, wakes and shearing layers and their relationship to the principal of marginal instability, *J. Fluid Mech* 88: 535–540.

List, E. J. (1982). Turbulent jets and plumes, *Annual Review of Fluid Mechanics* 14: 189–212.

Morton, B. R. (1965). Modelling fire plumes, *10th Int. Symposium on Combustion, 844-859.*

Panchapakesan, N. R. & Lu (1993). Turbulence measurements in axisymmetric jets of air and helium. part 2. helium jet, *J. Fluid Mechanics* 246: 225–247.

Panchapakesan, N. R. & Lumley, J. L. (1993). Turbulence measurements in axisymmetric jets of air and helium. part 1. air jet, *J. Fluid Mechanics* 246: 197–223.

Papantoniou, D. & List, E. J. (1989). Large-scale structure in the far field of buoyant jets, *J. Fluid Mechanics* 209: 151–190.

Rajaratnam, N. (1976). *Turbulent Jets*, Elsevier Science, Place of publication.

Ricou, F. & Spalding, D. B. (1961). Measurements of entrainment by axisymmetrical turbulent jets, *J. Fluid Mechanics* 8: 21–32.

Rooney, G. G. & Linden, P. F. (1996). Similarity considerations for non-boussinesq plumes in an unstratified environment, *J. Fluid Mechanics* 318: 237–250.

Schefer, R. W., Houf, W. G., Bourne, B. & Colton, J. (2006). Spatial and radiative properties of an open-flame hydrogen plume, *Int. J. Hydrogen Energy* 31: 1332–1340.

Schefer, R. W., Houf, W. G. & Williams, T. C. (2008a). Investigation of small-scale unintended releases of hydrogen: buoyancy effects, *Int. J. Hydrogen Energy* 33: 4702–4712.

Schefer, R. W., Houf, W. G. & Williams, T. C. (2008b). Investigation of small-scale unintended releases of hydrogen: momentum-dominated regime, *Int. J. Hydrogen Energy* 33: 6373–6384.

Shabbir, A. & Georg, W. K. (1994). Experiment on a round turbulent plume, *J. Fluid Mechanics* 275: 1–32.

Spiegel, E. A. & Veronis, G. (1960). On the boussinesq approximation for a compressible fluid, *Astrophys. J.* 131: 442–447.

Steward, F. R. (1970). Prediction of the height of turbulent diffusion buoyant flames, *Combust. Sci. Technol.* 2: 203–212.

Swain, M. R., Filoso, P. A. & Swain, M. N. (2007). An experimental investigation into the ignition of leaking hydrogen, *Int. J. Hydrogen Energy* 32: 287–295.

Takeno, K., Okabayashi, K., Kouchi, A., Nonaka, T., Hashiguchi, K. & Chitose, K. (2007). Dispersion and explosion field tests for 40 mpa pressurized hydrogen, *Int. J. Hydrogen Energy* 32: 2144–2153.

Tennekes, H. & Lumley, J. L. (1972). *A First Course in Turbulence*, MIT Press, Cambridge, MA.

Thomas, P. H. (1963). The size of flames from natural fires, *9th Symposium on Combustion, 844-859.*

Townsend, A. A. (1966). The mechanism of entrainment in free turbulent flows, *J. Fluid Mechanics* 26: 689–715.

Turner, J. S. (1973). *Buoyancy effects in fluids*, Cambridge Univ. Press, London.

Turner, J. S. (1986). Turbulent entrainment: The development of the entrainment assumption, and its application to geophysical flows, *J. Fluid Mech* 173: 431–471.

Venkatakrishnan, L., Bhat, G. S. & Narasimha, R. (1999). Experiments on plume with off-source heating: Implications for cloud fluid dynamics, *J. Geophys. Res.* 104: 14271–14281.

Woods, A. W. (1997). A note on non-boussinesq plumes in an incompressible stratified environment, *J. Fluid Mechanics* 345: 347–356.

Yimer, I., Campbell, I. & Jiang, L. Y. (2002). Estimation of the turbulent schmidt number from experimental profiles of axial velocity and concentration for high-reynolds-number jet flows, *Canadian Aeronautics Space Journal* 48: 195–200.

3

Controlled Mixing and Transport in Comb-Like and Random Jet Array Stirring Systems

S. Delbos[1], E. Chassaing[1], P. P. Grand[2],
V. Weitbrecht[3] and T. Bleninger[4]

[1]*Institute for Research and Development of Photovoltaic Energy (IRDEP), UMR 7174 EDF - CNRS - Chimie-ParisTech*
[2] *Nexcis*
[3] *VA f. Wasserbau/Hydrologie/Glaziologie ETH Zürich*
[4]*Institute for Hydromechanics, Karlsruhe Institute of Technology (KIT)*
[1,2]*France*
[3]*Switzerland*
[4]*Germany*

1. Introduction

Electrodeposition of alloys from low concentrated solutions or solutions whose elements have very different redox potentials is often strongly dependent on mass-transfer rate of one of the electroactive species. For such alloy systems, electrodeposition is carried out potentiostatically because the deposition potential has to be controlled in order to monitor good alloy composition. The applied potential is such that one of the species is deposited at its diffusion limiting current, and the other is deposited under activation-controlled regime. The diffusion layer's thickness controls the limiting current, and its uniformity controls the uniformity of the deposited film (Bard & Faulkner, 2001).
In electronical engineering, the paddle-cell system has been used for a long time to enhance mass-transfer rate and homogeneity (Powers & Romankiw, 1972). It can be used for different electrochemical systems (Datta & Landolt, 2000). The paddle-cell system, where a paddle-like object is moved back and forth continuously through the solution, has been used for a long time in electronical engineering to enhance mass transfer rate and homogeneity. It can be used for different electrochemical systems such as Pb-Sn (Datta & Landolt, 2000) or Fe-Ni (Powers & Romankiw, 1972). The paddle system is widely used in industry (McHugh et al., 2005; Keigler et al., 2005), but few studies deal with improving the process conditions in relationship with mean and local mass transfer.
A comb-like stirring system has been developed at IRDEP. It is similar to the paddle system and the shear plate system (Wu et al., 2005). The comb can be considered as an improvement of the paddle system by having several paddles one after another in a comb like arrangement. The comb is moved back and forth continuously through the solution. It is used for enhancing the homogeneity of electrodeposited Cu-In-Se layers that are the main component of CIS thin-film photovoltaic devices (Lincot et al., 2004).
Another system, the jet array, can also be used to plate Ni-Fe (permalloy) static wafers (Tzanavaras & Cohen, 1995). Hereby no moving parts pass through the solution, but the

solution itself is pumped through small openings causing jet mixing in the system. An improvement of this system was investigated.

The mixing and transport processes are described for each stirring system. Firstly, the paddle system characteristics are described based on a literature review. Secondly, the comb system will be introduced by describing the performed experiments and results, as well as proposing characteristic non-dimensional numbers for design purposes. Thirdly, the randomly firing jet array system will by analysed and parametrized based on the experimental results obtained from (Delbos et al., 2009b;a), which also describe the experimental details of their experiments, but which will not be repeated here.

2. Paddle-cell system

The paddle-cell, which is quite similar to the comb-like system, is well documented and can provide insight for comb-like system development. The paddle-cell was first developed by IBM (Powers & Romankiw, 1972) to provide laminar agitation and to improve mass transfer during copper electrodeposition. A paddle goes back and forth (reciprocates) above a horizontal cathode (fig. 1). Experimental data (Schwartz et al., 1987; Rice et al., 1988) and numerical simulations (Wilson & McHugh, 2005; Mandin et al., 2007) show that geometrical parameters of the paddle, size and aspect ratio of the cell and paddle shape have a strong influence on mass-transfer phenomena within the electrochemical cell.

In literature a relationship has been proposed between the diffusion layer thickness and paddle geometrical parameters. The Sherwood number, which can be defined as the ratio of the advection lengthscale and the diffusion lengthscale, was shown to be equal to:

$$Sh = \frac{\text{Advection lengthscale}}{\text{Diffusion lengthscale}} = \frac{g+h}{\delta} = \alpha Re^m Sc^{1/3} \tag{1}$$

where m and α are coefficients that depend on the geometry of the system, $Re = V \cdot (h+g)/\nu$, $Sh = (h+g)/\delta$, $Sc = \nu/D$. In the case of the paddle-cell, the advection lenghtscale is $g+h$, because the size of the eddies is determined by the smallest dimension available to them. The two other dimensions (heigth of the fluid and dimension of the cell in the direction parallel to the comb movement) are clearly bigger.

When Sherwood, Reynolds and Schmidt numbers can be linked by such a relationship, the diffusion layer thickness δ, through the Sherwood number, is totally controlled by the geometrical parameters of the tank. The value of m depends on turbulence intensity : $m \sim 0.5$ denotes laminar flow, $m \sim 0.75$ denotes turbulent flow (Bard & Faulkner, 2001; Cussler, 1997;

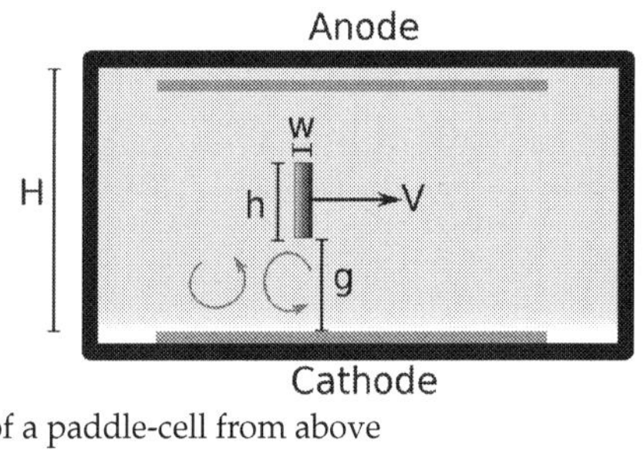

Fig. 1. Schematic view of a paddle-cell from above

Tennekes & Lumley, 1972; Brumley & Jirka, 1988). This type of analysis was also used for other systems, for example the shear flow system (Wu et al., 2005).
Numerical simulations were performed by (Wilson & McHugh, 2005) proposing a modification of equation 1 by including further characteristical numbers to describe the physical meaning of the previous used coefficients. In Wilson and McHugh's paper, three adimensional numbers have been introduced : the blockage ratio h/H, the proximity ratio h/g, the Strouhal number $f_v h/u_B$, where f_v is the vortex shedding frequency and u_B is the eddy velocity:

$$Sh = 0.566Re^{0.583}\left(\frac{h}{g}\right)^{0.151}\left(\frac{h}{H}\right)^{0.168}St^{0.283}Sc^{1/3} \tag{2}$$

The insight provided by these studies was useful for chosing relevant geometric parameters and designing characteristic non-dimensional parameters.

3. Comb-like system

3.1 Electrochemical reactor

A comb-like system is used for stirring. It is a piece of chemically inert material such as PP, PTFE or PVDF that has the shape of a comb. It is located in front of the cathode and it reciprocates in the direction parallel to the cathode, thus creating turbulence near the cathode (fig. 2). It has the advantages of simplicity and easy maintenance, but the drawback of a built-in anisotropy: the teeths create vertical patterns on the electrodeposited layer.
A modular tank was built to analyze the influence of the hydrodynamical parameters on a Cu-Ni electrodeposition process. In this tank local flow velocities have been measured in high temporal resolution using Laser Doppler Velocimetry (LDV). In addition, electrodeposition of Cu-Ni alloy films has been carried out on 5 x 5 cm^2 glass–Mo substrates.
The experimental tank allowed the study of a wide range of parameter variations (table 1). Results from studies on the paddle-cell (Schwartz et al., 1987; Rice et al., 1988; Wilson & McHugh, 2005) suggest that the most important parameters should be the distance between the comb and the cathode (g, see figure 2), the stirring frequency (f) and stroke (S, maximum amplitude of stirring), the width of the comb (h), the shape of the teeth section, and the dimensions of the cell (L and H). Prior unpublished studies also showed that the mesh of the comb (M) could have an influence on the hydrodynamical conditions.

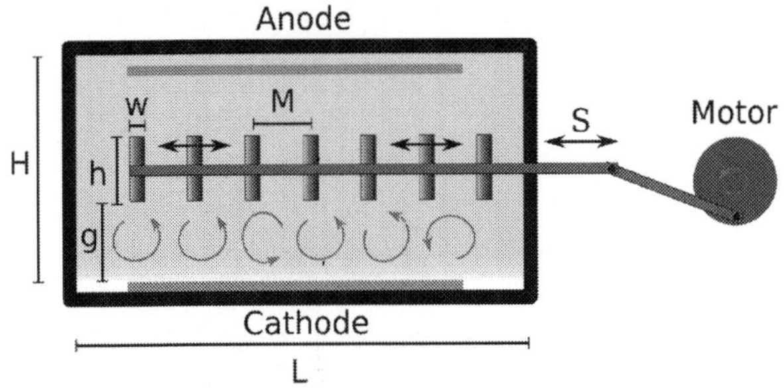

Fig. 2. Schematics of a comb-like system seen from above.

Parameter	Range
g (mm)	4 – 500
M (mm)	3 – 16
w (mm)	1 – 8
h (mm)	1 – 20
S (mm)	≤ 40
f (Hz)	≤ 7
Tooth section	rectangular circular triangular double triangle
H	10 – 500 mm
L	150 – 700 mm

Table 1. Hydrodynamical conditions of the comb-like system.

3.2 Electrodeposition

In order to investigate mass-transport phenomena, an electrochemical model system was developed in the laboratory (Ollivier et al., 2009). This system, the Cu-Ni system is an appropriate model: copper deposition is controlled by mass transfer phenomena, while nickel deposition is controlled by charge-transfer. Its deposition kinetics is a simplification of the Cu-In-Se codeposition system that is used for synthetizing solar cells in the laboratory, but easier to handle.

For the preparation of large volumes of Cu-Ni electrolyte ($\sim$ 40 L), the chemicals ($NiSO_4$ 0.165 M, $CuSO_4$ 0.025 M, Sodium Citrate 0.25 M) were first dissolved in smaller volumes (2 x 5 L), and immediately diluted into the whole water volume, to prevent the caking (mass precipitation) of the 40 L-electrolyte.

The electrodeposition takes place on a glass substrate covered with a sputtered Mo layer. The resistivity of the Mo layer is ρ_{ca} = 13 – 17 $\mu\Omega$.cm, its thickness e = 0.38 – 0.52 μm and its sheet resistance is = 0.27 – 0.44 Ω. This is the usual substrate for thin layer solar cells (Lincot et al., 2004).

The substrate is placed in an adapted basket that allows the substrate to be carried through the cleaning process. The substrate is first rinsed with high-purity water (18.2 MΩ), cleaned with soaped water and thoroughly rinsed with high-purity water. Finally it is dipped for 10 minutes into a NH_3 solution (25 %) to deoxidize the Mo (mainly to remove MoO_2). The substrate is afterwards rinsed with flowing water, and dried with argon.

The electrodeposition is performed in potentiostatic mode. The counter-electrode is a vertical 5 x 15 cm^2 Ti mesh covered with IrO_2 (Dimensionally stable Anode, DSA). A saturated Mercurous Sulfate Electrode (MSE, +0.65 V / Normal Hydrogen Electrode) is used as a reference electrode. The potentiostat is an Autolab PGSTAT30 controlled by the GPES software running under MS Windows.

A Pt wire is applied on the substrate to achieve the electrical contact, the substrate is then rinsed with flowing water, just before being dipped in the electrolyte. The substrate is then immersed and the potential of -1.60 V/MSE is applied, and the total charge is -2.8 C/cm^2 The resulting electrodeposited layers are composed of 90 - 99 % of copper and 1 - 10 % of nickel. This composition is the same as in similar experiments performed in a jet firing array system (Delbos et al., 2009a).

3.3 Chemical analysis

The thickness and chemical composition of the layers were analysed by Energy-Dispersive Spectroscopy X-Ray Fluorescence Fischerscope Xray Xan controlled by the WinFTM software running under Windows. The uncertainty of the measurement was 0.26 % for the thickness, 0.04 % for the copper content, 5.73 % for the nickel content. The local electrodeposited copper quantity $n_{Cu}(x,y)$ (in mol/cm^2) is linked to the local diffusion layer thickness $\delta(x,y)$ by this relationship:

$$n_{Cu(x,y)} = \frac{i_{Cu}(x,y)t}{\rho_{Cu} z_e F} = \frac{C_{Cu} D_{Cu} t}{\delta(x,y)} \quad (3)$$

where $i_{Cu}(x,y)$ is the local copper partial current density, t is the electrodeposition time, $z_e = 2$ is the number of electrons involved in the reaction, F is the Faraday's constant, ρ_{Cu} is the per volume ratio of Cu, C_{Cu} is the concentration of Cu^{II} species in the electrolyte, and D_{Cu} is the diffusion coefficient of Cu^{II} species.

The copper quantity is therefore locally dependent on the local diffusion layer thickness. In this set of experiments, mappings of $n_{Cu(x,y)}$ were performed: 5 rows of 20 measurements. Each row (parallel to the x direction) has a resolution of 2.5 mm, and each row is 10.75 mm from its neighbors. The standard deviation of these measurements is therefore a good indication on the spatial variations of the diffusion layer thickness.

Because of flaking on the edges of the plates, the standard deviation is calculated on the three central rows of measurements.

3.4 Laser Doppler Velocimetry

Velocity measurements have been performed using a 2-D Laser Doppler Velocimetry (LDV) system to determine the mean and turbulent flow characteristics. Such a device was already used to correlate flow velocity with electrodeposition patterns in a jet-firing plating cell (Delbos et al., 2009a;b). A 5 W Ar-Ion laser, a Dantec data acquisition apparatus (FiberFlow 60*81 BSA 55X) and the BSA Flow software running under MS Windows were used for data acquisition. ZrO_2 particles were used as seeding particles, and they were approximately 3 μm in diameter.

A backscatter probe with a focal length of 310 mm was used. The system provides a temporal resolution of the order of 10^{-2} s and the size of the measurement volume is about 0.1 mm x 0.1 mm x 1.6 mm.

The laser probe was located above the measurement volume, firing in the y-direction, and the laser had to cross the free air-water interface. In order to keep the interface horizontal, a small piece of glass is kept at the air-water interface above the measurement volume. The lack of this piece of glass, or the presence of a drop of water on the piece of glass could lead the acquisition data frequency to decrease by a factor of 100.

Each measurement series lasted 120 - 150 s, leading to approximately 2,000 - 3,000 velocity signals at each point before moving to another point of measurement. Measurements were taken on a horizontal line at a distance $z_m = 2$ mm away from the deposition electrode and 70 mm from the bottom of the tank (total fluid depth: 120 mm). With the help of a 2-D traversing system the LDV probe was positioned at the different measurement locations. The set-up is schematized on figure 3.

u and w, the velocity components on the x- and z-directions were measured. Simple treatment was applied to the data: for each point of measurement, in both measured velocity directions (u in the x-direction and w in the z-direction), the mean velocity and the Root Mean Square

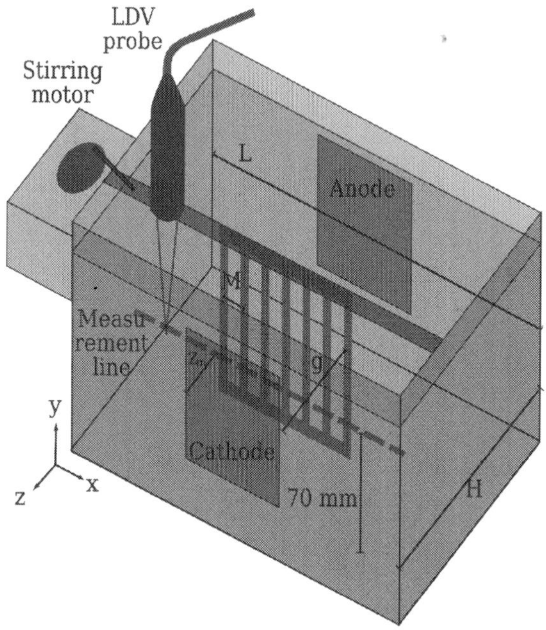

Fig. 3. Schematics of the LDV measurement set-up in the comb experiment.

(RMS) velocity were calculated, e.g. for w:

$$\overline{w} = \frac{1}{N}\sum_i w_i \text{ is the mean velocity} \quad (4)$$

$$w_{\text{RMS}} = \sqrt{\frac{1}{N}\sum_i (w_i - \overline{w})^2} \text{ is the RMS velocity} \quad (5)$$

where each w_i is one measurement in a series of N measurements. In the following, the RMS velocity is also called the *turbulent fluctuations*.
The uncertainty of the mean and RMS velocity measurements was determined by measuring the velocity at the same point 30 times and then calculating the normalized standard deviation of the measurements ($\sigma_{\text{norm}} = \sigma/\mu$, see section 3.3). The uncertainty of measurement is high for the w component ($\sigma_{\overline{w}}$ = 13.2 %), but it is more acceptable for the u component ($\sigma_{\overline{u}} < 5\%$). These values are summarized in table 2.

4. Results : control parameters of Cu-Ni electrodeposition in comb-like systems

The objective was to define two control parameters for the comb-like system: one that controls the limiting deposition current i_L on the whole cathode, the other that controls the standard deviation of electrodeposited copper σ_{Cu} on a horizontal line (x direction) on the cathode.

Measurement	$\overline{u}$	$\overline{w}$	u_{RMS}	w_{RMS}
σ	4.2 %	13.2 %	4.1 %	7.1 %

Table 2. Uncertainty of LDV measurements in the modular comb electrolyser.

S (mm)	10	14	24
σ_{Cu}	13.32	8.87	3.73

Table 3. Stroke – Cu standard deviation.

Therefore, the effect of the variation of geometrical parameters on the limiting current and the homogeneity of the electrodeposited layer was studied. From these results the two control parameters were defined.
A set of experiments, previously described in (Delbos, 2008), were performed in order to quantify the effect of each geometric parameter on copper deposition and flow parameters. Geometric parameters f, S, M, g, H, and L were tested for relevance (see fig.2). In contrary to the paddle-cell, the flow is confined between the comb and the cathode, leading to a characteristic size of flow vortices (eddy size) $u_B \sim g$. Consequently the parameters H and L do not seem to have an influence on flow velocity and electrodeposition pattern. Only f, S, M, g and teeth shape are found as relevant parameters (Delbos, 2008).

4.1 Effect of the variation of the stroke on electrodeposition and flow

In this section, the effect of the variation of the stroke of the comb is shown. In this set of experiments all parameters but the stroke remain constant ($H = 66$ mm, $L = 310$ mm, $M = 7$ mm, $h = 4$ mm, $g = 7$ mm, $f = 4$ Hz, $w = 1$ mm, the shape of the teeth was rectangular). The stroke S is varied from 10 mm to 24 mm. Fig. 4 shows the results: (a) presents the copper quantity measured by XRF along the horizontal axis of the cathode, (b) and (c) present the mean flow and the turbulent fluctuations measured at $z_m = 2$ mm from the cathode on an axis parallel to the axis were the copper quantity was measured.
The increase of the stroke at constant frequency increases the velocity of the comb, and therefore the energy input. Consequently the turbulent fluctuations increase with comb stroke (fig 4(c)), leading to increased mass transfer and increased deposited copper quantity (fig 4(a)). Increasing the stoke also increases the variations of mean flow (fig 4(b), but apparently the strong variations of the mean flow do not lead to inhomogeneous diffusion layer thickness: the copper quantity is more homogeneous for high values of S (table 3 and fig. fig 4(a)).

4.2 Non-dimensional parameters

From the same kind of experiments that are shown in section 4.1 and the literature (Wilson & McHugh, 2005; Schwartz et al., 1987; Rice et al., 1988), relevant non-dimensional parameters were chosen to characterize the comb-like system:

- non-dimensional stroke S/M
  In this system, each tooth of the comb behaves as a simple paddle, but the vortices shed by each tooth interact with the vortices shed by the neighboring tooth. The non-dimensional stroke S/M is therefore of primary importance for mass-transfer phenomena.
- Proximity ratio h/g
  For grid stirring devices, the turbulent fluctuations (RMS) decay proportionally to the inverse of the distance from the stirring device (at the cathode surface, $u_{RMS} \propto g^{-1}$) (Thompson & Turner, 1975). For the comb system, the energy input is proportionnal to the volume of fluid displaced by the comb, which is proportional to h, and its decay should be proportional to the distance between comb and cathode g. The proximity ratio (h/g) was therefore chosen as a relevant parameter for the system.
- Reynolds number
  For the paddle cell, $Re = (g + h)fS/\nu$. The same definition was chosen for the comb system.

In our experiments, $100 < Re < 4000$. Nevertheless LDV measurements showed that the flow was turbulent even for low Reynolds numbers ($u_{RMS}/\overline{u} \sim 2$, see fig. 4).

- Solidity ratio M/w
  For grid-stirring devices, a solidity ratio can be computed, it is equal to the surface of the grid divided by the surface of the holes (Thompson & Turner, 1975). Similarly the solidity ratio of a comb-like system is M/w: the solution volume displaced by the comb depends on this mesh size over tooth thickness ratio. The larger this ratio, the bigger the volume displaced by the comb, and the more energy is introduced in the system.
- The Schmidt number $Sc = \nu/D$
  In these experiments, the diffusion coefficient is $D = 3 \cdot 10^{-10}\mathrm{m}^2/\mathrm{s}$ and the kinematic viscosity is $\nu = 1.62 \cdot 10^{-6}\mathrm{m}^2/\mathrm{s}$ (Henninot, 1999), leading to $Sc = 5.4 \cdot 10^3$.

4.3 Deposition current control

In case of diffusion-controlled reactions, the current can be linked to the diffusion layer thickness through the classical relationship:

$$J_c = \frac{n\mathrm{F}DC_B}{\delta} \tag{6}$$

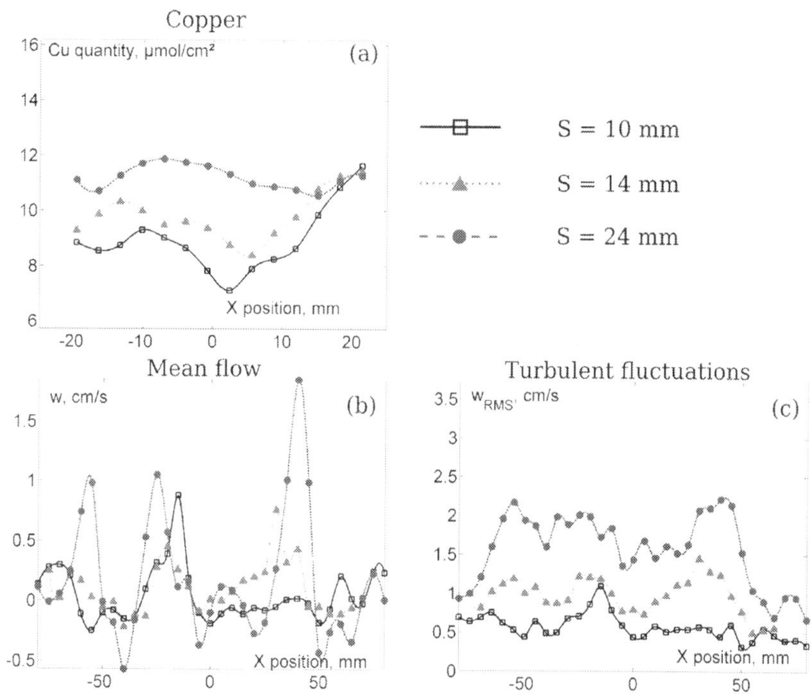

Fig. 4. Effect of the stroke on copper deposition and flow parameters

δ is itself linked to the Sherwood number $Sh = l_B/\delta$, where l_B is the typical size of the eddies. For this system, we assumed that $l_B = g$ because the eddies are confined between the comb and the cathode, therefore the largest possible size for the eddies is g.
The Sherwood number is linked to the Reynolds and Schmidt numbers (equation 7):

$$Sh = \alpha Re^{1/2} Sc^{1/3} \tag{7}$$

$$Sh = \left(\frac{M}{w}\right)^{1/4} \cdot \left(\frac{Sf(h+g)}{\nu}\right)^{1/2} \cdot \left(\frac{\nu}{D}\right)^{1/3} \tag{8}$$

Combining eq. 6 with eq. 8 and taking $l_B = g$ yields:

$$J_c = n\mathrm{F}C_B \left(\frac{M}{w}\right)^{1/4} \cdot \frac{[Sf(h+g)]^{1/2}}{g} \cdot D^{2/3} \cdot \nu^{-1/6} \tag{9}$$

The experimental deposition current is plotted as a function of the geometric deposition current J_C in fig. 5. The measured current density is indeed proportional to the geometric current density. That means that the number $L = (M/w)^{1/4} \cdot [Sf(h+g)]^{1/2}/g$ is a geometric control parameter for the deposition current: for higher values of L, the current density gets higher.

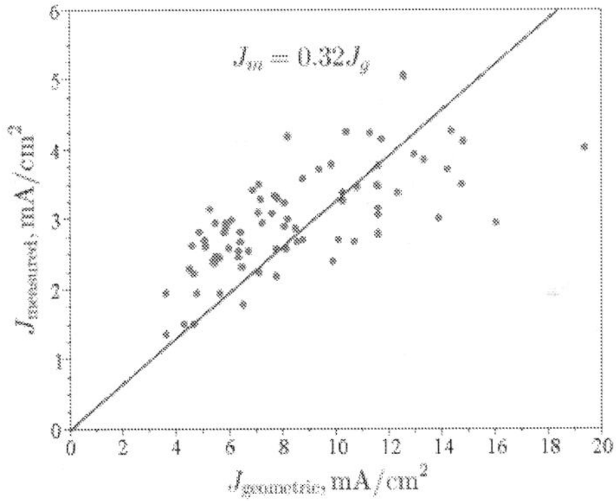

Fig. 5. Measured deposition current J_{measured} is plotted as a function of the geometric deposition current $J_{\mathrm{geometric}}$

4.4 Homogeneity control

To define a control parameter for the homogeneity of the electrodeposited layer the standard deviation of the electrodeposited layer versus each relevant non-dimensional parameter has been plotted separately. This allowed the design of a control parameter depending on the non-dimensional stroke, the proximity ratio and a modified Reynolds Number

$$K = \frac{S}{M} \cdot \frac{g}{h} \cdot \frac{fSh}{\nu} \tag{10}$$

Figure 6 shows the standard deviation of copper in Cu-Ni layers σ_{Cu} versus the non-dimensional parameter K. The variation is monotonous: the standard deviation σ_{Cu}

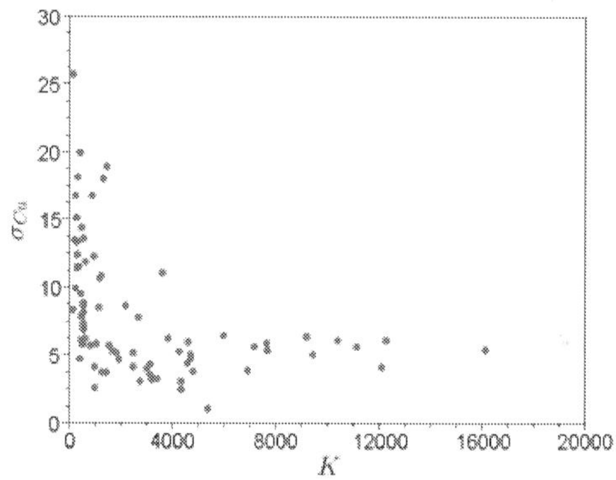

Fig. 6. σ_{Cu} in Cu-Ni electrodepositions vs. non-dimensional control parameter K.

decreases when K increases, meaning that the homogeneity of the electrodeposition increases when K increases.

The minimum standard deviation of Cu quantity obtained for this comb-like system was 1.07 % on 5 x 5 cm^2. The experimental conditions for this electrodeposition were $M = 9$ mm, $h = 4$ mm, $w = 1$ mm, $g = 25$ mm, $S = 24$ mm, $f = 5.4$ Hz, and the tooth section was rectangular. This value can be compared to the homogeneity of Fe-Ni layers synthesized in paddle-cell industrial electrolysers: Andricacos et al. (1994) report that the normalized standard deviation on circular wafers (diameter 30 cm) can be as low as 1.38 % for the Fe-Ni thickness and 0.86 % for the Fe content of the Fe-Ni layer. The homogeneity of electrodeposited layers performed in comb-like reactors are therefore comparable to the ones performed in industrial paddle-cell electrolysers used for electronic applications.

4.5 Interactions between the two control parameters

The two previously defined two control parameters allow the optimization of Cu-Ni homogeneity and the deposition current. As the control parameters K and L are made out of the same geometrical parameter, their relashionship has been studied. Figure 7 shows the geometrical deposition current J versus the homogeneity control parameter K.

For $K > 5000$ corresponding to homogeneous deposition (see figure 6), J can take the whole range of possible values. That means that it is possible to choose any deposition current and still perform homogeneous electrodeposition, which is particularly interesting for industrialization of such processes.

5. Results: control parameters of Cu-Ni electrodeposition in jet systems

5.1 Mixing and transport in randomly firing jet systems

The turbulence-generating device using a randomly firing jet array has been presented in (Variano et al., 2004; Variano & Cowen, 2008), where an array of jets fires in the same direction at the same discharge (volume of fluid per unit of time), randomly switched on and off. 1D jet arrays are typically holed pipes, and 2D jet arrays are typically holed plates. A definition diagram of a jet array is shown in figure 8. The stirring tank used in this study and related electrochemical methods are described in (Delbos et al., 2009a;b).

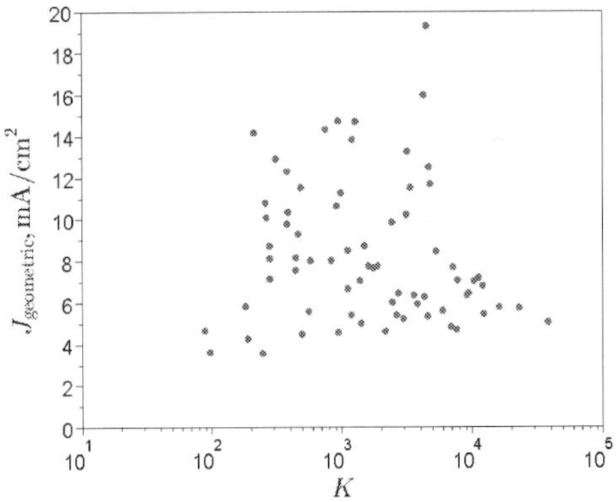

Fig. 7. Geometric deposition current $J_{\text{geometric}}$ vs homogeneity control parameter K

Previous studies Delbos et al. (2009b;a); Variano & Cowen (2008) showed that the important parameters for turbulence control in such systems were mainly the distance between the jets M, and the cathode z_d, as well as the momentum source fraction, which is the ratio of time of firing and time of sleep. However, no relationship between current density and geometrical parameters of the jet array has been reported yet. Furthermore, no information has been found related to the spatial variations of δ and its effects on the homogeneity of the electrodeposited layer in such systems.
The following findings are based on the experimental results obtained from (Delbos et al., 2009b;a), which also describe the experimental details of their experiments, which will not be repeated here.
The operating conditions of the jet systems are described in table 4

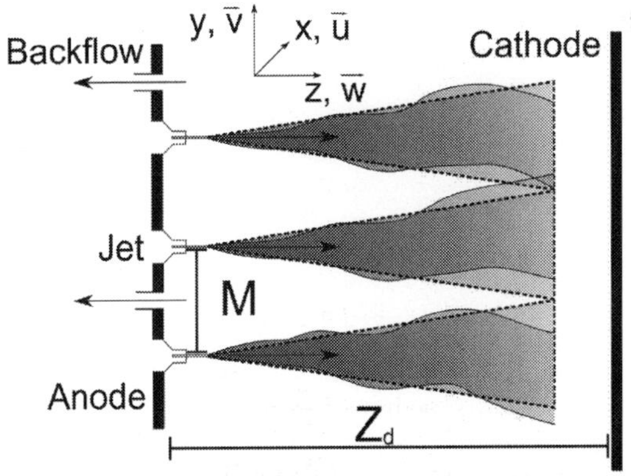

Fig. 8. Definition diagram of a jet array, showing the single jet discharges and their merging.

Parameter	Range
W_0 (m/s)	2.6, 7.4
d (mm)	0.7 – 1
z_m (mm)	60 – 230
M (mm)	18, 29, 45
ϕ_j(%)	50, 100

Table 4. Hydrodynamical conditions random jet array system. See List of symbols and figure 8

5.2 Non-dimensional parameters

As for the comb systems, first the non-dimensional geometrical parameters that control the flow characteristics are defined:

- Merging ratio
  As mentioned in (Delbos et al., 2009a;b), the merging length is a predominant factor for homogeneity: if the merging ratio z_d/L is smaller than 1, the jets are not merged and impinging patterns appear on electrodeposited layers. When the jet raster is different in horizontal and vertical directions, there are two merging lengths L_v and L_h, and therefore both corresponding merging ratios are used for the design of non-dimensional parameters.
- Reynolds number
  The Reynolds number of the individual jets was chosen to describe the turbulence intensity.

$$Re = \frac{W_0 d}{\nu} \tag{11}$$

In these set of experiments, the temperature could not be well controlled but was measured. The variation of the kinematic viscosity with temperature was therefore taken into account. The variation of water dynamic viscosity was taken from (Likhachev, 2003):

$$\eta = \eta_0 \exp \frac{E}{\mathrm{R}(T + T_0)} \tag{12}$$

where $\eta_0 = 2.414 \cdot 10^{-5}$ Pa.s, R = 8.314 J.K^{-1}.mol^{-1}, E = 2060 J.mol^{-1}, $T_0 = 133.15°$C, T is in °C. The variation of volumic mass is described in (McCutcheon et al., 1993):

$$\rho(T) = 1000 \cdot (1 - \frac{(T + 288.9414) \cdot (T - 3.9863)^2}{508923 \cdot (T + 68.12963)}) \tag{13}$$

where T is in °C

- Nondimensional turbulence decay ratio
  For perforated plates, the turbulence decay is proportional to the inverse of the distance from the turbulence generating device to the power 3/2 ($u_{\mathrm{RMS}} \propto z^{-3/2}$)) (Thompson & Turner, 1975). As the turbulence intensity decreases with the distance from the turbulence generation system, the stirring efficiency decreases with the distance from the nozzle. We propose a nondimensional turbulence decay ratio $(\ell/z_d)^{3/2}$ (where ℓ is the size of the jet array) to quantify the decreased homogeneity due to the turbulence decay:

- Randomization coefficient
  Finally, the randomization had to be taken into account. The influence of the source fraction $\phi_j = \mu_{on}/(\mu_{off} + \mu_{on})$ on the turbulence generation was studied in (Variano & Cowen, 2008). This study shows that the optimal value of ϕ_j for turbulence generation is 12.5 %. In the case of this study, the source fraction was 50 % in the random case, and 100 % in the continuous case. A coefficient k was designed so as to quantify the influence of random and non-random injection, and the most fitting results, resulting from euristic methods, are:
  k = 0.45 when the injection is continuous
  k = 0.80 when the injection is random

5.3 Deposition current control

For the jet system, eq.8 did allow us to describe a geometric control parameter of the system. A similar relationship is proposed, which was found through heuristic methods:

$$Sh = \frac{l_B}{\delta} = \alpha Re^{0.1} Sc^{-0.9} \tag{14}$$

where $l_B = \ell$: the eddy size l_B is the dimension of the jet array ℓ (a result described in (Delbos et al., 2009b)), $\alpha = [(L_h L_v)/z_d^2]^{0.1}$
Combining eq. 6 with eq.14 yields:

$$J_c = nFC_B \left(\frac{L_h L_v}{z_d^2} \right)^{0.1} \cdot \frac{(W_0 d)^{0.1}}{\ell} \cdot \frac{D^{1.9}}{\nu(T)} \tag{15}$$

The experimental deposition current is plotted as a function of the geometric deposition current J_C in fig. 9. The measured current density is indeed proportional to the geometric current density. That means that the number $L = \left(\frac{L_h L_v}{z_d^2} \right)^{0.1} \cdot \frac{W_0 d}{\ell}$ is a geometric control parameter for the deposition current: for higher values of L, the current density gets higher. Surprisingly, the type of injection (random or continuous) does not play any role in the value of the deposition current. In (Delbos et al., 2009b), we showed that the random injection increased the turbulent velocity but decreased the mean flow. It could be that these effects balance each other, causing random and continuous injection to yield similar values of current density.

5.4 Homogeneity control

This control parameter was built with the same method used for the comb system. The proposed nondimensional control parameter is:

$$K = \underbrace{k}_{(1)} \underbrace{\frac{z_d}{L_v} \frac{z_d}{L_h}}_{(2)} \underbrace{\left(\frac{\ell}{z_d} \right)^{3/2}}_{(3)} \underbrace{\frac{W_0 d}{\nu(T)}}_{(4)} \tag{16}$$

where (1) is the randomization coefficient, (2) are the merging ratios, (3) is the turbulence decay ratio, and (4) is the Reynolds number. The choice of 1 as the exponent for all the member except the turbulence decay ratio yielded the best results in term of monotony of the $K - \sigma$ plot.
Figure 10 shows the standard deviation of copper in Cu-Ni layers σ_{Cu} versus the nondimensional parameter K. The variation is monotonous: the standard deviation σ_{Cu}

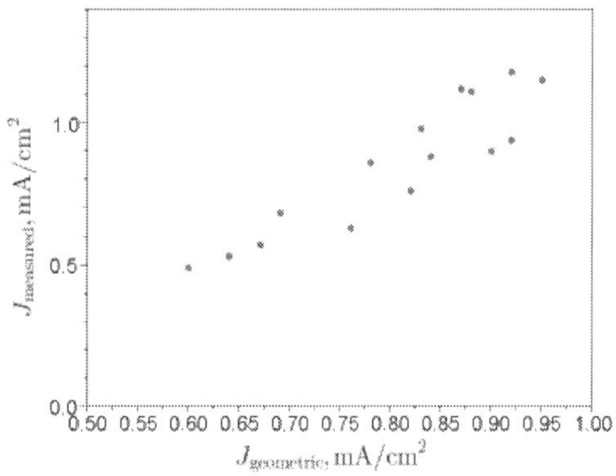

Fig. 9. Measured deposition current is plotted as a function of the geometric deposition current J_C

decreases when K increases, indicating that the homogeneity of the electrodeposited layer increases when K increases. Random injection generates more homogeneous deposition, high values of distance nozzle-to-cathode, initial velocity, and low values of merging distance (and therefore low values of jet mesh) are also helpful for homogeneous depositions.

5.5 Interactions between the two control parameters

The methodology used for the comb was also used for the jet array. The figure 11 shows the geometric current as a function of K. It shows that for high values of K yielding the best homogeneity, a wide range of values of J can be chosen.

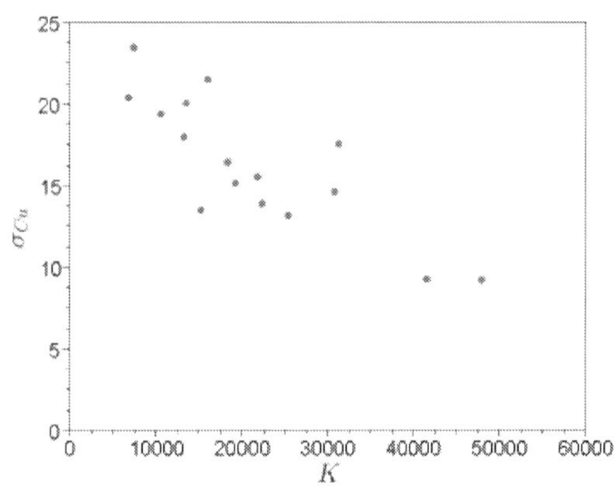

Fig. 10. σ_{Cu} in Cu-Ni electrodepositions vs. nondimensional control parameter K

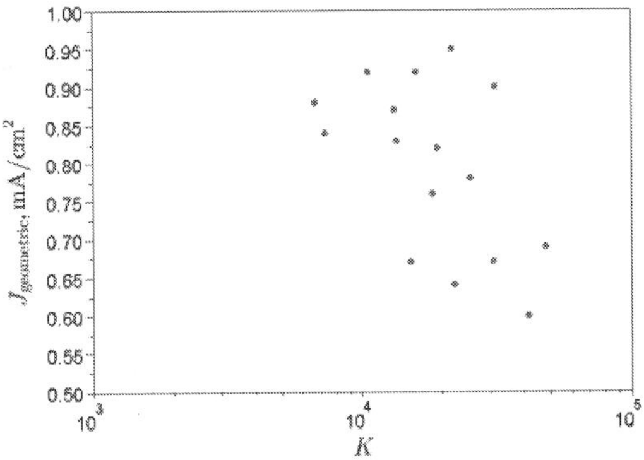

Fig. 11. J vs. K for the jet array system

6. Conclusion

In the microelectronics industry, the paddle-cell is widely used for enhancing mass transfer in diffusion controlled plating of alloys. In this paper two alternative stirring sytems, the comb-like system and the random jet array, are studied in details. In particular two control parameters are developed. Electrodeposited layers mappings were linked to LDV measurements of the flow velocity in order to the find geometrical parameters relevant to electrodeposition control. From these experiments two control parameters for each system were designed. They allow comb and random jet array systems operators to control the mean deposition current and the homogeneity of the electrodeposited layer. More generaly, these parameters control the mass transfer in these systems.

7. List of symbols

δ	diffusion layer thickness (mm)
ν	kinematic viscosity (m^2/s)
ϕ_j	momentum source ratio
ρ_{ca}	resistivity of the cathode (Ω.m)
σ	standard deviation
C	concentration (mol/L)
D	Molecular diffusion coefficient (m^2/s)
d	diameter of the jet nozzles (mm)
H	dimension of the cell in the direction perpendicular to the cathode (mm)
f	stirring frequency (Hz)

f_v	vortex shedding frequency (Hz)
g	distance between cathode and paddle or comb (mm)
h	height of paddle or height of comb teeth (mm)
i_{Cu}	copper partial current (A/cm^2)
i_L	diffusion-limited current (A/cm^2)
K	Homogeneity control parameter
L	Current control parameter
ℓ	dimension of the jet array (mm)
L	dimension of the cell in the direction parallel to the cathode (mm)
L_h	horizontal merging length (mm)
L_v	vertical merging length (mm)
LDV	Laser Doppler Velocimetry
l_B	advection lengthscale (mm)
M	distance between the teeth of the comb or distance between the nozzles (mm)
n_Cu	Copper quantity in the electrodeposited layer (mol/mm^2
Re	Reynolds number
RMS	Root Mean Square deviation
S	Stroke (mm)
Sc	Schmidt number
Sh	Sherwood number
St	Strouhal number
u,w	Flow velocity in the directions horizontal, parallel to the cathode and perpendicular to the cathode (mm/s)
u_B	typical eddy size (mm)
$\overline{u},\overline{w}$	Mean flow velocity (mm/s)
u_{RMS}, w_{RMS}	turbulent fluctuations (mm/s)
V	velocity of the paddle or velocity of the comb (mm/s)
W_0	initial velocity of the jets (mm/s)
w	width of paddle or width of comb teeth (mm)
XRF	X-Ray Fluorescence
x,y,z	coordinates, horizontal, parallel to the cathode, vertical, and perpendicular to the cathode (mm)
z_d	distance between nozzles and cathode (mm)
z_e	number of electrons in the electrochemical reaction (mm)
z_m	distance between the cathode and the LDV measurements (mm)

8. References

Andricacos, P., Branger, M., Browne, R. M., Dukovic, J. O., Fu, B. W. B., Hitzfeld, R. W., Flotta, M., McKenna, D. R., Romankiw, L. T. & Sahami, S. (1994). Multi-compartment electroplating system. US patent No 5312532, IBM.

Bard, A. J. & Faulkner, L. R. (2001). *Electrochemical methods: fundamentals and applications*, John Wiley & Sons, Inc.

Brumley, B. H. & Jirka, G. H. (1988). Air water transfer of slightly solulble gases – turbulence, interfacial processes and conceptual models, *Physicochemical hydrodynamics* 10(3): 295–319.

Cussler, E. L. (1997). *Diffusion - Mass transfer in fluid systems, 2nd edition*, Cambridge University Press.

Datta, M. & Landolt, D. (2000). Fundamental aspects and applications of electrochemical microfabrication, *Electrochemical Acta* 45: 2535–2558.

Delbos, S. (2008). *Électrodépot de Cu-In-Se contrôlé par turbulence pour la production de cellules solaires*, PhD thesis, Université Pierre et Marie Curie Paris.

Delbos, S., Grand, P.-P., Chassaing, E., Weitbrecht, V., Bleninger, T., Jirka, G. H., Lincot, D. & Kerrec, O. (2009). Application of randomly firing jet arrays for electrodeposition, *Journal of the Electrochemical Society* 156(11): 161–166.

Delbos, S., Weitbrecht, V., Bleninger, T., Grand, P.-P., Chassaing, E., Lincot, D., Kerrec, O. & Jirka, G. H. (2009). Homogeneous turbulence at an electrodeposition surface induced by randomly firing jet arrays, *Experiment in Fluids* 46(6): 1105–1114.

Henninot, C. (1999). *Codéposition électrochimique d'un alliage*, PhD thesis, INPL ENSIC Nancy.

Keigler, A., Liu, Z., Harrell, J. & Wu, Q. (2005). Method and apparatus for fluid processing a workpiece. International patent No 2005/042804A3, NEXX Systems, Inc.

Likhachev, E. (2003). Dependence of water viscosity on temperature and pressure, *Technical Physics* 48(4): 514–515.

Lincot, D., Guillemoles, J.-F., Taunier, S., Guimard, D., Sicx-Kurdi, J., Chomont, A., Roussel, O., Ramdani, O., Hubert, C., Fauvarque, J.-P., Bodereau, N., Parissi, L., Panheleux, P., Fanouillère, P., Naghavi, N., Grand, P.-P., Benfarah, M., Mogensen, P. & Kerrec, O. (2004). Chalcopyrite thin film solar cells by electrodeposition, *Solar Energy* 77: 725–737.

Mandin, P., Cence, J.-M., Fabian, C., Gbado, C. & Lincot, D. (2007). electrodeposition process modeling using continuous and discrete scales., *Computer and Chemical engineering* 31: 980–992.

McCutcheon, S. C., Martin, J. L. & Barnwell, T. (1993).

McHugh, P. R., Wilson, G. J., Woodruff, D. J., Zimmerman, N. & Erickson, J. J. (2005). Reactor having multiple electrodes and/or enclosed reciprocating paddles, and associated methods. US patent No 2005000817A1, Semitool, Inc.

Ollivier, A., Muhr, L., Delbos, S., Grand, P. P., Matlosz, M. & Chassaing, E. (2009). Copper–nickel codeposition as a model for mass-transfer characterization in copper-indium-selenium thin-film production, *Journal of Applied Electrochemistry* 39(12): 2337–2344.

Powers, J. V. & Romankiw, L. T. (1972). Electroplating cell including means to agitate the electrolyte in laminar flow. US patent No 3652442, IBM.

Rice, D. E., Sundstrom, D., McEachen, M. F., Klumb, L. A. & Talbot, J. B. (1988). *J. Electrochem. Soc* 135(11): 2777–2780.

Schwartz, D. T., Higgins, B. G. & Stroeve, P. (1987). *J. Electrochem. Soc* 134(7): 1639–1645.

Tennekes, H. & Lumley, J. L. (1972). *A first course in turbulence*, The MIT Press.

Thompson, S. M. & Turner, J. S. (1975). Mixing across an interface due to turbulence generated by an oscillating grid, *Journal of Fluid Mechanics* 67: 349.

Tzanavaras, G. & Cohen, U. (1995). Precision high rate electroplating cell and method. US patent No 5421987.

Variano, E. A., Bondenschatz, E. & Cowen, E. A. (2004). A random synthetic jet array driven turbulence tank, *Experiments in fluids* 37(4): 613–615.

Variano, E. A. & Cowen, E. A. (2008). A random-jet-stirred turbulence tank, *Journal of Fluid Mechanics* 640: 1-32.

Wilson, G. & McHugh, P. (2005). Unsteady numerical simulation of the mass transfer within a reciprocating paddle electroplating cell, *J.Electrochem. Soc* 152(6): 356–365.

Wu, B. Q., Liu, Z., Keigler, A. & Harrell, J. (2005). Diffusion boundary layer studies in an industrial wafer plating cell, *journal of the Electrochemical Society* 152(5): 272–276.

4

Application of Airborne Sound Waves for Mass Transfer Enhancement

Sergey V. Komarov
Nippon Light Metal Company, Ltd.
Japan

1. Introduction

Efforts to reduce energy use and greenhouse gas emissions continue to place increasingly stringent requirements on the industrial sector of the economy. This tendency is especially pronounced in energy-intensive industries which use high temperature processing. Examples are chemical and metallurgical processes, power generation and waste treatment. In these areas, many chemical reactions involved in the material fabrication or treatment proceed under mass transfer control, meaning that rate-controlling step of the reactions is not a chemical reaction itself (termed kinetic control), but transport of the chemical reagents to or from the reaction zone. This is associated with the well-known fact that the rates of chemical reactions increase with temperature to a greater extent than those of mass transfer. In the mass transfer controlling regime, any enhancement of the mass transfer rate should raise productivity of the above-mentioned processes, that in turn provides an option to reduce the energy consumption and production cost. Another important point to note is that inadequate control of mass transfer between different regions of the reactors or furnaces often results in a local under- or overheating of the materials being processed. Both these phenomena are considered to be a major source of atmospheric pollutant emissions. Typical examples of such processes are solid waste incineration and fuel combustion.

In most of the above-mentioned processes gaseous, liquid and solid phases coexist in one process. Nevertheless, in this chapter, we shall restrict our consideration to the gas-phase mass transfer. In gas, mass transfer can occur through the following two mechanisms: (1) convection which typically occurs in the gas bulk and (2) molecular diffusion which is the dominant near interfaces. In many cases, the rate of convective mass transfer can be relatively easily enhanced by blowing a gas into the media bulk or by applying some other method of increasing the forced convection. Therefore, the main resistance to the mass transfer is located usually in the boundary layers adjacent to the interfaces between different phases. Under these conditions, there is a very limited choice of techniques available for controlling the mass transfer at the interfaces. This is especially true in regard to those processes which involve high temperatures.

In these circumstances, sound or ultrasound waves provide a unique tool making it possible to supply energy directly to the interfaces and, thus, to influence the interfacial mass transfer rates. Attractiveness of ultrasonics is associated with the following features of sound. First, sound waves have the ability to propagate through homogeneous elastic mediums including gas without significant losses, and thus to effectively transfer the acoustic energy from a sonic generator to the materials being processed. Second, when the waves are incident upon an interface, the scattering or reflecting of the waves from the interface is responsible for a number of phenomena that occur at the interfaces. Examples of these phenomena include acoustic streaming, radiation pressure, forced turbulence and capillary waves. Most of them have a direct influence on the interfacial mass transfer process. Such effects are unachievable by any other methods. Third, along with the technical efficiency, sonic/ultrasonic treatment should be competitive as regards cost, because it provides an effective transmission of acoustic energy at a relatively low cost for ultrasonic equipment.

The idea of using ultrasonics for improving process performance or for changing material structure is rather old. As early as the 1920s, Wood and Loomis (Wood & Loomis, 1927) had studied the effects of ultrasonics on the atomization of liquids, the emulsification of immiscible liquids, and the changes in the structure of crystallized organic substances. During the years which followed, ultrasonic effects have become the subject of numerous extensive studies. The studies have proved the benefits of using ultrasonics in a variety of technological processes. In the majority of them, sound waves of ultrasonic range are radiated into liquids through a submersible sonotrode under conditions in which acoustic cavitation occurs in the liquid. Ultrasonic sonochemistry, a relatively new field of chemistry, exploits the acoustic cavitation to influence chemical reactions. Acoustic cavitation is a phenomenon of nucleation, oscillation and implosion of countless bubbles in liquids. The last phenomenon, namely bubble implosion, is of great importance in many ultrasonic applications. Cumulative microjets and shock waves, which are generated at the last moment of bubble implosion are thought to be one of the main contributor to the effects observed during ultrasonic treatment. Also, the ultrasonic cavitation is a very efficient route to enhance mass transfer rates both in the bulk and at the interfaces in liquids (e.g. Margulis, 1995).

However, the liquid-phase sonoprocessing has a number of serious limitations. As ultrasound waves are scattered on gas-liquid interfaces, the surface of cavitation bubbles has a screening effect on propagating acoustic energy. This leads to a significant decrease in the size of the zone that can be effectively treated by ultrasonic. This limits the ultrasonic applications for many processes proceeding in large-size reactors. Moreover, at higher temperatures, sonoprocessing through liquid phases requires designing a cooling system, choosing special materials for the waveguide components, and optimizing them under conditions of sharp temperature drops. Unfortunately, this is not always possible

On the other hand, the gas-phase sonoprocessing possesses some advantages as compared to the treatment of liquid phases that make it very useful for applying to high-temperature processes. First, the sound generator can be positioned a safe distance away from the high temperature zone. Second, in gas, sound waves can propagate over a longer distance than that in liquids because the sound propagation in gas is not restricted by cavitation. The gas-phase sonoprocessing is especially appealing for the processes that use gas blowing or injection. Typical examples are burners, solid waste incinerators, converters for making steel and cooper. In these processes, a sound wave can be relatively easily produced by passing a part of blown gas through a pneumatic-type sound generator that does not require any additional consumption of energy. In addition, the gas can serve as a coolant of the sound generator parts.

The goal of this chapter is to give a deeper insight into the possibilities and limitations of airborne sound waves as a tool to enhance the rates of gas-phase mass transfer and its related phenomena. Following the introduction, the second section gives a brief theoretical consideration on the sound related fundamental phenomena that is necessary for a better understanding of the following sections. The next, third section presents a short review of some recent studies that examined sound or ultrasound effects on the gas-phase mass transfer around spherical drops and particles, followed by a brief survey of recent efforts in applying acoustic oscillations to in a number of technological processes. The forth section provides a discussion of our results and some considerations for the enhancement of mass treansfer rates in pyrometallurgical processes. Finally, the last section concludes the chapter with some general remarks.

2. Brief theoretical consideration

Sound is a wave that is created by oscillating objects and that travels through an elastic medium from one location to another. The simplest type of sound wave is a plane travelling wave. The wavefront in such a wave is a plane surface meaning that the oscillating energy is transmitted in the form of parallel beam. When the oscillation amplitude is small enough (linear regime), one can neglect any interaction between the medium and wave, and consider that the wave propagates adiabatically.
In actual practice this situation is very seldom realized. In the material sonoprocessing, sound waves of high intensity propagate inside reactor vessels filled with solid and/or liquid materials, and bounded by side walls and bottom. This suggests that the wave can be reflected or absorbed by surfaces belonging to the materials or vessels. In addition, processes often generate small particulates like solid particles, dust or liquid droplets. When the travelling wave impinge on their surfaces, a part of the wave energy can be scattered and lost. Moreover, propagation of high intense sound wave is non-adiabatic because a part of its energy is lost during the compression half cycle. This phenomenon is know as disspation. The dissipation and scattering are thought to be the main cause of the wave attenuation. Besides, scattering and dissipation of waves are responsible for so-called non-linear acoustic effects which have a direct bearing on the transfer of mass. The following sections introduce most important characteristics of sound waves and, then, briefly explain the non-linear effects to that extent which is necessary for understanding the matter of the present chapter.

2.1 General characteristics of sound waves

In the case of a harmonic travelling wave, the oscillatory motion can be expressed in terms of displacement of the particle,ξ relative to their equilibrium position as follows

$$\xi = \xi_0 \sin(\omega t - kx) \tag{1}$$

This equation is the basis for derivation of other equations which describe the wave propagation. Thus, the velocity of oscillations, V, and pressure, P can be given as

$$V = V_0 \cos(\omega t - kx) \tag{2}$$

$$P = P_0 \sin(\omega t - kx) \tag{3}$$

where x is the direction of wave propagation, angular frequency $\omega = 2\pi f$ and amplitudes ξ_0, V_o, P_o are related to each other as $V_0 = \xi_0\omega$ and $P_0 = \xi_0\omega\rho c$. The wave number k can be simply defined as $k = \omega/c$ or as $k = 2\pi/\lambda$, where c and λ are the sonic velocity and wave length, respectively. Frequency, f is commonly subdivided into audible sonic and inaudible ultrasonic ranges although physically there is no difference between sound and ultrasound waves. The boundary between these ranges lies at about 16 kHz. Note that if the travelling wave is plane and propagates without attenuation, all the amplitudes ξ_0, V_o, P_o are constant. Based on the above relationships, one can derive an expression of sound intensity, *J*, which is defined as the average rate of sound energy transmitted by a travelling wave through a unit area normal to the direction of sound propagation

$$J = \frac{1}{2}\rho c V_0^2 \tag{4}$$

In practice, since the sound intensity varies by several orders of magnitude, it is often expressed in logarithmic form known as sound intensity level,*SIL*

$$SIL = 10\log\frac{J}{J_{ref}} \tag{5}$$

where the reference intensity, J_{ref} is equal to 10^{-12} W/m². The unit of SIL is decibel, *dB*.
There is no universally accepted criterion for distinguishing between low and a high-intense sound waves. However, many experimental results suggest that the sonic/ultrasonic effects can be obtained only if wave amplitude exceeds some threshold value. For gases, the threshold value of V_0 can be estimated from the experimentally determined value of SIL at which ultrasonic effects become significant. This value is equal to 130 dB (Mednikov, 1965; Blinov, 1991). Thus, using Eqs.(4) and (5), one can estimate the threshold V_0 for air (c=340 m/s, ρ=1.2 kg/m³) to be approximately 0.2 m/s.

2.2 Sound wave reflection. Standing wave.

As mentioned above, in many applications, sound waves propagate inside vessels like combustion chambers, furnaces, etc. This results in an impingement of the waves upon the vessel walls and in a reflection from the wall surfaces. Moreover, the principle of sonoprocessing itself involves exposing the surfaces of treated materials to a sound wave which, in general, can reflect from the surfaces or pass through it. In this connection, it is important to know conditions of wave reflection. When a sound wave is incident from a medium 1 on a plate made of medium 2 at right angle, the reflection coefficient, R can be expressed by the following equation (Heuter & Bolt, 1966)

$$R = \frac{\frac{1}{4}\left(m - \frac{1}{m}\right)^2 \sin^2\frac{2\pi d}{\lambda_2}}{1 + \frac{1}{4}\left(m - \frac{1}{m}\right)^2 \sin^2\frac{2\pi d}{\lambda_2}} \tag{6}$$

Here, d is the plate thickness, λ_2 is the length of sound wave in medium 2, m is the ratio of acoustic impedances of the media 1 and 2. The acoustic impedance is a key parameter in understanding the behaviour of sound wave when incident on a surface. Generally, acoustic impedance implies resistance to the propagation of sound wave. Quantitatively, it can be expressed as the ratio of sound pressure generated in a point of medium to velocity of oscillations of the medium particles during sound propagation. Expressing the ratio in terms of the appropriate amplitudes, one obtains the following relationship for acoustic impedance, Z

$$Z = \frac{P_0}{V_0} = \frac{\rho c V_0}{V_0} = \rho c \quad (7)$$

where ρ and c have the same notations as above. A special feature of gas phases that their acoustic impedances are several orders of magnitudes smaller than those of liquid and solid phases. For example, values of Z for air, water and steel at 20°C are 406, 1.47×10^6 and 4.56×10^7 kg/m^2·c, respectively.

Insertion of these values into equations (6) and (7) gives the results illustrated in Figure 1. This figure shows the reflection coefficients as a function of d/λ_2 for two cases. In both the cases a sound wave propagates in air. However, in the first case the medium 2 is water while in the second one it is steel. As can be seen, decrease and following increase in R near $d/\lambda_2 = 0.5$ and 1.0 occurs so sharply that the corresponding descending and ascending lines are indistinguishable in the figure. The inset (b) shows an enlarged view of variation of R in the vicinity of $d/\lambda_2 = 0.5$. The data reveal that the sharp variation of R takes place within a very narrow range of d/λ_2. Besides, it is clearly seen that this variation is steeper for steel plate than for water layer. Another region of sharp change in R appears at very small d/λ_2 close to zero. Here, it can be seen again that R for steel plate increases with d/λ_2 significantly faster compared to that for water layer.

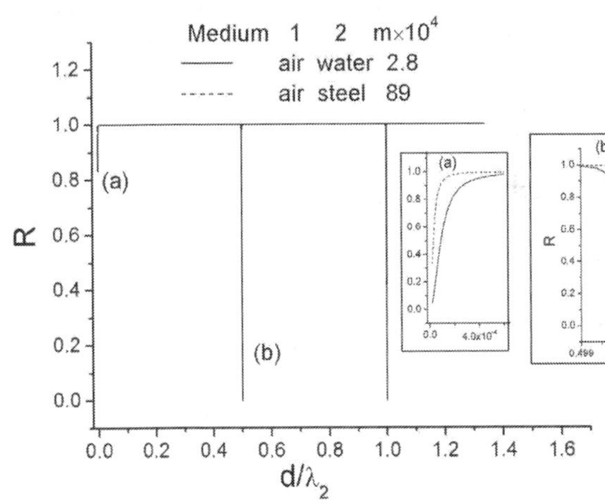

Fig. 1. Dependence of reflection coefficient on d/λ_2 for sound waves propagating in air and falling onto a water film (solid line) and a steel plate (dashed line).

Thus, one can conclude that, except for very thin films ($d/\lambda_2 \approx 0$), sound can pass through a bulk layer of liquid or solid only when the layer thickness is varied within a very narrow range around strictly fixed values. In practice, such cases, however, are few and far between. Therefore, it can be said that when a sound wave is introduced into a vessel filled with materials to be processed, the wave is completely reflected from all surfaces on which it is incident. The interference between incident and reflected waves may result in formation of so-called standing wave. Since the standing waves play an important role in the airborne sonoprocessing, it is necessary to consider this problem in some detail.

The simplest case of standing wave is one that forms when a travelling wave is incident upon a plane surface at right angle and completely reflected from the surface. In this standing wave, the oscillatory motion of fluid particles is expressed similarly to travelling waves in terms of displacement, oscillation velocity and sound pressure as given by Eqs.(8)~(10).

$$\xi = 2\xi_0 \cos(kx)\sin(\omega t) \tag{8}$$

$$V = 2V_0 \cos(kx)\cos(\omega t) \tag{9}$$

$$P = 2P_0 \sin(kx)\sin(\omega t) \tag{10}$$

where notations follow those of Eqs.(1)~(3).

It is remarkable that the values of amplitudes in the standing wave are double those in the travelling one. The solution of the above equations, although not shown here, reveals that, in the standing wave, there are points at which no oscillations occur. These points are referred to as node. The nodes are evenly spaced at intervals of one-half wave length. The points where the amplitude of the standing wave is a minimum are called antinodes. The antinodes occur midway between the nodes.

In standing wave, contrary to travelling one, there is no energy transfer. Instead, the energy is alternately transferred from kinetic to potential energy. By analogy with sound intensity level, *SIL*, the energy inside a standing wave can be evaluated by the sound pressure level, *SPL* in dB units as follows

$$SPL = 20\log\frac{P_0}{P_{ref}} \tag{11}$$

here the reference pressure, P_{ref} is equal to 2×10^{-5} Pa.

A special type of the standing wave occurs when it is formed between parallel surfaces, one of which is a sound-radiating surface while the other one is a reflecting surface. The longitudinal distribution of sound pressure, P in such a wave obeys the following equation (e.g. Rayleigh, 1945)

$$P = \xi_0 \omega c \rho_0 \frac{\cos k(L-x)}{\sin kL}\sin \omega t \tag{12}$$

where L is the distance between surfaces while the other notations are the same as above. Of most practical interest is the case when the distance L is an integer multiple of one-half wavelength according to Eq.(13)

$$L = \frac{n\lambda}{2},\ n = 1,2,3,... \tag{13}$$

Insertion of Eq.(13) into Eq.(12) gives $\sin kL=0$ meaning that the sound pressure P should increase to infinity. Of course it is physically impossible because of dissipative effects, the role of which increases with the oscillation amplitude. This kind of wave, called forced resonance standing wave, is very important for practical application because it allows to achieve very high levels of sound pressure with relatively small inputs of energy. In practice, the above resonance conditions can be realized either by fixing the sound radiator at a distance of L from the reflecting surface, or by adjusting the sound frequency in such a way to satisfy the relation (13).

It is to be noted that, in practice, geometry of vessels can be more complicated than the above case of parallel surfaces. Therefore, the resonance conditions are often determined experimentally by measuring the sound pressure inside the vessel as a function of the sound frequency and vessel geometric parameters.

2.3 Attenuation of sound waves.

As has been mentioned, all the above phenomena are well described in the framework of linear acoustics without considering any influence of the surrounding medium on wave propagation. Actually, as a sound wave is propagated through a medium, its intensity is attenuated through a number of mechanisms. As a matter of fact, the wave attenuation is the main reason of non-zero net mass flux that is, in turn, a cause of the mass transfer in high-intense sound wave.

The attenuation mechanisms include scattering and dissipation. Scattering refers to the reflection or refraction of sound waves when they impinge on an obstacle with dimension close to or less than the sound wavelength. Here, by an obstacle is meant any foreign substance on the second phase, for instance particulate, acoustic impedance of which differs greatly from that of the surrounding medium. Despite the fact that mechanism of scattering is very complicated, different theoretical considerations and experimental observations suggest similar relationships between the intensity of scattering and three key parameters: sound frequency, f, obstacle dimension, d and its mass fraction, C_0. Thus, Landau and Lifshits (Landau & Lifshits,1986) derived a formula which suggests that the intensity of scattering,I_s is proportional to the sixth power of d and fourth power of f for the case of rigid particles suspended in a gas. The reported dependences of I_s on C_o are linear for both the gaseous (Temkin, 1998) and liquid mediums (Carlson & Martinsson, 2002). Therefore, the intensity of scattering becomes significant only at higher frequencies, generally in the MHz range and/or at larger volume fractions of particulates. Since the commonly used frequency ranges in the airborne sonoprocessing do not exceed several tens kHz and because the fraction of particles, which can present in gas phases, is rather small, influence of scattering on the wave propagation will be neglected in the present discussion. However, it is to be noted that the scattering is of fundamental importance in considering the behavior of the particles themselves.

Sound energy dissipation, or absorption, is assumed to be the main cause of sound intensity attenuation in most of the applications considered in the present chapter. Generally, there are two main mechanisms of sound energy absorption in fluids. These are viscous dissipation, that occurs due to the normal and shear stresses induced in fluids on sound wave propagation, and heat conduction, that results from non-adiabatic nature of heat transfer during compression and rarefaction cycles in sound wave. Both these mechanism contribute the sound energy dissipation that can be expressed quantitatively in terms of the absorption coefficient, α, that is the sum of the viscosity-related term, α_V, and heat conduction-related term, α_T. For low-amplitude oscillations, α is expressed by Eq.(14)

$$\alpha = \alpha_V + \alpha_T = \frac{\omega^2}{2\rho c^3}\left(\frac{4}{3}\eta + \eta'\right) + \frac{\omega^2}{2\rho c^3}\kappa\left(\frac{1}{C_V} - \frac{1}{C_P}\right) \tag{14}$$

where $\omega = 2\pi f$, η is the shear viscosity, η' is the bulk viscosity, κ is the thermal conductivity, C_V and C_P are the specific heats at constant volume and pressure. In this case, attenuation of sound intensity due to absorption follows the exponential law according to Eq.(15).

$$J_x = J_o e^{-2\alpha x} \tag{15}$$

where J_0 and J_x are the intensities of sound waves at two points spaced each other by a distance of x in the direction of wave propagation. As readily seen from Eq.(14), energy absorption varies in proportion to the square of the frequency, f.

Attenuation of high-amplitude sound oscillations occurs more vigorously than that of low-amplitude oscillations. The stronger attenuation is associated with the conversion of a part of initial sinusoidal wave into the energy of arising high-frequency harmonics, rather than with absorption energy. Therefore, the attenuation coefficient of high-amplitude sound wave can exceed the absorption coefficient of low-amplitude sound wave by two orders of magnitude. There are also the other reasons but their detailed consideration is beyond the scope of this chapter. The relevant information on this topic can be found in books (Abramov, 1998; Hamilton & Blackstock, 1998).

2.4 Acoustic streaming.

The sound energy attenuation is responsible for a number of non-linear effects which serve as the basis for many ultrasonically based technologies. The main acoustic nonlinear effects arising in gas-phase systems are acoustic streaming, forced vorticity, and radiation pressure. There is a great body of literature devoded to investigations of the above effects. Detailed discussion on the results of these investigations can be found in a variety of books,e.g.(Rayleigh, 1945; Mason & Thurston, 1965; Hamilton & Blackstock, 1998; Lighthill, 2001) and reviews,e.g (Makarov & Ochmann, 1996,1997; Leighton, 2004). This section briefly outlines the acoustic streaming which is of primary importance in understanding the matter presented in the following sections.

Acoustic streaming is the steady flow that is generated in a fluid medium due to momentum transfer associated with the attenuation of a sound wave. The physical origin of acoustic streaming was best explained by Lighthill (Lighthill, 1978). Briefly, the explanation is as follows. Propagation of a sound wave causes fluctuations of the medium fluid particles at certain time scale and amplitude which are governed by the wave frequency and intensity. By applying the Reynolds stress approach to express the mean momentum flux due to the wave in the same manner as it does for turbuent pulsations, one can derive Reynolds stress, τ, in a sound wave as

$$\tau = \overline{\rho u_i u_j} \tag{16}$$

where $u_{i,j}$ are the fluctiating velocity in the sound wave. The bar signifies a mean value. The wave attenuation results in a spatial variaton of the Reynolds stress that can cause a non-zero net force exerted on the fluid. This force is capable of generating a steady flow termed acoustic streaming.

It is generally recognized that there three types of acoustic streamings. They differ each other by the spatial scale on which they can spread. The acoustic streaming of the first type is generated in an unbounded body of fluid. In this case, the streaming is a steady flow directed away from the sonic generator in the direction of wave oscillations. As its scale can be larger compared to the sound wavelength, it is called large-scale streaming. It is generally agreed that the large-scale streaming originates from the sound energy absorption briefly considered above. Under some conditions, the velocity of this type of streaming can be as high as several m/s (Lighthill, 1978).

The acoustic streamings of the second and third types are generated in the presence of solid obstacles (walls, particulates, etc.) placed in an acoustic field. In this case, the attenuation occurs because of frictional dissipation between an oscillating gas volumes and solid surface within the resulting boundary layer. The acoustic streaming of the second type is generated outside the boundaly layer. The scale of such an outer streaming is much smaller than that of the first type, and is equal approximately to the wavelength. The acoustic streaming of the third type is called inner small-scale streaming because it is induced within the bounday layer, the dimension of which is much smaller than wavelength. The effective thickness of the boundaly layer is about 5 times larger than that of acoustic boundaly layer which is given by the following expression

$$\delta = (\nu / \omega)^{0.5} \tag{17}$$

where ν is the kinematic viscosity of fluid, ω is the angular frequency of sound, $\omega=2\pi f$ (Zarembo, 1971).

3. Combustion and environmental control applications.

3.1 Mechanisms of mass transfer enhancements

As pointed in the above section, application of acoustic oscillations can lead to the occurrence of several phenomena responsible for improvements in the gas-phase mass transfer characteristics. The persistence of the acoustic effects and their magnitude should vary from process to process depending on the gas flow pattern inside the vessel, the presence of solid or liquid particulates as well as their size, temperature and so on. Numerous studies have shown that when acoustic oscillations are imposed on homogeneous flames or gas jets, the mass transfer enhancement is achieved through an intensification of turbulent mixing and improvement in entrainment characteristics of gas flames and jets. When a process involves chemical reactions proceeding at the surfaces of particulate or bulk materials of another phase, both the turbulent mixing in the gas bulk and flow at the surfaces have been found to be important in enhancing the mass transfer.

Experimental difficulties, encountered in high-temperature measurements, have motivated researchers to conduct cold model and numerical investigations. Other studies, although not focusing specifically on high temperature processes, have attempted to clarify the effects of acoustic oscillations on mass and heat transfer at room temperatures. The results of both groups of studies are of great importance for elucidating the mechanisms of mass transfer enhancement. Of special interest are studies that examine the mass transfer at curved surfaces like spheres and cylinders.

In one of the earlier study (Larsen & Jensen, 1977), evaporation rate for single drops of distilled water was measured under sound pressure of 132~152 dB and frequency of 82~734 Hz. The drops were suspended in dry air in upward motion and subjected to a horizontal standing wave sound field. The range of drop diameter was 0.8~2 mm. The authors used two dimensionless numbers, the drop diameter based Reynolds number, Re and the Strouhal number, S to correlate them with the experimentally determined Sherwood numbers, Sh defined on the basis of drop diameter, d. The expressions for the dimensionless numbers are given below.

$$\mathrm{Re} = \frac{V_0 d}{\nu} \tag{18}$$

$$S = \frac{d}{\xi_0} \tag{19}$$

$$Sh = \frac{k_M d}{D} \tag{20}$$

Here, k_M is mass transfer coefficient, D is the diffusion coefficient. The other notations are the same as above. The Strouhal number is a dimensionless number describing the oscillation flow around the sphere. At $S < 1$, Sherwood number was found to be increased proportionally to the 0.75 power of Re/S. Since, if V_0 is kept constant, the displacement ξ should increase with a decrease in frequency ω, these data suggest that lower frequencies are preferable for the mass transfer enhancement at $S < 1$. However, as $S > 1$, Sherwood number increased only with the 0.2 power of ReS^2. This suggests that higher frequencies are more desirable to enhance mass transfer rate at $S > 1$. The perhaps most interesting finding of this work was that flow around the drops at $S < 1$ and $S > 1$ is different. In the first case, gas flows completely around the sphere during a half-cycle of acoustic oscillation. If Re is high enough, the gas oscillations result in separation of boundary layer followed by buildup and shedding of eddy structures downstream of the separation points. It is assumed that these phenomena are the main cause of the mass and transfer enhancement at $S < 1$. On the other hand, when $S > 1$, the gas particles perform relatively small oscillations around drop causing an acoustic streaming to occur at its surface.

In one of the recent investigations (Kawahara et al.,2000), a small glass sphere (diam.1.6mm), covered by 0.4 to 0.6 mm thick layers of camphor or naphthalene, was positioned at a pressure node of a ultrasonic standing wave field to determine a distribution of the mass transfer rate over the sphere surface. Since the experiments were performed under a very high frequency of 58 kHz, a strong acoustic streaming was generated around the sphere that was found to be the main reason of the mass transfer enhancement. It was shown that the mass transfer due to the acoustic streaming is a strong function of the location on the surface being a maximum at the equator and a minimum at the poles. The authors derived the following expression to calculate the averaged Sherwood number, Sh as a function of the r.m.s. amplitude of gas particle velocity, B_{rms}, ω and D.

$$Sh = 1.336\,\mathrm{Re}; \quad \mathrm{Re} = \frac{B_{rms}}{\sqrt{\omega D}} \tag{21}$$

The other notations are the same as above. Notice that in their study the Reynolds number, Re is based on the acoustic streaming velocity. Here the Strouhal number, when defined by Eq.(19) , can be estimated to be much more than 1.

In another experimental study (Sung et al., 1994), the authors investigated mass transfer from a circular cylinder of 25 mm in diameter positioned in a steady flow on which acoustic pulsations were superimposed. The cylinder surface was precoated by a thin layer of naphthalene. Contrary to the above mentioned paper (Larsen & Jensen, 1977), in this study the directions of the steady and oscillatory flows were parallel to each other. The pulsation frequency was ranged from 10 to 40 Hz. The main conclusion from their results is that the enhancement of mass transfer rate is more effective at larger pulsation amplitudes and higher frequencies due the vortex shedding. Estimates show that the Strouhal number in these experiments is more then 1.

A detailed analysis, both experimental and theoretical, of evaporation from acoustically levitated droplets of various liquids was provided by Yarin et al. (Yarin et al., 1999). In their investigation, the frequency was 56 kHz. Therefore, the Strouhal number was assumed to be much more than unit. By plotting the Sherwood numbers against the Reynolds numbers, the authors showed that all data fall well on a straight line that is in agreement with their theoretical predictions. Both the Sherwood and Reynolds numbers were defined according to Eqs.(21). It was concluded that the effect of the acoustic field on droplet evaporation appears to be related to the acoustic streaming and squeezing of the drop by the acoustic radiation pressure.

A great body of experimental studies has been performed regarding the effects of acoustic oscillations on heat transfer from various geometries and surfaces. Taking into consideration the analogy between heat and mass transfers, a brief mention of some results of these studies will be made here. As before, our main interest is to clarify the effects of sound intensity and frequency on the heat/mass transfer characteristics.

One of the earlier study (Fand & Cheng, 1962) examined the influence of sound on heat transfer from a circular cylinder in the presence of a mean crossflow. In the experiments, air was blown onto the surface of the cylinder having a diameter of 3/4 inch. Simultaneously, the cylinder surface was exposed to high-intense acoustic oscillations at two frequencies, 1100 and 1500 Hz. The experimental data were presented as plots of $\alpha = Nu_v / Nu_0$ against the crossflow Reynolds number, Re_{cf} based on the cylinder diameter and crossflow velocity. Here, Nu_v and Nu_0 are the Nusselt numbers measured in the absence and presence of acoustic oscillations, respectively. They are given as follows

$$Nu_v = \frac{h_v d}{\lambda} \qquad Nu_0 = \frac{h_0 d}{\lambda} \tag{22}$$

where h_v and h_0 are the heat transfer coefficients from the cylinder in the absence and presence of acoustic oscillations, respectively, d is the cylinder diameter and λ is the thermal conductivity. Although the authors did not mention the Strouhal number, S in their paper, using the equations of the above sections, one can estimate S to be much more than 1.

The results of this study showed the following. At Re_{cf} about 1000, which was the lowest Re_{cf} examined, a 20 per cent augmentation of α was obtained at a SPL of 146 dB regardless of frequency. The augmentation mechanism was assumed to be an interaction similar to thermoacoustic streaming. As Re_{cf} increased to about 5000, α was reduced to 1. Then, α increased again with Re_{cf} reaching a maximum value at Re_{cf} = 8000~9500 on frequency of

1100 Hz and at Re_{cf} = 9000~11000 on frequency of 1500 Hz. In these ranges of Re_{cf}, the increase in heat transfer appears to be the result of two different interactions: (1) a resonance interaction between the acoustic oscillations and the vortices shed from the cylinder; (2) a modification of the flow in the laminar boundary layer on the upstream portion of the cylinder similar to the effect of free stream turbulence. Here, the augmentation of α was more pronounced for the case of lower frequency.

In a more recent heat transfer study (Gopinath & Harder, 2000), a preheated 5-mm cylinder was exposed to acoustically imposed low-amplitude zero-mean oscillatory flows to investigate the mechanisms of heat transfer from the cylinder at frequencies of 585 to 1213 Hz. Two distinct flow regimes were found to be important. The first one is the attached flow regime which show the expected square root dependence of the Nusselt number, Nu on the appropriate Reynolds number, Re. The second regime is predicted to be an unstable regime in which vortex shedding is prevalent, contributing to higher transfer rates so that the Nu number becomes proportional to $Re^{0.75}$. These findings are in good agreement with those reported in the above mass transfer studies.

The similar relationship between Nu and Re results has been obtained in another recent study (Uhlenwinkel et al., 2000) on much higher frequencies, 10 and 20 kHz. The heat transfer rate was determined by using cylindrical hot-film/wire probes positioned in the acoustic field of strong standing waves. The Nusselt number was found to increase as the 0.65 power of the Reynolds numbers. The experiments revealed a 25-fold increase in the heat transfer rate compared to that of free convection regardless of frequency in the range examined. Because the authors used rather high displacement amplitudes of sound waves and very small probes, the Strouhal number in their experiments should be more than unit suggesting the above- mentioned vortex formation and shedding. This is assumed to be the main reason of why the acoustic effect was so great in this study.

The above results can be summarized as follows. The amplitude of acoustic oscillations plays a crucial role in the enhancement of mass transfer from objects like particles and cylinders. That was the main reason why most of the above-mentioned effects were observed under the resonance acoustic oscillations. The mass transfer coefficient is increased in proportional to the 0.5~1.0 power of the velocity amplitude with a tendency for the power to become close to 0.5 at larger Strouhal numbers (acoustic streaming controlling regime) and to increase up to 1 at smaller Strouhal numbers (vortex shedding controlling regime). The effect of frequency was less pronounced. Moreover, there is a lack of agreement in the literature on the sign of this effect. There are reports showing increase, decrease and no effect of frequency on the mass- heat transfer rates.

It is to be noted that all the above studies dealt with the objects which were fixed in position in the acoustic fields. In actual practice, particles, no matter whether they are purposely added or generated during a process, can be entrained in the flow of the surrounding gas. Moreover, when the airborne particles are exposed to an acoustic field, they can be forced to oscillate on the same frequency as the acoustic field. Both types of particle motion can affect the mass transfer rate remarkably. However, because of the great experimental difficulties, to the best of our knowledge there have not been any experimental studies in this area.

3.2 Improvements in fuel combustion efficiency

These above-mentioned and other findings have motivated extensive research on the application of acoustic oscillations to improve the process performances in combustion, environmental and waste treatment technologies. The results obtained have strongly suggested that acoustic oscillations offer very attractive possibilities for designing novel processes with improved combustion efficiency and low pollutant emission. These findings would be of considerable interest for experts dealing with such energy intensive industrial processes as metallurgy, material recycling and waste treatment. Below is a brief survey of some recent results presenting the acoustic effects on the combustion efficiency and pollutant emission.

The results of one of the first study in this field (Kumagai & Isoda, 1955) revealed that an imposition of sound vibrations on a steady air flow yields about 15% augmentation of a single fuel droplet burning rate compared to the conventional one. The sound effects was found to be independent of the vibration frequency. More recently, Blaszczyk (Blaszczyk, 1991) investigated combustion of acoustically distributed fuel droplets under various frequencies. The conclusion was that about 14% increase in fuel combustion rate can be achieved at the 120~300 Hz frequency range despite the sound intensity was relatively low, 100~115 dB.

The influence of acoustic field on the evaporation/combustion rates of a kerosene single fuel droplet was investigated experimentally under standing wave conditions (Saito et al., 1994). The authors concluded that the rate increased by 2~3 times when the droplet was fixed at a velocity antinode position of the wave at frequencies < 100 Hz and relatively low sound pressure levels of 100 ~110 dB.

Effects of acoustic oscillations on evaporation rate of methanol droplets (diam.50~150 μm) at room temperatures were investigated in another study (Sujith et al., 2000). The authors found that a 100% increase in the evaporation rate can be obtained only in the presence of a high intense acoustic field at a SPL of 160 dB. There was a weak tendency toward an increase of the effect with frequency ranging from 410 to 1240 Hz. It is to be note that most of the above data support the mechanism in which the obtained enhancement of liquid fuel combustion occurs due to a better mixing between the fuel vapor and oxidant at the droplet interface.

Approximatelly the same effects of acoustics were found on the combustion of solid fuel particles. Yavuzkurt et al. (Yavuzkurt et.al., 1991a) investigated the effect of an acoustic field on the combustion of coal particles in a flame burner by injecting the particles of 20~70 μm into the burning gas stream and by monitoring the light intensity emitted from the flame. Averaged values of light intensity were 2.5~3.5 times higher at SPL of 145~150 dB and frequency of 2000 Hz compared to those without sound application. Additionally, the authors performed a numerical simulation of combustion phenomena of 100-μm coal particles, the results of which revealed 15.7 and 30.2 percent decreases in the char burn-out time at frequency of 2000 Hz and sound intensity levels of 160 and 170 dB, respectively (Yavuzkurt et.al., 1991b). The main reason for the char burning enhancement is that the high-intensity acoustic field induces an oscillating slip velocity over the coal particles which augments the heat and mass transfer rates at the particle surface.

Four loudspeakers were used to apply an acoustic field to 125-μm black liquor solid particles, injected into a reactor tube at a gas temperature of 550°C (Koepke & Zhu, 1998). The intensity of the field was 151 dB, frequency was ranged from 300 to 1000 Hz. The results revealed a 10 percent reduction of char yield compared with that obtained without acoustic field application. Besides, significantly increased yields of product gases CO and CO_2 were also observed with acoustic treatment. On the whole, the results revealed that the acoustic effects were more pronounced for the initial period of particle heat-up. The above two works (Yavuzkurt et.al., 1991a; Koepke & Zhu, 1998) also include brief overviews of earlier publications on the acoustically improved fuel combustion.

3.3 Reduction of combustion-related pollutant emission

In parallel with the combustion enhancement, forced acoustic oscillations provide a way to significantly reduce emission of such pollutions as NO_x,CO and soot particulates. Especially, a large body of literature has been published on suppressing the NO_x formation due to acoustically or mechanically imposed oscillations. Good reviews on this topic can be found in the relevant literature, for example (Hardalupas & Selbach, 2002; Mcquay et al., 1998; Delabroy et al., 1996). NO_x reduction level was found to be strongly dependent on the experimental conditions. The reported values are ranged from 100%(Delabroy et al.,1996) to 15% (Keller et al.,1994) decrease in NO_x emission rate as compared with that for steady flow conditions. The suppression mechanism has been well established. A sound wave, being propagated through a gas, can be thought as turbulent flow fluctuations of certain scale and amplitude which are governed by the wave frequency and intensity, respectively. Thus, imposing acoustic oscillations on flame front enhances the turbulent mixing resulting in reduced peak temperatures at the front that, in turn, is the reason of reduced emission of thermal NO_x. When acoustic field is imposed upon flame containing liquid/solid particles, oscillations of gas around the particles provide an additional mechanism of the peak temperature reduction due to convection. One more reason of low NO_x emission is that high amplitude acoustic oscillations induce a strong recirculation of flue gas inside the combustion chamber. This results in entrainment of the already formed NO_x into the flame zone where NO_x is reduced by hydrocarbon radicals homogeneously or heterogeneously on the surface of carbonaceous solid particles.

The same mechanism causes lowering of emission of CO and other gaseous pollutants although the literature on this subject is much less than that on the NO_x emission control. For example, a large decrease in NO and CO emissions was observed in the presence of acoustic oscillations imposed to an ethanol flame in a Rujke tube pulse combustor (Mcquay et al., 1998). Taking concentration values at steady conditions as a reference, the decreases were 52 ~100% for NO and 53~90% for CO depending on SPL (136 to 146 dB), frequency(80 to 240 Hz) and excess air (10 to 50%). Another example is the work (Keller et al., 1994) the authors of which obtained emissions levels of a premixed methane-air flame below 5 ppm for NO_x and CO.

Few studies examined the effect of forced acoustics on soot emission from different types of flame: a spray ethanol flame of a Rijke tube combustor (Mcquay et al., 1998), acetylene (Saito et al.,1998) and methane diffusion flames (Demare & Baillot, 2004; Hertzberg, 1997). The oscillation frequencies were also different: 200 Hz (Demare & Baillot, 2004), 40~240 Hz (Mcquay et al., 1998), < 100 Hz (Saito et al., 1998) and 40~1000 Hz (Hertzberg, 1997). In spite of such different conditions, all the authors reported full disappearance of soot emission from the flames with acoustic excitation. The results of these studies suggested that acoustic oscillations enhance the mixing of fuel and ambient gas that causes a re-oxidation of soot particles at the flame zone.

4. Pyrometallurgical applications

Another promising area of airborne sonoprocessing is pyrometallurgy. As has been mentioned in the introductory section, several important chemical reactions in pyrometallurgical processes occur at the interface between gas and molten bath under gas-phase mass-transfer control. An important feature of these processes is that many of them use a high speed gas jet to promote the chemical reactions between the gas and molten metal. Taken together, these features provide the basis for designing a low-cost and high-performance method of sonoprocessing.

The first attempt to use the energy of sound waves for enhancing the rates of pyrometallurgical processes was made in the former Soviet Union in the steelmaking industry. High-intense acoustic oscillations were applied to a basic oxygen converter, that is the most powerful and effective steelmaking process. For a better understanding of the following discussion, the main features of converter process will be explained in more details.

A schematic diagram of a converter process is shown in Figure 3. Iron-based solid scrap and molten pig iron containing 4%C, 0.2~0.8%Si, minor amount of P and S, are charged into a barrel-shaped vessel. Capacity of the vessel can be as large as 400 tons. Fluxes (burnt lime or dolomite) are also fed into the vessel to form slag, which absorbs impurities of P and S from scrap and iron. A supersonic jet of pure oxygen (1) is blown onto the molten bath (2) through a water-cooled oxygen lance (3) to reduce the content of carbon, dissolved in the molten metal, to a level of 0.3~0.6% depending on steel grade. For a high efficiency of the process, the oxygen flow rate must be very high, several normal cubic meters per minute per ton of steel. Impingement of such a high-speed jet upon the molten metal bath is attended by deformation of its surface producing a pulsating crater in the molten metal and causing splashing of the metal at the crater zone as schematically shown in Fig. 3. Typically, the process takes about 20 minutes.

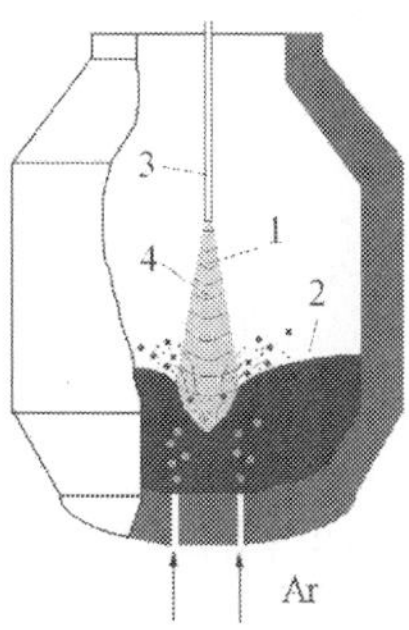

Fig. 3. A schematic representation of converter process.

The oxidation of carbon, which is often termed decarburization, is the main reaction in converter process. The decarburization reaction can proceed in two possible ways. The first one is the direct oxidation by gaseous oxygen according to

$$[C] + 0.5O_2 = CO \tag{23}$$

The second way is the indirect oxidation via formation of iron oxide according to

$$[Fe] + 0.5O_2 = (FeO) \tag{24}$$

$$(FeO) + [C] = [Fe] + CO \tag{25}$$

Here, parentheses and square brackets denote matters dissolved in the slag and metal, respectively. The reactions (23) occurs under the gas-phase mass transfer control. The reaction (24) is controlled by mass transfer of oxygen in both the gas and liquid phases. The

decarburization reaction occurs with a vigorous evolution of CO gas. As a results the slag is foamed and the lance tip becomes submerged into the metal-slag emulsion.
In an attempt to enhance the gas-phase mass transfer rate, an acoustically assisted converter process has been tested. In the process, a pneumatic sonic generator of the Hartmann type was built in the tip of a oxygen lance of a 10-t pilot converter (Blinov, 1991; Blinov & Komarov, 1994). Hence, the sound waves (4) propagated to the molten bath through the gas phase inside the converter as shown in Fig. 3. Design and operating principle of the Hartmann generators was briefly described in our previous review (Komarov, 2005). For the more details, the reader is referred to the earlier publications (Borisov, 1967; Blinov, 1991). The working frequency of sonic generator was 10 kHz. The intensity measured at a distance of 1 m from the generator was 150 dB.

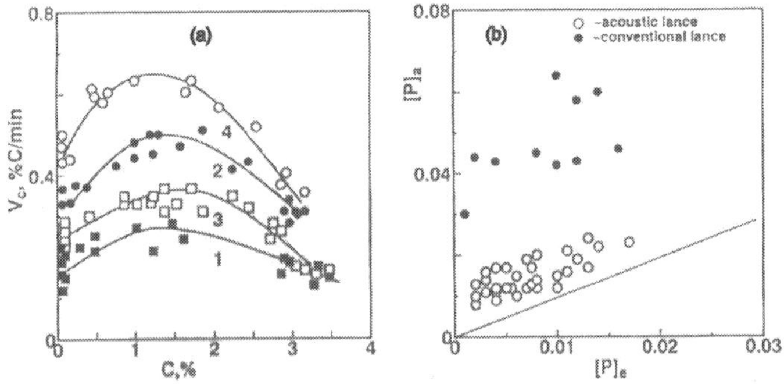

Fig. 4. Dependence of decarburization rate on carbon content (a) and relationship between actual and equilibrium content of phosphorus in the melt after completing the blowing operation.

Figure 4(a) presents the decarburization rate as a function of carbon content for two oxygen flow rates, 4(1,2) and 7(3,4) $Nm^3/min{\cdot}t$ and two oxygen lances: 1,2 - conventional lance, 3,4 - acoustic lance. The shape of the curves is typical for the decarburization rate in converter process: at the beginning, the rate increased as the carbon content reduced, passed through a maximum and then decreased. As can be seen from this figure, there is a significant effect of the acoustic oscillations on the decarburization rate. This effect seems to be stronger in the intermediate stage of the process while the carbon content is ranged from 0.5 to 2.5%. In the first and final stages of oxygen blowing operation, the effect of acoustic oscillations becomes less pronounced. The average enhancement of decarburization rate due to the acoustic lance application was about 40% under the given test conditions.
It is interesting to note that, in parallel with the enhancement in decarburization rate, there also has been a rise in the efficiency of phosphorus removal from the metal as well when the acoustic oscillations are applied. This reaction can be expressed as follows (Oeters, 1994)

$$[P] + 2.5(FeO)+1.5(CaO) = 0.5Ca_3(PO_4)_2 + 2.5[Fe] \quad (26)$$

The controlling mechanism of this reaction is more complicated compared to the decarburization reaction, however, it is well known that higher concentrations of FeO in slag is promote the reaction (26). Figure 4(b) is a plot of actual phosphorus concentration,

$[P]_a$ versus equilibrium one, $[P]_e$ for conventional and acoustically assisted process. The values of $[P]_a$ were measured by analyzing metal samples taken at the end of oxygen blow. The equilibrium values were determined according to the theory of regular solution based on the measurements of slag composition at the final stage of the blowing operation (Ban-Ya, 1993). Figure 4(b) shows that equilibrium for the phosphorous distribution between the metal and slag is not attained in the conventional process. This implies that a considerable amount of phosphorus remains in the metal. However, the use of the acoustic lance for blowing operation makes the phosphorous distribution closer to the equilibrium state, as can be seen from Fig. 4(b). Thus, acoustic oscillations were found to be capable of improving the efficiency of both the decarburization and phosphorus removal reactions.

To elucidate possible mechanisms of these improvements, two sets of laboratory scale experiments have been performed. In the first one, an effect of acoustic oscillations on the generation of drops in the above-mentioned crater zone was investigated by using cold models. The second set was aimed at clarifying the gas-phase mass transfer mechanism when the free surface of a liquid is exposed to acoustic oscillations. Below is some details on the experimental procedure and results.

4.1 Generation of drops

It has been known that the intensive drop formation occurs when a gas jet impacts with the gas-liquid interface. To investigate the drop formation a number of lances was designed to perform cold model experiments taking into consideration of the acoustic, aerodynamic and hydrodynamic similarity. In the experiments, the lances were installed vertically at a 0.1-m distance from the free surface of a 0.1-m depth water bath filled in a cylindrical vessel of 0.28 m in inner diameter. Air was blown onto the bath surface to cause a crater formation and drop generation. The drops were detached from the crater surface and carried away from the crater by the gas flow towards the vessel wall where they were trapped by a helical spout. Acoustic oscillations were generated using a specially designed small-scale pneumatic sonic generator of the Hartmann type operating at a frequency of 10 kHz. The generator was positioned above the water bath surface at such a distance that to obtain approximately the same sound pressure level at the crater as that during the pilot converter tests. Magnitude and frequency of turbulent oscillations was measured by using hot wire anemometry. The sensor was fixed close to the gas-liquid interface at the places free of the drop generation. Besides, a small hydrophone was used to measure frequency of oscillations generated in the water bath near the crater. The hydrophone was fixed in the water bath at a depth of 5 cm from the undisturbed free surface. More details on the experimental setup, procedure and results can be found in the following references (Blinov, 1991; Blinov & Komarov, 1994). Below is a brief description of the experimental results.

Magnitude of turbulent oscillations, ε_t was in direct proportion to the gas jet speed. The generation of drops began as ε_t reached a threshold value, irrespective of whether the acoustic oscillations are applied or not. There was a tendency for the threshold value to slightly reduce with the sound wave application. In either case, once begun, the drop generation continues with the rate rising proportionally to ε_t. On the whole, the application of acoustic oscillations caused the drop generation rate to increase by 20~50% depending on the lance design.

One possible explanation for the drop generation mechanism and the acoustic effect on it is as follows. A gas flow reflected from the interface enhances the horizontal component of flow velocity in the liquid near the impact zone. As the gas flow velocity is very high, a high level of turbulent oscillations are generated in the flow. The turbulent oscillations disturb the gas-liquid interface that results in the formation of capillary waves. The separation of a drop happens at the instant at which the wave amplitude exceeds a threshold value, A_c. This is schematically shown in Figure 5, where A denotes the amplitude of the first largest crest of the wave. This amplitude is the following function of kinematic viscosity, ν and wave length, λ (Tal-Figiel, 1990)

$$A = \frac{4\nu}{f\lambda} \tag{27}$$

Note that here f is the frequency of oscillations generated in water.
The drop formation becomes possible at a threshold amplitude of capillary wave, A_c

$$A_C \geq (4 \sim 7)A \tag{28}$$

The length of a capillary wave can be found from the following equation (Tal-Figiel, 1990)

$$\lambda = \left(\frac{2\pi\sigma}{f^2\rho}\right)^{\frac{1}{3}} \tag{29}$$

where σ is the surface tension of liquid. Substituting this expression into formula (27) Eq.(30) can be obtained.

$$A = 2.169\nu\left(\frac{\rho}{f\sigma}\right)^{\frac{1}{3}} \tag{30}$$

The frequency, *f* was measured by means of the above-mentioned hydrophone.
In the absence of acoustic oscillations, *f* varied over a wide spectrum from 12.5 to 230 Hz, with the fundamental frequency ranging from 135 to 200 Hz. It was found that the fundamental frequency increases twice and more under the application of acoustic oscillations. This phenomenon is assumed to be the main reason for the observed enhancement in drop generation rate due to acoustic oscillations.

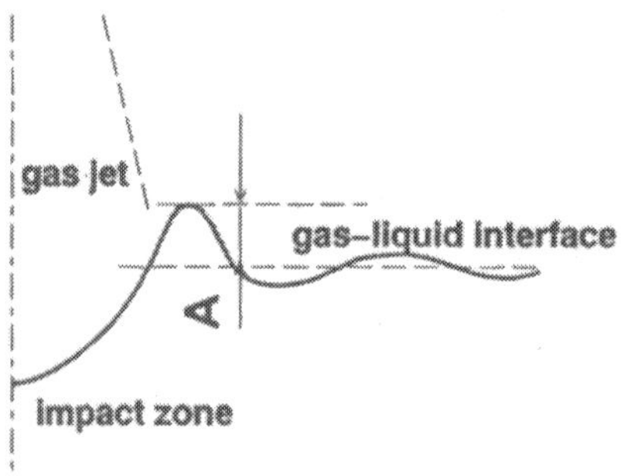

Fig. 5. A shematical representation of gas jet impact.

4.2 Mechanism of acoustically enhanced mass transfer

This section presents results of a cold model study concerning the possible effects of acoustic oscillations on the mass transfer characteristics with special emphasis on the influences of the oscillation frequency. In the experiments, the rates of the following gas-liquid absorption reactions were measured under different experimental conditions

$$2NaOH(aq) + CO_2 = Na_2CO_3(aq) + H_2O \quad (31)$$

$$2Na_2SO_3(aq) + O_2 = 2Na_2SO_4(aq) \quad (32)$$

$$O_2 = O_2(aq) \quad (33)$$

In these equations, aq denotes aqueous solution. A distinguishing characteristic of these reaction is that they proceed under different controlling regimes. The controlling mechanisms of these reactions were examined experimentally. The rate of the first reaction was found to be controlled by the interface mass transfer in both the liquid and gas phases. The second reaction proceeded under mixed control, the chemical reaction and interface gas-phase mass transfer. The rate of the third reaction, physical absorption of O_2 by water, was under the interface liquid-phase mass transfer control.

Figure 6 gives some details on the experimental setup used. This figure was reproduced from our previous paper (Komarov et al., 2007). The above-mentioned aqueous solutions or distilled water were filled into a cylindrical acrylic vessel 0.28 m in diameter and 0.47 m in height covered with an acrylic lid. The depth of the liquid bath was 0.2 m through all experiments. Gas, CO_2-N_2 mixture (reaction 31), air-N_2 mixture (reaction 32) or pure oxygen (reaction 33), was blown onto the liquid bath surface through a vertical tube (I.D.3 mm) fixed at the lid so that the distances between the axis lines of vessel and tube, and between the tube end and bath free surface was 0.02 and 0.13 m, respectively. The liquid bath was agitated by a 6 blade rotary impeller. The impeller was installed vertically at the vessel axis line. The flow rates of blown gas and the rotation speed of impeller were relatively low throughout these experiments. Hence, no drop generation ocurred and the area of the free surface of liquids was assumed to remain unchangeable regardless of the experimental conditions.

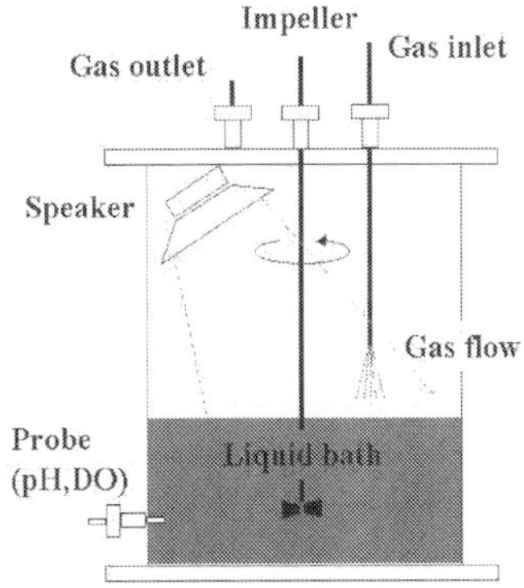

Fig. 6. Experimental setup for investigation of the acoustic effects on the mass transfer characteristics.

Sound waves were generated by using a powerful loudspeaker with the following characteristics: frequency range 70 ~ 18000 Hz, maximum input electrical power 50 W. The loudspeaker was fixed at the vessel lid so that the its vibrating diaphragm was inclined to the liquid free surface at an angle of about 27° as shown in Fig.6. Two broken lines show approximately the sound beam boundaries. The probe shown in the figure was used in order to measure the rates of reactions (31), pH probe, and (33), DO probe. Details on the measurement procedure can be found in our earlier publication (Komarov et al., 2007).
Exposing the gas-liquid interfaces to sound waves resulted in enhancement of the rates of all the above reactions, however the effect was different depending on the sound frequency, sound intensity, conditions of blowing gas and impeller rotation speed. For the reaction of CO_2 absorption, there was a tendency for the decrease in effect of acoustic oscillations as the velocity of blown gas increases and the rotation speed of impeller decreases. At the same gas velocity and rotation speed, the effect of sound for this reaction at a higher frequency was greater than that at a lower one. The largest enhancement in the mass transfer coefficient was 1.8 times at frequency of 15 kHz and gas velocity of 5 m/s. However, the frequency influence was rather complicated. There were frequencies at which the mass transfer coefficient peaked. Additional measurements of sound pressure level in the working space showed that the peaks originated from resonance phenomena occurring inside the vessel at certain frequencies.
The rate of Na_2SO_3 absorption was also enhanced with frequency. The measurements were performed at those frequencies where no resonance phenomena was observed. A 70% augmentation in the absorption rate was obtained within the frequency range of 0~7 kHz at a relatively high velocity of blowing gas, U_g =20 m/s. Therefore, the effect of sound application on this reaction appears to be stronger than that on the CO_2 absorption reaction.
The effect of acoustic oscillations on reaction (33) was significantly smaller as compared to those of reactions (31) and (32). The mass transfer coefficient rose appriximately by 20% as the oscillation frequency increased from 0 to 3 kHz. Notice that such a small effect was obtained at the velocity of blowing gas as low as 2.4 m/s.
Therefore, these three reactions can be arranged in order of increasing effect of sound on the reaction rates in the following way: physical absorption of oxygen by water, CO_2 absorption by NaOH aqueous solution and oxygen absorption by Na_2SO_3 aqueous solution. Thus, the above experimental results suggests definitely that the main reason of the increase in the absorption rates is an acoustically enhanced gas-phase mass transfer.
An analogy between turbulent and acoustic oscillations is thought to provide the best explanation for the mechanisms causing the observed mass transfer enhancement. Lighthill (Lighthill, 2001) was one of the firsts who noticed this analogy. In turbulent flows, fluid particles perform oscillations the amplitude and frequency of which are governed mainly by the flow velocity and the surrounding geometry. Propagation of a sound wave in a fluid medium causes fluctuations of the fluid particles too, with the only difference that they are oscillated at frequencies and amplitudes which are governed by the sound frequency and intensity (or pressure), respectively. This was confirmed by the following measurement results. Figure 7 shows the results of Fourier analysis of turbulent fluctuations generated near the air-water free surface exposed to a sound wave at a frequency of 880 Hz. This figure was reproduced from our previous paper (Komarov et al., 2007). The results make it clear that the fluctuation at the frequency of sound has much higher amplitude than fluctuations at the other frequencies. The measurements were performed at a distance of 2 mm above the surface using a highly sensitive hot-wire anemometry.

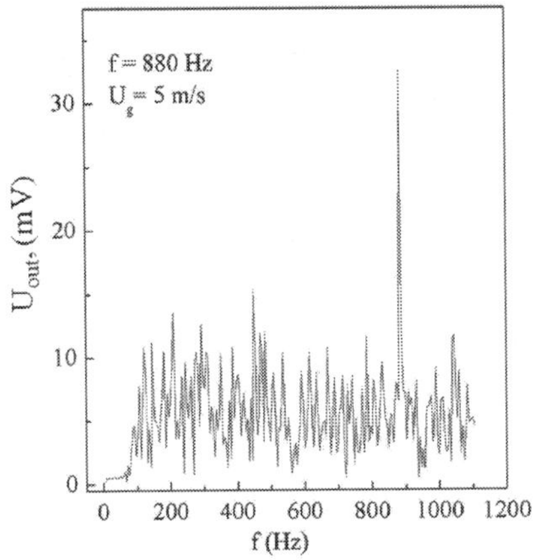

Fig. 7. Fourier analysis results of fluctuations imposed by a sound wave

Based on the above analogy, an attempt was made to explain the effects of sound frequency and intensity using relationships obtained for turbulent fluctuations. Omitting the intermediate transformations, the final expression for mass transfer coefficient, k can be written as follows

$$k \propto Sc^{-\frac{2}{3}} V_a^{\frac{1}{2}} \omega^{\frac{1}{4}} \tag{34}$$

where Sc is the Shmidt number (=ν/D), ν and D is the kinematic viscosity and diffusion coefficient, respectively, V_a is the average amplitude of velocity oscillations in acoustic boundary layer and ω is the angular frequency (=2πf). The thickness of acoustic boundary layer,δ is determined from Eq.(17). The underlying assumptions in deriving expression (34) were that, firstly, a viscous dissipation of acoustic oscillations occurs within the acoustic layer; secondly, the dissipation mechanism in gas phases at the gas-liquid interface is the same as that at the gas-solid interface. Readers interested in more detals are referred to the above-cited paper (Komarov et al., 2007)

Thus, expression (34) shows that the gas-phase mass transfer coefficient should increase with the one-forth power of the sound frequency. When this prediction is compared with the above experimental results, it becomes clear that the experiments show a little weaker frequency dependence. For example, according to the experimental results, the absorption rates of CO_2 and O_2 increased by 1.36 and 1.44 times within frequency ranges of 3 ~15 kHz and 1~7 kHz, respectively. However, for these frequency ranges, relationship (34) predicts enhancement in k by 1.5 and 1.63 times, respectively. The reason of why the experiments show less enhancement effect as compared with the predictions is that the above two reactions are controlled by the gas-phase mass transfer rate only in part as it has been explained in the previous sections.

Also, relationship (34) reveals that mass-transfer coefficient is proportional to the square root of oscillatory velocity amplitude, Va, that is considered to be a characteristic of sound field pressure. However, experimental verification of this prediction presents a considerable difficulty under the present experimental conditions. The reason is that, since sound waves propagate inside the vessel, they experience multiple reflections from both the free surface of liquid and vessel wall that causes resonance-like phenomena. This results in appearance of the above-mentioned maximums of mass transfer coefficient at certain frequencies, and makes it difficult to measure the sound pressure at the free surface of water bath.

As it has been briefly mentioned above, the effect of acoustic oscillations on mass transfer rate reduces with increasing the velocity of gas blown onto the free surface. This tendency is readily apparent from the finding that both the gas flows and sound waves produce turbulent oscillations at the gas-liquid interface. In high temperature processes, which use high velocity gas jets, the turbulence level in the gas phase should be very high. Under such conditions, the acoustic effects should be weak. Therefore, it would be interesting to estimate the threshold amplitudes of sound waves at which they are still effective in enhancing the gas-phase mass transfer for a given magnitude of the gas turbulent oscillations.

These estimates were made by considering two types of turbulent diffusion coefficients at the gas-liquid interface: that which originates from natural turbulent fluctuations of high speed gas flow, D_t and that which results from imposed acoustic oscillations, D_a. The expressions for these coefficients have been derived in the following form (Komarov et al., 2007)

$$D_t = \frac{0.4\rho v_t^3}{\sigma} z^2 \tag{35}$$

$$D_a = 0.8 V_0 k z^2 \tag{36}$$

where ρ is the density of gas, σ is the surface tension of liquid, v_t is the characteristic turbulent velocity, V_0 is the amplitude of oscillation velocity, k is the wave number of sound wave and z is the distance from the liquid surface.

By equating D_t with D_a, one can obtain a threshold velocity amplitude of sound wave, V_0^* at which the effects of natural turbulence and acoustically imposed oscillations on the gas phase mass-transfer rate are equal.

$$V_0^* = \frac{\rho v_t^3}{2k\sigma} \tag{37}$$

This expression suggests the following. First, in high intensity turbulent flows, the sound waves should have very high oscillation amplitudes in order to be effective. Second, the threshold velocity amplitude decreases with increasing the sound frequency if the other parameters are fixed. Recall that the wave number is proportional to sound frequency as described in the section 2.1.

Using Eq.(37), one can estimate V_0^* for conditions of the present cold model experiments and pilot converter tests. The estimate results are shown in Fig. 8 as plots of V_0^* versus v_t for the cold model at frequencies of 1 (line 1) and 10 kHz (line 2), and for the pilot converter at a frequency of 10 kHz (line 3), respectively. Two right vertical axes indicate the sound intensity level (SIL) in decibel units determined according to Eq.(5). This figure was reproduced from our previous paper (Komarov et al., 2007).

The following values of the physical properties were used in the estimates: ρ = 1.2 kg/m^3 and σ = 0.07 N/m for air-water system at 20°C, and ρ = 0.18 kg/m3 and σ = 1.4 N/m for the converter process including a CO gas atmosphere and molten steel at 1600°C. The sonic speeds were taken to be equal to 340 and 860 m/s for room and high temperatures, respectively. The shaded areas indicate the approximate ranges of the variation in v_t and V_0^*. The characteristic turbulent velocity was determined assuming a 5 % level of turbulence relative to the gas velocity at the nozzle exit. For the cold model conditions, the values of SIL were estimated to be 110~130 dB.

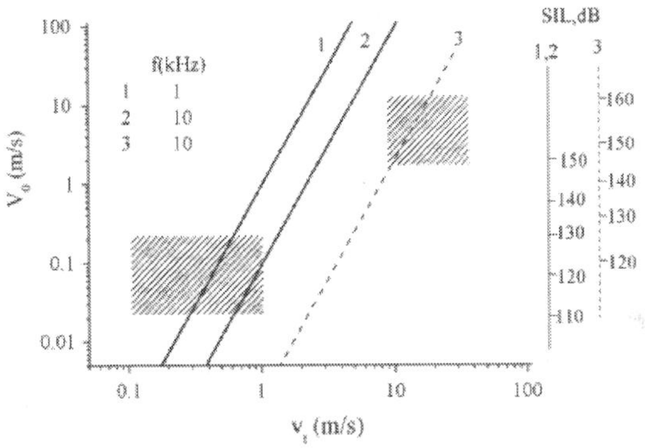

Fig. 8. A plot of V_0 versus v_t: 1 and 2 - cold model, 3 – pilot converter

Estimates of V_0^* for the pilot converter process, assuming that v_t is varied in the range of 10 to 35 m/s, suggest that the acoustic oscillations can be capable of enhancing the mass-transfer rate at sound pressure levels of 145~160 dB and frequency of 10 kHz. These estimates appear to be consistent with the above experimental observations.

Thus, the results presented in this section allow the following conclusions to be drawn. Application of high-intense acoustic oscillations causes the rate of decarburization reaction to enhance. A few mechanisms appear to exist that can result in this enhancement. The first one is the acoustically enhanced generation of drops at the crater zone where the high speed oxygen jet impact with the molten metal bath. The metal drops are oxidized by gaseous oxygen to FeO when flying through the crater zone. The acoustically imposed oscillations enhance the oxidation rate of drops through the above considered intensification of turbulent fluctuations and acoustic streaming at the drop surface. When such oxidized drops are delivered into the slag, its oxygen potential is assumed to become higher as compared with conventional blowing operation. As a result, the rates of decarburization and phosphor removal are enhanced.

4. Concluding remarks

Recently, considerable research efforts have been devoted to the investigation of acoustic oscillations for improving the performance of processes involving high and elevated temperatures. The research results have strongly suggested that the acoustic oscillations have the potential to enhance the efficiency of those processes, the rate of which is controlled by gas-phase mass transfer. Examples include, but not limited to, combustion of liquid and solid fuels, treatment of high temperature exhaust gas, steelmaking converter and Peirce-Smith converter for the refining of cooper. At higher temperatures, attractiveness of sonic and ultrasonic waves is associated with its ability to propagate through gas, and thus to transfer the acoustic energy from a gas or water cooled ultrasonic generator to a higher temperature area for material processing. Furthermore, if the wave intensity is high enough, its propagation initiates such non-linear phenomena as acoustic streaming, forced turbulence and capillary waves which are the prime causes of acoustic effects, especially at the gas-liquid or gas-solid interfaces.

A survey reveals that amplitude of acoustic oscillations plays a crucial role in the enhancement of gas-phase mass transfer from objects like particles and drops, while the effect of frequency is less pronounced. According the reported results, the mass transfer coefficient is increased in proportional to the 0.5~1.0 power of the velocity amplitude. Two controlling regimes were found to be important: 1- acoustic streaming controlling and 2 - vortex shedding controlling regime.

Experimental results have showed that the high-intense acoustic oscillations are capable of enhancing the rate of gas-phase mass transfer controlling reactions in steelmaking converter process. The following two mechanisms were found to play an important role in this enhancement: 1- acoustically enhanced generation of molten metal drops, 2 - acoustically intensified turbulent fluctuations and acoustic streaming at the drop surface.

Industrial competitiveness of the ultrasonic-based technologies is reinforced by relatively low cost of the power-generating equipment. In some special cases, the acoustic energy can be produced without any additional energy consumption by means of a comparatively simple device. An example is the pneumatic sonic generator applied to a process which uses gas blowing or injection.

5. References

Abramov, O. V. (1998). *High-Intensity Ultrasonics : Theory and Industrial Applications*, Gordon and Breach Science Publishers, ISBN : 9789056990411, Amsterdam.

Abramov, O. V; Khorbenko, S. G. & Shvelga, S. (1984). *Ultrasonic Processing of Materials*, Mashinostroenie, Moscow, [in Russian]

Bird, R. B.; Steart, W. E. & Lightfoot E.N. (2001). *Transport Phenomena*, John Wiley & Sons, Inc., ISBN 0-471-41077-2 , New York.

Blaszczyk, J. (1991). Acoustically disturbed fuel droplet combustion. *Fuel*, Vol.70,No.9, 1023-1025 , ISSN 0016-2361.

Blinov, K. A. (1991). *Application of Acoustic Oscillations for Steelmaking Processes*, Metallurgia, Chelyabinsk, Russia, [in Russian].

Blinov, K. A. & Komarov, S. V.(1994). Liquid dispersion and fine particle coagulation in depevopment of metallurgical processes. *Proceeding of the Second International Symposium on Metallurgical Processes for the Year 2000 and Beyond and the 1994 TMS Extraction and Process Metallurgy Meeting*.pp.413-427, ISBN 10 0873392418, Sxan Diego, California, September 20-23, TMS, Warrendale, Pa.

Borisov, Yu. Ya. (1967). Hartmann type gas-jet oscillators, In : Source of Power Ultrasound, Rozenberg L.D. (Ed.), 7-110,Nauka,Moscow [in Russian].

Carlson, J & Martinsson, P. E.(2002). A Simple Scattering Model For Measuring Particle Mass Fractions In Multiphase Flows. *Ultrasonics*, Vo.39,585-590, ISSN 0041-624X

David, J & Cheeke, N. (2003). *Fundamentals and Applications of Ultrasonic Waves*, CRC Press,ISBN 9780849301308, Boca raton.

Delabroy, O.; Lacas, F.; Poinsot, T.; Candel, S.; Hoffmann, T.; Hermann, J.; Gleis, S. & Vortmeyer, D .(1996). A study of NOx reduction by acoustic excitation in a liquid fueled burner. *Combustion Science and Technology,* Vol.119, 397-408, ISSN 0010-2202.

Demare, D. & Baillot, F. (2004). Acoustic enhancement of combustion in lifted nonpremixed jet flames. *Combustion and Flame,* Vol.139,No.4, 312-328, ISSN 0010-2180.

Fand,R.M. & Cheng,P.(1962). The Influence of Sound on Heat Transfer from a Cylinder in Crossflow. *Int.J.Heat Mass Transfer,* Vo.6, 571-596, ISSN 0017-9310.

Gopinath,A. & Mills, A.F.(1994). Convective Heat Transfer Due to Acoustic Streaming Across the Ends of Kundt Tube. *Journal of Heat Tranfer,*Vo.116, 47-53, ISSN 1528-8943

Hamilton, M. F & Blackstock, D. T. (1998). *Nonlinear Acoustics, Theory and Applications,* Academic Press, ISBN : 0123218608 , California

Hardalupas, Y. & Selbach, A. (2002). Imposed oscillations and non-premixed flames, Prog Energ Combust, Vol.28, 75-104, ISSN 0360-1285.

Hertzberg, J. (1997). Conditions for a Split Diffusion Flame. *Combustion and Flame,* Vol.109, 314-322, ISSN 0010-2180.

Hueter, T. F. & Bolt, R. H. (1966). *Sonics,* John & Sons, Inc., ISBN, New York

Kawahara, N.; Yarin, A. L.; Brenn, G.; Kastner, O. & Durst, F. (2000). Effect of acoustic streaming on the mass transfer from a sublimating sphere. *Physics of Fluids,* Vo.12.,No.4 ,912-923, ISSN 1070-6631.

Keller, J. O.; Bramlette, T. T.; Barr P.K. & Alvarez J. R.(1994). NO_x and CO emissions from a pulse combustor operating in a lean premixed mode. *Combustion and Flame,* Vol.99, 460-466, ISSN 0010-2180.

Koepke, S. A & Zhu, J. Y. (1998). Pyrolysis of Black Liquor in a High-intensity Acoustic Field. *Combustion Science and Technology,* Vol.140, 315-331, ISSN 0010-2202.

Komarov,S.V.; Kuwabara, M.& and Abramov, O.V. (2005). High Power Ultrasonics in Pyrometallurgy: Current Status and Recent Developments. *ISIJ International,* Vol. 45, 1765-1782, ISSN 0915-1559.

Komarov, S. V.; Noriki, N.; Osada, K.; Kuwabara, M. & Sano, M. (2007). Cold Model Study on Mass-Transfer Enhancement at Gas-Liquid Interfaces Exposed to Sound Waves. Metallurgical and Materials Transactions, Vol.38B, 809-818, ISSN 1073-5615

Kumagai, S. & Isoda, H. (1955). Combustion of Fuel Droplets in a Vibrating Air Field. *Proceeding the Fifth International Symposium on Combustion,* pp.129-132, Pittsburgh, May 1955, Reinhold Publishing, New York.

Landau, L. D & Lifshits, E. M. (1986). *Fluid Mechanics,* Nauka, Moscow,Russia [in Russian].

Larsen, P. S. & Jensen, J. W.(1977). Evaporation Rates of Drops in Forced Convection with Superposed Transverse Sound Field. *Int.J.Heat Mass Transfer,* Vo.21, 511-517, ISSN 0017-9310.

Leighton, T. G. (2004). From seas to surgeries, from babbling brooks to baby scans: The acoustics of gas bubbles in liquids. *International Journal of Modern Physics B.,* Vol.18, 3267-3314, ISSN 0217-9792.

Lighthill, J. (1978),Acoustic Streaming. *Journal of Sound and Vibration,* Vo.61, No.3, 391-418, ISSN 0022-460X.Lighthill,J. (2001), *Waves in Fluids,* Cambridge University Press, ISBN 0-521-01045-4, Cambridge.

Loh, B. G. & Lee D. R. (2004). Heat Trabsfer Characteristics of Acoustic Streaming by Longitudinal Ultrasonic Vibration. *J Thermophys Heat Tran,* Vo.18, No.1, 94-99, ISSN 0887-8722

Makarov, S. & Ochmann, M. (1996). Non-linear and thermoviscous phenomena in acoustics. Part I. *Acustica,* Vol.82, 579-606, ISSN 00017884 .

Makarov, S. & Ochmann,M.(1997). Non-linear and thermoviscous phenomena in acoustics. Part II. *Acustica,* Vol.83, 197-222, ISSN 00017884.

Margulis, M.A. (1995). *Sonochemistry and Cavitation,* Taylor & Francis, ISBN: 978-2881248498, Amsterdam.

Mason, W. P. & Thurston, R. N. (1965). *Physical Acoustics,* Vol.4, Part B, Academic Press Inc., New York.

Mcquay, M. Q.; Dubey, R. K. & Nazeer, W. A. (1998). An experimental study on the impact of acoustics and spray quality on the emissions of CO and NO from an ethanol spray flame. *Fuel,* Vol.77, 425-435, ISSN 0016-2361.

Mednikov, E. P. (1965). *Acoustic coagulation and precipitation of aerosols* (authorized translation from the Russian by Chas V. Larrick), USSR Academy of Sciences Press, Moscow.

Oeters, F. (1994). *Metallurgy of Steelmaking.* pp.47-54, Stahleisen, Dusseldorf

Rayleigh, J. W. S. (1945). *Theory of Sound,* Vol.1,2. Dover Publications, N.Y.

Saito, M.; Sato, M. & Suzuki, I. (1994). Evaporation and combustion of a single fuel droplet in acoustic fields. *Fuel,* Vol.73, No.3, 349-353 , ISSN 0016-2361.

Saito, M.; Sato, M. & Nishimura, A. (1998). Soot suppression by acoustic oscillated combustion. *Fuel,* Vol.77, No.9-10, 973-978, ISSN 0016-2361.

Sujith, R. I.; Waldherr, G. A.; Jagoda, J. I. & Zinn, B. T. (2000). Experimental Investigation of the Evaporation of Droplets in Axial Acoustic Fields, *Journal of Propulsion and Power,* Vo.16, 278-285, ISSN 0748-4658.

Sung, H. J.; Hwang, K. S. & Hyun, J.M. (1994). Experimental Stydy on Mass Transfer from a Circular Cylinder in Pulsating Flow. *Int.J.Heat Mass Transfer,* Vo.37, No.15, 2203-2210, ISSN 0017-9310.

Tal-Figiel, B. (1990). Conditions for instability of the liquid-liquid interface in a ultrasoinc field. *International Chemical Engineering,* Vo.30, No.3, 526-534, ISSN 1385-8947.

Temkin, S.(1998).Sound propagation in dilute suspensions of rigid particles. *J. Acoust. Soc.Am,* Vol.103, No.2,838-849, ISSN 0001-4966

Uhlenwinkel, V.; Meng, R. & Bauckhage, K. (2000). Investigation on heat transfer from circular cylinders in high power 10 kHz and 20 kHz acoustic resonant fields. *Int.J.Therm.Sci.,* Vo.39, 771-779, ISSN 1290-0729.

Vainstein, P.; Fichman, M. & Gutfinger, C.(1995). Acoustic Enhancement of heat transfer between two parallel plates. *Int.J.Heat Mass Transfer,* Vo.38, No.10, 1893-1899, ISSN 0017-9310

Yarin, A. L.; Brenn, G.; Kastner, O.; Rensink, D. & Tropea, C. (1999). Evaporation of acoustically levitated droplets. *J. Fluid Mech.,* Vol.399, 151-204, ISSN 0022-1120.

Yavuzkurt, S.; Ha, M. Y.; Koopman, K. & Scaroni, A. W. (1991,a). A Model of the Enhancement of Coal Combustion Using High-Intensity Acoustic Fields. *Journal of Energy Resources Technology,* Vol.113, No.4, 277-285, ISSN 0195-0738.

Yavuzkurt, S.; Ha, M. Y.; Reethof, G.; Koopmann, G. & Scaroni, A. W.(1991,b). Effect of an acoustic field on the combustion of coal particles in a flat flame burner. *Journal of Energy Resources Technology,* Vol.113, Vol.4, 286-293, ISSN 0195-0738.

Zarembo, L. K. (1971). Acoustic streaming. In : *High-Intensity Ultrasonic Fields,* Rozenberg L.D. (Ed.), 137-199, Plenum, New York.

5

Mass-transfer in the Dusty Plasma as a Strongly Coupled Dissipative System: Simulations and Experiments

Xeniya Koss, Olga Vaulina, Oleg Petrov and Vladimir Fortov
Institution of Russian Academy of Sciences
Joint Institute for High Temperatures RAS
Russia

1. Introduction

The problems associated with the mass-transfer processes in dissipative systems of interacting particles are of great interest in various fields of science (plasma physics, medical industry, physics and chemistry of polymers, etc.) (Frenkel, 1946; Cummins & Pike, 1974; Balescu, 1975; March & Tosi, 1995; Ovchinnikov et al., 1989; Dodd et al., 1982; Thomas & Morfill, 1996; Fortov et al., 1996; Fortov et al., 1999). Nevertheless, the hydrodynamic approaches can successfully describe these processes only in the case of the short-range interactions between particles. The main problem involved in studies of non-ideal systems is associated with the absence of an analytical theory of liquid. To predict the transport properties of non-ideal systems, the various empirical approaches and the computer simulations of dynamics of the particles with the different models for potentials of their interaction are used (Frenkel, 1946; Cummins & Pike, 1974; Balescu, 1975; March & Tosi, 1995; Ovchinnikov et al., 1989). The simulations of transport processes are commonly performed by methods of molecular dynamics, which are based on solving of reversible motion equations of particles, or Langevin equations taking into account the irreversibility of the processes under study.

The diffusion is the basic mass-transfer process, which defines the losses of energy (dissipation) in the system of particles and its dynamic features (such as the phase state, the conditions of propagation of waves and the formation of instabilities). When the deviations of the system from the statistical equilibrium are small, the kinetic coefficients of linear dissipative processes (constants of diffusion, viscosity, thermal conductivity etc.) can be found from Green-Kubo formulas that were established with the help of the theory of Markovian stochastic processes under an assumption of the linear reaction of the statistical system on its small perturbations. These formulas are the important results of the statistical theory of irreversible processes. According to these formulas, the diffusion coefficient D can be found from the following relationship:

$$D = \int_0^{\infty} \langle V(0)V(t) \rangle dt \, / \, m \, . \qquad (1)$$

Here $<V(0)V(t)>$ is the velocity autocorrelation function (VAF) of grains, t is the time, and m is the dimension of the system. The diffusion coefficient can be also obtained from the analysis of a thermal transfer of the grains through the unit area of the medium:

$$D = \lim_{t \to \infty} <(\Delta l)^2>/(2mt), \tag{2}$$

where $\Delta l = \Delta l(t)$ is the displacement of an isolated particle from its initial position during the time t. In both equations (1), (2), the brackets < > denote the ensemble and time averaging (the averaging over all time intervals with the duration t). As the relationships (1)-(2) were obtained without any assumptions on a nature of a thermal motion, they are valid for gases as well as for liquids and solids in the case of the small deviations of the system from its steady state condition. In the general case of non-ideal fluids, the analytical solutions of Eqs.(1)-(2) are unavailable wich makes impossible to find the diffusion coefficient. The simple solution, $D \equiv D_0 = T/(\nu_{fr}M)$, known as the Einstein relationship, exists only for the non-interacting ("brownian") particles; here M and T are the mass and the temperature of a grain, respectively, and v_{fr} is the friction coefficient.
Due to the existing level of experimental physics, it is necessary to go out of the bounds of diffusion approximation, and modern methods of numerical simulation (based on the theory of stochastic processes) allow one to make it. A description within the macroscopic kinetics may be insufficient for the analysis of mass-transfer processes on physically small time intervals. A study of the mass-transfer processes on short observation times is especially important for investigation of fast processes (e.g. the propagation of shock waves and impulse actions, or progression of front of chemical transformations in condensed matter (Ovchinnikov et al., 1989; Dodd et al., 1982)), and also for the analysis of transport properties of strongly dissipative media (such as colloidal solutions, plasma of combustion products, nuclear-induced high-pressure dusty plasma (Cummins & Pike, 1974; Fortov et al., 1996; Fortov et al., 1999)), where the long-term experiments should be carried out to measure the diffusion coefficients correctly.

2. Mass-transfer processes in non-ideal media

Consider the particle motion in a homogeneous dissipative medium. One can find a displacement of j-th particle in this medium along one coordinate, $x_j = x_j(t)$, under an action of some potential F and random F_{ran} forces from the Langevin equation

$$M\frac{d^2x_j}{dt^2} = -Mv_{fr}\frac{dx_j}{dt} + F + F_{ran}. \tag{3}$$

In a statistical equilibrium of system of particles ($M<(dx_j/dt)^2> = <MV_x(t)^2> \equiv T$) the mean value of the random force is zero, $<F_{ran}(t)> = 0$, and its autocorrelation function $<F_{ran}(0)F_{ran}(t)>= 2B\delta(t)$ corresponds to the delta-correlated Gaussian process, where $\delta(t)$ is the delta-function, and $B = T\nu_{fr}M$ (due to the fluctuation-dissipation theorem). Under these assumptions the Eq.(1) describes the Markovian stochastic process.
To analyze a dependence of mass-transfer on the time t, we introduce the following functions:

$$D_{G\text{-}K}(t) = \int_0^t \langle V_x(0)V_x(t)\rangle dt \tag{4a}$$

$$D_{msd}(t) = <(x_j)^2>/(2t) \tag{4b}$$

where $V_x(t)=dx_j/dt$ is the velocity of a j-th particle. With the small deviations of the system from the equilibrium state, both functions ($D_{G\text{-}K}(t)$ and $D_{msd}(t)$) with $t \to \infty$ should tend to the same constant $D = \lim_{t\to\infty} D(t)$, which corresponds to the standard definition of diffusion coefficient.

Neglecting the interparticle interaction ($F = 0$: the case of "brownian" particles), one can find the VAF, using the formal solution of Eq.(3) under assumption of $<F_{ran}(t)\ V_x(0)> = 0$ (Cummins & Pike, 1974):

$$\langle V_x(0)V_x(t)\rangle = \frac{T}{M}\exp\left(-\nu_{fr}t\right) \tag{5}$$

Then the mass-transfer evolution function $D_{G\text{-}K}(t)$, Eq.(4a), may be written as

$$D_{G-K}(t) = D_0\left(1-\exp\left(-\nu_{fr}t\right)\right) \tag{6a}$$

To find the mean-square displacement of a j-th particle, one should multiply Eq.(3) by x_j. Then, if there is no correlation between the slow particle motion and the "fast" stochastic impact ($<F_{ran}x_j>=0$), the simultaneous solution of Eqs.(3), (4b) in a homogeneous medium ($M<(dx_j/dt)^2> \equiv T$, $<(\Delta l)^2> = m<x_j^2>$) can be presented as (Ovchinnikov et al., 1989)

$$D_{msd}(t)/D_o = 1-\left(1-\exp\left(-\nu_{fr}t\right)\right)/\nu_{fr}t \tag{6b}$$

Thus, for the "brownian" case, when $t \to \infty$ and $\upsilon_{fr}t \gg 1$, we have $D_{G\text{-}K}(t) = D_{msd}(t) \to D_o$, and on small time intervals ($\upsilon_{fr}t \ll 1$) the motion of particles has a ballistic character: $<x^2>\equiv<x_j^2>\approx T\ t^2/M$ and $D_{msd}(t) = <x^2>/(2t) \propto t$.

The analytical solution of Eq. (3) may be also obtained for an ideal crystal under assumption that the restoring force $F = -M\omega_c^2 x_j$ acting on particles in lattice sites can be described by the single characteristic frequency ω_c (the case of harmonic oscillator). In this case we will have

$$M\frac{d^2x_j}{dt^2} = -M\upsilon_{fr}\frac{dx_j}{dt} - M\omega_c^2 x_j + F_{ran} \tag{7}$$

After multiplying both parts of this equation by $x = x_j$, rearranging and averaging, taking into account that $\langle F_{ran}x\rangle = 0$ and $M<(dx/dt)^2> =2<V_x(t)^2>\equiv T$), we will obtain (Vaulina et al., 2005b):

$$M\frac{d^2\langle x^2\rangle}{dt^2} = -M\upsilon_{fr}\frac{d\langle x^2\rangle}{dt} - 2M\omega_c^2\langle x^2\rangle + 2T \tag{8}$$

Then the simultaneous solution of Eq.(8) and Eq. (4b) can be written as

$$\frac{D_{msd}(t)}{D_o} = \frac{1-\exp(-\nu_{fr}t/2)\left(\cosh(\nu_{fr}t\psi)+\sinh(\nu_{fr}t\psi)/\{2\psi\}\right)}{2\xi_c^2\nu_{fr}t} \tag{9}$$

where $\psi=(1-8\xi_c^2)^{1/2}/2$, and $\xi_c = \omega_c/\nu_{fr}$. In the case of $(1-8\xi_c^2) < 0$, the ψ value is imaginary: $\psi = i\psi^*$, where $\psi^* = (8\xi_c^2-1)^{1/2}/2$. In this case, $\sinh(i\psi^*\nu_{fr}t)=i\sin(\psi^*\nu_{fr}t)$, $\cosh(i\psi^*\nu_{fr}t)=\cos(\psi^*\nu_{fr}t)$, and the expression for $D_{msd}(t)$ function will include the trigonometric functions instead of the hyperbolic functions.

To define VAF, $\langle V_x(0)V_x(t)\rangle \equiv \langle V_x(t_o)V_x(t_o+\tau)\rangle$, we will use the following designations: $V_x(t_o)\equiv V_0$, $V_x(t_o+\tau)\equiv V$, $X(t_o)=x$, $X(t_o+\tau)\equiv x+\Delta x$. The Eq.(7) can be presented as two expressions at two various instants of time (t = t_o and t= t_o+τ). Then, we can multiply the first of them by V, and the second by V_o, respectively. The sum of these two expressions, averaged on the particle's ensemble for all time intervals with the duration $t=\tau$ (taking into account, that $<x\Delta x>=0$, $\langle F_{ran}(t_o)V(t_o+\tau)\rangle = \nu_{fr}M\langle V(t_o)V(t_o+\tau)\rangle$, $\langle F_{ran}(t_o+\tau)V(t_o)\rangle = 0$ (Cummins & Pike, 1974; Ovchinnikov et al., 1989)), can be written as

$$\frac{d\langle V_oV\rangle}{dt} = -\upsilon_{fr}\langle V_oV\rangle - \omega_c^{\,2}\frac{d\langle x^2\rangle}{dt} \tag{10}$$

The solution of this equation is

$$<V_oV>= \frac{1}{2}\frac{d^2\langle x^2\rangle}{dt^2}, \tag{11}$$

and the Eq.(10) may be rewritten as

$$\frac{d^2\langle V_oV\rangle}{dt^2} = -\upsilon_{fr}\frac{d\langle V_oV\rangle}{dt} - 2\omega_c^{\,2}\langle V_oV\rangle. \tag{12}$$

Thus, in this case of harmonic oscillator, we will have for VAF

$$\langle V_x(0)V_x(t)\rangle = \frac{T}{M}\exp\left(-\nu_{fr}t/2\right)\left(\cosh(\nu_{fr}t\psi) - \sinh(\nu_{fr}t\psi)/\{2\psi\}\right) \tag{13}$$

and the simultaneous solution of Eq.(13) and Eq.(4a) can be written as

$$\frac{D_{G-K}(t)}{D_0} = \frac{\exp\left(-\nu_{fr}t/2\right)}{\psi}\cdot\sinh\left(\nu_{fr}t\psi\right). \tag{14}$$

When ψ is imaginary, the both expressions for VAF and $D_{msd}(t)$ functions (Eqs. (13), (14)) will include the trigonometric functions instead of the hyperbolic ones (see above).

The normalized VAF, $f(t) = M<V_x(0)V_x(t)>/T$, and mass-transfer evolution functions ($D_{G-K}(t)/D_o$, $D_{msd}(t)/D_o$) for various values of ξ_c are presented in Fig. 1, where the time is given in units of inverse friction coefficient (υ_{fr}^{-1}). It is easy to see that for short observation times a particle in a lattice site also has the ballistic character of motion ($<x^2> \approx T t^2/M$, $D_{msd}(t)$= $<x^2>/(2t) \propto t$). With the increasing time ($\upsilon_{fr}t \gg 1$) both evolution functions tend to zero: $D_{G-K}(t) = D_{msd}(t) \rightarrow 0$, because for the harmonic oscillator the mean-square declination $<(\Delta l)^2>$ is constant: $<(\Delta l)^2> = mT/(M\omega_c^2)$.

For liquid media, the exact analytic expression for $<V_x(0)V_x(t)>$, $D_{G-K}(t)$ and $D_{msd}(t)$ can't be obtained. Nevertheless, we should note some features referring to the relations between the mentioned functions, both in the case of "brownian" particles (see Eqs. (5)-(6b)) and in the case of harmonic oscillator (see Eq. (13)), which may take place for liquids:

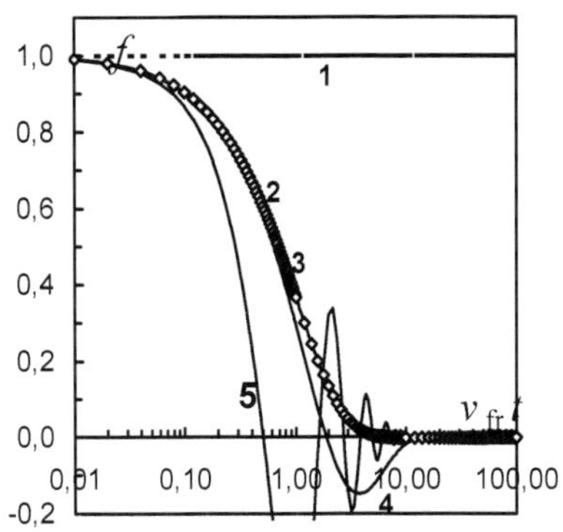

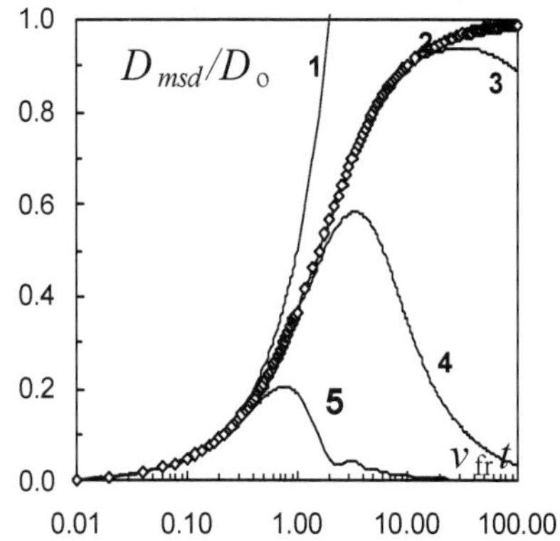

Fig. 1. Functions $f(t\nu_{fr})$ (**a**) and $D_{msd}(t\nu_{fr})/D_o$ (**b**) for : **1** - ballistic mode ($f(t)$=1, $D_{msd}(t) \propto t$); **2** - "brownian" case (Eqs.(5), (6b)); and for harmonic oscillator (Eqs.(13), (9)) with the different ξ_c: **3**- 0.033; **4** - 0.38; **5** - 2.

$$D_{G\text{-}K}(t) = \frac{d\{tD_{msd}(t)\}}{dt} \equiv \frac{1}{2}\frac{d\langle x^2\rangle}{dt}, \tag{15a}$$

$$\langle V_x(0)\,V_x(t)\rangle = \frac{d^2\{tD_{msd}(t)\}}{dt^2} \equiv \frac{1}{2}\frac{d^2\langle x^2\rangle}{dt^2}. \tag{15b}$$

The mean-square displacement evolution $D_{msd}(t)$ was studied numerically in (Vaulina et al., 2005b; Vaulina & Dranzhevski, 2006; Vaulina & Vladimirov, 2002) for non-ideal systems with a screened Coulomb pair interaction potential (of Yukawa type):

$$U = (eZ)^2\exp(-r/\lambda)/r\,. \tag{16}$$

Here r is a distance between two particles with a charge eZ, where e is an electron charge, λ is a screening length, $\kappa = l_p/\lambda$, and l_p is the mean interparticle distance, which is equal the inverse square root from the surface density of particles for two-dimensional (2d-) systems, and it is the inverse cubic root of their bulk concentration in three-dimensional (3d-) case. As a result of numerical simulation, the characteristic frequencies for the body-centered cubic (*bcc*-) lattice ($\omega_c = \omega_{Bcc} \approx 2eZ\exp(-\kappa/2)[(1+\kappa+\kappa^2/2)/(l_p^3M\pi)]^{1/2}$) and for hexagonal lattice ($\omega_c = \omega_h \approx 1.16\omega_{Bcc}$) were obtained (Vaulina et al., 2005b; Vaulina & Dranzhevski, 2006; Vaulina & Vladimirov, 2002). It was also shown that these frequencies are responsible for the mean time t_a of "settled life" of particles in liquid-like systems and their values define the evolution of mass-transfer for the observation times $t < t_a \approx 2/\omega_c$. Taking into account Eqs.(15a)-(15b), it is easy to assume that the behavior of VAF and $D_{G\text{-}K}(t)$ for the mentioned observation times in liquid-like Yukawa systems is also close to the behavior of these functions for harmonic oscillator .

The dynamics of 3d- systems of particles with the various types of pair isotropic potentials was numerically investigated in (Vaulina et al., 2004). Those potentials represented different combinations of power-law and exponential functions, commonly used for simulation of repulsion in kinetics of interacting particles (Ovchinnikov et al., 1989):

$$U = U_c\,[b_1 \exp(-\kappa_1\, r/l_p) + b_2\,(l_p/r)^n \exp(-\kappa_2\, r/l_p)]. \tag{17}$$

Here $b_{1(2)}$, $\kappa_{1(2)} = l_p/\lambda_{1(2)}$ and n are variable parameters, and $U_c = (eZ)^2/r$ is the Coulomb potential. In the context of investigation of dusty plasma properties, the screened Coulomb potential (16) (b_1=1, b_2=0, $\kappa_1 = l_p/\lambda$) is of particular interest. But it should be noted that the simple model (16) agrees with numerical and experimental results in a complex plasma only for short distances $r < \lambda$ between two isolated macro-particles in a plasma (Konopka et al., 1997; Daugherty et al., 1992; Allen, 1992; Montgomery et al., 1968). With increasing distance, the effect of the screening weakens, and the asymptotic character of the potential U for large distances $r >> \lambda$ can follow the power-law dependence: $U \propto r^{-2}$ (Allen, 1992) or $U \propto r^{-3}$ (Montgomery et al., 1968); thus, the parameters of the potentials (17) will be $\kappa_1 = l_p/\lambda$; $\kappa_2 = 0$; n =1-2; $b_1 >> b_2$, respectively.

It was noticed that the mass-transfer processes and spatial correlation of macroparticles in these 3d- systems are defined by the ratio of the second derivative U'' of a pair potential $U(r)$ in the point of the mean interparticle distance $r = l_p$ to the grains' temperature T, if the following empirical condition is met (Vaulina et al., 2004):

$$2\pi > |\,U'(l_p)\,|\, l_p \,/\, |\,U(l_p)\,| > 1. \tag{18}$$

In this case, the spatial correlation of particles didn't depend on the friction (ν_{fr}) and was defined by the value of effective coupling parameter $\Gamma^* = M l_p^2 U''/(2T)$ in the range between $\Gamma^* \sim 10$ and the point of crystallization of the system ($\Gamma^* \sim 100$), where for all considered cases the formation of *bcc*- structure was observed with the characteristic oscillation frequency of the grains (Vaulina et al., 2004):

$$\omega_c^2 = \omega_{bcc}^2 \equiv 2\,|\,U''(l_p)\,|\,/\,(\pi M). \tag{19a}$$

It is to expect that the characteristic oscillation frequency in the hexagonal lattice for the grains, interacting with the potentials (17), may be written similarly to the frequency found for the quasi-2d- Yukawa systems (Vaulina et al., 2005b; Vaulina & Dranzhevski, 2006):

$$\omega_c^2 = \omega_h^2 \equiv 2.7\,|\,U''(l_p)\,|\,/\,(\pi M). \tag{19b}$$

The behavior of VAFs and of the evolution functions, $D_{msd}(t)$, $D_{G\text{-}K}(t)$, are studied numerically in the next part of this paper for quasi-2d- and 3d- non-ideal systems with the different interaction potentials, which obey the Eqs.(17),(18).

3. Parameters of numerical simulation

The simulation was carried out by the Langevin molecular dynamics method based on the solution of the system of differential equations with the stochastic force F_{ran}, that takes into account processes leading to the established equilibrium (stationary) temperature T of macro-particles that characterizes kinetic energy of their random (thermal) motion. The simulation technique is detailed in Refs. (Vaulina et al., 2005b; Vaulina & Dranzhevski, 2006; Vaulina & Vladimirov, 2002; Vaulina et al., 2003). The considered system of N_p motion equations (N_p is a number of grains) included also the forces of pair interparticle interaction F_{int} and external forces F_{ext}:

$$M\frac{d^2\vec{l}_k}{dt^2} = \sum_j F_{int}(l)\big|_{l=|\vec{l}_k-\vec{l}_j|}\frac{\vec{l}_k-\vec{l}_j}{|\vec{l}_k-\vec{l}_j|} + \vec{F}_{ext} - Mv_{fr}\frac{d\vec{l}_k}{dt} + \vec{F}_{ran} \tag{20}$$

Here $F_{int}(l) = -\frac{\partial U}{\partial l}$, and $l = |\vec{l}_k - \vec{l}_j|$ is the interparticle distance. To analyze the equilibrium characteristics in the systems of particles interacting with potentials (17), the motion equations (20) were solved with various values of effective parameters that are responsible for the mass transfer and phase state in dissipative non-ideal systems. These parameters were introduced by analogy with the parameters found in (Vaulina & Dranzhevski, 2006; Vaulina & Vladimirov, 2002; Vaulina et al., 2004), namely the effective coupling parameter

$$\Gamma^* = a_1 l_p^2 U''(l_p)/(2T), \tag{21}$$

and the scaling parameter

$$\xi = \omega^*/v_{fr}, \text{ where } \omega^* = |a_2 U''(l_p)|^{1/2}(2\pi M)^{-1/2}. \tag{22}$$

Here $a_1 = a_2 \equiv 1$ for 3d- systems, and $a_1 = 1.5$, $a_2 = 2$ for quasi-2d- case. The calculations were carried out for a uniform 3d- system and for a quasi-2d- system simulating an extensive dusty layer. The scaling parameter was varied from $\xi \approx 0.04$ to $\xi \approx 3.6$ in the range typical for the laboratory dusty plasma in gas discharges; thus, the values of Z were varied from 500 to 50000, the particle mass, M, was 10^{-11} - 10^{-8} g, and the l_p values were ~100-1000 μm. The Γ^* value was varied from 10 to 120.

In the 3d- case the external forces were absent ($|\boldsymbol{F}_{ext}| \equiv 0$), and the periodical boundary conditions were used for all three directions x, y and z. The greater part of calculations was performed for 125 independent particles in a central calculated cell that was the cube with the characteristic size L. The length of the cell L (and the corresponding number of particles) was chosen in accordance with the condition of a correct simulation of the system's dynamics: $L >> l_p|U(l_p)|/\{|U'(l_p)|l_p - |U(l_p)|\}$, that satisfies the requirement of strong reducing of the pair potential at the characteristic distance L (Vaulina et al., 2003). So, for example, for Yukawa potential this conditions may be presented in the form $l_p/L << \kappa$ (Totsuji et al., 1996). The potential of interparticle interaction was cut off on the distance L_{cut} ~ $4\, l_p$, which was defined from the condition of a weak disturbance of electrical neutrality of the system: $U'(L_{cut})\, L_{cut}^2 << (eZ)^2$. To prove that the results of calculation are independent of the number of particles and the cutoff distance L_{cut}, the additional test calculations were carried out for 512 independent particles with $L_{cut} = 7\, l_p$ and $\Gamma^* = 1.5$, 17.5, 25, 49 and 92. The disagreement between the results of these calculations was within the limits of the numerical error and didn't exceed ± (1-3)%.

In the quasi-2d- case, the simulation was carried out for the monolayer of grains with periodical boundary conditions in the directions x and y. In z direction the gravitational force Mg, compensated by the linear electrical field $E_z = \beta z$ ($|\boldsymbol{F}_{ext}| \equiv F_{ext}^z = Mg - eZ\beta z$), was considered. Here β is the gradient of electrical field, and $F_{ext}^x = F_{ext}^y \equiv 0$. The number of independent particles in the central calculated cell was varied from 256 to 1024; accordingly, the cutoff distance of potential was changed from $5l_p$ to $25l_p$. The value of the gradient β of electrical field E_z, confining the layer in z direction, was varied from ~10^{-2} V/cm^2 to ~100 V/cm^2, and for the simulated monolayers of grains the β value was in an agreement with the criterion of formation of mono-layer's dust structures proposed in (Vaulina et al., 2005a): $eZ\beta < 2\sum_{i=1}^{N_p} U'(l_i)/l_i$. Under this condition we have not detected any considerable dependence of particles' dynamics on the values of β and N_p in our simulations.

4. Results of the numerical simulation and their discussion

The evolution of mass-transfer processes, obtained in the numerical experiments for quasi-2d- and 3d- systems with various interaction potentials for different values of ξ and Γ^* is illustrated in Figs. 2-5 where the normalized VAFs, $f(v_{fr}t) = M <V(0)V(v_{fr}t)>/(mT)$, and mass-transfer evolution functions ($D_{G\text{-}K}(v_{fr}t)/D_o$, $D_{msd}(v_{fr}t)/D_o$) are presented. In Figs. 2-3, the curves 1 are the solution of Langevin equation with neglecting of the interparticle interaction (see Eqs. (5)-(6 a,b)). It can be easily seen that in the presence of interparticle interactions, the behavior of $<V(0)V(t)>$, $D_{msd}(t)$, $D_{G\text{-}K}(t)$ on short observation times ($v_{fr}\ t \ll 1$) corresponds to the motion typical for "brownian" particles. With time, the functions $D_{msd}(t)$, $D_{G\text{-}K}(t)$ reach their maximums D_{msd}^{max} and $D_{G\text{-}K}^{max}$. However, neither the relative magnitude D_{msd}^{max}/D_o, $D_{G\text{-}K}^{max}/D_o$ nor the position $t_{max}v_{fr}$ of these maximums depend on Γ^* and are defined by the value of the scaling parameter ξ for 3d- problem as well as for the simulated 2d- system. This feature was noticed earlier for the functions $D_{msd}(t)$ (Vaulina et al., 2005b; Vaulina & Dranzhevski, 2006; Vaulina & Vladimirov, 2002; Vaulina et al., 2004).

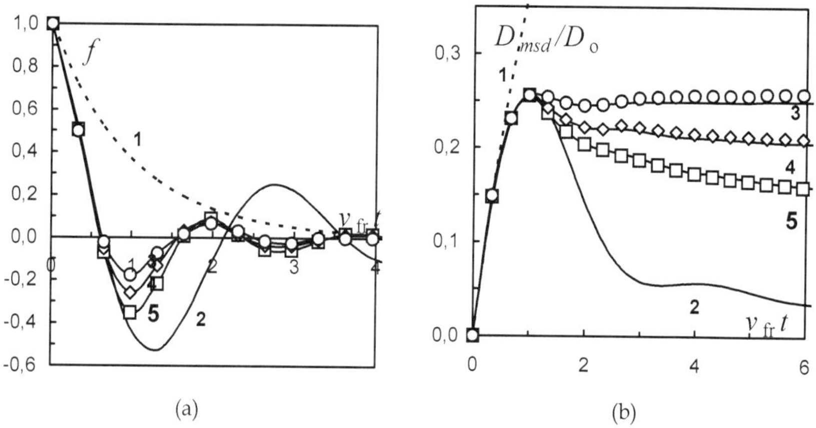

Fig. 2. Functions $f(tv_{fr})$ (**a**) and $D_{msd}(tv_{fr})/D_o$ (**b**) for: **1** - "brownian" case; **2** - harmonic oscillator with ξ_c =1.53. And the numerical results for quasi-2d- problem with ξ= 0.93 (ξ_c =1.53) and various Γ^*: **3** - 12; **4** - 27; **5** - 56; and for different potentials U: solid lines - $U/U_c = \exp(-4r/l_p)$; ○- $U/U_c = 0.1\exp(-2r/l_p) + \exp(-4r/l_p)$; ◇- $U/U_c = \exp(-4r/l_p) + 0.05\ l_p/r$; □- $U/U_c = 0.05(l_p/r)^3$.

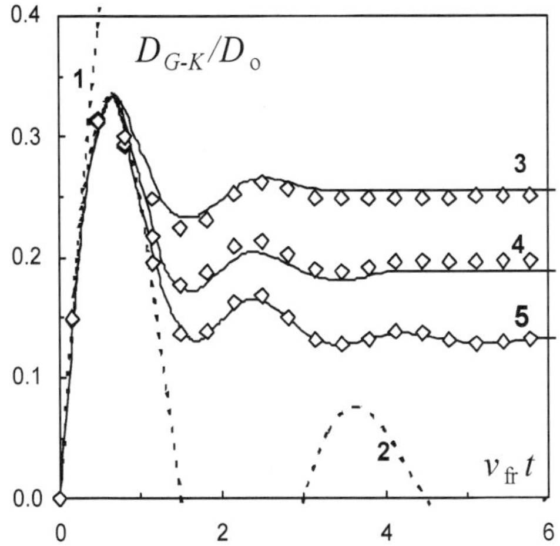

Fig. 3. Functions $D_{G-K}(t\upsilon_{fr})/D_o$ for: (**1**) - "brownian" case; (**2**) - for harmonic oscillator with ξ_c =1.53. And the numerical results for quasi-2d- problem (ξ = 0.93, ξ_c =1.53) with different Γ^*: **3** - 12; **4** - 27; **5** - 56. Fine lines - D_{G-K}/D_o (Eq. (4a)), ◇- D_{G-K}/D_o (Eq. (15a).

It is easy to see that the evolution of $<V(0)V(t)>$, $D_{msd}(t)$, $D_{G-K}(t)$ functions for the systems with the different pair potentials is defined by the particle temperature, T, the effective coupling parameter, Γ^*, the scaling parameter ξ (see Figs. 2-5) and also that the relation between $D_{msd}(t)$ and $D_{G-K}(t)$ is in accordance with Eq. (15a) (see Fig. 3).

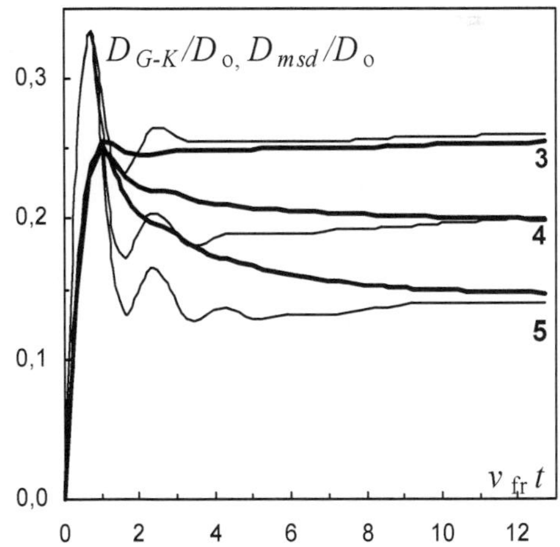

Fig. 4. Functions $D_{G-K}(t\upsilon_{fr})/D_o$ (fine lines, Eq. (4a)) and $D_{msd}(t\upsilon_{fr})//D_o$ (thick lines, Eq. (4b)) for quasi-2d- problem(ξ= 0.93, ξ_c =1.53) with different Γ^*: **3** - 12; **4** - 27; **5** - 56.

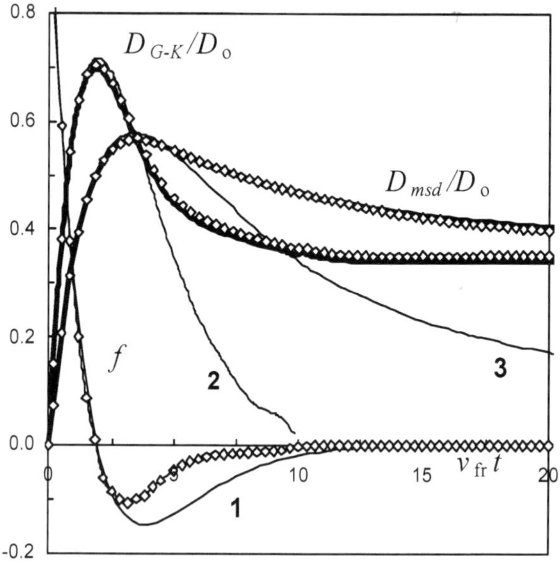

Fig. 5. Functions $f(tv_{fr})$ (**1**), $D_{G\text{-}K}(tv_{fr})/D_o$ (**2**) and $D_{msd}(tv_{fr})/D_o$ (**3**) for harmonic oscillator (fine lines) with ξ_c =0.19. And the numerical results for 3d- problem with ξ= 0.19 (ξ_c =0.38) and Γ^* = 27 for different potentials U: (thick lines) - U/U_c =exp(-2.4r/l_p);◇- U/U_c = exp(-4.8 r/l_p) + 0.05 l_p/r.

With $t \to \infty$, both $D_{msd}(t)$ and $D_{G\text{-}K}(t)$ functions tend to the same constant value D that corresponds to the diffusion coefficient. The normalized coefficients $D^*=D(v_{fr}+\omega^*)M/T$ vs. the Γ^* parameter for quasi-2d- systems with various pair potentials are shown in Fig. 6. It can be easily noticed that the D^* value for the systems under study is defined by the value of Γ^*. It was observed that the difference between the diffusion coefficients of weakly dissipative ($\xi > 0.3$) and weakly dispersive ($\xi < 0.25$) quasi-2d- structures with Γ^* between ~6 and ~97 is rather small; within the mentioned range of Γ^* the deviations of diffusion coefficients from their mean value don't exceed 7%. This difference increases noticeably with an increase of $\Gamma^* > 100$. These deviations were observed for quasi-2d- Yukawa systems as well as for 3d- systems with different types of pair potentials (Vaulina et al., 2005b; Vaulina & Dranzhevski, 2006; Vaulina & Vladimirov, 2002; Vaulina et al., 2004). Note that the obtained functions $D^*(\Gamma^*)$ have two critical points, one of which is a point of inflection (Γ^*~ 98-108) that, possibly, reflects a phase transition between the hexatic phase and the liquid. The second critical point (the point of a abrupt change of D) lies near Γ^*~ 153-165, where $D \to 0$, and the system under study is transforming into the solid with a perfect hexagonal lattice. The similar behavior of $D^*(\Gamma^*)$ was observed for quasi-2d Yukawa systems (Vaulina & Dranzhevski, 2006). The mean value of normalized diffusion coefficient, D^*, averaged for different values of Z, κ, v_{fr} and β) for the quasi-2d- systems with the different potentials is presented in Fig. 7, where the dependence $D^*(\Gamma^*)$ for 3d-structures (Vaulina & Vladimirov, 2002; Vaulina et al., 2004) is also shown.

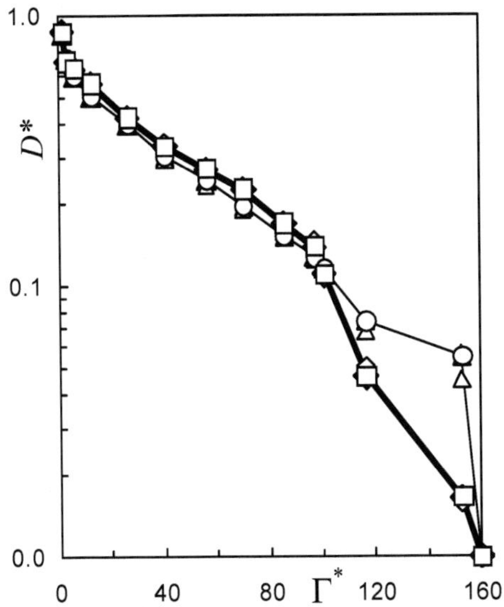

Fig. 6. Value of D^* vs. Γ^* for various potentials U: (black symbols) - $U/U_c = \exp(-4\, r/l_p) + 0.05\, l_p/r)$; (white symbols) - $U/U_c = 0.05(l_p/r)^3$; for different ξ: (◇, ◆) - 1.86; □ - 0.93, O - 0.23, (△, ▲) - 0.12. The solid lines are the averaged data of simulation for the quasy-2d- systems with $\xi < 0.25$ (thick line) and for $\xi > 0.3$ (fine line).

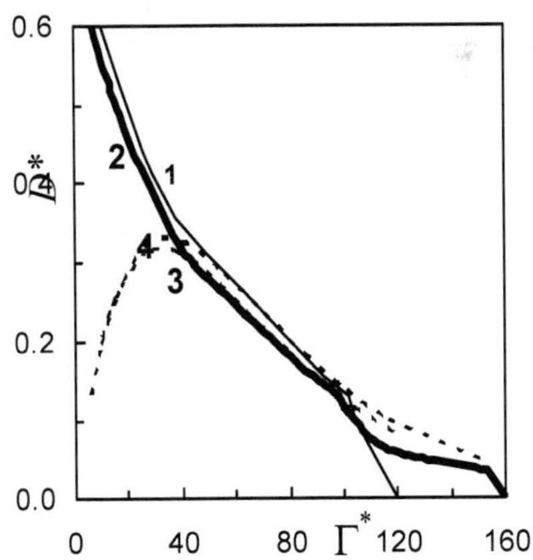

Fig. 7. Value of D^* vs. Γ^* for: **1 -** 3-d systems (see (Vaulina et al., 2004)); **2 -** quasi-2d- systems (averaged); **3 -** Eq.(23) for 2d- case with $\Gamma^*_c = 98$; **4 -** Eq.(23) for 3d- problem with $\Gamma^*_c = 102$.

The temperature dependence of the diffusion coefficient D of macroparticles in the 3d- systems with various types of potentials and for the quasi-2d- structures with the screened

Coulomb potential was found in (Vaulina & Dranzhevski, 2006; Vaulina & Vladimirov, 2002; Vaulina et al., 2004). There was shown that the diffusion coefficient for strongly coupled liquid-like systems can be presented as

$$D \approx \frac{T\Gamma^*}{12\pi(\xi+1)\nu_{fr}M}\exp\left(-3\frac{\Gamma^*}{\Gamma^*_c}\right), \tag{23}$$

where Γ^*_c is the crystallization point of the structure under study (Γ^*_c = 102 for the 3d-problem and Γ^*_c = 98 for the 2d- case). The approximation of the numerical results for obtained diffusion coefficients by Eq.(23) is shown in Fig. 7. The accuracy of this approximation, which for Γ^* >50 is just within 5%, decreases to 35% with Γ^* decreasing down to the value of $\Gamma^* \approx 30$ (Konopka et al., 1997; Daugherty et al., 1992; Allen, 1992). It should be noted that relation (23) is in accordance with the empirical "jumps" theory developed for molecular fluids that is based on the analogies between the liquid and the solid states of matter (Frenkel, 1946; March & Tosi, 1995); and it allows experimental determination of Γ^* in a strongly coupled system from measurements of the mean interparticle distance l_p, temperature T_p, and diffusion coefficient D without additional physical assumptions on the character of pair potential (Vaulina et al., 2003).

The comparison of evolution of mass-transfer processes in liquid-like 3d- and quasi-2d-systems with the behavior of analytical $D_{msd}(t)$, $<V_x(0)V_x(t)>$, $D_{G\text{-}K}(t)$ functions, obtained for harmonic oscillator, Eqs.(9),(13),(14), demonstrates a good agreement for observation times $\nu_{fr}\, t \le 1/\xi$ (see Figs. 2, 3, 5). Thus, in accordance with mentioned "jumps" theory the time of activation τ_o of "jumps" (the mean time of "settled life" of the particles) in the simulated systems practically doesn't depend on temperature and is defined by the oscillation frequency of the grains in "settled" condition: $\tau_o \approx 2/\omega_c$. The simulations also show that the system of interacting particles can be characterized by the constant values of transport coefficients only for intervals $t \gg \tau_o$, in contrast to the system of "brownian" particles, for which the evolution functions $D_{msd}(t)$ (or $D_{G\text{-}K}(t)$) tend to D_o for $t \gg \nu_{fr}^{-1}$.

The measurement of functions $<V(0)V(t)>$, $D_{msd}(t)$, $D_{G\text{-}K}(t)$ at the short observation times can be useful for the passive diagnostics of dust component in non-ideal plasma in the case of local statistical equilibrium of a dusty sub-system. As all the mentioned functions are connected by the relationships (15a)-(15b) and unambiguously depend on such parameters of grains as their temperature T, characteristic frequency ω_c and the friction coefficient ν_{fr}, it is possible to simultaneously determine all the mentioned parameters by measuring any of the functions $<V(0)V(t)>$, $D_{msd}(t)$ or $D_{G\text{-}K}(t)$ and using a procedure of best fitting of this chosen function by the respective analytical function for harmonic oscillator. Additionally, the information on T and ω_c allows one to estimate the value of the coupling parameter Γ^* of the system under study from the Eq. (21)-(22).

5. Mass transfer in the dusty plasma

The dusty plasma is an ionized gas containing micron-size charged grains (macroparticles) of solid matter (dust). This type of plasma is ubiquitous in nature (in space, in molecular dust clouds, in planetary atmospheres) and often appears in a number of technological processes (for example, fuel burning, an industrial processing of semiconductors etc) (Thoma et al., 2005; Morfill et al., 2003). The experiments with dusty plasma are carried out mostly in gas-discharges of various types. In gas-discharge plasma, micron-size grains acquire a significant (negative) electric charge, and can form dust structures similar to a liquid or to a solid. Dependent on the experimental conditions, these structures can be close to the uniform three-dimensional (3d-) or to the non-isotropic quasi- two-dimensional (2d-) systems, which consist of several (usually from one up to ten) horizontal layers of macroparticles (Nunomura et al., 2006; Ratynskaia et al., 2006; Nosenko & Goree, 2004; Gavrikov et al., 2005; Nunomura et al., 2005). Owing to their size, the dust particles may be video-filmed (Ratynskaia et al., 2006; Nosenko & Goree, 2004; Gavrikov et al., 2005; Nunomura et al., 2005; Vaulina & Vladimirov, 2002; Vaulina et al., 2004). It makes the laboratory dusty plasma a good experimental model, which can be used to study various physical phenomena in systems of interacting particles, that attract widespread interest in the physics of non-ideal plasmas as well as in other areas such as plasma chemistry, physics of the atmosphere, medicine, physics of polymers etc.

A study of the kinetic (transport) coefficients (constants of diffusion, viscosity, thermal conductivity etc.) for dusty plasma is of great interest (Nunomura et al., 2006; Ratynskaia et al., 2006; Nosenko & Goree, 2004; Gavrikov et al., 2005; Nunomura et al., 2005). These constants are fundamental parameters that reflect the nature of interaction potentials and a phase state of the system. When the deviations of the system from the statistical equilibrium are small, the kinetic coefficients can be found from Green-Kubo formulas that were established with the help of the theory of Markovian stochastic processes under an assumption of the linear reaction of the statistical system on its small perturbations (March & Tosi, 1995; Ovchinnikov et al., 1989). The diffusion is the basic mass-transfer process, which defines the losses of energy (dissipation) in the system. The collisions of grains with the neutral particles of surrounding gas have a dramatic effect on the dissipation of dust energy in weakly ionized laboratory plasma.

The measurement of the $<V(0)V(t)>$-, $D_{msd}(t)$-, $D_{G\text{-}K}(t)$- functions at the short observation times can be used for the diagnostics of dust component in plasma. As all the mentioned functions are connected by the Eqs. (15a)-(15b) and uniquely determined by the parameters of grains (by their temperature T, characteristic frequency ω_c and the friction coefficient ν_{fr}), it is possible to simultaneously determine all these parameters using the best fitting of any from the measured functions ($<V(0)V(t)>$, $D_{msd}(t)$, $D_{G\text{-}K}(t)$) by the respective analytical function for harmonic oscillator. The information on T and ω_c allows one to estimate the value of the effective coupling parameter Γ^* and the scaling parameter ξ, which determine the dynamics of grains in the system under study.

Nevertheless we should note that to apply the results of simulation of particle motions by LMDM for analysis of dynamics of grains in plasma and to prove the validity of measurements of diffusion constants with the help of Green-Kubo formula (4a), one needs to examine, whether the considered Langevin model is valid under experimental conditions (i.e. to prove the validity of the Markovian approach of the condition of local equilibrium of dust system and of the assumption of the linear reaction of this system on its perturbations). The main objections to the use of the Langevin model in a treatment of the results of real experiments are the possible influence of boundary conditions, external fields and strong interparticle interactions on the migration of particles and their energy exchange with the surrounding medium (thermostat) in real experiments (Ovchinnikov et al., 1989). A lot of special questions arise in an interpretation of dusty plasma experiments. These are the influence of the openness of dusty plasma system and of the irregular distribution of grains' stochastic energy on the degrees of freedom, including the question on a capability of the use of this energy as the main thermodynamical characteristic that describes the kinetic temperature of the dusty component and characterizes the exchange of its energy with a thermostat.

Thus, to verify a validity of the Langevin model, we should prove the diffusive character of the grains' migration (with $t \to \infty$: $<x^2> \propto t$, $D = const$), and also check the relations (15a), (15b), that result from the simulation of the dynamics of particles using the mentioned Langevin equations. (In the case when these relations are valid, the Green-Kubo formula (see Eq.(4a) with $t \to \infty$: $D_{G\text{-}K}(t) = D_{msd}(t) \equiv D$) is automatically true.)

6. The description of experiments and their results

The experiments were carried out for mono-disperse grains (with material density $\rho_p \approx 1.5$ g cm^{-3}, radiuses $a_p \approx 2.75$ μm and $a_p \approx 6.37$ μm) in the near-electrode area of RF-discharge in argon. The pressure in the discharge was $P = 0.03 - 0.5$ Torr, and its power - $W \approx 2\text{-}30$ W. The simplified scheme of the experimental setup is shown in Fig. 8. To visualize the cloud, a flat beam of He-Ne laser ($\lambda = 633$ nm) was used. The laser illumination had two regimes: in the first, the defocused laser beam illuminated the whole dusty structure inside the trap; this allowed to determine its dimensions. The second regime was used for the detailed observation of the horizontal cross-section of the dusty cloud. In this case the laser beam represented so called "laser knife" with the width ~ 2.5 cm and the waist size ~ 200 μm. The positions of grains were registered with a high-speed CMOS video camera (frame rate f_{vc} = 500 s^{-1}). The video-recording was processed by the special software, which allowed identifying the positions of each particle in the field of view of the video-system. Then the pair correlation functions $g(l)$, the mean inter-grain distances, the velocity autocorrelation functions ($<V(\delta t)V(t)> = (<V_x(\delta t)V_x(t)> + <V_y(\delta t)V_y(t)>)/2$), the mass-transfer functions ($D_{G\text{-}K}(t)$, $D_{msd}(t)$), and the diffusion coefficients D were obtained. The deviations of the measured parameters in two registered freedom degrees (x, y) were minor and didn't exceed ~ 0.5-3%: $<V_x((\delta t)V_x(t)> \cong <V_y(\delta t)V_y(t)>$, $<V_x^2> \cong <V_y^2>$, $<x^2> \cong <y^2>$, where $\delta t = f_{vc}^{-1}$; and the value of the average dust velocities $<V_x(t)> \cong <V_y(t)> \equiv 0$.

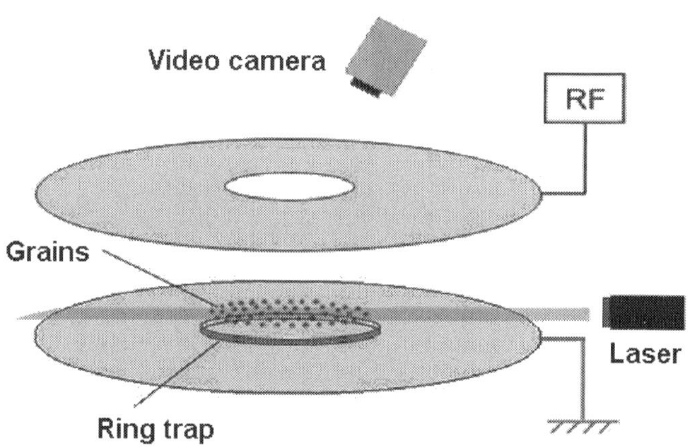

Fig. 8. The simplified scheme of experimental setup.

Under the experimental conditions, the dusty structures consisted of several (from 1 up to ~10) dusty layers, and the observed dusty structures were changing from the weakly correlated fluids to the dust crystals with the mean interparticle distance l_p from ~ 500 to ~1000 μm. The duration of one experiment under stable conditions was ~ 5-10 s. The experimental pair correlation functions $g(l)$ for the grains of different sizes, forming dusty mono-layers and multi-layer systems, are presented in Fig. 9.

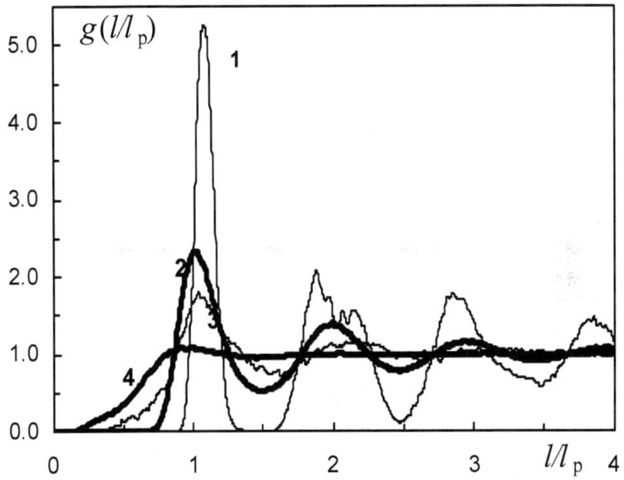

Fig. 9. The pair correlation function $g(l/l_p)$ for various experiments: **1** - a_p = 6.37 μm, mono-layer, P = 0.03 Torr; **2** - a_p = 2.755 μm, mono-layer, P = 0.35 Torr; **3** - a_p = 6.37 μm, multi-layer system, P = 0.11 Torr; **4** - a_p = 2.755 μm, multi-layer system, P = 0.04 Torr.

The magnitude of the maximum g_{max} of functions $g(l)$ and the mean interparticle distance l_p, which was obtained by analysis of the g_{max} position, are given in Table 1 for different experiments. The random errors of determination of the l_p and g_{max} values were less than 5% for all cases. Nevertheless, we should note that the magnitude of $g(l)$ peaks may be misrepresented because of: (i) limited number of grains in the field of view of the video camera; (ii) the probability of simultaneous registration of grains in the next layer in the case of the multi-layer structure.

The results of measurement of velocity autocorrelation functions, $<V(\delta t)V(t)>$, and mass-transfer functions ($D(t)=D_{G\text{-}K}(t)$ and $D(t)=D_{msd}(t)$) for various experiments are presented in Fig. 10 and Fig. 11 where the normalized values are shown: $f(t)= <V(\delta t)V(t)>/V_T^2$ and $D(t)/D_o$, where $V_T^2 = T/M$ is the mean square velocity of stochastic "thermal" motion of grains. The values of V_T^2 and ν_{fr} are presented in Table 1, and the methods of their determination are discussed in the next section.

For the most experiments, the motion of grains was diffusive: with the time increasing ($t\rightarrow \infty$), the mass-transfer functions tended to the same value: $D_{G\text{-}K}(t) \approx D_{msd}(t) \rightarrow D$, excluding the crystalline dusty structures, where with $t\rightarrow \infty$, the value of the mean square displacements of grain from its equilibrium position $<x^2> \cong <y^2> \rightarrow const$, and the D constant $\rightarrow 0$. The measured value of the diffusion coefficient D is given in Table 1. The difference between the diffusion constants determined from the Green-Kubo formula (4a), $D = D_{G\text{-}K}(t\to\infty)$, and from the analysis of the mean square displacements, $D = D_{msd}(t\to\infty)$ (4b), didn't exceed 5%, that corresponds to the random error. The systematic errors in determination of D due to the finite duration of measurements were the same (≤ 5%). The time t_D of establishing the constant value of D (±5%) was from ~ 1 s to 5 s in the relation to the gas pressure P (with the increase of P, the t_D value was increasing also). With the low pressure P ($\xi_c \geq 1$), the $D_{msd}(t)$ function achieves its constant value, D, faster than the $D_{G\text{-}K}(t)$ function (see Fig. 11 a,c,d); otherwise with the higher pressure P (ξ_c <<1, the case of non-oscillating $D_{G\text{-}K}(t)$), the $D_{G\text{-}K}(t)$ value tends to the constant faster than $D_{msd}(t)$, that allows to measure the diffusion coefficient D from the Green-Kubo formula in shorter time (see Fig. 11b).

P, Torr	l_p, mm	g_{max}	V_T^2, mm²/s²	D, mm²/s	ν_{fr}, s⁻¹	ω_c, s⁻¹	Γ_{3d}/Γ_{2d}	Z_{min}
R = 2.75 μm, mono-layer								
0.11	1	2.75	0.803	0.0025	30.5	9.7	92/102	5605
0.19	0.84	2.55	0.720	0.002	50.8	9.8	72/80	4303
0.35	0.92	2.35	0.949	0.002	98	8.8	54/60	4478
0.5	0.75	2.7	1.468	0.00135	143	16.3	80/89	6119
R = 2.75 μm, multi-layer system								
0.04	1.1	1.1	26.3	0.81	11	13.2	6.3/7	8800
0.06	1	1.08	20.1	0.58	15.5	10.23	4.1/4.6	5912
0.1	0.57	1.095	20.7	0.3	34	22.44	6.2/6.9	5581
0.14	0.6	1.05	43.1	0.66	44	15.62	1.6/1.8	4195
R = 6.37 μm, mono-layer								
0.03	1	5.25	0.45	→ 0	3.5	13.0	293/326	26311
0.05	1	2.45	1.78	0.022	6	12.5	69/77	25396
0.08	0.88	2.75	1.23	0.01	8.4	13.5	90/100	22642
0.42	0.57	4.9	0.73	→ 0	44	27.3	262/291	23852
R = 6.37 μm, multi-layer system								
0.05	0.92	1.28	12.94	0.449	7	14	10/11	24964
0.07	0.82	1.74	4.51	0.095	8.2	16.4	31.5/35	24742
0.08	0.85	2	2.90	0.048	8.25	15	44/49	23883
0.11	0.88	1.66	4.97	0.12	11.8	13.1	21/23	21968

Table 1. Parameters of dusty component for various experiments

We emphasize, that under conditions of our experiments we have not observed the anomalous-, or super- diffusions that were investigated in a set of works (Ratynskaia et al., 2006; Wen-Tau & I, 1998). With increase of time t ($t\to\infty$), the both mass-transfer functions tended to the same value: $D_{G\text{-}K}(t) \approx D_{msd}(t) \to D$; i.e. the mean square displacements of particles were proportional to the time t for all presented experiments (see curves 1, 2 in Figs. 11 a,b,c,d).

The results of examination of relation between the VAFs, $<V(\delta t)V(t)>$, mass-transfer functions ($D_{G\text{-}K}(t)$, $D_{msd}(t)$), and the mean-square displacement ($<\Delta l^2> = <x^2>+<y^2>$) are presented in Figs. 10 - 11. In all the cases there was obtained a good agreement between the direct measurements of functions $<V(\delta t)V(t)>$, $D_{G\text{-}K}(t)$ and those functions, calculated from the measurements of $<\Delta l^2>$ using the Eqs. (15a)-(15b). Thus, we can conclude that the stochastic model, given by the system of Langevin equations, allows the correct description of the dust motion under experimental conditions.

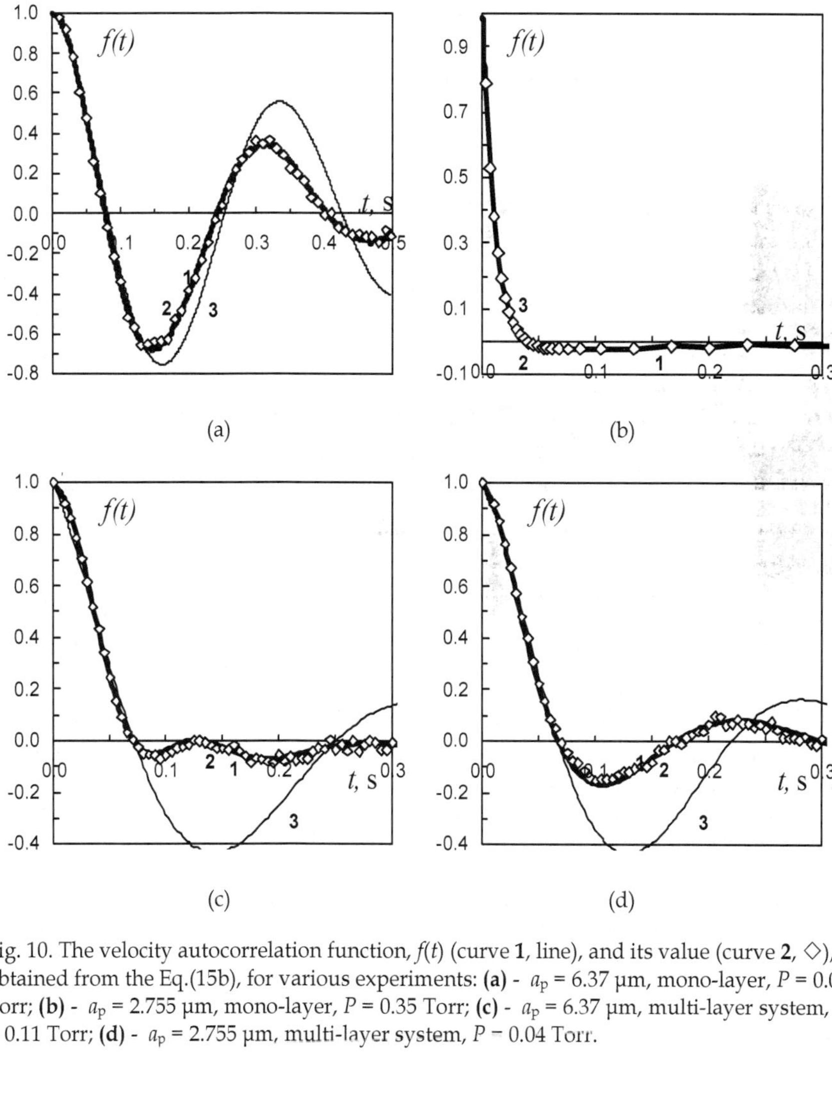

Fig. 10. The velocity autocorrelation function, $f(t)$ (curve **1**, line), and its value (curve **2**, ◇), obtained from the Eq.(15b), for various experiments: **(a)** - a_p = 6.37 µm, mono-layer, P = 0.03 Torr; **(b)** - a_p = 2.755 µm, mono-layer, P = 0.35 Torr; **(c)** - a_p = 6.37 µm, multi-layer system, P = 0.11 Torr; **(d)** - a_p = 2.755 µm, multi-layer system, P = 0.04 Torr.

Curve **3** is the $f(t)$ function for the harmonic oscillator, Eq.(13), with parameters indicated in Table 1.

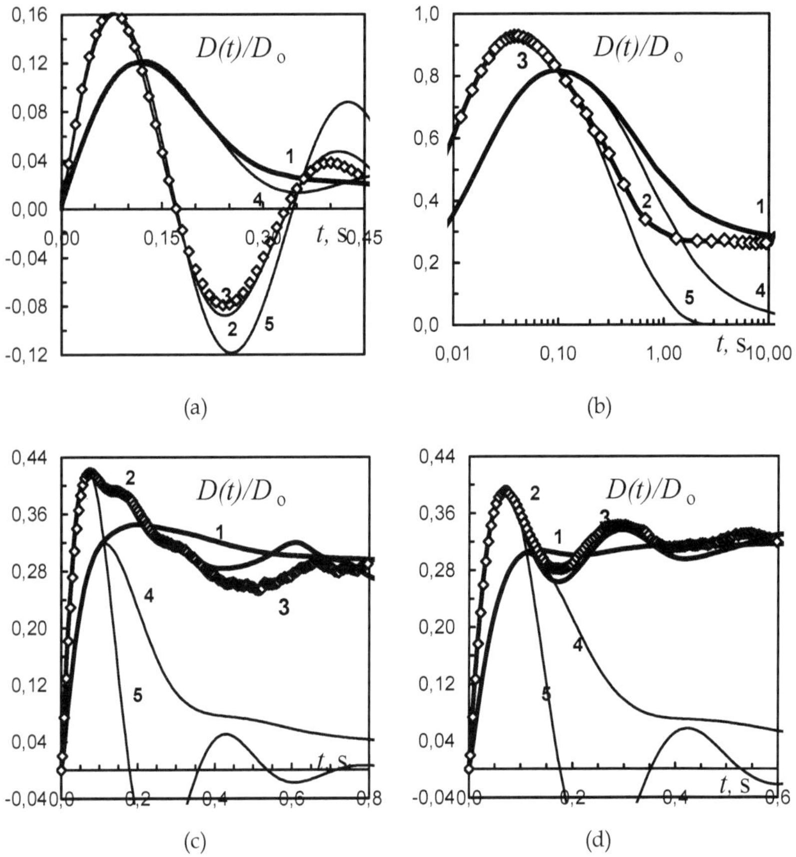

Fig. 11. The mass-transfer evolution functions $D(t)/D_o$: (curve **1**, line) – $D(t) \equiv D_{msd}(t)$; (curve **2**, line) - $D(t) \equiv D_{G\text{-}K}(t)$, Eq.(1b); (curve **3**, ◇) - $D(t) \equiv D_{G\text{-}K}(t)$, Eq.(15a), for various experiments: **(a)** - a_p= 6.37 μm, mono-layer, P=0.03 Torr; **(b)** - a_p= 2.755 μm, mono-layer, P = 0.35 Torr; **(c)** - a_p = 6.37 μm, multi-layer system, P=0.11 Torr; **(d)** - a_p= 2.755 μm, multi-layer system, P = 0.04 Torr.

Curve **4**, Eq.(9), and curve **5**, Eq.(14), are the corresponding functions for the harmonic oscillator with parameters indicated in Table 1.

7. The determination of parameters of dusty subsystem

The parameters of macroparticles (the mean-square velocity V_T^2 of their stochastic "thermal" motion, the characteristic frequency ω_c and the friction coefficient ν_{fr}) are presented in

Table 1. These parameters were obtained by the best fitting of the measured functions, $<V(\delta t)V(t)>$, $D_{msd}(t)$ and $D_{G\text{-}K}(t)$, and the corresponding analytical solutions, Eqs. (9, 13, 14), for the harmonic oscillator. The results of this procedure for various experiments are shown in Figs. 10-11. Note that the measured functions presented in these figures are relative, namely, they are the normalized values: $f(t)= <V(\delta t)V(t)>/V_T^2$ (see Fig. 10), and $D(t)/D_o$ (Fig. 11). The errors in determination of these relative functions for the same intervals of time, $t = (t'- t_o)$, were less than 2-3% for various initial time t_o which were chosen for processing experimental results. These random errors may be related to the fluctuations of dusty plasma parameters during experiments, and these errors were less than the errors in the determination of dusty plasma parameters (V_T^2 ,ν_{fr}, ω_c) by the best fitting of experimental data to the analytical solutions as discussed below.

The errors of the determined parameters were ~ 5-7% for V_T^2, less than 10% for the friction coefficient ν_{fr} , and less than 5% for ω_c. To optimize the procedure of fitting of experimental data, the initial values of ω_c and ν_{fr} were chosen in accordance with analytical approximations for the dependence of maximum D_{max} of function $D_{msd}(t)$ and the position of this maximum t_{max} on the parameter $\xi_c = \omega_c/\nu_{fr}$, that were proposed in (Vaulina & Dranzhevski, 2006; Vaulina et al., 2005b):

$$D_{max} \approx D_o/(1+ 2\xi_c). \tag{24}$$

$$t_{max}\nu_{fr} \approx 4\sqrt{2}\pi\,/(1+8\sqrt{2}\xi_c). \tag{25}$$

The accuracy of these approximations is about 5%. The Eq.(25) can be also used for choosing of the frame rate f_{vc} of video camera (registering the positions of grains) and the duration of measurement t_D that are necessary for the correct determination of dust parameters and diffusion coefficients D. So, the f_{vc} value should be much higher than $1/t_{max}$, and the duration t_D of measurement for determination of D should satisfy the condition $t_D >> t_{max}$. The preliminary estimation of values of ω_c and ν_{fr} under the experimental conditions may be performed using the existing theoretical models.

For all cases, the retrieved values of the mean-square velocity $V_T^2 = T/M$ were in accordance with the fitting of velocity distributions ($\varphi(V_x)$, $\varphi(V_y)$) by maxwellian functions (see Fig. 12). The difference between the values of V_T^2, obtained by two different methods, didn't exceed 5-7% and, in most cases, was within the limits of experimental error.

The retrieved values of ν_{fr} were in a good agreement with their theoretical estimations using the free-molecular approximation that under experimental conditions (for argon of the room temperature) may be presented as ν_{fr} [s^{-1}] $\approx$ 1144 P[Torr]/(a_p[µm]ρ_p[g/cm^3]) (Lifshitz & Pitaevskii, 1981; Raizer, 1991). The difference between the measured and theoretical values of ν_{fr} was less than 12% for all experiments.

For the direct examination of an accuracy of retrieved values for characteristic frequencies ω_c, we need an information on the form of pair potential U (see Eq.(19a,b)). Nevertheless, as the data on the dust parameters (ω_c ,V_T^2, l_p in Table 1) allows one to estimate the value of the effective coupling Γ^* and scaling ξ parameters (see Eqs.(19a,b, 21, 22)), we can compare the measured magnitudes of maximums, g_{max}, for pair correlation functions, and of diffusion coefficients, D, with existing numerical data. The values of Γ^*, obtained from solving Eqs.(19a,b,21), are presented in Table 1. For the case of mono-layer the value of $\Gamma^* = \Gamma_{2d}$ was determined in the 2-d approach ($a_o = 2.7$, $a_1 = 1.5$, $a_2 = 2$). For multi-layer systems the estimations were performed for the two limiting cases: for 3-d case, $\Gamma^* = \Gamma_{3d}$ (a_o = 2, $a_1 = a_2 \equiv$ 1), and for 2-d case, $\Gamma^* = \Gamma_{2d}$ (a_o=2.7, a_1 = 1.5, a_2 = 2).

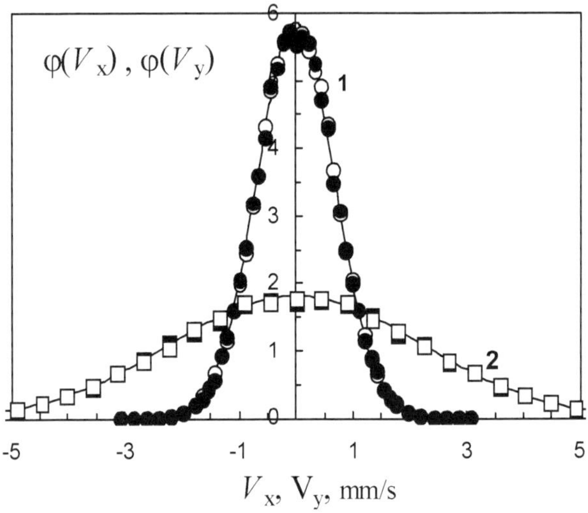

Fig. 12. The measured distribution of the velocities: $\varphi(V_x)$ -(●; ■) and $\varphi(V_y)$ -(○;□) - for the particles of the radius a_p = 6.37 μm and various experiments: (●;○) - mono-layer, P = 0.03 Torr; (■;□) - multi-layer system, P = 0.11 Torr. And their best-fit Maxwellian distributions with $V_T^2=T/M$ equal to: **1** - 0.46 mm²/s²; **2** - 4.8 mm²/s².

The comparison of the experimental values of maximums, g_{max}, and of the normalized diffusion coefficient $D^*= \nu_{fr}\ D(1+\xi)M/T$ with the results, obtained via the numerical simulation for 3-d problem and mono-layer (Vaulina et al., 2004; Vaulina & Dranzhevski, 2006), is shown in Figs. 13a - 13b. (Here we note that the 2d- and 3d-systems were simulated here and in (Vaulina et al., 2004) for a wide range of pair isotropic potentials, and in (Vaulina & Dranzhevski, 2006) for the quasi-2d-Yakawa systems.) It is easy to see that the measured dependence $g_{max}(\Gamma^*)$ is in a good agreement with numerical data (see Fig.13a). The difference between them is in the limits of experimental (~5%, see Section 2) and numerical (~5%) errors in the determination of g_{max}. The values of errors are shown in Fig.13a as a 5% confidence interval. The measured dependence $D^*(\Gamma^*)$ is also in a good agreement with the numerical data for all considered experiments (Fig.13b). The deviations between the experimental and numerical values of D^* don't exceed the experimental (~10%, see Section 2), and numerical error, δ_{cal}, in the determination of D; the δ_{cal} value increases from 7% to 15% when Γ^* varies from ~ 100 to ~ 5.

The value of the minimum charge $Z = Z_{min}$, that a grain can acquire in plasma, may be obtained under the assumption that the surrounding plasma doesn't screen the interaction between grains. Thus, in the case of Coulomb interparticle interaction, we obtain $Z_{min} \cong \omega_c \{\pi M\ l_p{}^3/5.4\}^{1/2}$ (see Eq. (19a,b)). The values of Z_{min} are presented in Table 1. The error of this estimation of Z_{min} is determined by the experimental errors of ω_c and l_p and is about 13%.

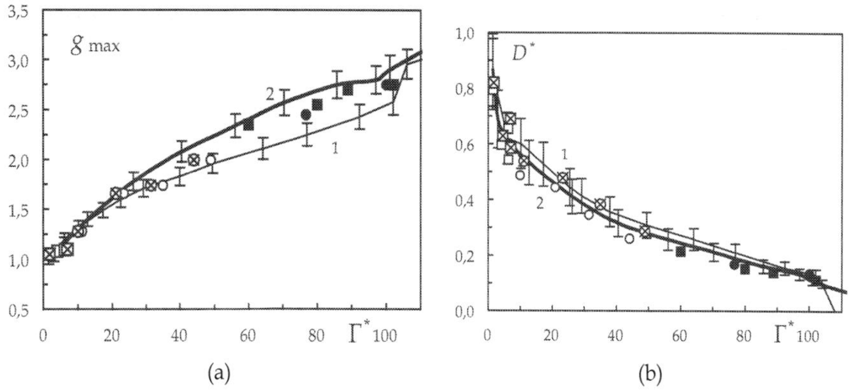

Fig. 13. Maximum g_{max} (**a**) of the $g(l)$ function and the normalized diffusion coefficient $D^*= D(\nu_{fr}+\omega^*)\, M/T$ (**b**) vs. Γ^* for: **1** - 3d- systems; **2** - mono-layer (Vaulina et al., 2004; Vaulina & Dranzhevski, 2006). The symbols are the experimental results (Table 1) for the grains of radius a_p: (○; ●; ⊗)- 6.37 μm; (□; ■; ⊠) - 2.755 μm, that form the dusty mono-layer (●; ■), or multi-layer structure (○; □; ⊗; ⊠). For the multi-layer structure the values of ω^* and Γ^* were determined for two cases: (○; □) - 3d- system $\Gamma^* = \Gamma_{3d}$; (⊗; ⊠) - 2d- system, $\Gamma^* = \Gamma_{2d}$.

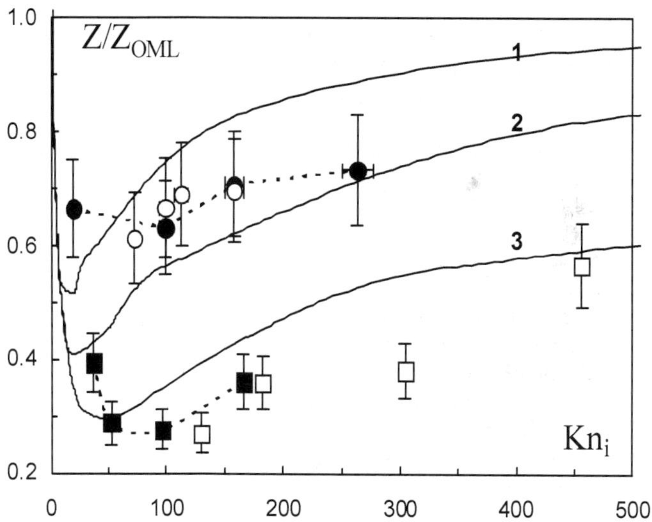

Fig. 14. Calculated dependence of Z/Z_{OML} (lines) on Kn_i for various a_p/λ: **1** - 0.07 ; **2** - 0.023 ; **3** - 0.007. Symbols are the Z_{min}/Z_{OML} value for the grains of radius a_p:: (○; ●)- 6.37 μm; (□; ■) - 2.755 μm; that form the dusty mono-layer (●; ■), or multi-layer structure (○; □).

The ratio of Z_{min} to the grain charge, $Z_{OML}\approx (2.7\, a_p T_e /e^2)$, obtained in the Orbital Motion Limited (OML) approach, are shown in Fig. 14 versus the Knudsen number $Kn_i = l_{in}/\, a_p$. Here T_e =3 eV is the electron temperature typical for rf-discharge in argon (Raizer, 1991;

Raizer et al., 1995), $l_{in} = (8\ T_i/\{\pi\ m_i\ \nu_{in}^2\})^{1/2}$ is an ion mean free path between their collisions with the gas neutrals, ν_{in} [s^{-1}] $\cong 8\ 10^6\ P$ [Torr] is the effective frequency of these collisions for single-charged argon ions (Raizer, 1991; Raizer et al., 1995), $T_i \cong 0.026$ eV is their temperature and m_i is the mass. The results of numerical simulations (Vaulina et al., 2006) of the charging of a single spherical particle in the weakly ionized argon-discharge plasma in the case of the neglecting of the drift ion velocity V_{id} are also presented in Fig. 14 as the ratio, Z/Z_{OML}, for the various values of a_p/λ = 0.07; 0.023; 0.007; here $\lambda \approx \lambda_{Di} = \{T_i/(4\pi e^2 n_i)\}^{1/2}$, and n_i is the concentration of ions (~10^7-5 10^8 cm^{-3}). A simple quantitative comparison of the presented numerical results, $Z(Kn_i)$, with the experimental data, $Z_{min}(Kn_i)$, is incorrect at least because of two reasons. First of them it is the neglecting of the dust screening, with a presence of which the obtained value of Z_{min} can be much lower than the real grain charge. The second reason is the neglecting of the drift ion velocity V_{id} in the numerical simulations (Vaulina et al., 2006) that may be not suitable for conditions of considered experiments in the near-electrode area of RF- discharge. In spite of that, we should note the good qualitative agreement between numerical $Z(Kn_i)$ and experimental dependencies $Z_{min}(Kn_i)$.

8. Conclusion

The results of numerical investigation of mass-transfer processes in extensive quasi-two-dimensional and three-dimensional non-ideal dissipative systems are presented. The particles in these systems were considered interacting with various isotropic pair potentials, which represented different combinations of power-law and exponential functions, commonly used for simulation of repulsion in kinetics of interacting particles. The calculations were performed in a wide range of parameters typical for the laboratory dusty plasma in gas discharges.

The evolution of mass-transfer processes, the velocity autocorrelation functions and the diffusion constants were studied. It was obtained that for the systems under study the particle temperature, the effective coupling parameter and the scaling parameter determine all mentioned characteristics. It was shown that the evolution of mean-square displacement of particles for the short observation times corresponds to the lattice oscillations with frequency proportional to the second derivative of pair potential of interparticle interaction. The estimations of the characteristic oscillation frequencies of particles (ω_c) in 3d- bcc-structures and in 2d- hexagonal lattice are presented. It is shown that these frequencies are responsible for the time of "settled life" of particles ($\tau_o \approx 2/\omega_c$) in non-ideal liquid systems and define the behavior of mass-transfer processes at the short observation times ($t < \tau_o$). The obtained results are in a good agreement with the "jumps" theory.

The results of the experimental study of mass-transfer processes are presented for the dust systems, forming in the laboratory plasma of radio-frequency (RF-) capacitive discharge. The experiments were carried out for macroparticles of various sizes ($a_p \approx 2.75$ μm and $a_p \approx 6.37$ μm) within a wide range of coupling parameters of dusty sub-system. The velocity autocorrelation functions, mass-transfer functions, the diffusion coefficients, pair correlation functions and the concentration were measured. For the most experiments, the motion of grains has the diffusive character. The diffusion coefficients, D, were determined from the Green-Kubo relation and from the mean-square displacement of particles. The difference between the values of D, obtained by two different methods, didn't exceed 5% and was within the limits of experimental error.

In all the cases there was obtained a good agreement between the direct measurements of the velocity autocorrelation and mass-transfer functions and those functions, calculated from the mean-square displacement of particles using the Eqs. (15a)-(15b). That means that the stochastic model, given by the system of Langevin equations, can be used for the correct description of the motion of dust under experimental conditions

The method of simultaneous determination of dusty plasma parameters, such as kinetic temperature of grains, their friction coefficient, and characteristic oscillation frequency is proposed. The parameters of dust were obtained by the best fitting the measured velocity autocorrelation and mass-transfer functions, and the corresponding analytical solutions for the harmonic oscillator. The coupling parameter of the systems under study and the minimal values of grain charges are estimated. The obtained parameters of the dusty sub-system (diffusion coefficients, pair correlation functions, charges and friction coefficients of the grains) are compared with the existing theoretical and numerical data.

9. Acknowledgement

This work was partially supported by the Russian Foundation for Fundamental Research (project no. 07-08-00290), by CRDF (RUP2-2891-MO-07), byNWO (project 047.017.039), by the Program of the Presidium of RAS, by the Russian Science Support Foundation, and by the Federal Agency for Science and Innovation (grant no. MK-4112.2009.8).

10. References

Allen, J.E. (1992). Probe theory - the orbital motion approach. *Physica Scripta,* 45, 497-503, ISSN 0031-8949

Balescu, R. (1975). *Equilibrium and Nonequilibrium Statistical Mechanics,* Wiley Interscience, ISBN 0-471-04600-0, Chichester

Photon Correlation and Light Beating Spectroscopy. (1974). Eds. by Cummins, H.Z. & Pike, E.R., Plenum, ISBN 0306357038, New York

Daugherty, J.E.; Porteous, R.K.; Kilgore, M.D. & Graves, D.B. (1992). Sheath structure around particles in low-pressure discharges. *J.Appl. Phys.,* 72, 3934-3942, ISSN 0021-8979

Dodd, R. K.; Eilbeck, J. C.; Gibbon, J. & Morris, H. C. (1982). *Solitons and Nonlinear Wave Equations,* Academic, ISBN 012219120X, New York

Fortov, V.E.; Nefedov, A.P.; Petrov, O.F.; *et al.* (1996). Particle ordered structures in a strongly coupled classical thermal plasma. *Phys. Rev. E,* 54, R2236–R2239, ISSN 1539-3755

Fortov, V.; Nefedov, A.; Vladimirov, V.; *et al.* (1999). Dust particles in a nuclear-induced plasma. *Physics Letters A,* 258, 305 - 311, ISSN 0375-9601

Frenkel, Ya. I. (1946). *Kinetic Theory of Liquid,* Oxford University Press, Oxford

Gavrikov, A.; Shakhova, I.; Ivanov, A.; *et al.* (2005). Experimental study of laminar flow in dusty plasma liquid. *Physics Letters A,* 336, 378-383, ISSN 0375-9601

Konopka, U.; Ratke, L. & Thomas, H.M. (1997). Central Collisions of Charged Dust Particles in a Plasma. *Phys. Rev. Lett.,* 79, 1269-1272, ISSN 0031-9007

Lifshitz, E.M. & Pitaevskii, L.P. (1981). *Physical Kinetics,* Pergamon Press, ISBN 0-7506-2635-6, Oxford

March, N.H. & Tosi, M.P. (1995). *Introduction to Liquid State Physics*, World Scientific, ISBN 981-02-4639-0, Singapore

Montgomery, D.; Joyce, G. & Sugihara, R. (1968). Inverse third power law for the shielding of test particles. *Plasma Phys.*, 10, 681-686, ISSN 0032-1028

Morfill, G.E.; Tsytovich, V.N. & Thomas, H. (2003). Complex plasmas: II. Elementary processes in complex plasmas. *Plasma Physics Reports*, 29, 1-30, ISSN 1063-780X

Nosenko, V. & Goree, J. (2004). Shear Flows and Shear Viscosity in a Two-Dimensional Yukawa System (Dusty Plasma). *Phys. Rev. Lett.*, 93, 155004, ISSN 0031-9007

Nunomura, S.; Samsonov, D.; Zhdanov, S. & Morfill, G. (2005). Heat Transfer in a Two-Dimensional Crystalline Complex (Dusty) Plasma. *Phys. Rev. Lett.*, 95, 025003, ISSN 0031-9007

Nunomura, S.; Samsonov, D.; Zhdanov, S. & Morfill, G. (2006). Self-Diffusion in a Liquid Complex Plasma. *Phys. Rev. Lett.*, 96, 015003, ISSN 0031-9007

Ovchinnikov, A.A.; Timashev, S.F. & Belyy, A.A. (1989). *Kinetics of Diffusion Controlled Chemical Processes*, Nova Science Publishers, ISBN 9780941743525, Commack, New York

Raizer, Yu.P.; Shneider, M.N.; Yatsenko, N.A. (1995). *Radio-Frequency Capacitive Discharges*, CRC Press, ISBN 0-8493-8644-6, Boca Raton, Florida

Raizer, Yu.P. (1991). *Gas Discharge Physics*, Springer, ISBN 0-387-19462-2, Berlin

Ratynskaia, S.; Rypdal, K.; Knapek C.; *et al.* (2006). Superdiffusion and Viscoelastic Vortex Flows in a Two-Dimensional Complex Plasma. *Phys. Rev. Lett.*, 96, 105010, ISSN 0031-9007

Thoma, M.H.; Kretschmer, M.; Rothermel, H.; *et al.* (2005). The plasma crystal. *American Journal of Physics*, 73, 420-424, ISSN 0002-9505

Thomas, H. M. & Morfill, G. E. (1996). Melting dynamics of a plasma crystal. *Nature*, 379, 806-809, ISSN 0028-0836

Totsuji, H.; Kishimoto, T.; Inoue, Y.; *et al.* (1996). Yukawa system (dusty plasma) in one-dimensional external fields. *Physics Letters A*, 221, 215-219, ISSN 0375-9601

Vaulina, O. S. & Dranzhevski, I.E. (2006). Transport of macroparticles in dissipative two-dimensional Yukawa systems. *Physica Scripta*, 73, №6, 577-586, ISSN 0031-8949

Vaulina, O. S. & Vladimirov, S. V. (2002). Diffusion and dynamics of macro-particles in a complex plasma. *Phys. Plasmas*, 9, 835-840, ISSN 1070-664X

Vaulina, O.S.; Petrov, O.F.; Fortov, V.E.; *et al.* (2003). Experimental studies of the dynamics of dust grains in gas-discharge plasmas. *Plasma Physics Reports*, 29, 642-656, ISSN 1063-780X

Vaulina, O. S.; Vladimirov, S. V.; Petrov, O. F. & Fortov, V. E. (2004). Phase state and transport of non-Yukawa interacting macroparticles (complex plasma). *Phys. Plasmas*, 11, 3234-3237, ISSN 1070-664X

Vaulina, O. S.; Adamovich, K. G. & Dranzhevskii, I. E. (2005a). Formation of quasi-two-dimensional dust structures in an external electric field. *Plasma Physics Reports*, 31, 562-569, ISSN 1063-780X

Vaulina, O. S.; Petrov, O. F. & Fortov, V. E. (2005b). Simulations of mass-transport processes on short observation time scales in nonideal dissipative systems. *JETP*, 100, No. 5, 1018–1028, ISSN 1063-7761

Vaulina, O. S.; Repin, A. Yu.; Petrov, O. F. & Adamovich, K. G. (2006). Kinetic temperature and charge of a dust grain in weakly ionized gas-discharge plasmas. JETP, 102, №6, 986 - 997, ISSN 1063-7761

Wen-Tau, Juan & I, Lin (1998). Anomalous Diffusion in Strongly Coupled Quasi-2D Dusty Plasmas. *Phys. Rev. Lett.*, 80, 3073-3076, ISSN 0031-9007

Part 2

Mass Transfer and Hydro-, Magnetohydro- and Electro- Dynamics

6

Forced Convection Mass-Transfer Enhancement in Mixing Systems

Rafał Rakoczy and Stanisław Masiuk
Institute of Chemical Engineering and Environmental Protection Process, West Pomeranian University of Technology al. Piastów 42, 71-065 Szczecin Poland

1. Introduction

The design, scale-up and optimization of industrial processes conducted in agitated systems require, among other, precise knowledge of the hydrodynamics, mass and heat transfer parameters and reaction kinetics. Literature data available indicate that the mass-transfer process is generally the rate-limiting step in many industrial applications. Because of the tremendous importance of mass-transfer in engineering practice, a very large number of studies have determined mass-transfer coefficients both empirically and theoretically.

Agitated vessels find their use in a considerable number of mass-transfer operations. They are usually employed to dissolve granular or powdered solids into a liquid solvent in preparation for a reaction of other subsequent operations (Basmadjian, 2004). Agitation is commonly used in leaching operations or process of precipitation, crystallization and liquid extraction.

Transfer of the solute into the main body of the fluid occurs in the three ways, dependent upon the conditions. For an infinite stagnant fluid, transfer will be by the molecular diffusion augmented by the gradients of temperature and pressure. The natural convection currents are set up owing to the difference in density between the pure solvent and the solution. This difference in inducted flow helps to carry solute away from the interface. The third mode of transport is depended on the external effects. In this way, the forced convection closely resembles natural convection expect that the liquid flow is involved by using the external force.

Mass-transfer process in the mixing systems is very complicated and may be described by the non-dimensional Sherwood number, as a rule is a function of the Schmidt number and the dimensionless numbers describing the influence of hydrodynamic conditions on the realized process. In chemical engineering operations the experimental investigations are usually concerned with establishing the mass-transfer coefficients that define the rate of transport to the continuous phase.

One of the key aspects in the dynamic behaviour of the mass-transfer processes is the role of hydrodynamics. On a macroscopic scale, the improvement of hydrodynamic conditions can be achieved by using various techniques of mixing, vibration, rotation, pulsation and oscillation in addition to other techniques like the use of fluidization, turbulence promotes or magnetic and electric fields etc.

In this work, the focus is on a mass-transfer process under various types of augmentation technique, i.e.: rotational and reciprocating mixers, and rotating magnetic field. According to the information available in technical literature, the review of the empirical equations useful to generalize the experimental data for various types of mixers is presented. Moreover, the usage of static, rotating and alternating magnetic field to augment the mass process intensity instead of mechanically mixing is theoretical and practical analyzed.

2. Problem formulation of mass diffusion under the action of forced convection

Under forced convective conditions, the mathematical description of the solid dissolution process may be described by means of the integral equation of mass balance for the component i, namely,

$$\int_V \frac{\partial \rho_i}{\partial \tau} dV = -\int_S \rho_i \cdot \left(\overline{w_i} \cdot \overline{n}\right) dS + \int_V j_i \, dV \tag{1}$$

where: j_i - volumetric mass source of component i, $kg_i \cdot m^3 \cdot s^{-1}$; $\overline{n}$ - normal component; w_i - vector velocity of component i, $m \cdot s^{-1}$; S - area, m^2; V - volume, m^3; ρ_i - concentration of component i, $kg_i \cdot m^{-3}$; τ - time, s.
The above equation (1) for homogenous mixture may be written in the following differential form

$$\frac{\partial \rho_i}{\partial \tau} + div\left(\rho_i \overline{w_i}\right) = j_i \tag{2}$$

The vector velocity of component i, $\overline{w_i}$, can be defined by using the averaged value of momentum, ($[\overline{q}_i]_{avg}$)

$$\overline{w_i} = \lim_{V \to 0} \frac{\left[\overline{q_i}\right]_{avg}}{m_i} \tag{3}$$

The velocity of component i, $\overline{w_i}$, in relation to the velocity of mixture or liquid, $\overline{w}$, is defined as follows

$$dif\,\overline{w_i} = \overline{w_i} - \overline{w} \Rightarrow dif\,\overline{w_i} = \overline{w_i} - \frac{1}{\rho}\sum_i \rho_i \overline{w_i} \tag{4}$$

Introducing the relation (4) in equation (2), gives the following relationship for the mass balance of i component

$$\frac{\partial \rho_i}{\partial \tau} + div\left[\rho_i\left(dif\,\overline{w_i} + \overline{w_i}\right)\right] = j_i \tag{5}$$

The above equation (5) can be rewritten as follows

$$\frac{\partial \rho_i}{\partial \tau} + div\left(\rho_i \overline{w}\right) = -div\left(\overline{j_i}\right) + j_i \tag{6}$$

The mass concentration of component i, c_i, may be defined in the following form

$$c_i = \frac{\rho_i}{\rho} \Rightarrow \rho_i = \rho c_i \tag{7}$$

Then, the differential equation of mass balance (equation 6) for the mass concentration of component i, c_i, may be given by:

$$\rho \frac{dc_i}{d\tau} + div\left(\overline{j_i}\right) - j_i = 0 \tag{8}$$

The term $\left(\rho \frac{dc_i}{d\tau}\right)$ in the above equation (8) expresses the local accumulation of relative mass and the convectional mass flow rate of component i, whereas the term $\overline{j_i}$ is total diffusion flux density of component i. The term (j_i) describes the intensity of the process generation of the volumetric mass flux of component i in the volume V due to the dissolution process. The resulting diffusion flux is expressed as a sum of elementary fluxes considering the concentration (c), temperature (T), thermodynamic pressure gradient (p), and the additional force interactions $(\overline{F})$ (i.e. forced convection as a result of fluid mixing) in the following form

$$\overline{j_i} = \overline{j_i}(c) + \overline{j_i}(T) + \overline{j_i}(p) + \overline{j_i}\left(\overline{F}\right) \tag{9}$$

A more useful form of this equation may be obtained by introducing the proper coefficients as follows

$$\overline{j_i} = -\rho D_i\, grad(c_i) - \rho D_i k_T\, grad(\ln T) - \rho D_i k_p\, grad(\ln p) - \rho D_i k_{\overline{F}} \overline{F} \tag{10}$$

where: D_i - coefficient of molecular diffusion, $m^2 \cdot s^{-1}$; k_T - relative coefficient of thermodiffusion, $kg_i \cdot kg^{-1}$; k_p - relative coefficient of barodiffusion, $kg_i \cdot kg^{-1}$; $k_{\overline{F}}$ - relative coefficient of forced diffusion, $kg_i \cdot m^2 \cdot kg^{-1} \cdot N^{-1}$.
In the case of the experimental investigations of mass-transfer process from solid body to its flowing surrounding dilute solution, the boundary layer around the sample is generated. This layer is dispersed in the agitated volume by means of the physical diffusion process and the diffusion due to the forced convection. Then, the differential equation of mass balance equation for the mass concentration of component i diffuses to the surrounding liquid phase is given as follows

$$\frac{\partial c_i}{\partial \tau} + \overline{w}\, grad\, c_i + div\left(\frac{\overline{j_i}}{\rho}\right) = \frac{j_i}{\rho} \tag{11}$$

Introducing relation (10) in the equation (11) gives the following relationship for the balance of the mass concentration of component i where force $\overline{F}$ is generated by the mixing process

$$\frac{\partial c_i}{\partial \tau} + \overline{w}\, grad\, c_i + div\left[-D_i\, grad\, c_i - a\left(\frac{\partial c_i}{\partial T}\right)_{\substack{p=const\\ p_{\overline{F}}=const}} grad\, T - \nu\left(\frac{\partial c_i}{\partial p}\right)_{\substack{T=const\\ p_{\overline{F}}=const}} grad\, p\right] + div\left(\frac{\overline{F}}{\overline{G}} c_i\right) = \frac{j_i}{\rho} \tag{12}$$

where: ν - kinematic viscosity, m^2·s^{-1}; $\dot{G}$ - mass flow rate, kg·s^{-1}.
The above relation (12) may be treated as the differential mathematical model of the dissolution process of solid body. The right side of the above equation (12) represents the source mass flux of component *i*. This expression may be represented by the differential kinetic equation for the dissolution of solid body as follows

$$
\begin{aligned}
&-\frac{dm_i(\tau)}{d\tau} = \beta_i dF_s(\tau)\frac{dc_i(\tau)}{d\tau}d\tau \Rightarrow -\frac{d\rho_i(\tau)}{d\tau} = \beta_i \frac{dF_s(\tau)}{dV_s(\tau)}\frac{dc_i(\tau)}{d\tau}d\tau \Rightarrow \\
&\Rightarrow -\frac{d\rho_i(\tau)}{d\tau} = \beta_i \frac{dc_i(\tau)}{dr\,d\tau}d\tau \Rightarrow \frac{d\rho_i(\tau)}{d\tau} = \beta_i\left(grad\,c_i(\tau)\overline{e_r}\right) \Rightarrow \\
&\Rightarrow \beta_i = \frac{d\rho_i(\tau)}{\left(grad\,c_i(\tau)\overline{e_r}\right)d\tau}
\end{aligned}
\tag{13}
$$

where: β_i - mass-transfer coefficient in a mixing process, kg·m^{-2}·s^{-1}; m_i - mass of dissolving solid body, kg$_i$; F_s - surface of dissoluble sample, m^2.
The above equation (13) cannot be integrated because the area of solid body, F_s, is changing in time of dissolving process. It should be noted that the change in mass of solid body in a short time period of dissolving is very small and the mean area of dissolved cylinder may be used. The relation between loss of mass, mean area of mass-transfer and the mean driving force of this process for the time of dissolving duration is approximately linear and then the mass-transfer coefficient may be calculated from the simple linear equation

$$
[\beta_i]_{avg} = \frac{\Delta m_i}{[F_s]_{avg}\,\Delta c_i \Delta\tau}
\tag{14}
$$

Taking into account the above relations (equations 13 and 14) we obtain the following general relationship for the mass balance of component *i*

$$
\begin{aligned}
&\frac{\partial c_i}{\partial\tau} + \overline{w}\,grad\,c_i + div\left[-D_i\,grad\,c_i - a\left(\frac{\partial c_i}{\partial T}\right)_{\substack{p=const\\ p_{\overline{F}}=const}} grad\,T - \nu\left(\frac{\partial c_i}{\partial p}\right)_{\substack{T=const\\ p_{\overline{F}}=const}} grad\,p\right] + \\
&+div\left(\frac{\overline{F}}{\dot{G}}c_i\right) = \frac{[\beta_i]_{avg}\left(grad\,c_i\right)\overline{e_r}}{\rho}
\end{aligned}
\tag{15}
$$

The obtained equation (15) suggests that this dependence may be simplified in the following form:

$$
\frac{\partial c_i}{\partial\tau} + \overline{w}\,grad\,c_i - div\left(D_i\,grad\,c_i\right) + div\left(\frac{\overline{F}}{\dot{G}}c_i\right) = \frac{[\beta_i]_{avg}\left(grad\,c_i\right)\overline{e_r}}{\rho}
\tag{16}
$$

The agitated vessels find their use in a considerable number of mass-transfer operations. Practically, the intensification of the mass-transfer processes may be carried out by means of the vertical tubular cylindrical vessels equipped with the rotational (Nienow et al., 1997) or the reciprocating agitators (Masiuk, 2001). Under forced convective conditions, the force $\overline{F}$ in equation (16) may be defined (Masiuk, et al. 2008):

- for rotational agitator $\overline{F_{rot}} = \rho V \frac{d\overline{w_{rot}}}{dt} \Rightarrow \overline{F_{rot}} = \rho V 2\pi^2 n^2 d_{rot} \overline{e_\varphi}$ (17)

- for reciprocating agitator $\overline{F_{rec}} = \rho V \frac{d\overline{w_{rec}}}{dt} \Rightarrow \overline{F_{rec}} = \rho V 4\pi^2 A f^2 \overline{e_z}$ (18)

where: A - amplitude of reciprocating agitator, m; d_{rec} - diameter of reciprocating agitator, m; d_{rot} - diameter of rotational agitator, m; f - frequency of reciprocating agitator, s^{-1}; n - rotational speed of agitator, s^{-1}; V - liquid volume, kg·m^{-3}; ρ - liquid density, kg·m^{-3}.
Introducing the proposed relationships (17) and (18) in equation (16), give the following relations for the agitated system by using the rotational and reciprocating agitator, respectively

$$\frac{\partial c_i}{\partial \tau} + \overline{w}\, grad\, c_i - div\left(D_i\, grad\, c_i\right) + div\left(\frac{\rho V n^2 d_{rot}}{\dot{G}_{rot}} c_i \overline{e_\varphi}\right) = \frac{\left[\beta_i\right]_{avg}\left(grad\, c_i\right)\overline{e_r}}{\rho} \quad (19)$$

$$\frac{\partial c_i}{\partial \tau} + \overline{w}\, grad\, c_i - div\left(D_i\, grad\, c_i\right) + div\left(\frac{\rho V A f^2}{\dot{G}_{rec}} c_i \overline{e_z}\right) = \frac{\left[\beta_i\right]_{avg}\left(grad\, c_i\right)\overline{e_r}}{\rho} \quad (20)$$

The mass flow rate for the rotational and the reciprocating agitator can be approximated by the following equations:

$$\dot{G}_{rot} = \rho V n \quad (21)$$

$$\dot{G}_{rec} = \rho V f \quad (22)$$

Taking into consideration the above relations (equations 21 and 22), we obtain the following relationships:

$$\frac{\partial c_i}{\partial \tau} + \overline{w}\, grad\, c_i - div\left(D_i\, grad\, c_i\right) + div\left(n d_{rot} c_i \overline{e_\varphi}\right) = \frac{\left[\beta_i\right]_{avg}\left(grad\, c_i\right)\overline{e_r}}{\rho} \quad (23)$$

$$\frac{\partial c_i}{\partial \tau} + \overline{w}\, grad\, c_i - div\left(D_i\, grad\, c_i\right) + div\left(A f c_i \overline{e_z}\right) = \frac{\left[\beta_i\right]_{avg}\left(grad\, c_i\right)\overline{e_r}}{\rho} \quad (24)$$

2.1 Definition of dimensionless numbers for mass-transfer process

The governing equations (23) and (24) may be rewritten in a symbolic shape which is useful for the dimensionless analysis. The introduction of the non-dimensional quantities denoted by sign (*) into these relationships yield:

$$\frac{c_{i_0}}{\tau_0}\left[\frac{\partial c_i^*}{\partial \tau^*}\right] + \frac{w_0 c_{i_0}}{l_0}\left[\overline{w}^*\, grad\, c_i\right] - \frac{D_{i_0} c_{i_0}}{l_0^2}\left[div^*\left(D_i^*\, grad^* c_i^*\right)\right] + \\ + \frac{n_0 d_{rot_0} c_{i_0}}{l_0}\left[div^*\left(n^* d_{rot}^* c_i^* \overline{e_\varphi}\right)\right] = \frac{\left[\beta_i\right]_{avg_0} c_{i_0}}{l_0 \rho_0}\left[\frac{\left[\beta_i\right]_{avg}^*\left(grad^* c_i^*\right)\overline{e_r}}{\rho^*}\right] \quad (25)$$

$$\frac{c_{i_0}}{\tau_0}\left[\frac{\partial c_i^*}{\partial \tau^*}\right]+\frac{w_0 c_{i_0}}{l_0}\left[\overline{w}^* \, grad\, c_i\right]-\frac{D_{i_0} c_{i_0}}{l_0^2}\left[div^*\left(D_i^* \, grad^* c_i^*\right)\right]+$$
$$+\frac{A_0 f_0 c_{i_0}}{l_0}\left[div^*\left(A^* f^* c_i^* \overline{e_z}\right)\right]=\frac{\left[\beta_i\right]_{avg_0} c_{i_0}}{l_0 \rho_0}\left[\frac{\left[\beta_i\right]_{avg}^*\left(grad^* c_i^*\right)\overline{e_r}}{\rho^*}\right] \tag{26}$$

The non-dimensional forms of these equations may be scaled against the convective term $\left(\frac{w_0 c_{i_0}}{l_0}\right)$. The dimensionless forms of the equations (25) and (26) may be given as follows:

$$\frac{l_0}{\tau_0 w_0}\left[\frac{\partial c_i^*}{\partial \tau^*}\right]+\left[\overline{w}^* \, grad\, c_i\right]-\frac{D_{i_0}}{l_0 w_0}\left[div^*\left(D_i^* \, grad^* c_i^*\right)\right]+$$
$$+\frac{n_0 d_{rot_0}}{w_0}\left[div^*\left(n^* d_{rot}^* c_i^* \overline{e_\varphi}\right)\right]=\frac{\left[\beta_i\right]_{avg_0}}{\rho_0 w_0}\left[\frac{\left[\beta_i\right]_{avg}^*\left(grad^* c_i^*\right)\overline{e_r}}{\rho^*}\right] \tag{27}$$

$$\frac{l_0}{\tau_0 w_0}\left[\frac{\partial c_i^*}{\partial \tau^*}\right]+\left[\overline{w}^* \, grad\, c_i\right]-\frac{D_{i_0}}{l_0 w_0}\left[div^*\left(D_i^* \, grad^* c_i^*\right)\right]+$$
$$+\frac{A_0 f_0}{w_0}\left[div^*\left(A^* f^* c_i^* \overline{e_z}\right)\right]=\frac{\left[\beta_i\right]_{avg_0}}{\rho_0 w_0}\left[\frac{\left[\beta_i\right]_{avg}^*\left(grad^* c_i^*\right)\overline{e_r}}{\rho^*}\right] \tag{28}$$

The equations (27) and (28) include the following dimensionless groups characterising the mass-transfer process under the action of the rotational or the reciprocating agitator:

$$\frac{l_0}{\tau_0 w_0} \Rightarrow \frac{D}{\tau_0 n_0 d_{rot_0}} \Rightarrow S_{rot}^{-1} \tag{29a}$$

$$\frac{l_0}{\tau_0 w_0} \Rightarrow \frac{D}{\tau_0 A_0 f_0} \Rightarrow S_{rec}^{-1} \tag{29b}$$

$$\frac{D_{i_0}}{l_0 w_0} \Rightarrow \left(\frac{\nu}{l_0 w_0}\right)\left(\frac{D_{i_0}}{\nu}\right) \Rightarrow \left(\frac{\nu}{w_0 D}\right)\left(\frac{D_{i_0}}{\nu}\right) \Rightarrow Re^{-1}\, Sc^{-1} \Rightarrow Pe_{mass}^{-1} \tag{30}$$

$$\frac{n_0 d_{rot_0}}{w_0} \Rightarrow \left(\frac{n_0 d_{rot_0}}{n_0 d_{rot_0}}\right) \Rightarrow 1 \tag{31a}$$

$$\frac{A_0 f_0}{w_0} \Rightarrow \left(\frac{A_0 f_0}{A_0 f_0}\right) \Rightarrow 1 \tag{31b}$$

$$\frac{[\beta_i]_{avg_0}}{\rho_0 w_0} \Rightarrow \left(\frac{[\beta_i]_{avg_0} d_s}{\rho_0 D_i}\right)\left(\frac{\nu}{w_0 d_{rot}}\right)\left(\frac{D_i}{\nu}\right)\left(\frac{d_{rot}}{d_s}\right) \Rightarrow$$
$$\Rightarrow \left(\frac{[\beta_i]_{avg_0} d_s}{\rho_0 D_i}\right)\left(\frac{\nu}{n d_{rot}^2}\right)\left(\frac{D_i}{\nu}\right)\left(\frac{d_{rot}}{d_s}\right) \Rightarrow Sh\, Re_{rot}^{-1}\, Sc^{-1}\left(\frac{d_{rot}}{d_s}\right) \tag{32a}$$

$$\frac{[\beta_i]_{avg_0}}{\rho_0 w_0} \Rightarrow \left(\frac{[\beta_i]_{avg_0} d_s}{\rho_0 D_i}\right)\left(\frac{\nu}{w_0 d_{rec}}\right)\left(\frac{D_i}{\nu}\right)\left(\frac{d_{rec}}{d_s}\right) \Rightarrow$$
$$\Rightarrow \left(\frac{[\beta_i]_{avg_0} d_s}{\rho_0 D_i}\right)\left(\frac{\nu}{A_0 f_0 d_{rec}}\right)\left(\frac{D_i}{\nu}\right)\left(\frac{d_{rec}}{d_s}\right) \Rightarrow Sh\, Re_{rec}^{-1}\, Sc^{-1}\left(\frac{d_{rec}}{d_s}\right) \tag{32b}$$

where: D - diameter of vessel, m.
Taking into account the proposed relations (29-32), we find the following dimensionless governing equations:

- for rotational agitator

$$S_{rot}^{-1}\left[\frac{\partial c_i^*}{\partial \tau^*}\right]+\left[\overline{w}^*\, grad\, c_i\right]-Pe_{mass}^{-1}\left[div^*\left(D_i^*\, grad^* c_i^*\right)\right]+ \left[div^*\left(n^* d_{rot}^* c_i^* \overline{e_\varphi}\right)\right]= Sh\, Re_{rot}^{-1}\, Sc^{-1}\left(\frac{d_{rot}}{d_s}\right)\left[\frac{[\beta_i]_{avg}^*\left(grad^* c_i^*\right)\overline{e_r}}{\rho^*}\right] \tag{33}$$

Name	Symbol	Definition	Significance	Interpretation and Remarks
Strouhal	S	$\frac{\tau_0 w_0}{D}$	$\frac{convection}{unsteadiness}$	Dimensionless number describing oscillating flow mechanism
Péclet (mass)	Pe_{mass}	$\frac{w_0 D}{D_i}$	$\frac{hydrodynamic\ convection}{mass\ diffusion}$	Dimensionless independent mass-transfer parameter. $Pe_{mass} = Re\, Sc$
Reynolds (rot)	Re_{rot}	$\frac{n_0 d_{rot_0}^2}{\nu_0}$	$\frac{interial\ force\ /\ convection}{viscous\ force}$	Reynolds number for the rotational agitator.
Reynolds (rec)	Re_{rec}	$\frac{A_0 f_0 d_{rec}}{\nu_0}$		Reynolds number for the reciprocating agitator.
Sherwood	Sh	$\frac{[\beta_i]_{avg} d_s}{\rho_0 D_i}$	$\frac{convective\ mass\ transport}{diffusive\ mass\ transport}$	mass-transfer Stanton number $St_m = Sh\, Sc^{-1}\, Re^{-1}$
Schmidt	Sc	$\frac{\nu_0}{D_i}$	$\frac{momentum\ diffusion}{molecular\ diffusion}$	$Sc = Re^{-1}\, Pe_{mass}$

Table 1. Dimensionless parameters in equations (33) and (34) and their physical role

-for reciprocating agitator

$$S_{rec}^{-1}\left[\frac{\partial c_i^*}{\partial \tau^*}\right]+\left[\overline{w}^* \, grad \, c_i\right]-Pe_{mass}^{-1}\left[div^*\left(D_i^* \, grad^* c_i^*\right)\right]+ \\ +\left[div^*\left(A^* f^* c_i^* \overline{e_z}\right)\right]=Sh\,Re_{rec}^{-1}\,Sc^{-1}\left(\frac{d_{rec}}{d_s}\right)\left[\frac{\left[\beta_i\right]_{avg}^*\left(grad^* c_i^*\right)\overline{e_r}}{\rho^*}\right] \tag{34}$$

Table 1 summarizes all essential and independent dimensionless parameters met in mass-transfer process under the action of the agitated systems.

From dimensionless form of equation (34) and (35) it follows that:

- for rotational agitator $Sh\,Re_{rot}^{-1}\,Sc^{-1}\left(\frac{d_{rot}}{d_s}\right)\sim 1 \Rightarrow Sh \sim Re_{rot}\,Sc\left(\frac{d_s}{d_{rot}}\right)$ (35)

- for reciprocating agitator $Sh\,Re_{rec}^{-1}\,Sc^{-1}\left(\frac{d_{rec}}{d_s}\right)\sim 1 \Rightarrow Sh \sim Re_{rec}\,Sc\left(\frac{d_s}{d_{rec}}\right)$ (36)

2.2 Mass-transfer characteristics

Under convective conditions a relationship for the mass-transfer similar to the relationships obtained for heat-transfer may be expected of the form (Incropera & DeWitt, 1996):

$$Sh = f\left(Re, Sc\right) \tag{37}$$

The two principle dimensionless groups of relevance to mass-transfer are Sherwood and Schmidt numbers. The Sherwood number can be viewed as describing the ratio of convective to diffusive transport, and finds its counterpart in heat transfer in the form of the Nusselt number (Basmadjian, 2004).

The Schmidt number is a ratio of physical parameters pertinent to the system. This dimensionless group corresponds to the Prandtl number used in heat-transfer. Moreover, this number provides a measure of the relative effectiveness of momentum and mass transport by diffusion.

Added to these two groups is the Reynolds number, which represents the ratio of convective-to-viscous momentum transport. This number determines the existence of laminar or turbulent conditions of fluid flow. For small values of the Reynolds number, viscous forces are sufficiently large relative to inertia forces. But, with increasing the Reynolds number, viscous effects become progressively less important relative to inertia effects.

Two additional dimensionless groups, the Péclet number and the Stanton number (see Table 1) are also used and they are composed of other non-dimensional groups.

Evidently, for relation (37) to be of practical use, it must be rendered quantitative. This may be done by assuming that the functional relation is in the following form (Kay & Crawford, 1980)

$$Sh = a\,Re^b\,Sc^c \tag{38}$$

The mass-transfer coefficients in the agitated systems can be correlated by the combination of Sherwood, Reynolds and Schmidt numbers. Using the proposed relation (38), it has been found possible to correlate a host of experimental data for a wide range of operations. The coefficients of relation (38) are determined from experiment.

For mass-transfer under natural-convection conditions, where the Reynolds number is unimportant, the mass-transfer may be described by using the following general expression (Bird et al., 1966)

$$Sh = f(Gr, Sc) \tag{39}$$

The Grashof number, Gr, plays the same role in free convection that the Reynolds number plays in forced convection. Recall that the Reynolds number provides a measure of the ratio of the inertial to viscous forces acting on a fluid element. In contrast, the Grashof number indicates the ratio of the buoyancy force to the viscous force acting on the fluid (Fox & McDonald, 1993).

Under forced convection conditions where the Grashof number is unimportant the boundary layer theory suggests the following form of the relation (38) (Garner & Suckling, 1958)

$$Sh = 0.023Re^{0.83}\,Sc^{0.44} \tag{40}$$

The exponent upon of the Schmidt number is to be 0.33 (Noordsij & Rotte, 1968; Jameson, 1964; Condoret et al., 1989; Tojo et al., 1981; Lemcoff & Jameson, 1975) as there is some theoretical and experimental evidence for this value (Sugano & Rutkowsky, 1968), although reported values vary from 0.56 (Wong et al., 1978) to 1.13 (Lemlich & Levy, 1961).

A dimensionless group often used in literature is the Colburn factor j for mass transfer, which is defines as follows (Geankoplis, 2003)

$$j_m = St_m Sc^{0.66} \tag{41}$$

3. Mass-transfer correlations in rotationally agitated liquid-solid systems

We have compiled a list of the most frequently used correlations and tabulated them in Table 2. The following correlations have been found useful in predicting transport coefficients in the agitated systems. These equations may be successfully applied to analyze the case of the dissolution of solid bodies in a stirred tank.

As it mentioned above, mass-transfer process is very complicated and may be described by the non-dimensional Sherwood number, as a rule is a function of the Schmidt number and the dimensionless numbers describing the influence of hydrodynamic conditions on the realized process. Use of the dimensionless Sherwood number as a function of the various non-dimensional parameters (see relation 37 or 39) yields a description of liquid-side mass-transfer, which is more general and useful. In majority of the works, both theoretical and practical (see Table 2), the correlations of mass-transfer process have the general form

$$Sh = 2 + aRe^{b}\,Sc^{c} \tag{42}$$

where the Sherwood number is a function of the Reynolds number, the Schmidt number, and differ by the fitting parameters a, b and c.

References	Correlation	Situation	Comments
(Hixon & Baum, 1941)	$Sh = 0.16\, Re^{0.62}\, Sc^{0.5}$	Solid-liquid agitated systems (turbine agitator with inclined blades)	$Re > 6.7 \cdot 10^4$
(Hixon & Baum, 1944)	$Sh = 2.7 \cdot 10^{-5}\, Re^{1.4}\, Sc^{0.5}$		$Re < 6.7 \cdot 10^4$
(Hixon & Baum, 1942)	$Sh = 3.5 \cdot 10^{-5}\, Re^{1.0}\, Sc^{0.5}$	Solid-liquid agitated systems (propeller agitator)	$Re = 3 \cdot 10^4 - 5 \cdot 10^6$
(Humphrey & Van Ness, 1957)	$Sh = 0.58\, Re^{0.58}\, Sc^{0.5}$	Solid-liquid agitated systems (propeller agitator, baffles)	$Re = 6.3 \cdot 10^3 - 3.3 \cdot 10^5$
	$Sh = 0.022\, Re^{0.87}\, Sc^{0.5}$	Solid-liquid agitated systems (turbine agitator, baffles)	$Re = 3.1 \cdot 10^4 - 9 \cdot 10^4$
(Barker & Treybal, 1960)	$Sh = 0.052\, Re^{0.833}\, Sc^{0.5}$	Solid-liquid agitated systems (turbine agitator, baffles)	$Re = 3.1 \cdot 10^3 - 3 \cdot 10^5$
(Harriot, 1962)	$Sh = 2 + 0.6\, Re^{0.5}\, Sc^{0.33}$	Solid particles suspended in agitated vessel with baffles	$Re = \frac{v_{slip} d_p}{\nu} \Rightarrow Re = \frac{v_T d_p}{\nu}$
(Weinspach 1967)	$Sh = 10.68 \cdot 10^2\, Re^{0.15}\, Sc^{0.08} \left(\frac{d_s}{d}\right)^{1.15}$	Correlation for various types of agitators and mixer with baffles	$Re \to$ turbulent region
	$Sh = 7.2 \cdot 10^{-2}\, Ar^{0.33}\, Sc^{0.56} \left(\frac{d_s}{d}\right)$		$Ar = \frac{d^3 g \Delta\rho \rho_c}{\eta_c^2}$ $Ar\left(\frac{d_s}{d}\right) = 10^2 - 10^7; Sc = 10^2 - 10^5$
(Levins & Gastonbury, 1972)	$Sh = 2 + 0.44\, Re^{0.5}\, Sc^{0.38}$	Solid particles with significant density difference	$Re = \frac{v_{slip} d_p}{\nu}$

Table 2. Summary of mass-transfer correlations for agitated systems

References	Correlation	Situation	Comments
(Liepe & Möckel, 1976)	$Sh = 2 + 0.67 Re_{\varepsilon}^{3/8} Sc^{1/3}$	Solid-liquid agitated systems	$d_s < d_{s,\lim}$ $d_{s,\lim} = 5.2\left(\frac{\Delta\rho}{\rho}\right)^{-5/6}\left(\frac{\nu_c^3}{\langle\bar{\varepsilon}\rangle}\right)^{1/4}$ d_s - size of solid grain, m
	$Sh = 2 + 0.35 Re_{\varepsilon}^{2}\left(\frac{\Delta\rho}{\rho}\right)^{1/3}$		$d_s > d_{s,\lim}$
(Herndl & Mersmann, 1982)	$Sh = 2 + 0.46 Re_{\varepsilon}^{2/3} Sc^{1/3}$	Solid-liquid agitated systems	$Re_{\varepsilon} = \left(\frac{d_s^4\langle\bar{\varepsilon}\rangle}{\nu_c^3}\right)^{1/3}$ $;1 < Re_{\varepsilon} < 10^3$ ε - energy dissipation rate, W·kg1
(Kirwan, 1987)	$Sh = 2 + 0.52\, Re^{0.52}\, Sc^{0.33}$	Small solid particles	$Re = \frac{E^{1/3} d_p^{4/3}}{\nu} < 1.0$
(Ditl et al., 1991)	$Sh = 2 + 0.43 Re_{\varepsilon}^{0.6} Sc^{0.39}\left(\frac{D}{d}\right)^{-0.25}$	Solid-liquid agitated systems	-
(Basmadjian, 2004)	$Sh = 1.46\, Re^{0.65}\, Sc^{0.33}$	Solid-liquid baffled vessel $Re = 10^4 - 10^6$	$Sh = \frac{k_c d_v}{D_i}; Re = \frac{d_i^2 n}{\nu}$ k_c - mass transfer coefficient, $m^2 \cdot s^{-1}$ $d_i n$ - dimensional velocity, where n - represents the number revolutions per unit time, $m \cdot s^{-1}$
(Hartmann et al. 2006)	$Sh = 2 + 0.6\, Re_p^{0.5}\, Sc^{0.33}$	Solid particles suspended in liquid	$Re_p = \frac{v_T d_p}{\nu}$ Re_p - particle Reynolds number

Table 2. Summary of mass-transfer correlations for agitated systems-continuation

Instead of Reynolds number, it would be more natural to express the Sherwood number in terms of the Pèclet number, which is the product of the Reynolds and Schmidt numbers (see Table 2). When this number is small, transport takes place due to mass diffusion, and the fluid velocity, density and viscosity do not affect the transport rates. In this case, the Sherwood number is a constant, dependent only on the configuration. For laminar flow past a spherical body, the limiting value of this dimensionless number is equal to 2.

4. Dissolutions of solid body in a tubular reactor with reciprocating plate agitator

4.1 Literature survey

Numerous articles concern with the effect of vibration on mass-transfer in a gas-liquid contradicting in a reciprocating plate column (Baird & Rama Rao, 1988; Baird et al., 1992; Gomma & Al Taweel, 2005; Gomma et al., 1991; Rama Rao & Baird, 1988,2000, 2003; Sundaresan & Varma, 1990). In these papers is stated that the oscillatory motion of the gas-liquid system in reciprocating plate columns assures much higher interfacial contacting areas than in conventional bubble column with the lower power consumption. The dissolution of solid particles into water and other solutions was investigated by Tojo et al. (Tojo et al., 1981), where agitation realized by using circular flat disc without perforation. The mass-transfer coefficient is calculated by measuring the slope of the concentration time curve in the first second of particle dissolution. The breakage of chalk aggregates in both the vibrating and rotating mixers and the analyse of the model of breakage which relates the pseudo-equilibrium aggregate size to the energy dissipation rate in the stirred vessel has been investigated by Shamlou et al. (Shamlou et al., 1996). The breakage of aggregates in both the vibrating and the rotating mixers occurs by turbulent fluid stresses, but it is not depended on the source generating the liquid motion.

4.2 Experimental details

The intensification of the dissolution under the action of reciprocating agitators were carried out by means of the reactor presented by (Masiuk, 2001). It is consisted of the vertical vessel, the system measuring mass of dissolving solid body (rock-salt), the device for measuring concentration in the bulk of mixed liquid (distilled water) and the arrangement for reciprocating plate agitator at various amplitudes and frequencies. The agitation was carried out with a single reciprocating plate with flapping blades oriented horizontally where it reciprocated in a vertical direction (Masiuk[a], 1999) and different number of the multihole perforated plates agitators. The detail schemes of the vibratory mixers are given in the relevant references (Masiuk, 1996; Masiuk[b], 1999; Masiuk, 2000). Additionally, the main geometrical dimensions of mixers and the operating ranges of the process parameters are collected in the Table 3.

The reciprocating agitator was driven by the electric a.c. motor coupled through a variable gear and a V-belt transmission turned a flywheel. A vertical oscillating shaft with the perforated plates and a hardened steel ring through a sufficiently long crank shaft were articulated eccentrically to the flywheel. An inductive transducer mounted inside the ring and a tape recorder was used to measure the inductive voltage directly proportional to the total force straining of the shaft.

The average mass-transfer coefficient was calculated from a mass balance between a dissolving solid cylindrical sample and no flowing surrounding dilute solution (see equation 14). Two conductive probes connected to a multifunction computer meter were used to measuring and recording of the concentration of the achieve solution of the salt. The mass of the rock salt sample decreasing during the process of dissolution is determined by an electronic balance that connected with rocking double-arm lever. On the lever arm the sample was hanging, the other arm connected to the balance. In the present investigation the change in mass of solid body in a short time period of dissolution is very small and the mean area of dissolved cylinder of the rock salt may be used. Than the mean mass-transfer

Parameters	Operating value
Geometrical dimensions of mixer	
Vessel diameter, m	0.205
Vessel height, m	1.033
Height of liquid level in the vessel, m	0.955
Geometrical dimensions of single reciprocating plate with flapping blades	
Plate diameter, m	0.036
Width of windows in the plate, m	0.039
Plate height, m	0.036
Clearance between the plate and exhaust of the blade, m	0.015-0.024
Hydraulic diameter, m	0.00386-0.01
Geometrical dimensions of multihole perforated plates	
Agitator diameter, m	0.204
Number of plates	1-5
Diameter of hole, m	0.005-0.06
Number of holes	2-650
Distance between plates, m	0.04
Operating conditions	
Mean diameter of sample, m	0.011-0.0316
Mean length of the sample, m	0.03-0.065
Amplitude, m	0.03-0.19
Frequency, s^{-1}	0.058-1.429
Time of dissolution, min	2
Loss of mass for 2 min dissolution, kg	$0.14 \cdot 10^{-3}$-$6.83 \cdot 10^{-3}$
Mean driving force, $kg_{NaCl} \cdot (kg_{solvent})^{-1}$	0.002-0.0137

Table 3. The main geometrical dimensions of mixer and operating conditions

coefficient may be calculated from the linear kinetics equation using the mean concentration driving force of the process determined from two time response curves.

Raw rock-salt (>98% NaCl and rest traces quantitative of chloride of K, Ca, Mg and insoluble mineral impurities) cylinders were not fit directly for the experiments because their structure was not homogeneous (certain porosity). Basic requirement concerning the experiments was creating possibly homogeneous transport conditions of mass on whole interfacial surface, which was the active surface of the solid body. These requirements were met thanks to proper preparing of the sample, mounting it in the mixer and matching proper time of dissolving. As an evident effect were fast showing big pinholes on the surface of the dissolved sample as results of local non-homogeneous of material. Departure from the shape of a simple geometrical body made it impossible to take measurements of its area with sufficient precision. So it was necessary to put those samples through the process of so-called hardening. The turned cylinders had been soaked in saturated brine solution for about 15 min and than dried in a room temperature. This process was repeated four times. To help mount the sample in the mixer, a thin copper thread was glued into the sample's axis. The processing was finished with additional smoothing of the surface with fine-grained abrasive paper. A sample prepared in this way had been keeping its shape during dissolving for about 30 min. The duration of a run was usually 30 sec. The rate of mass-transfer involved did not produce significant dimensional change in diameter of the cylinder. The time of a single dissolving cycle was chosen so that the measurement of mass loss could be made with sufficient accuracy and the decrease of dimensions would be relatively small (maximum about 0.5 mm).

Before starting every experiment, a sample which height, diameter and mass had been known was mounted in a mixer under the free surface of the mixed liquid. The reciprocating plate agitator was started, the recording of concentration changes in time, the weight showing changes in sample's mass during the process of solution, and time measuring was started simultaneously. After finishing the cycle of dissolving, the agitator was stopped, and then the loss of mass had been read on electronic scale and concentration of NaCl (electrical conductivity) in the mixer as well. This connection is given by a calibration curve, showing the dependence of the relative mass concentration of NaCl on the electrical conductivity (Rakoczy & Masiuk, 2010).

4.3 Results and discussion

In order to establish the effect of the Reynolds number on mass-transfer coefficient data obtained from the experimental investigations are graphically illustrated in a $\log\left(Sh\,Sc^{-0.33}\right)$ versus $\log(Re)$ systems for the in figure 1. These experimental results were obtained for the single reciprocating plate with flapping blades.

Figure 1 demonstrates that, within the limits of scatter among the plotted data represented by the points, the non-dimensional Sherwood number increase in the Reynolds number. The

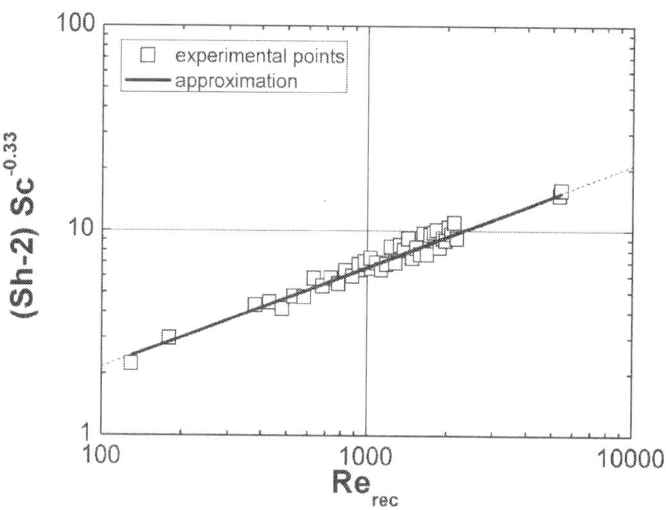

Fig. 1. Effect of the Reynolds number on the mass-transfer rate for the single reciprocating plate with flapping blades oriented horizontally

relation between the experimental results shown in this figure can be described by using the relation similar to relationship (37)

$$Sh = 0.17\, Re_{rec}^{0.33}\, Sc^{0.33}\left(1 + 0.65\, Re_{rec}^{0.22}\right) \tag{43}$$

The obtained correlation is valid for the following range of the process parameters: $Re_{rec} = 130 - 5400;\ Sc = 874 - 1244$. It is believed that the proposed dimensionless equation (44) is useful to generalize the experimental data in this work for the whole region of the reciprocating Reynolds number without the break of the correlating graphical line. Moreover, figure 1 presents a graphical form of equation (43), as the full curve, correlated the data very well with standard deviation $\sigma = 0.66$. The difference between the predicted and calculated values of the non-dimensional Sherwood number is less than $\pm 15\%$ for approximately 80% of the data points.
In present experimental investigations of the influence of the perforated plate reciprocating agitators on the mass-transfer process was additionally investigated for different number of plates varying from 1 to 5. The results of the mass-transfer experiments for different number of plates should be correlated using the relationship similar to the expression obtained for the reciprocating agitator with single perforated plate with flapping blades (equation 43).
In the present report the mass-transfer process is described by the similar somewhat modified relationship (37) between the dimensionless Sherwood number and the reciprocating Reynolds number

$$Sh = c_1\, Re_{rec}^{0.33}\, Sc^{0.33}\left(1 + c_2 Re_{rec}^{0.22}\right) \tag{44}$$

It is decided, that the constant c_2 in the relation (44) is not depended on the number of perforated plates N and equal 0.11. The constant c_1 has been computed as the function of parameter N employing the principle of least square method. Then the equation (44) for different number of the perforated plates may be rewritten in the following general dimensionless correlation

$$Sh = f^{Sh}(N)\, Re_{rec}^{0.33}\, Sc^{0.33}\left(1 + 0.11\, Re_{rec}^{0.22}\right) \tag{45}$$

where the function $f^{Sh}(N)$ is depended on the number of multihole plates.
The shape of the mass-transfer characteristics for the single plate as well as for different plates is similar and the function $f^{Sh}(N)$ may be described by means of the following relationship

$$f^{Sh}(N) = 0.17\, N \exp(-0.12\, N) \tag{46}$$

In the case of this work, the experimental data have been correlated using the new mean modified dimensionless Sherwood number, Sh^*, as a ratio of the dimensionless Sherwood, Sh, number and the dimensionless maximum density of mixing energy, $\rho^*_{E_{max}}$.
In the selection of the suitable agitator for the transfer process, it is not sufficient to take into consideration the power consumption and mixing time separately. The density of the maximum mixing energy, $\rho_{E_{max}}$, is defined as the product of the maximum power

consumption, $P_{\max}$, and the mixing time, t_{mix}, relate to the volume of mixed liquid, V_L (Masiuk & Rakoczy, 2007)

$$\rho_{E_{\max}} = \frac{P_{\max} t_{mix}}{V_L} \qquad [\text{J}\cdot\text{m}^{-3}] \tag{47}$$

The values of the density of the maximum mixing energy (47) may be determined by using the relationships describing the maximum power consumption

$$\begin{gathered} Po = f^P(N) Re_{rec}^{-1}\left(1 + 0.255 Re_{rec}^{0.95}\right) \Rightarrow \\ \Rightarrow Po = \left[0.89 N^{1.37} \exp(-0.15N)\right] Re_{rec}^{-1}\left(1 + 0.255 Re_{rec}^{0.95}\right) \Rightarrow \\ \Rightarrow \left(\frac{P_{\max} S^2}{\rho D^2 (2\pi A f)^3 (1 - S^2)}\right) = \left[0.89 N^{1.37} \exp(-0.15N)\right] Re_{rec}^{-1}\left(1 + 0.255 Re_{rec}^{0.95}\right) \end{gathered} \tag{48}$$

and the mixing time

$$\begin{gathered} \Theta = f^{t_{mix}}(N)\left(\frac{H_L}{d_h}\right) Re_{rec}^{-2}\left(1 + 0.021 Re_{rec}^{1.25}\right) \Rightarrow \\ \Rightarrow \Theta = \left[10.5 \exp(-1.06N)\right]\left(\frac{H_L}{d_h}\right) Re_{rec}^{-2}\left(1 + 0.021 Re_{rec}^{1.25}\right) \Rightarrow \\ \Rightarrow \left(\frac{t_{mix}\eta}{\rho H_L^2}\right) = \left[10.5 \exp(-1.06N)\right]\left(\frac{H_L}{d_h}\right) Re_{rec}^{-2}\left(1 + 0.021 Re_{rec}^{1.25}\right) \end{gathered} \tag{49}$$

where: H_L - height of liquid level in the vessel, m; P - power, W; S - fraction of open area of the reciprocating plate, m; η - liquid viscosity, kg·m^{-1}·s^{-1}.
The simple transformations gives the following equation describing the dimension density of the maximum mixing energy (Masiuk & Rakoczy, 2007)

$$\rho_{E_{\max}} = \psi f^E(N)\left(1 + 0.021 Re_{rec}^{1.25}\right)\left(1 + 0.255 Re_{rec}^{0.95}\right) \qquad [\text{J}\cdot\text{m}^{-3}] \tag{50}$$

where the function $f^E(N)$ is depended on the number of the multihole plates. The parameter ψ has the following form (Masiuk & Rakoczy, 2007)

$$\psi = \frac{4\left(1 - S^2\right)}{\pi S^2} \frac{H_L^{1.15}}{d_h^{3.15}} \frac{\eta^2}{\rho} \qquad [\text{J}\cdot\text{m}^{-3}] \tag{51}$$

where d_h - hydraulic diameter of perforated plate reciprocating agitators, m.
Hence, the dimensionless maximum density of maximum mixing energy, $\rho^*_{E_{\max}}$, may be calculated by means of the following equation

$$\rho^*_{E_{\max}} = \frac{\rho_{E_{\max}}}{\psi} \Rightarrow \rho^*_{E_{\max}} = f^E(N)\left(1 + 0.021 Re_{rec}^{1.25}\right)\left(1 + 0.255 Re_{rec}^{0.95}\right) \tag{52}$$

where:

$$f^E(N) = 9.35N^{1.37}\exp(-1.21N) \tag{53}$$

Taking into account the above equations (52) and (53), we find the following form of the modified dimensionless Sherwood number, Sh^*

$$Sh^* = \frac{Sh}{\rho^*_{E_{max}}} \Rightarrow Sh^* = \frac{(0.17N\exp(-0.12N))Re_{rec}^{0.33}Sc^{0.33}(1+0.11Re_{rec}^{0.22})}{(9.35N^{1.37}\exp(-1.21N))(1+0.021Re_{rec}^{1.25})(1+0.255Re_{rec}^{0.95})} \tag{54}$$

The graphical illustration of the equation (54) is given in the coordinates $(Sh^*Sc^{-0.33}, Re_{rec})$ log-log system in figure 2.

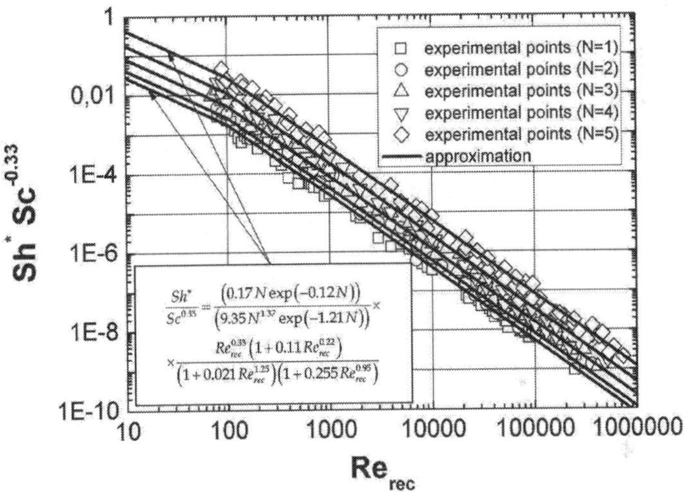

Fig. 2. Modified mass-transfer characteristics for reciprocating agitator with different number of plates

The rate of mass-transfer increases with increasing number of plates. The modified dimensionless Sherwood number, Sh^*, increases with increasing number of plates. Within the laminar region of the flow, the mass-transfer characteristics obtain the extreme (see figure 3). In the region of large Reynolds number, the curves of the modified dimensionless Sherwood number versus the reciprocating Reynolds number decreases sharply. Additionally, the change of plates of the reciprocating agitator has considerable profitable influence on the mass-transfer process.

5. Enhancement of solid dissolution process under the influence of transverse rotating magnetic field (TRMF)

The transverse rotating magnetic field (TRMF) is a versatile option for enhancing several physical and chemical processes. Studies over the recent decades were focused on application of magnetic field (MF) in different areas of engineering processes (Rakoczy &

Masiuk, 2009; Rakoczy, 2006). Static, rotating or alternating MFs might be used to augment the process intensity instead of mechanically mixing. Practical applications of RMF are presented in (Volz & Mazuruk, 1999; Melle et al., 1999; Walker et al. 2004; Nikrityuk et al. 2006; Yang et al., 2007; Fraňa et al. 2006; Spitzer, 1999).

Recently, TRMF are widely used to control different processes in the various engineering operations. This kind of magnetic field induces a time-averaged azimuthal force, which drives the flow of the electrical conducting fluid in circumferential direction. The magnetic filed lines rotate in the horizontal direction with the rotation frequency of the field, ω_{TRMF}. It is obvious that this parameter is equal to the frequency of alternating current. An electrical field, $\overline{E}$, is generated perpendicular to the magnetic field, $\overline{B}$. Perpendicular to the electric field, the Lorenz magnetic force, $\overline{F_m}$, is acting as the driving force for the liquid rotation.

The TRMF may be generated by cylindrical inductor resembling the stator of a three-phase asynchronous electrical engine. The similar inductor of TRMF was applied in (Lu & Li, 2000; Nekrassov & Chekin, 1961; Nekrassov & Chekin, 1962).

In the case of the electrical conducting fluid, the TRMF induces currents inside this liquid. These interact with the field of the inductor and generate electromagnetic force inside the electrical conducting liquid. The pattern of hydrodynamic flows due to TRMF in a cylindrical volume depends on the number of pole pair, p_{TRMF}, of TRMF inductor. The distribution of TRMF may be considered as the quasi-uniform or non-uniform for the number of pole pair equal to 1 and $p_{TRMF} > 1$, respectively. The application of TRMF may be generated the specific flow in a liquid cylindrical volume. The liquid flow is an azimuthal motion of medium around the central axis of the cylindrical volume. This flow is directed as the magnetic field rotates and the azimuthal velocity is maximal at the cylindrical vessel's walls. Obviously, this movement may determine the heat or mass-transfer inside the rotating liquid.

According to available in technical literature, research has not been focused on the mass-transfer during dissolution of a solid body to the surrounding liquid under the action of RMF. In the present report, the experimental investigations has been conducted to explain the influence of this kind of MF on the mass-transfer enhancement. Moreover, the influence of localisation of a NaCl-cylindrical sample localisation on the mass-transfer rate was experimentally determined.

5.2 Influence of transverse rotating magnetic field (TRMF) on mass-transfer process

The influence of TRMF on mass-transfer process is manifested by the magnetic force defined as the divergence of Maxwell's stress tensor and expressed by the following relation (Rakoczy & Masiuk, 2010)

$$\frac{\overline{F_m}}{V} = div\left[-\frac{1}{2}\mu_m \overline{H}^2 E + \mu_m\left(\overline{H}\overline{H}\right)\right] \Rightarrow \frac{\overline{F_m}}{V} = -\frac{1}{2\mu_m} grad\,\overline{B}^2 + \frac{1}{\mu_m} Div\left(\overline{B}\overline{B}\right) \tag{55}$$

where $\overline{B}$ - magnetic induction vector, kg·A^{-1}·s^{-1}; $\overline{H}$ - magnetic field strength vector, A·m^{-1}; V - liquid volume exposed to TRMF, m^3; μ_m - magnetic permeability, kg·m·A^{-2}·s^{-2}.

In the above equation (55), the term $\left(\frac{1}{\mu_m} Div\left(\overline{B}\overline{B}\right)\right)$ may be treated as the part of magnetic force and this expression describes the enhancement of flux density by the forced mass convection. Introducing this term in the equation (16), gives the following relationship

$$\frac{\partial c_i}{\partial \tau} + \overline{w}\, grad\, c_i - div\left(D_i\, grad\, c_i\right) + div\left(\frac{V c_i}{\dot{G}_{TRMF}\, \mu_m} Div\left(\overline{B}\overline{B}\right)\right) = \frac{\left[\beta_i\right]_{avg}\left(grad\, c_i\right)\overline{e}_r}{\rho} \tag{56}$$

The mass flow rate for this case may be approximated by using the following relation

$$\dot{G}_{TRMF} = \rho V \omega_{TRMF} \Rightarrow \dot{G}_{TRMF} = \frac{\rho V}{\tau_{TRMF}} \tag{57}$$

where τ_{TRMF} - time of magnetic field diffusion, s.
This parameter may be defined basing on the well-known advection-diffusion type equation (Möβner & Gerbeth, 1999; Guru & Hiziroğlu, 2004)

$$\frac{\partial \overline{B}}{\partial \tau} = rot\left(\overline{w} \times \overline{B}\right) + \frac{1}{\sigma_e \mu_m} \Delta \overline{B} \tag{58}$$

where w - velocity, m·s^{-1}; σ_e - electrical conductivity, A^2·s^3·kg^{-1}·m^{-3}; $\nu_m = \left(\sigma_e \mu_m\right)^{-1}$ - effective diffusion coefficient (magnetic viscosity), m^2·s^{-1}
The above equation (58) is also called the induction equation and it characterizes the temporal evolution of the magnetic field. The term, $rot\left(\overline{w} \times \overline{B}\right)$, in this equation dominates when the conductivity is large, and can be regarded as describing freezing of MF lines into the liquid. The term, $\frac{1}{\sigma_e \mu_m} \Delta \overline{B}$, in the *B*-field equation may be treated as a diffusion term. When the electrical conductivity, σ_e, is not too large, MF lines diffuse within the fluid.
The relation (58) may be expressed as follows

$$\frac{B_0}{\tau_0}\left[\frac{\partial \overline{B}^*}{\partial \tau^*}\right] = \frac{w_0 B_0}{l_0}\left[rot^*\left(\overline{w}^* \times \overline{B}^*\right)\right] + \frac{\nu_{m_0} B_0}{l_0^2}\left[\nu_m^* \Delta^* \overline{B}^*\right] \tag{59}$$

The dimensionless form of this equation may be used to examine the effect of liquid flow on the MF distribution. The non-dimensional forms of these equations may be scaled against the term $\left(\frac{\nu_{m_0} B_0}{l_0^2}\right)$. The dimensionless form of the equation (59) may be expressed by

$$\frac{l_0^2}{\nu_{m_0} \tau_0}\left[\frac{\partial \overline{B}^*}{\partial \tau^*}\right] = \frac{w_0 l_0}{\nu_{m_0}}\left[rot^*\left(\overline{w}^* \times \overline{B}^*\right)\right] + \left[\nu_m^* \Delta^* \overline{B}^*\right] \tag{60}$$

The introduced magnetic Fourier number

$$Fo_m = \frac{\nu_{m_0} \tau_0}{l_0^2} \tag{61}$$

and magnetic Reynolds number, directly analogous to the traditional Reynolds number,

$$Re_m = \frac{w_0 l_0}{\nu_{m_0}} \tag{62}$$

describes the relative importance of advection and diffusion of the MF.

Taking into account the above definitions of the non-dimensional groups, we obtain the following general relationship of the magnetic induction equation

$$Fo_m^{-1}\left[\frac{\partial \overline{B}^*}{\partial \tau^*}\right] = Re_m\left[rot^*\left(\overline{w}^* \times \overline{B}^*\right)\right] + \left[\nu_m^* \Delta^* \overline{B}^*\right] \tag{63}$$

It should be noticed that the time of magnetic field diffusion, τ_{TRMF}, may be defined as follows

$$Fo_m^{-1} \sim 1 \Rightarrow \frac{l_0^2}{\nu_{m_0}\tau_0} \sim 1 \Rightarrow \tau_0 \sim \frac{l_0^2}{\nu_{m_0}} \Rightarrow \tau_{TRMF} = \frac{D^2}{\nu_m} \tag{64}$$

Then, the equation (57) may be rewritten in the form

$$\dot{G}_{TRMF} = \frac{\rho V}{\tau_{TRMF}} \Rightarrow \dot{G}_{TRMF} = \frac{\rho \nu_m V}{D^2} \tag{65}$$

Taking into consideration the above relations (Eqs 56 and 65), we obtain the following relationships:

$$\frac{\partial c_i}{\partial \tau} + \overline{w}\, grad\, c_i - div\left(D_i\, grad\, c_i\right) + div\left(\frac{D^2 c_i}{\rho \nu_m \mu_m} Div\left(\overline{B}\overline{B}\right)\right) = \frac{\left[\beta_i\right]_{avg}\left(grad\, c_i\right)\overline{e_r}}{\rho} \tag{66}$$

The governing equation (66) may be rewritten in a symbolic shape which is useful for the dimensionless analysis. The introduction of the non-dimensional quantities denoted by sign $\left(^*\right)$ into these relationships yield:

$$\begin{aligned} &\frac{c_{i_0}}{\tau_0}\left[\frac{\partial c_i^*}{\partial \tau^*}\right] + \frac{w_0 c_{i_0}}{l_0}\left[\overline{w}^*\, grad\, c_i^*\right] - \frac{D_{i_0} c_{i_0}}{l_0^2}\left[div^*\left(D_i^*\, grad^* c_i^*\right)\right] + \\ &+ \frac{D^2 c_{i_0} B_0^2}{\rho_0 \nu_{m_0} \mu_{m_0} l_0^2}\left[div^*\left(\frac{D^* D^* c_i^*}{\rho^* \nu_m^* \mu_m^*} Div^*\left(\overline{B}^*\overline{B}^*\right)\right)\right] = \frac{\left[\beta_i\right]_{avg_0} c_{i_0}}{l_0 \rho_0}\left[\frac{\left[\beta_i\right]_{avg}^*\left(grad^* c_i^*\right)\overline{e_r}}{\rho^*}\right] \end{aligned} \tag{67}$$

The non-dimensional forms of the above equation (67) may be scaled against the convective term $\left(\frac{w_0 c_{i_0}}{l_0}\right)$ as follows

$$\begin{aligned} &\frac{l_0}{\tau_0 w_0}\left[\frac{\partial c_i^*}{\partial \tau^*}\right] + \left[\overline{w}^*\, grad\, c_i^*\right] - \frac{D_{i_0}}{l_0 w_0}\left[div^*\left(D_i^*\, grad^* c_i^*\right)\right] + \\ &+ \frac{D B_0^2}{\rho_0 \nu_{m_0} \mu_{m_0} w_0}\left[div^*\left(\frac{D^* D^* c_i^*}{\rho^* \nu_m^* \mu_m^*} Div^*\left(\overline{B}^*\overline{B}^*\right)\right)\right] = \frac{\left[\beta_i\right]_{avg_0}}{\rho_0 w_0}\left[\frac{\left[\beta_i\right]_{avg}^*\left(grad^* c_i^*\right)\overline{e_r}}{\rho^*}\right] \end{aligned} \tag{68}$$

The equation (68) includes the following dimensionless groups characterising the influence of the TRMF on the mass-transfer process

$$\frac{DB_0^2}{\rho_0 \nu_{m_0} \mu_{m_0} w_0} \Rightarrow \left(\frac{B_0^2 \sigma_e D}{\rho_0 w_0}\right) \Rightarrow \left(\frac{B_0^2 \sigma_e}{\rho_0 \omega_{TRMF}}\right) \Rightarrow \left(\frac{B_0^2 \sigma_e D^2}{\rho_0 \nu}\right)\left(\frac{\nu}{\omega_{TRMF} D^2}\right) \Rightarrow Q\, Re_\omega^{-1} \tag{69}$$

$$\frac{[\beta_i]_{avg_0}}{\rho_0 w_0} \Rightarrow \left(\frac{[\beta_i]_{avg_0} d_s}{\rho_0 D_i}\right)\left(\frac{\nu_0}{\omega_{TRMF} D^2}\right)\left(\frac{D_i}{\nu_0}\right)\left(\frac{D}{d_s}\right) \Rightarrow Sh\, Re_\omega^{-1}\, Sc^{-1}\left(\frac{D}{d_s}\right) \tag{70}$$

Taking into account the proposed relations (29-32, 69 and 70), we find the following dimensionless governing equations

$$S^{-1}\left[\frac{\partial c_i^*}{\partial \tau^*}\right] + \left[\overline{w}^* \, grad\, c_i^*\right] - Pe_{mass}^{-1}\left[div^*\left(D_i^*\, grad^* c_i^*\right)\right] +$$
$$+ Q\, Re_\omega^{-1}\left[div^*\left(\frac{D^* D^* c_i^*}{\rho^* \nu_m^* \mu_m^*} Div^*\left(\overline{B}^* \overline{B}^*\right)\right)\right] = Sh\, Re_\omega^{-1}\, Sc^{-1}\left(\frac{D}{d_s}\right)\left[\frac{[\beta_i]_{avg}^*\left(grad^* c_i^*\right)\overline{e}_r}{\rho^*}\right] \tag{71}$$

From dimensionless form of equation (71) it follows that

$$Sh\, Re_\omega^{-1}\, Sc^{-1}\left(\frac{D}{d_s}\right) \sim Q\, Re_\omega^{-1} \Rightarrow Sh \sim Q\, Sc\left(\frac{d_s}{D}\right) \tag{72}$$

The TRMF may be characterized by means of the following numbers:

- the dimensionless Chandrasekhar number

$$Q = \frac{magnetic\ force}{dissipative\ force} \Rightarrow Q = \frac{B_0^2 D \sigma_e}{\rho_0 \nu_0} \tag{73}$$

- the dimensionless field frequency based Reynolds number

$$Re_\omega = \frac{interial\ force\ /\ convection}{viscous\ force} \Rightarrow Re_\omega = \frac{\omega_{TRMF} D^2}{\nu_0} \tag{74}$$

where ω_{TRMF} - angular velocity of magnetic field, rad·s^{-1}.

The interaction between of the TRMF and the used liquid may be also described by means of the non-dimensional Hartmann number:

$$Ha = B_0 D \sqrt{\frac{\sigma_e}{\rho_0 \nu_0}} \tag{75}$$

It should be noticed that the Chandrasekhar number is the square of the Hartmann number $Q = Ha^2$. It should be noticed that the interaction between MF and the induced current in the electrically conducting liquid produces a Lorentz force:

A significant consequence of the above expressions (73-75) is that the fluid rotation under action of the TRMF may be completely characterized by means of a single dimensionless number which is denoted as the magnetic Taylor number, Ta_m. The dimensionless Hartman and field frequency based Reynolds numbers can be unified to this non-dimensional group as follows:

$$Ta_m = \frac{centrifugal\ force}{viscous\ force} \Rightarrow Ta_m = Q\,Re_\omega \Rightarrow Ta_m = Ha^2\,Re_\omega \Rightarrow Ta_m = \frac{B_0^2 D^4 \omega_{TRMF} \sigma_e \rho_0}{\eta_0} \tag{76}$$

The above dimensionless magnetic Taylor number (76) is a measure of the azimuthal liquid velocity induced by the TRMF. Moreover, this parameter describes the Lorentz-force amplitude of RMF action and scales the force that drives the convection due to the applied TRMF.

The above definitions of non-dimensional groups (73-75) lead to the final expression of the relationship (72)

$$Sh \sim Q\,Sc\left(\frac{d_s}{D}\right) \Rightarrow Sh \sim Ha^2\,Sc\left(\frac{d_s}{D}\right) \Rightarrow Sh \sim \left(Ta_m Re_\omega^{-1}\right)Sc\left(\frac{d_s}{D}\right) \tag{78}$$

5.3 Experimental details

All experimental measurements of mass-transfer process using the TRMF were carried out in a laboratory set-up including electromagnetic field generator. A schematic of the experimental apparatus is presented in figure 3.

This setup may be divided into: a generator of the rotating electromagnetic field (1), a glass container (2) with the conductivity probes (3-4), an electric control box (5) and an inverter (6) connected with multifunctional electronic switch (8) and a personal computer (7) loaded with special software. This software made possible the electromagnetic field rotation control, recording working parameters of the generator and various state parameters.

From preliminary tests of the experimental apparatus, the glass container is not influenced by the working parameters of the stator. The TRMF was generated by a modified 3-phase stator of an induction squirrel-cage motor, parameters of which are in accordance with the Polish Standard PN-72/E-06000. The stator is supplied with a 50 Hz three-phase alternating current. The transistorized inverter (4) was used to change the frequency of the rotating magnetic field in the range of $f_{TRMF} = 1 - 50\,s^{-1}$. The stator of the electric machine, as the RMF generator is made up of a number of stampings with slots to carry the three phase winding. The number of pair poles per phase winding, p, is equal to 2. The windings are geometrically spaced 120 degrees apart. The stator and the liquid may be treated as apparent virtual electrical circuit of the closed flux of a magnetic induction. The stator windings are connected through the a.c. transistorized inverter to the power source. The generator produces an azimuthal electromagnetic force in the bulk of the TRMF reactor with the magnetic field lines rotating in the horizontal plane.

For the experimental measurements, MF is generated by coils located axially around of the cylindrical container. As mentioned above, this field is rotated around the container with the constant angular frequency, ω_{TRMF}. The TRMF strength is determined by measuring a magnetic induction. The values of the magnetic induction at different points inside the glass container are detected by using a Hall probe connected to the personal computer.

In the present work, the local mass-transfer rates were correlated with the magnetic field measurements. The aim was to investigate a dissolution process of the NaCl-cylinder probe

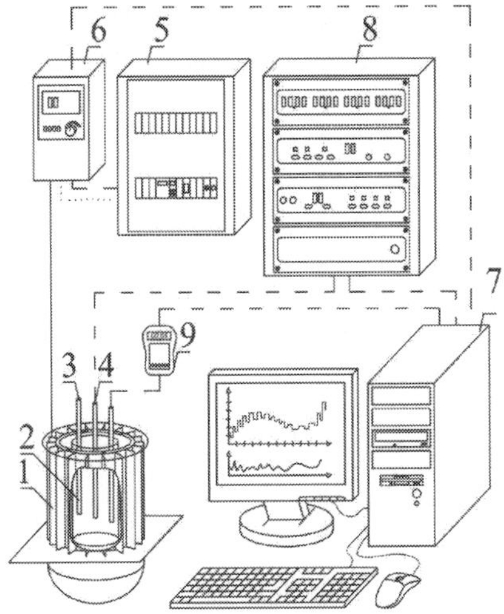

Fig. 3. Sketch of experimental set-up: 1 - generator of rotating magnetic field, 2 - glass container, 3,4 - conductivity probes, 5 - electronic control box, 6 - a.c. transistorized inverter, 7 - personal computer, 8 - multifunctional electronic switch, 9 - Hall probe

under the action of TRMF. Three measurement positions (I, II and III) were located between the wall and the centre of the glass container. All measurement positions (see Fig.4) were investigated only for a few TRMF levels. The exact localization of a NaCl cylinder sample is graphically presented in figure 4.

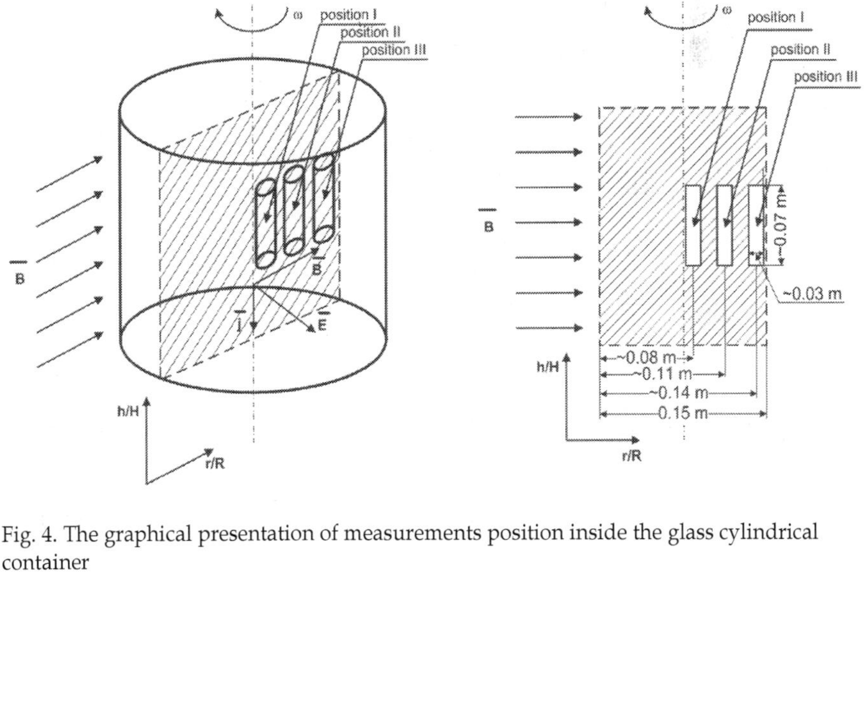

Fig. 4. The graphical presentation of measurements position inside the glass cylindrical container

The TRMF is produced by coils located around the cylinder, and the axes are directed along the radius. When the alternating current is supplied to the windings, the generated MF rotates around the cylinder column. An a.c. transistorized inverter controlled the rotating frequency (equal to the alternating frequency current in the range of $f_{TRMF} = 1 - 50 s^{-1}$).

A typical example of dependence relationship between the spatial distribution of the magnetic induction and various current values of level is presented in figure 5.

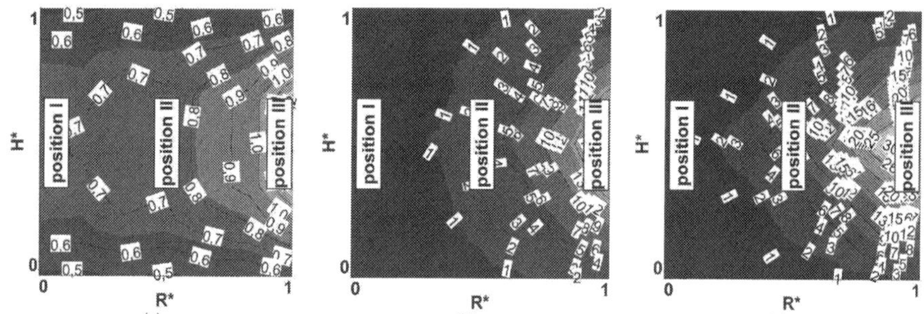

(a) (b) (c)

Fig. 5. The typical example of the dependence between the magnetic induction and the various values of current intensity: (a) $f_{TRMF} = 1 s^{-1}$, (b) $f_{TRMF} = 25 s^{-1}$, (c) $f_{TRMF} = 50 s^{-1}$

The TRMF with the magnetic induction, B, is controlled by using an a.c. alternating current frequency equal to the frequency of TRMF. As follows form the experimental measurements, the values of a magnetic induction are spatially distributed and time independent. It is seen that in the area occupied with this container, the TRMF distribution depends strongly on the spatial coordinates (see Fig.5). A local mass-transfer rate was

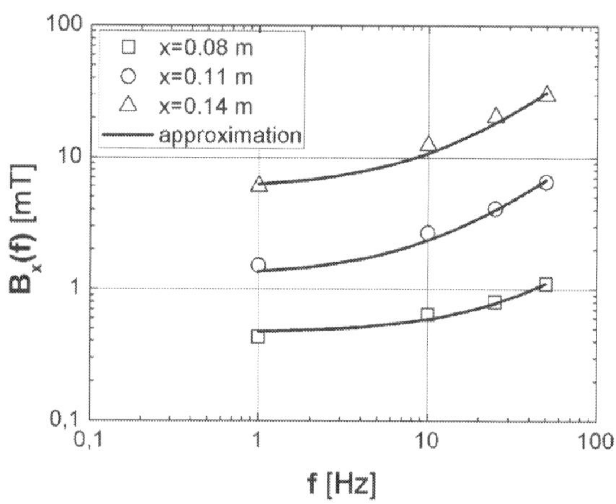

Fig. 6. The variations of the averaged magnetic induction in different localizations of NaCl-cylinder sample

obtained in three radial locations (I for x = 0.08 m; II for x = 0.11 m; III for x = 0.14 m) form the wall of the cylindrical container. It is seen that the magnetic induction at locations III and II is greater than an location I (see Fig.5). Typical profiles of the averaged magnetic induction at different locations of a NaCl-cylinder sample are graphically shown in figure 6. In order to establish the effect of a current level on the averaged values of a local magnetic induction, the obtained data shown in this figure may be analytically described by:

$$x = 0.08\,m \Rightarrow B_x(f) = 0.013f + 0.46 \tag{79a}$$

$$x = 0.11\,m \Rightarrow B_x(f) = 0.11f + 1.25 \tag{79b}$$

$$x = 0.14\,m \Rightarrow B_x(f) = 0.51f + 5.71 \tag{79c}$$

The enhancement of the dissolutions process under the action of TRMF was carried out by using the experimental set-up shown in Fig.3 and described in section 4.2.

5.4 Result and discussion

Under the TRMF conditions, a relationship for the mass-transfer can be described in the general form:

$$Sh = f\left[\left(Ta_m Re_\omega^{-1}\right), Sc\right] \tag{80}$$

In the presented report, the mass-transfer process under the action of TRMF is described by using the modified magnetic Taylor number, which defines the magnitude of the magnetic effect in the tested experimental set-up. The results of our experiments suggest that this non-dimensional number may be expressed as follows:

$$\left\{Ta_m\right\}_x = \frac{\left[B_x(f)\right]^2 x^4 \omega_{TRMF} \sigma_e \rho_0}{\eta_0} \tag{81}$$

The modified dimensionless field frequency based Reynolds number may defined by using the following relation

$$\left\{Re_\omega\right\}_x = \frac{\omega_{TRMF} x^2}{\nu_0} \tag{82}$$

In order to establish the effect of all important parameters on mass-transfer process in the tested set-up, in the wide range of variables data obtained in the present work have been analyzed, we propose the following relationship

$$Sh = 2 + a\left[\left\{Ta_m\right\}_x \left\{Re_\omega\right\}_x^{-1}\right]^b Sc^c \tag{83}$$

The effect of mass-transfer process under the action of RMF can be described using the variable $(Sh-2)Sc^{-c}$ proportional to the term $a\left[\left\{Ta_m\right\}_x \left\{Re_\omega\right\}_x^{-1}\right]^b$. The experimental results obtained in this work are graphically illustrated in $\log\left\{(Sh-2)Sc^{-c}\right\}$ versus $\log\left[a\left[\{Ta_m\}_x\{Re_\omega\}_x^{-1}\right]^b\right]$ in figure 7.

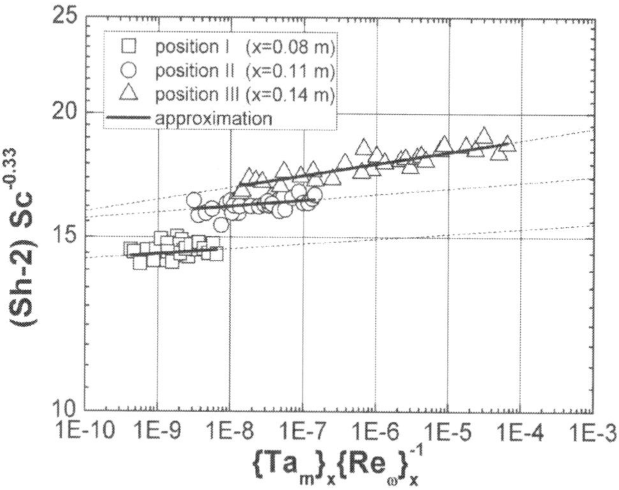

Fig. 7. The graphical presentation of mass-transfer data under the action of the TRMF

The exponent of the dimensionless Schmidt number is 0.33 as there is some theoretical and experimental evidence for this value. The experimental results in figure 7 suggest that the term $(Sh-2)Sc^{-0.33}$ versus the term $\left[\{Ta_m\}_x\{Re_\omega\}_x^{-1}\right]$ may be analytically described by a unique monotonic function. The constants and exponents are computed employing the Matlab software and the principle of least squares and the proposed relationship (19) may be rewritten in the following forms

$$x = 0.08\,m \rightarrow \frac{(Sh-2)}{Sc^{0.33}} = 16\left[\{Ta_m\}_x\{Re_\omega\}_x^{-1}\right]^{0.005} \tag{84a}$$

$$x = 0.011\,m \rightarrow \frac{(Sh-2)}{Sc^{0.33}} = 18\left[\{Ta_m\}_x\{Re_\omega\}_x^{-1}\right]^{0.006} \tag{84b}$$

$$x = 0.14\,m \rightarrow \frac{(Sh-2)}{Sc^{0.33}} = 21\left[\{Ta_m\}_x\{Re_\omega\}_x^{-1}\right]^{0.012} \tag{84c}$$

Figure 7 presents a graphical form of the above equation (84) as the full curves, correlated the experimental data very well with the standard deviation equal to $\sigma_I = 0.17$ (localization I), $\sigma_{II} = 0.21$ (localization II) and $\sigma_{III} = 0.58$ (localization III). The average percentage errors of these data are equal to $\overline{\delta}_I = 0.54\%(R = 0.89)$, $\overline{\delta}_{II} = -0.18\%(R = 0.86)$ and

$\overline{\delta}_{III} = 0.14\%(R = 0.75)$. Figure 7 gives an overview results, in the form of the proposed analytical description (see Eqs 84a-84c) for the experimental investigations. The first conclusion drawn from the inspection of this graph is that the proposed relationships fit the analysed experimental data very well.

Figure 7 demonstrates that, within the scatter limits among the plotted data represented by points, the dimensionless Sherwood number increases with increasing the magnetic local Taylor number. It was found that as the intensity of the MF increases, the relative velocity of liquid inside the cylindrical container increases. This liquid rotation may be defined by using the proposed term $\left[\{Ta_m\}_x \{Re_\omega\}_x^{-1}\right]$. The obtained relationships (see equation 84) indicate that the transfer rates would increase with this term for location I - $Sh \sim \left[\{Ta_m\}_x \{Re_\omega\}_x^{-1}\right]^{0.005}$, location II - $Sh \sim \left[\{Ta_m\}_x \{Re_\omega\}_x^{-1}\right]^{0.006}$ and location III - $Sh \sim \left[\{Ta_m\}_x \{Re_\omega\}_x^{-1}\right]^{0.012}$, respectively.

Moreover, the experimental results shown in figure 7 suggest the following expression:

$$\frac{(Sh-2)}{Sc^{0.33}} = fn\left[\left(\{Ta_m\}_x \{Re_\omega\}_x^{-1}\right), \left(\frac{x}{D}\right)\right] \tag{85}$$

where the Sherwood number is function of the adequate dimensionless numbers and the dimensionless location of a NaCl-cylindrical sample, $\left(\frac{x}{D}\right)$. The effect of location of sample may be described by means of the relatively simple unique dimensionless relationship as follows:

$$Sh = 2 + a\left[\{Ta_m\}_x \{Re_\omega\}_x^{-1}\right]^b Sc^c \left(\frac{x}{D}\right)^d \tag{86}$$

The constant a and exponent b, c and d in Eq.(86) are computed by using the principle of least square. Applying the software Matlab the analytical relationship may be obtained:

$$Sh = 2 + 22.5\left[\{Ta_m\}_x \{Re_\omega\}_x^{-1}\right]^{0.015} Sc^{0.33} \left(\frac{x}{D}\right)^{0.3} \tag{87}$$

The graphical presentation of Eq.(87) is given in figure 8.

The proposed form of Eq.(87), which is presented in figure 8 as the full curve, correlates the data very well with a standard deviation $\sigma = 0.05$. The average percentage error of all the data is $\overline{\delta} = -0.21\%(R = 0.92)$. Figure 8 demonstrate that, within the limits of scatter among the plotted data represented by the points, the mass-transfer rate (dimensionless Sherwood number) increase with increasing the term $a\left[\{Ta_m\}_x \{Re_\omega\}_x^{-1}\right]^b Sc^c \left(\frac{x}{D}\right)^d$. This figure shows a strongly increase in mass-transfer process when the RMF is applied. It was found that the

Sherwood number increase with increase in the local magnetic Taylor number. It should be noted that the mass-transfer rate is increased with the dimensionless location of a NaCl- cylindrical sample $\left(\frac{x}{D}\right)^{0.3}$.

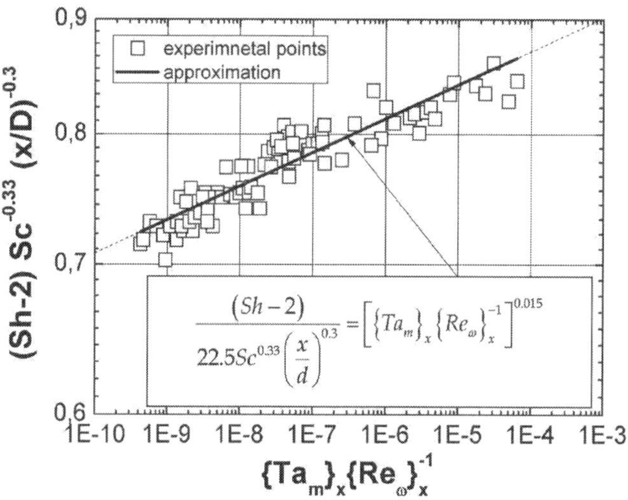

Fig. 8. The graphical presentation of equation (87)

6. Conclusion

The present considerations show interesting features concerning the effects of forced convection on the mass-transfer in the mixing systems. Based on the experimental results, the effects of the reciprocating agitators for the wide range of Reynolds number are correlated by the suitable equations. With the respect to the other very useful mass-transfer relationships given in the pertinent literature, the theoretical description of problem and the equations predicted in the present work is much more attractive because it generalizes the experimental data taking into consideration the various parameters, which defined the hydrodynamic state and the intensity of forced convection effects in the tested mixing systems. Moreover, the list of empirical equations for the prediction of mass-transfer coefficients for the rotational agitated systems has been presented. The tabulated correlations in Table 2 may be successfully applied to study the dissolution of solid bodies in the rotational stirred tank.

It should be noticed that the novel approach to the mixing process presented and based on the application of transverse rotating magnetic field (TRMF) to produce better hydrodynamic conditions in the case of the mass-transfer process. From practical point of view, the dissolution process of solid body is involved by using the turbulently agitated systems. In previous publications are not available data describing the mass-transfer operations of the dissolution process under the TRMF conditions. Therefore, the

experimental investigations have been conducted to explain the influence of this kind of magnetic field (MF) on the mass-transfer enhancement. Moreover, the influence of localization of NaCl-cylindrical sample on the mass-transfer was determined. The influence of the TRMF on this process is described using the non-dimensional parameters formulated on the base of fluid mechanics equations. These dimensionless numbers allow quantitative representation and characterization of the influence of hydrodynamic state under the TRMF conditions on the mass-transfer process. The dimensionless groups are used to establish the effect of TRMF on this operation in the form of the novel type dimensionless correlation.
The study of mass-transfer process under the action of TRMF results in significant enhancement of the solid dissolution rate. The mass-transfer measurements obtained in the presence of this kind of MF indicates a strong dependency on dissolution rate per localization of NaCl-cylindrical sample. In the present report, the mass-transfer under the action of TRMF is described by novel type of dimensionless group (the local magnetic Taylor number).

7. References

Baird, M.H.I. & Rama Rao, N.V. (1988). Characteristics of a counter current reciprocating plate bubble column. II. Axial mixing and mass transfer, *Canadian Journal of Chemical Engineering*, Vol. 66, 222-230. ISSN: 0008-4034

Baird, M.H.I., Rama Rao, N.V. & Vijayan S. (1992). Axial mixing and mass transfer in a vibrating perforated plate extraction column, *Canadian Journal of Chemical Engineering*, Vol. 70, 69-76. ISSN: 0008-4034

Barker, J.J. & Treybal, R.E. (1960). Mass transfer coefficients for solids suspended in agitated liquids, *AIChE Journal*, Vol. 6 (2), 289-295. ISSN 0001-1541

Basmadjian, D. (2004). *Mass transfer. Principles and applications*, CRC Press LLC, ISBN 0-8493-2239-1, USA.

Bird, R.B., Stewart, W.E. & Lightfoot, E.N. (1966). *Transport phenomena*, Wiley, ISBN 0-471-41077-2, USA.

Condoret, J.S., Riba, J.P. & Angelino, H. (1989). Mass transfer in a particle bed with oscillating flow, *Chemical Engineering Science*, Vol. 44, No. 10, 2107-2111. ISSN 0009-2509

Ditl, P., Rieger, F. & Roušar, I. (1991). The design of agitated dissolution tanks, *Proceedings of 7th Europ. Conf. on Mixing*, pp. 483-490, Brugge, 1991.

Fox, R.W. & McDonald A.T. (1990). *Introduction to fluid mechanics*, Wiley, ISBN 0-471-20231-2, USA.

Fraňa K., Stiller J., Grundmann R. (2006): Transitional and turbulent flows driven by a rotating magnetic field, *Magnetohydrodynamics*, Vol. 42, 187-197. ISSN 0024-998X

Garner, F.H. & Suckling, R.D. (1958). Mass transfer form a soluble solid sphere, *AIChE Journal*, Vol. 4, No. 1, 114-124. ISSN 0001-1541

Geankoplis, C.J. (2003). *Transport processes and separation processes. Principles*, Pearson Education Inc., ISBN 0-13-101367-X, USA.

Gomaa, H.G. & Al Taweel, A.M. (2005). Axial mixing in a novel pilot scale gas-liquid reciprocating plate column, *Chemical Engineering and Processing*, Vol. 44, 1285-1295. ISSN 0255-2701

Gomaa, H.G., Landau, J. & Al Taweel, A.M (1991). Gas-liquid contacting in reciprocating plate columns. I. Hydrodynamics, *Canadian Journal of Chemical Engineering*, Vol. 69, 228-239. ISSN: 0008-4034

Guru, B.S. & Hiziroğlu, H.R. (2004). *Electromagnetic field theory fundamentals*, Cambridge University Press.

Harriot, P. (1962). Mass transfer to particles: Part I. Suspended in agitated systems, *AIChE Journal*, Vol. 8, No. 93, 93-102. ISSN 0001-1541

Hartmann, H., Derksen, J.J., van der Akker, H.E.A. (2006). Numerical simulation of a dissolution process in a stirred tank reactor, *Chemical Engineering Science*, Vol. 61, 3025-3032. ISSN 0009-2509

Herdnl, G. & Mersmann, A. (1982). Fluiddynamik und Stoffübergang in gerührten Suspensionen, *Chemical Engineering and Technology*, Vol. 54, No. 3, 258-259. ISSN 1521-4125

Hixon, A. & Baum, S.J. (1941). Agitation. Mass transfer coefficients in liquid-solid agitation systems, *Industrial and Engineering Chemistry*, Vol. 33, No. 11, 1433-1438. ISSN 0888-5885

Hixon, A. & Baum, S.J. (1942). Agitation. Performance of propellers in liquid-solid systems, *Industrial and Engineering Chemistry*, Vol. 34, No. 1, 120-125. ISSN 0888-5885

Hixon, A. & Baum, S.J. (1944). Mass transfer and chemical reactions in liquid-solid agitation, *Industrial and Engineering Chemistry*, Vol. 36, No. 6, 528-531. ISSN 0888-5885

Humphrey, D.W. & Van Ness H.C. (1957). Mass transfer in a continuous-flow mixing vessels, *AIChE Journal*, Vol. 3, No. 2, 283-286. ISSN 0001-1541

Incropera, F.P. & DeWitt D.P. (1996). *Fundamentals of heat and mass transfer*, John Wiley & Sons Inc., ISBN 0-471-30460-3, USA.

Jameson, G.J. (1964). Mass (or heat) transfer form an oscillating cylinder, *Chemical Engineering Science*, Vol. 19, 793-800. ISSN 0009-2509

Kays, W.M. & Crawford M.E. (1980). *Convective heat and mass transfer*, McGraw-Hill, ISBN 0-9668461-4-1, USA.

Lemcoff, N.O. & Jameson, G.J. (1975). Solid-liquid mass transfer in a resonant bubble contractor, *Chemical Engineering Science*, Vol. 30, 363-367. ISSN 0009-2509

Lemlich R. & Levy, M.R. (1961). The effect of vibration on natural convective mass transfer. *AIChE Journal*, Vol. 7, 240-241. ISSN 0001-1541

Levins, D.M. & Glastonbury, J.R. (1972). Particle-liquid hydrodynamics and mass transfer in a stirred vessel II: Mass transfer, *Trans. Inst. Chem. Engr.*, Vol. 50, No. 32, 132-146, ISSN 0360-7275

Lu, X.S. & Li, H. (2000). Fluidization of $CaCO_3$ and Fe_2O_3 particle mixtures in a transverse rotating magnetic field, *Powder Technology*, Vol. 107, 66-78. ISSN 0032-5910

Masiuk, S. & Rakoczy, R. (2007). Power consumption, mixing time, heat and mass transfer measurements for liquid vessels that are mixed using reciprocating multiplates agitators. *Chemical Engineering and Processing*, Vol. 46, 89-98. ISSN 0255-2701

Masiuk, S. (1996). Heat transfer measurements in a liquid vessel that is mixed using vibratory agitator, *Chemical Engineering Journal*, Vol. 61, 107-112. ISSN 1385-8947

Masiuk, S. (1999)[a]. Mieszadło wibracyjne (Reciprocating agitator), Patent PL 175516.

Masiuk, S. (1999)[b]. Power consumption measurement in a liquid vessel that is mixed using vibratory agitator, *Chemical Engineering Journal*, Vol. 75, 161-165. ISSN 1385-8947

Masiuk, S. (2000). Mixing time for a reciprocating plate agitator with flapping blades, *Chemical Engineering Journal,* Vol. 79, 23-30. ISSN 1385-8947

Masiuk, S. (2001). Dissolution of solid body in a tubular reactor with reciprocating plate agitator. *Chemical Engineering Journal,* Vol. 83, 139-144. ISSN 1385-8947

Masiuk, S., Rakoczy, R. & Kordas, M. (2008). Comparison density of maximal energy for mixing process using the same agitator in rotational and reciprocating movements. *Chemical Engineering and Processing,* Vol. 47, 1252-1260. ISSN 0255-2701

Melle, S., Calderon, O.G., Fuller, G.G. & Rubio M.A. (2002): Polarizable particle aggregation under rotating magnetic fields using scattering dichroism, *Journal of Colloid and Interface Science,* Vol. 247, 200-209. DOI:10.1006/jcis.2001.8087

Möβner, R. & Gerbeth, G. (1999). Buoyant melt flows under the influence of steady and rotating magnetic fields, *Journal of Crystal Growth,* Vol. 197, 341-345. ISSN 0021-9797

Nekrassov, Z.I. & Chekin, V.V. (1961). The effect of an alternating magnetic field on a fluidized bed of ferromagnetic particles, *Izvestiya Akademii Nauk SSSR, Metallurgy Fuels,* Vol. 6, 25-29. (in Russian)

Nekrassov, Z.I. & Chekin, V.V. (1962). The effective viscosity of fluidized bed of polydisperse ferromagnetic solids in alternating magnetic field, *Izvestiya Akademii Nauk SSSR, Metallurgy Fuels,* Vol. 1, 56-59. (in Russian)

Nienow, A.W., Edwards, M.F. & Harnby, N. (1997). *Mixing in the process industries,* Butterworth-Heinemann, ISBN 0750637609, UK

Nikrityuk, P.A., Eckert, K. & Grundmann, R. (2006): A numerical study of unidirectional solidification of a binary metal alloy under influence of a rotating magnetic field, *International Journal of Heat and Mass Transfer,* Vol. 49, 1501-1515. ISSN 0017-9310

Noordsij, P. & Rotte, J.W. (1967). Mass transfer coefficients to a rotating and to a vibrating sphere. *Chemical Engineering Science,* Vol. 22, 1475-1481, ISSN 0009-2509

Rakoczy, R. & Masiuk, S. (2009). Experimental study of bubble size distribution in a liquid column exposed to a rotating magnetic field, *Chemical Engineering and Processing: Process Intensification,* Vol. 48, 1229-1240. ISSN 0255-2701

Rakoczy, R. & Masiuk, S. (2010). Influence of transverse rotating magnetic field on enhancement of solid dissolution process. *AIChE Journal,* Vol. 56, 1416-1433. ISSN 0001-1541

Rakoczy, R. (2006). Hydro-thermal behavior of liquid systems in a magnetic field, Doctro's Thesis, Szczecin University of Technology.

Rama Rao, N.V & Baird, M.H.I. (1988). Characteristics of a counter current reciprocating plate bubble column. I. Holdup, pressure drop and bubble diameter, *Canadian Journal of Chemical Engineering,* Vol. 66, 211-221. ISSN: 0008-4034

Rama Rao, N.V & Baird, M.H.I. (2000). Axial mixing and gas holdup with reciprocating doughnut plates, *Canadian Journal of Chemical Engineering,* Vol. 78, 261-264. ISSN: 0008-4034

Rama Rao, N.V & Baird, M.H.I. (2003). Gas liquid mass transfer in a 15 cm diameter reciprocating plate column, *Journal of Chemical Technology and Biotechnology.,* Vol. 78, 134-137. ISSN 1097-4660

Rousseau, R.W. (1987). *Handbook of separation process technology,* Wiley, ISBN 978-0-471-89558-9 , USA.

Shamlou, P.A., Gierczycki, A.T. & Tichner-Hooker, N.J. (1996). Breakage of flocs in liquid suspensions agitated by vibrating and rotating mixers, *Chemical Engineering Journal*, Vol. 62, 23-34. ISSN 1385-8947

Spitzer K.H. (1999): Application of rotating magnetic field in Czochralski crystal growth, *Crystal Growth and Characterization of Materials*, Vol. 38, 39-58. ISSN 0960-8974

Sugano, Y. & Rutkowsky, D.A. (1968). Effect of transverse vibration upon the rate of mass transfer for horizontal cylinder, *Chemical Engineering Science*, Vol. 23, 707-716, ISSN 0009-2509

Sundaresan, A. & Varma, Y.B.G. (1990). Interfacial area and mass transfer in gas-liquid cocurrent upflow and countercurrent flow in reciprocating plate column, *Canadian Journal of Chemical Engineering*, Vol. 68, 952-988. ISSN: 0008-4034

Tojo, K., Miyanami, K. & Minami, I. (1981). Vibartory agitation in solid-liquid mixing, *Chemical Engineering Science*, Vol. 36, 279-284. ISSN 0009-2509

Tojo, K., Miyanami, K. & Mitsui, H. (1981). Vibratory agitation in solid-liquid mixing, *Chemical Engineering Science*, Vol. 36, 279-284, ISSN 0009-2509

Treybal, R.E. (1980). *Mass transfer operations*, McGraw-Hill, ISBN 0070651760, USA

Volz, M.P. & Mazuruk, K. (1999): Thermoconvective instability in a rotating magnetic field, *International Journal of Heat and Mass Transfer*, Vol. 42, 1037-1045. ISSN 0017-9310

Walker, J.S., Volz, M.P. & Mazuruk K. (2004): Rayleigh-Bénard instability in a vertical cylinder with a rotating magnetic field, *International Journal of Heat and Mass Transfer*, Vol. 47, 1877-1887. ISSN 0017-9310

Weinspach, P.M. (1967). Der Lösevorgang im Fließbett und im Rührgefäß, *Chem.-Ing.-Tech.*, Vol. 39, No. 5/6, 213-236.

Wong, P.F.Y., Ko, N.W.M. & Yip, P.C. (1978). Mass transfer from large diameter vibrating cylinder, *Trans. Item*, Vol. 56, 214-216.

Yang, M., Ma, N., Bliss, D.F. & Bryant G.G. (2007): Melt motion during liquid-encapsulated Czochralski crystal growth in steady and rotating magnetic field, *International Journal of Heat and Mass Transfer*, Vol. 28, 768-776. ISSN 0017-9310

7

Unsteady Magnetohydrodynamic Convective Heat and Mass Transfer Past an Infinite Vertical Plate in a Porous Medium with Thermal Radiation, Heat Generation/Absorption and Chemical Reaction

Stanford Shateyi[1] and Sandile Motsa[2]
[1]*University of Venda*
[2]*University of Swaziland*
[1]*South Africa*
[2]*Swaziland*

1. Introduction

Convective flows with simultaneous heat and mass transfer under the influence of a magnetic field and chemical reaction arise in many transport processes both naturally and artificially in many branches of science and engineering applications. This phenomenon plays an important role in the chemical industry, power and cooling industry for drying, chemical vapour deposition on surfaces, cooling of nuclear reactors and petroleum industries.

Natural convection flow occurs frequently in nature. It occurs due to temperature differences, as well as due to concentration differences or the combination of these two, for example in atmospheric flows, there exists differences in water concentration and hence the flow is influenced by such concentration difference.

Changes in fluid density gradients may be caused by non-reversible chemical reaction in the system as well as by the differences in molecular weight between values of the reactants and the products. Chemical reactions can be modeled as either homogeneous or heterogeneous processes. This depends on whether they occur at an interface or as a single phase volume reaction. A homogeneous reaction is one that occurs uniformly throughout a given phase. On the other hand, a heterogeneous reaction takes place in a restricted area or within the boundary of a phase. In most cases of chemical reactions, the reaction rate depends on the concentration of the species itself. A reaction is said to be of first order, if the rate of reaction is directly proportional to the concentration itself, (Cussler, 1988). For example, the formation of smog is a first order homogeneous reaction. Consider the emission of nitrogen dioxide from automobiles and other smoke-stacks. This nitrogen dioxide reacts chemically in the atmosphere with unburned hydrocarbons (aided by sunlight) and produces peroxyacetylnitrate, which forms an envelop which is termed photo-chemical smog.

The study of heat and mass transfer with chemical reaction is of great practical importance in many branches of science and engineering. (Das et al., 1994) studied the effects of mass transfer flow past an impulsively started infinite vertical plate with constant heat flux and chemical reaction. (Anjalidevi & Kandasamy, 1999) studied effects of chemical reaction, heat and mass transfer on laminar flow along a semi-infinite horizontal plate. More recently, intensive studies have been carried out to investigate effects of chemical reaction on different flow types (see Seddeek et al. 2007; Salem & Abd El-Aziz 2008; Mohamed 2009; Ibrahim et al.,2008).

Convection flows in porous media has gained significant attention in recent years because of their importance in engineering applications such as geothermal systems, solid matrix heat exchangers, thermal insulations, oil extraction and store of nuclear waste materials. These can also be applied to underground coal gasification, ground water hydrology, wall cooled catalytic reactors, energy efficient drying processes and natural convection in earth's crust. Detailed reviews of flow through and past porous media can be found in (Nield and Benjan, 1999). (Abel et al., 2001) studied the two-dimensional boundary layer problem on mixed convection of an incompressible visco-elastic fluid immersed in a porous medium over a stretching sheet. (Ali, 2007) analyzed the effect of lateral mass flux on natural convection boundary layer induced by heated vertical plate embedded in a saturated porous medium.

The use of magnetic field that influences heat generation/absorption process in electrically conducting fluid flows has many engineering applications. For example, many metallurgical processes which involve cooling of continuous strips or filaments, which are drawn through a quiescent fluid. The properties of the final product depend to a great extent on the rate of cooling. The rate of cooling and therefore, the desired properties of the end product can be controlled by the use of electrically conducting fluids and the applications of the magnetic fields (Vajravelu & Hadjinicalaou 1997). Many works have been reported on flow, heat and mass transfer of electrically conducting fluids over semi-infinite/infinite plates/stretching surfaces in the presence of magnetic field (see for instance Chamkha & Khaled, 2000; Shateyi et al. 2007, 2010; Makinde & Sibanda 2008; Bég et al. 2009, Pal & Talukdar 2010, Makinde & Aziz 2010).

The study of heat generation or absorption in moving fluids is important in problems dealing with chemical reactions and those concerned with dissociating fluids. Heat generation effects may alter the temperature distribution and this in turn can affect the particle deposition rate in nuclear reactors, electronic chips and semi conductor wafers. Although exact modeling of internal heat generation or absorption is quite difficult, some simple mathematical models can be used to express its general behaviour for most physical situations. Heat generation or absorption can be assumed to be constant, space-dependent or temperature-dependent. (Crepeau & Clarksean 1997) have used a space-dependent exponentially decaying heat generation or absorption in their study on flow and heat transfer from a vertical plate. Several interesting computational studies of reactive MHD boundary layer flows with heat and mass transfer in the presence of heat generation or absorption have appeared in recent years (see for example Patil & Kulkarni, 2008; Salem & El-Aziz, 2008; Samad & Mohebujjaman, 2009; Mohamed, 2009; Mahdy, 2010).

When technological processes take place at higher temperatures thermal radiation heat transfer has become very important and its effects cannot be neglected (Siegel & Howel, 2001). The effect of radiation on MHD flow, heat and mass transfer become more important industrially. Many processes in engineering areas occur at high temperature and a knowledge of radiation heat transfer becomes a very important for the design of the pertinent equipment. The quality of the final product depends to a great extent on the heat controlling factors, and the knowledge of radiative heat transfer in the system can lead to a desired product with sought qualities. Different researches have been forwarded to analyze the effects of thermal radiation on different flows (Cortell, 2008; Bataller, 2008; Ibrahim et al. 2008; Shateyi, 2008; Shateyi and Motsa, 2009; Aliakba et al., 2009; Shateyi et al., 2010, Hayat, 2010; Cortell, 2010; among other researchers).

In spite of all the previous studies, the unsteady MHD free convection heat and mass transfer for a heat generation/absorption with radiation absorption in the presence of a reacting species over an infinite permeable plate has received little attention. Hence, the main objective of this chapter is to investigate the effects of thermal radiation, chemical reaction, heat source/sink parameter of an electrically conducting fluid past an infinite vertical porous plate subjected to variable suction. The plate is assumed to be embedded in a uniform porous medium and moves with a constant velocity in the flow direction in the presence of a transverse magnetic field. The governing equations are transformed into a system of nonlinear ordinary differential equations by using suitable similarity transformations. Numerical calculations are carried out using the spectral homotopy analysis method (SHAM). The SHAM is a modified version of the homotopy analysis method (HAM), (Liao, 2003) and the Chebyshev spectral collocation method (Trefethen, 2000). The SHAM has been successfully used by (Motsa et al. 2010a,b). One strength of the SHAM is that it removes restrictions of the HAM such as the requirement for the solution to conform to the so-called rule of solution expression and the rule of coefficient ergodicity. Also, the SHAM inherits strengths of the HAM, for example, it does not depend on the existence of a small parameter in the equation to be solved, it avoids discretization, and the solution obtained is in terms of an auxiliary parameter $\hbar$ which can be conveniently used to determine the convergence rate of the solution.
Graphical results for the velocity, temperature and concentration profiles based on the numerical solutions are presented and discussed. We also discuss the effects of various parameters on the skin-friction coefficient and the rate of heat and mass transfer at the surface.

2. Mathematical formulation

We consider an unsteady two-dimensional flow of an incompressible and electrically conducting viscous fluid, along an infinite vertical permeable plate embedded in a porous medium, with thermal radiation, heat generation/absorption and chemical reaction. The $x-$ axis is taken on the infinite plate, and parallel to the free-stream velocity which is vertical and the $y-$ axis is taken normal to the plate. Since the motion is two dimensional and the length of the plate is large enough, all the physical variables are independent of x. A magnetic field B_0 of uniform strength is applied transversely to the direction of the flow. The transversely applied magnetic field and magnetic Reynolds number are assumed to be very small so that the induced magnetic field and the Hall effect are negligible. In this chapter assumption is also made that there is no applied voltage which implies the absence of an electric field. The plate is maintained at constant temperature T_w and concentration C_w, higher than the ambient temperature T_∞ and concentration C_∞, respectively.
The analysis considers a homogeneous first-order chemical reaction with constant rate k_c between the diffusing species and the fluid. The fluid is assumed to have constant properties except that the influence of the density variations with temperature and concentration which are considered only in the body force term. The governing equations for the problem considered in this chapter are based on the balances of mass, linear momentum, energy and concentration species. These equations are as given below:

Continuity equation:

$$\frac{\partial v}{\partial y} = 0, \tag{1}$$

Momentum equation:

$$\frac{\partial u}{\partial t}+v\frac{\partial u}{\partial y} = \nu\frac{\partial^2 u}{\partial y^2}+g\beta_t(T-T_\infty)+g\beta_c(C-C_\infty)-\frac{\sigma B_0^2}{\rho}u-\frac{\nu}{K}u-\frac{b}{K}u^2, \tag{2}$$

Energy equation:

$$\frac{\partial T}{\partial t}+v\frac{\partial T}{\partial y}=\alpha\frac{\partial^2 T}{\partial y^2}-\frac{1}{\rho c_p}\frac{\partial q_r}{\partial y}-\frac{Q_0}{\rho c_p}(T-T_\infty)+Q_1^*(C-C_\infty), \tag{3}$$

Mass diffusion equation:

$$\frac{\partial C}{\partial t}+v\frac{\partial C}{\partial y}=D\frac{\partial^2 C}{\partial y^2}-k_c(C-C_\infty). \tag{4}$$

Where x, y are the dimensional distance along and perpendicular to the plate, respectively. u and v are the velocity components in the x, y directions respectively, g is the gravitational acceleration, ρ is the fluid density, β_t and β_c are the thermal and concentration expansion coefficients respectively, K is the Darcy permeability, b is the emperical constant, B_0 is the magnetic induction, T is the thermal temperature inside the thermal boundary layer and C is the corresponding concentration, σ is the electric conductivity, α is the thermal diffusivity, c_p is the specific heat at constant pressure, D is the diffusion coefficient, q_r is the heat flux, Q_0 is the dimensional heat absorption coefficient, Q_1^* is the coefficient of proportionality of the radiation and k_c is the chemical reaction parameter.
The boundary conditions for the velocity, temperature and concentration are:

$$u=U_0,\quad v=v(t),\quad T=T_w,\quad C=C_w \text{ at } y=0 \tag{5}$$

$$u=0,\quad T=T_\infty,\quad C=C_\infty, \text{ as } y\to\infty. \tag{6}$$

By using the Rosseland diffusion approximation (Hossain *et al.* 1999) and following (Raptis, 1999) among other researchers, the radiative heat flux, q_r is given by

$$q_r=-\frac{4\sigma^*}{3K_s}\frac{\partial T^4}{\partial y}, \tag{7}$$

where σ^* and K_s are the Stefan-Boltzman constant and the Roseland mean absorption coefficient, respectively. We assume that the temperature differences within the flow are sufficiently small such that T^4 may be expressed as a linear function of temperature.

$$T^4\approx 4T_\infty^3T-3T_\infty^4. \tag{8}$$

Using (7) and (8) in the last term of equation (3) we obtain

$$\frac{\partial q_r}{\partial y}=-\frac{16\sigma^* T_\infty^3}{3K}\frac{\partial^2 T}{\partial y^2}. \tag{9}$$

In order to obtain a local similarity solution (in time) of the problem under consideration we introduce a time dependent length scale δ as

$$\delta=\delta(t). \tag{10}$$

A convenient solution of equation (1) in terms of this length scale is considered to be in the following form

$$v = v(t) = -\frac{\nu}{\delta} v_0, \tag{11}$$

where $v_0 > 0$ is the suction parameter.
These non-linear partial differential equations are then transformed by a similarity transformation into a system of ordinary differential equations given as;

$$f'' + (2\eta + v_0)f' + Gr\theta + Gc\phi - Mf - \frac{1}{Da}f - \frac{ReFs}{Da}f^2 = 0, \tag{12}$$

$$\left(\frac{1+R}{Pr}\right)\theta'' + (2\eta + v_0)\theta' - \Omega\theta + Q_2\phi = 0, \tag{13}$$

$$\frac{1}{Sc}\phi'' + (2\eta + v_0)\phi' - \gamma\phi = 0, \tag{14}$$

where primes denote differentiation with respect to η and $Da = \frac{K}{\delta^2}$ is the local Darcy number, $Fs = \frac{b}{\delta}$ is the local Forchheimer number, $Re = \frac{U_0\delta}{\nu}$ is the local Reynolds number, $Pr = \frac{\nu}{\alpha}$ is the Prandtl number, $\Omega = \frac{Q_0}{\rho c_p}$ is the heat absorption parameter, $Q_2 = \frac{Q_1^*(C_w - C_\infty)}{(T - T\infty)}$ is the absorption of radiation parameter, $\gamma = \frac{k_c\delta^2}{\nu}$ is the chemical reaction parameter, $Sc = \frac{\nu}{D}$ is the Schmidt number, $M = \frac{\sigma\delta^2 B_0^2}{\nu\rho}$ is the magnetic field parameter, R is the thermal radiation parameter, $Gr = g\beta_t(T_w - T_\infty)\delta^2/\nu U_0$ is the local Grashof number, $Gc = g\beta_c(C_w - C_\infty)\delta^2/U_0$ is the modified Grashof number.
The corresponding boundary conditions for $t > 0$ are transformed to:

$$f = 1, \quad \theta = 1, \quad \phi = 1, \quad \text{at } \eta = 0, \tag{15}$$

$$f = 0\,, \quad \theta = 0, \quad \phi = 0 \quad \text{as } \eta \to \infty. \tag{16}$$

The resultant system of coupled and non-linear ordinary differential equations are then solved by the Spectral homotopy analysis method (SHAM). The effects of various parameters such as local Darcy number, Schmidt number, Hartmann, radiation parameter, reaction rate parameter and Grashof numbers on the velocity, temperature, concentration, skin-friction coefficient, Nusselt number and Sherwood numbers are shown in figures and tables and analyzed in detail.

3. Solution methods

To solve the equation system (12) - (14) we first use the Chebyshev pseudospectral method (CPM) to solve equation (14) for $\phi(\eta)$ and use the result in (13) to solve for $\theta(\eta)$ using the Chebyshev pseudospectral method again. The solutions for $\theta(\eta)$ and $\phi(\eta)$ are substituted in (12) to obtain a nonlinear equation in $f(\eta)$. To solve the resulting nonlinear equation we use the spectral homotopy analysis method (SHAM) which is based on a blend of the Chebyshev pseudospectral method with the standard homotopy analysis method (HAM).
Before applying the CPM and SHAM we use the domain truncation approach to approximate the domain of the problem $[0, \infty)$ by the numerical computational domain $[0, L]$, where L is a fixed length that is taken to be sufficiently larger than the thickness of the boundary layer.

We then transform the domain $[0,L]$ to the domain $[-1,1]$, on which the Chebyshev spectral method can be implemented, using the algebraic mapping

$$\xi = \frac{2\eta}{L} - 1, \quad \xi \in [-1,1]. \tag{17}$$

To solve (13) and (14) we use the Chebyshev pseudospectral method and approximate the unknown functions $\theta(\xi)$ and $\phi(\xi)$ as a truncated series of Chebyshev polynomials of the form

$$\theta(\xi) \approx \sum_{k=0}^{N} \theta_k T_k(\xi_j), \quad \phi(\xi) \approx \sum_{k=0}^{N} \phi_k T_k(\xi_j), \qquad j = 0,1,\ldots,N \tag{18}$$

where T_k is the kth Chebyshev polynomial, and $\xi_0, \xi_1, \ldots, \xi_N$ are Gauss-Lobatto collocation points (see Canuto et al. 1988) defined by

$$\xi_j = \cos\frac{\pi j}{N}, \quad j = 0,1,\ldots,N. \tag{19}$$

and $N+1$ is the number of collocation points. Derivatives of the functions $\theta(\eta)$ and $\phi(\eta)$ at the collocation points are represented as

$$\frac{d^r\theta}{d\eta^r} = \sum_{k=0}^{N} \mathbf{D}_{kj}^r \theta(\xi_j), \quad \frac{d^r\phi}{d\eta^r} = \sum_{k=0}^{N} \mathbf{D}_{kj}^r \phi(\xi_j) \tag{20}$$

where r is the order of differentiation and $\mathbf{D} = \frac{2}{L}\mathcal{D}$ with $\mathcal{D}$ being the Chebyshev spectral differentiation matrix (see for example, Canuto et al. 1988; Trefethen 2000) whose entries are defined as

$$\begin{aligned}
\mathcal{D}_{jk} &= \frac{c_j}{c_k}\frac{(-1)^{j+k}}{\xi_j - \xi_k} \qquad j \neq k; j,k = 0,1,\ldots,N, \\
\mathcal{D}_{kk} &= -\frac{\xi_k}{2(1-\xi_k^2)} \qquad k = 1,2,\ldots,N-1, \\
\mathcal{D}_{00} &= \frac{2N^2+1}{6} = -\mathcal{D}_{NN}.
\end{aligned} \tag{21}$$

Substituting equations (18) - (21) in (13) yields

$$\mathbf{C\Phi} = 0 \tag{22}$$

subject to the boundary conditions

$$\phi(\xi_N) = 1, \quad \phi(\xi_0) = 0, \tag{23}$$

where

$$\mathbf{C} = \frac{1}{Sc}\mathbf{D}^2 + (2\mathbf{J} + v_0\mathbf{I})\mathbf{D} - \gamma\mathbf{I}, \tag{24}$$

$$\mathbf{\Phi}_0 = [\phi(\xi_0), \phi(\xi_1), \ldots, \phi(\xi_N)]^T, \tag{25}$$

$$\mathbf{J} = diag([\eta_0, \eta_1, \ldots, \eta_N]). \tag{26}$$

In the above definitions T stands for transpose, $\mathbf{I}$ denotes an identity matrix of order $(N+1)$ and *diag* denotes a diagonal matrix of size $(N+1) \times (N+1)$. The boundary conditions (23) are imposed on the first and last rows of the matrices $\mathbf{C}$ and $\mathbf{\Phi}$ as follows

$$\begin{pmatrix} 1 & 0 & \cdots & 0 & 0 \\ & & & & \\ & & \mathbf{C} & & \\ & & & & \\ 0 & 0 & \cdots & 0 & 1 \end{pmatrix} \begin{pmatrix} \phi(\xi_0) \\ \phi(\xi_1) \\ \vdots \\ \phi(\xi_{N-1}) \\ \phi(\xi_N) \end{pmatrix} = \begin{pmatrix} 0 \\ 0 \\ \vdots \\ 0 \\ 1 \end{pmatrix} \tag{27}$$

The solution for $\mathbf{\Phi}$ is then obtained from solving

$$\mathbf{\Phi} = \mathbf{C}^{-1}\mathbf{K}_2, \tag{28}$$

where $\mathbf{K}_2$ is the right hand side of equation (27). Applying the Chebyshev spectral method on equation (13) to solve for $\theta(\eta)$ gives

$$\mathbf{B}\mathbf{\Theta} = -Q_2\mathbf{\Phi} \tag{29}$$

subject to the boundary conditions

$$\theta(\xi_N) = 1, \quad \theta(\xi_0) = 0, \tag{30}$$

where

$$\mathbf{B} = \frac{1+R}{Pr}\mathbf{D}^2 + (2\jmath + v_0\mathbf{I})\mathbf{D} - \Omega\mathbf{I}, \tag{31}$$

$$\mathbf{\Theta}_0 = [\theta(\xi_0), \theta(\xi_1), \ldots, \theta(\xi_N)]^T, \tag{32}$$

The boundary conditions (30) are imposed on the first and last rows of the matrices in equation (29) as follows;

$$\begin{pmatrix} 1 & 0 & \cdots & 0 & 0 \\ & & & & \\ & & \mathbf{B} & & \\ & & & & \\ 0 & 0 & \cdots & 0 & 1 \end{pmatrix} \begin{pmatrix} \theta(\xi_0) \\ \theta(\xi_1) \\ \vdots \\ \theta(\xi_{N-1}) \\ \theta(\xi_N) \end{pmatrix} = \begin{pmatrix} 0 \\ -Q_2\phi(\xi_1) \\ \vdots \\ -Q_2\phi(\xi_{N-1}) \\ 1 \end{pmatrix} \tag{33}$$

The solution for $\mathbf{\Theta}$ is then obtained from solving

$$\mathbf{\Theta} = \mathbf{B}^{-1}\mathbf{K}_1, \tag{34}$$

where $\mathbf{K}_1$ is the right hand side of equation (34). Once the solutions for $\phi(\eta)$ and $\theta(\eta)$ have been found using equations (28) and (34), respectively, the SHAM approach is then used to solve then nonlinear equation (27) for $f(\eta)$.

In applying the SHAM, it is convenient to make the boundary conditions homogeneous by making use of the transformation

$$F(\xi) = f(\eta) + f_0(\eta), \quad f_0(\eta) = e^{-\eta} \tag{35}$$

where $f_0(\eta)$ is an initial guess chosen to satisfy the boundary conditions for $f(\eta)$. Substituting (35) in the governing equation (12) and the associated boundary conditions gives

$$F'' + a_1(\eta)F' + a_2(\eta)F - \beta_1 F^2 = \psi(\eta), \tag{36}$$

subject to

$$F(0)=0, \quad F(\infty)=0, \tag{37}$$

where

$$a_1 = 2\eta + v_0, \; a_2 = -\left(M + \frac{1}{Da} + \frac{2ReFsf_0(\eta)}{Da}\right), \; \beta_1 = \frac{ReFs}{Da}, \tag{38}$$

$$\psi = -\left[f_0'' + (2\eta + v_0)f_0' + Gr\theta + Gc\phi - Mf_0 - \frac{1}{Da}f_0 - \frac{ReFs}{Da}f_0^2\right] \tag{39}$$

We look for an initial approximation, $F_0(\eta)$, which is the solution of the linear part of the governing equation (36) given by

$$F_0'' + a_1(\eta)F_0' + a_2(\eta)F_0 = \psi(y), \tag{40}$$

subject to

$$F_0(0) = F_0(\infty). \tag{41}$$

The problem (40) - (41) is first transformed, using (17), to the interval [-1,1] then solved using the Chebyshev pseudospectral method. The unknown function $F_0(\xi)$ is approximated as

$$F_0(\xi) \approx \sum_{k=0}^{N} F_0(\xi_k)T_k(\xi_j), \quad j = 0,1,\ldots,N \tag{42}$$

After transforming the derivatives and substituting equations (42) in (40) - (41) we obtain

$$\mathbf{AF}_0 = \mathbf{\Psi} \tag{43}$$

subject to the boundary conditions

$$F_0(\xi_N) = F_0(\xi_0) = 0, \tag{44}$$

where

$$\mathbf{A} = \mathbf{D}^2 + \mathbf{a}_1\mathbf{D} + \mathbf{a}_2, \tag{45}$$

$$\mathbf{F}_0 = [F_0(\xi_0), F_0(\xi_1), \ldots, F_0(\xi_N)]^T, \tag{46}$$

$$\mathbf{\Psi} = [\psi(\xi_0), \psi(\xi_1), \ldots, \phi(\xi_N)]^T, \tag{47}$$

$$\mathbf{a}_s = diag([a_s(\eta_0), a_s(\eta_1), \ldots, a_s(\eta_{N-1}), a_s(\eta_N)]), \quad s = 1,2, \tag{48}$$

The boundary conditions (44) are imposed on the first and last rows of the matrices **A** and **Ψ** as follows

$$\begin{pmatrix} 1 & 0 & \cdots & 0 & 0 \\ & & & & \\ & & \mathbf{A} & & \\ & & & & \\ 0 & 0 & \cdots & 0 & 1 \end{pmatrix} \begin{pmatrix} F_0(\xi_0) \\ F_0(\xi_1) \\ \vdots \\ F_0(\xi_{N-1}) \\ F_0(\xi_N) \end{pmatrix} = \begin{pmatrix} 0 \\ \psi(\xi_1) \\ \vdots \\ \psi(\xi_{N-1}) \\ 0 \end{pmatrix} \tag{49}$$

The solution for $\mathbf{F}_0$ is then obtained from solving

$$\mathbf{F}_0 = \mathbf{A}^{-1}\mathbf{K}_0, \tag{50}$$

where $\mathbf{K}_0$ is the right hand side of (49).
To find the SHAM solutions of (36) we begin by defining the following linear operator

$$\mathcal{L}[\tilde{F}(\eta;q)] = \frac{\partial^2 \tilde{F}}{\partial \eta^2} + a_1 \frac{\partial \tilde{F}}{\partial \eta} + a_2 \tilde{F} \tag{51}$$

where $q \in [0,1]$ is the embedding parameter, and $\tilde{F}(\eta;q)$ is an unknown function. The zeroth order deformation equation is given by

$$(1-q)\mathcal{L}[\tilde{F}(\eta;q) - F_0(\eta)] = q\hbar\{\mathcal{N}[\tilde{F}(\eta;q)] - \psi(\eta)\} \tag{52}$$

where $\hbar$ is the non-zero convergence controlling auxiliary parameter and $\mathcal{N}$ is a nonlinear operator given by

$$\mathcal{N}[\tilde{F}(\eta;q)] = \frac{\partial^2 \tilde{F}}{\partial \eta^2} + a_1 \frac{\partial \tilde{F}}{\partial \eta} + a_2 \tilde{F} - \beta_1 \tilde{F}^2 \tag{53}$$

Differentiating (52) m times with respect to q and then setting $q = 0$ and finally dividing the resulting equations by $m!$ yields the mth order deformation equations

$$\mathcal{L}[F_m(\eta) - \chi_m F_{m-1}(\eta)] = \hbar R_m(\eta), \tag{54}$$

subject to the boundary conditions

$$F_m(0) = F_m(\infty) \tag{55}$$

where

$$R_m(\eta) = F''_{m-1} + a_1 F'_{m-1} + a_2 F_{m-1} - \beta_1 \sum_{n=0}^{m-1} F_n F_{m-1-n} - \psi(y)(1-\chi_m) \tag{56}$$

and

$$\chi_m = \begin{cases} 0, & m \leq 1 \\ 1, & m > 1 \end{cases}. \tag{57}$$

Using the transformation (17) and applying the Chebyshev pseudospectral transformation on equations (54)-(56) gives

$$\mathbf{A}\mathbf{F}_m = (\chi_m + \hbar)\mathbf{A}\mathbf{F}_{m-1} - \hbar(1-\chi_m)\mathbf{\Psi} + \hbar P_{m-1} \tag{58}$$

subject to the boundary conditions

$$F_m(\xi_N) = F_m(\xi_0) = 0, \tag{59}$$

where A and $\mathbf{\Phi}$, are as defined in (45) and (47), respectively, and

$$\mathbf{F}_m = [F_m(\xi_0), F_m(\xi_1), \ldots, F_m(\xi_N)]^T, \tag{60}$$

$$P_{m-1} = -\beta_1 \sum_{n=0}^{m-1} \mathbf{F}_n \mathbf{F}_{m-1-n} \tag{61}$$

To implement the boundary conditions (59) we delete the first and last rows of P_{m-1} and $\mathbf{\Psi}$ and delete the first and last row and column of $\mathbf{A}$ This results in the following recursive formula for $m \geq 1$.

$$\mathbf{F}_m = (\chi_m + \hbar)\mathbf{F}_{m-1} + \hbar\mathbf{A}^{-1}[\mathbf{P}_{m-1} - (1-\chi_m)\mathbf{\Psi}], \tag{62}$$

Thus, starting from the initial approximation, which is obtained from (50), higher order approximations $F_m(\xi)$ for $m \geq 1$, can be obtained through the recursive formula (62).

4. Results and discussion

In this section we give numerical results obtained by the spectral homotopy analysis method for the main parameters affecting the flow. Tables 1-6 display results for the skin friction $f'(0)$, Nusselt number, $-\theta'(0)$ and the Sherwood number $-\phi(0)$ when different parameters are varied. Table 1 gives the values of the local skin-friction for different values of the Hartmann number M. Analysis of the tabular data shows that magnetic field strength enhances the local skin friction as highlighted with increases in the absolute values of the skin friction as M increases. Physically, this implies that the plate surface exerts a drag force on the fluid. Table 2 shows the effect of the local Darcy number on the local skin friction. From this table we observe that the skin friction decreases in absolute values as the Darcy parameter increases

M	2nd order	4th order	6th order	8th order	10th order
1	-1.84655375	-1.84671328	-1.84671464	-1.84671465	-1.84671465
5	-2.78956864	-2.78968689	-2.78968758	-2.78968758	-2.78968758
10	-3.65197128	-3.65204866	-3.65204895	-3.65204896	-3.65204896
15	-4.34317940	-4.34323240	-4.34323255	-4.34323255	-4.34323255
20	-4.93446135	-4.93449949	-4.93449957	-4.93449957	-4.93449957

Table 1. Values of the $f'(0)$ for different values of M at different SHAM orders

Da	2nd order	3rd order	4th order	6th order	8th order
1	-1.84655375	-1.84670017	-1.84671328	-1.84671464	-1.84671465
5	-1.43582267	-1.43582436	-1.43582440	-1.43582440	-1.43582440
10	-1.37843023	-1.37843045	-1.37843046	-1.37843046	-1.37843046
15	-1.35893671	-1.35893678	-1.35893678	-1.35893678	-1.35893678
20	-1.34911952	-1.34911955	-1.34911955	-1.34911955	-1.34911955

Table 2. Values of the $f'(0)$ for different values of Da at different SHAM orders

The effect of thermal radiation on the skin friction is depicted on Table 3. We clearly observe in this table that the absolute values of the skin friction are reduced as the thermal radiation parameter R increases. Table 4 has been prepared to show the effect of wall suction velocity on the skin friction. We observe that the magnitude of the local skin friction significantly increases with increasing values of the wall suction velocity.

R	2nd order	4th order	6th order	8th order	10th order
5	-1.71772583	-1.71778800	-1.71778832	-1.71778833	-1.71778833
10	-1.64867944	-1.64870680	-1.64870690	-1.64870690	-1.64870690
15	-1.60822374	-1.60823854	-1.60823858	-1.60823858	-1.60823858
20	-1.58025980	-1.58026899	-1.58026901	-1.58026901	-1.58026901
30	-1.54255011	-1.54255482	-1.54255483	-1.54255483	-1.54255483

Table 3. Values of the $f'(0)$ for different values of R at different SHAM orders

v_0	2nd order	4th order	6th order	8th order	10th order
1	-1.84422963	-1.84438345	-1.84438474	-1.84438475	-1.84438475
2	-2.56784574	-2.56822755	-2.56823124	-2.56823128	-2.56823128
3	-3.38334080	-3.38395995	-3.38396583	-3.38396590	-3.38396590
4	-4.26060301	-4.26139618	-4.26140302	-4.26140309	-4.26140309
5	-5.17790146	-5.17878811	-5.17879476	-5.17879482	-5.17879482

Table 4. Values of the $f'(0)$ for different values of v_0 at different SHAM orders

Table 5 has been included to show the effects of varying the wall suction velocity v_0, thermal radiation parameter R, heat absorption parameter Ω, absorption of radiation parameter Q_2 and the Prandtl number Pr on the local Nusselt number. It can be clearly observed that the rate of heat transfer between the wall and the fluid increases for increasing values of v_0, high values of Q_2 as well as Pr. The Nusselt number is observed to be reduced by increasing values of R and small values of Q_2.

v_0	R	Ω	Q_2	Pr	$-\theta'(0)$
1.0	1.0	1.0	1.0	0.7	0.95840112
2.0	1.0	1.0	1.0	0.7	1.22084437
3.0	1.0	1.0	1.0	0.7	1.50119641
4.0	1.0	1.0	1.0	0.7	1.79526126
5.0	1.0	1.0	1.0	0.7	2.09985847
1.0	2.0	1.0	1.0	0.7	0.76503772
1.0	4.0	1.0	1.0	0.7	0.58038779
1.0	5.0	1.0	1.0	0.7	0.52676177
1.0	10.0	1.0	1.0	0.7	0.38340898
1.0	1.0	2.0	1.0	0.7	1.12989889
1.0	1.0	4.0	1.0	0.7	1.41719168
1.0	1.0	5.0	1.0	0.7	1.54193396
1.0	1.0	10.0	1.0	0.7	2.05653970
1.0	1.0	1.0	2.0	0.7	0.81994800
1.0	1.0	1.0	5.0	0.7	0.40458864
1.0	1.0	1.0	10.0	0.7	-0.28767696
1.0	1.0	1.0	15.0	0.7	-0.97994256
1.0	1.0	1.0	20.0	0.7	-1.67220816
1.0	1.0	1.0	1.0	0.1	0.33842173
1.0	1.0	1.0	1.0	0.7	0.95840112
1.0	1.0	1.0	1.0	1.0	1.17519349
1.0	1.0	1.0	1.0	10.0	5.39718685
1.0	1.0	1.0	1.0	50.0	20.81046541

Table 5. Values of the $-\theta'(0)$ for different values of v_0, R, Ω,Q_2 and Pr

Table 6 depicts the effects of varying v_0, γ and Sc on the local Sherwood number. We observe from this table that all these three parameters cause the local Sherwood number to increase. Final results are computed for the main physical parameters which are presented by means of graphs. The influence of the thermal Grashof number Gr, the magnetic field parameter M, the Darcy number Da, absorption radiation parameter Q_2, thermal radiation parameter R, heat absorption Ω, and the suction parameter v_0 on the velocity profiles can be analyzed from Figs. (a)-(g). Fig (a) shows the influence of thermal buoyancy force parameter Gr on the velocity. As can be seen from this figure, the velocity profile increases with increases in the values of the thermal buoyancy. We actually observe that the velocity overshoot in the boundary layer region. Buoyancy force acts like a favourable pressure gradient which accelerates the fluid within the boundary layer therefore the solutal buoyancy force parameter Gm has the same effect on the velocity as Gr. From Fig (b) we observe that the effect of magnetic field is to decrease the value of velocity profile through out the boundary layer which result in the thinning of the boundary layer thickness. Fig (c) displays the influence of the Darcy number Da on the velocity profile. Increasing the Darcy number increases the velocity. The effect of increasing the value of the absorption parameter on the velocity is shown in Fig (d). We observe in this figure that increasing the value of the absorption of the radiation parameter due to increase in the buoyancy force accelerates the flow rate.

Fig (e) depicts the effect of varying thermal radiation parameter R on the flow velocity. We observe that the thermal radiation enhances convective flow. Fig (f) illustrates the influence of heat absorption coefficient Ω on the velocity. Physically, the presence of heat absorption (thermal sink) effect has the tendency in resulting in a net reduction in the flow velocity. This behaviour is seen from Fig (f) in which the velocity decreases as Ω increases. The hydrodynamic boundary layer decreases as the heat absorption effects increase. The effects of v_0 on the velocity field are shown in Fig (g). It is clearly seen from this figure that the velocity profiles decrease monotonically with the increase of suction parameter indicating the usual fact that suction stabilizes the boundary layer growth.

The influence of heat absorption, radiation absorption, thermal radiation and suction on the temperature distribution is respectively, shown on Figs (h) - (k). Fig (h) depicts the effects of heat absorption Ω on the temperature distribution. It is observed that the boundary layer absorbs energy resulting in the temperature to fall considerably with increasing values of Ω. This is because when heat is absorbed, the buoyancy force decreases the temperature profile. The effect of absorbtion of radiation parameter on the temperature profile is shown on Fig (i). It is seen from this figure that the effect of absorption of radiation is to increase temperature in the boundary layer as the radiated heat is absorbed by the fluid which in turn increases the temperature of the fluid very close to the porous boundary layer and its effect diminishes far away from the boundary layer. From Fig (j) we observe that the effect of thermal radiation is to enhance heat transfer as thermal boundary layer thickness increases with increase in the thermal radiation. We observe that the effect of R is to increase the temperature distribution in the thermal boundary layer. This is because the increase of R implies increasing of radiation in the thermal boundary layer, and hence increases the values of the temperature profiles in thermal boundary layer. Lastly the effect of suction parameter v_0 on the temperature field is displayed in Fig (k). We see that the temperature profiles decrease with increasing values of v_0. This is because sucking decelerates fluid particles through the porous wall reducing the growth of the fluid boundary layer as well as thermal and concentration boundary layers.

v_0	γ	Sc	$-\phi'(0)$
1	1	0.6	1.53261796
2	1	0.6	1.97076626
3	1	0.6	2.44776350
4	1	0.6	2.95336689
5	1	0.6	3.47988849
1	0	0.6	1.28843084
1	5	0.6	2.25555303
1	10	0.6	2.90467826
1	20	0.6	3.87130625
1	50	0.6	5.84331065
1	1	0.6	1.53261796
1	1	1.0	2.13711784
1	1	2.0	3.45301022
1	1	5.0	6.90551441
1	1	10.0	12.23149717

Table 6. Values of the $-\phi'(0)$ for different values of v_0, γ and Sc

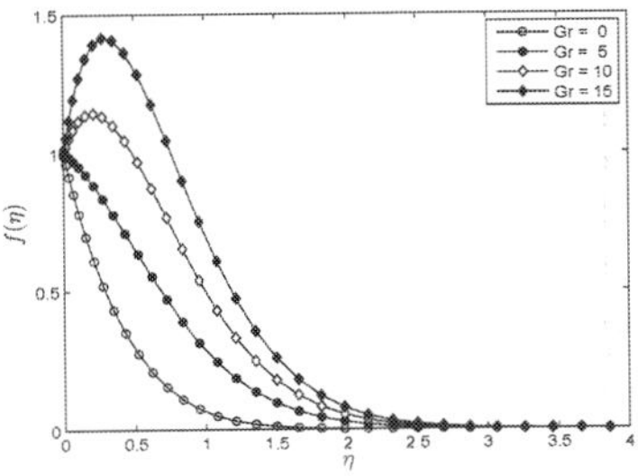

(a) Plot of $f'(\eta)$ for varying Gr

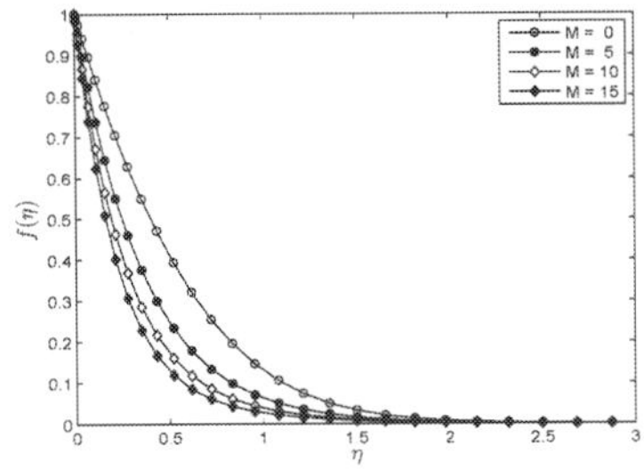

(b) Plot of $f'(\eta)$ for varying M

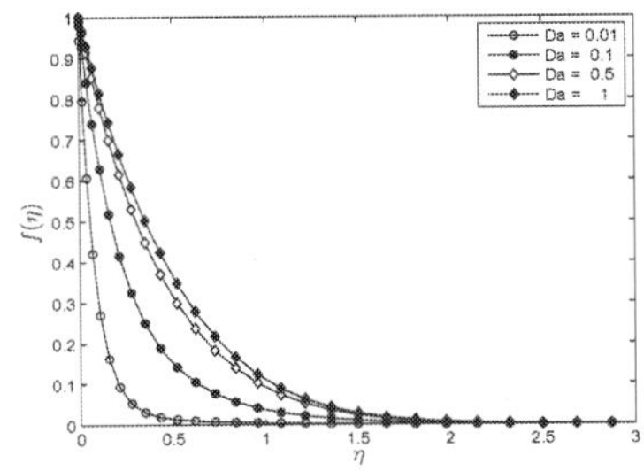

(c) Plot of $f'(\eta)$ for varying Da

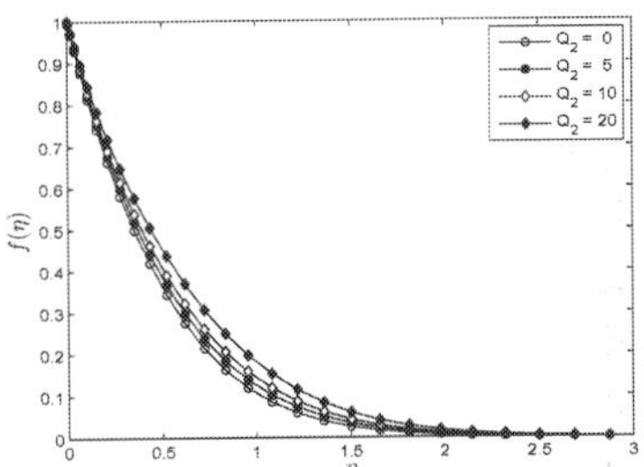

(d) Plot of $f(\eta)$ for varying Q_2

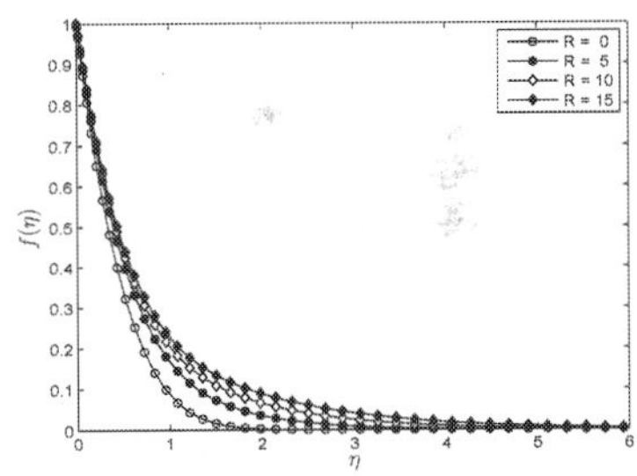

(e) Plot of $f(\eta)$ for varying R

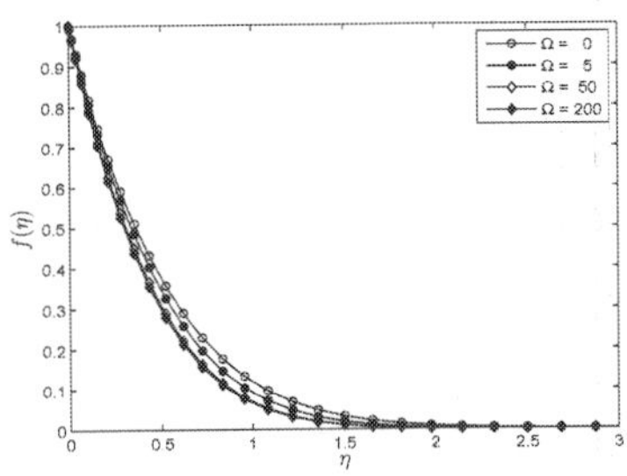

(f) Plot of $f(\eta)$ for varying Ω

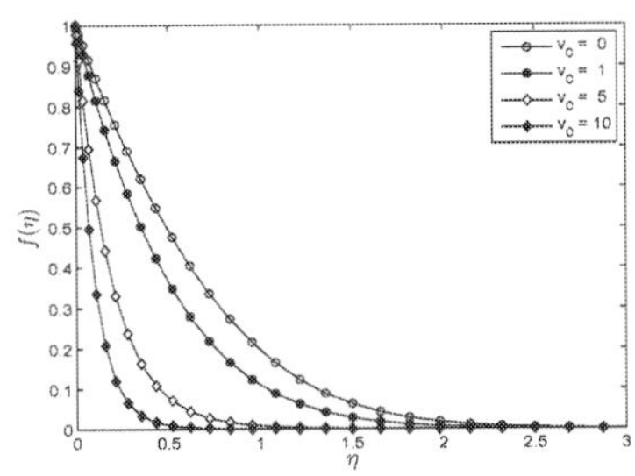

(g) Plot of $f(\eta)$ for varying v_0

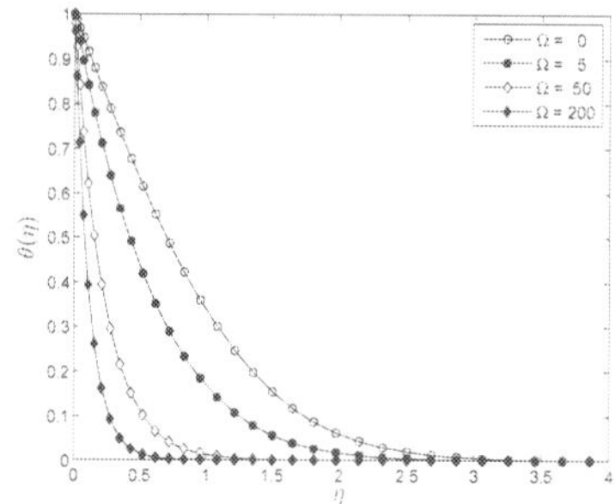

(h) Plot of $\theta(\eta)$ for varying Ω

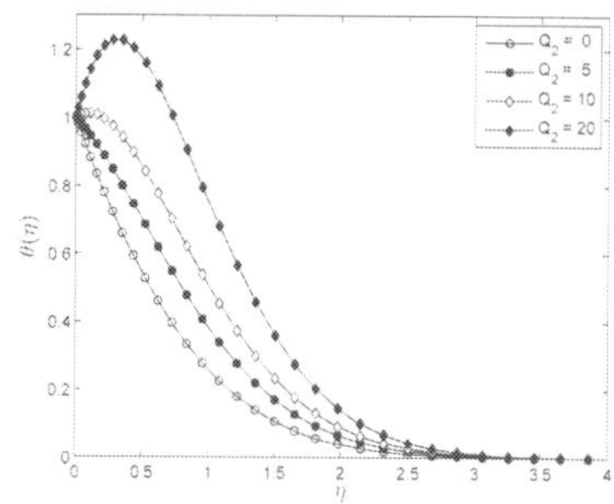

(i) Plot of $\theta(\eta)$ for varying Q_2

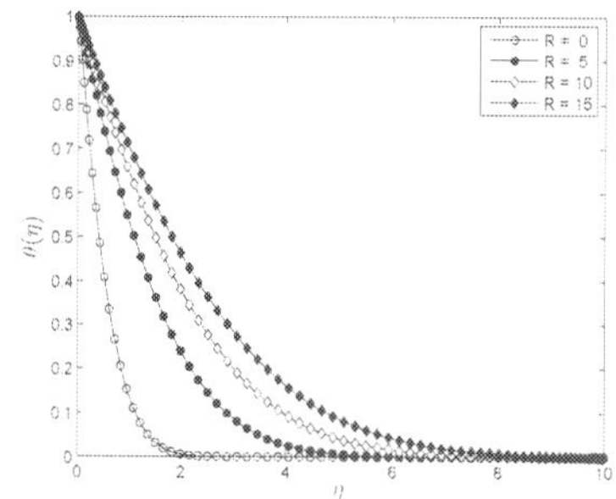

(j) Plot of $\theta(\eta)$ for varying R

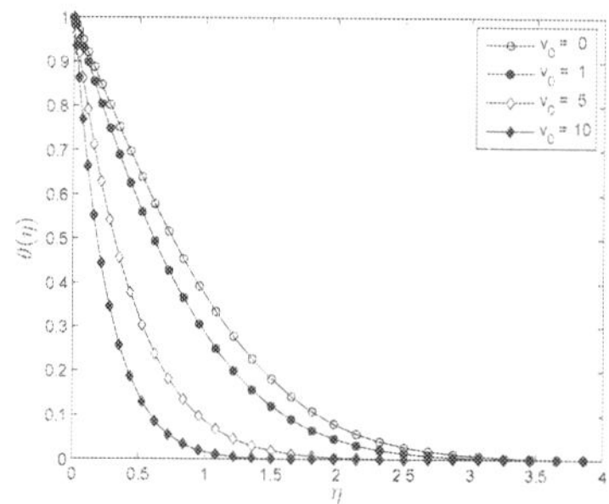

(k) Plot of $\theta(\eta)$ for varying v_0

Figs (l) and (m) depict the influence of the non-dimensional chemical reaction parameter γ and the suction parameter on concentration profiles, respectively. The effect of chemical reaction parameter is very important in the concentration field. Chemical reaction increases the rate of interfacial mass transfer. Reaction reduces the local concentration, thus increases its concentration gradient and its flux. In Fig (m) we see that the concentration profiles decrease with increasing values of the suction parameter.

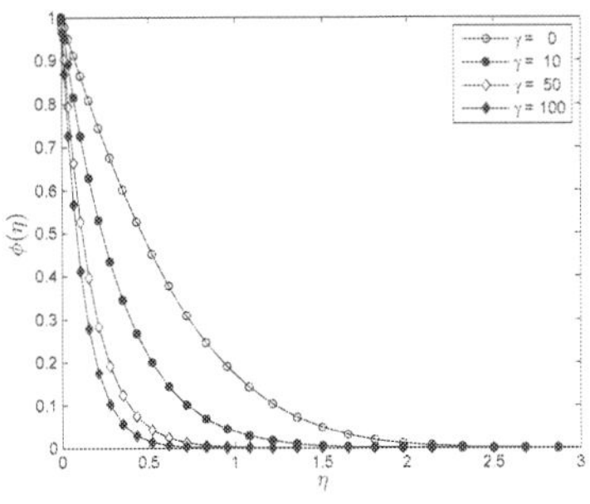

(l) Plot of $\phi(\eta)$ for varying γ

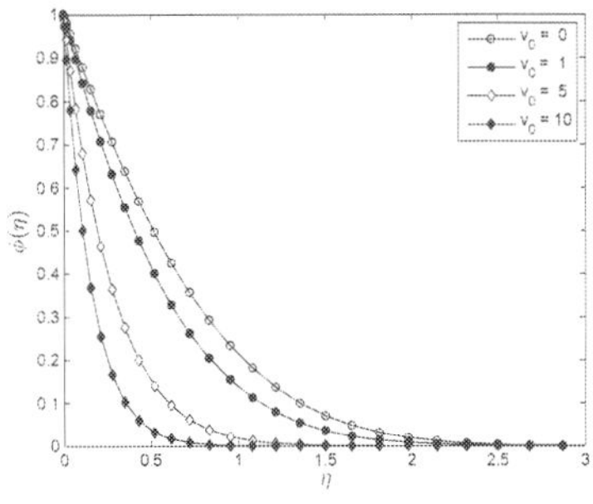

(m) Plot of $\phi(\eta)$ for varying v_0

5. Conclusions

In this chapter, a new improved numerical technique to solve the problem of unsteady magnetohydrodynamic convective heat and mass transfer past an infinite permeable vertical plate in porous medium with thermal radiation, heat absorption and chemical reaction has been employed. The highly non-linear momentum, energy and species boundary layer equations are converted into ordinary differential equations using similarity transformations before being solved using the spectral homotopy analysis method. The effects of various physical parameters like buoyancy parameter, Hartmann number, Darcy number, suction, thermal radiation and chemical reaction on velocity, temperature and concentration profiles are obtained.
The following main conclusions can be drawn from the present chapter:

1. Wall suction stabilizes the velocity, thermal as well as concentration boundary layer growth.
2. Boundary layer flow attain minimum velocity values for large Hartmann numbers.
3. Buoyancy parameter is to increase the velocity distribution in the momentum boundary layer.
4. The presence of heat absorption effects cause reductions in the fluid temperature which resulted in decreases in the fluid velocity.
5. The concentration decreases with increasing the chemical reaction parameter.
6. Both the velocity and temperature profiles increase with increasing values of radiation absorption parameter.

These results might find wide applications in engineering, such as geothermal system, heat exchangers, and nuclear waste depositors.

6. References

[1] Cussler, E.L. (1988). *Diffusion Mass Transfer in Fluid Systems,* Cambridge University Press, London.

[2] Das, U.N, Deka, R., & Soundalgekar, V.M. (1994). Effects of mass transfer on flow past an impulsively started infinite vertical plate with constant heat flux and chemical reaction. *J. Forshung Im Ingenieurwesen-Engineering Research Bd,* 60, 284-287.

[3] Anjalidevi, S.P. & Kandasamy, R. (1999). Effects of chemical reaction, heat and mass transfer on laminar flow along a semi infinite horizontal plate. Heat and Mass Transfer, 35, 465-467.

[4] Seddeek, M.A., Darwish, A.A. & Abdelmeguid, M.S. (2007). Effects of chemical reaction and variable viscosity on hydromagnetic mixed convection heat and mass transfer for Hiemenz flow through porous media with radiation, *Commun Nonlinear Sci. Numer. simulat.,* 15, 195-213.

[5] Salem, A.M. & Abd El-Aziz, M. (2008). Effect of Hall currents and chemical reaction on hydromagnetic flow of a stretching vertical surface with internal heat generation/absorption. *Applied Mathematical Modelling,* 32, 1236-1254.

[6] Mohamed, R. A., (2009). Double-Diffusive Convection-radiation Interaction on Unsteady MHD Flow over a Vertical Moving Porous Plate with Heat Generation and Soret Effects, *Applied Mathematical Sciences,* 3, (13), 629-651.

[7] Ibrahim, F.S., Elaiw, A.M., & Bakr, A.A. (2008). Effect of chemical reaction and radiation absorption on the unsteady MHD free convection flow past a semi infinite vertical permeable moving plate with heat source and suction. *Commun Nonlinear Sci. Numer. simulat.,* 13, 1056-1066.

[8] Mahdy, A. (2010). Effect of chemical reaction and heat generation or absorption on double-diffusive convection from a vertical truncated cone in a porous media with variable viscosity. *International Communications in Heat and Mass Transfer,* 37, 548-554.

[9] Nield, D. A., & Bejan, A., (1999). *Convection in Porous Media, 2nd edition.* Springer-Verlag, New York .

[10] Abel, M.S., Kan, S.K., Prasad, K.V. (2001). Convective heat and mass transfer in a visco-elastic fluid flow through a porous medium over a stretching sheet. *International Journal of Numerical Methods for Heat and Fluid flow,* 11(8), 779-792.

[11] Ali, M.E. (2007). The effect of lateral mass flux on the natural convection boundary layers induced by a heated vertical plate embedded in a saturated porous medium with internal heat generation. *Int. J. Thermal Sci.,* 46, 157-163.

[12] Vajravelu, K., & Hadjinicolaou, A., (1997). Convective heat transfer in an electrically conducting fluid at a stretching surface with uniform free stream.*Int. J. Eng. Sci,* 35, 1237-1244.

[13] Chamkha, A. J., & Khaled, A. R. A., (2000). Similarity solutions for hydromagnetic mixed convection heat and mass transfer for Hiemenz flow through a porous medium. *Int.J.Numerical Methods for Heat and Fluid Flow,* 10, 94-115.

[14] Shateyi, S., Sibanda, P., & Motsa, S.S., (2007). Magnetohydrodynamic flow past a vertical plate with radiative heat transfer, *Journal of Heat Transfer,* 129, 1708-1714.

[15] Shateyi, S. Motsa, S.S., & Sibanda, P. (2010). The Effects of Thermal Radiation, Hall Currents, Soret, and Dufour on MHD Flow by Mixed Convection over a Vertical Surface in Porous Media, *Mathematical Problems in Engineering,* Volume 2010, Article ID 627475, 20 pages doi:10.1155/2010/627475.

[16] Makinde, O.D & Sibanda, P. (2008). Magnetohydrodynamic mixed convective flow and

heat and mass transfer past a vertical plate in a porous medium with constant wall suction. *ASME - Journal of Heat Transfer* , 130 (11260), 1-8.

[17] Bég, O.A., Zueco, J., Bhargava, R., & Takhar, H.S. (2009). Magnetohydrodynamic convection flow from a sphere to a non-Darcian porous medium with heat generation or absorption effects: network simulation. *International Journal of Thermal Sciences,* 48, 913-921.

[18] Pal, A. & Talukdar, B. (2010). Perturbation analysis of unsteady magnetohydrodynamic convective heat and mass transfer in a boundary layer slip flow past a vertical permeable plate with a thermal radiation and chemical reaction, *Commun Nonlinear Sci. Numer. simulat.*, 15, 1813-1830.

[19] Makinde, O.D & Aziz, A. (2010). MHD mixed convection from a vertical plate embedded in a porous medium with a convective boundary condition. *International Journal of Thermal Sciences,* 49, 1813-1820.

[20] Crepeau, J.C. & Clarksean, R. (1997). Similarity solutions of natural convection with internal heat generation *J. Heat Transfer,* 119, 183-185.

[21] Patil P.M., & Kulkarni, P.S. (2008). Effects of chemical reaction on free convective flow of a polar fluid through a porous medium in the presence of internal heat generation. *International Journal of Thermal Sciences,* 47, 1043-1054.

[22] Salem, M.A., & Abd El-Aziz, M. (2008). Effect of Hall currents and chemical reaction on hydromagnetic flow of a stretching vertical surface with internal heat generation/absorption. *Applied Mathematical Modelling,* 32, 1236-1254.

[23] Samad, M.A., & Mohebujjaman, M. (2009). MHD Heat and Mass Transfer Free Convection Flow anong a Vertical Stretching Sheet in Presence of Magnetic Field with Heat Generation. *Research Journal of Applied Science, Engineering and Technology,* 1(3), 98-106.

[24] Siegel, R., & Howell, J.R. (2001). *Thermal Radiation Heat Transfer,* Speedy Hen, London, CA, United Kingdom , (ISBN: 1560328398 / 1-56032-839-8).

[25] C. Canuto, M. Y. Hussaini, A. Quarteroni, and T. A. Zang (1998). *Spectral Methods in Fluid Dynamics*, Springer-Verlag, Berlin, 1988.

[26] Cortell, R., (2008). Effects of viscous dissipation and radiation on the thermal boundary layer over a nonlinearly stretching sheet. *Phys Lett A,* 372, 631-636.

[27] Bataller, R.C., (2008). Radiation effects for the Blasius and Sakiadis flows with a convective surface boundary condition. *Appl Math Comput,* 206: 832-840.

[28] Ibrahim, F.S., Elaiw, A.M., & Bakar, A.A. (2008). Effect of the chemical reaction and radiation absorption on the unsteady MHD free convection flow past a semi infinite vertical permeable moving plate with heat source and suction. *Communications in Nonlinear Science and Numerical Simulation,* 13, 1056-1066.

[29] Shateyi, S. (2008). Thermal Radiation and Buoyancy Effects on Heat and Mass Transfer over a Semi-Infinite Stretching Surface with Suction and Blowing, *Journal of Applied Mathematics,* Volume 2008, Article ID 414830, 12 pages doi:10.1155/2008/414830.

[30] Shateyi, S., & Motsa, S.S. (2009). Thermal Radiation Effects on Heat and Mass Transfer over an Unsteady Stretching Surface. *Mathematical Problems in Engineering,* Volume 2009, Article ID 965603, 13 pages doi:10.1155/2009/965603.

[31] Aliakbar, V., Alizadeh-Pahlavan, A. & Sadeghy, K., (2009). The influence of thermal radiation on MHD flow of Maxwellian fluids above stretching sheets. *Commun Nonlinear Sci Numer Simulat,* 14(3): 779-794.

[32] Hayat, T., & Qasim, M. (2010). Influence of thermal radiation and Joule heating on MHD

flow of a Maxwell fluid in the presence of thermophoresis. *Int. J. Heat Mass Transfer,* doi:10.1016/j.ijheatmasstransfer.2010.06.014.

[33] Cortell, R, (2010). Suction, viscous dissipation and thermal radiation effects on the flow and heat transfer of a power-law fluid past an infinite porous plate, *Chem Eng Res Des,* doi:10.1016/j.cherd.2010.04.017

[34] S.J. Liao (2003), Beyond perturbation: Introduction to homotopy analysis method. Chapman & Hall/CRC Press, 2003.

[35] S. S. Motsa, P. Sibanda and S. Shateyi (2010a), A new spectral-homotopy analysis method for solving a nonlinear second order BVP. *Commun. Nonlinear Sci. Numer. Simulat.* 15 (2010a) 2293-2302.

[36] S. S. Motsa, P. Sibanda, F. G. Awad, S. Shateyi (2010b), A new spectral-homotopy analysis method for the MHD Jeffery-Hamel problem, *Computer & Fluids* , 39 (2010b) 1219-1225.

[37] M. A. Hossain, M.A. Alim, and D. A. S. Rees (1999), The effect of radiation on free convection from a porous vertical plate. *Int. J. Heat Mass Transfer,* 42, (1999), 181 - 191.

[38] A. Raptis (1998), Flow of a micropolar fluid past a continuously moving plate by the presence of radiation, *Int. J. Heat Mass Transfer,* 41, (1998), 2865-2866.

[39] L. N. Trefethen (2000), Spectral Methods in MATLAB, SIAM, 2000.

8

Mass Transfer, and Effects of Magnetic Fields on the Mass Transfer in Close Binary System

Davood Manzoori
Department of Physics, University of Mohaghegh Ardabili, P. O.Box. 179, Ardabil Iran

1. Introduction

The gravitational potential of a binary system is described by the Roche model where each star dominates the gravitational potential inside regions called Roche lobes.The two Roche lobes meet at the inner Lagrange point along the joining of the two stars. Figure 1 shows equipotential surfaces in the orbital plane for a binary system. As may be seen from Fig. 1, there are five equilibrium points(i.e. $\nabla\phi = 0$) three of which i.e., L_1, L_2, L_3,are along the line centers. Two are peripheral to the masses and lie as the critical points along equipotentials that envelope both stars. two other points ,i.e., L_4, L_5 lie opposite to each other, perpendicular to the line of centers. These are quasi-equilibrium points for which local orbits are possible because of coriolis acceleration. If either star fills its Roche lobe, matter will stream from the Roche lobe filling star through the inner Lagrange point to the other star in a process known as Roche lobe overflow(RLOF). This actually occurs before the photosphere reaches the Roche lobe radius in the absence of magnetic fields or other constrains on the mass flow. This mass transfer affects both the evolution of the components of the binary as well as the binary properties such as orbital period and eccentricity. Roche lobe overflow can be triggered by the evolution of the binary properties or by evolution of the component stars. On the one hand, the orbital separation of the binary can change so that the Roche lobe can shrink to within the surface of one of the stars. On the other hand, stellar evolution may eventually cause one of the stars to expand to fill its Roche lobe. When both stars in the binary are main-sequence stars, the latter process is more common. Since the more massive star will evolve first, it will be the first to expand and fill its Roche lobe. At this stage, the mass exchange can be conservative (no mass is lost from the binary) or non-conservative (mass is lost). Depending on the details of the mass exchange and the evolutionary stage of the mass-losing star there are several outcomes that will lead to formation of a relativistic binary. The primary star can lose its envelope, revealing its degenerate core as either a helium, carbon-oxygen, or oxygen-neon white dwarf, it can explode as a supernova, leaving behind a neutron star or a black hole, or it can simply lose mass to the secondary so that they change roles. Barring disruption of the binary, its evolution will then continue. In most outcomes, the secondary is now the more massive of the two stars and it may evolve off the main sequence to fill its Roche lobe. The secondary can then initiate mass transfer or mass loss with the result that the secondary also can become a white dwarf (WD), neutron star (NS), or black hole (BH). The relativistic binaries that result from this process fall into a number of observable categories. A WD-MS or WD-WD binary may eventually become a cataclysmic variable once the white dwarf begins to accrete material from its companion. If the companion is a Main Sequence star RLOF can

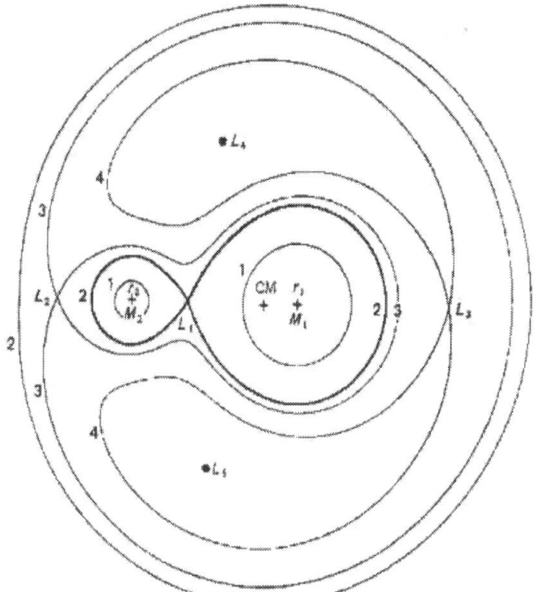

Fig. 1. Roche surface for q = 0.4. The Lagrangian points are indicated

be triggered by evolution of the companion. If the companion is another white dwarf, then RLOF is triggered by the gradual shrinking of the orbit through the emission of gravitational radiation. WD-WD cataclysmic variables are also known as AM CVn stars. If the total mass of the WD-WD binary is above the Chandrasekhar mass, the system may be a progenitor to a type I supernova. The orbit of a NS-MS or NS-WD binary will shrink due to the emission of gravitational radiation. At the onset of RLOF, the binary will become either a low-mass X-ray binary(if the donor star is a WD or MS with $M_1 \leq 2M_\odot$), or a high-mass X-ray binary(if the donor is a more massive main-sequence star).These objects may further evolve to become millisecond pulsars if the NS is spun up during the X-ray binary phase[1]. A comprehensive table of close binary types that can be observed in electromagnetic radiation can be found in Hilditch (2001). The type of binary that emerges depends upon the orbital separation and the masses of the component stars. During the evolution of a $10M_\odot$ star, the radius will slowly increase by a factor of about two as the star progresses from zero age main sequence to terminal age main sequence. The radius will then increase by about another factor of 50 as the star transitions to the red giant phase, and an additional factor of 10 during the transition to the red supergiant phase. These last two increases in size occur very quickly compared to the slow increase during the main-sequence evolution of the star. Mass transfer can be divided into three cases see (Thomas 1977) related to the timing of the onset of RLOF.

Case A: If the orbital separation is small enough (usually a few days), the star can fill its Roche lobe during its slow expansion through the main-sequence phase while still burning hydrogen in its core.

[1]See the website, http://www.livingreviews.org

Case B: If the orbital period is less than about 100 days, but longer than a few days, the star will fill its Roche lobe during the rapid expansion to a red giant with a helium core. If the helium core ignites during this phase and the transfer is interrupted, the mass transfer is case B.

Case C: If the orbital period is above 100 days, the star can evolve to the red supergiant phase before it fills its Roche lobe. In this case, the star may have a CO or ONe core.

Case A mass transfer occurs during the slow growth, case B during the first rapid expansion, and case C during the final expansion phase. The nature of the remnant depends upon the state of the primary during the onset of RLOF and the orbital properties of the resultant binary depend upon the details of the mass transfer.

Wood (1950) studied period variations of binaries and suggested that mass ejections could be a cause of period change. Huang (1963) revised the problem of mass transfer in binary systems, he modified the Jeans(1924-1925) mode of mass ejection, through suggesting two modes of mass transfer i.e., slow and intermediate modes, in each case he was able to derive the following equations:

$$\frac{\Delta P}{P} = -3(\frac{M_2 - M_1}{M_2})\frac{\Delta M_1}{M_1} + \frac{3e\Delta e}{1-e^2} \qquad for\ slow\ mode \tag{1}$$

$$\frac{\Delta P}{P} = (1+3\gamma_2)(\frac{M_2 + M_1}{M_1 + M_2}) - 3\frac{\Delta M_1}{M_1} + \frac{\Delta M_2}{M_2} + \frac{3e\Delta e}{1-e^2} \qquad for\ intermediate\ mode \tag{2}$$

Where

$$\gamma_2 = \frac{(M_1 + M_2)^2}{M_1 M_2}[\frac{a_e}{a(1-e^2)}]^{1/2}$$

P, denotes the orbital period, $M_i s$, the masses of the components, e, eccentricity, a, semi-major axis, and Δ implies the variations of respective parameters. A true review of mass and angular momentum transfer and their consequences on the evolution of binary stars may be find by Thomas (1977).

2. Conservative mass transfer

When no ejected matter leaves a binary system, the mass transfer is said to be conservative. During consevative mass transfer, the orbital elements of the binary can change due to transfer of angular momentum from one star to the companion. Consider a system with a totat mass $M = M_1 + M_2$, semi-major axis, a, eccentricty, e, and the total orbital angular momentum, J,

$$J = M_1 M_2 \sqrt{\frac{Ga(1-e^2)}{M_1 + M_2}} \tag{3}$$

will also be conserved, where G, is universal grvitational constant. Hence: $\dot{J} = 0, \dot{M} = 0\ and\ \dot{M}_1 = -\dot{M}_2$. And We may also write

$$a = c^2 (M_1 M_2)^{-2} J^2 \tag{4}$$

Where:

$$c = \sqrt{\frac{M}{G(1-e^2)}}$$

, c is a constant
Differentiating eq. 3 yields

$$\frac{\dot{a}}{a} = 2\frac{\dot{J}}{J} - 2\frac{\dot{M_1}}{M_1} - 2\frac{\dot{M_2}}{M_2} + \frac{\dot{M}}{M} \tag{5}$$

Using the requirements for conservative mass transfer (mentioned just after the eq. 3) in eq. 5 we get

$$\frac{\dot{a}}{a} = 2\dot{M_1}(\frac{M_2 - M_1}{M_1 M_2}) \tag{6}$$

If M_1 is the mass losing star $\dot{M_1}$ will be negative, therefore we conclude that if the mass donor star is more massive, then the orbit will shrink and hence period decreases. But if the mass donor is less massive than the accretor component then orbit will be widened and period increases.
By using the Kepler's third law ($\frac{P^2}{a^3} = \frac{4\pi^2}{GM}$) in eq. 6, the period variation with time $\dot{P}$ due to mass transfer can be written as,

$$\dot{P} = \frac{3M_1 \dot{(}M_1 - M_2)}{M_1 M_2} P \tag{7}$$

Where P, is the orbital period.

3. Non conservative mass transfer

In case of non conservative mass transfer both mass and angular momentum can be removed from the system. Following Demircan et al. (2006) Orbital Angular Momentum (OAM) of a two body system is given by

$$J = (\frac{M_1 M_2}{M_1 + M_2})a^2\Omega = (\frac{q}{(1+q)^2})Ma^2\Omega \tag{8}$$

Where,

$$(\frac{M_1 M_2}{M_1 + M_2})a^2 = I$$

is moment of Inertia and $\Omega = \frac{2\pi}{P}$ is angular speed, P, is orbital period and $M = M_1 + M_2$, $q = \frac{M_2}{M_1}$ are total mass and mass ratio, respectively. If we assume isotropic mass loss from the surface of the components, then,

$$\dot{J} = (\frac{q}{(1+q)^2})\dot{M}a^2\Omega \tag{9}$$

Since the dynamics of a two body system obey Kepler's third law therefore one expects transfer of mass would change the, a, P and M accordingly,

$$3\frac{\dot{a}}{a} + 2\frac{\dot{\Omega}}{\Omega} = \frac{\dot{M}}{M} \tag{10}$$

assume q as a constant, differentiation of equation 8 with respect to "t", will give

$$\frac{\dot{J}}{J} = \frac{\dot{M}}{M} + 2\frac{\dot{a}}{a} + \frac{\dot{\Omega}}{\Omega} \tag{11}$$

if $\frac{\dot{a}}{a}$ is substituted from equation 10 to equation 11 we get,

$$\frac{\dot{J}}{J} = \frac{5}{3}\frac{\dot{M}}{M} - \frac{1}{3}\frac{\dot{\Omega}}{\Omega} = \frac{5}{3}\frac{\dot{M}}{M} + \frac{1}{3}\frac{\dot{P}}{P} \tag{12}$$

But substitution of $\frac{\dot{\Omega}}{\Omega}$ from 10 in to 11 will give

$$\frac{\dot{J}}{J} = \frac{3}{2}\frac{\dot{M}}{M} + \frac{1}{2}\frac{\dot{a}}{a} \tag{13}$$

Therefore the last two equations i.e. 12 and 13, give how the loss of OAM and mass are caused the orbital period, P, and orbital radius, a, to change.
According to equations 8 and 9 the isotropic mass loss implies

$$\dot{J} = \frac{\dot{M}}{M}J \tag{14}$$

This eq. implies that the only source of OAM loss is mass loss, which were assumed to be isotropic. Substituting $\frac{\dot{J}}{J}$ from eq 14 to 12 will give,

$$\dot{P} = -2\frac{\dot{M}}{M}P \tag{15}$$

In this last eq. since the $\frac{\dot{P}}{P}$ is measurable observationally, therefore $\frac{\dot{M}}{M}$ can be calculated from eq. 15.
Now if we use the following eqs. taken from Stepien (1995) to calculate relative angular momentum lost from a system as,

$$J = G^{2/3}M^{5/3}\Omega^{-1/3}q(1+q)^{-2} \tag{16}$$

$$J_{lost} = \dot{J} = -\frac{1}{3}G^{2/3}M^{5/3}\Omega^{-4/3}q(1+q)^{-2}\dot{\Omega} \tag{17}$$

and hence the relative OAM lost only by magnetized star wind

$$\frac{J_{lost}}{J} = -\frac{1}{3}\frac{\dot{\Omega}}{\Omega} \tag{18}$$

Where

$$\dot{\Omega} = 1.8 \times 10^{-8}\left(\frac{k^2(M_1R_1^2 + M_2R_2^2)(1+q)^2}{qG^{2/3}M^{5/3}}\right)\Omega^{7/3}e^{-1.3\Omega} \tag{19}$$

Using the value of $\frac{J_{lost}}{J}$ in eq. 14, the relative mass lost from the system can be estimated. Since eq.15 may give the total mass ejected from the system. Hence the transferred mass from one component to the other easily can be estimated.

4. Effects of magnetic fields on the mass loss and mass transfer

Now it is generally accepted among the astrophysicists that close binary systems with a cool F-K type star display enhanced magnetic activities. Short period close binaries i.e., those having an orbital period $P < 5 - 6$ days, possess the just mentioned characteristics, due to rapid rotation (Richard & Albright 1993). Many authors (see e.g., Richards 1993, 1992; Hall 1989 & Olson 1981) have discussed that the secondary and/or primary in the close binaries having a late type component show a variety of time dependent magnetic properties, which causes brightness variations in the light curve, radiations of X-ray, Ultraviolet, infrared, and cyclic variations in the orbital period of the binary.

Generations of Polidal and Toroidal magnetic Fields: It is a well known fact that the dynamo mechanism is likely cause of large scale magnetic field productions in the stars having convective layers (e.g., see parker, 1955). The differential rotation between the radiative core and convective envelope, winds up the field and causes a deformation (shearing) of the poloidal field which, in turn, generates an additional toroidal field component and thus creates a Lorentz force which counteracts the shear due to the poloidal field.

The effects of magnetic fields on the mass and angular momentum transfer and / or loss of both the companions are quiet appreciable. In a close binary, where the spin and orbital angular momenta are strongly coupled. This stellar spindown, forces a decrease in the orbital period of the system even without mass transfer.The coupling constant depends on the magnetic field strength and can be important if the field is strong enough (on the order or few mega gauss). Another effect of the magnetic field is to alter the spin through torquing of the star by mass outflow (see next section).

5. Magnetic braking

The net results of the mass exchange is a mass transfer from the primary to secondary during whole contact phase (Huang et al. 2007). In addition the magnetic braking is a common phenomenon to all the contact binary stars (see Bradstreet and Guinan 1994; Huang et al. 2007).

According to Bradstreet and Guinan (1994), Stepien (1995, 2006) the role of AM loss is crucial in the formation and evolution of the contact low mass binary stars. The magnetized star winds move outward from the active star, but are twisted due to rapid rotation of the star. Charged particles in the star wind get trapped in the magnetic field of the star and are dragged along the field lines. The result is Angular Momentum (AM) transfer from the star by magnetic field to the charged particles. As the winds leave the star surface they are dragged by the magnetic field which, in turn, slows down the rotation of star. For close binaries in which synchronization of rotational and orbital period is expected, loss of rotational angular momentum occurs at the expense of orbital AM. As a result, the period decreases (consistent with the observations) i.e., the components spin up and approach one another to form a single rapid rotating star (see Stepien 1995; Skumanich 1972). As stated by Stepien (1995, 2006) contact binary stars are magnetically very active and it is generally accepted that they lose mass and Angular Momentum (AM) via magnetized wind. Moreover the separation of the components is relatively low. Therefore, one expects the magnetic field interactions between the two components to be intensified and consequently its effect on the AM loss to be enhanced, due to the formation of magnetic loops between the surface magnetic fields of the components (see Fig 2 & 3). This statement is consistent with the Bradstreet and Guinan (1994) that the magnetic torque produced by magnetic field in the wind depends on the strength of

magnetic field. But the details of this idea and its quantitive formulations and experimental verifications will remain a challenge for the future.
The particles in the stellar winds that leave the star radially, at the stellar surface, their tangential velocity components are equal to the rotational surface velocity of the star. (e.g., for the Sun this velocity $v_{surf} = 2km/s$). When the same particle are traveled to outer space, it is expected to slow down to much lower velocities at large distances (e.g. for the Sun particles at earth distance, $a = 1.5 \times 10^8 km$, $v_{surf} = \frac{R_\odot}{a} \times 2 \simeq 1 \times 10^{-2}$ provided the angular momentum (of the star is conserved). But in the case the Sun, the particles velocities measurements by spacecrafts (e.g. Helios) is in order of 1-10 km/s. That is $10^2 - 10^3$ times faster than the expected velocities, the cause is that charged particles travel along the twisted open field lines and not just radially outwards. Therefore the magnetic energy of the field per unit volume must be much larger than the particle's kinetic energy and thus its trajectory is dominated by direction of magnetic field lines rather than gravitional field (see Strassmier 2001) and references there in). If some of the stellar magnetic field lines are open and reconnect, with companion star, then the particles either may collide with the parties which were guided by the companion star along the field lines in similar fashion to the primary star, or may fall in to the atmosphere of the companion star. (see the Figs. 2 & 3). These charged particles would carry the angular momentum of the star with themselves.
With this picture in mind, as explained earlier, the magnetic field lines are bent due to rapid rotation of the star, their curvature cause a counteract force on the surrounding stellar plasma if we assume that magnetic poles are coinciding with the rotation poles, then the dissipated angular momentum is very small and braking is almost negligible. But if field is anchored at or near the equatorial plane then the braking would be strongest and therefore maximum angular momentum is removed. Observations indicate that magnetic braking must be very effective for the observable surface, is larger in the late F-K type stars.
Due to tidal interaction between the components in binary systems, the component stars rotate much faster (10-100 times) as compared to a single star, therefore one expects the magnetic fields production in the surfaces (subsurface) of the components in binaries containing one or both components as late (F-K) type stars to be much stronger as compared with a single star (Yuan & Quian, 2007, eq. 7) the magnetic force is sensitive to (a^{-4}), the central separation between the components, but face to face separation between two components is much smaller as compared to central separation (i.e., a). The two magnetic fields as pictured above are superposed (see Fig. 4)
There are some observational evidences to support the above picture. Lestrade (1996) detected radio emissions from intra - binary region of the stars UX Ari and 2 CrB, which he attributed to gyro synchrotron process associated with large scale magnetic fields. Siarkowski (1996) used X - ray light curves of RS CVn binary AR Lacertae to map spatial structure of its corona, and found regions of enhanced X - ray emission and extended structures that interconnect the two stars. Gunn et. al. (1999) presented radio interferometeric observations of Algol- type binary V 505 Sagittarii and modulated radio flux density levels with evidence of eclipses of the emission regions of both conjunctions of the binary. The form of the light curve obtained implied that the radio source involves at least some enhanced emission in the intra - binary active regions. Gunn et. al. (1994), made EUV radio observations of active RS CVn binary CF Tucanae, their observations indicated an, active intra - binary region and field interaction in active close binary (see Figs. 2 & 3)
Uchida and Sakurai (1985) discussed the formation of corona and origin of flares in RS CVn binaries having starspots, and interpreted in terms of reconnection of the magnetic flux tubes

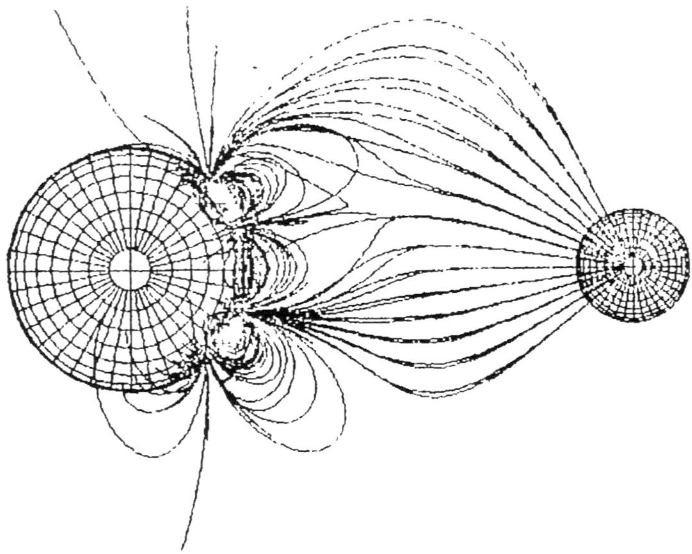

Fig. 2. Formation of large scale magnetic loops between two components of a binary star. (Taken from Uchida & Skurai 1985)

of the companion star, they found that the injection of the hot plasma in to the large scale Pole - to - spot connections was required to explain the extended corona with large emission measure they attributed this to the sweeping - pinch mechanism (see Uchida & Shibata, 1984)

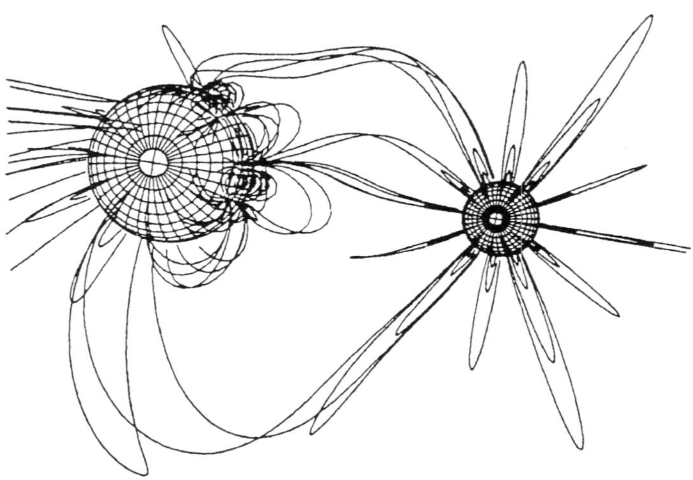

Fig. 3. Formation of large scale magnetic loops between two components of a binary star. (Taken from Uchida & Skurai 1985)

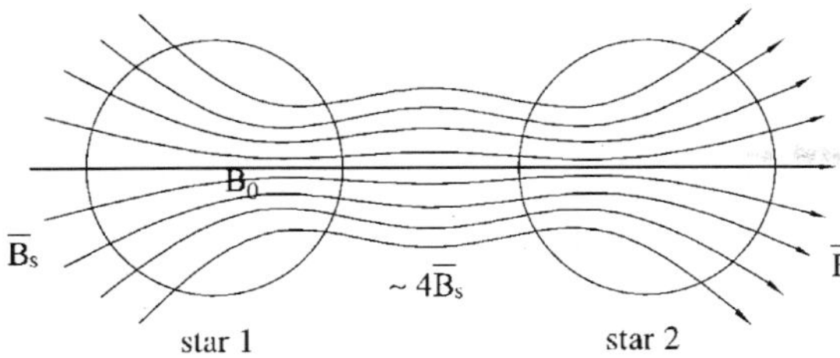

Fig. 4. Schematic sketch of magnetic field structure for a RS CVn binary star when both the components are magnetically active (Taken from Yuan & Quian 2007)

associated in the twisted magnetic flux tubes which emerge and reconnect with the flux tubes connecting poles and spots. Uchida & Shibata (1985) proposed a jet formation mechanism in which a packet of the toroidal component of magnetic field ($B\phi$) plays a role. Such a packet of toroidal field produced by the action of rapid rotation of the star in the region where $\beta = \frac{P_g}{P_m} \gg 1$. They showed the jet can be accelerated when toroidal component of the field emerges from the $\beta >> 1$ region to the $\beta \ll 1$ region. When this happens a progressive pinch is produced through the relaxation of the field and mass is ejected out by pinch near the magnetic axis and also by $\mathbf{J} \times \mathbf{B}$ force in the twisted field region surrounding the axis.

6. Conclusions

In Algol type binaries one outcome of conservative mass transfer through Roche lobe filling mechanism of late secondary component is circumbinary disk, which can significantly influence the orbital evolution and cause the orbit to shrink on a sufficiently long time scale. Rapid mass transfer in Algols with low mass ratio can also be accounted for by this scenario. The observed decrease in orbital periods of some Algol binaries (see Manzoori, 2007, 2008 manzoori & Ghozaliasl 2007) suggests OAM loss and magnetic braking mechanism, during the binary evolution.

One of the interesting evolutionary consequences of mass transfer discovered in recent years, is the formation of double contact binaries. It seems the concept of double contact binaries first was introduced by Wilson (1979), as type of close binary which occurs only for nonsynchronism, the binaries in which both components fill their limiting lobes, naming them as double contact may be misguiding, because they do not even have a single contact point. This can occur, when at least one of the components rotates faster than synchronously, so that its limiting lobe is smaller than the respective Roche lobe (and the limiting lobe would be bigger than Roche lobe if it rotates slower than synchronously). The situation can occur as a natural consequence of mass transfer, because transfer normally spins -up an accreting star by converting orbital to rotational angular momentum.

As stated earlier in section 1, one of the main characteristics of (semidetached, Algol- type) binaries is mass transfer due to Roche lobe filling of late secondary component. Such mass - exchange will spin -up the mass accreting star (most of whose mass is transferred material, by the end of rapid phase of mass transfer). But according to Wilson et al. (1985) this rapid rotation will be damped quickly, as soon as mass transfer is stopped. Among semidetached systems there are few members which show asynchronous rotation, see table 1 of the Wilson and Twigg (1980). Thus, near to or after the end of rapid phase of mass transfer, the mass losing star would normally be found to fill its Roche lobe (rotating synchronously), while the accreting star (or perhaps only its outer envelope) may have accommodated all of its capacity of the angular momentum which can hold, and so fill its limiting "rotational" lobe. According to Wilson and Caldwell (1978) the β- Lyrae and V356 Sgr and also U Cep and some of the Algol type systems are most likely candidates of double contact binaries. A main evolutionary consequence of their finding of double contact is that, since the accreting star after filling the limiting lobe will not be able to accommodate any more of the new high angular momentum material transferred by donor star, therefore such a system must find an alternate way for accommodation of the transferred matter, as first noted by Wilson and Caldwell (1978); this is a probable explanation for the thick circumstellar disks seen in β Lyr and V356 Sgr, and is perhaps related to numerous strange effects seen is U Cep (see Manzoori 2008). Notice that the existence of double contact systems introduce a symmetry in to the morphology of close binaries, in that stars may fill lobes exactly not only as a result off mass loss, but also as a result of mass gain.

The disk formed as mentioned, reduces a substantial fraction of the primary's light and therefore the star appears relatively under luminous, therefore it should make primary eclipse shallower than expected (see Manzoori 2006, 2008). For the same reason it would reduce also the irradiation of the secondary light by the primary, (reduces reflection effect), if the disk emits relatively little light of its own, the effect of its eclipse by the secondary star will be slight, but the eclipse of the secondary by the disk should be appreciable. Thus secondary eclipse should be wider than primary eclipse, and should also be deeper than it would be without an eclipse by the disk. One of the main conditions of the existence of double contact phase in the evolution of binary stars is nonsynchronous rotation of mass accreting component reaching the centrifugal limit. Naturally a star can not exceed its limiting rotational lobe because its equatorial matter would then be centrifugally unbound (see Wilson, 1983). The discovery of only very few members of this type of binaries, might be due to relatively brief time scale for binary to spend in double contact phase according Li et al. (2006).

According to Bradstreet and Guinan (1994) in close binary systems with period $P < 0.5d$, where tidal forces are strong enough to cause synchronization of the components, the role of spin -orbit coupling becomes important in evolution of the binary and angular mentum loss from the system through magnetic braking spin up with time, and orbital period decreases. Consequently magnetic activity of the stars increases. This process can cause a detached binary composed of late type components to become a short period contact W UMa - type binary. Webbink (1976, 1985) stated that in short period contact binaries with period $P < 0.5d$, after formation the loss of angular momentum through gravitational radiations play an important role in the evolution of the system and loss of angular momentum can ultimately lead to the coalescence of the binary system in to a single star. It is believed now the rapidly rotating F-K Com G - type gaint stars with intense H_{α} emissions are formed in this way. According to Stepien (1995; 2006) W UMa stars are magnetically very active and it is generally accepted that they lose mass and Angular Momentum (AM) via magnetized wind. Gazeas and Stepien (2008), analyzing over hundred cool contact binaries, found several correlations among the geometrical and physical parameters and they stated that "the coalescence of both components in to a single fast rotating star, is the final fate of contact binaries". This statement is not agreed with that of the Li et al. (2005). Li et al. in a series of three papers (2004a; 2004b; 2005) discussed the structure and evolution of low mass contact binaries, with and without spin and orbital angular momentum loss, and taking in to account effects of energy transfer, with negligible gravitational radiation. They presented an ultimate model of W UMa systems exhibiting cyclic evolution around a state of marginal contact, in time scales of about 6×10^6 yr, without loss of contact, if spin angular momentum of both components are included; 9×10^6 yr, if spin angular momentum of both components are neglected. According to them, the ratio of spin angular moment of both components to the orbital angular momentum of the system becomes larger as the evolution proceeds, so that the system would then coalesce in to a rapidly rotating star, when the spin angular momenta of both components become more than one third of the orbital angular momentum of the system. Hence the cyclic evolution mentioned for the contact binary systems can not last forever. It only lasts about 7×10^9 yr, in good agreement with the observational result of old cluster NGC188; containing at least four W UMa type stars.

7. References

Demircan, O. Eker, Z., Karatas, y., & Bilir, S., 2006, MNRAS, 366, 1511.

Bradstreet, D. H.; Guinan, E. F, 1994 ,ASPC, 56, 228

Gazeas, K., & Stepien, K., 2008, MNRAS, 390, 1577.

Hall, D. S., 1989, SSRV, 50, 219.

Hilditch, R. W. An introduction to close binary stars / Cambridge University Press, Cambridge, X+381 pp., 2001,.

Huang, Su. Shu., 1963, ApJ, 138, 417.

Huang, H. Q., Song,H. F., Bi, S.L., 2007, Chin J., Astron & Astrophs 2007, vol 7, No. 2, 235.

Jeans J. H. 1924, M. N., 85, 2.

Jeans J. H. 1925, ibid, P 912.

Lestrade, J. F, 1996, IAUS.176, 173

Li, L., Han, Z., Zhang, F., 2004a, MNRAS ,351 137

Li, L., Han, Z., Zhang, F., 2004b, MNRAS ,355, 1383

Li, L., Han, Z., Zhang, F., 2005, MNRAS, 360, 272.

Manzoori, D., & Ghozaliasl, Gh., 2007, AJ, 133, 1302.

Manzoori, D., 2007b, PASA, 24, 6168.

Manzoori, D., APSS, 2008, 318, 57 .

Olson, E.C., 1981, ApJ, 250, 704.

Parker, E. N., 1955, ApJ, 122, 293.

Richards, M. T., 1992, ApJ, 387, 329.

Richards, M. T., 1993, ApJS, 86, 255.

Richards, M.T. & Albright, G.E., 1993, ApJS, 88, 199.

Skumanich, A., 1972, ApJ, 171, 565.

Stepien, K., 1995, MNRAS, 274, 1019.

Stepien, K., 2006, MNRAS, 274, 1019.

Strassmeier, K.G. (2001) Lecture Notes in Physics, vol 563, 48.

Thomas H.C., (1977). Ann. rev. Astron. Astrophs., 15, 127, 51.

Uchida, Y., Shibata, K. (1984). Pub. Astron. Soc. Japan, 36, 118.

Uchida, Y., Sakurai, T. (1985). IAUS, 107,281.

Webbink, R. E., 1976, ApJ, 311, 829.

Webbink, R. E., 1985, In Interacting Binary Stars, Ed. R.E., Pringle (Cambridge Univ Press Cmbridge)

Wilson, R. E., 1979, ApJ, 234, 1054.

Wilson, R. E., APJ, 1983, 234, 1054

Wilson, R. E., & Caldwell, C. N., 1978, Apj, 221, 917.

Wilson, R. E., & Twigg, L. W., 1980, IAUS, 88, 263.

Wood, F. B., 1983, APJ, 234, 1054
Yuan, J., Qian, S. 2007, ApJ, 669, L93-L95.

A Mass Transfer Study with Electrolytic Gas Production

Eudésio O. Vilar[1], Eliane B.Cavalcanti[2] and Izabelle L.T. Albuquerque[3]
[1,3]*Federal University of Campina Grande-PB*
[2]*Tiradentes University/ITP-SE*
Brazil

1. Introduction

In general, tall vertical electrolyzers are used industrially to produce only gases like chlorine, hydrogen and oxygen, or gases and products such as soda and chlorine. Moreover, these electrolyzers usually have a very short cathode-anode distance and often operate under forced convection. For many electrochemical processes mass transfer in electrolytic cells, in particular to electrodes, must be optimized to operate economically. Many electrochemical reactions involve a gaseous component and a great deal of research has been devoted to the study of the specific features of these reactions. Three main areas have been investigated such as: the bubble formation (Chirkov & Psenichnikov, 1986), the mass transfer and hydrodynamic instabilities at gas-evolving surfaces (Kreysa & Kuhn, 1985), and the behavior of gas in porous electrodes in fuel cells (White & Twardoch, 1988). Many works were developed up to the present about gas-evolving electrodes (St-Pierre & Wragg, 1993a, 1993b; Vogt, 1979, 1984a, 1984b, 1984c, 1989a, 1989b, 1992, 1994, 1997; Czarnetzki & Janssen, 1989; Boissonneau & Byrne, 2000; Ellis et al., 1992; Janssen et al. 1984; Lastochkin & Favelukis, 1998; Wongsuchoto et al., 2002; Buwa & Ranade, 2002; Gabrielli et al., 2002; Correia & Machado, 1998; Lasia, 1998; Iwasaki et al., 1998; Fahidy & Abdo, 1982; Lasia, 1998, 1997; Barber et al., 1998; Eigeldinger & Vogt, 2000; Solheim et al., 1989; Elsner & Coeuret, 1985; Dykstra et al., 1989; Khun & Kreysa, 1989; Lubetkin, 1989; Martin & Wragg, 1989; Lantelme & Alexopoulos, 1989; Gijsbers & Janssen, 1989; Chen, 2001; Lasia & Rami, 1990; Kienzlen et al., 1994; Saleh, 1999; Janssen, 1978) but, few data on mass transfer with different cathode geometries under flow-by or flow-through electrolyte conditions with gas-evolving have been studied (Fouad & Sedahmed, 1974; Rousar et al., 1975;; Janssen & Barendrecht, 1979, Mohanta & Fahidy, 1977; Sedahmed, 1978; Sedahmed & Shemilt, 1981; Elsner & Marchiano, 1982; Albuquerque et. al., 2009). This chapter does not intend to explore in detail the mechanism during the bubbles formed electrolytically but only show an comparative study about the effect of hydrodynamic condition over mass transfer gas-electrodes for two cathodes geometries, during the hydrogen production at chlor-alkali cell by diaphragm process in laboratory scale.

2. Mass transfer

In general, it is necessary to consider three basic mechanism to the mass transfer in electrochemical systems, : migration, convection, diffusion and reaction.

Migration is the movement of charged species through the electrolyte due to a potential gradient; the current of electrons through the external circuit must be balanced by the passage of ions through the solution between the electrodes (both cations to the cathode and anions to the anode). It is, however, not necessarily an important form of mass transport for the electroactive species, even if it is charged. The forces leading to migration are purely electrostatic and, hence, do not discriminate between types of ions. As a result, if the electrolysis is carried out with a large excess of an inert electrolyte in the solution, this carries most of the charge, and little of the electroactive species Ox (oxidized specie) is transported by migration, i.e. the transport number (Bockris & Reddy,1977) O_x is low.
Convection is the movement of a species due to fluid dynamic forces. In practice, these forces can be induced by stirring or agitating the electrolyte solution or by flowing it through the electrochemical cell. Sometimes the electrode can be moved (e.g., rotating disk electrodes). When such forms of forced convection are present, they are normally the predominant mode of mass transport. By the other hand, natural convection can arises from small differences in density, temperature or gases caused by the chemical change at the electrode surface. The treatment of mass transport, highlights the differences between laboratory experiments and industrial-scale electrolysers. As is pointed out by (Pletcher & Walsh, 1993), the need in an industrial cell is only to promote the desired effect within technical and economic restraints and this permits the use of a much wider range of mass transport conditions. In particular, a diverse range of electrode-electrolyte geometry and relative movement are possible.
Diffusion and reaction. Diffusion is the movement of a species down a concentration gradient and it occurs whenever there is an electrical charge exchange at a surface. An electrode reaction (generally fast reaction) converts starting material to product, e.g.;

$$O_x + e^- \rightarrow R_{ed} \tag{1}$$

where O_x and R_{ed} are the oxidized and reduced species respectively, hence close to the electrode surface there is a (concentration) boundary layer (up to 0.01mm thick) in which the concentration of Ox is lower at the surface than in the bulk solution while the opposite is the case for R_{ed} and, hence, O_x will diffuse towards and R_{ed} away from the electrode.
Fundamental mass transport studies in industrial electrolytic cells are dependent of the fluid dynamic or by the inertial and viscous forces. This ratio is given by the well-known Reynolds number *Re,* calculated from the Equation:

$$Re = \frac{\rho L \nu}{\mu} = \frac{L \nu}{\upsilon} \tag{2}$$

where p is the density of the solution, μ its dynamic viscosity, υ its kinematic viscosity, ν a mean flow velocity and L a characteristic length (for example, the length of a flat plate electrode). At higher Reynolds number, the viscous damping is no longer predominant and turbulence increase, by the other hand, any obstacles to fluid flow, or roughness in the electrolytic cell will can cause the commencement of turbulence or micro-turbulence at lower Reynolds number. In a particularly case of electrolytic cell with gas production, the Reynolds number can be obtained by the following Equation:

$$Re^* = \frac{V_g . d}{A . \upsilon} \tag{3}$$

where d is the bubble detachment diameter (m), A the electrode area (m^2) and V_g is the volumetric flow rate of gas bubbles ($m^3.s^{-1}$), defined as;

$$V_g = RTj/nFP \tag{4}$$

where R (8.314 $JK^{-1}mol^{-1}$) is the gas constant, T the absolute temperature (K), j the current density of electrolysis ($A.m^{-2}$), n the estequiometric number of electrons, F the Faraday constant (96,485 $C.mol^{-1}$) and P the pressure (atm).
Natural or forced turbulence in electrolytic cells is usually advantageous since the eddies both increase mass transport of the electroactive species to the electrode surface and promote the exchange of species between the bulk solution and the boundary layer, minimizing local pH and other concentration changes due to the electrode reaction. It is not uncommon to introduce insulating nets, bars or other structural features into the cell to act as turbulence promoters. By the other hand, the morphology of the electrode surface can act as turbulence auto-promoter (e.g. mesh, reticulated metal, particulate bed, fibrous material).

2.1 Sherwood number

The Sherwood number is a measure of the rate of mass transfer, k_d , which is usually calculated in electrolytic cells from the limiting current density j_L for several cells and electrodes configurations under specific hydrodynamic conditions, i.e., the potential of the electrode is held at a value where all the electroactive species reaching the surface undergo the electrode reaction. The Sherwood number can be obtained using the relationship:

$$Sh = \frac{k_d L}{D} = \frac{j_L L}{nFC^{\infty} D} = f \left| \begin{array}{l} \text{laminar or turbulent flow} \\ \text{fluid properties} \\ \text{temperature} \\ \text{cell configuration} \\ \text{structure and active area of the electrode} \end{array} \right. \tag{5}$$

where L (m) is a characteristic dimension of the system, n the number of electrons involved in reaction, F the Faraday constant, C^{∞} the bulk concentration ($mol.m^{-3}$) of the specie electrochemically active and D ($m^2.s^{-1}$) its diffusivity. In Equation (5) the hydrodynamic condition (laminar or turbulent flow) can be evaluated by the choice of the Equation (2) or (3) and the Schmidt number by Equation (6), where it represents the relationship between the resistivity of momentum and mass diffusivities.:

$$Sc = \frac{\mu}{\rho D} = \frac{\upsilon}{D} \tag{6}$$

In general, mass transport in electrolytic cells with flow may be expressed in terms of the following expression:

$$Sh = kRe^{a}Sc^{b} \tag{7}$$

In general, to Newtonian fluids, it is assumed 0.333 for the constant b. The constants "k" and "a" many be obtained from the logarithmic linearization of equation (7). The current limiting density, j_L is generally determined from the choice of a extremely fast reaction, for example, the electroreduction of the ferricyanide-ion in alkaline solution providing a diffusional control under various flow rate conditions. A more detailed approach on obtaining experimental current density limit may be found in specialized publications (Coeuret & Storck, 1984; Walsh, 1993; Bockris & Reddy, 1977).

2.2 The gas evolution mechanism

Gas evolution occurs on an electrode through several phenomena. The gas produced by electrochemical reactions on the electrode dissolves in the electrolyte and is transported by diffusion (boundary layer concentration) and convection towards the bulk of the solution. The mechanism of growth and detachment of bubbles from electrode surface can develops in two or three steps depending on the size and its configuration geometry- for example, perforated plate, meshes or expanded electrodes. As presented by (Gabrielli et al.,1989), the first correspond the transient step or the bubble´s radius variation with time and its depend of the electrolyte density. During the bubble growth the second step can be limited by diffusion of the dissolved molecular gas in the solution or by the kinetics of the production of the gas. When the bubble is larger than the electrode, it is assumed that the gas produced in molecular form is all transformed to the gaseous from which increases the bubble size. The last stage of the bubble evolution, i.e. its detachment from the surface, occurs when the balance between the forces which tend to maintain it on the electrode and the forces which tend to release it is broken. These various forces include the weight of the bubble, the buoyancy, the superficial tension, the pressure, the inertia and the electrostatic forces.

2.3 Electrolytic gas production from chlorine-alkali cell

The Figure 1 has shown the well-known electrolytic diaphragm process to produce chlorine and soda products (Almeida Filho et al., 2010; Abdel-Aal & Hussein, 1993; Abdel-Aal et al., 1993). The saturated aqueous sodium chloride (saturated brine) feeds the anodic

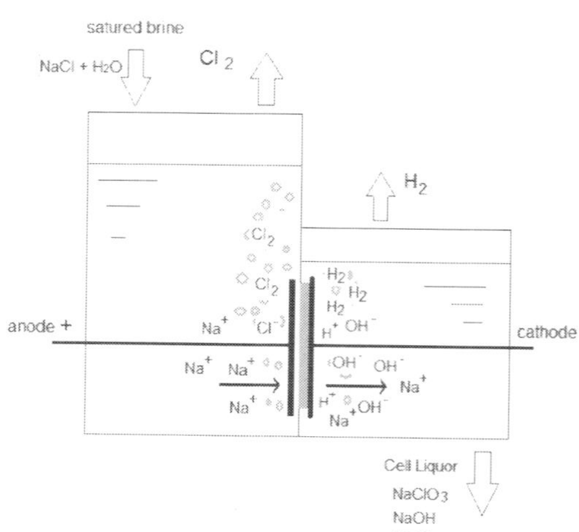

Fig. 1. Basic schematic of an electrolytic cell to produce chlorine and soda by the diaphragm process (Almeida Filho et.al, 2010)

compartment. The chlorine gas produced by the anodic reaction leaves the semi-cell, while the brine diffuses to the cathode compartment through the diaphragm due to the hydrostatic pressure drop between the two compartments. Hydrogen and hydroxyl ions are produced in the cathode compartment, which together with the sodium ions (Na^+) present in brine (anodic compartment) form sodium hydroxide (NaOH) at the same time that chlorine and hydrogen gas flow outside the cell. The part of the NaCl that did not react in the anodic compartment to produce chlorine gas diffuses into the cathode compartment through the diaphragm, joining the NaOH to form an aqueous solution of NaCl and NaOH called cell liquor.

The main reactions that occur in the process are as follows:

$$2NaCl \rightarrow 2Na^+ + Cl_2 + 2e^- \quad \text{(anode)} \tag{8}$$

$$2H_2O + 2e^- \rightarrow H_2 + 2OH^- \quad \text{(cathode)} \tag{9}$$

$$2Na^+ + 2OH^- \rightarrow 2NaOH + Cl_2 + H_2 \quad \text{(global reaction)} \tag{10}$$

In the electrolytic production of chlorine-soda, high current density produces bubbles that can cover some parts of the electrode surfaces, causing an undesirable decrease in mass transfer. These limitations can be minimized through proper tuning of the cathode geometry and the electrolytic cell configuration (St-Pierre & Wragg, 1993). In industrial electrolytic operations that involve gas production, perforated plate or expanded electrodes are traditionally used to increase the reactive area per unit volume of the cell. However, the accumulation of generated bubbles on the surface of the cathode can block the electrochemically active area. This reduces efficiency by increasing ohmic drops in the layer of electrolyte adjacent to the electrode surface. Thus, the increase in the volume of bubbles adsorbed per unit area causes a decrease in mass transfer at the electrode surface (Vogt, 1984; Albuquerque, 2006, 2009). For these reasons, there has been increased interest in finding electrode geometries that promote the detachment of gas bubbles in order to increase mass transfer and ultimately efficiency.

An effective method for increasing the rate of mass transfer is to induce electrolyte turbulence near the surface to prevent the accumulation of bubbles. The behavior of this type of system was studied with expanded metal electrodes in which the electrochemical reaction on the electrode surface is controlled by diffusion and detachment of gas bubbles. (Elsner,1984) concluded that the mechanism that drives the resulting increase in mass transfer varies based on the type and orientation of the expanded metal electrode geometry and the volumetric flow direction of the electrolyte. In general, we can assume that forced convection and detachment of bubbles will improve mass transfer when the geometry of the electrode does not inhibit the release of the bubbles generated electrochemically. A strong correlation between the mass transfer coefficient and gas production has been shown in mass transfer studies. (Fouad & Sedahmed, 1973) studied this relationship for electrodes oriented vertically and horizontally, concluding that the average mass transfer coefficient is greater for horizontal electrodes. (Nishiki et al., 1987) found that generated gas bubbles decrease the conductivity between electrodes by increasing the resistance of the solution. This affected overall cell performance by increasing the potential (energy consumption of the cell). It is evident that appropriate choice of electrode material and geometry may help to mitigate such problems. (Hine et al., 1984) studied perforated plate electrodes, concluding that variation in electrolyte resistance and overvoltage is a function of both the porosity and distance between the electrode/diaphragm interface. The porosity appears to be an important parameter for reducing the cell potential. (Jorne & Louvar, 1980) and (Jansen et al.,1984) concluded that expanded metal electrodes with a three-dimensional texture can help to prevent generated gas from accumulating on the electrode surface, thereby decreasing the ohmic drop.

3. Un example of experimental study of mass transfer with gas production

The relevance and main contribution of this study was to compare and analyze the influence of the flow perpendicular to two geometries of cathode used in electrochemical industry, on the mass transfer associated with the electrolytic production of hydrogen. The electrolytic cell used in this study is a prototype for laboratory-scale production of chlorine-soda via an electrolytic diaphragm process (see Fig. 1). The reactor has two compartments of plexiglas with 1.45 L and 0.316 L to the anode and cathode electrodes respectively, separated by an asbestos-coated diaphragm (deposited on the cathode) like shows the Fig. 2. The Fig. 3 shows the two geometric shapes to the cathode - perforated plate and mesh geometry both with 7.0 x 8.0 cm made from commercial SAE 1020 alloy. The reduction of potassium ferricyanide in alkaline medium was used for the mass transfer study with NaOH as the electrolyte support. A PAR (Princeton Applied Research)-VMP3 potentiostat, was utilized for this purpose. Table 3 lists the properties of the electrolyte solution to 27°C.

Composition	$K_3Fe(CN)_6$ = 0.005 N $K_4Fe(CN)_6$ = 0.05 N NaOH = 1.0 N
ν ($m^2.\ s^{-1}$)	0.9648×10^{-6}
D^a ($m^2.\ s^{-1}$)	6.0×10^{-10}

[a] The diffusion coefficient was calculated from the Stokes-Einstein equation: $D\mu/T = 2{,}49 \times 10^{-15}$ [kg.m.s^{-2}.K^{-1}]

Table 3. Composition and properties of the electrolyte solution.

The reduction of potassium ferricyanide in alkaline medium was used for the mass transfer study with NaOH as the electrolyte support. A PAR (Princeton Applied Research)-VMP3 potentiostat, was utilized for this purpose. Table 3 lists the properties of the electrolyte solution to 27°C.

The experimental procedure was performed at the following conditions: volumetric flow rate between 0.03 and 0.13×10^{-3} L/s. A procedure found in the literature (Elsner,1984) was used to determine the average mass transfer coefficient with gas production. This procedure consisted of measuring the concentration variation of the reduced electroactive species ($Fe(CN)_6^{3-}$) with respect to time. The electro-reduction of ferricyanide ions in alkaline solution occurs under diffusional control. The electrochemically generated current intensity from controlled diffusion in the presence of hydrogen bubbles can then be determined from the following equation:

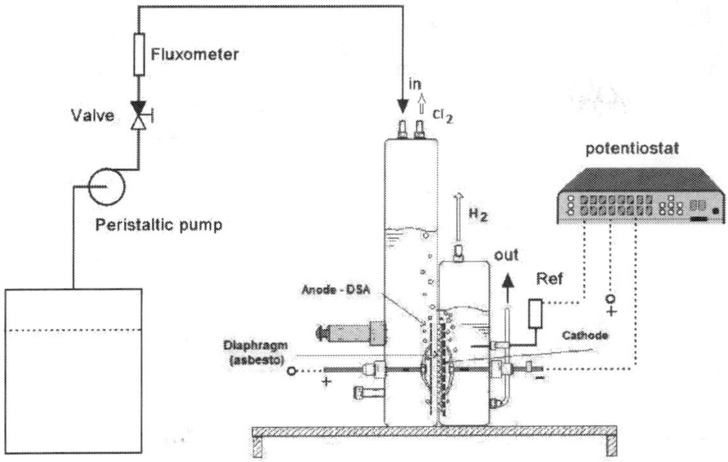

Fig. 2. Experimental set-up. Ref-reference electrode (Ni), out – soda produced (Albuquerque et. Al, 2009)

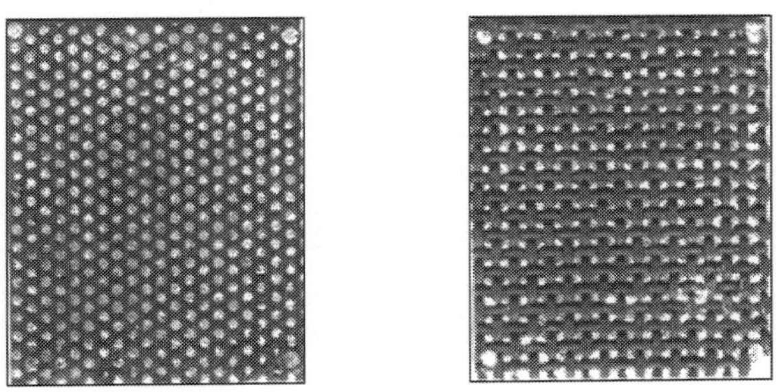

(a) (b)

Fig. 3. Cathode geometries (a) perforated plate and (b) mesh geometry.

$$j_d^g = \frac{nFV_c\Delta C}{t} \tag{11}$$

Where j_d^g (A) is the current intensity from diffusion in the presence of bubbles produced electrochemically, ΔC (mol.m^{-3}) is the gradient concentration (ferricyanide ion concentration before and after electrolysis), V_c is the volume of the cathode compartment (m^3), n is the number of electrons involved in the and t is the time of electrolysis (s). From the electrolytic current intensity, the average mass transfer coefficient was determined from the following expression:

$$\overline{k_d^*} = \frac{j_d^g}{nFA\overline{C}} \tag{12}$$

where $\overline{k_d^*}$ is the combined average mass transfer rate ($m.s^{-1}$), *A* the active area of the cathode (m^2) and $\bar{C}$ is the average concentration of ferricyanide ions during electrolysis ($mol.m^{-3}$). The ferricyanide concentration was determined by amperometric titration (Vilar, 1996) using a cobalt chloride (0.0339M) solution like agent in a three-electrode cell setup consisting of a Hg/HgO reference electrode, a working rotatable platinum electrode (1,000.0 rpm, 2.0 mm diameter) controlled by CTV101 speed control unit, both - Radiometer analytical and 1x1cm sheet of platinum as counter electrode. The experimental setup was controlled by potentiodynamic technique using a PAR (Princeton Applied Research)-VMP3 Potentiostat.

3.1 Modeling

The following correlation was determined to best represent the chlorine-soda electrochemical reactor used in the present work (Zlokarnik, 2002):

$$\overline{Sh} = aRe^bSc^{1/3} = \frac{\overline{k_d^*}L}{D} = a\left(\frac{\upsilon L}{\nu}\right)^b\left(\frac{\nu}{D}\right)^{1/3} \tag{13}$$

where *L* (the characteristic dimension) is given by the following relationship between the porosity of the electrode ε and the specific area A_s (m^{-1}):

$$L = \frac{\varepsilon}{A_s} \tag{14}$$

and

$$A_s = \frac{A_g}{V_s} \tag{15}$$

where ε is the porosity (0.51 and 0.75 to perforated plate and mesh geometry respectively) Ag is the geometric area ($57.0x10^{-4}m^2$ both), and Vs the volume of solid electrode.

3.2 Results and discussion

Figure 4 shows the effect of the percolation rate of electrolyte through the diaphragm on the average mass transfer coefficient. The percolation rate ($m.s^{-1}$) was calculated as the ratio between the feed flow and the open cathode area (0.00287 m^2 for perforated plate and 0.0042 m^2 for mesh geometry).

It can be observed in Figure 4 that for the perforated plate geometry, the combined average mass transfer coefficient decreases with increasing percolation rate of electrolyte. The opposite behavior is observed for the mesh geometry. There are also two distinct regions in both curves, highlighted by the inflection points. This is characteristic of areas of hydrodynamic transition phenomena, probably due to laminar flow with rippling. Figure 5 helps to describe this behavior. The geometric influences are illustrated by the vector velocity of percolation (black arrows) and the direction of micro-convection (white arrows) caused by the rise of the bubbles. For the perforated plate geometry, Figure 5 (A) and (B) illustrates the supposition that the layer of micro-convection caused by the rise bubbles is pushed away from electrode surface when the cross-percolation velocity of electrolyte is

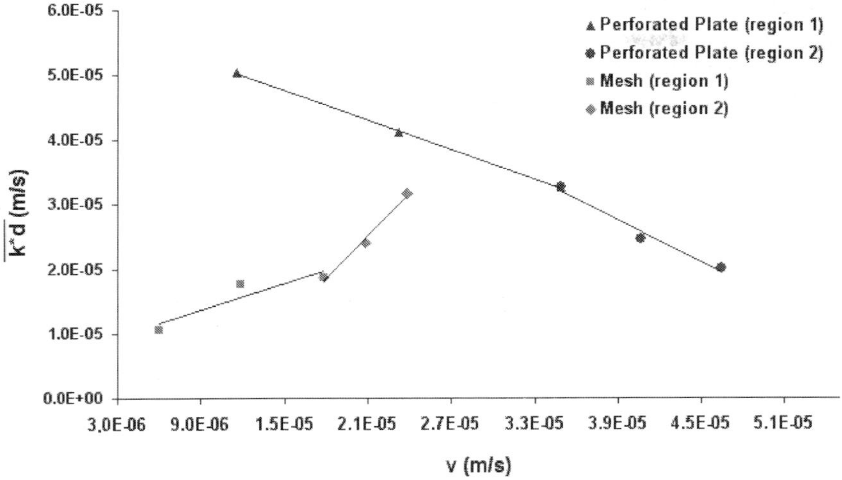

Fig. 4. Combined average mass transfer coefficient with respect to percolation rate of electrolyte.

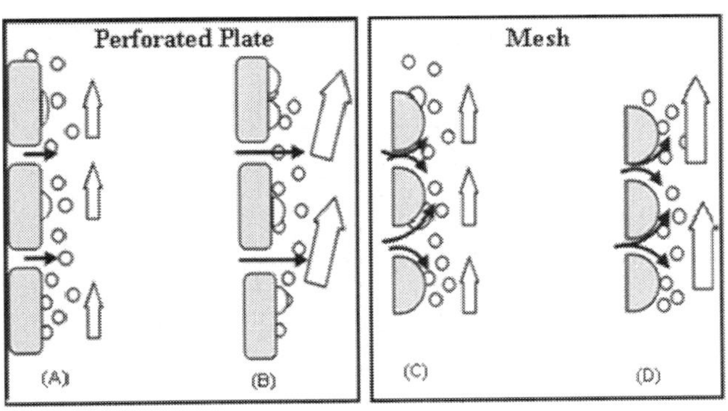

Fig. 5. Hypothesis of the situation between the change of cross-velocity percolation vectors with the rise of bubbles for; -Perforated plate, (A) -low and (B) - high velocities, -Mesh geometry, (C) -low and (D) - high velocities.

increased. This phenomena can hinder the detachment of the bubbles adhered to the cathode surface, causing a decrease in the rate of mass transfer with increasing percolation velocity. For the mesh geometry, a contrary phenomenon is illustrated. The Figure 5 (C) and (D) illustrates the same vector representation, but in this case the curved surface promotes increasing the velocity in the Prandtl hydrodynamic layer (Coeuret & Storck, 1984; Walsh, 1993), which enhances the detachment of gas bubbles. The increasing turbulence facilitates the detachment of the bubbles and the micro-convective movement reduces the Nernst boundary layer, and thereby increases the combined average mass transfer coefficient.

The Figure 5 can be explained by the supposition that turbulence can be more pronounced at the surface of the mesh electrodes than the surface of the perforated plate electrodes. For the mesh geometry, this mechanism is more significant at high percolation rates (see region 2 of Figure 4). Furthermore, this result indicates that for low percolation rates, the turbulence caused by micro-convection is not strong enough to detach the bubbles trapped in the mesh holes. This is probably due to greater bubble surface adhesion in this geometry.

With respect to dimensionless correlation, the constants a and b were determined from logarithm function applied to Equation (13). The results are shown in Figure 6 and the Table 4 list all the correlations and Reynolds numbers domains studied in this study.

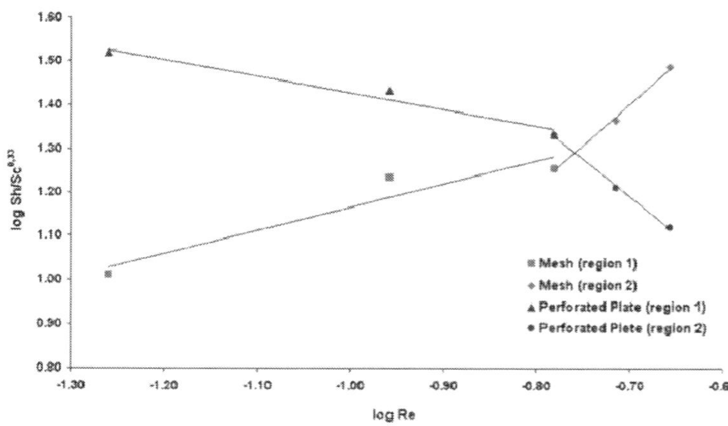

Fig. 6. The relationship between log (Sh/Sc$^{1/3}$) and log Re for mesh and perforated plate cathode geometries.

Region	Mesh	Perforated Plate	Reynolds
1	$\overline{Sh} = 11.01\,\mathrm{Re}^{0.38}.Sc^{1/3}$	$\overline{Sh} = 49.11\,\mathrm{Re}^{0.53}.Sc^{1/3}$	0.055<Re<0.165
2	$\overline{Sh} = 486.41\,\mathrm{Re}^{1.84}.Sc^{1/3}$	$\overline{Sh} = 1.01\,\mathrm{Re}^{-1.70}.Sc^{1/3}$	0.165<Re<0.220

Table 4. Empirical correlations for both geometries and Reynolds number domains studied.

These correlations were compared with those found in the literature. (Stephan & Vogt, 1974) proposed a model expressed by Equation (17), which correlates the mass transfer in various systems with gas evolution. This model was evaluated for 32 experiments, as shown in Figure 7.

$$Sh = \frac{\overline{k_d^*}.d}{D} = \frac{3.385}{C_\phi^{0.33}}\left(Re^*Sc\right)^{0.487}(1-\theta)^{0.5} \tag{17}$$

where d is the bubble detachment diameter (d = 40 μm for bubbles of hydrogen in alkaline solution), θ is the fraction of area covered (θ = 0.2 for semi-spherical bubbles and 0.3 for spherical bubbles) and C_ϕ is the sphere diameter (C_ϕ= 8 for bubble and 4 semi-spherical bubbles). The Reynolds number Re^* was determined by Equations (3) and (4) and the combined average mass transfer rate, k_d^* by Equation (12). The results of the present study were compared with the experimental data compiled by (Stephan & Vogt, 1974) as shows by the Figure 7. These data were obtained from acidic or alkaline solutions using various electrode materials such as platinum, copper and graphite. The data are valid for the following domain: 0° C < T < 80°C; 3 $A.m^{-2}$ < j <10^5 $A.m^{-2}$; 160.00 < Sc < 23,000.00 and $3.0\ 10^{-6}$ < Re^* < $9.0x10^{-1}$

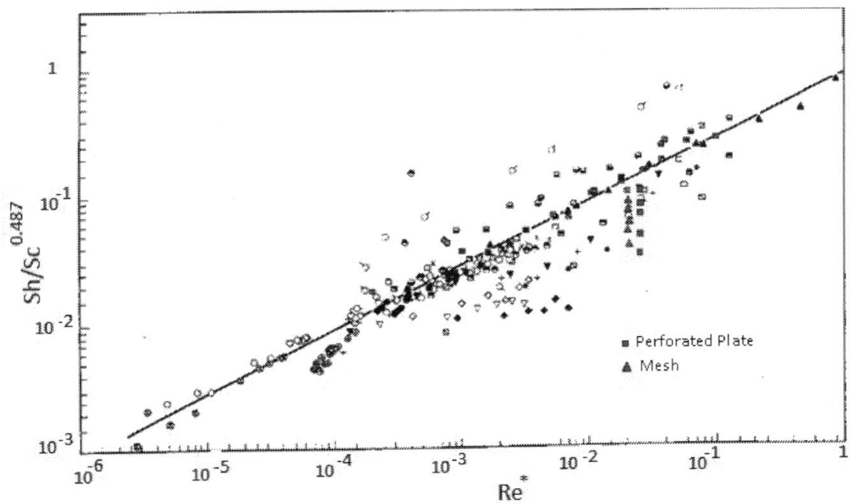

Fig. 7. Comparison of the experimental data of mass transfer with gas production (Stephan & Vogt, 1974) and the experimental data from this study.

According to Equations (3) and (4), the value of the Reynolds number is related to both the bubble velocity and the current density. For the purpose of comparison, only current density was used in this work so there is only one value of the Reynolds number.

4. Conclusions

It was shown that the mechanism controlling the average combined mass transfer coefficient during hydrogen production in electrochemical processes is dependent on the electrode geometry. The perforated plate geometry with deposited asbestos showed a slight advantage, compared with the mesh geometry, due to the detachment of bubbles from the active surface at low percolation velocities. Furthermore, increasing the percolation velocity resulted in a decrease of the average mass transfer coefficient, due to displacement of the micro-convective layer away from the electrode surface. For the mesh geometry, increasing the percolation velocity leads to an increase in the average mass transfer due to combined micro-convective effects. Specifically, rising bubbles associated with increased flow velocity over the curved wire surface, contribute to the displacement of bubbles blocked by adhesion. Finally for the chlorine-soda diaphragm process, a particularly operational industrial condition utilizing a percolation rate between $5.32\ 10^{-6}$ and $6.16\ 10^{-6}\ m.s^{-1}$, the present study showed that the perforated plate geometry is plus advantageous.
To improve the electrochemical cells with electrolytic gas production it is very important for the mass transfer researches with new electrodes materials and geometries for cathodes and/or anodes. Thus it may be possible to achieve low energy consumption in a high efficiency process and low residues production.

5. References

Abdel-Aal, H.K & Hussein, I.A. (1993), Parametric study for saline water electrolysis: Part I-Hydrogen production, *International Journal of Hydrogen Energy*,18,6,pp.485-489.

Abdel-Aal, H.K; Sultant, S.M. & Hussein, I.A. (1993), Parametric study for saline water electrolysis:Part II-Chlorine evolution, selectivity and determination, *International Journal of Hydrogen Energy*,18,7,pp.545-551.

Albuquerque, I. L. T. (2006); Influence of cathode geometry on the mass transfer and cathode potential during the water electrolysis in alkaline environment, MsC, Federal University of Campina Grande-PB, Brazil.

Albuquerque, I.L.T.; Cavalcanti, E.B. & Vilar, E.O. (2009), Mass transfer study of electrochemical processes with gas production, *Chemical Engineering and Process: Process Intensification*,48,pp.1432-1436.

Almeida Filho, E.M.; Vilar, E.O. & Feitoza, A.C.O. (2010), Phisical-chemical characterization and statistical modeling applied in a chlor-alkali diaphragm-cell process, *Chemical Engineering and Design*, article In press, doi: 10.1016/j.cherd.2010.08.007.

Barber, J.; Morin, S. & Conway,B.E. (1998), Specificity of the kinetics of H_2 evolution to the structure of single-crystal Pt surfaces, and the relation between opd and upd H, *Journal of Electroanalytical Chemistry*, 446, pp.125-138.

Bockris, J.O'M. & Reddy, A.K.N. (1977), *Modern Electrochemistry*, Vol.1, Plenum Press/Rosetta Edition, New York.

Boissonneau, P.& Byrne P. (2000) An experimental investigation of bubble - induced free convection in a small electrochemical cell, *Journal of Applied Electrochemistry*, 30, pp. 767-775.

Buwa, V. V. & Ranade,V. V. (2002), Dynamics of gas–liquid &ow in a rectangular bubble column: experiments and single/multi-group CFD simulations, *Chemical Engineerign Science*, 57,pp.4715-4736.

Chen, Wein-X.(2001), Kinetics of hydrogen evolution reaction on hydrogen storage alloy electrode in alkaline solution and effects of surface modification on the electrocatalytic activity for hydrogen evolution reaction, *International journal of Hydrogen Energy*, 26, pp.603-608.

Chirkov, Yu.G. & Psenichnikov, A.G. (1986). 37th Meeting of the International Society of Electrochemistry, *Soviet Electrochem.*,21,114,p.283, August (1986), USSR, Vilnius

Coeuret, F. & Storck, A. (1984) *Elements de genie electrochimique*, Tec. & Doc. Lavoisier, Paris, France.

Correia, A.N. & Machado, A.S. (1998), Hydrogen evolution on electrodeposited Ni and Hg ultramicroelectrodes, *Electrochimica ActaI*,43,3-4, pp. 367-373.

Czanetzki, L.R. & Janssen, L.J.J. (1989), Electrode current distribution in a hypochlorite cell, *Journal of Applied Electrochemistry*,19,pp.630-636.

Dykstra, P.A.; Kelsall, G.H.; Liu, X. & Tseung, A.C.C. (1989), New electrodes for oxygen evolution in acidic solution, *Journal of Applied Electrochemistry*, 19,pp.697-702.

Eigeldinger, J. & Vogt, H. (2000), The bubble coverage of gas-evolving electrodes in a flowing electrolyte, *Electrochimica Acta*, 45,pp.4449-4456.

Ellis, K.A.; Fahidy, T.Z.& Pritzker, M.D. (1992) Application of computer vision to bubble detection at a gas-evolving electrode, *Chemical Engineerig Science*, 47, 13/14, pp.3623-3630.

Elsner, C. & Marchiano, S.L. (1982) The effect of electrolytically formed gas bubbles on ionic mass transfer at a plane vertical electrode, *Journal of Applied Electrochemistry*, 12, pp. 735-742.

Elsner, C.; (1984) Transfert de matière et distribution du potentiel et du courant sur des électrodes de métal deployé, en présence de bulles électro-engendrées, Cap. IV, Dr. Thesis, ENSCR, Université de Rennes I, France.

Elsner, C. & Coeuret, F., (1985), Potential distribution along gas evolving electrodes, *Journal of Applied Electrochemistry*,15,4,pp.567-574

Fahidy, T. Z. & Abdo, S. E. (1982), On the induction time of the electrolytic evolution of hydrogen bubbles, *Electrochimica Acta*, 27,10,pp.1521-1523.

Fouad, M.G. & Sedahmed, G.H. (1973), Mass transfer at horizontal gas-evolving electrodes, *Electrochimica Acta*, 18, pp. 55-58.

Fouad, M.G. & Sedahmed, G.H. (1974), Mass transfer at cylindrical gas evolving electrodes, *Electrochimica Acta*, 19, p.861

Gabrielli, C.; Huet, F.& Nogueira, R.P. (2002), Electrochemical impedance of H_2-evolving Pt electrode under bubbleinduced and forced convections in alkaline solutions, *Electrochimica Acta*, 47, pp.2043-2048.

Gabrielli, C; Macias, A.;Keddam, M.; Huet, F. &. Sahar, A. (1989), Potential drops due to an a gas-evolving electrode, *Journal of Applied Electrochemistry*, 19, pp. 617-629.

Gijsbers, H.F.M. & Janssen, L.J.J. (1989), Distribution of mass transfer over a 0.5-m-tall hydrogen-evolving electrode, *Journal of Applied Electrochemistry*,19, pp. 637-648.

Hine, F.; Yasuda, M.; Ogata, Y. & Hara, K. (1984) Hydrodynamic studies on a vertical electrolyzer with gas evolution under forced circulation, *Journal of Electrochemistry Society*, 83, p. 131.

Iwasaki, A.; Kaneko, H.; Abe, Y. & Kamimoto, M. (1998), Investigation of electrochemical hydrogen evolution under microgravity condition, *Electrochimica Acta*,46,5-6,pp.509-514.

Janssen, L.J.J.; Sillen C.W.M.P.; Barendrecht, E. & Stralen, S.J.D.(1984) Bubble behaviour during oxygen and hydrogen evolution at transparent electrodes in KOH solution, *Electrochimica Acta*, 29, pp. 630 - 642.

Jassen, L.J.J. (1978), Mass transfer at gas evolving electrodes, *Electrochimica Acta*,23,p.81

Jassen, L.J.J. & Barendrecht, E.(1978), The effect of electrolytic gas evolution on mass transfer at electrodes, *Electrochimica Acta*,24,p.735

Jorne, J. & Louvar, J.F. (1980) Gas diverting electrodes in the chlor-alkali membrane cell. Journal of Electrochemistry Society, 2, p.127.

Kienzen, V.; Haaf, D. & Schnurnberger, W. (1994), Location of hydrogen gas evolution on perforated plate electrodes in zero gap cells, *International Journal of Hydrogen Energy*, 19,9,pp.729-732.

Kreysa, G. & Kuhn, M. (1985), *Journal of Applied Electrochemistry*, 15, p.283.

Khun, M. & Kreysa, G. (1989), Modelling of gas-evolving electrolysis cells. II Ivestigarion of the flow field around gas-evolving electrodes, *Journal of Applied Electrochemistry*, 19, pp.677-682.

Lantelme, F. & Alexopoulos, H. (1989), Role of gas bubbles adsorbed on a carbon electrode: electroreduction of chlorine gas in fused salts, *Journal of Applied Electrochemistry*, 19,pp.649-656.

Lasia, A. (1998), Hydrogen evolution:oxidation reactions on porous electrodes, *Journal of Electroanalytical Chemistry*, 454, pp.115-121.

Lasia, A. (1997), Porous electrodes in the presence of a concentration gradient, *Journal of Electroanalytical Chemistry*, 428,pp.155-164.

Lasia, A. & Rami, A. (1990), Kinetics of hydrogen evolution on nickel electrodes, *Journal of Electroanalytical Chemistry*, 294, pp.123-141.

Lastochkin, D. & Favelukis, M. (1998) Bubble growth in a variable diffusion coefficient liquid, *Chemical Engineering Journal*, 69, pp.21-25.

Lubetkin, S.D. (1989), Measurement of bubble nucleation rates by an acoustic method, *Journal of Applied Electrochemistry*, 19, pp.668-676.

Martin, A.D. & Wragg, A.A. (1989), The vertical distribution of current in a gas-evolving membrane cell, *Journal of Applied Electrochemistry*,19,pp.657-667.

Mohanta, S. & Fahidy, T.Z. (1977), The effect on anodic bubble formation on cathodic mass transfer under natural convection condition, *Journal of Applied Electrochemistry*,7,p.235

Nishik, Y. i, K.; Tokuda, Aoki, K. & Matsuda, H., (1987) Current distribution in a two-dimensional narrow gap cell composed of a gas evolving electrode with an open part, *Journal of Applied Electrochemistry*, 17, pp. 67-76.

Pletcher, D. & Walsh, F.C. (1993), *Industrial Electrochemistry*, 2ªedition, Black Academic & Professional, Chapman&Hall, London.

Rousar, I.; Kacin, J.; Lippert, E.; Smirous, F. & Cezner, V. (1975), Transfer of mass or heat to an electrode in the region of hydrogen evolution II. Experimental verification of mass and heat transfer equations, *Electrochimica Acta*,20,p.295

Sedahmed, G.H. (1978), Mass transfer enhancement by the counter electrode gases in a new cell design involving a three-dimensional gauze electrode, *Journal of Applied Electrochemistry*,8,p.399

Saleh, M.M. (1999), Mathematical modelling of gas evolving flow-through porous electrodes, *Electrochimica Acta*,45,pp.959-967.

Solheim, A.; Johansen, S.T.; Rolseth, S. & Thonstad, J. (1989), Gas induced bath circulation in aluminium reduction cells, *Journal of Applied Electrochemistry*, 19,pp.703-712.

Stephan, K. & Vogt, H. (1979) A model for correlating mass transfer data at gas evolving electrodes, *Electrochimica Acta*, 24, pp. 11-18.

St-Pierre, J. & . Wragg, A.A, (1993a) Behaviour of electrogenerated hydrogen and oxgen bubbles in narrow gap cells–Part I. Experimental, *Electrochimica Acta*, 38, 10, pp. 1381–1390.

St-Pierre, J. & . Wragg, A.A, (1993b) Behaviour of electrogenerated hydrogen and oxgen bubbles in narrow gap cells–Part II. Application in chlorine production, *Electrochimica Acta*, 38, 13, pp. 1705–1710.

Vilar, E.O. (1996) Transfer de matière entre um fritté métallique et un liquid - application aux electrodes poreuses percolées. Dr. Thesis, ENSCR, Université de Rennes I, France.

Vogt, H. (1979), On the supersaturation of gas in the concentration boundary layer of gas evolving electrodes, *Electrochimica Acta*,25,pp.527-531.

Vogt, H. (1984a) The rate of gas evolution at electrodes - I. An Estimate of efficiency of gas evolution of the basis of bubble grwth data, *Electrochimica Acta*, 29,2, pp.175 - 180

Vogt, H. (1984b), The rate of gas evolution at electrodes - II. An Estimate of efficiency of gas evolution from the supersaturation of electrolyte adjacent to a gas-evolving electrode, *Electrochimica Acta*, 19, pp. 167 - 173.

Vogt, H. (1984c) Studies on gas-envolving electrodes: The concentration of dissolved gas in electrolyte bulk, *Electrochimica Acta*, 30,2,pp.265-270.

Vogt, H. (1997), Contribution to the interpretation of the anode effect, *Electrochimica Acta*, 42,17,pp.2695-2705.

Vogt, H.(1989a), The problem of the departure diameter of bubbles at gas-evolving electrodes, *Electrochimica Acta*, 34,10, pp.1429-1432.

Vogt, H. (1989b), Mechanism of mass transfer of dissolved gas from a gas-evolving electrode and their effect on mass transfer coefficient and concentration overpotential, *Journal of Applied Electrochemistry*,19,pp.713-719.

Vogt,H.(1992) The role of single-phase free convection in mass transfer at gas evolving electrodes - I. Theoretical, *Electrochimica Acta*, 28, pp. 1421-1426.

Vogt,H. (1994) The axial hypochlorite distribution in chlorate electrolyzers, *Electrochimica Acta*,39,4,pp.2173-2179.

Walsh,F. (1993) *A first course in electrochemical engineering*, The Electrochemical Consultancy, Romsey, England.

White, S.H. & Twardoch, U.M.(1988), *Journal of Electrochemical Society*, 135, p. 893.

Wongsuchoto, P.; Charinpanitkul,T. & Pavasant, P. (2003), Bubble size distribution and gas–liquid mass transfer in airlift contactors, *Chemical Engineering Journal*, 92, pp. 81-90.

Zlokarnik, M. (2002) *Scale-up in Chemical Engineering*, Wiley-VCH Verlag GmBH & Co. KGaA ISBNs: 3-527-30266-2 (Hardback); 3-527-60056-6 (Electronic).

10

Mass Transfer Equation and Hydrodynamic Effects in Erosion-Corrosion

A. Yabuki
Hiroshima University
Japan

1. Introduction

Localized corrosion frequently occurs near the inlet of copper alloy heat exchanger tubes in seawater. Localized corrosion occurs when protective corrosion-product film that forms on the surface of the copper alloy is broken away by shear stress and turbulence causing the underlying metal surface to come into direct contact with the corrosive liquid. This phenomenon is known by several different terms: erosion-corrosion, flow-induced localized corrosion, flow-accelerated corrosion, or flow assisted corrosion (FAC), etc. (*Chexal et al.*, 1996; *Murakami et al.*, 2003). Damage by erosion-corrosion largely depends on hydrodynamic conditions such as the flow velocity of a liquid. Thus, this type of corrosion is characterized by the "breakaway velocity" at which the surface protective film is destroyed as the flow velocity increases (*Syrett*, 1976). To predict the extent of damage to copper alloys under a flowing solution, it is imperative to elucidate the relationships between damage to the materials and the hydrodynamic characteristics of the corrosive solution. Erosion-corrosion of copper alloys often proceeds via a diffusion-controlled process, and the mass-transfer equation for an oxidizing agent over the surface of a material is generally adopted. To apply the mass transfer equation to erosion-corrosion damage, mass transfer in both the concentration boundary layer and in the corrosion-product film on the material need to be considered, because the corrosion-product film that forms on the material confers a resistance to corrosion (*Mahato et al.*, 1980; *Matsumura et al.*, 1988). Flow velocity is generally used as the hydrodynamic parameter to predict erosion-corrosion damage, because it is quite simple. However, flow velocity is not sufficient to accurately predict damage, since erosion-corrosion frequently occurs in a turbulent region where the direction of flow changes, such as in a pipe bend, an elbow and or tee pipe fittings. Several papers have reported that the Sherwood number, a dimensionless number used in mass transfer operations, is useful as the mass transfer coefficient in the concentration boundary layer (*Sydberger et al.* 1982; *Poulson*, 1983, 1993, 1999; *Wharton*, 2004). Poulson reported that the Sherwood number in many flow conditions can be estimated through electrochemical measurements (*Poulson*, 1983). However, the Sherwood number also might inaccurately describe the condition of a corrosion-product film. Nešićet et al. conducted a numerical simulation of turbulent flow when a rust film was present, and found that fluctuations in turbulence affected both mass transfer through the boundary layer and the removal of the film (*Nešić et al.*, 1991). A numerical simulation of pipe flow has also been used to investigate erosion-corrosion (*Ferng et al.*, 2000; *Keating et al.*, 2001; *Postlethwaite et al.*, 1993; *Wharton et al.*, 2004).

This chapter on erosion-corrosion damage will discuss use of both the mass transfer equation as it relates to damage of materials and near-wall hydrodynamic effects to predict damage. Erosion-corrosion tests of copper alloys were conducted in a corrosive solution under various flow velocities using a jet-in-slit testing apparatus. A damage profile for each specimen was prepared using a surface roughness meter to evaluate local damage. The depth of the damage, calculated using the mass transfer equation, was related to the experimental data to confirm the applicability of the equation. Using the mass transfer coefficient of the corrosion-product films obtained from the mass transfer equation, the condition of the film and the breakaway properties were compared for each material. In addition, the near-wall hydrodynamic conditions at the material surface in the apparatus were measured using pressure gauges. The measured hydrodynamic conditions were applied to the equation used to predict the corrosion damage. The relationship between the near-wall hydrodynamic effects on the material surface and the corrosion of metallic materials under a flowing solution was investigated.

2. Erosion-corrosion damage

2.1 Experimental

The jet-in-slit testing apparatus used in the erosion-corrosion test is shown in Fig. 1. The testing apparatus consisted of a test solution tank, a pump, a flow meter and a test cell. Figure 2 shows a detailed schematic rendering of the test cell.

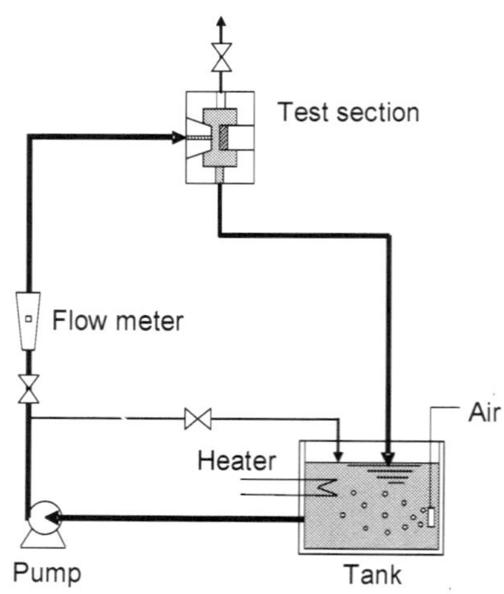

Fig. 1. Schematic diagram of a jet-in-slit testing apparatus.

In this apparatus, the test solution was allowed to flow from the nozzle into the slit between the specimen and the nozzle. The diameter of the specimen was 16 mm. The nozzle was made of a polymethyl-methacrylate resin with a bore diameter of 1.6 mm. The gap between the nozzle top and the specimen was 0.4 mm. As the solution was injected from the nozzle mouth into the slit, the solution filled the slit and flowed radially over the specimen surface. As the solution approached the periphery of the specimen, the cross-sectional area of the flow increased, and, consequently, the flow velocity decreased. The rapid reduction in flow velocity created a shear stress and an intense turbulence in the flow, similar to what is expected downstream of orifice plates (*Matsumura et al.*, 1985). As a result, localized corrosion damage in the jet-in-slit test can be accounted for primarily by shear stress and the turbulence of the flow.

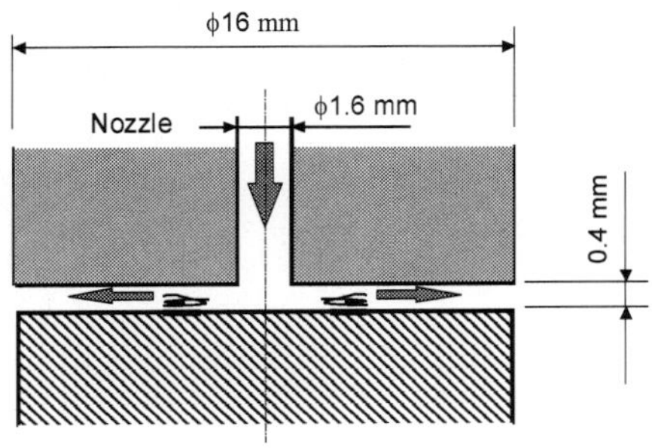

Fig. 2. Test section in the jet-in-slit corrosion-testing apparatus.

A 1 wt% $CuCl_2$ solution saturated with air was used as the test solution. Cu^{2+} was used as the oxidizing agent to accelerate the corrosion reaction. The temperature of the test solution was maintained at 40 °C. The flow velocities at the nozzle outlet were varied from 0.2 to 7.5 $m \cdot s^{-1}$. At a flow rate of 0.4 $L \cdot min^{-1}$, the fluid velocity at the nozzle outlet was 3.3 $m \cdot s^{-1}$ and the Reynold's number at that point was 8100. The test duration was 1 h.

The materials used in the investigation were pure copper (Cu) and three copper alloys, namely a beryllium copper alloy (BeCu) and two types of copper nickel alloys (70CuNi and 30CuNi). The chemical compositions of the test materials are shown in Table 1.

Symbol	Primary chemical composition / wt%
Cu	99.99Cu
70CuNi	30.2Ni-Cu
30CuNi	31.6Cu-Ni
BeCu	1.85Be-Cu

Table 1. Chemical composition of the copper alloys used in the tests.

Damage depth was determined by comparing the difference in the specimen surface profile before and after the test using a surface roughness meter and by determining the mass loss of the specimen. The damage depth rate was obtained by converting the maximum damage depth into $mm \cdot y^{-1}$.

2.2 Measurement of damage profiles

Cross-sectional profiles of the Cu and BeCu specimens after the test at a flow rate of 0.8 $L \cdot min^{-1}$ are shown in Fig. 3. The dotted line indicates the profile before the test as determined by the following: volume loss as calculated using measurement of the mass loss and the density of a specimen. The same position was used to measure the profile pre- and post-test, and then the difference between the two profiles was cylindrically integrated to obtain the volume loss. Then, the position was shifted vertically, and the procedure was repeated to determine if the results coincided. Both the Cu and BeCu specimens were significantly damaged in the central region of the specimen (A) and in an area approximately 2 mm from the center of the specimen (B). The damage in region A was due to shear stress, while that in region B was due to turbulence, as described above (*Matsumura et al.*, 1985). The ratio of the damage in the central region A to that in region B was approximately two-thirds for the Cu specimen. On the other hand, the ratio for the BeCu specimen was approximately one-half. Thus, the damage to the BeCu specimen was much greater in the central region. This result indicates that the corrosion resistance of a corrosion-product film depends on the hydrodynamic conditions of a flowing solution. To evaluate in detail the role of the hydrodynamic effect in erosion-corrosion damage, the 1-mm radii of spots in the center regions of the damaged areas of the specimens and disturbed regions 2 to 3 mm from the center regions were chosen, and the maximal damage depths at both locations were measured under various velocities.

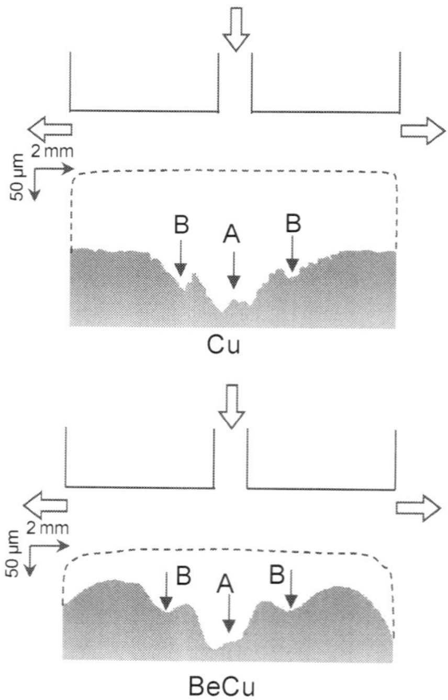

Fig. 3. Cross-sectional profile of a copper specimen (upper panel) and a BeCu specimen (lower panel) tested in a solution flowing at 0.8 L · min^{-1} for 1 h. The dotted line is the profile before the test. A and B are the central and disturbed regions.

3. Mass transfer equation in erosion-corrosion

3.1 Mass transfer equation

Various hydrodynamic parameters have been proposed to control the occurrence and extent of erosion-corrosion. The mass transfer coefficient is a parameter that relates the rate of a diffusion-controlled reaction to the concentration driving force, and includes both diffusional and turbulent transport processes. Erosion-corrosion of copper alloys mainly proceeds under cathodic control because the rate-controlling step in corrosion is the transport of the oxidizing agent from the bulk of the fluid to the metal surface. When the surface of the copper alloy is exposed to a flowing fluid, a concentration boundary layer is formed in the bulk of the fluid outside of the corrosion-product film, as shown in Fig. 4.

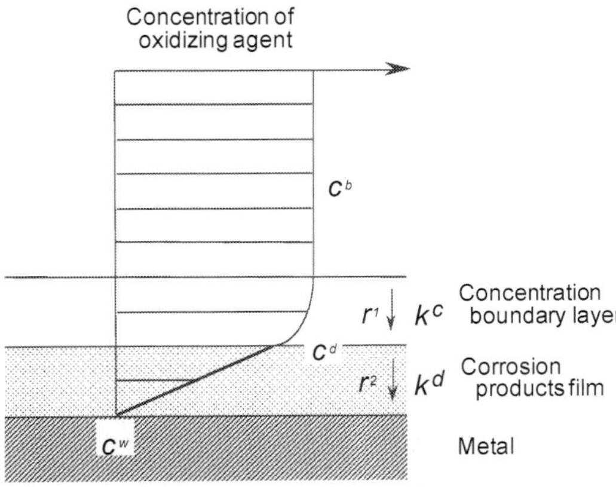

Fig. 4. Distribution of the oxidizing agent concentration in a solution flowing over a metal surface.

The diffusion rates of the oxidizing agent r_1 and r_2 in the concentration boundary layer and in the corrosion-product film, respectively, can be determined, as follows:

$$r_1 = k_c\,(c_b - c_d) \tag{1}$$

$$r_2 = k_d\,(c_d - c_w) \tag{2}$$

where c_b , c_d and c_w (mol · L^{-1}) are the oxidizing agent concentrations in the bulk of the flowing fluid, at the outside surface of the corrosion-product film, and at the metal surface, respectively. k_c and k_d (mm · y^{-1}) are the mass transfer coefficients in the concentration boundary layer and in the corrosion-product film, respectively.
The corrosion rate should be proportional to the diffusion rate of the oxidant. In the steady state, the mass transfer rates in the concentration boundary layer are equal to that in the corrosion-product film. Accordingly, the corrosion rate, R_c (mm · y^{-1}), can be given by the following reaction by using the conversion factor K (L · mol^{-1}):

$$R_c = K\,r_1 = K\,r_2 \tag{3}$$

The concentration of the oxidizing agent at the metal surface, c_w, may be zero (=0), since a very rapid electrochemical reaction is assumed. Equations (1)-(3) are combined to give:

$$R_c = Kc_b \,/\, (1/k_c + 1/k_d) = c_b \,/\, (1/Kk_c + 1/Kk_d) \tag{4}$$

Equation (4) indicates that the corrosion rate is directly proportional to the concentration of the oxidizing agent, c_b, and inversely proportional to the combined resistance to mass

transfer, $1/Kk_c+1/Kk_d$. The concentration of the oxidizing agent, c_b , is 0.075 mol · L^{-1}, which corresponds to a $CuCl_2$ concentration of 1 wt%.
The issue of whether the mass transfer equation can be applied to the experimental results was examined. The problem is how to determine the mass transfer coefficients, k_c and k_d. According to the definition of the mass transfer coefficient, the coefficient in the concentration boundary layer, k_c, is inversely proportional to the thickness of the concentration boundary layer. It was previously determined that the thickness is dependent on the flow velocity and is inversely proportional to the velocity to the power of 0.5 for laminar flow and of 0.8 for turbulent flow (*Bird et al.*, 1960). Accordingly,

$$k_c \propto u^{0.5} \quad \text{(for laminar flow, Re<2300)} \tag{5}$$

$$k_c \propto u^{0.8} \quad \text{(for turbulent flow, Re>2300)} \tag{6}$$

It may be assumed that the mass transfer coefficient in the corrosion-product film, i.e., k_d, is also inversely proportional to the thickness of the corrosion-product film, but is initially independent of flow velocity, because the thickness of the corrosion-product film is nearly constant. After the increase in the corrosion rate, it is assumed that k_d depends on the flow velocity to the power, i.e., k_c. This is because the surface after the breakaway of the corrosion-product film consisted of a completely naked area, while at the same time the area was still covered with residual corrosion product (*Matsumura et al.*, 1985). Accordingly,

$$Kk_d = \alpha \text{ (constant, < breakaway velocity)} \tag{7}$$

$$Kk_d = \beta u^n \text{ (> breakaway velocity)} \tag{8}$$

where α, β and n are constants. Under these assumptions, Kk_c and Kk_d were determined and fitted the experimental data.

3.2 Damage depth rate and fitting by mass transfer equation

Figure 5 shows the relationship between the damage depth rate at the central and disturbed regions of a Cu specimen and the flow velocity. The solid curves in the figure were calculated using the mass transfer equation and fitted to the experimental data. Using the same procedure, the experimental data and the fitted lines for BeCu, 70CuNi and 30CuNi are shown in Figs. 6, 7 and 8, respectively. The coefficient in the concentration boundary layer, Kk_c, was determined and used to fit the experimental data. The constants α, β and n in equation (7) and (8), which are related to the mass transfer coefficient in a corrosion-product film, are listed in Table 2. These parameters are discussed below.
The damage depth for the Cu specimen increased slightly with increasing flow velocity at lower velocities (Fig. 5). The damage depth increased rapidly at a certain velocity, namely the breakaway velocity (*Syrett*, 1976). The breakaway velocity at the central region was 2 m · s^{-1} and at the disturbed regions it was 0.8 m · s^{-1}. The damage depth at a velocity less than the breakaway velocity for the central and disturbed regions fit the same curve. The damage depth doubled at the breakaway velocity in both regions, and further increased at higher flow velocities. This result indicates that the corrosion-product film formed on the Cu was easily broken away by the turbulence that occurred in the disturbed regions, and was not due to shear stress. It was confirmed that the damage depth, determined by the mass transfer equation, was well fitted to experimental damage depth for Cu, although the damage varied at different regions of the specimen.

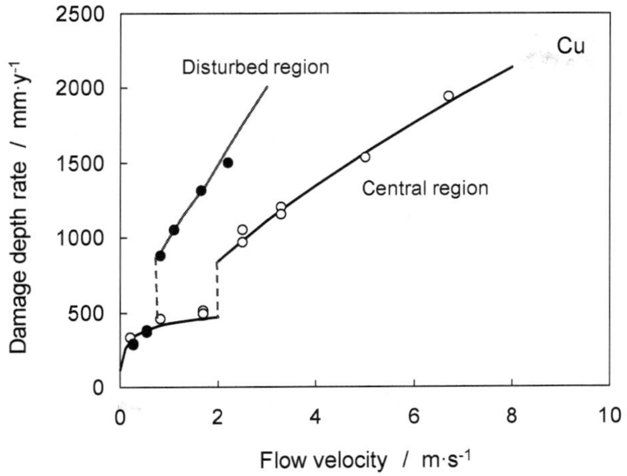

Fig. 5. Relationship between flow velocity and damage depth rate at the central and disturbed regions of a pure copper (Cu) specimen tested in a jet-in-slit testing apparatus. The curves were calculated using the mass transfer equation as fitted to the experimental data.

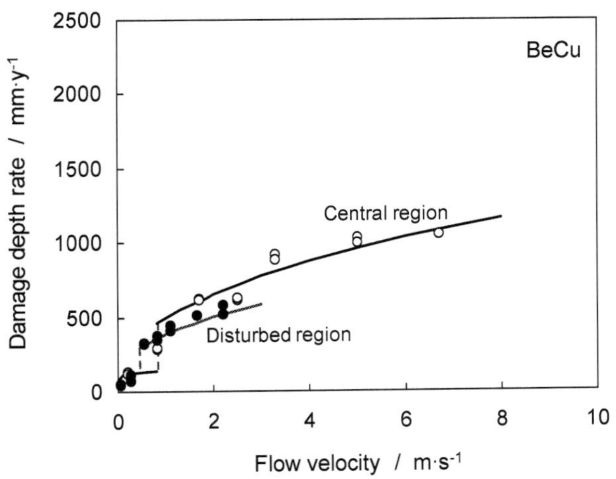

Fig. 6. Relationship between the flow velocity and damage depth rate in the central and disturbed regions of a beryllium copper alloy (BeCu) specimen tested in a jet-in-slit testing apparatus. Curves were calculated by using the mass transfer equation as fitted to the experimental data.

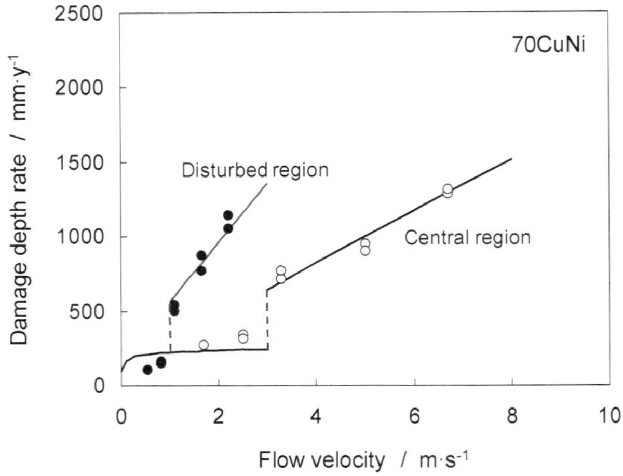

Fig. 7. Relationship between the flow velocity and damage depth rate in the central and disturbed regions of a copper nickel alloy (70CuNi) specimen tested in a jet-in-slit testing apparatus. Curves were calculated using the mass transfer equation as fitted to the experimental data.

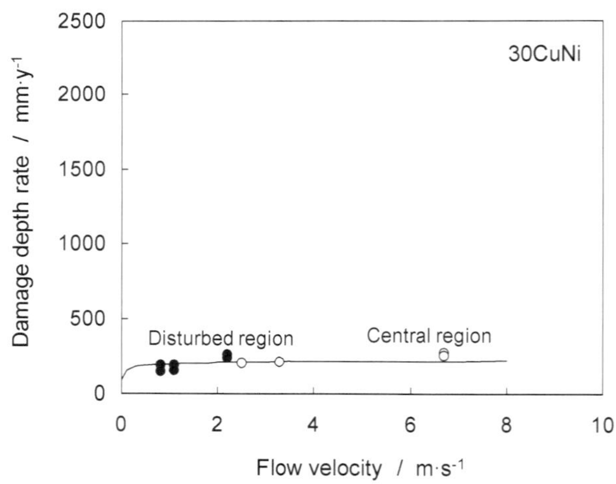

Fig. 8. Relationship between the flow velocity and damage depth rate in the central and disturbed regions of a copper nickel alloy (30CuNi) specimen tested in a jet-in-slit testing apparatus. Curves were calculated using mass transfer equation as fitted to the experimental data.

Symbol	Region	V_b / m · s^{-1}	α	β	n
Cu	Central	2	8000	12000	0.6
	Disturbed	0.8	↑	40000	↑
BeCu	Central	0.8	2000	10000	0.3
	Disturbed	0.5	↑	7000	↑
70CuNi	Central	3	3500	4000	0.9
	Disturbed	1	↑	12000	↑
30CuNi	Central	-	3000	-	-
	Disturbed	-	↑	-	-

Table 2. Constants for the mass transfer equation determined as fitted to the damage depth rate of each copper alloy.

The damage depth rate for the BeCu specimen increased with increasing flow velocity, but was relatively low, compared to the rate for the Cu specimen (Fig. 6). The breakaway velocity at the central region was 0.8 m · s^{-1}, while that in the disturbed region was 0.5 m · s^{-1}. Moreover, the breakaway velocity was lower than that for the Cu specimen. The damage depth rate was very low at velocities less than the breakaway velocity, compared to the rate for the Cu specimen. At velocities greater than the breakaway velocity, the damage depth rate increased slightly with increasing flow velocity. However, the behavior of the damage depth rate was similar in both the central and disturbed regions. The corrosion behavior of the BeCu specimen was different from that of the Cu specimen. The damage depth rate, calculated from the mass transfer equation, could also be fitted to the experimental data for the BeCu sample.

For the 70CuNi alloy, the breakaway velocity in the central regions was 3 m · s^{-1}, while that in the disturbed region was 1 m · s^{-1} (Fig. 7). The damage depth rate at the breakaway velocity was increased three-fold, and also increased with an additional increase in the flow velocity. Although the damage was low compared to the damage to the Cu specimen, the corrosion behavior was similar to that of the Cu specimen. This result was attributed to the formation of a good quality anti-corrosion film due to the addition of nickel. The damage depth rate at a velocity lower than the breakaway velocity was linear, but the curve calculated from the mass transfer equation did not coincide with the experimental damage rate. This result was apparently caused by a slight breakaway of the film, although the damage was not fatal. The damage depth rate at a velocity higher than the breakaway velocity simulated the experimental rate.

The damage depth rate for the 30CuNi alloy was constant at both the central and disturbed regions under all velocities (Fig. 8). This result shows that 30CuNi is an excellent film for protecting against erosion-corrosion, even when the flow velocity is high. The damage depth rate calculated from the mass transfer equation was well-fitted to the experimental results.

The damage depth rate at the central region differed from that at the disturbed region, and was dependent on hydrodynamic conditions. However, it was confirmed that the mass transfer equation and the assumptions concerning the corrosion-product film as described by equations (7) and (8) can be applied to erosion-corrosion damage.

3.3 Characterization of film

The relationships between the flow velocity and the mass transfer coefficient in a corrosion-product film (Kk_d) in the central regions of each specimen are shown in Fig. 9. The mass transfer coefficient in the concentration boundary layer, Kk_c, is shown by the dotted line in the figure. The dashed lines in the Kk_d curves show the breakaway velocities for each material.

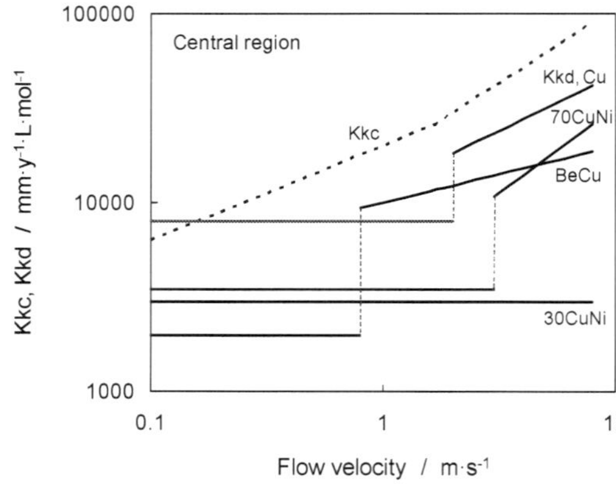

Fig. 9. Values of Kk_c and Kk_d determined by fitting to the damage depth rate in the central region of a specimen.

The mass transfer coefficient in the corrosion-product film, Kk_d, of Cu was very similar to the mass transfer coefficient of the concentration boundary layer, Kk_c, at a velocity less than the breakaway velocity. Therefore, the damage rate for Cu depends on the concentration boundary layer at lower velocities. However, the damage rate of the other copper alloys is determined by the condition of the corrosion-product film, because the Kkd of the other copper alloys was very low compared with Kk_c. This is equivalent to α, as shown in Table 2, and the corrosion resistance of the films that formed on BeCu, 70CuNi and 30CuNi was enhanced more than two-fold compared to the film that formed on pure Cu. At a velocity higher than the breakaway velocity, the Kk_d for the copper alloys was always lower than Kk_c. Consequently, the damage rate was mostly dependent on the mass transfer rate in the corrosion-product film. In other words, the damage rate is determined only by the corrosion-product film. At velocities higher than the breakaway velocity, the slope of Kk_d , listed in Table 2 as a power of n, was quite different for each material. The constant n appears to be the breakaway property of the film that formed on each material. The constant n for Cu was 0.6, which was similar to the change in the thickness of the concentration boundary layer and the same as Kk_c. The film that formed on BeCu was resistant to breakaway, since the constant for BeCu was as low as 0.3. The breakaway property of the film that formed on 70CuNi was nearly proportional to the flow velocity, since the constant, n, was 0.9. Thus, these results confirm that the breakaway property of each material was different at the central region, where the shear stress was dominant.

The relationships between the flow velocity and the mass transfer coefficient in the corrosion-product film (Kk_d) at the disturbed regions of each specimen are shown in Fig. 10. At velocities less than the breakaway velocity, the behavior observed in the disturbed regions was the same as that observed in the central region. At velocities higher than the breakaway velocity, the Kk_d for Cu exceeded the value of Kk_c. This result indicates that mass transfer in the concentration boundary layer was dominant, although small amounts of the corrosion-product film might remain on the specimen surface. The damage to the 70CuNi alloy affected both mass transfer coefficients, since the Kk_d for 70CuNi was comparable to Kk_c. The Kk_d for BeCu was so low that mass transfer in the corrosion-product film was dominant, as in the central region. The power, n, of the exponential equation for Kk_d at velocities higher than the breakaway velocity was different for each material, however, it was the same as that in the central region. Thus, although the extent of breakaway of the films that formed on each material was dependent on the hydrodynamic conditions, the increasing ratio for breakaway of the film with flow velocity was not dependent on hydrodynamics, but on the type of material.

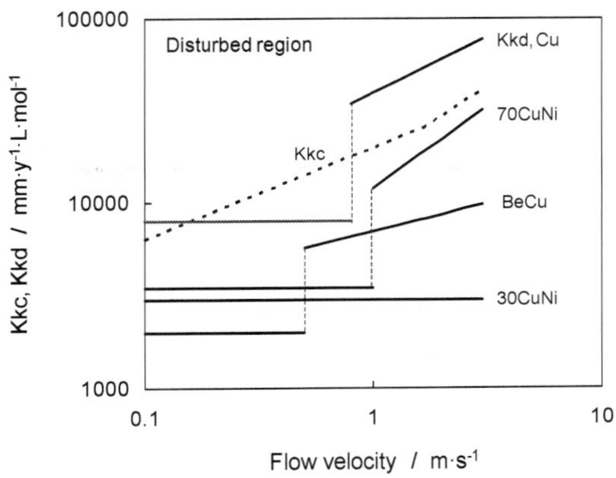

Fig. 10. Values of Kk_c and Kk_d determined by fitting to the damage depth rate in the disturbed region of the specimens.

Consequently, to predict erosion-corrosion damage for a copper-based material, the mass transfer equation can be used as a fundamental equation. However, the flow velocity does not adequately express various hydrodynamic conditions such as turbulence or shear stress. The Sherwood number seems to be more suitable than the flow velocity. Of course, the mass transfer coefficient in the concentration boundary layer can be predicted. However, prediction of the mass transfer coefficient in a corrosion-product film appears difficult, because the Sherwood number is almost propotional to the flow velocity at a Reynold's number of less than 10,000 in an impinging jet testing apparatus similar to the apparatus used in the present study (*Sydberger et al.*, 1982). An alternative parameter to describe hydrodynamic conditions rather than the flow velocity or the Sherwood number is desirable for prediction of the breakaway properties of a corrosion-product film. The corrosion-product film that formed on pure Cu tested at a lower flow velocity consisted of numerous particles, which were approximately 5 μm in diameter. Thus, breakaway of the corrosion-product film was equivalent to particle removal due to the hydrodynamic action of the flowing solution. Concerning the removal of the particles, it is important to investigate the relationships between particle morphology and adhesive force. Thus, selection of a hydrodynamic parameter related to erosion-corrosion was the most important issue initially. The mass transfer coefficient can be determined by electrochemical measurements, but it was thought that determination of the force acting on the surface of the material, for instance, a pressure measurement, was also useful. It is reasonable to use the mass transfer equation as a basic equation. A more complex equation should be developed for prediction of erosion-corrosion damage in an actual machine, along with a numerical simulation (*Ferng et al.*, 2000; *Keating et al.*, 2001; *Postlethwaite et al.*, 1993).

4. Hydrodynamic effects

4.1 Measurement of near-wall hydrodynamic conditions

The near-wall hydrodynamic conditions of each specimen were measured in the jet-in-slit corrosion testing apparatus using two pressure gauges and a wire. The set-up for the system is shown in Fig. 11. The fluid in the tank flows to a nozzle through a flow meter with a pump, and then returns to the tank. Three types of nozzles were prepared. One was the same size as the nozzle of the corrosion testing apparatus, and the others were 2-fold and 5-fold scale-ups. The fluid velocity at the nozzle equaled that of the corrosion testing apparatus. Two holes, each 0.3 mm in diameter, were bored into the surface of the measurement plate, and a wire 0.05 mm in diameter was set between the holes, as shown in Fig. 12.

Pressure gauges (PGM-02KG, Kyowa Electronic Instruments Co., Ltd.) were connected to the holes. The signal from each pressure gauge was input to a personal computer through a sensor interface (PCD-300, Kyowa Electronic Instruments Co., Ltd.). The horizontal velocity V_x (m · s^{-1}) and the vertical velocity V_y (m · s^{-1}) were calculated by the pressure differential $\Delta P = P_1 - P_2$ (Pa) and the wall pressure upstream of the wire, P_1 (Pa), respectively. The measured pressure was converted into velocity using equations (9) and (10), which are given by Bernoulli's law,

$$V_x = (2\Delta P / \rho)^{0.5} \tag{9}$$

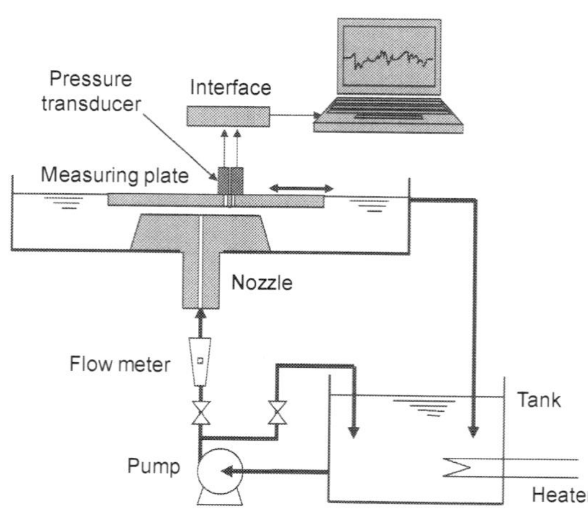

Fig. 11. Set-up for the system to measure the near-wall hydrodynamic conditions of a specimen in the jet-in-slit corrosion testing apparatus.

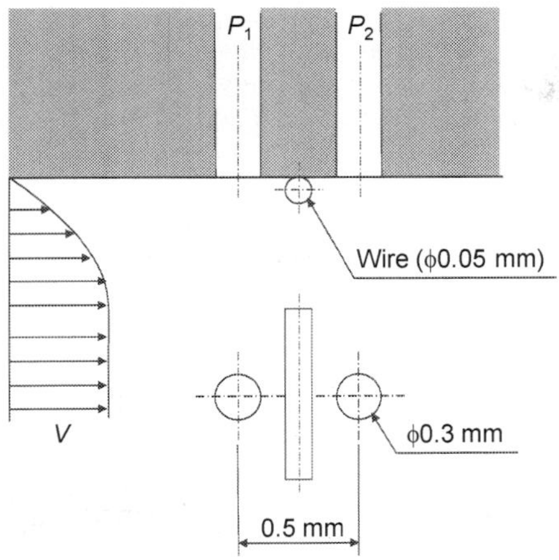

Fig. 12. Dimensions of the two holes connected to the pressure gauges and the wire set on the measuring plate for determination of near-wall hydrodynamic conditions.

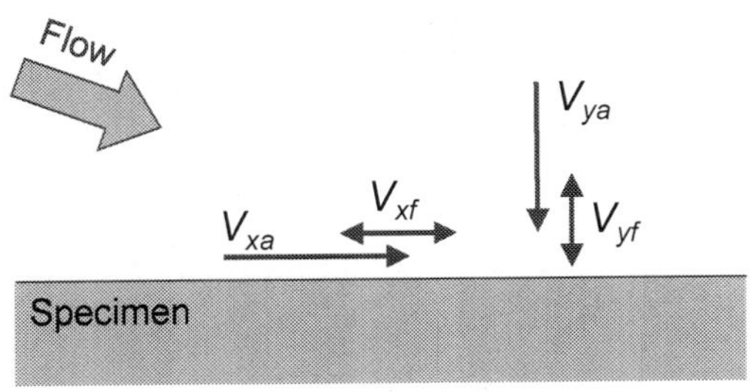

Fig. 13. The near-wall average velocity and the fluctuation in the horizontal (x) and vertical (y) directions relative to the specimen surface, V_{xa}, V_{xf}, V_{ya} and V_{yf}.

$$V_y=(2P_1/\rho)^{0.5} \quad (10)$$

where ρ ($kg \cdot m^{-3}$) is the density of the solution. The measuring plate was moved from the center of the nozzle mouth to the periphery to measure each point on the specimen surface.

Only the distributions in the right-half of the apparatus were measured, since the left-half distributions should be nearly identical to those of the right side. The sampling time was 10 s and the sampling frequency was 2 kHz. The average velocities and the fluctuations in the horizontal (x) and vertical (y) directions relative to the specimen surface, V_{xa}, V_{xf}, V_{ya} and V_{yf} (as shown in Fig. 13), were calculated from the time dependence of the horizontal velocity V_x and the vertical velocity V_y, respectively. V_{xa} and V_{ya} are equal to the wall shear stress and the hydrodynamic energy density, respectively (*S. Nešić et al., 1991; Bozzini et al., 2003*).

4.2 Near-wall hydrodynamic conditions

Near-wall hydrodynamic conditions in the jet-in-slit apparatus were measured using three types of nozzles. The distribution of the hydrodynamic conditions, which were measured using the same nozzle as was used in the corrosion testing apparatus, did not appear smooth, because the holes for the pressure gauge and wire were too large. However, the distributions of the hydrodynamic conditions that were measured using the nozzles that were scaled-up 2- and 5-fold, were very smooth and nearly identical. The measurement points were easily adjusted using the nozzle that was scaled-up 5-fold. Therefore, that nozzle was used to measure the near-wall hydrodynamic conditions. To compare the hydrodynamic conditions with the corrosion damage, the region in which the hydrodynamic conditions were measured corresponds to the nozzle in the corrosion testing apparatus. Thus, the value of that region was divided by 5.

The time-dependence of the pressure differential was located 1.5 mm from the center of the specimen (ΔP), and the wall pressure was at the center of the specimen, P_1, and both are shown in Fig. 14. Fluctuations in ΔP and P_1 were detected, and the frequencies were approximately 200 Hz and 130 Hz, respectively. The frequencies were not the same in both areas, so the fluctuations must have been due to the vibration of the fluid rather than to the pulsation of the pump.

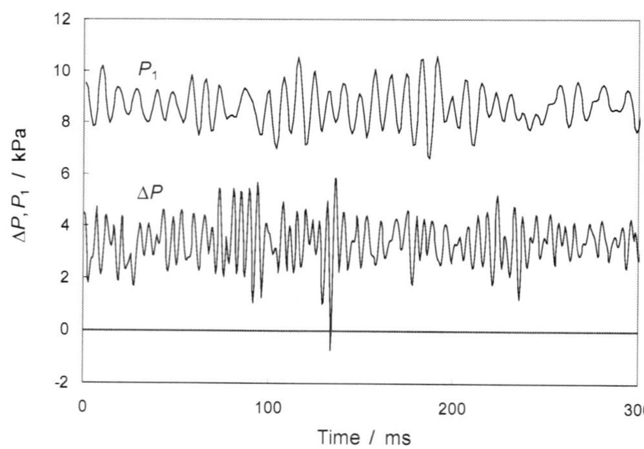

Fig. 14. Time-dependence of the pressure differential located 1.5 mm from the center of the specimen (ΔP) and the wall pressure at the center of the specimen (P_1).

ΔP and P_1, which were measured by pressure gauges were converted into horizontal velocity (V_x) and vertical velocity (V_y) using equations (9) and (10), respectively. Furthermore, the horizontal velocity V_x was calculated from the average velocity V_{xa} and the fluctuation V_{xf}. V_{xf} was 3 times the standard deviation of the velocity. The vertical velocity and the horizontal velocity were also used to calculate V_{ya} and V_{yf}. The distributions of the velocities and the fluctuations in the horizontal and vertical directions were obtained by moving the measuring plate.

Fig. 15 shows the distributions of the near-wall hydrodynamic conditions, V_{xa}, V_{xf}, V_{ya}, and V_{yf}, of the specimens. Only the distributions in the right-half of the apparatus are shown in the figure. The average vertical velocity, V_{ya}, was highest at the center of the specimen, since the solution was injected vertically and toward the specimen. V_{ya} showed a negative pressure in the area outside of 5 mm, which was due to boundary-layer separation in the area. The average horizontal velocity V_{xa} was highest 5 mm from the center of the specimen, which corresponded to the edge of the nozzle mouth, because the flow direction changed at this location. The fluctuations in the vertical and horizontal velocities V_{ya} and V_{yf} were highest in the area 10 mm from the center. As a result, the hydrodynamic parameters had different distributions, and the region showing the maximum differed among these parameters. Thus, we successfully measured the various hydrodynamic parameters in the jet-in-slit corrosion testing apparatus.

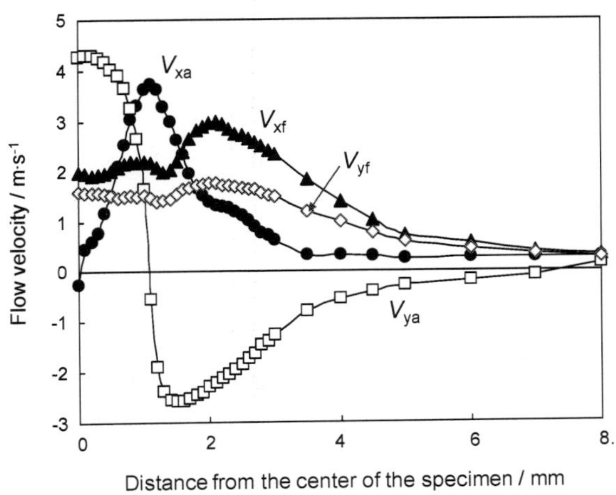

Fig. 15. The distributions of the near-wall hydrodynamic conditions, V_{xa}, V_{xf}, V_{ya} and V_{yf}, of the specimens. Only the distributions in the right-half of the apparatus are shown in the figure.

To relate the hydrodynamic conditions to the corrosion damage, V_{xa}, V_{xf} and V_{ya} were selected as follows. Only positive pressure was used for V_{ya}, because negative pressure in the area outside the region 5 mm from the center was caused by boundary-layer separation and was related to the fluctuations V_{xf} and V_{xf}. The vertical velocity fluctuation V_{yf} was similar to the horizontal velocity fluctuation V_{xf}, so only V_{xf} was used to predict the corrosion damage.

4.3 Prediction of damage using hydrodynamic effects and mass transfer equation

The jet-in-slit testing apparatus shown in section 2.1 was used for the corrosion tests performed under a flowing solution. Brass (58.3Cu-38.2Zn-3.10Pb-0.17Fe-0.23Sn) and 70CuNi (30.2Ni-Cu) were used as model materials for the corrosion tests. A 3% NaCl solution and a 1 wt% $CuCl_2$ solution saturated with air were used as the corrosive test solutions. The concentration of oxygen dissolved in the solutions was approximately 6.45 ppm. Cu^{2+} in the 1 wt% $CuCl_2$ solution acted as the oxidizing agent to accelerate the corrosion reaction. The test duration was 24 h in the 3% NaCl solution and 1 h in the 1 wt% $CuCl_2$ solution.

Cross-sectional profiles of the brass and 70CuNi specimens after flow-induced localized corrosion tests at a flow rate of 0.4 L · min^{-1} are shown in Fig. 16. The thin line indicates the profile before the test, which was determined as volume loss calculated from mass loss and the density of a specimen coinciding to the volume loss calculated from the surface profile before and after the test, as explained in section 2.2.

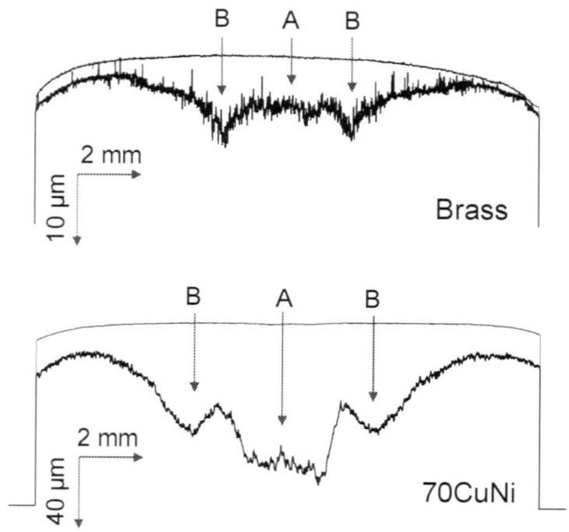

Fig. 16. Cross-sectional profiles of the brass (upper) and 70CuNi (lower) specimens after flow-induced corrosion tests at a flow rate of 0.4 L min^{-1}. The thin line is the profile before the test.

Brass was significantly damaged in the central region of the specimen (region A) and in the area approximately 2 mm from the center of the specimen (region B). The damage to the periphery was relatively low. This result was obviously due to the hydrodynamic effects of the corrosive solution. Similar to the brass specimen, the 70CuNi sample was significantly damaged in the central area (regions A and B), compared with the damage to the periphery. The damage in region B was deeper than that in region A for brass, but the damage in region B was less than that in region A for 70CuNi. The result is likely due to differences in the corrosion-product films that formed on the surface of each specimen under a flowing solution, because the hydrodynamic conditions near the surface were very similar for the two materials. The corrosion rate for both materials differed with the oxidizing agent concentration. The flow-induced localized corrosion of copper alloys proceeded through both the initiation step, which occurred following the mechanical destruction of the corrosion-product film, and the propagation step, which occurred following the repeated formation and breakaway of the products due to the local hydrodynamic effect (*S. Nešić et al., 1991; Bozzini et al., 2003*). Since wet-polished specimens were tested in the present study, most of the damage to the copper alloys occurred during the propagation step, which was equal to the steady state. The porous corrosion-product film was detected by direct observation of the specimen surface after the corrosion test.

The corrosion damage to the copper alloy in a 1 wt% $CuCl_2$ solution was much greater than the damage that resulted from the 3% NaCl solution, as shown in Fig. 16. This result indicated that the corrosion damage was mainly caused by the cathodic reaction. Thus, the flow-induced localized corrosion of copper alloys mainly proceeds under cathodic control, so that the rate-controlling step in corrosion is the transport of the oxidizing agent from the bulk of the fluid to the metal surface. The corrosion rate should be proportional to the diffusion rate of the oxidant. In the steady state, the corrosion rate, R_c ($m \cdot s^{-1}$), can be given by equation (4). Accordingly, the local damage d (μm) can be estimated using the testing time t (h), as follows:

$$d = 3.6\times10^9\, K c_b t \,/\, (1/k_c + 1/k_d) \tag{11}$$

Assuming that the corrosion-product film that formed on the copper alloys was either thicker than the concentration boundary layer or thinner but more dense, it had a structure that resisted diffusion of the oxidizing agent. Therefore, the diffusion of the oxidizing agent in the corrosion-product film was relatively low for the rate-controlling step, namely $k_c >> k_d$. Thus equation (11) gives

$$d = 3.6\times10^9\, K k_d\, c_b\, t \tag{12}$$

The corrosion-product film that formed on the surface was in a steady state, and its thickness and structure were determined by repeated formation and breakaway due to the hydrodynamic effects. In this process, the condition of the film is determined by the mechanical force acting on the film. When the velocity was high, the force acting on the film was large, resulting in thinning of the film, so that the mass transfer coefficient in the corrosion-product film became larger. Hence, the mass transfer coefficient in the corrosion-product film k_d ($m \cdot s^{-1}$) was assumed to be proportional to the velocities at the near-wall, as follows:

$$k_d = \gamma_{xa} V_{xa} + \gamma_{xf} V_{xf} + \gamma_{ya} V_{ya} \tag{13}$$

where each γ is a material-specific constant that corresponds to the contributing ratio for each hydrodynamic condition.

The damage depth profile for the copper alloys was calculated using equations (12) and (13) as fitted to the experimental damage profile using a trial-and-error method. The calculated and experimental profiles for brass and 70CuNi are shown in Figs. 17 and 18. The figures show only the right-half of the damage profile for each specimen. The data used for brass that was tested in a 3% NaCl solution were as follows: $K = 3.6\times10^{-6}\ m^3 \cdot mol^{-1}$ (= $63.5/2/8.9\times10^6$, where 63.5 $g \cdot mol^{-1}$ is the molecular weight of copper, 2 is the number of ion-exchanges in anodic and cathodic reactions, and $8.9\times10^6\ g \cdot m^{-3}$ is the density of copper), $c_b = 0.20\ mol \cdot m^{-3}$, which corresponds to dissolved oxygen of 6.45 ppm, t=24 h, and the γ values obtained by fitting to the measured data were: $\gamma_{xa} = 2.3\times10^{-7}$, $\gamma_{xf} = 4.6\times10^{-5}$ and $\gamma_{ya} = 1.2\times0^{-6}$, respectively.

The data used for 70CuNi that was tested in a 1 wt% $CuCl_2$ solution were as follows: $K = 7.1 \times 10^{-6}\ m^3 \cdot mol^{-1}$ ($=63.5/1/8.9\times10^6$, where 63.5 $g \cdot mol^{-1}$ is the molecular weight of copper, 1 is the number of ion-exchanges in anodic and cathodic reactions, and $8.9\times10^6\ g \cdot m^{-3}$ is the density of 70CuNi). $c_b = 74\ mol \cdot m^{-3}$, which corresponds to the Cu^+ concentration of the 1 wt% $CuCl_2$ solution, t = 1 h, and the γ values obtained for each material from fitting to the measured data were: $\gamma_{xa} = 3.8\times10^{-6}$, $\gamma_{xf} = 8.8\times10^{-6}$, $\gamma_{ya} = 3.5\times10^{-6}$, respectively. The calculated profiles were consistent with both the experimental data and the areas of maximal damage for both copper alloys—2 mm for brass and 1 mm for 70CuNi. Comparing the contributing ratio (γ) to each hydrodynamic condition, the fluctuation in horizontal velocity, γ_{xf}, was

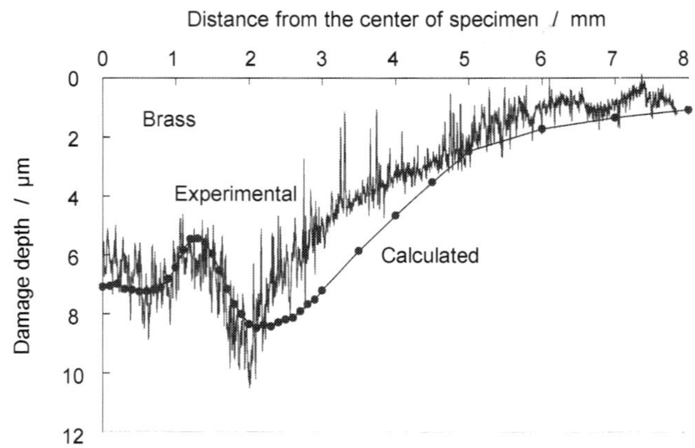

Fig. 17. The experimental and calculated damage depth profiles for brass tested in the 3% NaCl solution for 24 h. The figures show only the right half of the specimen.

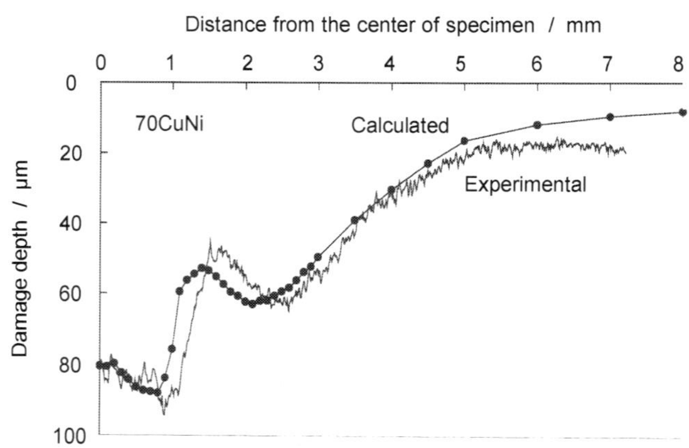

Fig. 18. The experimental and calculated damage depth profiles for 70CuNi tested in the 1 wt% $CuCl_2$ solution for 1 h. The figures show only the right half of the specimen.

dominant in the corrosion damage of brass. On the other hand, average vertical velocity, γ_{xa}, also affected the corrosion damage of 70CuNi, in addition to the fluctuation of horizontal velocity, γ_{xf}. Thus, the hydrodynamic effect on the corrosion damage of copper alloys was not attributed to a single parameter, such as flow velocity or Sherwood number, but instead was related to multiple effects from both horizontal and vertical force and fluctuation. The material-specific constant γ in equation (13) is related to the mechanical properties of the corrosion-product films that formed on the surface of copper alloys. Consequently, these properties are particularly important for the prediction of corrosion damage under a flowing solution.

7. Conclusion

Erosion-corrosion tests were carried out using a jet-in-slit testing apparatus, and the following results were obtained. The damage depth rate of Cu, BeCu and 70CuNi increased with increasing flow velocity, and the breakaway velocity was clearly evident. The films that formed on Cu and 70CuNi were significantly damaged by turbulence, such that the hydrodynamic conditions of the flowing solution affected the breakaway property of the corrosion-product film. Damage depth rate, calculated by the mass transfer equation, which involved mass transfer in the concentration diffusion layer and in the corrosion-product film, could to be fitted to the experimental damage rate. Thus, we confirmed that the mass transfer equation can be applied to erosion-corrosion damage to copper and copper alloys. The film condition remained nearly constant at a velocity lower than the breakaway velocity, and the films that formed on BeCu, 70CuNi and 30CuNi were enhanced more than two-fold compared with the film that formed on pure Cu, as evidenced by the analysis of the mass transfer coefficient in the corrosion-product film. The corrosion-product film was exponentially broken away at velocities greater than the breakaway velocity. The breakaway property of the corrosion-product film differed among the materials, since the power of the exponential equation was different for each material.
The relationship between the near-wall hydrodynamic effects on the material surface and the corrosion damage of material under a flowing solution was investigated. The near-wall hydrodynamic conditions on the surface of the test specimens were measured using pressure gauges, and both the distribution of the near-wall velocity and the velocity fluctuations were determined. Three types of hydrodynamic conditions were applied as parameters that are related to the mass transfer coefficient in a corrosion-product film in the equation used to predict corrosion damage. The damage profiles calculated from the equation could be fitted to that obtained from the corrosion test using the three hydrodynamic parameters. The determined material-specific constant agreed with the mechanical resistance of the corrosion-product film to the hydrodynamic effects in a corrosive liquid. Consequently, the mechanical properties of the corrosion-product films that formed on the surface of the copper alloys are particularly important for the prediction of corrosion damage under a flowing solution.

8. References

Bird R.B.; Stewart W.E. & Lightfoot E.N. (1960). *Transport phenomena,* John Wiliey & Sons, INC., pp. 647, ISBN 0-471-07392-X, New York.

Bozzini B.; Ricotti M.E.; Boniardi M. & Mele C. (2003). Evaluation of erosion–corrosion in multiphase flow via CFD and experimental analysis, *Wear,* Vol.255, No.1-6, 237-245, ISSN 0043-1648.

Chexal B.; Horowitz J.; Jones R.; Dooley B.; Wood C.; Bouchacourt M.; Remy F.; Nordmann F. & Paul P.St. (1996). *Flow-Accelerated Corrosion in Power Plants,* EPRI Distribution Center, ISBN TR-106611, Pleasant Hill, CA.

Ferng Y.M.; Ma Y.P. & Chung N.M. (2000). Application of local flow models in predicting distributions of erosion-corrosion locations, *Corrosion,* Vol.56, No.2, 116-126, ISSN 0010-9312.

Keating A. & Nešić S. (2001). Numerical prediction of erosion-corrosion in bends, *Corrosion,* Vol.57, No.7, 621-633, ISSN 0010-9312.

Mahato B.K.; Cha C.Y. & Shemilt L.W. (1980). Unsteady state mass transfer coefficients controlling steel pipe corrosion under isothermal flow conditions, *Corrosion Science*, Vol.20, No.3, 421-441, ISSN 0010-938X.

Matsumura M.; Oka Y.; Okumoto S. & Furuya H. (1985). *Laboratory Corrosion Tests and Standards STP 866*, ASTM, pp.358-372, ISBN 0-8031-0443-X, Philadelphia.

Matsumura M.; Noishiki K. & Sakamoto A. (1988). Jet-in-slit test for reproducing flow-induced localized corrosion on copper alloys, *Corrosion*, Vol.54, 79-88, ISSN 0010-9312.

Murakami, M.; Sugita K.; Yabuki A & Matsumura M. (2003). Mechanism of so-called erosion-corrosion and flow velocity difference corrosion of pure copper, *Corrosion Engineering (Zairyo-to-Kankyo)*, Vol.52, No. 3, 155-159, ISSN 0917-0480.

Nešić S. & Postlethwaite J. (1991). Hydrodynamics of disturbed flow and erosion-corrosion. Part I-Single-phase flow study, The Canadian Journal of Chemical Engineering, Vol.69, No. 3, 698-703, ISSN 00084034.

Postlethwaite J.; Nešić S.; Adamopoulos G. & Bergstrom D.J. (1993). Predictive models for erosion-corrosion under disturbed flow conditions, *Corrosion Science*, Vol.35, No.1-4, 627-633, ISSN 0010-938X. Poulson B. (1983). Electrochemical measurements in flowing solutions, *Corrosion Science*, Vol.23, No.4, 391-430, ISSN 0010-938X.

Poulson B. (1993). Advances in understanding hydrodynamic effects on corrosion, *Corrosion Science*, Vol.35, No.1-4, 655-665, ISSN 0010-938X.

Poulson B. (1999). Complexities in predicting erosion corrosion, *Wear*, Vol.233-235, 497-504, ISSN 0043-1648.

Sydberger T. & Lotz U. (1982). Relation Between Mass Transfer and Corrosion in a Turbulent Pipe Flow, *J. Electrochem. Soc.*, Vol.129, No.2, 276-283, ISSN 0013-4651.

Syrett B.C. (1976). Erosion-corrosion of copper-nickel alloys in seawater and other aqueous environments - a Literaturee review, *Corrosion*, Vol.32, 242-252, ISSN 0010-9312.

Wharton J.A. & Wood R.J.K. (2004). Influence of flow conditions on the corrosion of AISI 304L stainless steel, *Wear*, Vol.256, No.5, 525-536, ISSN 0043-1648.

11

Hydrodynamics and Mass Transfer in Heterogeneous Systems

Radmila Garić-Grulovic[1], Nevenka Bošković-Vragolović[2], Željko Grbavčić[2] and Rada Pjanović[2]
[1]*Institute for Chemistry, Technology and Metallurgy, University of Belgrade*
[2]*Faculty of Technology and Metallurgy, University of Belgrade*
Serbia

1. Introduction

Research of transport phenomena in liquid - particles systems, in past years, had a more theoretical then practical importance (Coudrec, 1985; Lee et al., 1997; Schmidt et al., 1999). For industrial use, especially with fast development of bio and water cleaning processes, better knowing of these systems become more important. An industrial application of these systems requires determination of transfer characteristics, especially mass transfer. Frequently mass transfer is the rate determining step in the whole process. However, in the real systems, it is not always easy to differentiate the limitation due to the mass transfers from that due to the hydrodynamic.

Mass transfer in liquid-solid packed and fluidized beds has been widely investigated in terms of particle–fluid mass transfer by dissolution, by electrochemical and by ion-exchange methods (Damronglerd et al. 1975; Koloini et al. 1976; Dwivedi & Upadhyay, 1977; Chun & Couderc, 1980; Kumar & Upadhyay, 1981; Rahman & Streat, 1981; Yutani et al. 1987). Some of the results of mass transfer in packed and fluidized beds have been obtained as the transfer between an immersed surface and the liquid (Riba et al. 1979; Bošković et al. 1994; Bošković-Vragolović, et al., 1996&2005). Liquid fluidization of particulate solids has a history which predates the now more commonly applied gas fluidization. The broad range of operations to which liquid fluidization has found applications are: classification of particles by size and density, a special case being sink-and-float separation by density; backwashing of granular filters and washing of soils; crystal growth; leaching and washing; adsorption and ion exchange; electrolysis with both inert and electrically conducting fluidized particles; liquid-fluidized bed heat exchangers and thermal energy storage; and bioreactors. Fluidized-bed bioreactors, which have received much attention during the past thirty years, are usually characterized by the catalytic use of enzymes or microbial cells that are immobilized by attachment, entrapment, encapsulation or self-aggregation. The most common application of such bioreactors is probably in wastewater treatment and, as in the case of the other operations mentioned above, liquid fluidization must in each case be weighed against competing schemes for achieving the same objective before it is adopted commercially (Epstein, 2003). In contrast to fluidized beds data, there are no published data on mass transfer in vertical and horizontal hydrotransport of particles.

Liquid–solid packed and fluidized beds have usually been investigated separately (Kumar & Upadhyay, 1981; Yutani et al. 1987; Comiti & Renaud, 1991; Schmidt et al., 1999), but many authors have noticed the similarity between the two systems (Dwivedi & Upadhyay, 1977).

2. Background of heterogeneous systems

There are a number of different types of systems designed for fluid-solid heterogeneous operations. Two-phase flow systems of fluid (gas or liquid) and particles, the way of realizing contact between the phases, namely how the introduction of fluid in the bed of particles can be (Fig. 1):

systems with a fixed bed of particles;

- packed bed (a),

systems with a moving bed of particles;

- Fluidized bed (b),
- spouted bed and spout-fluid beds (c),
- spouted bed and spout-fluid beds with draft tube (d),
- vertical two-phase flow systems (fluid-particle), i.e. transport systems (e).

Packed beds are an essential part of chemical engineering equipment. A packed bed is a column filled with a support material, in which one or more fluids flow (Fig. 1, a). In chemical processing, a packed bed is a hollow tube, pipe, or other vessel that is filled with a packing material. The packing can be randomly filled with small objects like Raschig rings or else it can be a specifically designed structured packing. Differently shaped packing materials have different surface areas and void space between the packing. Both of these factors affect packing performance. The purpose of a packed bed is typically to improve contact between two phases in a chemical or similar process. Packed beds are widely used in industry to contact two or more fluid phases at relatively low pressure drops. For process design purposes, it is essential that pressure drop be estimated for its proper operation. They are readily used in industry for catalytic reactions, combustion, gas absorption, distillation, drying, and separation processes.

A fluidized bed is a packed bed through which fluid flows at such a high velocity that the bed is loosened and the particle-fluid mixture behaves as though it is a fluid (Fig. 1, b). Thus, when a bed of particles is fluidized, the entire bed can be transported like a fluid, if desired. Both gas and liquid flows can be used to fluidize a bed of particles. In fluidized beds, the contact of the solid particles with the fluidization medium (a gas or a liquid) is greatly enhanced when compared to packed beds. The most common reason for fluidizing a bed is to obtain vigorous agitation of the solids in contact with the fluid, leading to excellent contact of the solid and the fluid and the solid and the wall. Fluidized beds are widely used in industry for mixing solid particles with gases or liquids. In most industrial applications, a fluidized bed consists of a vertically-oriented column filled with granular material, and a fluid (gas or liquid) is flow upward through a distributor at the bottom of the bed.

Fluidized bed technology has become a common method to increase heat and mass transfer in physical and chemical operations. Although a great amount of studies have been performed to characterize the physical properties of these systems, only a few satisfying models have been derived. This is especially true for liquid-solid fluidized beds, which usually play a minor role in practical applications. Furthermore, studies dealing with gas-solid or gas-liquid-solid fluidized beds may not be applied to liquid-solid fluidized beds, because of the very different physical behaviour of these systems (Schmidt et al., 1999).

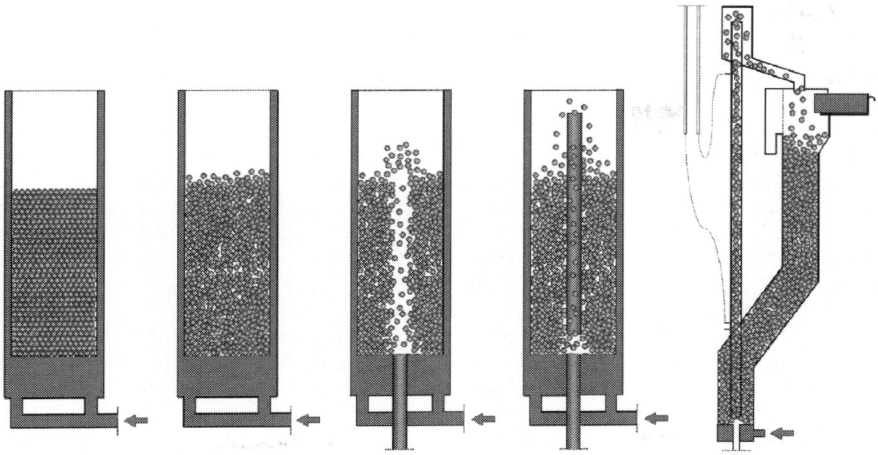

(a) (b) (c) (d) (e)

Fig. 1. Schematic diagram of heterogeneous systems:

a. Packed bed,
b. Fluidized bed,
c. Spout/Spout-fluid bed,
d. Spout/Spout-fluid bed with draft tube,
e. Vertical two-phase flow system (transport system).

A spouted bed can be formed in a vertical column where the fluid jet blows vertically upwards along the centre line of the centre line of the column thus forming a spout in which there is a fast moving fluid and entrained particle (Fig. 1, c). The remainder of the bed, the annulus between the spout and the column wall, is densely packed with particles moving slowly downwards and radially inwards. Fluid percolates through these particles from the spout.

The spouted bed technique, originally developed by Mathur & Gishler, 1955, has been successfully applied to a variety of diffusional, thermal, mechanical and other processes. Various modifications of the classical spouted bed have been proposed to improve its operability and provide better agreement between bed characteristics and process requirements. By introducing external annular fluid flow through the annulus bottom, spout-fluid bed can be formed (Nagarkatti & Chatterjee, 1974; Vuković et al., 1984; Sutanto et al., 1985). The draft tube in these systems is essentially a pneumatic or hydraulic riser which is relatively short even in the industrial scale units.

The addition, a draft tube in a spout and spout-fluid bed changes some of basic their characteristics (Fig. 1, d). The most important of these are:

- there is no bed height limitation imposed by a maximum spout and spout-fluid bed height.
- the minimum spout and spout-fluid (or circulating) velocity is lower in a bed with a draft tube.
- the draft tube forces all of the particles to travel through the entire annulus section surrounding the draft tube before re-entering the spout in the entry section thereby narrowing the particle residence time distribution in the annulus.

The spout and spout-fluid bed with draft tube is a very flexible fluid-particle contacting system because the fluid flowrates through the draft tube and annulus, and the solids circulation rate can be very easily controlled and adjusted in accordance with process needs. As a kind of high-performance reactor for fluid-solid particle reactions, the spouted bed technology is applied to a variety of chemical processes such as dryness, prilling, coating, gasification, combustion, and pyrolysis, etc. During the past decades, numerous modified spouted bed designs have been developed to overcome some of the limitations of the conventional spouted beds, to accommodate the diverse properties of the materials handled and enhance the operability, heat and mass transfer characteristics, and gas-solid or fluid-solid contacting efficiency, for example, multiple spouted beds, spouted fluidized beds, conical spouted beds, draft tube spouted beds, pulsed spouted beds, and rotating spouted beds et al. In addition, many experimental and theoretical studies, which aimed at grasping the more useful flow characteristics of the spouted beds, have also been performed (Hu et al., 2008).
The vertical two-phase fluid-solids flow (Fig. 1, e), is a modified of draft tube spout and spout-fluid bed. In these systems the fluid jet penetrates upward along the central line, forming a spout and entraining the particles up the spout. The tube was mounted in a modified top closed spouted bed in order to obtain non-fluctuating controlled flow of particles. The annulus between the spout and the hopper wall is a densely packed bed with particles moving slowly downwards and radially inward toward the spout. These systems have complex hydrodynamics and the magnitude of the solids flowrate through the transport tube is a major consideration in all applications (Garić, et al., 1995&1996; Garić-Grulović, et al., 2005). All these systems, except packed and classical fluidized bed (Fig. 1), contain elements of a vertical two-phase fluid-particle flow. Design and practical realization of any of these systems that contain elements of a vertical two-phase flow, requires knowledge of the laws that connect the fluid velocity, particle velocity, voidage and pressure gradient as heat and mass transfer coefficients. The vertical two-phase flow systems are important because of their desirable characteristics for use in chemical and biochemical reactors. Unlike many studies of heat and mass transfer in the vertical flow of gas-particle, that include a wide area of application (Matsumoto et al., 1978; Duduković et al., 1996; Mansoori et al., 2002), in the literature are much less data for momentum, heat and mass transfer in systems with the vertical flow of liquid-coarse particles.

3. Investigation in heterogeneous systems

Wall-to-fluid mass transfer in packed beds, fluidized beds and hydraulic transport of spherical glass particles and in single-phase flow regime has been studied experimentally using adsorption and dissolution method. Experiments were performed using spherical glass particles from 1.2 to 3.06 mm in diameter with water in a 25.4 and 40mm I.D. column (Bošković-Vragolović, et al., 2007 & 2009; Garić-Grulović, et al., 2009).
The adsorption method was introduced as a relatively simple mass transfer measurement technique, applicable for liquid flow investigations (Mitrović, et al., 1989). The adsorption method is based on the dynamic adsorption of an organic dye onto a surface covered with a thin layer of a porous adsorbent. In our experiments, very diluted solution of methylene blue ($c_0 = 2.5 \cdot 10^{-3}$ g/dm^3) was used as a fluid in the presence of inert glass particles. The foils of silica gel were used as adsorbent. ("Merck", DC-Alufolien Kie-selgel). Concentration profiles of methylene blue were measured in the flow of water through investigated beds, i.e. packed and fluidized beds (Bošković-Vragolović, et al., 2009). Colour intensity of the surface was determinate by "Sigma Scan Pro" software. Besides basic usage, for

determination of local and average mass transfer coefficients between fluid and solid surface, adsorption method is also, suitable for fluid flow visualization.

The rates of the dye adsorption are supposed to be diffusion controlled. The quantity of the dye transferred during a fixed time period is a function of diffusion boundary layer conditions. The rate of adsorption decreases from high values at the beginning to a constant value for investigated dye concentration. This constant value is the actual mass transfer rate through the completely formed boundary layer. For high concentration of the dye the concentration at the film's surface is $c_0 \neq 0$ and the adsorption process is controlled by mixed kinetics. Also, for high mass transfer rates mixed kinetics occur at lower concentration.

Adsorption from a much diluted solution and far from equilibrium conditions is a very fast process. If the exposure time is short (less than 10 min) the mass transfer rate depends only on the diffusion through the boundary layer. The concentration of adsorbate just above the adsorbent's surface is $c_i = 0$.

Mass flux is,

$$N_A = k\,(c_0 - c_i) = k\,c_0 \tag{1}$$

and for exposure time ,

$$N_A = \frac{c_p}{t} \tag{2}$$

where c_p, is the surface concentration of organic dye on adsorbent layer, c_0 is the bulk concentration.

If the induction period can be short (in the case of thin boundary layers) a simplified expression to calculate mass transfer coefficient is obtained (Končar-Djurdjević, 1953):

$$k = \frac{c_p}{t\,c_0} \tag{3}$$

For this method, value of the surface concentration c_p, is necessary for data quantification. This method is used since 1953, but determination of this parameter, c_p which now can be done easily, using suitable software (Sigma Scan Pro), gives absorption method new actuality.

The dissolution method is based on mass transfer from tube segment of benzoic acid to water. The investigated systems (packed bed, fluidized bed and the transport tube in hydraulic transport) are equipped with a tube segment prepared from benzoic acid (Bošković-Vragolović, et al., 2007; Garić-Grulović, et al., 2001&2009).

The mass transfer coefficient is calculated from the equation:

$$k_c = \frac{\Delta m}{A\,t\,\Delta c} = \frac{\Delta m}{(D_t \pi h)\,t\,\Delta c} \tag{4}$$

The transferred mass, Δm, is determined by measuring the weight loss of benzoic acid. The mass transfer area A is calculated from the mean value of tube segment diameter after and before dissolution. Since the weight loss of benzoic acid Δm is small, bulk concentration is negligible so equilibrium concentration c^* was taken as a driving force Δc. The benzoic acid solubility and diffusivity in water were taken from the literature (Kumar, et al., 1978; Kumar & Upadhyay, 1980).

In each run the average fluid/water temperature was recorded and corresponding values of the diffusion coefficients, fluid viscosity, fluid density and equilibrium solubility were considered in calculations. The water temperature was ranged from 14 to 16°C. Important properties of particles and bed are summarized in table 1.

Particles (glass spheres)					
d_p (mm)	1.2	1.94	2.06	2.98	3.04
ρ_p (kg/m^3)	2641	2507	2461	2509	2465
ε_{mF}	0.392	0.422	0.424	0.462	0.447
U_{mF} (m/s)	0.0129	0.0255	0.0268	0.0435	0.0409
U_t (m/s)	0.208	0.299	0.3025	0.371	0.3680

Table 1. Particle characteristics

4. Hydrodynamics and mass transfer

4.1 Hydrodynamics in heterogeneous systems

For design of heterogeneous systems, used for the mass transfer is important to know hydrodynamics characteristics of systems (i.e. liquid velocity, particle velocity and voidage). Investigation of hydrodynamics in liquid fluidized beds are covered by a large number of research papers, where are presented: the minimum fluidization velocity, the expansion of fluidized bed, the fluid-particle friction coefficient, the drag coefficient of bed and movement of particles (Makkawi & Wright, 2003; Yang & Renken, 2003).

For liquid-particle systems, the two most important characteristics are: for any liquid velocity of the fluidized bed voidage is uniform and the bed voidage is increasing function. Many authors have noted that on the diagram $U = f(\varepsilon)$ in the logarithmic scale is present a change of slope for $\varepsilon \approx 0.85$. Both Garside & Al Dibouni 1977, and Riba & Couderc 1977, have observed a change in the slope of their data on such a plot at $\varepsilon \approx 0.85$ and suggested correlations for low ε ($\varepsilon < 0.85$) and high ε ($\varepsilon > 0.85$) systems. Carlos & Richardson, 1968, and Kmiec, 1978, concluded, analyzing the movement of labelled particles in fluidized beds, that probably there is a change in the mechanism of momentum transfer at voidage $\varepsilon \approx 0.85$.

Our investigations in liquid fluidized beds showed agreement with mentioned literature results (Fig. 2).

Our investigation have shown that in the vertical liquid-particle flow, at voidage $\varepsilon \approx 0.85$ there is also change in mechanism of momentum transfer, which represents the transition from turbulent to parallel flow of particles (Fig 3&4).

In the hydraulic transport of coarse particles, the two characteristic flow regimes are defined as:

a. Turbulent flow, where the particles move vertically but with some radial movement. This regime is characteristic of the lower fluid and particle velocities and appears very much like the inverse settling of a particle suspension. At the same time, the flowing suspension is very much like a particulate fluidized bed, where the whole fluidized suspension flows relative to the tube walls.

b. Parallel flow, where the particles move vertically along a parallel streamline. This regime is characteristic of the higher fluid and particle velocities.

The 2D-photographic illustration of these two regimes is shown in Fig. 3, for glass particles, 5 mm in diameter. The water was coloured with methylene blue. Particle tracks were recorded using a strong light behind the column and an exposure of 0.5 s.

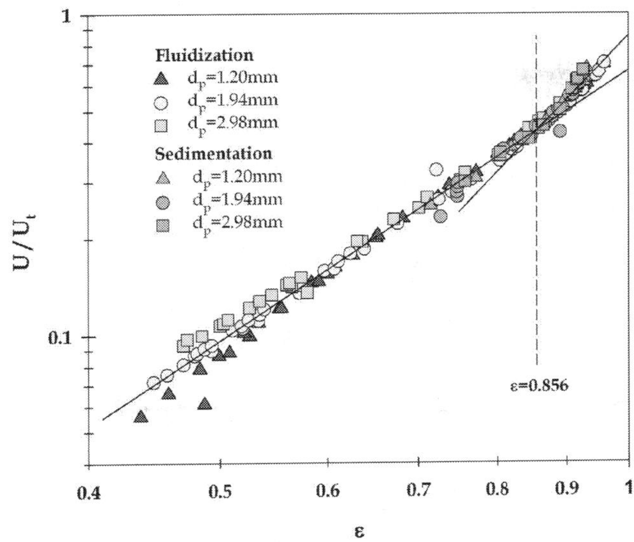

Fig. 2. The relationship between velocity and voidage for water-fluidized and sedime-nting beds of spherical glass particles (D_t = 40mm)

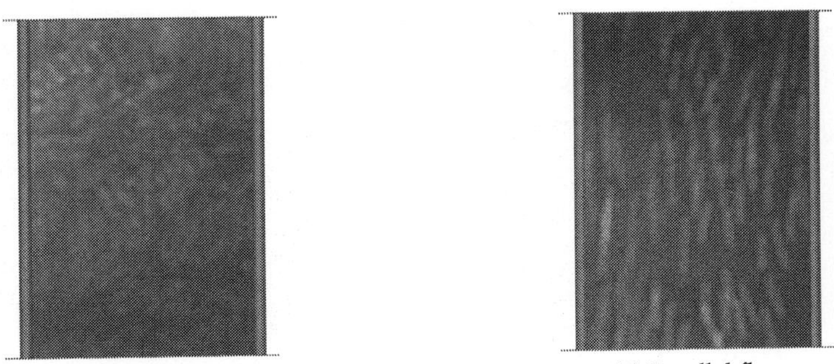

(A) Turbulent flow (B) Parallel flow

Fig. 3. Illustration of turbulent (A) and parallel (B) particle flow. Twodimensional column 60x8 mm, glass spheres 5 mm in diameter, water coloured by methylene blue

In our previous work (Grbavčić et al., 1992; Garić-Grulović et al., 2004&2008), we found that the choking criterion for vertical pneumatic transport lines of Day et al., 1990, is also valid as a criterion for regime designation in the hydraulic transport of coarse particles. Day et al., 1990, studied choking phenomena in vertical gas-solids flow. Using the steady state one-dimensional suspension momentum equation, with constant fluid properties including frictional effects between the bed and the wall (Leung, 1980, Grbavčić et al., 1992), could be expressed as,

$$-\frac{dP}{dz} = (\rho_p - \rho_f)g(1-\varepsilon) + F_w + \gamma\frac{d\varepsilon}{dz} \tag{5}$$

where

$$\gamma = \rho_p v^2 - \rho_f u^2 \tag{6}$$

They proposed $\gamma=0$ at the inlet of the transport tube as their choking criterion. This criterion together with a semi-theoretical relationship for the slip velocity at the inlet leads to an equation for predicting the choking velocity that agrees quite well with all available experimental data for vertical gas-particle flow. Fig. 4 shows the relationship between dimensionless parameter γ^* and transport line voidage, where:

$$\gamma^* = \frac{\rho_p v^2 - \rho_f u^2}{\rho_f U_t^2} \tag{7}$$

As can be seen, from Fig. 4, the critical voidage for regime transition in vertical transport is about $\varepsilon \approx 0.85$. The value of voidage $\varepsilon \approx 0.85$ is shown as a typical value in particulately fluidized and sedimenting beds (Fig. 2). Since in the turbulent flow regime in vertical transport, the frequency of particle collisions with the tube wall is much higher, it is reasonable to expect higher mass transfer rates in this flow regime.

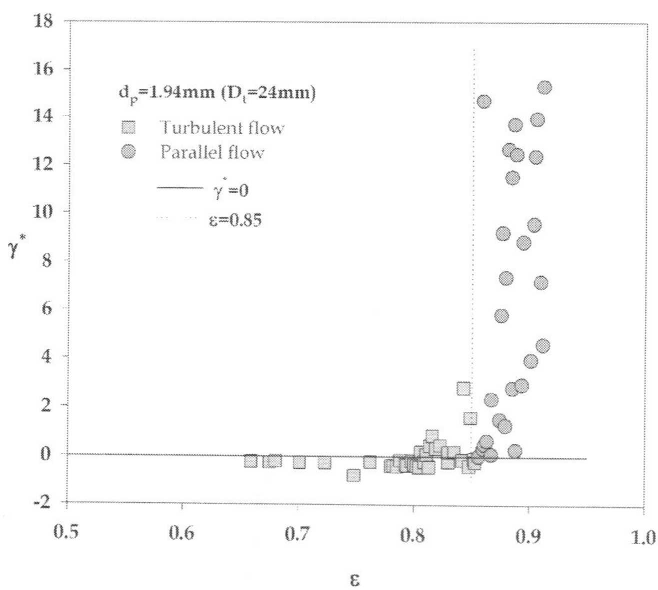

Fig. 4. Relationship. γ^* vs. ε for hydraulic transport (Grbavčić et al., 1992; Garić-Grulović et al., 2004&2008)

4.2 Mass transfer in heterogeneous systems

Mass transfer in heterogeneous systems is influenced by basic hydrodynamics parameters: fluid velocity, bed voidage and particles concentration. Fig. 5, shows relationship among mass transfer coefficient, liquid velocity and voidage in all investigated systems. Increasing liquid velocity increases mass transfer in packed bed. In fluidized bed increasing liquid velocity leads to expansion of bed (ε> 0.7) i.e. reduction of the concentration of particles in the bed, which contributes to reducing the mass transfer coefficient. In hydraulic transport, increasing liquid velocity and slightly changes of voidage have positive influence on mass transfer. Finally, as can be seen in Fig. 5, the influence of particle concentration for all system is slightly greater then the influence of liquid velocity.

Fig. 6, shows relationships between mass transfer coefficient and superficial liquid velocity. The highest mass transfer coefficient was at minimum fluidization velocity because of high concentration of moving particles.

Constant movement of particles in the bed causes mixing of fluid near the wall reduces the thickness of the boundary layer and increases the mass transfer (Yutani et al., 1987; Schmidt et al., 1999). In addition, investigation of Schmidt et al., 1999, indicate existence of a maximum of Sherwood number Sh = f (k), at bed voidage about 0.8. Results of our investigations (Fig. 5&6), show existence of maximum of mass transfer coefficient at transition from packed to fluidized bed, but there is no maximum for fluidized bed. In addition, with increasing fluid velocity fluidized bed is expands (ε>0.8, Fig. 6&7).

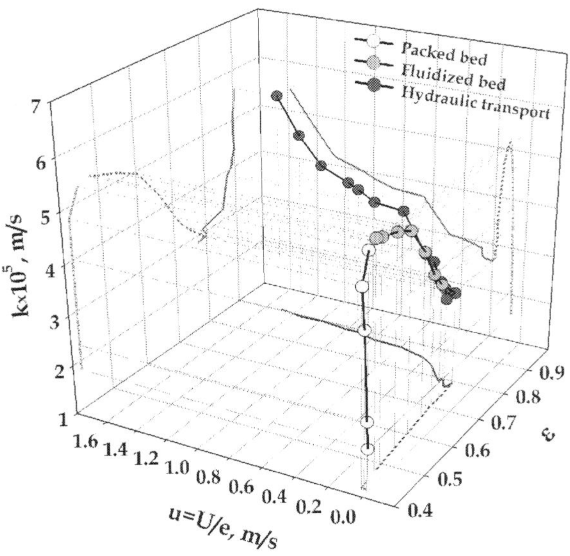

Fig. 5. The relationship between mass transfer coefficient, interstitial fluid velocity and voidage, for packed beds, fluidized beds and vertical transport (dissolution method; d*p*=1.94mm)

Chromatogram on Fig. 7, gives clear visualization of fluid flow around particles in the packed bed. Colour intensity is proportional to local mass transfer rates. Also, on Fig. 7, are shown chromatograms for fluidized beds at minimal velocity and for highly expanded fluidized bed. Uniform colour intensity could be observed in both cases, with higher intensity for minimal fluidized bed velocity.

Fig. 8, presents mass transfer coefficient as a function of superficial fluid velocity for packed beds, fluidized beds, vertical transport and single phase flow. With increasing liquid velocity in packed beds, mass transfer coefficient increases for all bed voidage. With increasing superficial liquid velocity in fluidized beds mass transfer coefficient slightly decreases, tends to the value of mass transfer coefficient for single phase flow. In the vertical transport for low transport velocities mass transfer coefficient is constant, and it is higher than mass transfer coellicient for single liquid flow. For higher transport velocities mass transfer coefficient increases, but there no significant difference between transport and single liquid flow.

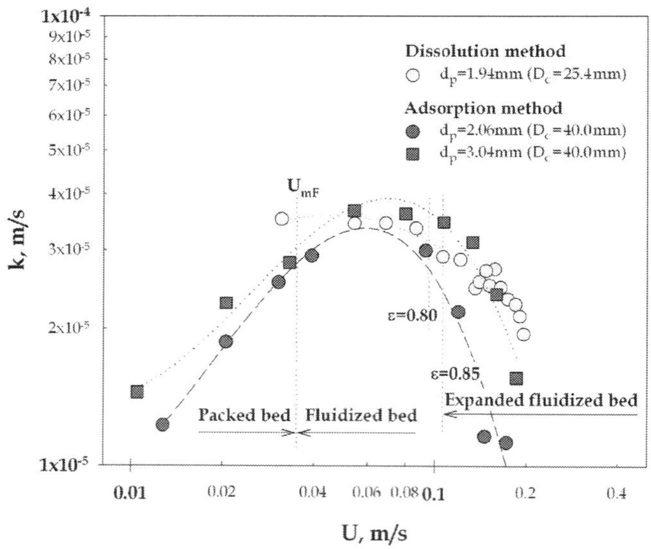

Fig. 6. Mass transfer coefficient vs. superficial fluid velocity for packed and fluidized beds

Fig. 9, present the relationship between particle Sherwood number and particle Reynolds number for different particle diameter in fluidized beds and vertical transport. The data are separated into tree groups with particle diameter. The experimental results show that with increasing Reynolds number (liquid velocity), Sherwood number (mass transfer coefficient), slightly decreases in fluidized beds.

With increasing liquid velocity the Sherwood number for vertical transport is constant for low transport velocities, but for higher transport velocities mass transfer increases. It implies that because of low particle concentration in transport column the influence of particles on diffusivity boundary layer is not significant for higher transport velocities.

Comparison of the data for fluidized bed (d_p=1.94mm) with several literature correlations (Upadhyay & Tripathi, 1975; Tournie et al., 1977; Bošković, et al., 1994), show significant difference between our data and the available correlations. The data calculated with it could be seen that are quit different.
The following correlation proposed by Upadhyay & Tripathi, 1975, for mass transfer in packed and fluidized bed,

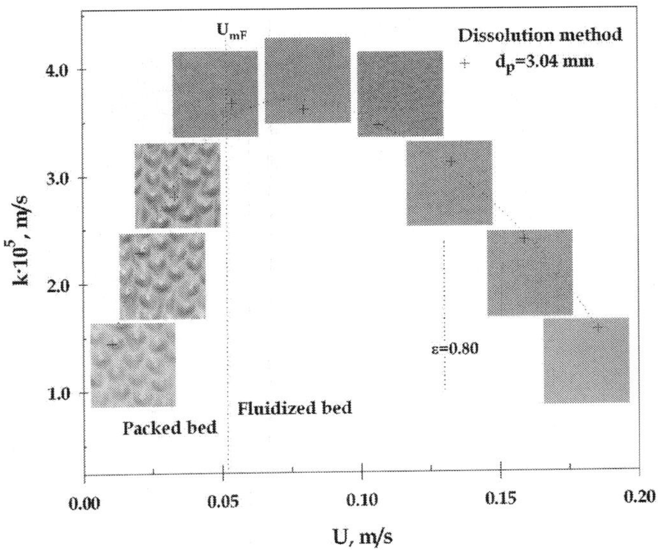

Fig. 7. Relationship between mass transfer coefficient and superficial liquid velocity in packed and fluidized beds

$$Sh_p = 3.8155 \cdot Re^{''0.2687} Sc^{1/3} \quad \text{for} \quad Re^{''} < 20 \tag{8}$$

$$Sh_p = 1.6218 \cdot Re^{''0.5553} Sc^{1/3} \quad \text{for} \quad Re^{''} > 20 \tag{9}$$

where $Re''=Re_p/(1-\varepsilon)$, in the following range of variables: $0.01<Re''<12000$, $572<Sc<70000$ and $0.268<\varepsilon<0.9653$. The data calculated with correlation (eq. 9) show the maximum which has not been confirmed by our experimental data (Fig. 9).
Also, Tournie et al., 1977, have given the correlation for mass transfer particle-fluid in fluidized bed,

$$Sh_p = 0.253 \cdot Re_p^{0.004} Ga^{0.319} Mv^{0.299} Sc^{0.4} \tag{10}$$

where $Mv=(\rho_p-\rho_f)/\rho_f$, and the equation (10) is recommended in the following range of variables: $1.6<Re_p<1320$, $2470<Ga<4.42 \cdot 106$, $0.27<Mv<1.14$, $305<Sc<1595$. The data calculated with Tournie et al., 1977, correlation for mass transfer particle-fluid in fluidized bed (eq. 10), show significant difference our experimental values (Fig. 9).

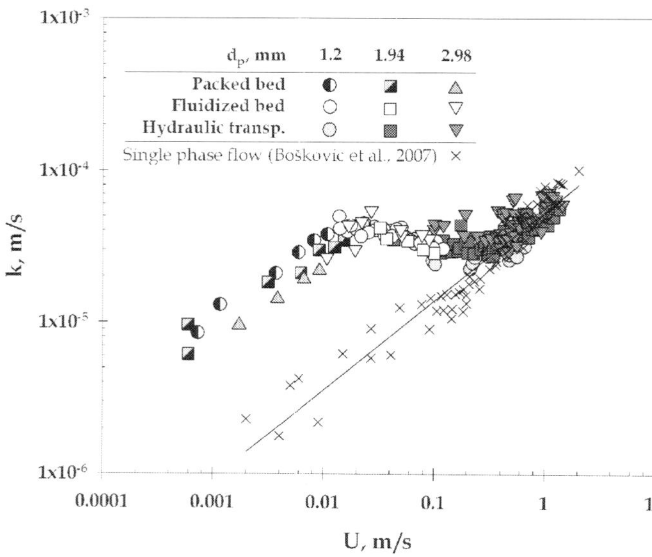

Fig. 8. Mass transfer coefficient vs. superficial fluid velocity for packed beds, fluidized beds, vertical transport and for single liquid flow by dissolution method

The mass transfers in liquid fluidized bed, Bošković et al., 1994, are correlated by the equation,

$$Sh_p = 0.261 \cdot Re_p^{0.03} Ga^{0.324} Sc^{1/3} \tag{11}$$

in the following range of variables: Re_p=15÷400, Sc=1361÷1932. This equation is derived for mass transfer between fluid and an immersed sphere in fluidized beds of spherical inert particles. The predicted values using the correlation by Bošković, et al., 1994 (eq. 11), are in good agreement with our experimental data (Fig. 9).

Fig. 10, present the relationship between mass transfer factor and particle Reynolds number in fluidized beds obtained by adsorption and dissolution methods. Also this picture gives the comparison of our experimental data with several literature correlations (Chu et al., 1953; Fan et al., 1960; Upadhyay & Tripathi, 1975; Dwivedi & Upadhyay, 1977; Bošković, et al., 1994), expressed through the mass transfer factor (j_D), which is defined as:

$$j_D = \frac{Sh}{Sc^{1/3} Re} \tag{12}$$

The data are calculated with correlation proposed by Chu et al., 1953, as follows,

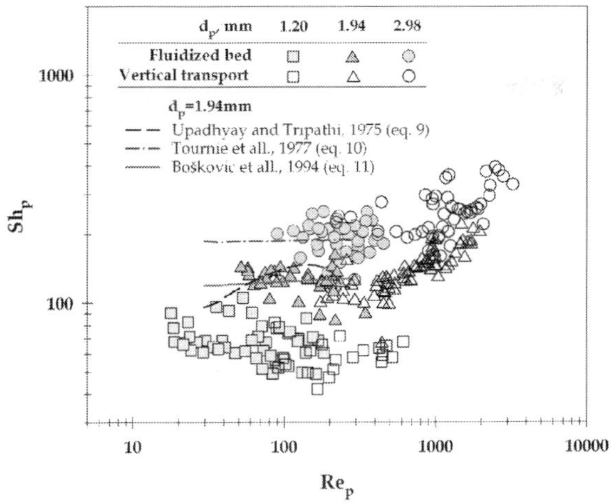

Fig. 9. Relationship between Sherwood number and Reynolds number for particles (dissolution method)

$$j_D = 5.7 \cdot Re''^{-1.22} \quad \text{for} \quad 1 < Re'' < 30 \tag{13}$$

$$j_D = 1.77 \cdot Re''^{-1.56} \quad \text{for} \quad 30 < Re'' < 1000 \tag{14}$$

where $Re'' = Re_p / (1-\varepsilon)$, show good agreement with our experimental data (Fig. 10).

Fan et al., 1960, on the basis of adsorption tests of phenol on activated carbon particles in liquid fluidized bed came to the following correlation,

$$j_D = 1.865 \cdot Re'^{\,0.48} \tag{15}$$

where $Re' = Re_p / \varepsilon$. The predicted values from the correlation (15) and our experimental data have shown significant difference (Fig. 10).

The data calculated with correlation proposed by Upadhyay and Tripathi, 1975, for mass transfer in packed and fluidized bed (from eq. 9&12),

$$j_D = 1.6218 \cdot Re''^{0.4447} \quad \text{for} \quad Re'' > 20\text{)(eq. 9)} \tag{16}$$

where $Re''=Re_p/(1-\varepsilon)$, show good agreement with the our experimental data (Fig. 10).

Also Fig. 10, gives the comparison of our experimental data with correlations of Dwivedi & Upadhyay, 1977, who proposed the following correlations,

$$j_D \cdot \varepsilon = 1.1068 \cdot Re_p^{-0.72} \quad \text{for} \quad Re_p < 10 \tag{17}$$

$$j_D \cdot \varepsilon = 0.4548 \cdot Re_p^{-0.4069} \quad \text{for} \quad Re_p > 10 \tag{18}$$

based on the many different experimental data for mass transfer of liquid in packed and fluidized beds. As can be seen, our data show very good agreement with correlation for lower values of Reynolds number for particles in the fluidized beds.

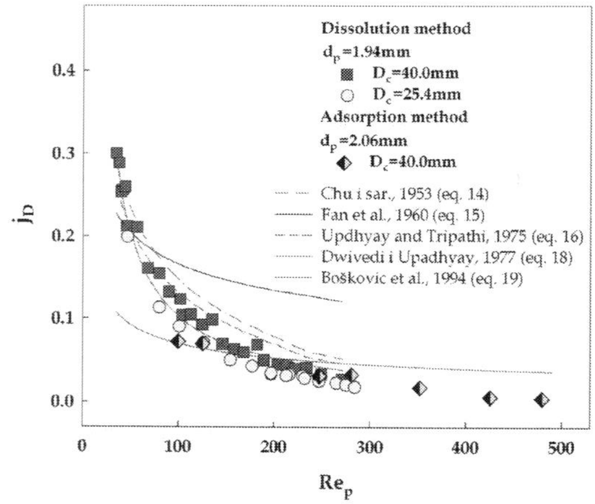

Fig. 10. Variation of mass transfer factor with Reynolds number for particles

And finally, the correlation of Bošković et al., 1994, has been derived for the mass transfer between fluid and an immersed sphere in fluidized beds of spherical inert particles (from eq. 11&12),

$$j_D = 0.261 \cdot Ga^{0.324} Re_p^{-0.97} \tag{19}$$

shows excellent agreement for the mass transfer wall-to-fluid (Fig. 10).

Fig. 11, presents the mass transfer factor as a function of Reynolds number in packed, fluidized beds and hydraulic transport for different experimental techniques. It could be seen that there is no significant difference between mass transfer factors obtained by this two methods (Bošković, et al. 1994; Bošković-Vragolović, et al., 2007). Often used dissolution method is very reliable, and agreement of data shows that the adsorption method gives good results, also. Advantage of adsorption method is possibility to obtain local mass transfer coefficients.

5. Conclusion

The basic hydrodynamic parameters of packed bed, fluidized beds and vertical transport are analyzed.

The influence of different parameters as liquid velocity, particles size and voidage on mass transfer in packed beds, fluidized beds and hydraulic transport are presented. The data for mass transfer in all investigated systems are shown using Sherwood number (Sh) and mass transfer factor-Colburn factor (j_D) as a function of Reynolds number (Re) for particles and for column.

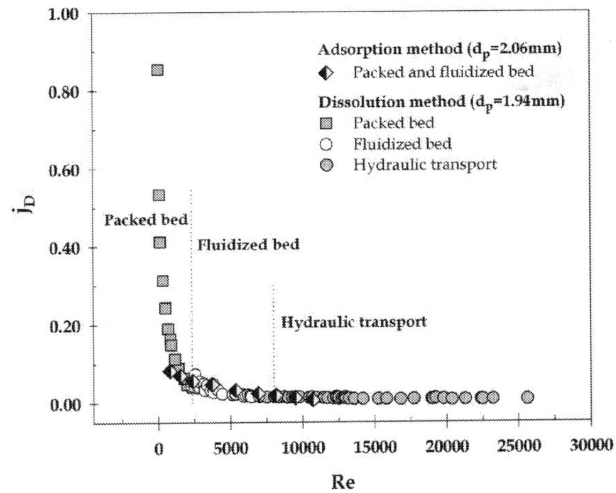

Fig. 11. The relationship between mass transfer factor and Reynolds number

With increasing liquid velocity in packed beds mass transfer coefficient increases while in fluidized beds mass transfer coefficient decrease. The highest mass transfer coefficient was at minimum fluidization velocity because of high concentration of moving particles.
In the vertical transport for low transport velocities mass transfer coefficient is constant, and it is higher than mass transfer coel1icient for single liquid flow. For higher transport velocities mass transfer coefficient increases, but there no significant difference between transport and single liquid flow.

6. Nomenclature

A	mass transfer area, m^2
c^*	equilibrium concentration, kg/m^3
c_i	concentration of adsorbate, kg/m^3
c_0	concentration of the bulk, kg/m^3
cp	surface concentration, kg/m^3
Δc	concentration difference (driving force), kg/m^3
D_{AB}	molecular diffusion coefficient, m^2/s
d_p	particle diameter, m
D_t	diameter of the column/transport tube, m
F_w	pressure gradient due to the suspension-wall friction, Pa/m
g	gravitational acceleration, m/s^2
Ga	Galileo number ($=(d_p^3\rho_f^2 g)/\mu^2$)
h	high of tube segment, m
j_D	mass transfer factor, $(k/U)\cdot Sc^{1/3}$
k	mass transfer coefficient, m/s
Δm	transferred mass, kg

Mv	density number (=$(\rho_p-\rho_f)/\rho_f$)
N_A	mass flux, kg/(m^{2}s)
P	dynamic pressure, Pa
Re	Reynolds number for column (=$D_t\rho_f U/\mu$)
Re_p	Reynolds number for particles (=$d_p\rho_f U/\mu$)
Sc	Schmidt number (=$\mu/(\rho_f D_{AB})$)
Sh	Sherwood number for column (=kD_t/D_{AB})
Sh_p	Sherwood number for particles (=kd_p/D_{AB})
t	exposure time, s
u	mean interstitial fluid velocity in the transport tube (=U/ε), m/s
U	superficial fluid velocity in the spout and in the transport tube, m/s
U_{mF}	minimum fluidization velocity, m/s
U_t	particle terminal velocity in an infinite medium, m/s
v	radially averaged particle velocity in the transport tube, m/s
z	vertical distance coordinate, m
	Greek letters
γ	defined by Eq. (6)
γ^*	defined by Eq. (7)
ε	voidage
ε_{mF}	voidage at minimum fluidization
ρ_f	fluid density, kg/m^3
ρ_p	particle density, kg/m^3
μ	viscosity of the fluid, Ns/m^2

7. Acknowledgments

Acknowledgment Financial support of the Research Council of Serbia is gratefully acknowledged.

8. References

Bošković, N.; Grbavčić, Ž. B.; Vuković, D. V. & Marković-Grbavčić, M. (1994). *Powder Techn. 79,* pp. 217.

Bošković-Vragolović, N.; Grbavčić, Ž.; Janković, D. & Minić, V. (1996). Mass and Momentum Transfer in Packed Beds of Spherical inert Particles, *J. Serb. Chem. Soc., 61(4-5),* pp.401-409.

Bošković-Vragolović, N.; Brzić, D. & Grbavčić., Ž. (2005). Mass transfer between a fluid and an immersed object in liquid-solid packed and fluidized beds, *J. Serb. Chem. Soc. 70 (11),* pp. 1373-1379

Bošković-Vragolović, N.; Garić-Grulović, R. & Grbavčić., Ž. (2007). Wall-to-liquid mass transfer in fluidized beds and vertical transport of inert particles, *J.Serb.Chem.Soc., 72(11),* pp. 1103-1113.

Bošković-Vragolović, N.; Garić-Grulović, R.; Grbavčić, Ž. & Pjanović, R. (2009). Mass Transfer and Fluid Flow Visualization in Packed and Fluidized Beds by the Adsorption Method, *Russian Journal of Physical Chemistry A, Vol. 83, No. 9,* pp. 1550-1553.

Carlos, C.R. & Richarrdson, J. F. (1968). Solids movement in liquid fluidized beds-I Particle velocity distribution, *Chem. Engng. Sci., 23 (8),* pp. 813-824.

Chu, J.C.; Kalil, J. & Wetteroth, W.A. (1953). Mass Transfer in a Fluidized Bed, *Chem. Enger. Prog., 49 (3),* pp. 141-149.

Chun, T. & Couderc, J. P. (1980). High flux mass transfer in liquid fluidization, *Chem. Eng. Sci. 35(8),* pp. 1707-1715.

Comiti, J. & Renaud, M. (1991). Liquid-solid mass transfer in packed beds of parallelepipedal particles: energetic correlation, *Chem. Eng. Sci.* 46(1), pp. 143-154.

Coudrec J.P. (1985). *Fluidization*, 2nd Edition, Editors Academic Press, London, pp. 1-46.

Damronglerd, S., Couderc, J.P. & Angelino, H. (1975). Mass transfer in particulate fluidization, *Trans. Inst. Chem. Engrs, 53,* pp. 175-180.

Day, J.Y.; Littman, H. & Morgan III, M.H. (1990). A New Choking Velocity Correlation for Vertical Pneumatic Conveying, *Chem.Eng.Sci., 45,* pp. 355-360.

Duduković, A.; Milošević, V. & Pjanović, R. (1996). Gas-solid and gas-liquid mass transfer coefficients, AIChE J., 42, pp. 269-270.

Dwivedi, P. N. & Upadhyay, S. N. (1977). Particle-fluid mass transfer in fixed and fluidized beds, *Ind. Eng. Chem. Proc. Des. Dev. 16(2),* pp. 157-165.

Epstein, N. (2003). Applications of Liquid-Solid Fluidization, *Int. J. of Chem. Reactor Eng., vol. 1,* R1.

Fan, L.S. ; Yang, C. & Wen, C.Y. (1960). Mass transfer in semifluidized beds for solid-liquid system, *AIChE J., 6,* pp. 482–487.

Froment, G.F. & Bischoff, K.B. (1990). *Chemical Reactor Analysis and Design (2nd Ed.),* Wiley Series in Chemical Engineering, John Wiley & Sons, ISBN-10: 0471510440.

Garić, R.V.; Grbavčić, Ž.B. & Jovanović, S.Dj. (1995). Hydrodynamic modeling of vertical non-accelerating gas-solids flow, *Powder Technol., 84 (1),* pp.65-74.

Garić, R.; Grbavčić, Ž. & Rožić, Lj. (1996). Indirect determination of the particle-wall friction coefficient in the vertical pneumatic and hydraulic conveying of coarse particles, *J.Serb.Chem.Soc., 61,* pp.391-400.

Garić-Grulović, R. V.; Grbavčić, Z.B. & Arsenijević, Z.Lj. (2001). Mass Transfer in Vertical Two-Phase Flow of Liquid Coarse Particles, *The 3rd Internacional Symposium for South-east European Countries (SEEC) on Fluidized Beds in Energy Production, Chemical and Process Engineering and Ecology,* Sinaia, Romania, Proc., pp.23-30.

Garić-Grulović, R. V.; Grbavčić, Ž. B. & Arsenijević, Z. Lj. (2004). Heat Transfer and Flow Pattern in Vertical Liquid-Solids Flow, *Powder Technol., 145 (3),* pp.163-171.

Garić-Grulović, R.V.; Grbavčić, Ž.B. & Arsenijević, Z.Lj. (2005). A pseudo-fluid representation of vertical liquid-coarse solids flow, *J.Serb.Chem.Soc., 70(5)* pp.775-784.

Garić-Grulović, R.; Bošković-Vragolović, N.; Grbavčić, Ž. & Arsenijević, Z. (2008). Wall-to-bed heat transfer in vertical hydraulic transport and in particulate fluidized beds, *International Journal of Heat and Mass Transfer, 51,* pp.5942–5948.

Garić-Grulović, R.; Grbavčić, Ž.; Bošković-Vragolović, N. & Arsenijević, Z. (2009). Mass transfer in vertical liquid–solids flow of coarse particles, Powder Technology, Volume 189, Issue 1, 25, pp. 130-136.

Garside, J. & Al-Dibouni, M. R.(1977). Velocity-Voidage Relationships for Fluidization and Sedimentation in Solid-Liquid Systems, Ind. Eng. Chem. Process Des. Dev., pp. 16 (2), pp 206–214.

Grbavčić, Ž.B.; Vuković, D.V.; Garić, R.V.; Hadžismajlović, Dž.E.; Littman, H.; Jovanović, Dj.S.; Littman, H. & Morgan III, M.H. (1992). Hydrodynamic Modeling of Vertical Liquid-Solids Flow, *Powder Technol., 72 (2),* pp. 183-191.

Hu, G. ; Gong, X. ; Wei, B. & Li, Y. (2008). Flow patterns and transitions of a novel annular spouted bed with multiple air nozzles, *Industrial and Engineering Chemistry Research, 47(3),* pp. 9759-9766.

Kmiec, A. (1978). Particle distribution and dynamics of particle movement in solid-liquid fluidized beds, *Chem. Engng. J., 15 (1),* pp. 1-12.

Koloini, T.; Sopčič, M. & Žumer, M. (1976). Mass transfer in liquid-fluidized beds at low reynolds numbers, *Chem. Eng. Sci. 32(6),* pp. 637-641.

Končar-Djurdjević, S. (1953). Application of a New Adsorption Method in the Study of Flow of Fluids, *Nature, 172(4384),* pp. 858.

Kumar, S.; Upadhyay, S.N. & Mathur, V.K. (1978). On the Solubility of Benzoic Acid in aqueous Carboxy Methyl Cellulose Solutions', *J. Chem. Eng. Data*, 23, pp.139-141.

Kumar S. & Upadhyay, S. N. (1980). Laminar Flow Dissolution and Diffusion in Non-Newtonian Fluid, *Ind. Eng. Chem. Fundamen.*, 19(1), pp. 75-79.

Kumar, S. & Upadhyay, S. N. (1981). Mass and momentum transfer to Newtonian and non-Newtonian fluids in fixed and fluidized beds, *Ind. Eng. Chem. Fundamen., 20,* pp. 186-195.

Lee, J.; Chun, H. & Shemilt, L. (1997). Wall-to-liquid mass transfer in liquid-solid and gas-liquid-solid fluidized beds, J.Chem.Eng.Japan, 30(2), pp. 246-252.

Leung, L.S. (1980). Fluidization, *in J.R.Grace and J.M.Matsen (Eds.), Plenum Press, New York,* pp. 25-68.

Makkawi, Y.T. & Wright, P.C. (2003). The voidage function and effective drag force for fluidized beds, *Chem. Engng. Sci., 58 (10),* pp. 2035-2051.

Mansoori, Z.; Saffar-Avval, M.; Tabrizi, H.B. & Ahmadi, G. (2002). Modelling of heat transfer in turbulent gas-solid flow, *Int. J. Heat Mass Transfer, 45,* pp. 1173-1184.

Mathur, K.B. & Gishler, P.E. (1955). A technique for contacting gases with coarse solid particles, *AIChE Journal, 1,* pp. 157-164.

Matsumoto, S.; Ohnishi, S. & Maeda, S. (1978). Heat Transfer to Vertical Gas-Solids Flows, *J. Chem. Engng. Japan, 11,* pp. 89-95.

Mitrović, M.; Končar-Đurđević, S.; Cvijović, S.; Voronjec, D. & Duduković, A. (1989). Fluid flow visualisation and analysis by the adsorption method, *J.Serb.Chem.Soc, 54 (8),* pp. 383-406.

Nagarkatti, A. & Chatterjee, A. (1974). Pressure and Flow Characteristics of a Gas Phase Spout-Fluid Bed and the Minimum Spouifluid Condition, *Can. J. Chem. Eng., 52,* pp. 185-195.

Rahman, K. & Streat, M. (1985). Mass transfer in fixed and fluidized beds, *Chem. Eng. Sci. 40 (9),* pp. 1783-1785.

Riba, J. P. & Couderc, J. P. (1977). Expansion de couches fluidisées par des liquids, *Can. J. Chem. Eng. 55,* pp. 118.

Riba, J. P.; Routie, R. & Couderc, J. P. (1979) *Fluidization,* J. F. Davidson, D. L. Keairns, Eds. Cambridge, University Press, Cambridge, England, pp. 157.

Schmidt, S.; Buchs, J.; Born, C. & Biselli, M. (1999). A new correlation for the wall-to-fluid mass transfer in liquid-solid fluidized beds, *Chem. Eng. Sci., 54,* pp. 829-839.

Sutanto, W.; Epstein, J. &. Grace, R. (1985). Hydrodynamics of Spout-Fluid Beds, *Powder Technology, 44,* pp. 205-212.

Vuković, D.V.; Hadžismajloviž, Dž.E.; Grbavčić, Ž.B.; Garić, R.V. & Littman, H. (1984). Flow Regimes for Spout-Fluid Beds, *Can.J.Chem.Eng. 62(6),* pp. 825-829.

Yang, Y. & Renken, A. (2003). A generalized correlation for equilibrium of forces in liquid-solid fluidized beds, *Chem. Engng. J., 92 (1-3),* pp. 7-14.

Yutani, N.; Ototake, N. & Fan, L. T. (1987). Research Article Statistical analysis of mass transfer in liquid-solids fluidized beds, *Ind. Eng. Chem. Res.*, 26, pp. 343-347.

12

Flow and Mass Transfer inside Networks of Minichannels

Florian Huchet
LCPC
France

1. Introduction

The process miniaturization constitutes a challenge for the Chemical Engineering domain. The particular benefit in term of the increase of the ratio between the transfer surface area and the fluid volume inside microfluidic system is really promising for the conception of efficient apparatus such as microreactors, micromixers and microseparators allowing a better chemical reaction control and heat and mass transfer intensification in order to realize sustainable industrial equipments. On other hand, a proper design of a microstructured platform where miniaturized reactors, mixers and separators are implemented with integrated sensors is crucial for the fabrication of new materials, chemical or bio-chemical products and testing new catalyst and reagent (Gunther & Jensen, 2006). Flow and mass transfer characterization inside these new tools of development and production is fundamental for their optimal design.

Yet, these new pieces of equipment are made in stainless steel integrating about hundreds of microchannels either about several tens of microexchangers. These microstructured exchangers can operate at high pressure and present three-dimensional geometries. Hessel *et al.* (2005) report the order of magnitude of the flow rate in various microstructured reactors. The flow rates range between 10 and 10000 $l.h^{-1}$ and the flow regime is usually transitional or turbulent. In spite of new experimental methods (Sato *et al.*, 2003; Natrajan & Christensen, 2007), it remains difficult to measure simultaneously a scalar quantity (concentration, temperature, velocity) at different locations of the microstructured reactors. Thus, a lot of difficulties occur in the prediction of wall transfer phenomena (heat, mass, momentum) in the microstructured reactors in view of their integration in chemical manufacture. A characterization of the flow behaviour and of heat and mass transfer performance is needed in order to develop and improve these microsystems for their application in process engineering. A large number of studies dealing with flow through microsystems of different shapes and flow configurations is available in the literature since a few years. Among them, T-microchannel (Bothe *et al.*, 2006) or hydrodynamics focusing (Wu & Nguyen, 2005) are some promising classes of flow configurations for microfluidics apparatus applications. Various complex geometries are usually studied by using numerical approaches or global measurements to characterize transfer phenomena in heat exchangers (Brandner *et al.*, 2006) or microreactors (Commenge *et al.*, 2004). The flow inside these microreactors or microexhangers are usually in the transitional or turbulent regime and the experimental description of all the hydrodynamics scales become more difficult than in classical macrodevices. In the mixing research area, the characterization of the mixing scales is nevertheless fundamental for the design and the optimization of the microscale devices. The fluid flow at the microscale level is mainly connected to the characteristics of flow in the transitional and turbulent regimes. The conditions of stationarity, homogeneity and isotropy cannot be assumed in confined turbulent flow in microsystems. Thus, it is of some importance, from both academic and practical points of view, to study confined flow and mixing with particular attention given to the small scale motion. In spite of the recent work dealing with local hydrodynamics analysis inside microchannels, in particular by µPIV (Li and Olsen, 2006), very few paper are dedicated to both hydrodynamics and mixing at the small scales especially in the near-wall vicinity. A high sampling frequency is required to adequately describe a confined turbulent flow characterized by non-Gaussian and high level fluctuations. Recently, it constitutes an important challenge for classical turbulence investigations techniques. (Natrajan & Christensen, 2007, Natrajan *et al.*, 2007).

The objective of the experimental research work presented in this chapter is to use several methods in order to characterize flow and mass transfer inside networks composed of crossing minichannels. The cells are some geometric model to study a complex confined flow such as those met in certain mini-heat exchangers or mini-catalytic reactors. The originality is to apply proper experimental methods in order to describe the transfer phenomena at several scales. The global approaches are relevant in the frame of the flow regimes identification and the comparison with other geometries in term of liquid-solid mass transfer performed at three large nickel electrodes and pressure drop measurements. The local approaches are performed in the frame of a multi-scales diagnostic of the flow by PIV (Particle Image Velocimetry) and by using electrochemical microsensors. The electrochemical method constitutes the originality of the used experimental tools. The high potential of electrochemical techniques (Yi *et al.*, 2006; Martemianov *et al.*, 2007) has recently attracted a significant attention in the microfluidic area due to its ability to detect a large range of species (chemical or biochemical) and the low cost instrumentation compared to optical methods for instance. Integration of multiple microelectrodes allows simultaneous measurements at different locations inside the microexchanger. The electrodiffusion probes are used for the mapping of wall shear rates in the flow cell. An array of 39 microelectrodes allows us to characterize the flow regimes, longitudinal and lateral evolutions of the flow structures and flow behaviour at the channels crossings. In other hand, the use of the electrochemical microsensors method is also adapted to the characterization of the mixing state in different geometries of minireactor composed of networks of minichannels.

Thus, this chapter is organized in several sections:

- the next section is dedicated to the presentation of the electrochemical diagnostics based on the condition of the diffusional limitation at the wall microprobes. Two methods allowing the assessment of the instantaneous wall shear rate determination are compared by using an adapted signal processing tools,
- the third section is dedicated to the materials and the calibration methods with a special attention given to the experimental cell composed of a network of crossing minichannel.
- the fourth section presents the local flow results obtained using PIV measurements and the electrodiffusion diagnostics,
- the fifth section deals with the global characterization by using liquid-solid mass transfer and pressure drop measurements,
- the sixth section is dedicated to the mixing performance characterization inside two differents networks of minichannels,
- conclusion and outlooks are finally drawn.

2. Electrochemical method and post-processing tools

2.1 Electrodiffusion technique

The technique is based on the wall shear rate measurement (Hanratty & Campbel, 1983) consisting in using mass transfer probes flush-mounted in the wall. A potential difference is applied between the microprobes acting as cathodes and a large area anode. A fast electrochemical reduction reaction takes place at the microprobes surface allowing the diffusion boundary layer development as drawn in figure 1.

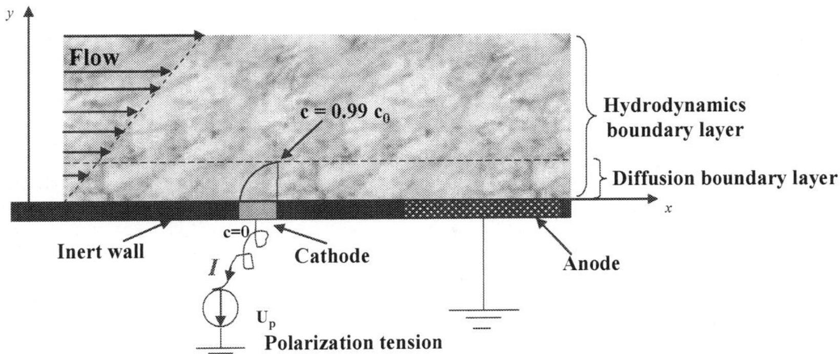

Fig. 1. Electrochemical method principle

The electrochemical reaction employed in the frame of our work is the reduction of ferricyanide ions on a circular platinum cathode:

$$Fe(CN)_6^{3-} + e^- \rightarrow Fe(CN)_6^{4-} \tag{1}$$

The principle involves the measurement of a current under diffusional limiting conditions at the microprobes in such a way that the reaction rate is diffusion-controlled through the mass transfer boundary layer, δ_d, and that the ionic migration can be neglected due to the presence of a supporting electrolyte. The measured intensity varies with the applied voltage between the anode and the cathode until it reaches a constant value, I_{lim}, corresponding to the limiting diffusion conditions. The mass transfer coefficient, k, can then be calculated by the Faraday's expression:

$$k = I_{lim} / v_e \Im A c_0 \tag{2}$$

where v_e is the number of electrons involved in the red-ox reaction, $\Im$ is the Faraday's constant, A is the surface area of the microelectrode and c_0 is the bulk concentration of the reacting species.
In the case of large active surface of the electrode, the measured mean current correspond to the global mass transfer at the wall, k_{mt}.
By working with microelectrodes, the mean measured limiting current is controlled locally by convective diffusion and the well-known Lévêque formula (Lévêque, 1928) can be applied to determine the mean wall shear rate, $\bar{s}$. The stationary equation has been solved (Reiss & Hanratty, 1963) for a circular microelectrode:

$$\bar{s} = 3.22\left(\bar{I}_{lim}/v_e \Im c_0 d_e^{5/3} D^{2/3}\right)^3 \quad (3)$$

where d_e is the diameter of the circular electrode, and D is the diffusion coefficient of the active species in the solution.

The analytical quasi-steady state interpretation solution of the measured current correctly describes the time response of the mass transfer rates, and the instantaneous wall shear rate, $s_q(t)$, can be related to the instantaneous mass transfer rates by the same equation as for steady flow (equation 2):

$$s_q(t) = 3.22(v_e \Im c_0)^{-3} D^{-2} d_e^{-5} I_{lim}^{\ 3}(t) \quad (4)$$

For high frequency fluctuations of the wall shear rate the filtering effect of the mass boundary layer damps the fluctuations of the mass transfer rate and the quasi-steady solution is not yet representative. The cut-off frequency under which the quasi-steady-state can be considered to be valid is rather low owing to the large value of the Schmidt number in the electrolyte (Sc≈1700). Two methods are currently used in order to evaluate the wall shear rate fluctuations.

From the power spectra density (PSD) of the instantaneous current fluctuations, $W_{ii}(f)$, the transfer function, $H(f)$, allows the assessment to the power spectra density of the wall shear rate fluctuations, $W_{ss}(f)$. Thus, the frequency response of the electrochemical probes is taken into account to restore the shear rate fluctuations spectrum from the current fluctuations one by using the following relationship:

$$W_{ss}(f) = W_{ii}(f) \ / \left|H(f)\right| \quad (5)$$

A correct use of equation 12 supposes two conditions:

i. the transfer function must be accurately known in the whole frequency range,
ii. the homogeneity condition with time-depending fluctuations and the average value must be uniform over the whole probe surface.

Concerning circular probes, several forms of $|H(f)|$ have been the objective of several studies (Nakoriakov *et al.*, 1986, Deslouis et al., 1990) that we have presently applied to the limiting current obtained after applying Fourier's transform. This function allows the determination of the wall shear stress dynamics.

The second method, based on the Sobolik's correction (Sobolik *et al.*, 1987), takes into account the calculation of the wall shear stress time-evolution.

2.2 Sobolik correction

This method is based on a correction with respect to the probe dynamic behaviour by using the diffusion-convection equation solution (Sobolik *et al.*, 1987). These authors solved the mass balance equation assuming that the concentration field is a similar function of three variables:

$$c(x, y, t) = c_0 f(\eta) \quad (6)$$

$$\eta = y\, f(t)^{1/3} / \delta_d(x) \quad (7)$$

where *f(t)* is a general time function which takes into account the time shifting of the wall shear rate. The resolution of the diffusion-convection equation in the whole mass boundary layer leads to a general expression of the time history of the wall shear rate, *s*(*t*):

$$s(t) = s_q(t) + \frac{2}{3} t_0 \left(\frac{\partial s_q}{\partial t}\right) \quad (8)$$

where t_0 is the characteristic time of the probe defined as a dynamic behaviour parameter of the electrodiffusion probe:

$$t_0(t) = 0.426 \; d_e^{2/3} D^{-1/3} s_q(t) \quad (9)$$

This relationship was used by several authors in different flow configurations (Labraga *et al.*, 2002 ; Tihon *et al.*, 2003) who found it relevant in unsteady flow conditions, even by comparison with inverse method (Rehimi *et al.*, 2006).

2.3 Power spectra density assessments

The comparison of the electrochemical transfer function and the corrected solution (Sobolik et al., 1987) can be made using a frequential representation of the signals. The procedure of the signal treatment is given in the present under-section according to the methodology presented in figure 2.

It corresponds to the steps generally applied in order to obtain the experimental characterization of a turbulent flow from passive scalars or from one of two components of the velocity.

The unsteady variations of the current measured from the microelectrodes correspond to the fluctuations of the concentration of the active species into the diffusive boundary layer. There are strongly connected to the flow fluctuations developed into and outside the hydrodynamical boundary layer. The recording of a random signal such as the current needs a one-dimensional Fourier transform in order to obtain the repartition of the energy in the frequencies space. It gives a physical sense to the temporal signal that appears as a noise. The present methodology is inspired from literature (Max, 1985). The first step consists in extracting the fluctuating value, i(t), of the recorded signal, $I_{lim}(t)$, defined by :

$$I_{lim}(t) = \overline{I_{lim}} + i(t) \quad (10)$$

The resulting signal is divided in several blocs, N, having the same number of points, N_e. Each bloc recovers the half part of the previous one. Each part of the signal, $i(t)_N$, is treated independently. This averaging method allows to remove perturbations (as ambient noise or electromagnetism wave) and to conserve the physical phenomenon representation. The number of points, N_e, of each bloc depends on the temporal resolution of the studied phenomenon. The sampling frequency is adjusted as a function of the turbulence level in order to describe all the physical information in the various sub-ranges of the spectra: from the scale of energy containing eddies to the smallest scale depending on the ratio of diffusivities, the Schmidt number.

In the other hand, each bloc is multiplied by a temporal window with the same size N_e. This function allows eliminating of the lobe phenomenon which occurs when a Fourier transform is applied to a finite signal.

This truncation effect can be reduced by several kinds of windows (Hanning, Blackman). Among them, the Hanning's window has been retained, which is defined by:

$$F_{ha}(t) = 0{,}5 \times (1 + \cos\frac{\pi t}{N_p}) \quad (11)$$

The power spectral density of the current, $W_{ii}(t)_N$, is obtained by discrete Fourier transform of the focused parts of the signal and their integration gives rise to the DSP, W_{ii}:

$$W_{ii}^{i} = \int_0^{N_e} (i(t)_N \times F_{ha}(t)) \times \exp(-j2\pi ft)dt \quad (12)$$

$$W_{ii} = \frac{\sum_{i=1}^{N} W_{ii}^{i}}{N} \quad (13)$$

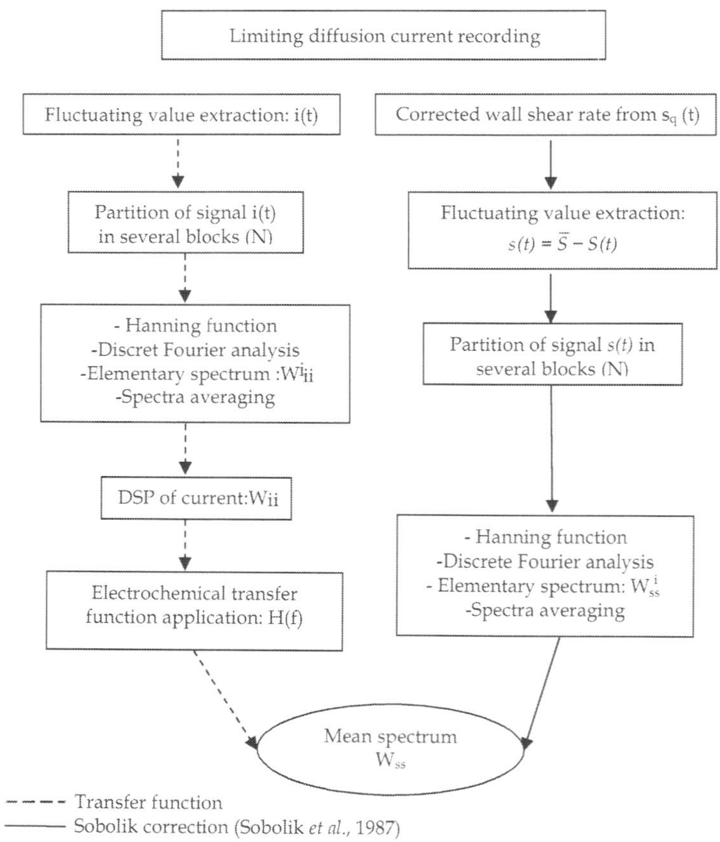

Fig. 2. Different steps of the signal processing

3. Materials & calibration methods

3.1 Experimental set-up

The experimental cell is shown in Fig. 3. It is made of Altuglas and composed of crossing minichannels. The individual square-cross sections of the channels (1.5 mm in side) intersect at right angles. The whole test section has a length of H=105 mm and a width of L=52 mm. At the inlet, there is a calming section containing glass spheres 2 mm in diameter, which allow better distribution of the fluid. Two bottom plates were successively used in order to perform two measurements techniques. One includes thirty-nine circular platinum microelectrodes flush-mounted to the wall allowing an electrodiffusion diagnostics of the wall-flow. The second one is a transparent plate required for the visualization in the frame of PIV measurements.

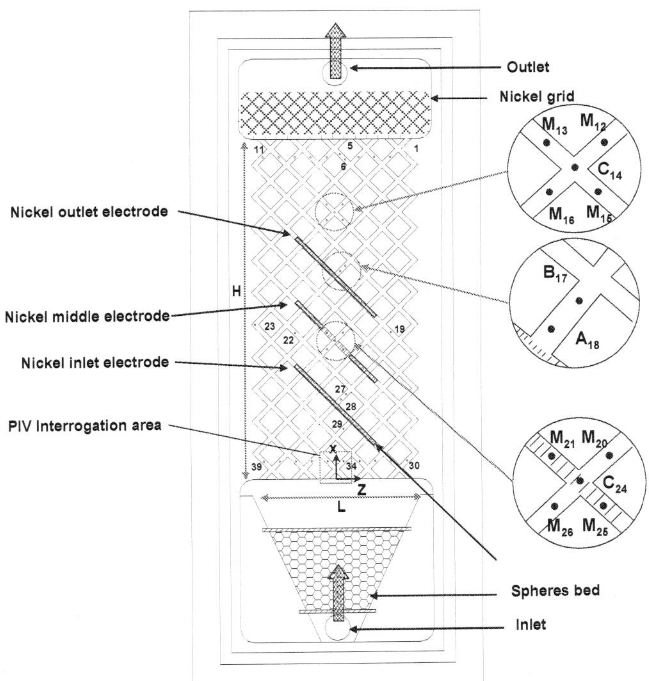

Fig. 3. Scheme of the experimental cell

The microelectrodes have a nominal diameter of 0.25 mm working as cathodes. The anode is made of a nickel grid located at the cell outlet section. As seen in Fig.3, the microelectrodes are numbered from right to left and from top to bottom. The microelectrode positions with respect to the individual minichannel sections are designated with four different labels: M (at the middle of a channel section), A (just after channel crossing), B (just before crossing), and C (at the center of a channel crossing). Two dimensionless parameters are used to determine the position of each probe inside the network of minichannels. The axial position is represented by the parameter $X=x/H$ and the lateral one by $Z=2z/L$. The large nickel electrodes (strips with dimensions of $d_h \times l_e=1.5\times3.65$ mm) are placed at three flow cell positions in order to study the global mass transfer inside the flow cell. The exact surface area of each electrode is obtained by image analysis.

A suitable electrochemical system is provided by an addition of 0.025 M equimolar potassium ferro/ferricyanide and 0.05 M potassium sulphate into water. The polarization voltage of -0.8 V is applied to ensure limiting diffusion current measuring conditions. A home-built electrodiffusion analyser is used to set the polarization voltage to the microelectrodes, to convert the measured currents into voltages and to amplify the resulting signals. A PC computer controlled the analyser operation and data recording. Data records (ranging from 30000 to 80000 samples, depending on the Reynolds number value) from eight current signals are provided at a sampling frequency ranging from 3 kHz to 8 kHz.

The experiments were performed at Reynolds numbers Re ranged from 50 to 3000. The Reynolds number, $Re=u_c.d_h/\nu$, is based on the channel hydraulic diameter, d_h, ν being the kinematic viscosity of the working fluid and the mean velocity inside individual channel sections u_c is defined by:

$$u_c = \frac{Q_0}{nA_c} \tag{14}$$

where n is the number of minichannels at the inlet and at the outlet (n=10), A_c, the section of an individual minichannel. By assuming a uniform repartition of the flow rate at inlet and outlet, the flow rate inside the minichannel, Q_c, is calculated from the total flow rate, Q_0, by:

$$Q_c = \frac{Q_0}{n} \tag{15}$$

All measurements have been carried out at the room temperature.

3.2 Calibration method

The electrochemical probes are made from a platinum wire 250 μm in diameter, but the real active surface area at which mass transfer occurs can be different of the geometrical one due to the manufacturing process. Thus, the calibration technique used in this work is based on the study of the transient response of the microelectrode to polarization switch-on (Sobolik *et al.*, 1998). This current response is described by the well-known solution of unsteady diffusion in stagnant fluid:

$$I(t) = \tfrac{1}{4}\nu_e F C_0 \pi\, d_e^2 \sqrt{D/\pi t} \tag{16}$$

with D, the molecular diffusion coefficient of the reacting species.

The transient current measured consecutively to a voltage step is used to determine the individual effective diameter, d_e, of each microelectrode. In spite of shape deformation during the process of microelectrode fabrication, the effective diameter values is found equal to 0.25 mm with a mean deviation of 0.01 mm for ten repetitions of the calibration. These values is found to be close to the platinum wire nominal diameter as shown in the table 1.

Electrode	1	2	3	4	5	6	7	8	9	10
d_e (mm)	0,251	0,249	0,252	0,262	0,267	0,275	0,298	0,250	0,267	0,288
Electrode	**11**	**12**	**13**	**14**	**15**	**16**	**17**	**18**	**19**	**20**
d_e (mm)	0,242	0,245	0,240	0,229	0,253	0,268	0,246	0,225	0,291	0,270
Electrode	**21**	**22**	**23**	**24**	**25**	**26**	**27**	**28**	**29**	**30**
d_e (mm)	0,246	0,258	0,256	0,256	0,241	0,282	0,289	0,256	0,264	0,230
Electrode	**31**	**32**	**33**	**34**	**35**	**36**	**37**	**38**	**39**	
d_e (mm)	0,243	0,249	0,250	0,258	0,254	0,266	0,224	0,262	0,242	

Table 1. Recapitulative of the microelectrode diameter

3.3 Molecular diffusion coefficient

The measurements of the diffusion coefficient, D, of the ferricyanide ions inside the working solution were obtained by the classical Levich method (Coeuret & Stork, 1984) which uses a rotating disc electrode system. The advantage of this device deals with the possibility to use a working electrode with a well defined surface area (in our experiments $S = 3.14 \times 10^{-6}$ m^2) and to work with small electrolyte volume in well-controlled hydrodynamic conditions. The diffusion coefficient, D, is obtained from the experimental dependence of the limiting diffusion current, I_L, versus the angular rotation speed of the disc, ω:

$$I_L = 0.621 \nu_e C_0 F S D^{2/3} \nu^{-1/6} \omega^{1/2} \tag{17}$$

where υ is the kinematic viscosity of the solution.

A set of experiments performed in a large range of temperature values ($285 < T(K) < 305$) are significant in order to check the Stokes-Einstein relationship between dynamic viscosity, diffusivity and absolute temperature :

$$\frac{\mu D}{T} \cong 2{,}18 \times 10^{-15} \mathrm{m}^2.\mathrm{Pa.K}^{-1} \tag{18}$$

4. Local flow diagnostics

4.1 PIV measurements

The experimental testing bench included a laser (Nd-Yag, 15 Hz, 120 mJ), a double image recorder camera (Kodak megaplus ES 1.0, 1008 × 1016 pixels) which is joined to a 28 mm lens and three macroscopic sleeves. The dedicated processor (PIV 2100) and Flowmanager V 4.5 software is used to perform the calculations of the flow fields using the cross-correlation method. The seeding material is spherical polyamide particles from Dantec (density = 1.03, d_p=20 μm). Interrogation areas are squares of 32 × 32 pixels. The laser, the CCD camera and the cell are placed on an individual moving system. The water pump are preceded by a mixer and the working cell is placed on a stiff table mounted on slender screws in order to reduce the vibrations induced by the pump. Micrometric moving systems are used to align the laser beam in the fluid plane and to accurately focalize the camera on the measurement plane. By moving the laser, the thickness of the laser sheet crossing the network cell has a minimum value less than 1 mm.

In those conditions, the magnification ratio is closed to 1:1 and the investigated visualisation field measured is 1 cm × 1 cm. The field depth of the image is measured by a diffraction grating and is approximately equal to 300 μm. The seeding concentration is adjusted to get between 5 and 10 particles in each interrogation window. The statistical averaging of the data was performed on a series of 1000 instantaneous velocity fields and the statistical convergence is checked on mean velocity, and second-order moments of fluctuating velocity.

The measurements are focused on eight zones corresponding to the location of the electrochemical probes. The experiments are performed at Reynolds number, *Re*, ranged from 145 to 1620. The results are limited to one zone at the inlet of the network in order to present the PIV results. The whole of the results are available in the publication of Huchet *et al.* (2008)a.

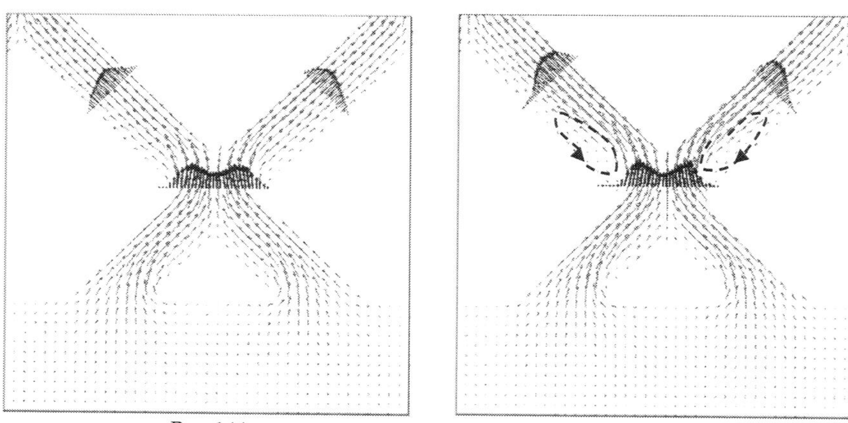

Re= 144 Re=1270

Fig. 4. Velocity profiles and mean flow fields in zone 1 at the inlet of the network for two Reynolds numbers.

The results of mean flow fields are presented in Fig. 4 at low Reynolds number (*Re* = 144) and higher Re value (Re=1270) at the inlet. For Re=144, no instability and no significant detachment appear after the crossing channels. Three velocity profiles are plotted, one of them is located at the crossing and is characterized by two peaks corresponding to the laminar velocity profile of each incoming channel. The velocity increases on both sides of the crossing centre and depicts the symmetrical distribution of the flow in the two outlet branches. The mean velocity at the crossing junction is found 1.7 times higher than in the outlet branches. Normally, the ratio between the velocity at the crossing section and the incoming channel velocity should be $\sqrt{2}$. The lack of resolution in the near wall region tends to overestimate the experimental values.

For *Re*=1270, and as previously observed, the velocity decreases in the crossing centre. A large recirculation zone is observed on the opposite side of the rear location of the crossing and is associated to a preferential flow which presents more important momentum transfer at this location. The recirculation extends over half of the length between two successive crossings and covers half of its width. The velocity fields are quite similar in each channel after the crossing and the flow structure is non-established after the impact of the two incident streams in comparison to the parabolic profile obtained in fully developed laminar flow.

4.2 Wall shear rate fluctuations measurements from microelectrodes
4.2.1 Power spectrum density analysis
4.2.1.1 PSD of the current : W_{ii}

Figure 5 shows the PSD of the current, W_{ii}, for a large range of Reynolds numbers (317<Re<3535) at three different positions in the network either at the inlet (electrode M34), at the middle (A28) and close the outlet (M15) of the system.

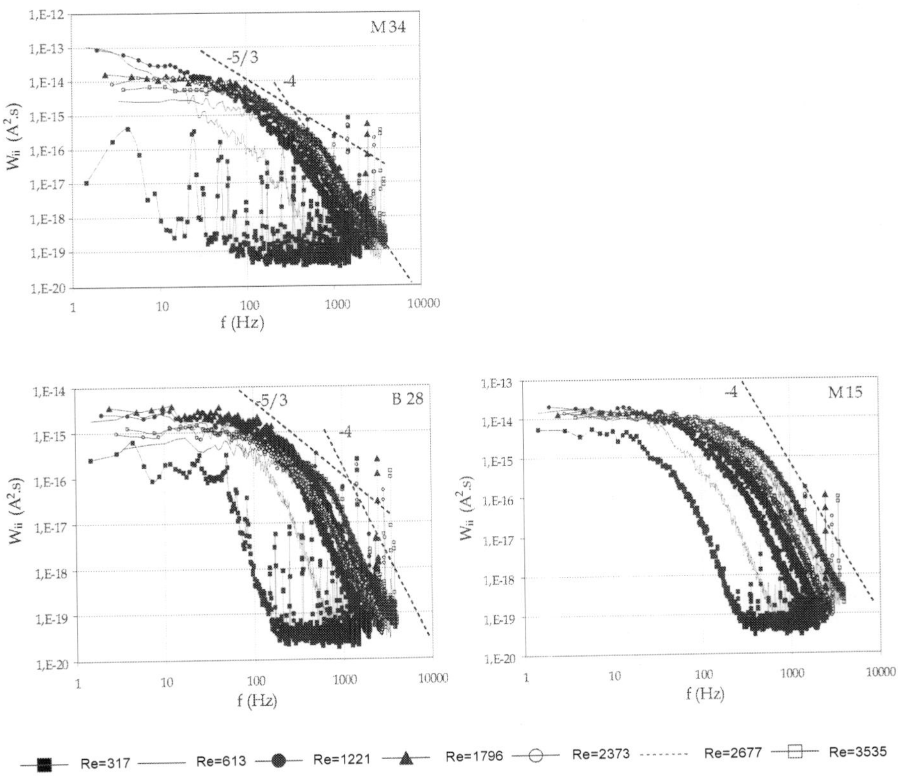

Fig. 5. PSD of the limiting current fluctuations for three electrodes (M_{34}, B_{28}, M_{15}) at different Reynolds numbers.

As previously proven by Tennekes & Lumley (1972), the cascades of the spectra correspond quite well to the standard concentration spectra of the electrochemical species dynamically mixed in the diffusion boundary layer. For the electrodes located in the non-established flow area (M34 to A28), the low frequency part of the spectra gives rise to a decrease until 30 Hz. From 30 Hz to 130 Hz, the logarithmic trends are characterized by a slope equal to $f^{-5/3}$ which corresponds to the inertial-convective sub-range where the large scale structures govern the transport of the passive scalar. Above f=130 Hz the spectra fall-down with a slope equal to f^{-4}. For M15 location, where the flow is fully developed, the spectra are characterized by a low frequency plateau and the downward slope reaches f^{-4} at a frequency (20<f(Hz)<120) which is increasing with Re.

4.2.1.2 PSD of the wall shear rate : W_{ss}

The PSD of the wall shear rate fluctuations obtained from using the electrochemical transfer function are presented in figure 6 and compared with the PSD calculated from the wall shear rate fluctuations from the Sobolik correction (equation 8).

Each spectrum is characterized by a low frequency plateau and a decrease in the high frequency area. The slope in this last part of the spectra differs according to the position of the probe in the network. The value of the slope depends on the nature of the flow. It has to be noticed in the present case that the first hypothesis is not respected. The transfer function does not manage to represent properly the whole range of the fluctuations as a cut-off frequency appears in each spectrum (500<f(Hz)<1500). Therefore, the linearization theory of the transfer function is not yet valid and the direct correction of the electrochemical signals seems to be a more attractive method to solve the issue of the dynamic behaviour of the electrochemical probe. The shape of the PSD of the corrected wall shear rate is similar to that calculated by the transfer function in the range of frequency corresponding to the intermediate and large flow structures (1<f(Hz)<1000). Above a frequency value equal to f≈1500Hz, the PSD of the corrected wall shear rate gives some information in the dissipative area of the spectra, which are impossible to obtain with the transfer function method.

It may be noticed that the level of the wall shear rate fluctuations calculated using the transfer function remains correct since this part of the spectra does not influence the integrate value of the PSD (Huchet *et al.*, 2007). Nevertheless, it is important to select the corrected method of Sobolik *et al.* (1987) to analyse mixing phenomenon from the electrochemical method application. The reason is that the flow structures associated to the micromixing phenomena which allow the best conditions for the reaction are located in the dissipative part of the spectra. Thus, we propose to apply in the following section this method for the characterization of the flow regimes by spectra integration to access the fluctuating rate of the wall shear rate:

$$FR = \sqrt{\overline{s'^2}} \Big/ \bar{s} = \sqrt{\int W_{ss} df} \Big/ \bar{s} \, , \tag{19}$$

with

$$s(t) = \bar{s} + s'(t) \tag{20}$$

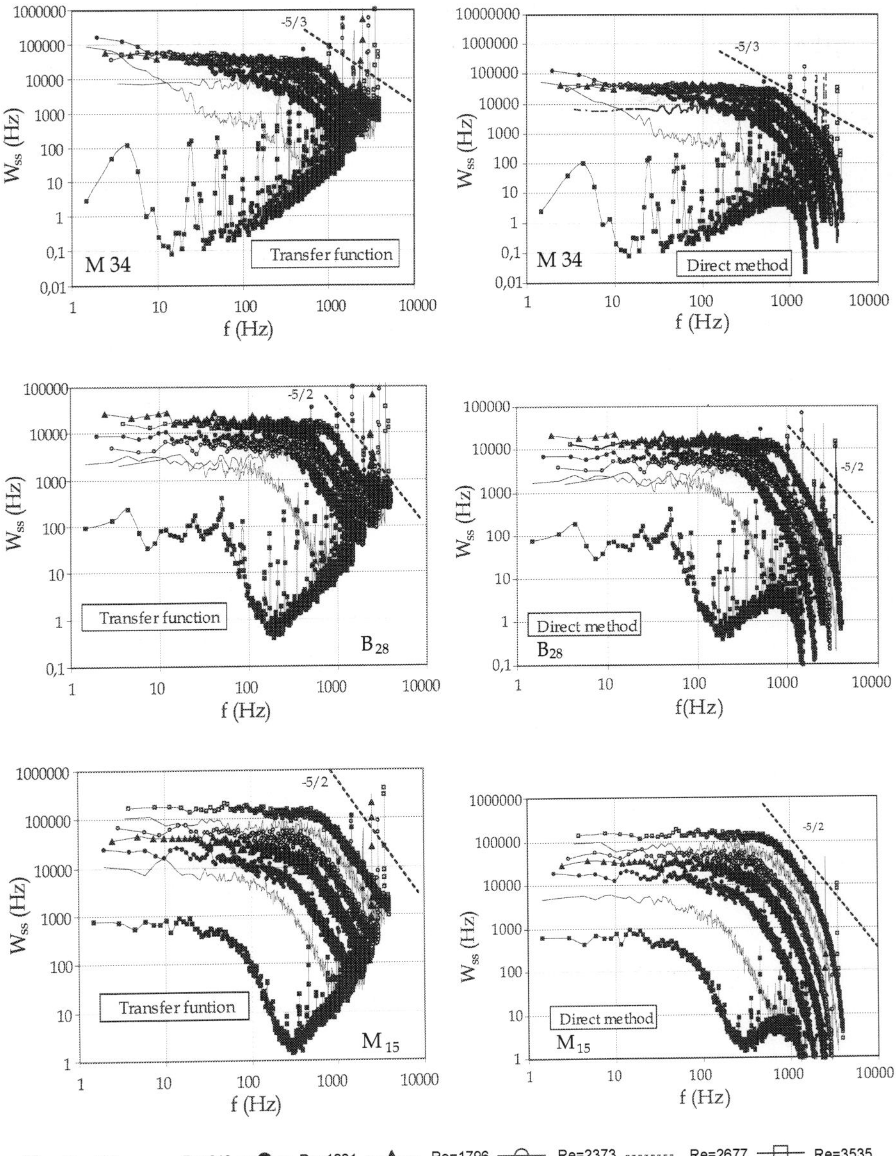

Fig. 6. Comparison of the PSD of the wall shear rate fluctuations between transfer function method and Sobolik's solution (direct solution) for three working microelectrodes and several Reynolds numbers.

4.2.2 Characterization of the flow regimes and the unsteady flow structures

4.2.2.1 Flow regimes

An example of wall shear rate courses measured inside the flow cell is shown in Fig.7.

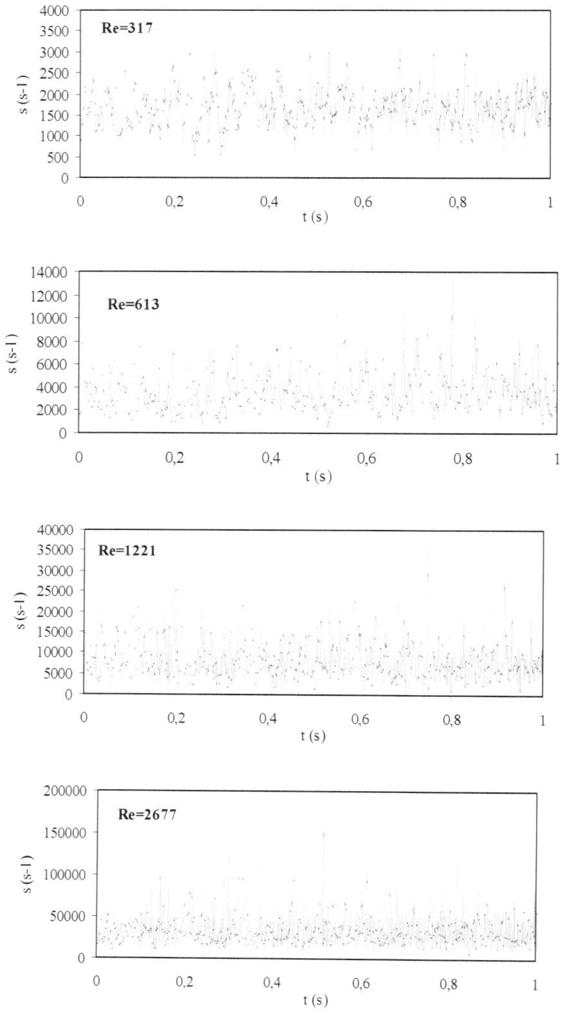

Fig. 7. Time-traces of the corrected wall shear rate recorded on electrode M15 for different Reynolds numbers

For these measurements carried out near the cell outlet, the first flow fluctuations are observed at the flow rate corresponding to Re~200. As the flow rate increases, the level of fluctuations is enhanced and the wall shear rate reaches values which are higher than those observed in flow channels of common sizes.

It is interesting to analyze the results obtained for the microelectrodes located at the different axial positions. The study of fluctuation rates *FR*, which is presented in Figure 8, illustrates the typical flow regimes achieved in the flow cell and also the gradual evolution of fluctuations along the cell.

The variation of *FR* with *Re* is characterized by an initial plateau at *FR*~0 (laminar flow regime), then *FR* increases sharply (transient flow regime) up to a practically constant level (regime of developed flow fluctuations), whose actual value is depending on the specific probe location. The critical value of *Re* corresponding to the onset of fluctuations varies from 560 (for the probe M34 located close to the cell inlet) to 200 (for all the electrodes behind the third channel crossing). Therefore, the flow inside the network of channels can be considered as established for the axial locations of *X*>0.3.

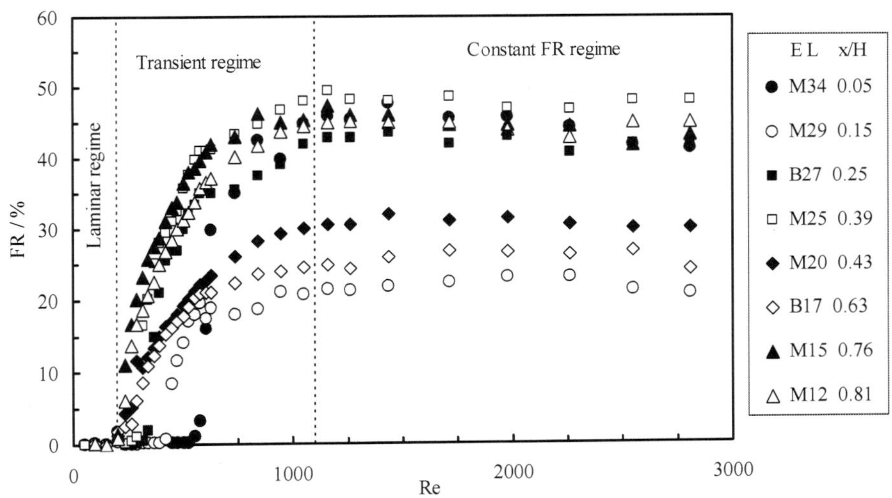

Fig. 8. Variations of fluctuation rates $FR = \sqrt{\overline{s'^2}}/\overline{s}$ with the Reynolds number obtained for the different axial locations $X = x/H$.

Only at low flow rates (for *Re*<200), the stable laminar regime is observed throughout the flow cell. On the other hand, the stabilization of near-wall flow fluctuations is reached at high flow rates and is observed anywhere in the cell for *Re*>1100. As the flow pattern is very complex (3D with recirculation zones behind the crossings), the final value of *FR* is very sensitive to the exact position of the microelectrode (especially with respect to the channel centerline). The fluctuation rates are found to stabilize at relative values ranging from 20 % to 50 %. The channel crossings have an effect on the enhancement of flow fluctuations and also on their earlier stabilization than in straight channels. The shape of *FR* versus *Re* curves is very similar to that observed for the evolution of fluctuations in packed beds of particles. The packed bed flow configurations exhibit also practically the same critical Re values characterizing the onset (Re~180) (Seguin *et al.*, 1998a) and the stabilization (Re~900) of flow fluctuations (Seguin *et al.*, 1998b).

4.2.2.2 Flow structures

In complement to the preliminary interpretation regarding the flow structure by PIV (see section 4.1) analysis and wall shear stress measurements (Huchet *et al.*, 2007), the wall turbulent eddies are classically assessed by the dimensionless autocorrelation of fluctuating velocity gradient and can then be calculated:

$$R_{ss}^{*} = \frac{R_{ss}}{R_{ss}(0)} \tag{21}$$

With

$$R_{ss}(t,dt) = \frac{s'(t)s'(t+dt)}{\sqrt{s'(t)^2}\sqrt{s'(t+dt)^2}} \tag{22}$$

Calculation of the autocorrelation provides information regarding the Taylor microscale, λ, and the integral length scale, Λ, by supposing the Taylor hypothesis, which is defined by:

$$\lambda = \tau_\lambda U_c \text{ with } R_{ss}^{*}(t = \tau_\lambda \to 0) \tag{23}$$

Performing a Taylor series expansion of the autocorrelation $R_{SS}^{*}(t)$, the osculating parabola thus obtained is supposed to yield the derivative covariance:

$$R_{ss}^{*}(t = \tau_\lambda) = 1 - \frac{t^2}{\tau_\lambda{}^2} \tag{24}$$

The integral scale is given by:

$$\Lambda = U_c \int_0^\infty R^{*}{}_{ss}(t)dt \tag{25}$$

Integral length scales estimate the size of the largest turbulent eddies and can also be defined as the size of the large energy containing eddies, i.e. eddies containing most of the turbulent kinetic energy. Taylor microscale is a measurement of the dimension of eddies which transfer the kinetic energy at the scale of dissipation where the viscous phenomena predominate. The Taylor microscales represent the small-scale motion which are of significant interest in term of molecular mixing or micromixing.

In Figs. 9, the autocorrelation curve of the wall shear rate signals is plotted as a function of time at several locations between the inlet and the outlet of the network corresponding to the position of the microelectrodes in the minichannels.

It shows a large range of characteristic times corresponding to the convective time of the structures in the near location of the probes. Thus, the turbulence macroscales and microscales were calculated according to equations 15 and 14 for Re=2950 by using the Taylor hypothesis.

The sizes of the integral length scale, Λ, and the Taylor microscales, λ, are gathered in Table 2 according to their locations.

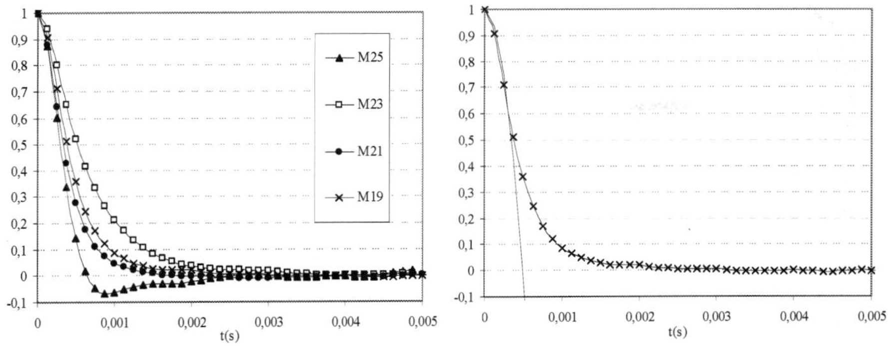

Fig. 9. Left: Dimensionless autocorrelation of the wall shear rate for different positions in the radial direction at X/H≈ 0.4 for *Re*=2950. Right: Autocorrelation function (solid line and crossing) at M19 location and osculating parabola (solid line) function (τ_λ=0.0005 s).

	Probes	*X/H*	*l/L/2*	τ_λ *(ms)*	λ *(mm)*	τ_Λ *(ms)*	**Λ (mm)**
Lateral evolution	M25	0.39	0.088	0.44	0.87	0.47	0.92
	M23	0.46	-0.959			3.32	6.54
	M21	0.43	-0.102	0.44	0.87	0.98	1.93
	M19	0.43	0.733	0.5	0.98	1.67	3.28
Axial evolution	M34	0.05	0.088	0.375	0.74	0.88	1.73
	M15	0.76	0.088	0.375	0.74	0.49	0.97
	M4	0.95	0.279	0.375	0.74	0.62	1.22
Crossing junction	C24	0.41	-0.008	0.44	0.87	0.81	1.59
	C14	0.79	-0.008	0.44	0.87	0.72	1.42
Before a crossing	B27	0.25	0.12	0.45	0.89	1.46	2.87
	B17	0.63	0.12	0.45	0.89	1.35	2.66
	B5	0.96	0.057	0.3	0.59		
After a crossing	A28	0.23	0.057	0.3	0.59	0.35	0.69
	A18	0.61	0.057	0.3	0.59	0.78	1.54

Table 2. Comparison of the macroscales and the Taylor microscales at different location in the network for Re=2950.

- Axial evolution
  The evolution between x/H=0.05 and x/H=0.95 at three locations (M34, M15 and M4) in the axial direction (0.088<z/(L/2)<0.279) is quite constant. The sizes of the Taylor microscale is equal to 0.74 mm and the sizes of the larger eddies is ranged between 0.97 mm and 1.73 mm which are of the same order of magnitude than the characteristic length of a square minichannel equal to 1.5 mm in side.

- Lateral evolution
  Regarding to the lateral evolution between $z/(L/2)$=-0.959 and $z/(L/2)$=0.733 at $x/H \approx$ 0.4, the results are scattered according to the position in the network. For M25 and M21 position, the sizes of the micro and macroscale respectively equal in average to 0.87 mm and 1.5 mm confirm the previous positions given that their location corresponding to the middle of the network ($z/L/2 \approx 0$) and both to the middle of a channel.
  At the lateral positions of the network (M19 and M23) larger integral scales, respectively equal to 3.28 mm and 6.54 mm are observed.
  It corresponds to the length of a minichannel between two successive crossings equal to 5.5 mm involving that the shape of these larger vortices are spread out in the longitudinal direction and convected on average along the streamwise direction by the flow. At M23, the correlation coefficient presents a very restricted zone with a parabolic behaviour and can not satisfy the calculation of the Taylor microscale.
- At the channels crossings ("A, B, C" probes)
  In the crossing location, at a smaller scale, the Taylor microscales are of the same order of magnitude at the crossing junction and before the crossing ($\bar{\lambda} \approx$ 0.88 mm) excepted for B5 position where a lower value is reported. This value is assumed to be due to the outlet effects which especially disturb the flow at this location. This particularity was already mentioned in the section 3.2 where the turbulent intensity field at the outlet is not representative of the remaining of the crossing flow inside the network.
  For microelectrodes located after the crossing (A), the low values of Taylor microscales ($\bar{\lambda} \approx$ 0.59 mm) are maintained due to the recirculation zone observed by PIV.
  The sizes of the macroscales at "A" and "C" positions equal in average to 1.5 mm are lower than at position "B". In the first two locations, the sizes are controlled by the junction at the crossing involving the reduction of the flow section, while "A" position presents smaller turbulent scales induced by the recirculation zone. The mean size of the macroscales ($\bar{\lambda} \approx$ 2.77 mm) found at the "B" position corresponds to the length between the crossing and the recirculation zone.

Finally, results regarding characteristic length scales of the turbulence (integral scales and Taylor microscales) show very different trends according to the geometry. Thus, from the PIV and electrochemical measurements, a general pattern describing the different scales inside a crossing of two minichannels transposable to the whole network of crossing minichannels in the constant fluctuations flow regime can be proposed. The integral scales are clearly dependent on the position in the network and particularly influenced by elbows and crossings.

In the present study, the confined geometry addicted by the crossing effects induces anisotropic spatial scales characteristics depending on the position in the channel and in the whole network.

4.2.2.3 Statistical flow properties

As seen in the previous section, the spatial heterogeneity of the flow structures need to investigate deeper the temporal anisotropy. Thus, statistics calculations are performed in order to characterize the hydrodynamics in term of degree of intermittency or anisotropy.

Experimental and numerical data dealing with intermittent turbulent flow are often characterized by statistical properties which are really different than in homogeneous and isotropic turbulence (Xu *et al.*, 2006; Portelli *et al.*, 2003). Most of the works concerning the study of the turbulent boundary layer has shown that this region is characterized by the

presence of ejection of coherent structures called "burst phenomena". These near-wall characteristics are linked by the presence of high velocity gradient and the measurements of the fluctuating quantities such as concentration or temperature are characterized by strong and rare fluctuations. In particular, the probability density function is deformed and the dimensionless fourth order moment, i.e. the flatness factor, F, is very different from the value 3 calculated in the case of Gaussian fluctuations. Moreover, the dimensionless third order moment, i.e. the skewness factor, S_k, can significantly increase above zero, which reveals intermittent and strong fluctuations. In the present work, we studied the small scales statistics issued of the electrochemical signals which are linked to the instantaneous limiting diffusion current. The fluctuations of the current correspond to the fluctuations of the concentration of the electrochemical species in the diffusion boundary layer.
Time-evolution of the fluctuating value of the wall shear rate measured at the location M15 is given in Fig. 7 for *Re*=2677. The temporal shifting around the mean value is the first criteria of the intermittency characteristics. More important positives fluctuations are noticed. The normalized histogram of the data at Re= 2950 is compared with a Gaussian distribution in the figure 10. It confirms the intermittency of the electrochemical signals characterized by an asymmetrical and non-Gaussian distribution of the fluctuations corresponding to a flatness factor of 8.3. Moreover, the skewness factor, S_k, which reveals the intensity of the fluctuations when its value is greater than 0, reaches 1.55 at this location.

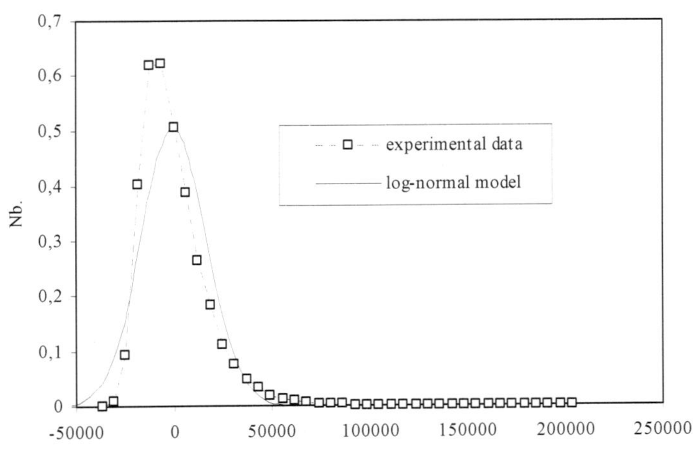

Fig. 10. Normalised distribution of the fluctuating wall shear rate at the location M15 for *Re*=2950 and comparison with a log-normal model.

For the purpose of our work, flatness and skewness factors are gathered in the Figure 11 for few locations corresponding to the positions of various microelectrodes. The calculation of the statistical properties of the electrochemical current have been performed by Adolphe *et al.* (2007) for transitional and turbulent straight channel flow which have described three different regimes such as:

- at Re<2000, in the laminar regime, the skewness values are $S_k \approx 0.3$ and flatness values are $F \approx 1.5$ revealing a range of fluctuations containing very low frequencies,

- for 2000<Re<4000, in the transient regime, the skewness factor is $S_k \approx 0.75$, which reveals the existence of strong and rare positive fluctuations associated with burst phenomena,
- in the turbulent regime (Re>10000), $S_k \approx 0$ and $F \approx 3$.

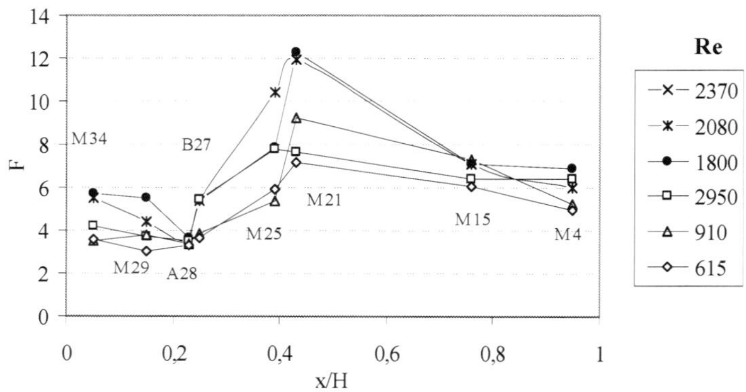

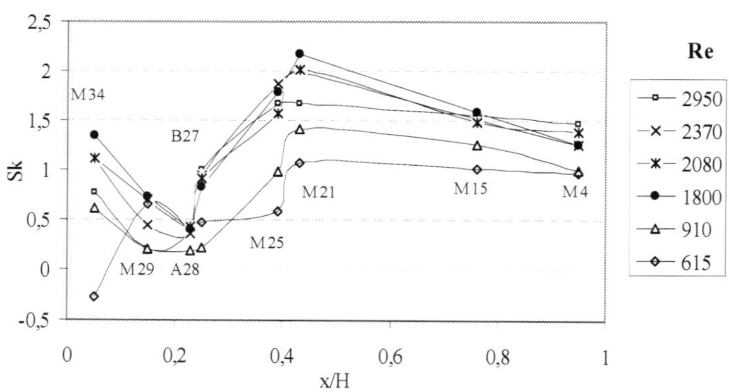

Fig. 11. Evolution of the flatness factor and the skewness factor in the network of crossing minichannels for various Reynolds numbers.

For most of the positions, the variation of F and S_k according to the Reynolds number are really different than in the straight channel configuration. Flatness factor varies between 3 and 11 until Re≈1800. From Re≈1800 to Re≈2800, the values decrease until to reach a constant value shifting between 5 and 8. The values of the skewness factor are ranged between 1 and 2 from the constant fluctuation flow regime (Re>1000). Thus, these statistical properties confirm a transient flow regime in the network of crossing minichannels. The transient flow regime is expanded above a Reynolds number equal to 1000 in spite of a constant fluctuation rate of the wall shear rate calculated by spectral analysis. Moreover, the fluctuations are noticeably stronger and the intermittent level is larger than in the case of a straight channel in transient regime.

In addition to the Reynolds number on the intermittency level, one can also notice about the influence of the axial location. The locations corresponding to the microelectrodes M29, A28 and B27 are characterized by lower values of the flatness and skewness factors. This zone ($0<x/H<0.3$) (section 4.2.2.1) is found as being the establishing length of the flow inside the network of crossing minichannels. The high values found at M34 are attributed to some inlet effects already mentionned in the PIV discussion in the section 3.2. At $x/H\approx0.4$, the intermittency level rises and is found as be maximum at this location in the network whatever the Reynolds number considered. Beyond $x/H\approx0.4$, the intermittency level is stable whatever the Reynolds number until the outlet ($x/H\approx0.95$).

The statistical properties of the fluctuating values of the wall shear rate provide information regarding the intermittency characteristics of the flow in the near wall region at several locations inside the network of crossing minichannels. This statistical analysis allows us to verify the results discussed previously about the anisotropic scales observed by PIV inside a confined flow. Nevertheless, in the present case, it seems that the crossings could be also responsible of the intermittency by the decrease of the fluctuation rate of the wall shear rate at the junction.

5. Liquid-solid mass transfer and pressure drop characterization

5.1 Liquid-solid mass transfer

The experimental results on wall mass transfer obtained for three large electrodes are plotted in Figure 12.

These data are represented in the dimensionless form $ShSc^{1/3}$ = f (Re) by using the Sherwood ($Sh = k_{mt}dh/D$), Schmidt ($Sc=\nu/D$) and Reynolds numbers.

If the results are presented in logarithmic coordinates, two almost linear regions can be distinguished. Two corresponding correlations cover almost the whole range of experimental data:

$$ShSc^{-1/3} = 0.94\,\mathrm{Re}^{0.41} \tag{26}$$

(for 15<Re<100, in laminar regime)

$$ShSc^{-1/3} = 0.29\,\mathrm{Re}^{0.66} \tag{27}$$

(for 200<Re<3500, fluctuating regime)

The transition between these two mass transfer regimes is observed around Re≈100–200, i.e. at the flow rate for which the inception of near-wall flow fluctuations is expected. The mass transfer coefficient measured in the flow cell is significantly higher than that observed in straight square channels. As seen in Figure 12 the experimental data are clearly above the lines representing the Lévêque equation (Lévêque, 1928) (for laminar regime):

$$ShSc^{-1/3} = 1.85\left(\frac{d_h}{l_e}\right)^{1/3}\mathrm{Re}^{1/3} \tag{28}$$

and the equation obtained by the Chilton–Colburn analogy and cited by Coeuret & Storck (1984) (for turbulent regime):

$$ShSc^{-1/3} = 0.023\,\mathrm{Re}^{0.8} \tag{29}$$

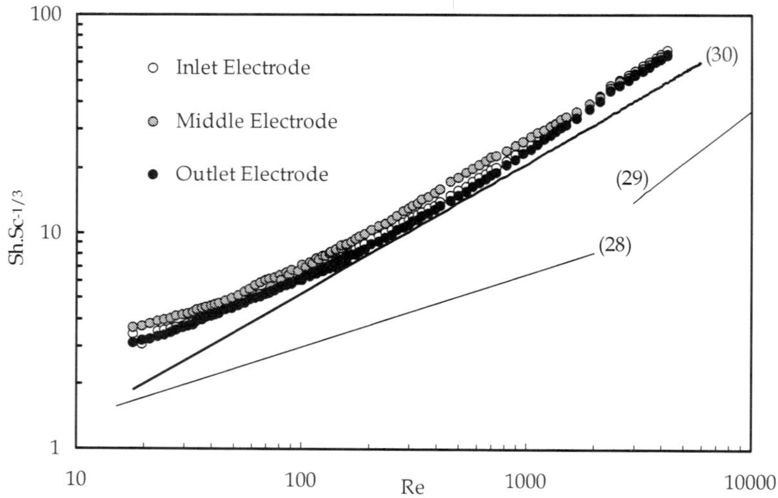

Fig. 12. Comparison of the measured wall mass transfer data with the correlation previously obtained for porous media and straight channels

For comparison, the classical correlation, which is proposed for mass transfer in fixed beds of spheres (Wakao & Funazkri, 1978), is also shown in Figure 12. After transformation of the particle dimensionless numbers into those for pore and application of the capillary model (Comiti & Renaud, 1989) for description of a microchannel network between individual spheres, the original Wakao correlation can be expressed in the form:

$$ShSc^{-1/3} = 1/3\,\mathrm{Re}^{0.6} \tag{30}$$

Where Re is identical to the Re_{pore}.

The analogy with a porous media takes into account three parameters. The specific surface, a_v, the porosity, ε, and the tortuosity, τ depend on the geometrical characteristics of the network such as represented in the figure 13:

a_v=1400 m^{-1} - ε=0,36

$$\tau = \frac{2(l + d_h)}{(l + d_h)\sqrt{2}} = \sqrt{2} \tag{31}$$

From the structural parameters from the capillary model, it is possible to determine the pore diameter of about 1.6 mm, close to the hydraulic diameter channels.

From the capillary representation of the network, the correlation of Wakao & Funazkri (1978) is, especially for Re>200, in relatively good agreement with the present mass transfer data (note the good agreement between the Reynolds number exponents in Equations 27 and 30). Consequently, from a mass transfer point of view, the studied flow cell can be compared to a porous medium rather than to a set of straight channels as demonstrated previously in the subsection 4.2.2.1 in the frame of the identification of the flow regimes from fluctuations rate of wall shear rate.

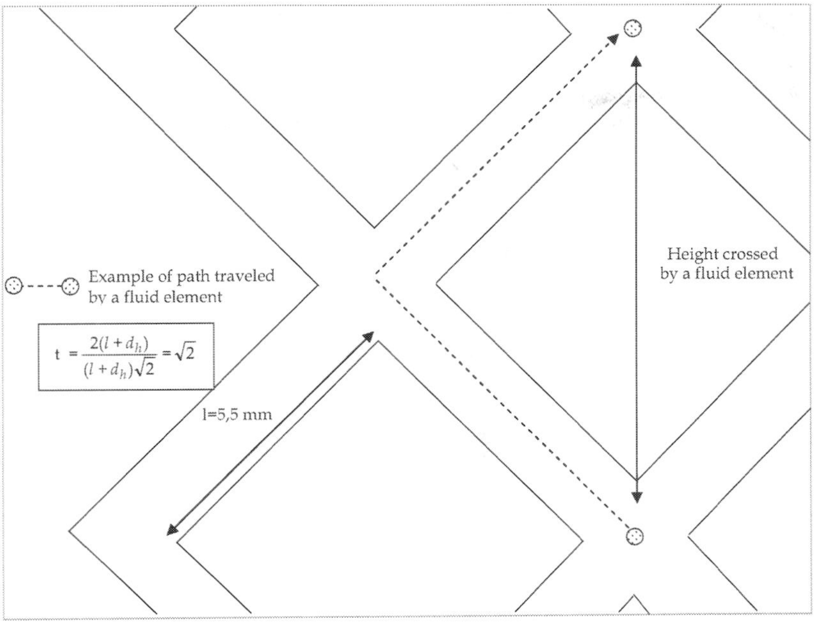

Fig. 13. Representation of motion of fluid particles used in calculating the tortuosity parameter through a capillary representation of the network of minichannels.

5.2 Pressure drop measurements

The measurements of pressure drop across the networks of height, H, are carried out by means of differential pressure valve located at the inlet and the outlet of the cell.

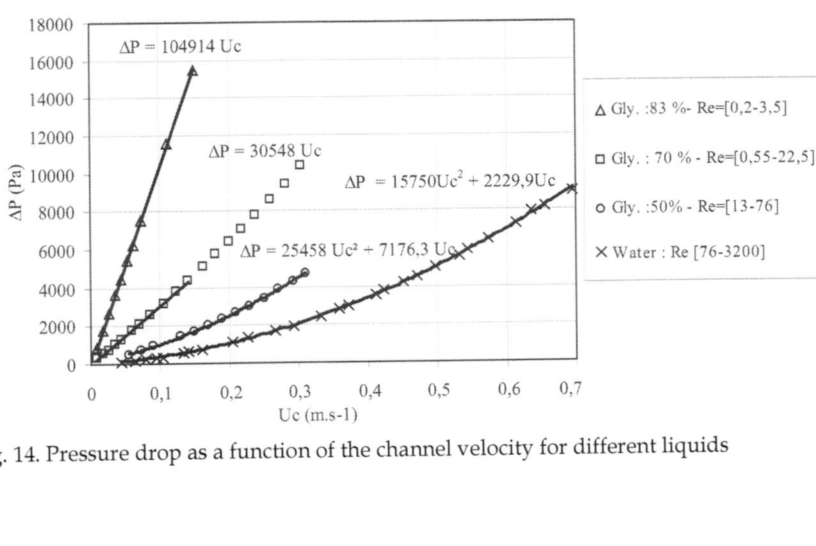

Fig. 14. Pressure drop as a function of the channel velocity for different liquids

The representation of the pressure drop as a function of the velocity is shown in Figure 14. Several aqueous solutions of glycerine are used in order to exhibit the linear part of the curve ΔP vs U_c and to identify the flow regime where the viscous force are preponderant.
The laminar flow regime can be divided in two parts:
-for Re<6~10, a linear part can be distinguished in the curves. This zone corresponds to a Reynolds number ranged between 0.5 and 10.
-for Re>10, a parabolic shape is representative of the pressure drop with a quadratic term similar to the Forchheimer type equation used for porous media:

$$\Delta P = M\eta U_0 + N\rho U_0^2 \tag{32}$$

Where U_0 is the superficial velocity, η the dynamic viscosity, and M and N depends on the porous medium structure.

6. Mixing performance in different networks of minichannels

We propose in the present section to use the electrochemical microsensors method in order to characterize the mixing performance in different geometries of network of minichannels.

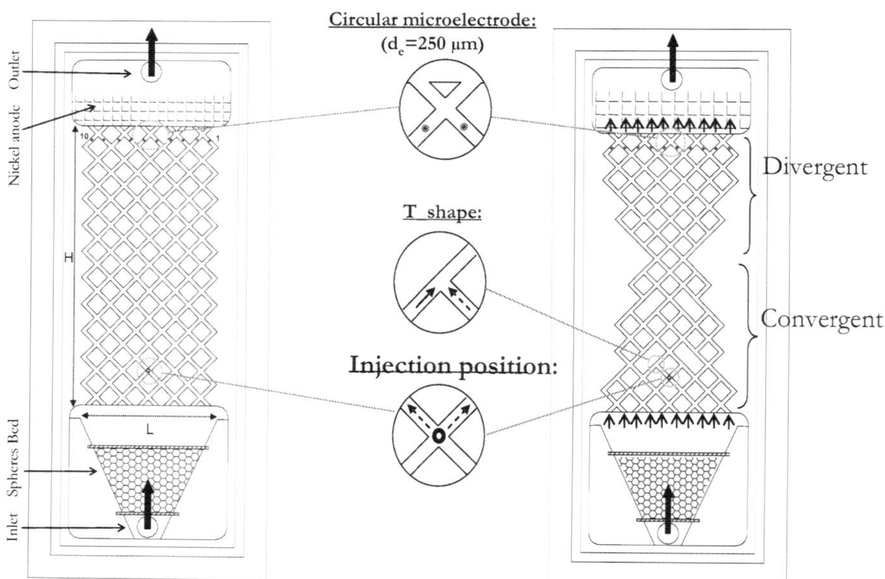

Fig. 15. Scheme of the experimental cells. Left: ×_network Right: T_network.

The working cells are presented in Fig. 15. One of them is the same network studied in the previous sections otherwise noticed ×_network. They are made in two altuglas plates and feature square minichannels in the upper plate. The second geometry, called T_network, corresponds to a better design for a mixing process.

The T_network is composed of a converging part in the first half of the cell followed by a diverging one in the remaining part of the network in order to induce a better distribution of the fluid until the outlet. The decrease of the flow section at the middle of the network reduces the distance between the streamlines and the path leading to the mixing process. Moreover, few T_shaped parts are integrated in the converging zone in order to enhance mixing.

The individual square-cross sections of the channels are the same either 1.5 mm in side. The whole test section has a length of H=105 mm and a width of L=52 mm. A third bottom plate is used for the two cells in order to perform electrochemical measurements. The location of the injection of the tracer in the flow cells is chosen at the second crossing in the median axis of the cells.

The limiting diffusion current on each microelectrode is registered at Reynolds numbers, Re, ranging from 50 to 250. The Reynolds number is based on the hydraulic diameter of the minichannels (d_h=1.5 mm) and the mean velocity inside an individual channel, U_c:

6.1 Measurements of the mixing level

6.1.1 Principle

The principle of mixing investigation is based on the simultaneous measurement of the limiting diffusion current at the channels outlets in order to quantify the mass amount of a tracer (the ferricyanide ions) injected upstream of the cells. The concentration of the electrochemical tracer, C_T, larger than the concentration in the electrochemical solution, C_0, flowing inside the cells is detected by the microelectrodes implemented at the wall of the ten outlet branches (from E1 to E10).

The network of crossing minichannels is primed by a minipump (Cole-Parmer Instrument) delivering a suitable electrochemical solution, kept at constant temperature (20 °C). The electrolyte is an aqueous solution of 0.002 M equimolar potassium ferro/ferricyanide and 0.057 M potassium sulphate working as supporting electrolyte. The physical properties of the solution are measured in the experimental conditions at constant temperature equal to 20 °C ($\nu=1.016\times10^{-6}$ m^2.s^{-1}, $D=7.98\times10^{-10}$ m^2.s^{-1}). The electrochemical tracer is an aqueous solution of potassium ferricyanide of concentration, C_T, ranged between 0.025 and 0.15 M. Injection, realized at the second crossing minichannel after the inlet in the central axis of the network, is performed by using two methods:

- the first method is a pulse injection by using a syringe of volume equal to V_{inj} = 5.75 ml,
- the second method is a step injection by using a micro-pump (Armen instrument) delivering a constant flow rate equal to q_{inj} = 5 ml.min^{-1}.

All the experimental parameters are gathered in table 1. The ratio between Péclet and Reynolds numbers for the different investigated flow rates is greater than 1000, allowing to satisfy a forced convective flow regime.

Injection type	q_{inj} (ml.min^{-1})	V_{inj} (ml)	C_0 (mol.m^{-3})	C_T (mol.m^{-3})	U_c (m.s^{-1})	Re	Pe=$U_c d_h/D$
					0.038	57	75188
Pulse		5.75			0.076	114	144737
			2	25-50-100-150	0.107	161	204887
Step	5				0.165	247	313909

Table 3. Experimental parameters for the two methods of injection.

6.1.2 Mixing criteria

A home built electrodiffusion analyser is used to set the polarization voltage to the microelectrodes, to convert the simultaneously measured currents into voltages, and thus amplify the resulting signals. A polarization voltage of -0.8 V is applied to ensure limiting diffusional conditions. A PC controls the analyser operation and data recording. Data records (60000 samples) for 10 current signals are provided at a sampling frequency of 1 kHz.
Thus, at the beginning of recording, the injection of the electrochemical tracer is performed and examples of the signals recorded are shown in Fig 16 for the T_network in the case of the pulse injection and in the case of the step-shaped one for the ×_network. Others examples of recording of the limiting diffusion current are presented in the paper of Huchet *et al.* (2008)b.
From the measurements of the current intensity $I_i(t)$ obtained on each electrochemical sensor (i=1 to 10), the molar percentage of the tracer in each outlet minichannel is obtained, taking into account the base-line value, I_0, due to the limiting diffusion current induced by the bulk solution. Applying equation 2, one can write for the probe i:

$$\frac{\Delta C_i}{C_0} = \frac{C_i(t) - C_0}{C_0} = \frac{I_i(t) - I_{i0}}{I_{i0}} \tag{33}$$

$\Delta C_i(t)$ is the increase of concentration of ferricyanide ions on the probe i due to the tracer injection.
For the pulse injection, the molar percentage $\%n_i$ of the tracer in the outlet branch i is given by the following equation:

$$\%n_i = \frac{Q_c C_0 \int_0^\infty \left[\frac{I_i(t)}{I_{i0}} - 1\right] dt}{\sum_{i=1}^{10} Q_c C_0 \int_0^\infty \left[\frac{I_i(t)}{I_{i0}} - 1\right] dt} 100 = \frac{\int_0^\infty \left[\frac{I_i(t)}{I_{i0}} - 1\right] dt}{\sum_{i=1}^{10} \int_0^\infty \left[\frac{I_i(t)}{I_{i0}} - 1\right] dt} 100 \tag{34}$$

For the step-shaped injection, I_i and ΔC_i are constant:

$$\%n_i = \frac{Q_c C_0 \Delta C_i}{\sum_{i=1}^{10} Q_c C_0 \Delta C_i} 100 = \frac{\Delta C_i}{\sum_{i=1}^{10} \Delta C_i} 100 \tag{35}$$

Calculation of the molar percentage in each minichannel informs about the distribution of the tracer at the cells outlet. Thus, qualitative information is obtained regarding the mixing process inside the working systems.
In the other hand, a criterion of deviation is as well chosen in order to compare the mixing effectiveness for each geometry of minireactor. This is a classical criterion for evaluation of the mixing state. It compares the observed mixing with a perfect mixing state. In the present work, the measured molar percentage of the tracer is compared to a perfect distribution of the tracer which corresponds to a value of 10 % in the case of an ideal radial distribution. Thus, the mean relative deviation, σ, is given by:

$$\sigma = \sqrt{\sum_{i=1}^{10} (\%n_i - 10\%)^2 \Big/ 10} \tag{36}$$

The highest values of σ advise on poor performance in terms of mixing effectiveness and the smallest values of σ reveal a good degree of mixing.

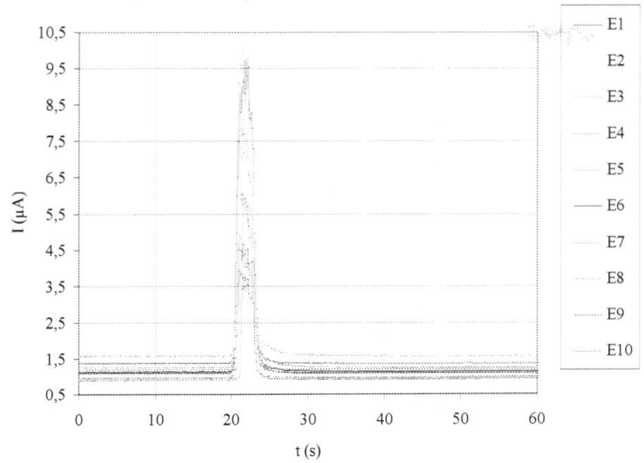

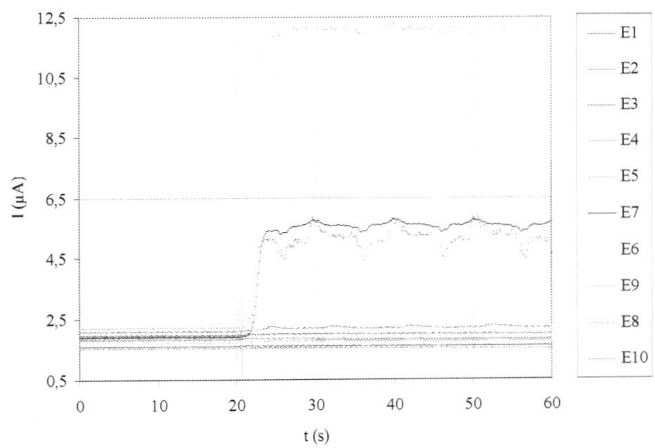

Fig. 16. Up : Time traces of the diffusion limiting current recorded during the pulse injection for the T_network. Down: Time traces of the diffusion limiting current recorded during the step-shaped injection for the ×_network.

6.2 Mixing performance of each network

6.2.1 Mixing quality

The radial distribution of the mass amount of tracer detected by the electrochemical sensors at the outlet of the two cells is presented in Figure 17 and 18 for Re=161 and C_T=50 mol.m^{-3}. The two injection techniques are representative of the tracer dispersion for each geometry of the cells. For the T_network, the radial dispersion is more important than in the ×_network and shows a better mixing state in this experimental condition.

Figures 19 and 20 synthesizes the whole experimental results which take into account the different tracer concentrations and the different Reynolds numbers studied. The relative deviation as a function of Reynolds number is plotted, emphasizing the preceding qualitative observation.

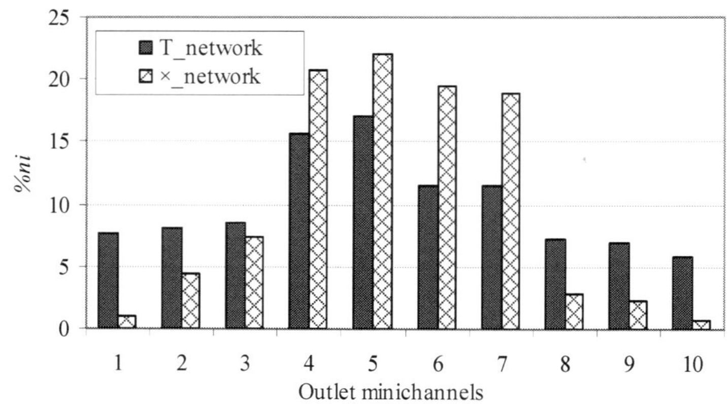

Fig. 17. Radial dispersion of the electrochemical tracer (C_T=50 mol.m^{-3}) at Re=160 for a pulse injection.

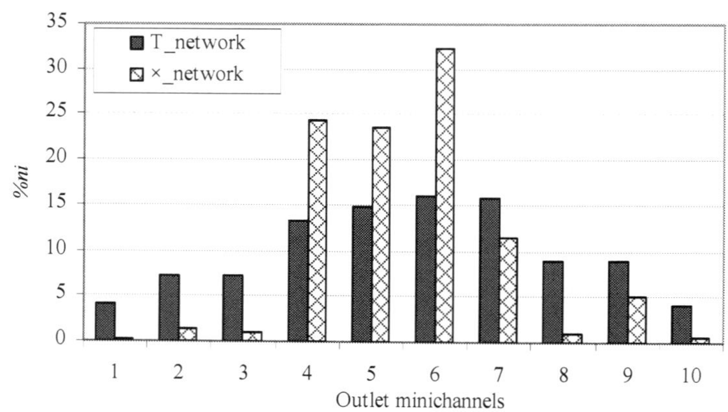

Fig. 18. Radial dispersion of the electrochemical tracer (C_T=50 mol.m^{-3}) at Re=160 for a step-shaped injection.

Except for *Re*≈57, the results reported in Fig. 20 show that the variances for the T_network are higher than for the ×_network, thus emphasizing a better effectiveness in terms of mixing whatever the Reynolds number. The zone of restriction between the convergent and the divergent contributes to decrease the path between the streamlines to force the mixing of the fluids. The values found at higher concentrations (C_T=100 mol.m^{-3} and C_T=150 mol.m^{-3}) for *Re*=57 are due to the dominating diffusion regime compared to the convective regime. Indeed, the concentration of the supporting electrolyte (K_2SO_4) is locally and briefly decreased near the injection point. Furthermore, the pulse injection, especially for the small Reynolds number values induces a local increase of the flow rate due to the duration of the pulse injection which is not negligible compared to the entrance flow rate in the cell.

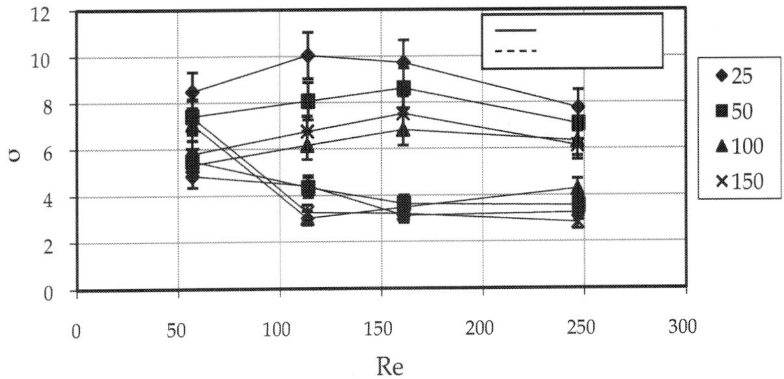

Fig. 19 Mixing effectiveness as a function of the Reynolds number at different lectrochemical tracer concentrations with the pulse injection method.

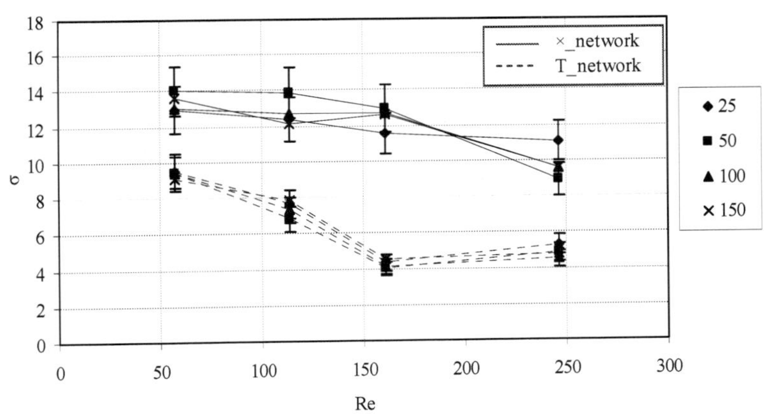

Fig. 20. Mixing effectiveness as a function of the Reynolds number at different electrochemical tracer concentrations with the step-shaped method.

For the step-shaped method (Fig. 20), the results confirm those previously obtained with the pulse method regarding the effectiveness of the mixing between the two geometries. Oppositely to the pulse injection, no dependence of the tracer's concentration is noticeable on the results for the two geometries. The results are not altered by the hydrodynamics generated by the injection, even at low Re values. The step-shaped technique can be considered as a better method in the mixing characterization for this kind of confined geometry.

For the ×_network, σ is found to be constant and equal in average to 13 until Re=161. Above Re=161, σ decreases to reach a value close to 10. This increase in terms of mixing performance is correlated to the first instabilities measured inside the network of crossing minichannels since the first fluctuations of the wall shear rate are quantified from $Re\approx200$.
For the T_network, we observe a sharp decrease from Re≈57 until Re≈161. From $Re\approx161$, σ is found to be constant at a value close to 4.5. This improvement of the mixing state is due to the geometry of the T_network inducing for a given flow rate several flow regimes in the network. For example, at Re≈114 at the inlet (calculation based on eq. 14), the Reynolds number varies in the channels from the converging zone until the restriction zone. At this last location for example, it remains only two channels for the fluid circulation and the velocity in each channel corresponds to a velocity equal to about 0.26 $m.s^{-1}$. Thus, the Reynolds number reaches a value equal to 400 at this position and thus contributes to the destabilization of the flow and an increase of the mixing capacity.

6.2.2 Energetic dissipation

The energy consumption of the working systems has to be evaluated in order to compare their efficiency as minimixer. Thus, measurements of pressure drop across the networks of height, H, are carried out by means of differential pressure valve located at the inlet and the outlet of the cells.

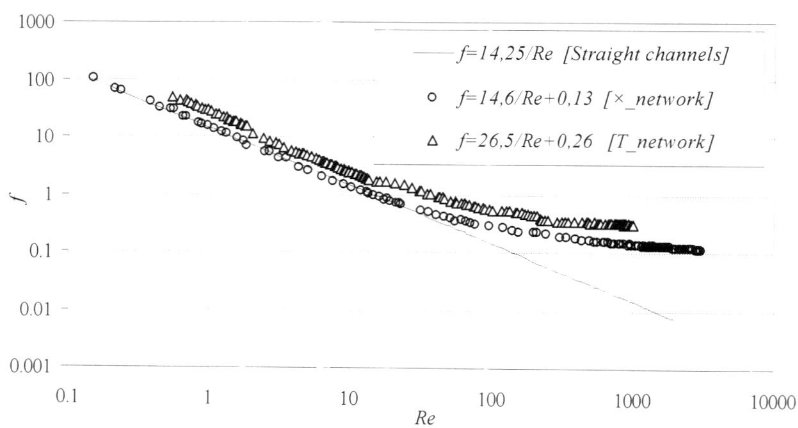

Fig. 21. Evolution of friction factor as a function of Reynolds number for the different geometries of network.

The experimental values of the pressure drop for all the used fluids and for all the Reynolds numbers are presented in Fig. 21 by using the friction coefficient, f, given by:

$$f = \frac{\Delta P}{H} \frac{d_h}{2\rho U_c^2} \tag{37}$$

This representation is useful to compare the different studied geometries (T_network and ×_network) with a set of ten straight parallel channels. Until $Re\approx10$, the obtained values for

×_network and a set a straight channels are similar. For Re>10, an inertial contribution appears for the ×_network and the T_network due to the different shapes (crossing, bend or T_shaped).
It is important to notice that whatever the flow cell studied in the present work, the experimental results are quite well correlated by an equation of type:

$$f = \frac{A}{\mathrm{Re}} + B \tag{38}$$

With A=14.6 and B=0.13 in the ×_network case and A=26.5 and B=0.26 in the T_network. This equation, validated in a large range of Reynolds numbers (1<Re<1000) is similar to that employed in porous media confirming the analogy with the packed bed of particles proposed previously. For the T_network, the pressure drop is approximatively two times larger than for the ×_network for a given Reynolds number value. But T_network clearly ensures better mixing quality even at low Reynolds number, thus its use at lower Reynolds numbers allows energy saving for an identical or a better mixing quality. For example, in order to obtain the same mixing effectiveness corresponding to σ equal to about 9 (Fig.20: Re=55 for T_network, Re=250 for ×_network) it is necessary to bring 6 times more energy into the ×_network. From Re=114 in the T_network, the mixing performance increase to reach a value equal to 6 while the energy consumption remains 2 times less important than in the ×_network at Re=250.

7. Conclusion and outlooks

Hydrodynamics and mass transfer inside networks of minichannels of square section is studied by using different experimental techniques. Each technique is performed in order to characterize the flow and the mass transfer at several levels. Global measurements from pressure drop measurements and liquid-solid mass transfer has given a comprehensive approach and the comparison with known geometries from semi-empirical correlations. The local diagnostics of the flow is determined from Particle Image Velocimetry and electrodiffusional method in order to characterize the flow scales.
The main part of the research work has concerned a network of minichannels or so called ×_network. The sobolik's method was compared with the transfer function method from the assessment of the power density spectra (PSD) of the fluctuation of the wall shear rate. Shear rate fluctuations are observed for Re>200. Under transitional flow conditions, the fluctuation rate increases rapidly and reaches a locally constant level above a Reynolds number of about 1000. The onset and the stabilization of the wall shear rate fluctuations were found similar than those already measured in the porous media. Moreover, the flow is found to be laminar and the pressure drop governed by viscous dissipation only at low Reynolds number (Re<10). For Re>10 the flow regime remains laminar but the pressure drop is described by a Forcheimer type equation due to the kinetic energy degradation. A capillary model based on the geometrical parameter of the network (specific surface, porosity and tortuosity) has been used in order to compare the global liquid-solid mass transfer results with the correlation of Wakao & Funazkri (1978). It confirms that the network of crossing minichannels can rather be compared to a porous media than to a set of straight channels.

Mean velocity fields by PIV and wall shear rate fluctuations analysis from electrochemical microsensors have been focused in order to analyse the characteristic scales of the flow. PIV velocity data emphasize the developing behaviour of the flow between two successive crossings in the whole network. The inhomogeneous characteristics of the flow are confirmed by using wall electrochemical microsensors at different locations in the network. The wall turbulence eddies are studied by direct high frequency measurements of the wall shear rate. It shows unusually small scales characteristics: integral scales and Taylor microscales are relatively scattered and tend to be dependent on the cell geometry. Higher sizes of the integral scales are found in the lateral part than in the central one of the network. The contribution of the crossings as regards the size of the mixing eddies is less important in the lateral zone ($-0.73>z/L/2>0.96$) than in the axial zone ($-0.3<z/L/2<0.3$). The Taylor microscales are controlled by the confinement effects which induce a scattered spatial distribution between two successive crossings. This spatial heterogeneous distribution of the small scales is added to a temporal intermittency analysed by means of the statistical properties of the fluctuating wall shear rate quantities. We generally noted that the flatness and the skewness factors are respectively greater than 5 and greater than 1. Thus, the intermittency level is clearly larger than in the case of a straight channel in transient regime. The constant regime fluctuation defined from $Re\approx1000$ corresponds to a high level of intermittency which starts at $x/H\approx0.3$, is maximum at $x/H\approx0.4$ then reaches a stabilization beyond $x/H\approx0.4$.

Figure 22 summarizes the results from different methods inside this cell model.

The last item of the chapter has concerned the mixing performance characterization for two geometries of network: ×_network & T_network. Among the two tracer response techniques investigated (step-shaped and pulse), the results showed that the step-shaped method is more appropriated for a confined flow system. The diffusion-convective phenomenon at high concentration of tracer is limited due to a better hydrodynamics control at the injection location.

The results in the Reynolds numbers range, $55<Re<250$, concerning the T_network seem to be promising in terms of mixing effectiveness compared to the ×_network. In spite of pressure loss two times larger in the T_network than in the ×_network, it is possible to obtain the same or a better mixing effectiveness and to minimize energy consumption by working at clearly lower Reynolds number values in the T_network.

The results regarding the ×_network confirm preceding local hydrodynamics study and liquid-solid mass transfer measurements which exhibit similar behaviour than that observed in porous media in the Reynolds numbers range $1<Re<1000$. Evaluation of pressure drop in the T_network is as well represented by a Forchheimer equation as met in porous media.

The final ambition is to build some mixing model from these experimental data in one node of the network until the whole of the cell. It will help to design new microreactors with complex geometry (network of microchannels, porous media, constructal geometry). In other hand, some improvements of electrochemical microsensors are essential for characterization of transfer phenomena considering the high potential of this technique for the understanding of hydrodynamics and mass transfer phenomena inside the microsystems. Integration of double rectangular strip microelectrodes will inform about the directional characteristics of the wall-flow and the flow microscales in order to characterize the micromixing inside the mini and microreactors.

In conclusion, the flow inside Microsystems is generally not established with a predominant effect of the walls. The electrochemical method is the only one able to analyze developing unsteady flow characteristics in order to understand the unsteady phenomenology of ultra-confined flow developed inside micromixers and microreactors. Our next investigations are devoted in the mixing analysis inside different shapes of micromixers (Huchet *et al.*, 2008c)

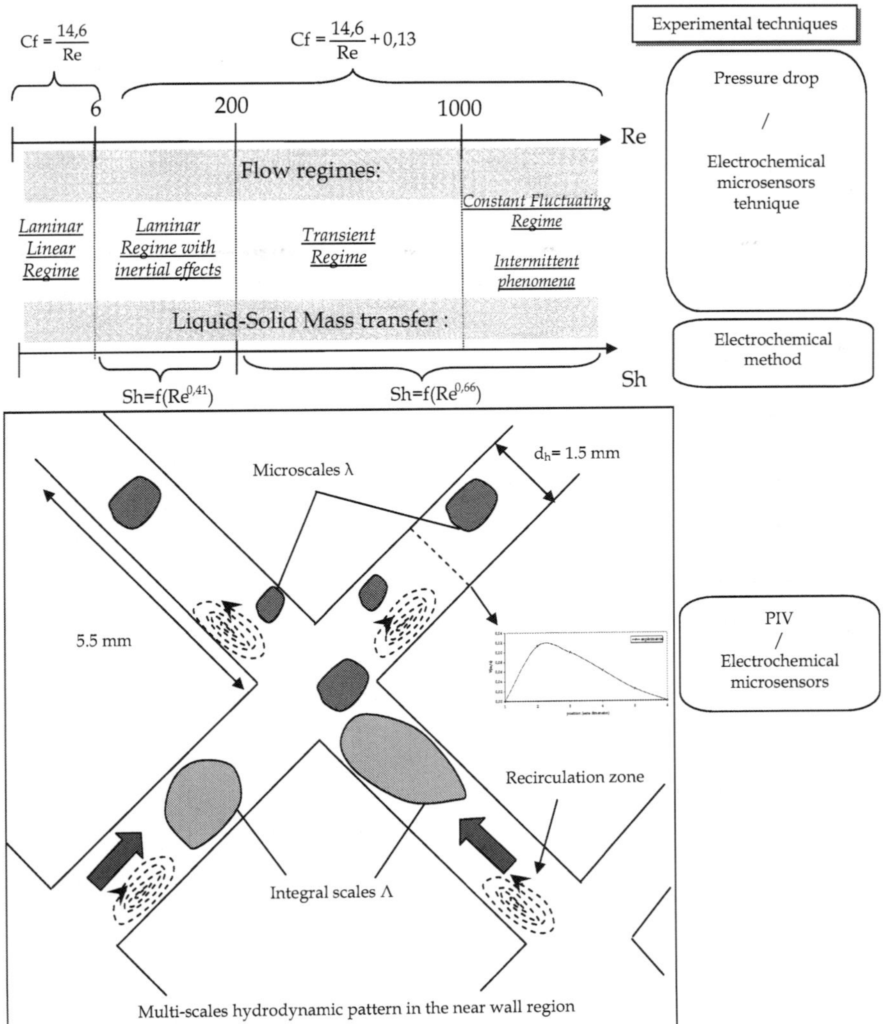

Fig. 22. Summary of results obtained from various experimental techniques in the ×_network

8. Acknowledgement

This work has been performed at the laboratory GEPEA -UMR CNRS 6144

9. References

Adolphe, X., Danaila, L. & Martemianov, S. (2007). On the small-scale statistics of turbulent mixing in electrochemical systems, *J. Electroanal. Chem.* 600: 119-130.

Bothe, D., Stemich, C. & Warnecke, H.J. (2006). Fluid mixing in a T-shaped micro-mixer, *Chem. Eng. Sci.* 61: 2950-2958.

Brandner, J.J., Anurjew, E., Bohnn, L., Hansjosten, E., Henning, T., Schygulla, U., Wenka, A. & Shubert, K. (2006). Concepts and realization of microstructure heat exchangers for enhanced heat transfer, *Exp. Therm. Fluid Sci.* 30: 801-809.

Coeuret, F. & Stork, A., (1984). *Eléments de génie électrochimique* Tec. Doc., Paris, 134-135.

Comiti, J. & Renaud, M. (1989). A new model for determining mean structure parameters of fixed beds from pressure drop measurements, *Chem. Eng. Sci.* 44: 1539-1545.

Commenge, J.M., Falk, L., Corriou, J.P. & M., Matlosz (2004). Intensification des procédés par microstructuration, *C. R. Physique* 5: 597-608.

Deslouis, C., Gil, O. & Tribollet, B. (1990). Frequency response of electrochemical sensors to hydrodynamic fluctuations, *J. Fluid Mech.* 215: 85-100.

Gunther, A., Jensen, K.F. (2006). Multiphase microfluidics: from flow, characteristics to chemical and materials synthesis, *Lab chip* 6: 1487-1503.

Hessel, V., Löwe, H. & Schönfeld, F. (2005). Micromixers - a review on passive and active mixing principles, *Chem. Eng. Sci.* 60: 2479-2501.

Lévêque, M.A. (1928). *Les lois de la transmission de la chaleur par convection*, Ann. Mines, Paris, 13: 201-239.

Hanratty, T.J. & Campbell, J.A. (1983). Measurement of wall shear stress in *'Fluid mechanics measurements'* (Eds.) J.R. Goldstein, Washington Hemisphere, 559.

Huchet, F., Comiti, J., Tihon, J., Montillet, A., & Legentilhomme, P. (2007). Electrodiffusion diagnostics of the flow and mass transfer inside a network of crossing minichannels, *J. App. Electrochem.* 37 1: 49-55.

Huchet, F., Legentilhomme, P., Comiti, J., Solliec, C., Legrand, J. & Montillet, A. (2008)a. Multi-scale analysis of near-wall turbulence intermittency inside a network of crossing minichannels by using electrochemical method and PIV measurements, *Int. J. Heat Fluid Flow* 29 5: 1411-1421.

Huchet, F., Comiti, J., Legentilhomme, P. & Bennadji, H. (2008)b. Mixing characterization and energetic dissipation in different networks of minichannels, *Chem. Eng. R. D.* 86 10A: 1135-1142.

Huchet, F., Havlica, J., Legentilhomme, P., Montillet, A., Comiti, J. & Tihon, J. (2008)c. Use of electrochemical microsensors for hydrodynamics study in crossing microchannels, Microfluid. Nanofluid. 5 1: 65-64.

Labraga, L., Lagraa, B., Mazouz, A. & Keirsbulck, L. (2002). Propagation of shear-layer structures in the near-wall region of a turbulent boundary layer, *Exp. Fluids* 33: 670-676.

Li, H. & Olsen, M. (2006). Examination of large-scale structures in turbulent microchannel flow, *Exp. Fluids* 40: 733-743.

Martemianov, S., Evdokimov, Y.K. & Adolphe, X. (2007). Electrodiffusion diagnostics of laminar flow using the delay-time method, J. Appl. Electrochem. 37: 1321-1328.

Max, J. (1985). *Méthodes et techniques de traitement du signal et applications aux mesures physiques*, ed. Masson, 4th edn.

Nakoryakov V. E., Budukov A. P., Kashinsky O. N. & Geshev P. I. (1986). Electrodiffusion method of investigation into the local structure of turbulent flows, ed. V.G. Gasenko, Novosibirsk.

Natrajan, V.K. & Christensen, K.T. (2007). Microscopic particle image velocimetry measurements of transition to turbulence in microscale capillaries, Exp. Fluids 43, 1-16.

Natrajan, V.K., Yamaguchi, E. & Christensen, K.T. (2007). Statistical and structural similarities between micro- and macroscale wall turbulence, Microfluid. Nanofluid. 3: 8-100.

Portelli, B., Holdsworth, P.C.W. & Pinton, J.F. (2003). Intermittency and non-Gaussian fluctuations of the global energy transfer in fully developed turbulence, *Phys. Rev. Lett.* 90 104501.1-104501.4.

Rehimi, F., Aloui, F., Ben Nasrallah, S., Doubliez, L. & Legrand, J. (2006). Inverse method for electrodiffusional diagnostics of flow, *Int. J. Heat Mass Transfer* 49: 1242-1254.

Reiss, L.P. & Hanratty, T.J. (1963). An experimental study of the unsteady nature of the viscous sublayer, *AIChE J.* 8: 154-160.

Sato, Y., Irisawa, G., Ishizuka, M., Hishida, K. & Maeda, M. (2003). Visualization of convective mixing in microchannel by fluorescence imaging, *Meas. Sci. Technol.* 14: 114-121.

Seguin, D., Montillet, A. & Comiti, J. (1998). Experimental characterization of flow regimes in various porous media-I: Limit of laminar flow regime, *Chem. Eng. Sci.* 53: 3751-3761.

Seguin, D., Montillet, A., Comiti, J. & Huet, F. (1998). Experimental characterization of flow regimes in various porous media II: Transition to turbulent regime, *Chem. Eng. Sci.* 53: 3897-3909.

Sobolik, V., Tihon, J., Wein, O. & Wichterle, K. (1998). Calibration of electrodiffusion friction probes using a voltage-step transient, *J. Appl. Electrochem.* 28: 329-335.

Sobolik, V., Wein, O. & Cermak, J. (1987). Simultaneous measurement of film thickness and wall shear stress in wavy flow of non-Newtonian liquids. *Coll. Czech Chem. Com.* 52: 913-928.

Tennekes, H. & Lumley, J.L. (1972). *A first course in turbulence*, Cambridge, MA: MIT Press, 8: 286.

Tihon, J., Tovchigrechko, V., Sobolik, V. & Wein, O. (2003). Electrodiffusion of the near-wall reversal in liquid films at the regime of solitary waves, *J. Appl. Electrochem.* 33: 577-587.

Wakao N. & Funazkri, T. (1978). Effect of fluid dispersion coefficients on particule-to-fluid mass transfer coefficients in packed beds. Correlation of Sherwood numbers, *Chem. Eng. Sci.*, 33, 1375-1384.

Wu, Z. & Nguyen, N. T. (2005). Hydrodynamic focusing in microchannels under consideration of diffusive dispersion: theories and experiments. *Sensors Actuators B* 107: 965-974.

Xu, C.X., Li, L., Cui, G.X. & Zhang, Z.S. (2006). Multi-scale analysis of near-wall turbulence intermittency. *J. Turbulence* 7: 1-15.

Yi, C., Zhang, Q., Li, C.W., Yang, J., Zhao, J. & Yang, M. (2006). Optical and electrochemical technique for cell based microfluidic systems, *Anal. Bioanal. Chem.* 384: 1259-1268.

13

Simulation of Hydrodynamics and Mass Transfer in a Valve Tray Distillation Column Using Computational Fluid Dynamics Approach

A. Jafari[1], S.M. Mousavi[2*] H. Moteshaffi[2],
H. Roohian[2] and H. Hamedi Sangari[3]
[1]*Department of Planning and Development,*
National Petrochemical Company,Tehran,
[2]*Biotechnology Group, Chemical Engineering Department,*
Tarbiat Modares University,Tehran,
[3]*R & T Management Department,*
National Iranian Oil Refining and Distribution Company, Tehran,
Iran

1. Introduction

Distillation is a widely used method to separate liquid mixtures into their components and has been applied in many separation processes such as those in petroleum, petrochemical, chemical and related industries. It shares a large portion of the capital investment, and is the largest consumer of energy in those industries. It is also commonly recognized that distillation is a very important process in today's industry and will continue to be in the future (You, 2004). Tray column is widely used as distillation column in chemical and petrochemical industry. It has some advantages compared with packed column, such as easy maintainability, low cost, convenient feed and side-stream withdrawal, and reliability for high pressure and high viscosity liquid. Numerous studies were carried out on tray columns and many different types of tray have been developed. Among them, valve tray is one of the most effective ones, and it plays an important role in commercial production due to its flexibility in handling a wide range of vapour throughputs. As the gas horizontally blows out of the valves and has a long hold-up time, the valve has low entrainment and higher average operating efficiency than sieve trays which had been in use before (Lianghua et al., 2008; Li *et al.*, 2009).

Even if valve tray columns are widely used in distillation processes separating liquid mixtures, a bottleneck that impedes the further improvement of these column internals is the fact that little is known about the detailed flow field through the valves on the tray for a given geometry under different operation conditions. The main reason is the complex flow

* Corresponding author at: Biotechnology Group, Chemical Engineering Department, Tarbiat Modares University, Jalal Al Ahmad highway, Tehran, Iran. Tel.: +98 21 82884917. Fax: 82884931,
E-mail address: mousavi_m@modares.ac.ir (S.M. Mousavi).

behaviours on the tray deck. For example, although it is well known that the liquid flow pattern, or velocity distribution, is a very important factor in distillation tray design, the details of the extremely complex hydrodynamic processes on trays still do not seem to be completely understood (Hirschberg et al., 2005; Lianghua et al., 2008). In contrast with the importance and common use of valve tray columns, due to the complex geometry of their valves, there is lack of report about simulating the hydrodynamics of valve trays by using computational fluid dynamics (Li *et al.*, 2009).

In the recent years, computational fluid dynamics (CFD) has attracted the attention of researchers as a powerful tool for fluid-flow phenomena in various devices. It is possible to use CFD method to simulate the complex fluid flow performance on trays. Compared to experimental method, CFD has the flexibility of testing different flow geometry and system conditions without suffering appreciable cost. At present, numerical simulation is mainly focused on the liquid flow field (Li *et al.*, 2009; Zhang & Yu, 1994; Liu *et al.*, 2000) or the gas-liquid two-phase flow (Delnoij et al., 1999; Deen et al., 2001; Buwa & Ranade 2002; Van Baten & Krishna, 2000) on a tray.

Since the liquid distribution on a valve tray is complex in its structure influenced by the hydrodynamics behaviour of gas through the bubble element, it is very important to study the hydrodynamics behaviour of the gas using CFD (Lianghua et al., 2008). The tray design based on computational fluid dynamics foundation is obviously superior to that based on experience or rough estimation, especially for the case of scaling up a column to large diameter (You, 2004). The idea of using CFD to incorporating the prediction of tray efficiency relies on the fact that the hydrodynamics is an essential influential factor for mass transfer in both interfacial and bulk diffusions, which could be understood by the effect of velocity distribution on concentration profile. This, in fact, opens an issue on the computation for mass transfer prediction based on the fluid dynamics computation (Sun *et al.*, 2007). Several recent publications have established the potential of CFD for describing tray hydraulics and tray performance (Wijn, 1996; Mehta et al., 1998; Fischer & Quarini, 1998; Yu *et al.*, 1999; Krishna & van Baten, 1999; Liu *et al.*, 2000; Ling *et al.*, 2004; Rahbar *et al.*, 2006; Li *et al.*, 2009; Alizadehdakhel, 2010).

The experts cited transport phenomena such as fluid flow, heat and mass transfer, and multi-phase flow as subjects that are insufficiently understood. In recent years there has been considerable academic and industrial interest in the use of computational fluid dynamics (CFD) to model two-phase flows in process equipment. The use of CFD models for gas–liquid bubble columns has also evoked considerable interest in recent years and both Euler–Euler and Euler–Lagrange frameworks have been employed for the description of the gas and liquid phases (Van Baten & Krishna, 2000). Mehta *et al.* (1998) have analysed the liquid phase flow patterns on a sieve tray by solving the time-averaged equations of continuity of mass and momentum only for the liquid phase. Interactions with the vapour phase are taken account by use of interphase momentum transfer coefficients determined from empirical correlations. Yu *et al.* (1999) attempted to model the two-phase flow behaviour using a two-dimensional model, focussing on the description of the hydrodynamics along the liquid flow path, ignoring the variations in the direction of gas flow along the height of the dispersion. Fischer and Quarini (1998) have attempted to describe the three-dimensional transient gas–liquid hydrodynamics. An important key assumption made in the simulations of Fischer and Quarini concerns the drag coefficient;

these authors assumed a constant drag coefficient of 0.44, which is appropriate for uniform bubbly flow.

As a result of the development of computer technology beginning in the sixties of last century, the computational fluid dynamics and heat transfer were initiated on the basis of closure of the differential equations of momentum and heat transfer respectively by the fluid dynamic researchers and mechanical engineers. Nevertheless, in the chemical engineering work, the prediction of concentration field is equally important, and the task of such development, which may be regarded as computational mass transfer, is naturally relied on the investigation by the chemical engineers. Recently, works on computational mass transfer have been reported covering its basic ground and various applications. The early research on the numerical simulation of concentration distribution was reported in 1960s, almost at the same time when CFD was developed (McFarlane et al., 1967; Farouqali & Stahl, 1969), but the simulation of concentration and temperature distributions depended largely on the pattern of velocity distribution by CFD. Yu et al. studied the concentration field on a sieve tray by applying the CFD method to solve the flow and mass transfer equations with the assumption of constant equilibrium ratio of the separated substance in the mixture. They obtained the mass transfer efficiency in terms of effectiveness of the tray. But their assumption of constant equilibrium ratio would lead to a large deviation in the case of having wide range of tray concentration. Therefore, a variable equilibrium ratio is adopted to recalculate the concentration field on the tray in this study (Yu et al., 1999). You analysed quantitatively the dependence of mass transfer performance of sieve distillation tray on the physical properties of fluid and tray flow patterns and discussed the ways to improve the efficiency of distillation sieve tray (You, 2004). Wang et al. presented a three-dimensional CFD model for describing the liquid-phase flow and concentration distribution on a distillation column tray. They considered both volume fractions of gas and liquid as well as the interfacial forces and the interphase mass transfer item. Simulations of three-dimensional liquid flow and the concentration distribution were carried out on both a single tray and all trays of a distillation column with 10 sieve trays under total reflux and the prediction by the present model for liquid flow on a single tray was confirmed by the experimental measurements (Wang et al., 2004). Sun et al. demonstrated the feasibility of the simplified computational mass transfer model for distillation column simulation, and compared the simulation results with the experimental data taken from literatures. They showed that by applying the modified model to the simulation of a commercial scale distillation tray column, predicted concentration at the outlet of each tray and the tray efficiency were satisfactorily confirmed by the published experimental data (Sun et al., 2007). The computational mass transfer under investigation now aims to the prediction of concentration distribution of complex fluid systems with simultaneous mass, heat and momentum transports and/or chemical reactions in chemical processes (Xigang & Guocong, 2008).

The purpose of this research was to develop a three-dimensional CFD modeling, within the two-phase Eulerian framework, to understand the hydrodynamics and mass transfer of valve trays in a distillation column. Velocity distributions, clear liquid height, phases volume fractions, and separation of materials under unsteady state conditions were predicted. The main objective of this work was to consider gas-liquid mass transfer (species distribution for each phase) for cyclohexane-n-heptane separation in a single valve tary distillation column.

2. Mathematical formulation

2.1 Governing equations

In this research Eulerian method was used to predict the behaviour of gas-liquid in a valve tray. The governing equations were solved sequentially. The non-linear governing equations were linearized to produce a system of equations for the dependent variables in every computational cell. The resulting linear system was then solved to yield an updated flow-field solution. A point implicit (Gauss-Seidel) linear equation solver was used in conjunction with an algebraic multigrid method to solve the resulting scalar system of equations for the dependent variable in each cell.

The continuity equation for phase q is:

$$\frac{\partial}{\partial t}(\alpha_q \rho_q) + \nabla.(\alpha_q \rho_q \vec{v}_q) = 0 \tag{1}$$

Where α, ρ and $\vec{v}$ shows volume fraction, density and velocity, respectively. The description of the multiphase flow as an interpenetrating continuum incorporated the concept of volume fraction, denoted by α_q. The volume fractions represent the space occupied by each phase, and the laws of conservation of mass and momentum are satisfied by each phase individually. The derivation of the conservation equations can be done by calculating the ensemble average of the local instantaneous balance for each of the phases or by using the mixture theory approach. It is assumed that density is constant, then the continuity equation (Eq. (1)) can be written as Eq. (2), and the volume fraction of each phase can be calculated via this equation:

$$\frac{\partial}{\partial t}(\alpha_q) + \nabla.(\alpha_q \vec{v}_q) = 0 \tag{2}$$

The momentum balance for phase q yields:

$$\frac{\partial}{\partial t}(\alpha_q \rho_q \vec{v}_q) + \nabla.(\alpha_q \rho_q \vec{v}_q \vec{v}_q) = -\alpha_q \nabla P + \nabla.\overline{\overline{\tau}}_q + \alpha_q \rho_q \vec{g} + \sum_{p=1}^{n}(\vec{R}_{pq}) + \alpha_q \rho_q (\vec{F}_q) \tag{3}$$

where t, P, $\vec{g}$, $\overline{\overline{\tau}}$, $\vec{R}_{pq}$, and $\vec{F}_q$ represents time, pressure, gravity, stress-strain tensor, interphase force, and body force, respectively. Here n is number of phases. The q phase stress-strain tensor ($\overline{\overline{\tau}}_q$) was defined as follow:

$$\overline{\overline{\tau}}_q = \alpha_q \mu_q (\nabla \vec{v}_q + \nabla \vec{v}_q^T) + \alpha_q (\lambda_q - \frac{2}{3}\mu_q) \nabla.\vec{v}_q \overline{\overline{I}} \tag{4}$$

where μ and λ_q are dynamic viscosity and bulk viscosity of phase q, respectively. The interphase force, $\vec{R}_{pq}$, depended on the friction, pressure, cohesion, and other effects and was subjected to the conditions that $\vec{R}_{pq} = -\vec{R}_{qp}$, $\vec{R}_{qq} = 0$ and that

$$\sum_{p=1}^{n} \vec{R}_{pq} = \sum_{p=1}^{n} K_{pq} (\vec{v}_p - \vec{v}_q) \tag{5}$$

where K_{pq} (K_{qp}) is the interphase momentum exchange coefficient; and it tends to zero whenever the primary phase is not present within the domain.

In simulations of the multiphase flow, the lift force can be considered for the secondary phase (gas). This force is important if the bubble diameter is very large. It was assumed that the bubble diameters were smaller than the distance between them, so the lift force was insignificant compared with the other forces, such as drag force. Therefore, there was no reason to include this extra term.

The exchange coefficient for these types of gas-liquid mixtures can be written in the following general form:

$$K_{pq} = \frac{\alpha_q \alpha_p \rho_p f}{\tau_p} \tag{6}$$

where f and τ_p are the drag function and relaxation time, respectively. f can be defined differently for the each of the exchange-coefficient models. Nearly all definitions of f include a drag coefficient that is based on the relative Reynolds number. In this study the basic drag correlation implemented in FLUENT (***Schiller-Naumann***) was used in order to predict the drag coefficient.

In comparison with single-phase flows, the number of terms to be modeled in the momentum equations in multiphase flows is large, which complicates the modeling of turbulence in multiphase simulations.

In the present study, standard k-ε turbulence model was used. The simplest "complete models" of turbulence are two-equation models in which the solution of two separate transport equations allows the turbulent velocity and length scales to be independently determined. Eeconomy, and reasonable accuracy for a wide range of turbulent flows explain popularity of the standard k-ε model in industrial flow and heat transfer simulations. It is a semi-empirical model, and the derivation of the model equations relies on phenomenological considerations and empiricism (FLUENT 6.2 Users Guide, 2005).

The equations k and ε that describe the model are as follows:

$$\frac{\partial}{\partial t}(\rho_m k) + \nabla.(\rho_m \vec{v}_m k) = \nabla.(\frac{\mu_{t,m}}{\sigma_k}\nabla k) + G_{k,m} - \rho_m \varepsilon \tag{7}$$

and

$$\frac{\partial}{\partial t}(\rho_m \varepsilon) + \nabla.(\rho_m \vec{v}_m \varepsilon) = \nabla.(\frac{\mu_{t,m}}{\sigma_\varepsilon}\nabla \varepsilon) + \frac{\varepsilon}{k}(C_{1\varepsilon} G_{k,m} - C_{2\varepsilon} \rho_m \varepsilon) \tag{8}$$

where k and ε shows turbulent kinetic energy and dissipation rate, respectively; and $C_{1\varepsilon}$, $C_{2\varepsilon}$, σ_k and σ_ε are parameters of the model. The mixture density and velocity, ρ_m and $\vec{v}_m$, are computed from:

$$\rho_m = \sum_{i=1}^{N} \alpha_i \rho_i \tag{9}$$

and

$$\vec{v}_m = \frac{\sum_{i=1}^{N} \alpha_i \rho_i \vec{v}_i}{\sum_{i=1}^{N} \alpha_i \rho_i} \tag{10}$$

The turbulent viscosity $\mu_{t,m}$ is computed from:

$$\mu_{t,m} = \rho_m C_\mu \frac{k^2}{\varepsilon} \tag{11}$$

The production of turbulent kinetic energy $G_{k,m}$ is computed from:

$$G_{k,m} = \mu_{t,m}(\nabla \vec{v}_m + (\nabla \vec{v}_m)^T) : \nabla \vec{v}_m \tag{12}$$

2.2 Species transport equations

To solve conservation equations for chemical species, software predicts the local mass fraction of each species, Y_i, through the solution of a convection-diffusion equation for the ith species. This conservation equation takes the following general form:

$$\frac{\partial}{\partial t}(\rho Y_i) + \nabla \cdot (\rho v Y_i) = -\nabla \cdot j + R_i + S_i \tag{13}$$

where R_i is the net rate of production of species i by chemical reaction and here it is zero. S_i is the rate of creation by addition from the dispersed phase plus any user-defined sources. An equation of this form will be solved for N -1 species where N is the total number of fluid phase chemical species present in the system. Since the mass fraction of the species must sum to unity, the Nth mass fraction is determined as one minus the sum of the N - 1 solved mass fractions.

In Equation (13), J_i is the diffusion flux of species i, which arises due to concentration gradients. By default, FLUENT uses the dilute approximation, under which the diffusion flux can be written as:

$$Ji = -\rho D_{i,m} \nabla Y_i \tag{14}$$

Here $D_{i,m}$ is the diffusion coefficient for species i in the mixture.

3. Numerical implementation

3.1 Simulation characteristics

In the present work, commercial grid-generation tools, GAMBIT 2.2 (FLUENT Inc., USA) and CATIA were used to create the geometry and generate the grids. The use of an adequate number of computational cells while numerically solving the governing equations over the solution domain is very important. To divide the geometry into discrete control volumes, more than 5.7×10^5 3-D tetrahedral computational cells and 37432 nodes were used. Schematic of the valve tray is shown in figure 1.

The commercial code, FLUENT, have been selected for simulations, and Eulerian method implemented in this software; were applied. Liquid and gas phase was considered as continuous and dispersed phase, respectively. The inlet flow boundary conditions of gas and liquid phase was set to inlet velocity. The liquid and gas outlet boundaries were specified as pressure outlet fixed to the local atmospheric pressure. All walls assumed as no slip wall boundary. The gas volume fraction at the inlet holes was set to be unity.

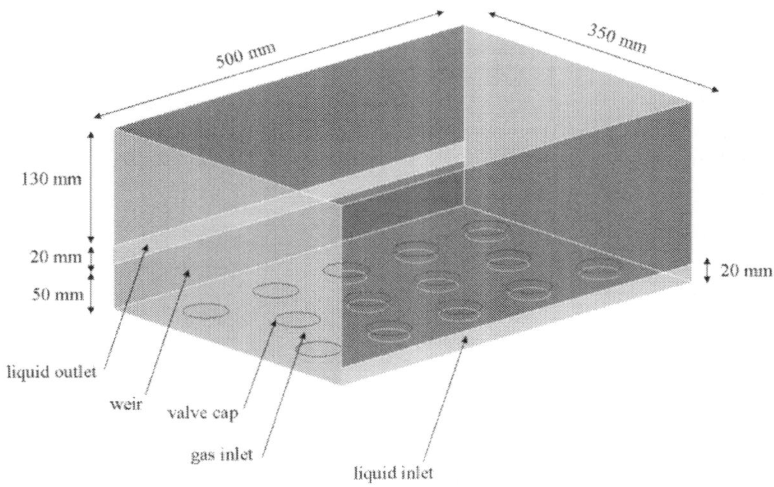

Fig. 1. Schematic of the geometry

The phase-coupled simple (PC-SIMPLE) algorithm, which extends the SIMPLE algorithm to multiphase flows, was applied to determine the pressure-velocity coupling in the simulation. The velocities were solved coupled by phases, but in a segregated fashion. The block algebraic multigrid scheme used by the coupled solver was used to solve a vector equation formed by the velocity components of all phases simultaneously. Then, a pressure correction equation was built based on total volume continuity rather than mass continuity. The pressure and velocities were then corrected to satisfy the continuity constraint. The volume fractions were obtained from the phase continuity equations. To satisfy these conditions, the sum of all volume fractions should be equal to one.

For the continuous phase (liquid phase), the turbulent contribution to the stress tensor was evaluated by the k-ε model described by Sokolichin and Eigenberger (1999) using the following standard single-phase parameters: $C_{\mu} = 0.09$, $C_{1\varepsilon} = 1.44$, $C_{2\varepsilon} = 1.92$, $\sigma_{K} = 1$ and $\sigma_{\varepsilon} = 1.3$.

The discretization scheme for each governing equation involved the following procedure: PC- SIMPLE for the pressure-velocity coupling and "first order upwind" for the momentum, volume fraction, turbulence kinetic energy and turbulence dissipation rate. The under-relaxation factors that determine how much control each of the equations has in the final solution were set to 0.5 for the pressure and volume fraction, 0.8 for the turbulence kinetic energy, turbulence dissipation rate, and for all species.

Using mentioned values for the under-relaxation factors, a reasonable rate of convergence was achieved. The convergence was considered to be achieved when the conservation equations of mass and momentum were satisfied, which was considered to have occurred when the normalized residuals became smaller than 10^{-3}. The normalization factors used for the mass and momentum were the maximum residual values after the first few iterations.

3.2 Confirmation of grid independency

The results are grid independent. To select the optimized number of grids, a grid independence check was performed. In this test water and air were used as liquid and gas phase, respectively. The flow boundary conditions applied to each phase set the inlet gas velocity to $0.64\,ms^{-1}$, and the inlet liquid velocity to $0.195\,ms^{-1}$. Four mesh sizes were examined and results have been represented in table 1. The data were recorded at 15 s, which was the point at which the system stabilized for all cases. Outlet mass flux of air was considered to compare grids. As the difference between numerical results in grid 3 and 4 is less than 0.3%, grid 3 was chosen for the simulation. Figure 2 shows the grid.

Grid	Number of elements	Outlet mass flux of air (g/s)
1	411×10^3	5.24
2	554×10^3	7.08
3	575×10^3	7.4
4	806×10^3	7.42

Table 1. Results of grid independency

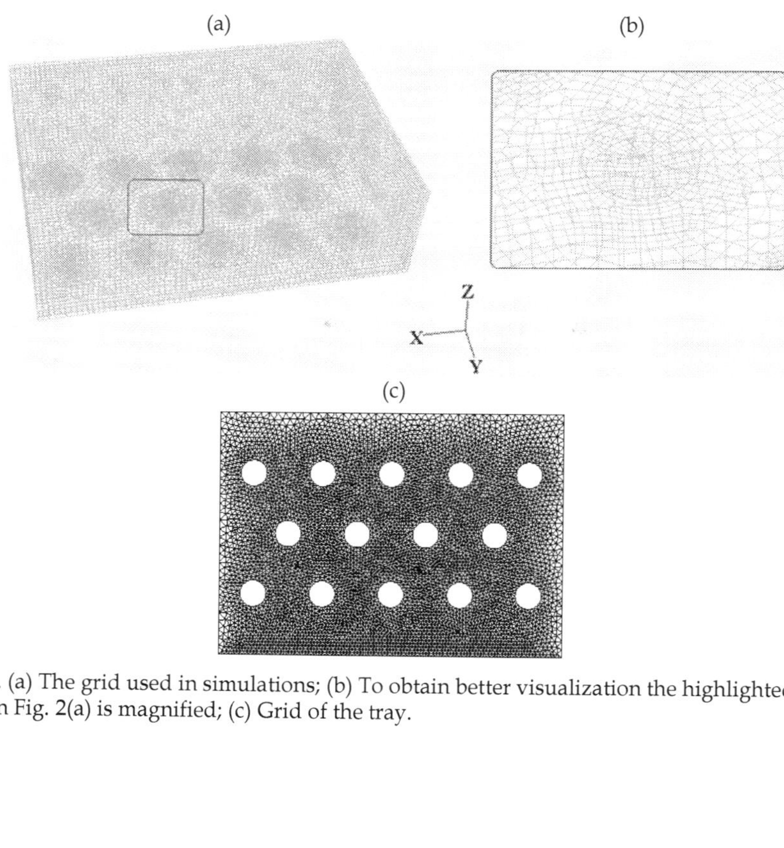

Fig. 2. (a) The grid used in simulations; (b) To obtain better visualization the highlighted part in Fig. 2(a) is magnified; (c) Grid of the tray.

4. Results and discussion

Here hydrodynamics and mass transfer of a distillation column with valve tray is studied. Two-phase, newtonian fluids in Eulerian framework were considered

4.1 Hydrodynamics behviour of a valve tray

Firstly water and air were used as liquid and gas phase. During the simulation, the clear liquid height, the height of liquid that would exist on the tray in the absence of gas flow, was monitored, and results have been presented in figure 3. As this figure shows, after a sufficiently long time quasi-steady state condition has established. The clear liquid height has been calculated as the tray spacing multiplied by the volume average of the liquid-volume fraction.

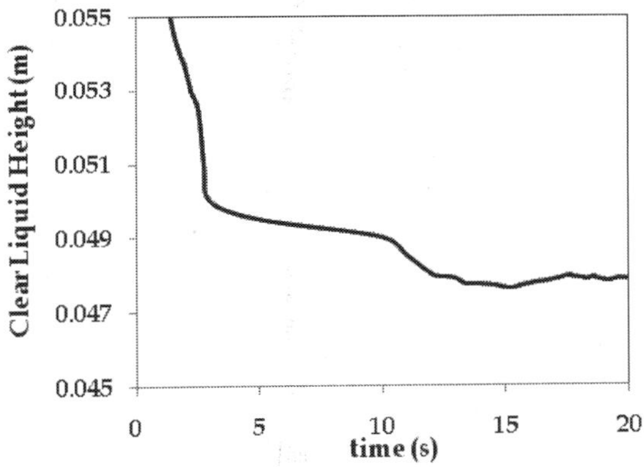

Fig. 3. Clear liquid height versus time.

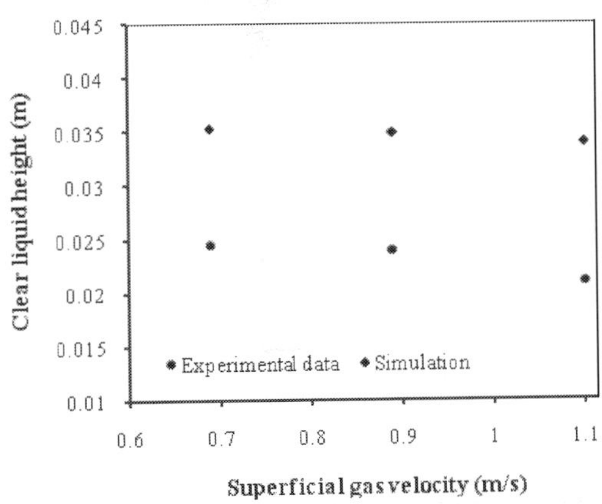

Fig. 4. Clear liquid height as a fuction of superficial gas velocity

In order to valid simulations, results (clear liquid height) were compared with semi-empirical correlations (Li et al., 2009). As figure 3 illustrates, around 15 s steady-state condition is achieved and the clear liquid height is about 0.0478 m. Simulation results are in good agreement with those predicted by semi-empirical correlations and the error is about 2%.

To investigate the effect of gas velocity on clear liquid height, three different velocities (0.69, 0.89 and 1.1 m/s) were applied. The liquid load per weir length was set to 0.0032 $m^3s^{-1}m^{-1}$, and clear liquid height was calculated for the air-water system. Results have been shown in figure 4 and they have been compared with experimental data (Li et al., 2009). As this figure represents, trend of simulations and experimental data are similar.

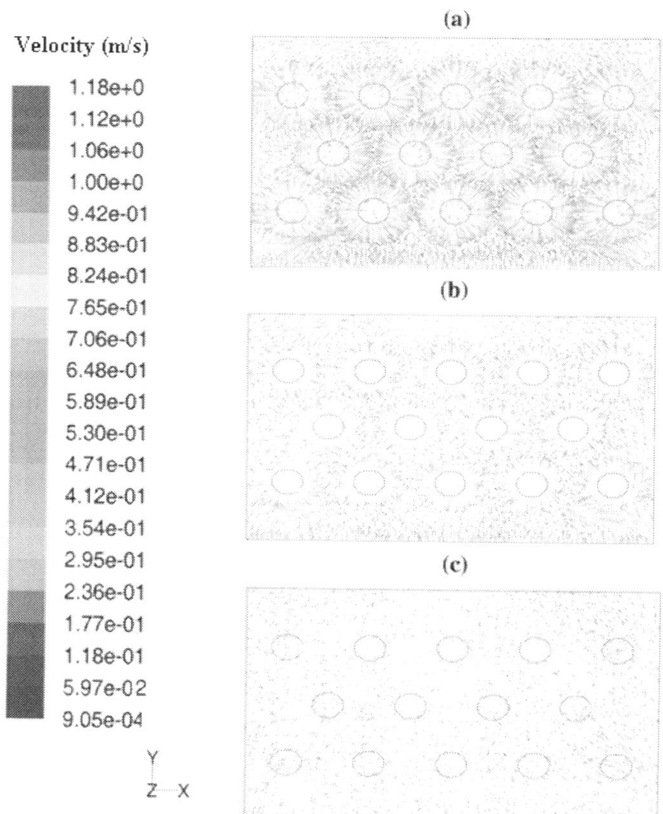

Fig. 5. Top view of liquid velocity vectors after 6 s at (a) z=0.003m, (b) z=0.009m and (c) z=0.015m.

Simulations continued and two phase containing cyclohexane (C_6H_{12}) and n-heptane (C_7H_{16}) were assumed. Numerical approach has been conducted to reach the stable conditions.

Because flow pattern plays an important role in the tray efficiency; numerical results were analysed and Liquid velocity vectors after 6 s have been represented in figure 5. As this figure illustrates, the circulation of liquid near the tray wall have been observed, confirmed experimentally by Yu & Huang (1980) and also Solari & Bell (1986). In fact, as soon as the liquid enters the tray, the flow passage suddenly expands. This leads to separation of the boundary layer. In turbulent flow, the fluids mix with each other, and the slower flow can easily be removed from the boundary layer and replaced by the faster one. The liquid velocity in lower layers is greater than that in higher layers, thus the turbulent energy of the former is larger, and this leads to the separation point of lower liquid layers moving backward toward the wall. Finally circulation produces in the region near the tray wall.
Gas velocity vectors have been shown in figure 6. As this figure represents, the best mixing of phases happens around caps. Such circulations around valves also have been reported elsewhere (Lianghua et al., 2008). Existence of eddies enhances mixing and has an important effect on mass transfer in a distillation column.

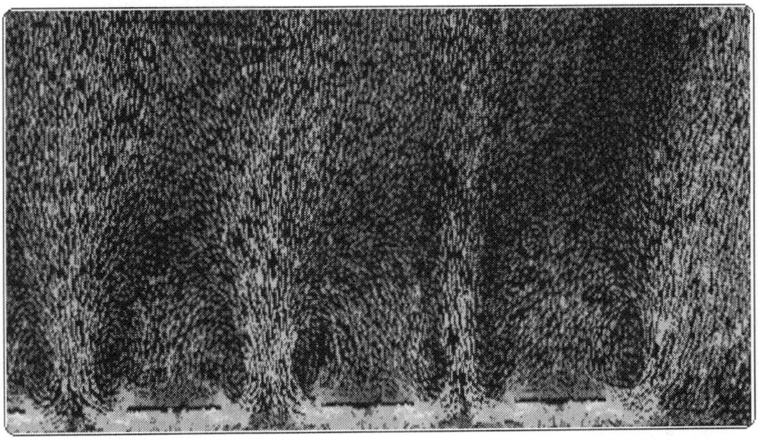

Fig. 6. Gas velocity vectors around caps after 6 s.

4.2 Mass transfer on a valve tray

It is assumed that cyclohexane transfers from liquid to the gas phase, and initial mass fraction of C_7H_{16} in both phases is about 0.15. Concentration is simulated by simultaneously solving the CFD model and mass-transfer equation. Mass fraction of C_7H_{16} in liquid phase versus time has been presented in figure 7. As this figure shows, with passing time mass fraction of n-heptane in liquid increases. In other words, the concentration of the light component (C_6H_{12}) in the gas phase increases along time (figure 8) and the C_7H_{16} concentration in this phase decreases. In addition, C_6H_{12} concentration in gas phase at higher layers increases. Figure 9 shows mass frcation of the light component at three different z and after 6 s. As contours (figure 8 and 9) illustrate, concentrations are not constant over the entire tray and they change point by point. This concept also has been found by Bjorn et al. (2002).

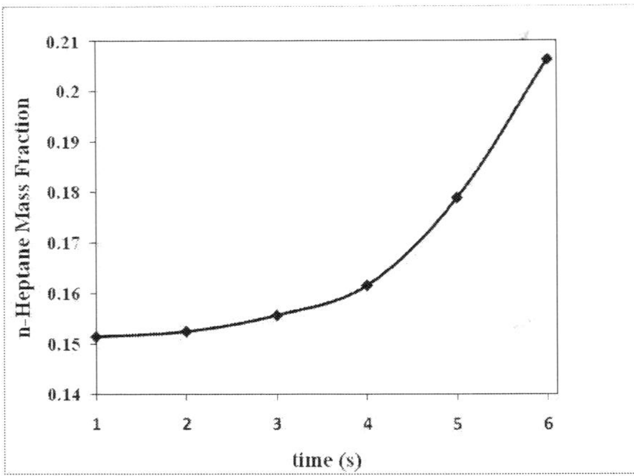

Fig. 7. Changes of n-heptane mass fraction in liquid phase with time.

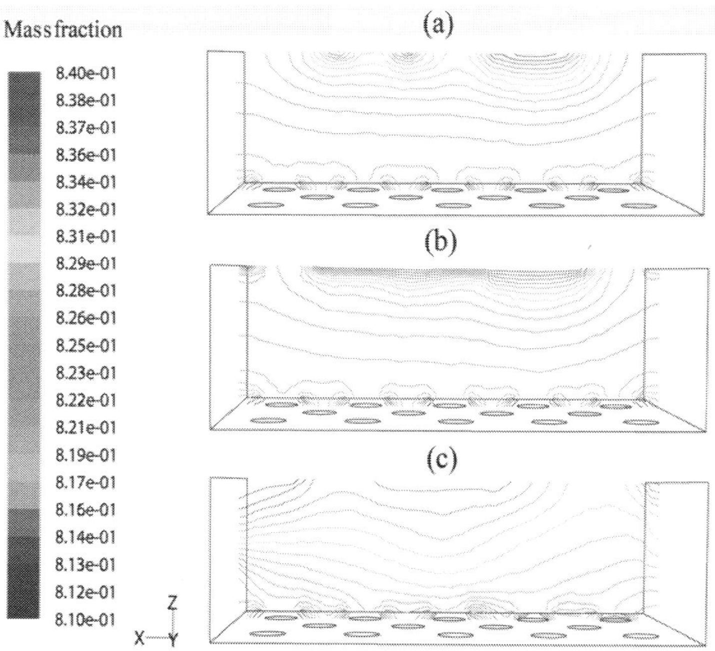

Fig. 8. Mass fraction contours of C_6H_{12} in the gas phase on an x-z plane after (a) 0.25 s, (b) 0.4 s, and (c) 0.6 s.

As mentioned in figure 5, fluid circulation happenes near the tray wall. Therefore, the liquid residence time distribution in the same zone is longer than that in other zones. With the increase of the liquid residence time distribution, mass transfer between gas and liquid is more complete than that in other zones. As the liquid layer moves up, the average concentration of C_6H_{12} in gas phase increases (figure 9) or C_6H_{12} concentration in liquid phase decreases.

A simulation test with high initial velocities of phases were done, and it was found that hydrodynamics have a significant effect on mass transfer. Results of the simulation have been presented in figure 10. Again liquid circulation were observed near the tray wall, and the maximum velocity can be seen around z=0.009m (figure 10 (b)). At this z, C_6H_{12} concentration is in the maximum value and after that the mass fraction becomes constant (figure 10 (c)).

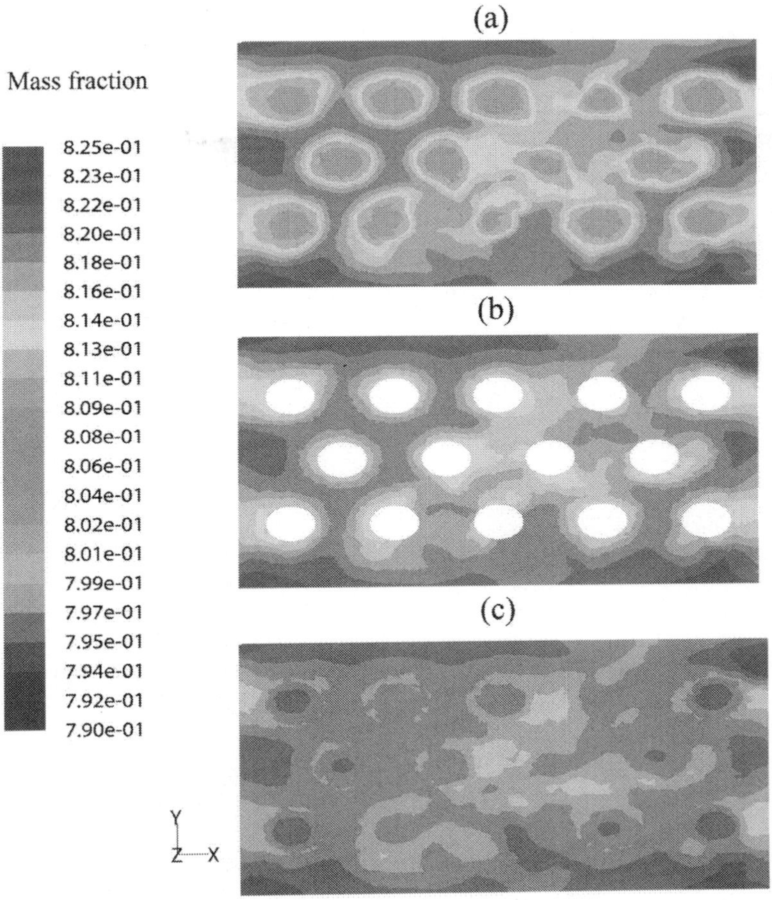

Fig. 9. Mass fraction contours of C_6H_{12} in the gas phase after 6 s at (a) z=0.003m; (b) z=0.009 m and (c) z=0.015m.

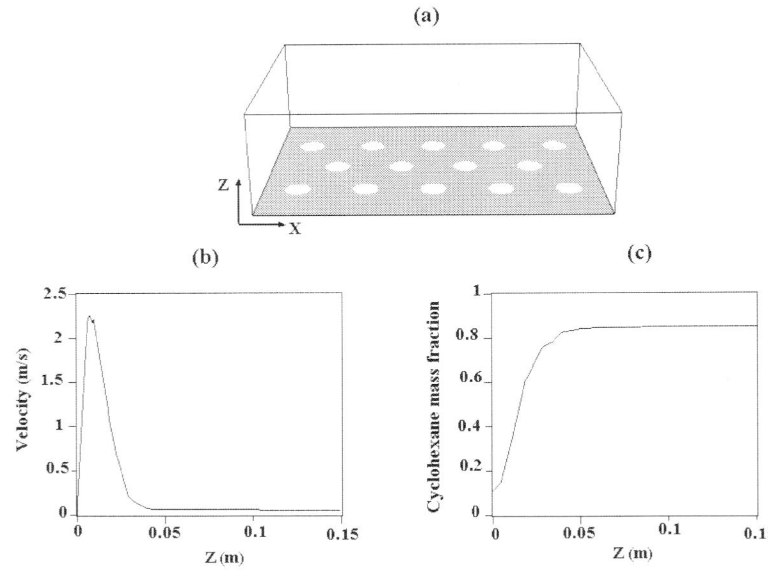

Fig. 10. Results at high initial velocities of gas and liquid afterr 6s. (a) The geometry; (b) Liquid velocity versus z and (c) Changes of C_6H_{12} mass fraction with z.

Figure 11 represents snapshots of gas hold-up at z=0. Fluid hold-up was calculated as the phase volume fraction. Near the tray, gas is dispersed by the continued liquid, and liquid hold-up decreases as height increases.

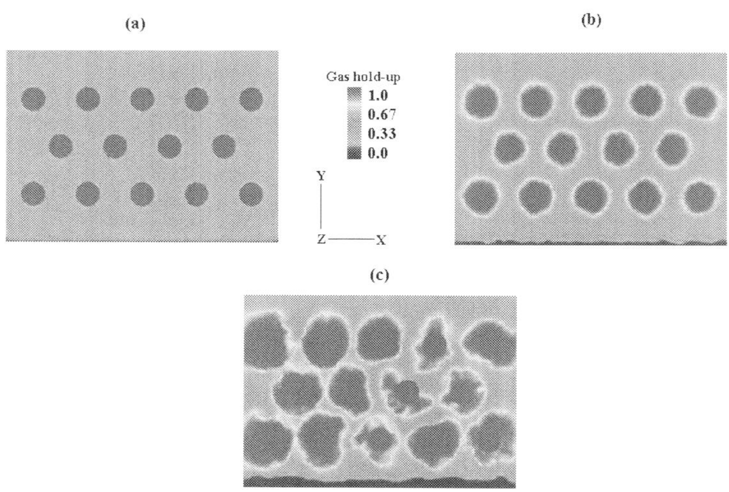

Fig. 11. Gas hold-up at z=0 at (a) 1s,(b) 3s, and (c) 6s.

5. Conclusion

A three dimensional two phase flow and mass transfer model was developed for simulation of hydrodynamics behaviour and concentration distribution in a valve tray of a distillation column. CFD techniques have been used and governing equations simultaneously were solved by FLUENT software. Eulerian method was applied in order to predict the behaviour of two phase flow. Clear liquid height for the system of air-water was calculated and results were compared with experimental data.

System of cyclohexane-n-heptane also were considered and its mass transfer were investigated. Eddies near the caps have been observed, and it is found that such circulations enhances mixing and has an important effect on mass transfer in a distillation column. Results show that the concentration of the light component (C_6H_{12}) in the vapour phase increases along time and the C_7H_{16} concentration in the vapour phase decreases. In addition, concentrations are not constant over the entire tray and they change point by point. This research showed that CFD is a powerful technique in design and analysis of mass transfer in distillation columns and the presented model can be used for further study about mass transfer of valve trays.

6. Acknowledgements

The authors thank the National Iranian Oil Refining & Distribution Company (NIORDC) because of financial support of this research (contract No. 88-1096). Special thanks to Eng. Mohammad Reza Mirian for his kind cooperation in this work.

7. References

Alizadehdakhel, A.; Rahimi, M. & Abdulaziz Alsairafi, A. (2010). CFD and experimental studies on the effect of valve weight on performance of a valve tray column. *Comp. Chem. Eng.*, 34, 1–8.

Bjorn, I.N.; Gren, U. & Svensson, F. (2002). Simulation and experimental study of intermediate heat exchange in a sieve tray distillation column. *Comp. Chem. Eng.*, 26, 499-505.

Buwa, V.V. & Ranade, V.V. (2002). Dynamics of gas-liquid flow in a rectangular bubble column: Experiments and single/multi-group CFD simulations. *Chem. Eng. Sci.*, 57, 22, 4715-4736.

C.H. Fischer & G.L. Quarini, 1998. Three-dimensional heterogeneous modelling of distillation tray hydraulics. Paper presented at the AIChE Annual Meeting, Miami Beach, FL, 15-19.

Deen, N.G.; Solberg, T. & Hjertager, B.H. (2001). Large eddy simulation of the gas-liquid flow in a square cross-sectioned bubble column. *Chem. Eng Sci.*, 56, 6341-6349.

Delnoij, E.; Kuipers, J.A.M. & van Swaaij, W.P.M. (1999). A three-dimensional CFD model for gas-liquid bubble columns. *Chem. Eng. Sci.*, 54, 13/14, 2217-2226.

Fluent 6.2 Users Guide (2005). Fluent Inc., Lebanon.

Hirschberg, S.; Wijn, E.F. & Wehrli, M. (2005). Simulating the two phase flow on column trays. *Chem. Eng. Res. Des.*, 83, A12, 1410–1424.

Krishna, R.; Van Baten, J.M.; Ellenberger, J.; Higler, A.P. & Taylor, R. (1999). CFD simulations of sieve tray hydrodynamics. *Chem. Eng. Res. Des.*, 77, 639–646.

Li, X.G.; Liuc, D.X.; Xua, Sh.M. & Li, H. (2009). CFD simulation of hydrodynamics of valve tray. *Chem. Engin. Proc.*, 48, 145–151.

Lianghua, W.; Juejian, C. & Kejian, Y. (2008). Numerical simulation and analysis of gas flow field in serrated valve column. *Chin. J. of Chem. Eng.*, 16, 4, 541-546.

Ling Wang, X.; Jiang Liu, C.; Gang Yuan, X. & Yu, K.T. (2004). Computational fluid dynamics simulation of three-dimensional liquid flow and mass transfer on distillation column trays. *Ind. Eng. Chem. Res.*, 43, 2556-2567.

Liu, C.J.; Yuan, X.G.; Yu, K.T. & Zhu, X.J. (2000). A fluid-dynamics model for flow pattern on a distillation tray. *Chem. Eng. Sci.*, 55, 12, 2287-2294.

McFarlane, R.C.; Muller, T.D. & Miller, F.G. (1967). Unsteady-state distribution of fluid composition in two-phase oil reservoirs undergoing gas injection. *Soc. Petrol. Eng. J.*, 7, 1, 61-74.

Mehta, B.; Chuang, K.T. & Nandakumar, K. (1998). Model for liquid phase flow on sieve trays. *Chem. Engin. Res. and Des.*, 76, 843-848.

Rahbar, R.; Rahimi, M.R.; Shahraki, F. & Zivdar, M. (2006). Efficiencies of sieve tray distillation columns by CFD simulation. *Chem. Eng. Technol.*, 29, 3, 326-335.

Sanyal, J.; Marchisio, D.L.; Fox, R.O. & Dhanasekharan, K. (2005). On the comparison between population balance models for CFD simulation of bubble columns. *Ind. Eng. Chem. Res.*, 44, 14, 5063-5072.

Solari, R.B.; Bell, R.L. (1986). Fluid flow patterns and velocity distribution on commerical-scale sieve trays. *AICHE J.*, 32, 4, 640-649.

Sokolichin, A.; Eigenberger, G.; (1999). Applicability of the standard turbulence model to the dynamic simulation of bubble columns. Part I. Detailed numerical simulations, *Chem. Eng. Sci.*, 54, 2273–2284.

Sun, Z.M.; Yu, K.T.; Yuan, X.G.; Liu, C.J. & Sun, Z.M. (2007). A modified model of computational mass transfer for distillation column. *Chem. Eng. Sci.*, 62, 1839 - 1850.

Van Baten, J.M. & Krishna, R. (2000). Modelling sieve tray hydraulics using computational fluid dynamics. *Chem. Eng. J.*, 77, 3, 143-151.

We, C.; Farouqali, S.M. & Stahl, C.D. (1969).Experimental and numerical simulation of two-phase flow with interface mass transfer in one and two dimensions. *Soc. Petrol. Eng. J.*, 9, 3, 323-337.

Wijn, E.F. (1996). The effect of downcomer layout pattern on tray efficiency. *Chem. Eng. J.*, 63, 167-180.

Xigang, Y. & Guocong, Y. (2008). Computational mass transfer method for chemical process simulation. *Chin. J. of Chem. Eng.*, 16, 4, 497-502.

YOU, X.Y. (2004). Numerical simulation of mass transfer performance of sieve distillation trays. *Chem. Biochem. Eng. Q.*, 18, 3, 223–233.

Yu, K.T.; Huang, J. (1980). Simulation of large tray and tray efficiency. *Paper presented at the AIChE Spring National Meeting*, Philadelphia.

Yu, K.T.; Yuan, X.G.; You, X.Y. & Liu, C.J. (1999). Computational fluid-dynamics and experimental verification of two-phase two dimensional flow on a sieve column tray. *Chem. Eng. Res. Des.*, 77A, 554-558.

Zhang, M.Q. & Yu, K.T. (1994). Simulation of two-dimensional liquid flow on a distillation tray. *Chin. J. Chem. Eng.*, 2, 2, 63-71.

Part 3

Mass Transfer in Food

14

Modeling Moisture Movement in Rice

Bhagwati Prakash[1] and Zhongli Pan[1,2]
[1]*University of California, Davis,*
[2]*Western Regional Research Center, ARS, USDA*
United States

1. Introduction

Rice is one of the leading food crops in the world with total annual production being about 448 million metric tons on milled rice basis in 2008/09 year (USDA, 2010). Rice is found in marketplace in different forms depending on level of its subsequent processing. Rough rice (or paddy rice) is the rice that is obtained just after harvest. After removal of its outer husk (or hull), it becomes brown rice. Brown rice after milling, where the bran layer and embryo is removed become whiter in color and is called white rice (or milled rice) that is favored form of human consumption in most countries.

Rough rice is generally harvested at 18-24% moisture contents on wet basis and requires drying down to 12-14% for safe storage. At commercial scale, drying is carried out by blowing heated air over grains causing them to lose moisture rapidly. In addition to drying, moisture movement inside rice kernels occurs when rice is exposed to dry or humid environments causing desorption or adsorption of moisture, respectively, during any of pre-harvest or post-harvest stages.

During any of moisture adsorption or desorption processes, the surface of kernel reaches the equilibrium moisture content in surrounding environmental conditions very rapidly, however, at center of kernel moisture changes slowly, developing moisture gradients within the kernel. Higher magnitudes of such moisture gradients are believed to be one of the major reasons causing fissures or cracks in rice, which result in broken rice on milling. Milled rice kernels that are three-fourths or more of the unbroken kernel length are called head rice while the rest are called broken rice (USDA, 1994). Since the full-length grain is preferred form of rice, broken rice has typically half market value than that of head rice (Mossman, 1986; Thompson & Mutters, 2006). Therefore, reducing rice fissuring has been an important goal in rice drying research.

In last five decades, many researchers have pursued mathematical modeling of rice drying process. Key objective of such model development was to determine the moisture of the drying rice sample after certain drying period. Development of models also assisted in understanding the impact of factors affecting drying process such that drying air temperature and speed of drying air and optimizes them for reducing drying time, without performing a large number of experiments. Mathematical models were also used to determine the moisture gradients within the rice kernels that might affect rice fissuring. In addition to improve drying process, mathematical models can also help in making decisions on whether rice at particular moisture can be exposed to certain environmental conditions for certain period of time without significant fissuring.

Among different moisture adsorption and desorption processes, drying has attracted most attention of researchers. From modeling perspective, there is very little difference between these processes except the magnitude of moisture movement is very rapid in case of drying. Drying process would be the main focus in this chapter, however, important information on other sorption processes will also be described when necessary. Modeling of two types of drying: convective air drying by heated air and radiative drying by infrared drying will be mainly covered in this chapter.

The purpose of this chapter is to illustrate different approaches pursued in modeling of drying processes in rice. Development of both empirical and theoretical models based upon principles of mass and heat transfer is described in this chapter. Near the end of chapter, brief discussion on determination of some of the key hygroscopic and thermal properties is provided. Our goal is to expose the reader to variety of options available in rice drying modeling literature and assist them to make well-informed choices for successful development of rice drying models.

2. Mechanism of moisture movement

During most of agricultural products drying, initially moisture is quickly removed, then is followed by progressively slower drying rates (Allen, 1960). Fig. 1. shows the typical rice drying curve of rice. It clearly shows that drying rate, which is slope of the drying curve, becomes smaller with progress of drying. Such drying is referred as falling rate drying.

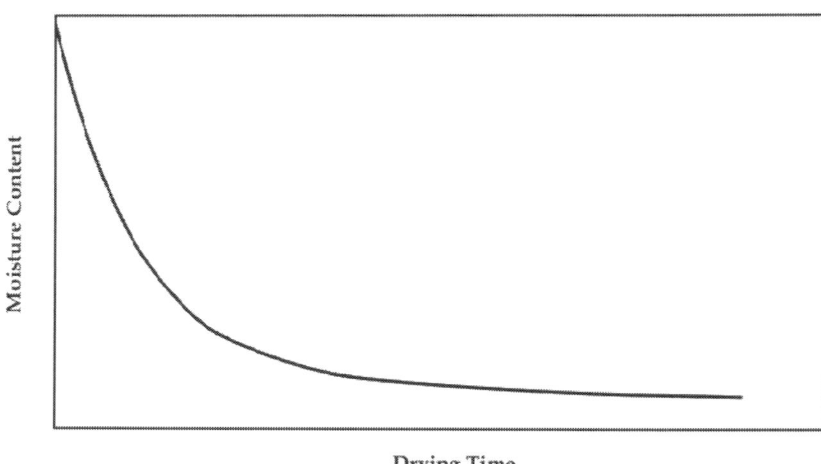

Fig. 1. Typical drying curve of rough rice

The cause of falling rate drying is inability of internal moisture movement to convey moisture to the surface at a rate comparable to that of its removal from the surface. Many theories were proposed to explain mechanism of internal moisture movement in such falling rate drying behavior. Some of them are: difference in vapor pressure, liquid diffusion, capillary flow, pore flow, unimolecular layer movement, multimolecular layer movement, concentration gradient and solubility of the absorbate. Each of these theories explains some aspects of drying in some materials but no universally applicable theory has been substantiated by experiments (Allen, 1960). Srikiateden & Roberts (2007) reviewed these mechanisms as reported in different solid foods and described the mathematical equations involved in such mechanisms.

Despite the uncertainty about the actual mechanism of moisture movement, most researchers (Mannapperuma, 1975; Steffe & Singh, 1980a; Aguerre et al., 1982; Lague, 1990; Sarker et al. 1994; Igathinathane & Chattopadhyay, 1999a; Meeso et al., 2007; Prakash & Pan, 2009) have described moisture movement in rice drying by Fick's laws of diffusion. Using such diffusion mechanism, Lu & Siebenmorgen (1992) have modeled moisture movement during adsorption and desorption processes. Their models predicted average moisture of the grain reasonably well. However, it should be noted that this alone does not establish diffusion as the mechanism of moisture movement in rice. A true criterion for the validity of the mechanism would be accurate prediction of moisture distribution within the grain (Hougen et al., 1940), which has not been fully considered in rice drying.

3. Mathematical models

Drying of any material normally involves both heat and mass transfer (or moisture transfer). Pabis & Henderson (1962) measured the center and surface temperature of yellow shelled maize kernels during heating by free convection in an oven at 71°C (i.e. 160°F). They observed that the surface and center temperatures differed significantly during the first 3 to 4 minutes only and therefore, grains can be considered isothermal during drying process. Citing this work as their basis, many researchers have assumed grains to be isothermal during the drying process and neglected heat transfer within the rice kernel. However, in case of infrared drying, where heating period is very short (typically less than two minutes), rice kernel cannot be considered isothermal and heat transfer within the rice kernel must be considered.

Different approaches were taken to mathematically model the moisture changes during the drying period. Based on the size of sample, these models can be broadly categorized into three: thin layer (or single layer) drying models, deep bed drying models and single kernel drying models. Thin layer drying models were mostly empirical or semi-empirical in nature while most single kernel drying models were based on mechanism of Fickian diffusion. Deep bed drying models are generally based on thin layer drying models. It is important to discuss the concept of equilibrium moisture content before we describe development of different models in detail.

At any fixed environmental conditions, a wet rice sample continues to lose moisture until an equilibrium state is reached. This moisture content is called equilibrium moisture content (EMC). In addition to rice variety, this EMC of a rice sample also depends on the temperature and humidity of ambient air. When a dry rice sample is exposed to humid environments, it gains moisture content and equilibrates to a value of EMC, which may be different than the EMC obtained during the moisture desorption. The difference in EMCs obtained during adsorption and desorption experiments are due to the different condition of the grains in their approach to equilibrium conditions (Allen, 1960). They cited Simmonds et al. (1953), who described the discrepancy between the EMC values obtained during adsorption and desorption to be due to the living nature of grain, which changes its chemical and physical nature according to environment.

Allen (1960) described another concept of EMC called dynamic EMC that is different than from the EMC value discussed previously, which they referred as static EMC. They considered that use of static EMC is inappropriate for drying process, because physical and chemical changes in the kernel during such process are very fast compared to the conditions used to determine static EMC. During drying, kernel surface becomes very dry while inner kernel has higher moisture content creating high moisture gradients. In such conditions, kernel surface that is exposed to drying air is not representative of the whole grain. This causes grain moisture to approach a value that is higher than its static EMC during drying. This moisture content value is called its dynamic EMC. If drying is conducted through a large range of moisture contents, this approach may reveal existence of more than one value of dynamic EMCs. They considered dynamic EMCs to be the logical choice to describe moisture loss during drying process while for gentle moisture movement processes such as exposure to humid or dry conditions, use of static EMC was more appropriate.
Bakker-Arkema & Hall (1965) dried alfalfa wafers and found that use of static EMC as boundary equations in second order differential equation of moisture transfer predicted drying behavior successfully. When static EMC was used, they found the moisture diffusivity to be almost constant during all but initial stages of drying. On the other hand, using dynamic EMC resulted in diffusivity value that changed rapidly with moisture content. Based on this work, they concluded that use of static EMC is more suitable for some biological products. In rice drying, most of recent works have used static EMCs in their models. Unless specified, EMC in this chapter refers to static equilibrium moisture content.

3.1 Thin layer drying model

Utilizing the analogy between heat transfer and drying process (Allen, 1960), drying rate of grains can be expressed as:

$$\frac{\partial M}{\partial t} = -k(M - M_e) \tag{1}$$

where, M (kg water/ kg dry solids) is moisture content of grain, t (s) is the period of drying, M_e (kg water/ kg dry solids) is the equilibrium moisture content (EMC) of the drying grain and k (s^{-1}) is the drying constant. This equation was found to be convenient, if dynamic EMC was used for M_e (Allen, 1960). Integrated form and solution of this equation is given by:

$$\log(M - M_e) = -\frac{k}{2.303}t + C \tag{2}$$

$$MR = \frac{M - M_e}{M_o - M_e} = e^{-kt} \tag{3}$$

where, C is the integration coefficient, MR is dimensionless moisture ratio, and M_0 (kg water/kg dry solids) is the initial moisture content of grains. The value of dynamic EMC used in such equations are determined using trial and error method by plotting, $\log(M-M_e)$ versus time for different assumed values of M_e. For the correct value of M_e, such plot would be a straight line. The slope of such straight line is used to determine the drying constant k.
Allen (1960) applied this drying equation to thin layer rice drying experimental data and found that the M_e and drying constant k, depended upon initial moisture content,

temperature, humidity and quantity of drying air. The rice drying experimental data also revealed that initial drying period i.e. up to 30 minutes needs different treatment, since it is dependent upon M_e and k values, which is different from those used in later drying periods. However this transition was smooth and knowledge of precise discontinuity was not required.

In last fifty years, there has been considerable interest among researchers to use empirical or semi-empirical models, to describe the drying curves during thin-layer grain drying. Some of these models are described in Table 1.

Reference	Model Equation
Allen (1960)	$MR = e^{-kt}$
Henderson (1974)	$MR = Ae^{-k_1 t} + Be^{-k_2 t}$
Agrawal & Singh (1977)	$MR = Ae^{-kt^n}$
Wang (1978)	$MR = 1 + At + Bt^2$
Midilli et al. (2002)	$MR = Ae^{-kt^n} + Bt$

Table 1. Thin layer drying model equations

In these models, k, A, B, k_1, k_2, n are constants dependent upon drying conditions such as ambient humidity, temperature of drying air and air flow rate through drying column.

Hacihafizoglu et al. (2008) reviewed twelve such models to describe thin layer drying of rice. The number of parameters in these models varies from one to four. They conducted drying experiments on long grain rough rice to determine the parameters of these models and determined the statistical fit of these models to the drying experimental data. In all but one model, the correlation coefficient was found to be higher than 0.98, which means that all of these can describe the thin layer drying of rough rice satisfactorily. As expected, model developed by Midilli et al. (2002) that has four parameters gave the best fit with the experimental results.

Advantages of such thin layer drying models are in their simplicity and ease of development. However, parameters in these models depend on specific drying conditions and therefore must be experimentally determined for each drying condition.

3.2 Deep bed drying models

In deep bed drying, conditions of drying air and grain vary at different depths making it difficult to use single value of drying constant and equilibrium moisture content, required in thin-layer drying equations. Two approaches were taken to model such deep bed drying systems. First approach required determination of mean bed temperature and then, use the thin layer drying equation with drying constants at that mean temperature. In second approach, deep drying bed was considered as series of thin layer drying beds, each having different temperature and therefore, different drying constants.

Allen (1960) conducted experiments to study deep bed drying of maize and rice and used principle of dimensional analysis, to determine mean bed temperature. Their predicted drying times in case of rice and maize drying were close to the actual values, the difference

being less than 10% of total drying periods. Nelson (1960) applied similar dimensional analysis and theory of similitude to study drying in deep bed grain dryers.
Boyce (1965) conducted enthalpy balances to thin layers of barley during drying, accounting both sensible heat and latent heat of vaporization and conducted layer by layer calculations to determine temperature of grains at different depths in the deep bed. Despite involving extensive computation in this approach, it has considerable merit as it uses the fundamentals principles of heat transfer, compared to empirical approach of Allen (1960).
Detailed description of deep bed drying models was considered beyond the scope of this chapter. For detailed information on these models and their implementation on computer reader can refer to works by Bakker-Arkema et al. (1967), Spencer (1969), Henderson & Henderson (1968) and Parry (1985).

3.3 Single kernel models

In last three decades, Fick's laws of diffusion have been extensively used to model moisture movement within the rice kernel in different forms namely white rice, brown rice and rough rice. In addition to average kernel moisture, such models also describe moisture distribution within the rice kernels that can be used to estimate moisture gradients and fissuring in rice. Some studies (Lague, 1990; Yang et al., 2002; Meeso et al., 2007; Prakash & Pan, 2009) have also considered heat transport within the kernel in their models.
Different rice varieties have their different physical, thermal and hygroscopic properties. Depending upon the variety of interest, researchers have determined these properties and developed appropriate models to describe drying. Detailed description of these modeling efforts is described in the next section.

4. Single kernel models

4.1 Kernel geometry

Rice kernel has an irregular shape. In addition to its shape, structure and thickness of husk also varies in the kernel (Fig. 2). Developing mathematical model for irregular shapes is computer power intensive and hence, most of the times rice kernel is approximated to simpler shapes such as sphere, cylinder, prolate spheroid and ellipsoid. Depending upon length-width ratio of milled rice, rice varieties are classified into three grain types: long, medium and short. Length to width ratio for long grain rice is larger than 3.0, medium grain rice is 2.0 to 2.9 and short grain rice is lower than 2.0 (USDA, 1994). Selecting the shape of model depends upon geometry of rice variety under study and computational tools available to solve the model.
Steffe & Singh (1980a) assumed spherical shape to model short grain rice forms. In rough rice model, they considered endosperm layer to have spherical shape that is surrounded by spherical shells of bran and husk. Their brown rice model consisted of endosperm and bran layers while white rice model consisted endosperm only. Assuming spherical shape made drying a one-dimensional transport process and easy to solve mathematically.
Prediction of moisture gradients accurately demanded more resemblance to the true shape of kernel. Lu & Siebenmorgen (1992), Sarkar et al. (1994), Igathinathane & Chattopadhyay (1999a) and Yang et al. (2002) assumed prolate spheroid shape to model medium and long grain rices while, Ece and Cihan (1993) considered short cylinder shape to model the short grain rice kernel. Due to their choice of model geometry, these studies have considered transport processes in two directions.

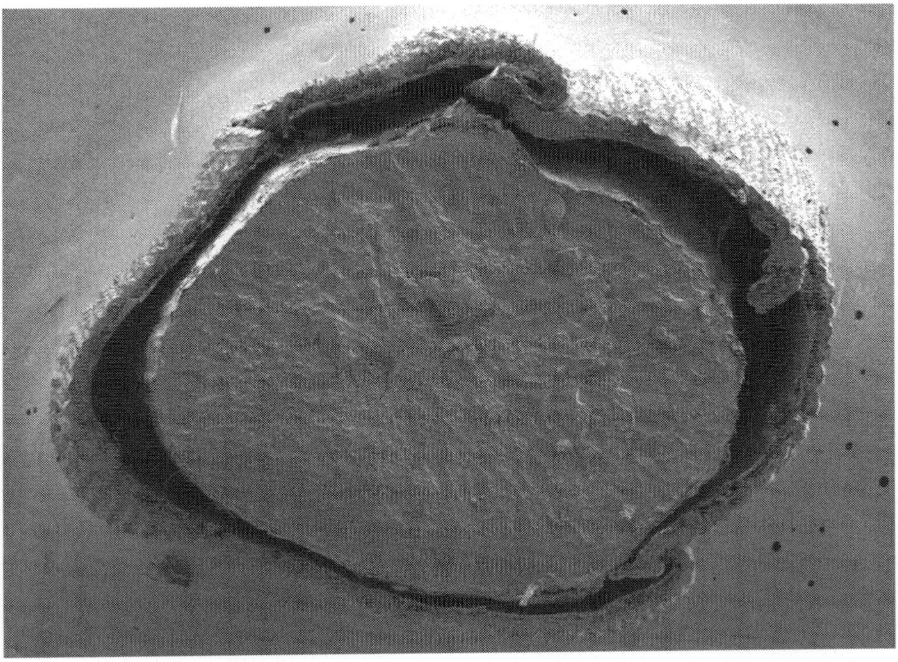

Fig. 2. Scanning electron micrograph of transverse section of rough rice (Lemont variety)

Use of prolate spheroid shape is suitable when rice kernel cross-section is circular. In case of medium grain rice variety Californian M206, kernel cross-section is not circular with one axis about 40% longer than the other. Table 2 shows the three dimensions of this rice variety. In this case, we have considered an ellipsoid shape with three unequal axes to represent the kernel. Here, moisture transport within the kernel is three-dimensional phenomenon.

		Dimensions (mm)			Weight (mg)
		Length	Width	Thickness	
White Rice	Mean	5.78	2.66	1.85	22.3
	(Std. Dev.)	(0.23)	(0.11)	(0.08)	(0.6)
Brown Rice	Mean	5.82	2.78	1.97	25.0
	(Std. Dev.)	(0.33)	(0.09)	(0.08)	(0.7)
Rough Rice	Mean	6.97	3.16	2.19	31.1
	(Std. Dev.)	(0.32)	(0.19)	(0.08)	(0.5)

Table 2. Kernel dimensions and weights of medium grain rice variety Californian M206 at 18% moisture on wet basis

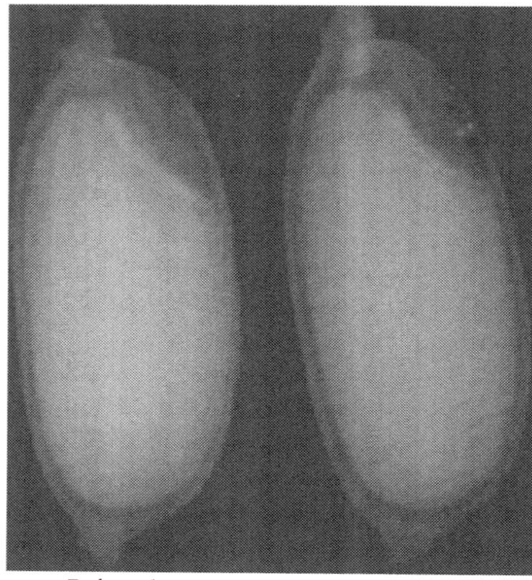

Fig. 3. X-ray images of two rough rice kernels before and after drying

In rough rice model, the kernel consists of three isotropic regions namely endosperm (or white rice), bran and husk. In addition to these regions, small air gap is also present between bran and husk regions in actual rough rice (Fig. 3). Using X-ray imaging, we observed size of this air gap to increase with progress of drying. In most models, existence of this air layer is not considered and moisture transfer resistance due to husk represents the equivalent resistances of this air gap and the husk region. Moisture transfer resistances present in the rough rice model are shown in Fig. 4.

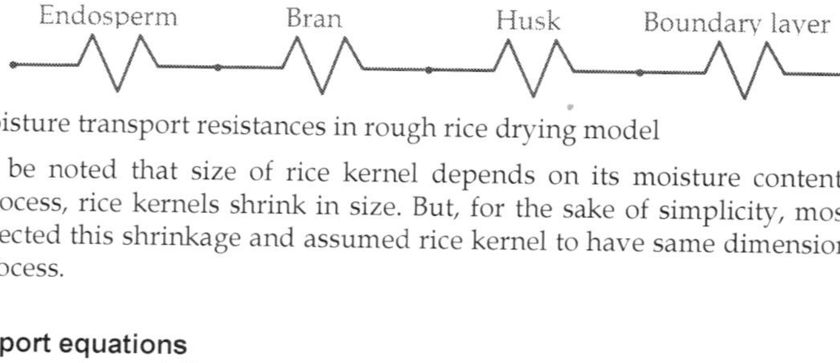

Fig. 4. Moisture transport resistances in rough rice drying model

It should be noted that size of rice kernel depends on its moisture content. During the drying process, rice kernels shrink in size. But, for the sake of simplicity, most researchers have neglected this shrinkage and assumed rice kernel to have same dimensions during the drying process.

4.2 Transport equations

Single-phase moisture diffusion and heat conduction is commonly used as the mechanism of moisture transfer and heat transfer, respectively, within the rice kernel. Fick's law of diffusion and Fourier's law of conduction are commonly used to describe these transport processes during rice drying, respectively. General equations corresponding to these laws are given by:

$$\frac{\partial M}{\partial t} = \nabla \bullet (D \nabla M) \tag{4}$$

$$\rho c_p \frac{\partial T}{\partial t} = \nabla \bullet (k \nabla T) + Q \tag{5}$$

where, M is the moisture content on dry basis (kg water/kg dry matter), t is time (s), ∇ is divergence operator, D is moisture diffusivity (m^2/s), ρ is density of rice components (kg/m^3), c_p is specific heat (J.kg^{-1}.°C^{-1}), T is temperature (°C), k is thermal conductivity (W.m^{-1}.°C^{-1}) and Q is volumetric heat generation (W/m^3).

Depending upon the assumed kernel shape, these transport equations can be applied in spherical, cylindrical or cartesian co-ordinates. Igathinathane & Chattopadhyay (1999a) have used prolate spheroid co-ordinates.

4.3 Boundary and initial conditions

Two kind of boundary conditions are commonly found in the moisture transport modeling in the rice drying literature: Dirichlet boundary condition (i.e. instant moisture equilibration of kernel surface to the ambient environment) and Newmann boundary condition. Mannappeeruma (1975), Steffe & Singh (1980a), Lu & Siebenmorgen (1992) and Meeso et al. (2007) considered Dirichlet boundary condition and assumed kernel surface moisture M_s (kg water/kg dry matter) to have the same moisture as the equilibrium moisture content M_e (kg water/kg dry matter) of rice in the ambient environmental conditions. This can be written as:

$$M_s = M_e \tag{6}$$

Sarker et al. (1994) and Yang et al. (2002) equated the outward moving moisture flux to moisture taken away by convective air and obtained the following Newmann boundary condition:

$$-D \frac{\partial M}{\partial n} = h_m (M_s - M_e) \tag{7}$$

where, h_m (m/s) is the surface moisture transfer coefficient and n is the outward normal at the kernel surface. It should be noted that Eqn. 6 and Eqn. 7 become identical for larger values of h_m.

Due to existing uncertainty on mechanism of moisture movement in the rice kernel, some researchers have assumed evaporation of liquid moisture to occur both at surface and inside the kernel (Yang et al., 2002) while others have assumed evaporation to occur only at surface of kernels (Meeso et al., 2007; Prakash & Pan, 2009).

Assuming evaporation to occur only at kernel surface during drying, heat transfer by convective air to the rice surface can be equated to the conductive heat entering the surface and change in enthalpy of evaporating moisture. This enthalpy balance represents the heat transfer boundary condition and can be described by:

$$-k \frac{\partial T}{\partial n} = h(T_s - T_a) - \frac{\rho \lambda V}{A} \bullet \frac{\partial M_{av}}{\partial t} \tag{8}$$

where, h (W.m^{-2}.°C^{-1}) is the convective heat transfer coefficient, T_s (°C) is the temperature of rice kernel surface, T_a (°C) is the temperature of drying air, λ (J/kg) is latent heat of vaporization, V (m^3) is the volume of the kernel, A (m^2) is the total surface area of the kernel and M_{av} (kg water/kg dry solids) is average moisture content of kernel at any given time on dry basis.

At internal boundaries i.e. bran-husk interface and endosperm-bran interface, moisture and heat fluxes leaving one region is equated to the respective fluxes entering other region. Rice is considered to have uniform moisture and temperature throughout the kernel before drying. This is set as the initial conditions in the models.

4.4 Volumetric heat generation

Assuming moisture evaporation to take place both at surface and inside the kernel, the heat generation due to such evaporating liquid moisture can be expressed as:

$$Q = \frac{\rho\lambda}{1+M} \bullet \frac{\partial M}{\partial t} \tag{9}$$

where, λ (J/kg) is latent heat of vaporization. Here $\partial M/\partial t$ must represent the rate of change of moisture content in liquid form. Determining this rate would require multiphase modeling that describes moisture in liquid and vapor phases separately. Such multiphase approach is not pursued in rice modeling yet. When moisture evaporation is assumed to occur only at kernel surface then there is no such volumetric generation.

Like other radiations, infrared radiation has power to penetrate the surfaces and generate heat. Heat transfer from infrared radiations to rice kernels can be modeled in two ways: assuming penetration of heat inside the kernel surface or assuming no heat penetration i.e. all of radiation heating the surface only.

Heat generated at certain depth below the kernel surface can be modeled as an exponential decay (Ginzburg, 1969). Datta & Ni (2002) described the heat generation due to infrared radiations as following:

$$Q = \frac{P_s}{d_p} e^{-\frac{d}{d_p}} \tag{10}$$

where, P_s is the infrared radiation power at the surface (W/m^2), d is the depth from the surface (m), and d_p is the penetration depth of infrared radiation (m).

Though assuming heat penetration is more fundamental approach, it requires knowledge of penetration depth that is still to be determined accurately. Some studies (Meeso, 2007; Prakash & Pan, 2009) have considered penetration depth of of 1-2 mm that was reported for grains by Ginzberg (1969) and Nindo et al. (1995). If penetration of heat is not assumed, then there is no heat generation due to infrared radiation. In such case, all radiation heat falling over the rice surface should be considered in the heat transfer boundary condition, which can then be rewritten as:

$$-k\frac{\partial T}{\partial n} = h(T_s - T_a) - \frac{\rho\lambda V}{A} \bullet \frac{\partial M_{av}}{\partial t} - P_s \tag{11}$$

4.5 Solution of models

Analytical and numerical solutions have been used to solve the transport equations in rice drying. Crank (1979) described analytical solution to Fick's law of diffusion for regular shapes such as sphere, cylinder and slab for different boundary conditions.

Aguerre et al. (1982) have assumed rough rice kernel as a homogeneous sphere and considered instant moisture equilibration of kernel surface to the ambient environment.

They have used the following analytical solution to predict average moisture content (M) of the kernel at any given time:

$$MR = \frac{M - M_e}{M_0 - M_e} = \frac{6}{\pi^2} \sum_{n=1}^{\infty} \frac{1}{n^2} \exp\left(-\frac{n^2 \pi^2 D t}{R^2} \right) \tag{12}$$

where, M_0 is the initial moisture content, M_e is the equilibrium moisture content to rice kernel at ambient drying conditions and R (m) is the equivalent radius of the rice kernel. Equivalent radius was determined by equating the volume of rice kernel (V, m^3) to that of sphere and is given by:

$$R = \left(\frac{3}{4\pi} V \right)^{1/3} \tag{13}$$

It should be noted that neglecting third and higher terms of series in Eqn. 12 result in the thin-layer drying equation reported by Henderson (1974) in Table 1.

Ece & Cihan (1993) assumed rice as a homogeneous short cylinder with Dirichlet boundary conditions at surface and used following analytical solution to determine average moisture constant at any given time:

$$MR = \frac{M - M_e}{M_0 - M_e} = \frac{8R^2}{L^2} \sum_{n=1}^{\infty} \sum_{m=1}^{\infty} \frac{1}{\alpha_n^2 \beta_m^2} e^{-(\alpha_n^2 + \beta_m^2)\frac{D}{R^2}} \tag{14}$$

where, R (m) is radius of cylinder, L (m) is height of cylinder. α_n (n=1,2,…) are the roots of Bessel function of zero order $J_0(x)$, and β_m (m=1,2,…) are defined as:

$$\beta_m = \frac{(2m-1)\pi R}{2L} \tag{15}$$

Aguerre et al. (1982) and Ece & Cihan (1993) did not consider presence of multiple components such as endosperm, bran and husk inside their models. If these components are considered in the model, known analytical solutions cannot be used to solve moisture transport equations. In such cases, numerical methods such as finite difference and finite element methods are used to solve heat and moisture transport equations.

Steffe & Singh (1980a) and Meeso et al. (2007) used finite difference methods in their one dimensional spherical shaped rice models. Igathinathane & Chattopadhyay (1999a) have used finite difference method in prolate spheroid co-ordinate system to model rice drying in two dimensions. Advantage of the finite difference method lies in its simplicity of implementation. However, it is not well suited to solve two or three-dimensional problems and/or problems consisting of material discontinuity. In such problems, finite element method is more suitable.

Many general-purpose finite element software packages such as Comsol Multiphysics or pdetool in MATLAB can be used to solve heat and mass transfer equations involved in the drying model. Lu & Sibenmorgen (1992), Sarker et al. (1994), Yang et al. (2002) and Prakash & Pan (2009) have used finite element method in their rice models. The representative meshed model geometry of rough rice in the three-dimensional model is shown in Fig. 5. As seen in this figure, only one-eighth of the actual rough rice volume was considered in this model. This was due to the existence of symmetry about the three axes in the rice kernel.

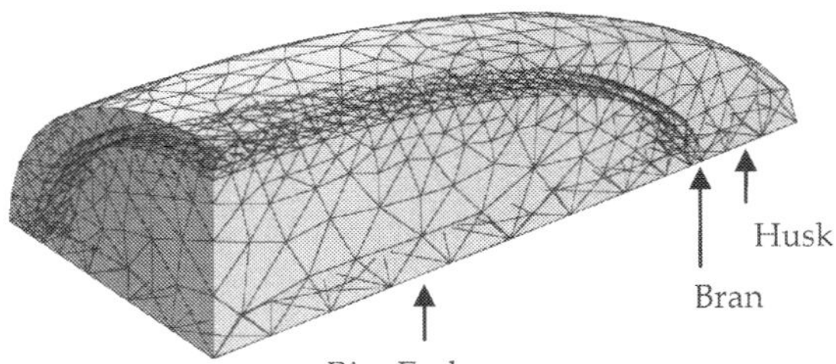

Fig. 5. Meshed rough rice model visualization in a finite element software program

4.6 Model validation

Measuring moisture distribution within the rice grain is very difficult due to its small size. Hence, most studies have measured the average kernel moisture and compared it with the model predicted average moisture to validate the model. Similarly, the center temperature of grain is measured and compared with its model predicted value to validate the heat transfer component in the models.

4.7 Moisture and temperature gradients

Single kernel models have the capability to determine moisture distribution and moisture gradients within the rice kernel. Fig. 6 shows distribution of moisture gradient produced inside the rough rice kernel after 20 minutes of drying at air temperatures of 45°C. Highest moisture gradients are observed along the shortest axis on both sides of bran layer. Among the three rice components, bran has least moisture diffusivity, which slows the movement of moisture across it and thus, produces such high gradients on bran-endosperm and bran-husk interface. Such high moisture gradients along the shortest axis, near the center of kernel may explain the occurrence of most fissures along the axial direction of the kernel.

The maximum temperature gradient inside rice kernel appeared within 20 s of drying period and the entire temperature gradients disappeared within 2 to 3 minutes when rough rice was dried by heated air at 60°C temperatures (Yang et al., 2002). Based on this observation, they concluded that impact of temperature gradients on fissuring is minimal. On the other hand, moisture gradients were found to increase with drying time up to about 15 to 30 minutes after which they started to decline (Sarker et al., 1996; Yang et al., 2002).

In infrared radiation drying, we determined the moisture gradients at two points located on bran-husk interface (P1) and bran-endosperm interface (P2), both along the shortest axis and close to center of kernel. These two points are expected to have highest moisture gradients. To understand the relationship between these moisture gradients and rice fissuring, drying experiments were performed for different drying periods and fissuring in rice was measured. Generally, rice fissuring is quantitatively described by head rice yield that is defined as the mass percentage of rough rice, which remains as head rice after milling. Lower values of HRY suggest more fissuring. Fig. 7 describes head rice yields obtained after infrared drying experiments conducted for different time periods and corresponding moisture gradients at points P1 and P2 obtained from the model.

Moisture gradients at bran-husk interface were higher than at the bran-endosperm interface due to rapid drying of husk. After 150 s of drying, moisture gradients at bran-husk interface remained almost constant, as husk was already very dry by this period. Significant reduction in head rice yield is observed only after 120 s of infrared heating period. While correlating fissuring and moisture gradient values, it should be noted that at higher moistures, rice kernels are more elastic and can endure higher moisture gradients while the same magnitude of moisture gradients in low moisture rice can cause it to fissure (Kunze & Calderwood, 2004).

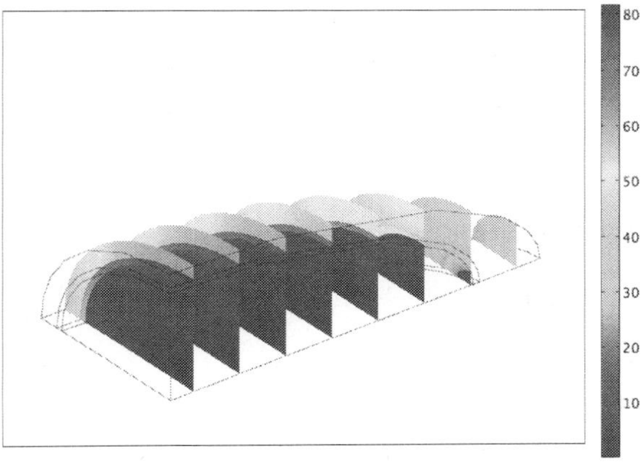

Fig. 6. Moisture gradient (% dry basis per mm) distribution in the rough rice kernel (Californian M206 variety) after 20 minutes of drying at 45°C air temperature

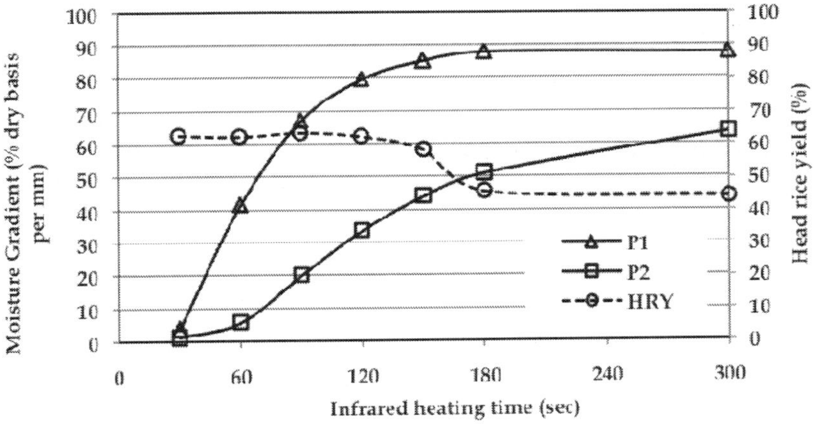

Fig. 7. Head rice yield (HRY) and moisture gradients at bran-husk interface (P1) and bran-endosperm interface (P2) during infrared drying for different time periods

5. Thermal and hygroscopic properties

Drying modeling demands knowledge of moisture diffusivity, thermal conductivity, specific heat, density, equilibrium moisture content and convective heat and mass transfer coefficients. Lague (1990) and Kunze et al. (2004) have described studies on physical, thermal, mechanical and hygroscopic properties of rice. These properties depend upon the rice variety, moisture content and sometimes temperature. Some of these properties such as equilibrium moisture content, moisture diffusivity, convective mass transfer coefficients and convective heat transfer coefficient, are described in detail in this section.

5.1 Equilibrium moisture content

At equilibrium, different components of rough rice have different moisture contents (Table 3). In addition to the rice form, variety and temperature, equilibrium moisture content (EMC) also depends upon whether rice is adsorbing or desorbing. Even among the equilibrated rice samples, the individual rice kernels were observed to have a standard deviation of 0.5% in 10% average MC rice to 4.5% in 26% average moisture rice (Siebenmorgen et al., 1990). The causes of such non-uniformity among equilibrated samples are not well understood.

R.H.	White rice		Brown rice		Rough rice		Husk	
%	ads.	des.	ads.	des.	ads.	des.	ads.	des.
20	7.5	10.2	9.5	9.8	9.9	11.1	10.5	11.3
40	10.2	13.6	9.8	11.7	10.6	13.3	13.0	14.3
60	15.2	17.4	12.7	15.0	12.7	16.5	15.8	17.6
80	19.4	21.0	16.8	18.7	17.3	20.0	18.6	21.0
98	30.9		27.2		26.7		26.3	

Table 3. Equilibrium moisture content (EMC, % dry basis) of different forms of Californian variety M206 rice and its husk during adsorption (ads.) and desorption (des.) at different relative humidities (RH, %)

At any given temperature, EMC of rice samples can be plotted versus the corresponding equilibrium relative humidity (ERH) to obtain the moisture isotherm. This ERH is also equal to the water activity (a_w) of the rice sample. There were many theoretical and empirical mathematical equations developed to describe the relationship between ERH or water activity and EMC and each equation had success in predicting the equilibrium moisture content data for a particular foodstuff at a given range of equilibrium relative humidities and temperatures. Among these numerous equations, the Guggenheim-Anderson-de Boer (GAB) equation, showed good fit to the isotherms in a broad range of a_W and was used by by several researchers to model sorption isotherms of rice (Gencturk et al., 1986; Reddy & Chakraverty, 2004; Iguaz & Virseda, 2007). It is given by:

$$M = \frac{M_0 C K a_w}{(1 - K a_w)(1 - K a_w + C K a_w)} \tag{16}$$

where, M_0, C and K are constants. A literature survey carried out by Sun (1999) for the selection of suitable isotherm model from 18 different authors' 763 data points indicated that when the sorption experiments were conducted at different temperatures then the Strohman-Yoerger equation is most appropriate to describe sorption isotherms of rice. It is expressed by:

$$a_w = \exp\left[C_1 e^{-C_2 M} \ln(P_s) - C_3 e^{-C_4 M}\right] \tag{17}$$

where, P_s is saturated pressure over liquid water and C_i (i =1 to 4) are different constants.

5.2 Moisture diffusivity

Moisture transfer within the kernel could occur in both liquid and vapor phase. Differentiating these two would require a multiphase modeling of heat and mass transfer. Such studies have not been pursued so far in rice drying. To account for both liquid and vapor phase diffusivity, usually an effective diffusivity, D_{eff} (m^2/s) is used. Effective diffusivity of different rice components namely endosperm, bran and husk are determined by minimizing the sum of squared differences between actual and predicted drying curves of white rice, brown rice and rough rice, respectively.

First, rice endosperm diffusivity is assumed. Then the white rice model is run to predict the moisture contents at different times. Error is determined by comparing the average values of model predicted moisture and experimentally measured white rice moisture. Then, the sum of squared errors (*SSE*) is computed as follows:

$$SSE = \sum_{i=1}^{n} \left(M_{ei} - M_{pi} \right)^2 \tag{18}$$

where, M_{ei} and M_{pi} are experimentally measured and model predicted moisture (%, dry basis) at a given time period, respectively, and n is the number of data points used to calculate *SSE*. Different diffusivity values are assumed in the white rice model and for each of them the corresponding *SSE* is computed. Diffusivity value that gives the lowest *SSE* is considered as the white rice diffusivity. Once rice endosperm diffusivity is known, brown rice model is run assuming different values of bran diffusivity. By comparing the *SSE* between the brown rice model predictions and the brown rice experimental data, diffusivity value of bran is determined. Similarly, diffusivity value of husk is determined.

In addition to diffusivity determinations of individual rice components, brown rice and rough rice models may be run assuming that they are composed of single homogeneous material and then these model predictions can be compared with the experimental drying data to determine the diffusivities of brown rice and rough rice as a whole.

Most researchers have described the temperature dependence of effective diffusivity by Arrhenius type exponential relationships. Table 4 shows some of such relationships. To compare predictions of moisture diffusivity from these relationships, moisture diffusivity (D, m^2/s) was calculated for one certain drying conditions i.e. drying air temperature of 43°C and drying air velocity (V) of 3 m/s. This diffusivity value is described in the last column of the Table 3.

Diffusivity predictions from these relationships vary significantly; for example endosperm diffusivity varied from 2.3E-11 to 1.7E-10 m^2/s. These differences can be explained on the basis of biological variations in rice kernels of different varieties, the specific assumptions such as kernel shape used in rice modeling and the drying conditions such as drying bed thickness, drying air velocity used in different studies. For a given volume, sphere has minimum surface area and therefore, models assuming spherical shape of rice grain would predict highest diffusivity values for same set of experimental data.

Rice endosperm has highest diffusivity among the three rice components. Between husk and bran, some researchers (Steffe & Singh, 1980a; Lu & Siebenmorgen, 1992) determined that husk had lower diffusivity but others (Sarker et al., 1994; Igathinathane & Chattopadhyay, 1999b) found that bran had lower diffusivity. For medium grain rice variety Californian M206, we have found diffusivity of bran to be smallest among rice components. Cause of such differences is not well understood.

Process, Rice type and Reference	Rice Component	Model	D (m^2/s)#
Drying Short grain (S6) Steffe & Singh (1980a, 1982)	 Endosperm Bran Hull Brown Rice Rough Rice	$D = \frac{Ae^{-(B/T)}}{3600}$ A = 2.57E-3 B = 2880 A = 0.797 B = 5110 A = 484 B = 7380 A = 0.141 B = 4350 A = 33.6 B = 6420	 7.9E-11 2.1E-11 9.7E-12 4.1E-11 1.4E-11
Drying Rewetted Short grain (S6) Steffe & Singh (1980b)	 Endosperm Bran	$D = \frac{Ae^{-(B/T)}}{3600}$ A = 1.29E-2 B = 3430 R^2 = 0.84 A = 1.82 B = 5400 R^2 = 0.99	 6.9E-11 1.9E-11
Drying Long grain (Bluebelle) Steffe & Singh (1982)	Rough Rice	$D = \frac{Ae^{BM}}{3600}$ $A = (9.8787E-3)e^{-(4295/T)}$ $B = 0.159T - 37.88$	3.2E-11
Adsorption Long grain (Lemont) Lu & Siebenmorgen (1992)	 Endosperm Bran Hull Brown Rice Rough Rice	$D = \frac{Ae^{-(B/T)}}{3600}$ A = 1.6163 B = 5289.5 R^2 = 0.91 A = 110.969 B = 7042.5 R^2 = 0.98 A = 3.0101 B = 6000.5 R^2 = 0.99 A = 79.2446 B = 6639.4 R^2 = 0.98 A = 5.9835 B = 6101.4 R^2 = 1.00	 2.4E-11 6.5E-12 4.7E-12 1.7E-11 6.8E-12
Adsorption Long grain (NewBonnet) Lu & Siebenmorgen (1992)	 Endosperm Bran Hull Brown Rice Rough Rice	$D = \frac{Ae^{-(B/T)}}{3600}$ A = 1.1855 B = 5206.1 R^2 = 0.91 A = 3.2867 B = 6013.5 R^2 = 0.99 A = 1.4641 B = 5842.8 R^2 = 0.98 A = 12.8355 B = 6080.8 R^2 = 0.97 A = 2.968 B = 5917.4 R^2 = 0.98	 2.3E-11 5.0E-12 3.8E-12 1.6E-11 6.1E-12
Drying Rocco Ece & Cihan (1993)	Rough rice	$D = \frac{0.05915V^{-0.85}e^{\left(\frac{-4706}{TV^{0.076}}\right)}}{3600}$	7.3E-12
Drying Long grain (Lemont) Sarker et al. (1994)	 Endosperm Bran Hull	$D = \frac{A \times 10^{-(B/T)}}{3600}$ A = 3.768E-3 B = 1213.3 R^2 = 0.95 A = 3.822E-2 B = 1941.1 R^2 = 0.91 A = 2.277E-1 B = 2046.6 R^2 = 0.96	 1.5E-10 7.6E-12 2.1E-11
Drying Medium grain parboiled rice (Pankaj) Igathinathane & Chattopadhyay (1999a, 1999b)	 Endosperm Bran Hull	$D = Ae^{-(B/T)}$ A = 1.08E-7 B = 2036 R2 = 0.99 A = 3.99E-6 B = 4096 R2 = 0.99 A = 1.66E-5 B = 4230 R2 = 0.99	 1.7E-10 9.4E-12 2.6E-11

Diffusivity value (D) calculated at drying conditions: T = 316 K (43°C) and V = 3 m/s.

Table 4. List of relationships to determine diffusivity (m^2/s) of rice fractions

It is well known that diffusivity of starch varies with change in its moisture content. We determined moisture diffusivity of rice components during adsorption and desorption process in different moisture contents ranges. We found that diffusivity of rice endosperm was higher during desorption than adsorption at moisture contents higher than 8% on dry basis. Between 8% and 19% moisture, white rice diffusivity increased with increase in moisture. On the other hand, diffusivity of husk decreased with increase in moisture content at all moisture contents. Bran diffusivity remained almost constant with change in moisture.

5.3 Convective heat transfer coefficient

Convective heat transfer coefficient determines the flux of heat input from the drying air to rice grains. It depends upon properties of drying air, porosity of rice bed and the shape of rice grains. Deep bed drying can be considered as packed bed of grains, while single layer or single kernel drying can be considered as grain being immersed in fluid.

Bird et al. (1960) described convective heat transfer coefficient h ($W.m^{-2}.^{\circ}C^{-1}$) in packed beds by the following correlation:

$$h = \left(\frac{\rho v' c_p}{\varepsilon}\right) \times \left(A\psi \,\mathrm{Re}^{-B}\right) \times \left(\frac{\mu c_p}{k}\right)_f^{\frac{-2}{3}} \tag{19}$$

$A = 0.61$ and $B = 0.41$ for $Re > 50$

$A = 0.91$ and $B = 0.51$ for $Re < 50$

where, ρ is density (kg/m^3), v' is superficial velocity (m/s), c_p ($J.kg^{-1}.^{\circ}C^{-1}$) is specific heat, ε is the void fraction, Re is dimensionless Reynolds number, μ ($kg.s^{-1}.m^{-1}$) is viscosity, k ($W.m^{-1}.^{\circ}C^{-1}$) is the thermal conductivity of the flowing gas and ψ depends upon shape of particles in bed. Value of ψ is 1 for sphere, 0.91 for cylinder and 0.86 for flakes. The subscript f indicates properties evaluated at the film temperature with others at the bulk temperature.

Taking a theoretical approach, Geankoplis (2000) has described a packed bed column as bundle of crooked tubes of varying cross-sectional area. For heat transfer in the bed of spheres and Reynolds number (Re) range of 10-10000, convective heat transfer coefficient is expressed by:

$$h = \left(\frac{\rho v' c_p}{\varepsilon}\right) \times \left(\frac{2.876}{\mathrm{Re}} + \frac{0.3023}{\mathrm{Re}^{0.35}}\right) \times \left(\frac{\mu c_p}{k}\right)_f^{\frac{-2}{3}} \tag{20}$$

For sphere submerged in a fluid surface heat transfer coefficient can be described as (Geankoplis, 2000):

$$Nu = \frac{hd}{k} = 2 + 0.6\,\mathrm{Re}^{1/2}\,\mathrm{Pr}^{1/3} \tag{21}$$

where, Nu is the dimensionless Nusselt number, d (m) is the characteristic dimension, and Pr is the dimensionless Prandtl number. Using this correlation, Bal (1968) found the value of surface heat transfer coefficient to vary between 57-85 $W.m^{-2}.^{\circ}C^{-1}$ over a wide range of temperatures.

Smith et. al. (1971) measured impact of shape, air velocity and air temperature on convective heat transfer coefficient of ellipsoid bodies. They defined geometry index, G to accommodate impact of shape as follows:

$$G = \frac{1}{4} + \frac{3c^2}{8a^2} + \frac{3c^2}{8b^2} \tag{22}$$

where, a, b, and c are the three dimensions of the ellipsoid. They allowed the test objects to cool down from a defined temperature and measured the change of temperature with time. With this temperature time data, they determined Nusselt number for different conditions and then it was fitted to a regression equation that is described below:

$$Nu = (0.32 - 0.22G)\mathrm{Re}^{(0.44+0.23G)} \tag{23}$$

Correlation coefficient for this model varied between 0.90 and 0.99 for values of G between 0.3 and 1.0. Mannapperuma (1975) used this equation to determine convective heat transfer coefficient for brown rice. For geometry index of 0.526, he found the coefficient to vary very small with drying temperatures. For air speed varying from 0.25 to 2.03 m/s the coefficient varied from 17-68 $\mathrm{W.m^{-2}.^{\circ}C^{-1}}$.

Sokhansanj and Bruce (1987) used following empirical equation for use in grain drying simulation:

$$\begin{aligned} h &= 3.27cG\,\mathrm{Re}^{-0.65}\,\mathrm{Pr}^{-0.67} \\ \mathrm{Re} &= \frac{Gd}{\mu} \end{aligned} \tag{24}$$

where, G is mass velocity of air ($\mathrm{kg.m^{-2}.^{\circ}s^{-1}}$), d (m) is the particle diameter and μ ($\mathrm{kg.s^{-1}.m^{-1}}$) is the viscosity of the air.

Apart from the above described methods, the convective heat transfer coefficient for a given drying process can also be measured experimentally by enthalpy balance as follows:

$$hA_s(T_a - T)dt = mc_p dT + \lambda dm \tag{25}$$

where, A_s is kernel surface area ($\mathrm{m^2}$), T_a is air temperature in drier (°C), T is temperature of kernel (°C) at drying time t (s), m is mass of wet kernels ($\mathrm{kg/m^3}$), c_p ($\mathrm{J.kg^{-1}.^{\circ}C^{-1}}$) is specific heat of grains and λ (J/kg) is latent heat of vaporization. Assuming, no moisture loss in the thin layer of kernel during heating i.e. period when $M = M_e$, Pabis & Henderson (1962) simplified this equation to determine convective heat transfer coefficient as follows:

$$h = \frac{\rho_g V c_p}{A_s t}\ln\left(\frac{T_a - T_0}{T_a - T}\right) \tag{26}$$

where, ρ_g ($\mathrm{kg/m^3}$) is density of wet grain, V is the kernel volume ($\mathrm{m^3}$), c_p ($\mathrm{J.kg^{-1}.^{\circ}C^{-1}}$) is specific heat of grains, T_0 is the initial temperature of grain (°C). When logarithm of temperature ratio is plotted versus time, the slope of curve can be used to evaluate the convective heat transfer coefficient.

5.4 Convective mass transfer coefficient

Surface mass transfer coefficient is dependent upon shape of kernels, air speed flowing over the kernels and properties of air in immediate environment. Patil (1988) and Sarker et al. (1994) have determined surface mass transfer coefficient, h_m (m/s) using following relationship:

$$Sh = \frac{h_m d_k}{D_{wa}} = 2 + 0.522\,\mathrm{Re}^{0.5}\,Sc^{0.33} \qquad (27)$$

where, Sh is the dimensionless Sherwood number, D_{wa} is diffusivity of water in air (m^2/s), d_k is equivalent diameter of the kernel (m) and Sc is dimensionless Schmidt number.

6. Scope for future research

Describing moisture movement accurately for small and complex biological material such as rough rice is very challenging. Despite decades of modeling work, still many assumptions in these models remain untested and they provide a wide horizon for future research.

Moisture distribution within the rice kernel has rarely been validated by experimental techniques, thus the uncertainty over the actual mechanism is still unresolved. Nagato et al. (1964) investigated moisture distribution in the rice kernel by taking hardness measurements at different points in the kernel. Recently, there has been interest in using magnetic resonance imaging (MRI) technique to determine moisture distribution within the rice kernels (Frias et. al., 2002; Ishida et al., 2004; Hwang et al., 2009). However, due to small moisture present in the kernels, MRI signal is weaker in rice kernels than in other fruits and vegetables making it difficult to determine moisture distribution quantitatively. We performed MRI experiments to determine moisture distributions within the rough rice kernel and found that moisture removal within the rice kernel was not uniform during drying. Moisture removal is faster near the embryo (also called germ) region than other parts of kernel. Such moisture removal was also observed by Nagato (1964). Cause of such non-uniformity of moisture removal from the kernel is yet to be understood and implemented in the model. Similarly, impact of the changing air-layer between bran and husk is yet to be investigated.

There has been progress in determining impact of moisture gradients on rice fissuring. Still, more research is required to determine limiting values of moisture gradients that must be avoided to minimize fissuring. As discussed earlier, such gradients would depend on moisture content of rice.

7. Conclusion

In this chapter, we have discussed the important advances made in development of mathematical models to describe moisture movement in rice. With progress of computation technologies, the modelling studies have shifted from empirical or semi-empirical approach to more rigorous and mechanistic approach. In this chapter, three such approaches to model rice drying namely thin-layer models, deep-bed models and single kernel drying models are discussed. Among these, single kernel models are based upon diffusion mechanism and can determine moisture distribution within the rice kernel. Since moisture gradients within the grain were found to affect fissuring in kernels, such models have found widespread use in recent times. Using robust computational tools, recent models have used two-dimensional and three-dimensional models of prolate spheroid and ellipsoid shape, respectively, to model moisture movement in rice kernel. These shapes resemble the actual kernel geometry more closely and therefore, such models are expected to predict moisture distribution more accurately. In drying processes such as infrared drying, heating of kernels is very rapid and therefore, temperature dependent properties such as moisture diffusivity would vary within the rice kernel and vary with time. In such cases, both heat transfer and moisture transfer processes needs to be solved simultaneously in the model. Knowledge of physical, thermal and hygroscopic properties of the rice variety under study are very important for accurate model predictions.

8. References

Aguerre, R.; Suarez, C. & Viollaz, P. E. (1982). Drying kinetics of rough rice grain. *Journal of Food Technology*, 17, 679-686

Agrawal, Y. C. & Singh, R. P. (1977). Thin-layer drying studies on short-grain rice. *ASAE Paper*, No. 77-3531, St. Joseph, U.S.

Allen, J. R. (1960). Application of grain drying theory to the drying of maize and rice. *Journal of Agricultural Engineering Research*, 5, 363-385

Bakker-Arkema, F. W. & Hall, C. W. (1965). Static versus dynamic moisture equilibria in the drying of biological products. *Journal of Agricultural Engineering Research*, 10, 4, 308-311

Bakker-Arkema, F. W.; Bickert, W. G. & Patterson, R. J. (1967). Simultaneous heat and mass transfer during the cooling of a deep bed of biological products under varying inlet air conditions. *Journal of Agricultural Engineering Research*, 12, 4, 297-307

Bal, S. (1968). *Analytical study of radiant heating of rice grain*, Unpublished MS thesis. Louisiana State University, Baton Rogue, U.S.

Bird, R. B.; Stewart, W. E. & Lightfoot, E. N. (1960). *Transport phenomena*, John Wiley and Sons, U.S.

Boyce, D. S. (1965). Grain moisture and temperature changes with position and time during through drying. *Journal of Agricultural Engineering Research*, 10, 4, 333-341

Crank, J. (1979). *The mathematics of diffusion*. Oxford University Press, Oxford, U.K.

Datta, A. K., & Ni, H. (2002). Infrared and hot-air-assisted microwave heating of foods for control of surface moisture. *Journal of Food Engineering*, 51, 355-364

Ece, M. C. & Cihan, A. (1993). A liquid diffusion for drying rough rice. *Transactions of the ASAE*, 36, 837-840

Frias, J. M.; Foucat, L.; Bimbenet, J. J. & Bonazzi, C. (2002). Modeling of moisture profiles in paddy rice during drying mapped with magnetic resonance imaging. *Chemical Engineering Journal*, 86, 173–178

Geankoplis, C. J. (2000). *Transport processes and unit operations*, 3rd ed., Prentice-Hall, Eaglewood Cliffs, U.S.

Gencturk, M. B.; Bakshi, A. S.; Hong, Y. C. & Labuza, T. P. (1986). Moisture transfer properties of wild rice. *Journal of Food Process Engineering*, 8, 243-261

Ginzburg, A. S. (1969). *Application of infra-red radiation in food processing*, Translated by A. Grochowski. Leonard-Hill, London, U.K.

Hacihafizoglu, O.; Cihan, A., & Kahaveci, K. (2008). Mathematical modelling of drying of thin layer rough rice. *Food and Bioproduct Processing*, 86, 268-275

Henderson, J. M. & Henderson, S. M. (1968). A computational procedure for deep-bed drying analysis. *Journal of Agricultural Engineering Research*, 13, 2, 87-95

Henderson, S. M. (1974). Progress in developing the thin layer drying equation. *Transactions of the ASAE*, 17, 1167-1168 & 1172

Hougen, O. A.; McCauley, H. J. & Marshall Jr, W. R. (1940). Limitations of diffusion equations. *Transactions of the American Institute of Chemical Engineers,* 36, 183-206

Hwang, S.; Cheng, Y.; Chang, C.; Lur, H. & Lin, T. (2009). Magnetic resonance imaging and analysis of tempering processes in rice kernels. *Journal of Cereal Science,* 50, 36-42

Igathinathane, C. & Chattopadhyay, P. K. (1999a). Moisture diffusion modelling of drying in parboiled paddy components. Part I: Starchy endosperm. *Journal of Food Engineering,* 41, 2, 79-88

Igathinathane, C., & Chattopadhyay, P. K. (1999b). Moisture diffusion modelling of drying in parboiled paddy components. Part II: Bran and Husk. *Journal of Food Engineering,* 41, 2, 89-101

Iguaz, A. & Virseda, P. (2007). Moisture desorption isotherms of rough rice at high temperatures. *Journal of Food Engineering,* 79, 794–802

Ishida, N.; Naito, S. & Kano, H. (2004). Loss of moisture from harvested rice seeds on MRI. *Magnetic Resonance Imaging,* 22, 871-875

Kunze, O. R. & Calderwood, D. L. (2004). Chapter 9: Rough rice drying - moisture adsorption and desorption. In *Rice: Chemistry and Technology,* E. T. Champagne, ed., 223-268, American association of cereal chemists, St. Paul, U.S.

Kunze, O. R.; Lan, Y. & Wratten, F. T. (2004). Chapter 8: Physical and mechanical properties of rice. In *Rice: Chemistry and Technology,* E. T. Champagne, ed., 191-221, American association of cereal chemists, St. Paul, U.S.

Lague, C. (1990). *Modeling pre-harvest stress-cracking of rice kernels.* Unpublished PhD diss. University of California, Davis, U.S.

Lu, R. & Siebenmorgen T. J. (1992). Moisture diffusivity of long-grain rice components, *Transactions of the ASAE,* 35, 6, 1955-1961

Mannapperuma, J. D. (1975). *Analysis of thermal and moisture stresses caused during drying of brown rice.* Unpublished MS thesis. Louisiana State University, Baton Rogue, U.S.

Meeso, N.; Nathakaranukule, A.; Madhiyanon, T. & Soponronnarit, S. (2007). Modeling of far-infrared irradiation in paddy drying process. *Journal of Food Engineering,* 78, 1248-1258

Midilli, A.; Kucuk, H. & Yapar, Z. (2002). A new model for single layer drying. *Drying Technology,* 20, 7, 1503–1513

Mossman, A. P. (1986). A review of basic concepts in rice-drying research. *Critical Reviews in Food Science and Nutrition,* 25, 1, 49-71

Nagato, K.; Ebata, M. & Ishikawa, M. (1964). On the formation of cracks in rice kernels during wetting and drying of paddies. *Proceedings of the Crop Science Society of Japan,* 33, 82-89

Nelson, G. L. (1960). A new analysis of batch grain-drier performance. *Transactions of the ASAE,* 3, 2, 81-85 & 88

Nindo, C. I.; Kudo, Y. & Bekki, E. (1995). Test model for studying sum drying of rough rice using far-infrared radiation. *Drying Technology,* 13, 225-238

Parry, J. L. (1985). Mathematical modelling and computer simulation of heat and mass transfer in agricultural grain drying: A review. *Journal of Agricultural Engineering Research,* 32, 1-29

Pabis, S. & Henderson, S. M. (1962). Grain drying theory III. The air/grain temperature relationship. *Journal of Agricultural Engineering Research,* 7, 1, 21-26.

Patil, N. D. (1988). Evaluation of diffusion equation for simulating moisture movement within an individual grain kernel. *Drying Technology,* 6, 1, 21-42

Prakash, B. & Pan, Z. (2009). Heat and mass transfer modeling of rough rice under convective and infrared drying. *ASAE Paper,* No. 096420, St. Joseph, U.S.

Reddy, B. S. & Chakraverty, A. (2004). Equilibrium moisture characteristics of raw and parboiled paddy, brown rice, and bran. *Drying Technology*, 22, 4, 837-851

Sarker, N. N.; Kunze, O. R. & Strouboulis, T. (1994). Finite element simulation of rough rice drying. *Drying Technology*, 12, 4, 761-775

Sarker, N. N.; Kunze, O. R. & Strouboulis, T. (1996). Transient moisture gradients in rough rice mapped with finite element model and related to fissures after heated air drying. *Transactions of the ASAE*, 39, 2, 625-631

Siebenmorgen, T. J.; Banaszek, M. M. & Kocher, M. F. (1990). Kernel moisture content variation in equilibrated rice samples. *Transactions of the ASAE*, 33, 6, 1979-1983

Simmonds, W. H. C.; Ward, G. T. & McEwen, E. (1953). The drying of wheat grain. *Transactions of the Institute of Chemical Engineers*, 31, 265, cited in Allen (1960).

Smith, R. E.; Bennet, A. H., & Vacinek, A. A. (1971). Convection film coefficients related to geometry for anomalous shapes. *Transactions of the ASAE*, 14, 1, 44-47

Sokhansanj, S. & Bruce, D. M. (1987). A conduction model to predict grain temperatures in grain drying simulation. *Transactions of the ASAE*, 30, 4, 1181-1184

Spencer, H. B. (1969). A mathematical simulation of grain drying. *Journal of Agricultural Engineering Research*, 14, 3, 226-235

Srikiateden, J. & Roberts, J. S. (2007). Moisture transfer in solid food materials: A review of mechanisms, models, and measurements. *International Journal of Food Properties*, 10, 739-777

Steffe, J. F., & Singh, R. P. (1980a). Liquid diffusivity of rough rice components. *Transactions of the ASAE*, 23, 3, 767-774 & 782

Steffe, J. F., & Singh, R. P. (1980b). Diffusivity of starchy endosperm and bran of fresh and rewetted rice. *Journal of Food Science*, 45, 356-361

Steffe, J. F., & Singh, R. P. (1982). Diffusion coefficients for predicting rice drying behaviour. *Journal of Agricultural Engineering Research*, 27, 6, 489-493

Sun, Da-W. (1999). Comparison and selection of EMC/ERH isotherm equations for rice. *Journal of Stored Products Research*, 35, 249-264

Thompson, J. F. & Mutters, R. G. (2006). Effect of weather and rice moisture at harvest on milling quality of California medium-grain rice. *Transactions of the ASABE*, 49, 2, 435-440

USDA (1994). Chapter 5. Inspection of Milled Rice *in Rice Inspection Handbook*. USDA Federal Grain Inspection Service. Washington D. C., U.S.

USDA (2010). *Rice Area, Yield, and Production*. USDA Foreign Agricultural Service. Available at www.fas.usda.gov. Accessed 11 May 2010

Verma, L. R.; Bucklin, R. A.; Endan, J. B. & Wratten, F. T. (1985). Effects of drying air parameters on rice drying models. *Transactions of the ASAE*, 28, 296-301

Wang, C. Y. (1978). *Simulation of thin-layer and deep-bed drying of rough rice*. Unpublished PhD diss. University of California, Davis, U.S.

Yang, W.; Jia, C. C.; Siebenmorgen, T. J.; Howell, T. A. & Cnossen, A. G. (2002). Intra-kernel moisture responses of rice to drying and tempering treatments by finite element simulation. *Transactions of the ASAE*, 45, 4, 1037-1044

15

Mass Transfer Mechanisms during Dehydration of Vegetable Food: Traditional and Innovative Approach

Derossi A.[1], Severini C.[1] and Cassi D.[2]
[1]*Department of Food Science, University of Foggia*
[2]*Department of Physics, University of Parma*
Italy

1. Introduction

All along people well know that through the reduction of water content it is possible to preserve food for a long time. Among foods, fresh fruits and vegetables show an average water content ranged between 90% and 98% which greatly affect their perishability. So, back in the past sun dehydration was the first drying method used to assure fruits and vegetables during long period of drought, winters, etc. In terms of stability, drying processes not only inhibit microbial growth but also several biological and chemical degradation reactions; nevertheless, they also affect sensorial and nutritional characteristics promoting the collapse of vegetable tissues and the degradation of vitamins and antioxidants (Ibarz and Barbosa-Canovas, 2003). Nowadays, it is well known that the "state" of water rather than its mass fraction is responsible of microbial growth and degradation reactions. Moreover, in the last 30 years the need of new technologies allowed to develop several dehydration methods such as hot air dehydration, osmotic dehydration, microwave dehydration, infrared (IR) dehydration, ultrasonic dewatering, hybrid technologies, etc. The introduction of these technologies in food industries has increased the quality of dried vegetables leading to an exponential increase of the market of these products. For instance, a report published from Research and Markets showed that the total West European dehydrated food market was worth euros 11.5 bilion in 2009.

Strictly speaking, during drying processes, the key factor of all traditional and innovative techniques is the mass transfer from vegetable tissues to its surrounding and *vice versa*. In general, water is the component that moves from vegetable tissues toward the surrounding air but this transfer may occurs through several mechanisms such as capillary flow, diffusion of water due to concentration differences, surface diffusion, vapor diffusion in the pores due to pressure gradient, water vaporization-condensation (Ibarz and Barbosa-Canovas, 2003), etc.; moreover, some of these mechanisms may affect each other, making the drying a very complex phenomenon. Furthermore, in the case of osmotic dehydration a countercurrent mass transfer occurs: a. water flows from vegetables to hypertonic solution; b. osmotic agents move toward vegetable tissues. On these bases, the possibility to obtain safe dried vegetables with high nutritional and sensorial quality, also maintaining a high production output and low energy costs, is strictly related to the ability in controlling mass transfer. To reach these purposes several questions need to be satisfied: *which are the mass transfers mechanisms involved during each dehydration technologies? how is possible to increase the dehydration rate? which mathematical models may be used to predict mass transfer? how the mass transfer may affect the quality of dehydrated vegetable food?*. Even though much has been done to give correct responses, much more remains to be explained. For instance, Fickean diffusion is widely recognized from food scientists as the predominant internal mass transfer mechanism during drying processes. Under this approach, water moves randomly from a region with high concentration toward a region at low concentration assuming that the moisture gradient inside vegetables is the only driving force of the motion. On this basis several empirical equations have been proposed to predict mass transfer. Although these models significantly increased the knowledge of dehydration, much more must be explained. For instance, Fick's laws which are widely used to model water diffusion during drying require several assumptions and simplifications that are often unrealistic. Some of these are: food are homogeneous and isotropic media; diffusion coefficients are independent of moisture concentration; samples are approximately considered as spheres, cylinders or slabs; heat transfer during drying is ignored; collapse of vegetable tissues during dehydration (easily detected by a visual aspect) is neglected (Saguy et al., 2005). The effects of microscopic structure on mass transfer has been completely dropped and only in the last years few pioneering papers were published on this topics. It should be considered that most food structures generally exhibit a disordered geometry, often due to percolation phenomena, which in many cases can be described in terms of fractal geometry. This feature dramatically affects diffusion phenomena, giving rise to anomalous laws (subdiffusion), involving universal geometrical parameters such as fractal dimension and spectral dimension.

This chapter serves to provide at the readers the conventional and emerging theories on mass transfer during traditional and innovative drying technologies of fruits and vegetables. After a first evaluation of the basic principles concerning the relation between water and food quality, the most important mechanisms of molecular motion, the variables affecting the rate of the most important drying technologies and their effects on kinetics are reviewed and discussed. Also, the most used mathematical models will be reported analyzing their advantages and the related assumption and limitation with the aim to give the basis for the modeling of mass transfer during vegetable dehydration. Moreover, the application of statistical-physic approach to study the random movement of water and solutes inside food three-dimensional microstructure will be reported.

2. The importance of dehydration treatments for safety and quality of vegetable food

2.1 Basic principles on the relation between water and food quality

Water is the most abundant component in foods. Among these, fruits and vegetables show a mass fraction of water in the range of 90% and 98%. Its amount and its peculiar chemical and physical characteristics make it the key factor for biological and chemical degradation reactions. In particular, water is the most important medium in which chemical and biological reagents may move, collide and react. Moreover, water may act as reagent and co-reagent of several degradation reactions (i.e. hydrolysis of lipids which degrades vegetable oils or the fat content of vegetables) and it stabilizes the most important biological structures such as enzymes, proteins, DNA, cellular membranes, etc. by its ability to produce a large number of hydrogen bounds. On these basis, taking into account the safety and quality of vegetable food, it is possible to state that water controls the growth of pathogen and/or alterative microorganisms, chemical and enzymatic reactions such as enzymatic browning (EB), lipid oxidation, Maillard reaction (NEB), vitamin degradation, texture degradation, etc. which are able to make food not safe and/or organoleptically or nutritionally unacceptable for the consumers. Furthermore, water significantly modify physical and chemical properties of vegetables such as thermal conductivity, heat capacity, dielectric properties, electrical conductivity, boiling and freezing point, firmness, etc., which are key factors for all dehydration technologies as well as for others important industrial processes. So, the knowledge regarding the correlation between both the above reactions, food properties and water is crucial for the correct production of dried vegetables. However, these correlations not always are linear but often very complex. For these reasons the fundamental aspects concerning the water in food are summarized hereafter.

In general, all people well know that reducing the amount of water in food it is possible to significantly increase its shelf life. So, for many years the water concentration in food was considered the only factor affecting their perishability. Nevertheless, Scott (1957) highlighted the importance of the "state" of water in food rather than its mass fraction stating that it is a key factor for degradation reactions. Scientific literature reports many terms to define the "state" of water in food some of which cannot be used as synonymous: "free" and "bound" water, "unfreezable water", "structured water", "rotationally water", "water mobility". These are often related to the analytical techniques used to detect them as in the case of "unfreezable water" estimated by differential scanning calorimeter (DSC) or "rotationally water" that is measured by nuclear magnetic resonance (NMR) which allows to measure the rate of rotational mobility of water molecules. However, all above terms may be considered similar if the follows general meaning it is taken into account: *the state of water means the availability of water molecules to be freely used for all biological and chemical reactions.* Significant different levels of water availability may be detected in vegetable food with the same water concentration. On the other hand foods with different water content may show the same water availability. So, vegetable foods with the same water content may show significantly different shelf life. These behaviours are caused by different interactions between water molecules and chemical compounds of vegetables such as sugars, proteins, salts, fats, vitamins, etc. Under this consideration "free" water may be considered as an unperturbed systems (also called bulk water) in which water molecules may interact with each other keeping constant their chemical and physical characteristics such as bound angle, internuclear distance, etc. Instead, "bound water" may be considered as a perturbed system in which water preferentially interacts with other molecules. Fennema (1999) defined bound water as: "*water that exists in the vicinity of solutes and other non aqueous constituents and that exhibits properties that are significantly altered from those of bulk water in the same system*". "Free" water may be freely used from microorganisms for their intracellular reactions or as reagent for chemical degradation reactions; instead, "bound" water is chemically bounded to solid matrix (solutes) and it is unavailable for microbial growth and chemical reactions. In 1957 Scott defined a mathematical parameter to measure the water availability in food: water activity (a_w). This index has thermodynamic basis:

$$\mu = \mu_0 + RT \ln (f/f0) \tag{1}$$

where μ and μ_0 are respectively the chemical potential of water and those of pure water (the reference state); f is the fugacity of the system at given conditions and f_0 is the fugacity at the reference state (Van der Berg & Bruin, 1981); *T* and *R* are the temperature (K) and gas constant (8.314 J K^{-1} mol^{-1}) respectively. Since in practical condition fugacity may be considered equal to partial vapour pressure (p) it is possible to obtain the following equation:

$$\mu = \mu + \mu_0 RT \ln (P/P_0) \quad (2)$$

if thermodynamic equilibrium is reached it is possible to use the following:

$$\mu = \mu + \mu_0 RT \ln (a_w) \quad (3)$$

So, water activity may be defined as the ratio between vapor pressure surrounding food (P) and the vapor pressure of pure water (reference state, P_0) both measured at the same temperature and pressure values:

$$a_w = P/P_0 \quad (4)$$

It is worth nothing that this parameter is valid only if thermodynamic equilibrium is reached (no water molecules move inside food due to different chemical potential) although in practical application this condition is often not satisfied. Since 1975, water activity is the most important parameter used to relate the water availability with the rate of degradation reactions and several books have been published on this topics (Ruand & Chen, 1998; Rockland & Beuchat, Roos et al., 1999).

2.2 Water activity and degradation reaction of vegetable food

In 1970, Labuza reported a food stability map in terms of water activity values (figure 1). In particular, the figure shows the rate of some degradation reactions as a function of a_w values. Moreover, a general moisture sorption isotherm of food is reported. According to Frank & Wen (1957), the map is subdivided in three regions. In the first one (0 - 0.2) water is strictly bounded with solid matrix of food and it cannot be used as reagent or co-reagent. Moreover, in this region viscosity is so high that mobility of chemical reagents is restricted and the rate of degradation reactions is reduced. In the second region, characterized from water activity values ranged between 0.2 and 0.7, water molecules progressively becomes available for some biological and chemical reactions such as Maillard reaction (non enzymatic browning) and enzyme activities but they are still unavailable for microbial growth with the exception of some molds. In the third region (0.7 - 1.0) water molecules are weakly bounded to the solutes or restrained by capillary forces; here, water molecules are freely usable for both microbial growth and chemical reactions which may quickly occur. In terms of microbial growth it is possible to observe a direct relation: as lower the water activity, as slower the growth rate of molds, yeasts or bacteria.

When microbial cells are exposed to a system with a high osmotic pressure (low water activity) they lose water because cellular membranes are semi-permeable. This phenomenon, called plasmolisys, proceeds until the osmotic equilibrium is reached but, in this condition microorganisms are unable to reproduce and they will die or remain dormant (Sperber, 1983; Troller, 1985; Amezaga-Johnston et al., 1999). In general, the reduction of a_w values has two inhibitory effects on the growth rate of microorganisms: 1. the increase of lag phase; 2. the reduction of generation time (Troller, 1985). Moreover, from figure 1 it is possible to observe that the resistances among the microorganisms are very different and may be graded as follows: molds > yeasts > bacteria. Furthermore, among these, different microorganisms may show significant variation in term of water activity resistance. For

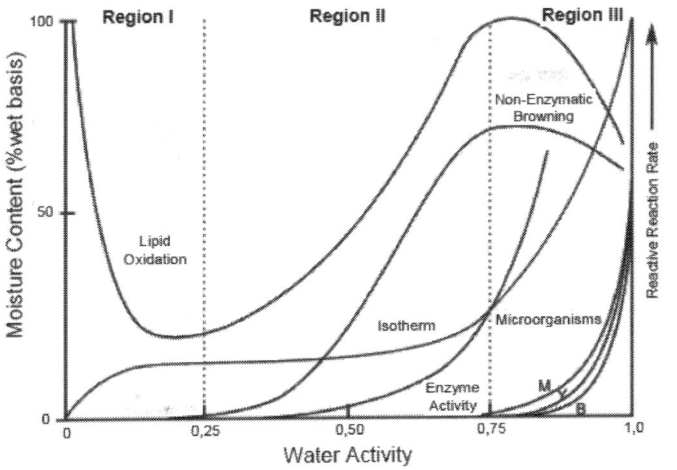

Fig. 1. Stability map of food as a function of water activity (from Schmidt, 2004)

instance, the out-growth of *Clostridium botulinum* spores, one of the most dangerous pathogen microorganisms, is inhibited at a_w values < 0.935 (Clavero et al., 2000). The generation times of *Listeria monocytogenes* at a_w values of 0.98 and 0.92 were reported respectively as 6.1 h and 12.7 h in systems at pH 6.2. Colatoni & Magri (1997) reported that the minimum value for the *Clostridium perfringens* spore germination is 0.935. Beuchat (1985) reported that *Staphylococcus aureus,* which is very tolerant to a_w values, shows a minimum for toxin production of 0.90 and a minimum of 0.86 for growth. In table 1 are reported the minimum a_w values for some of the most important microorganisms for food degradation.

It should be taken into account that the threshold reported in table 1 are often measured in model system with specific environmental conditions. In vegetable foods, the minimum a_w values for growth or toxin production of each microorganism must be considered as range of ranges of a_w values inside which the "true" limiting value may change when other inhibitory factors are applied. In fact, it should be considered that the resistance of microorganisms on a_w values is affected by several parameters such as pH, temperature, oxygen concentration, preservatives, solutes, etc. For instance, the type and the amount of solutes in food (their chemical composition) greatly affect the tolerance of microorganisms on water activity. This is because microorganisms have different levels of ability to adapt at an hypertonic environment regaining turgor pressure and excluding certain incompatible solutes (Kang et al., 1969; Gould & Measures, 1977; Christian, 1981; Buchanan & Bagi, 1997: Lenovich, 1985 Gould, 1985). For instance, it was reported that the limiting a_w values for several microorganisms is lower when sodium chloride rather than glycerol is used (Gould and Measures, 1977). Since a comprehensive analysis of the effects of water availability on microbial degradation is out of the principal aim of this chapter we report the most important papers and books where the effect of water activity on safety and quality of food was studied in details. (Dukworth, 1975; Labuza 1980; Rockland & Nishi, 1980; Rockland & Stewart, 1981; Troller, 1985; Simato & Multon, 1985; Troller, 1985; Drapon, 1985; Gould, 1985).

Water activity (a_w)	Bacterial	Yeast	Mold
0.98	*Some Clostridium pseudomonas*		
0.97	*Some Cl. perfringens and Cl. botulinum E*		
0.95	*Bacillus, Cl.botulinum A and B, Escherichia, Pseudomonas, Salmonella, Serratia*		
0.94-0.92	*Lactobacillus, Streptococcus, Pediococcus, Microbacterium, Vibrio*	*Rhodotorula, Pichia*	*Rizhopus, Mucor*
0.91-0.88	*Staphylococcus, Streptococcus, Lactobacillus*	*Hansenula, Saccharomyces, Candida, Torulopsis*	*Cladosporium*
0.87-0.80	*Staphylococcus*	*Saccharomyces*	*Paecilomyces, Aspergillus, Penicillium, Emericella*
0.80-0.62		*Saccharomyces*	*Eurotium, Monascus,*

Table 1. Limiting a_w values for the growth of some microorganisms of interest for vegetable food (adapted from Leistern and Radel, 1978)

As in the case of microbial growth, enzymatic activity increases with increasing of a_w values. Nevertheless, figure 1 shows that enzymes may catalyze biological reactions also at a_w values very low (close to 0.2). This is the case of lipase and lipoxygenase enzymes which are responsible of the vegetable oils degradation even if the a_w value is around of 0.025 and 0.05 respectively (Drapon, 1985; Brokmann & Acker, 1977). This behavior was explained by the action of lipid as plasticizing medium which may increase the mobility of reagents.

Non enzymatic browning (NEB) is a complex chemical reaction between reducing sugars, like glucose, and amino groups such as amino acids. NEB is responsible to modify the appearance, the taste and the nutritional value of food (Maillard, 1912; Martins et al., 2001). As reported from Martins et al. (2001) Maillard reaction consists in consecutive and parallel reaction steps affected by several parameters. For the production of dried vegetables, NEB is among the most important degradation reactions because the high temperature promotes the production of brown melanoidin pigments making the vegetables brown. Moreover, other changes such as degradation of the nutritive value of the involved proteins, production of volatile (Fors, 1983) and antioxidant compounds (Griffith & Johnson, 1957; Brands et al., 2000; Martins et al., 2001) are involved during NEB reaction. In terms of water availability the Maillard reaction has been reviewed from several authors (Mauron, 1981; Baltes, 1982; Yaylayan, 1997; Martins et al., 2001) which showed no direct relation between its rate and the a_w values. Figure 1 shows a bell-shaped trend in which the rate of NEB increases until a maximum value almost at 0.7 and then significantly decreases. The trend of NEB is a consequence of two different effects: 1.the changes of the mobility of chemical reagents; 2. the dilution of the system (Eichner, 1975; Labuza & Saltmarch, 1981). By increasing a_w values, water molecules progressively become free and may act as plasticizing media imparting mobility to the chemical reagents which produce the melonoidin pigments. This effect progressively increases until the maximum a_w values of ~ 0.7, after which the high water concentration dilutes the chemical species reducing their probability to interact each other (Labuza & Saltmarch, 1981; Maltini et al., 2003).

Moisture plays an important effect on lipid degradation in particular at medium-high temperature as in the case of drying processes. Perhaps the trend of lipid oxidation as a function of water activity is the most complex among the others degradation reactions. Lipid oxidation is a chemical auto-catalytic reaction during which unsatured fatty acids and oxygen react producing off-flavors and increasing rancidity. Several factors may influence oxidation rate: moisture content, type of fatty acids, the amount on metal ions, light, temperature, oxygen concentration, antioxidants, etc. From the figure 1 it is possible to observe an inverse correlation as a_w increase until values of 0.2, then a direct correlation till 0.75 and again a decrease of reaction rate as a_w value increases. This behavior was well studied from Labuza (1975) and Karel (1980) which hypothesized a combined effect of pro- and antioxidant factors. In comparison with other degradation reactions when water activity value is very low (~ 0.1) lipid oxidation rate is very high. This is because when solid matrix is dry a maximum contact between oxygen and lipid exists. Instead, as water activity increases until a_w value of 0.25, water molecules obstacle the collision between fatty acids (on solid matrix) and oxygen leading to a reduction of oxidation rate. Instead, between a_w values of ~ 0.2 and ~ 0.75 the pro-oxidant factors such as the increased mobility of chemical reagents, the solubilization of chemical species inside water, makes reaction rate and water activity directly correlated.

Others quality indexes are affected by water activity such as carotenoid content and texture properties of dehydrated food. Carotenoids are lipid soluble pigments responsible of the color of many fruits and vegetables such as tomato, carrot, grape, orange, cherry, etc. During drying they may undergo the same degradation reactions of lipids (Stephanovic & Karel, 1982). Chlorophylls were shown to be more resistant to degradation at lower a_w value, probably because the mobility of reagents is restricted and the probability to react is low due to the high viscosity of the system (Lajollo and Marquez, 1982). Betalaines, the major pigments of red beet showed a high stability in model system at low a_w values (Saguy et al., 1980; Saguy et al., 1984). As reported from Cohen and Saguy (1983) a reduction of a_w values from 0.75 to 0.32 produced an increase of half life of betanine in beet from 8.3 to 133 days.

3. Dehydration technologies for vegetable food: mass transfer mechanisms and process variables

3.1 The behavior of vegetables during drying processes

With the aim to correctly deal mass transfer mechanisms during dehydration processes it is important to briefly remind the importance of chemical composition of vegetables on drying. As previously reported, vegetables with the same water content may show significant different a_w values because of their chemical composition lead to different affinity for water. Solutes such as sugars, salts, ions, proteins, lipids and their relative concentration in fresh fruits or vegetables, interact with water molecules through different chemical bounds, making the water molecules more or less easily removable from biological tissues. In this way, each vegetable shows a unique behavior in terms of equilibrium between its water content and water activity values. This equilibrium may be described by sorption or desorption isotherms which respectively refer to the case in which vegetable food is under the process of increasing or decreasing its water content (Wolfe et al., 1972; Slade & Levine, 1981; van den Berg, 1985; Slade & Levin, 1985; Kinsella and Fox, 1986; van der Berg, 1986; Slade & Levine, 1988a; Slade & Levine, 1988b; Karel & Lund, 2003). So, it should be always considered that knowledge of the isotherms are a basic requirement to plan a correct drying process with the aim to maximize its advantage and minimize degradation reactions. In figure 2 are reported typical sorption and desorption isotherms of food. Usually they show a general S-shaped trend in which three regions may be clearly observed. These exactly reflect the regions of food stability map previously discussed (figure 1). For instance, the first section of desorption isotherms, in a range of a_w values between 0 and 0.2 - 0.3, is called "monolayer" and it is characterized from water molecules strictly absorbed on hydrophilic, charged and polar molecules such as sugars and proteins (Kinsella & Fox, 1986; Lahsasni et al., 2002; Hallostrom et al., 2007; Okos et al., 2007). Usually, the water in this region is considered as "unfrozen" and it is not available for chemical reactions and it cannot act as plasticizer. Also, as known, sorption and desorption isotherms are not overlapped, stating a completely different behavior in the cases in which water is removed from or added to vegetables. Moreover, during drying foods show higher water activity values in comparison with rehydration process.

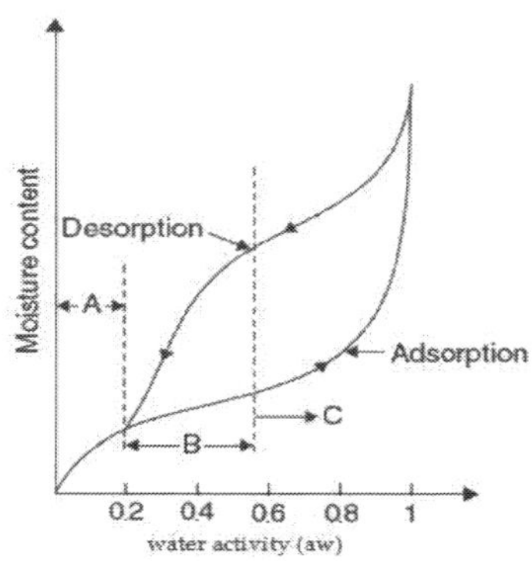

Fig. 2. Typical sorption and desorption isotherm of food (adapted from Okos et al., 2007)

This different behavior is called hysteresis and it may have different intensity and/or different shape as a consequence of several factors among which the chemical composition of food is one of the most important (Okos et al., 2007). So, vegetables with high content of sugars/pectins show an hysteresis in the range of monolayer while in starchy food the hysteresis occurs close to a_w of 0.7 (Wolf et al., 1972; Okos et al., 2007). Nevertheless, Slade & Levine (1991) stated that other factors such as temperature, physical structure (i.e. amorphous or crystalline phases), experimental history (i.e. previous desorption/sorption cycles) and sample history (i.e. pretreatments, thermal history during storage before drying, etc) may greatly affect the shape of hysteresis. For instance, it is commonly accepted that as temperature increases, the moisture content decreases leading to a reduction of isotherms; this effect is greater in desorption that on adsorption, producing a reduction of hysteresis.

3.2 Dehydration techniques and mass transfer mechanisms

Dehydration is one of the most important unit operation in Food Science. The terms dehydration and drying are generally used as synonymous but they not are exactly the same. Dehydrated vegetables are considered to have a mass fraction of water lower than

2.5%; instead dried vegetables may contain more than the 2.5% (Ibarz & Barbosa-Canovas, 2003). A complete and correct analysis of dehydration or drying technologies is an hard work. Traditionally, conventional and innovative dehydration techniques are the two most important classes considered in scientific literature. In the first group sun dehydration, hot air dehydration, spray drying, osmotic dehydration, freeze drying, fluidized bed drying, are the most important; instead, microwave drying, infrared drying, ultrasonic dehydration, electric and magnetic field dewatering, solar drying, are among the most studied innovative techniques. Another usual classification is based on the analysis of dryer plant. Craspite & Rotstein (2007) analyzing the design and the performance of several dryers stated that it is possible to classify them on the basis of supplying heat, type of drying, equipment, method of the product transporting, nature and state of feed, operating conditions and residence time. In the same way, Okos et al. (2007) analyzed several different drying techniques on the basis of a classification of dryer design. However, since many dehydration techniques may be combined and/or several methods to increase the dehydration rate may be used, the number of drying technologies available or in development stage is very high. For instance, Chua and Chou (2005) well reviewed new hybrid drying technologies classifying them in three groups: 1. Combined drying technology; 2. Multiple-stage drying; 3. Multiple-process drying. Moreover, the use of new methods to increase the mass transfer of the above technologies, may promote new dehydration techniques. This is the case in which the use of vacuum pressure was combined with osmotic dehydration giving two innovative techniques: vacuum osmotic dehydration (VOD) and pulsed osmotic dehydration (PVOD).

However, with the aim to study and classify the drying techniques on the basis of mass transfer mechanisms it is necessary to take into account the following factors: the physical state in which water molecules leave vegetable tissues; the physical state in which water molecules move inside vegetable pieces; the location from which water molecules leave the vegetables. Water may leave vegetables or move inside it as liquid and/or vapor; also, water molecules may leave the vegetables from their surface and/or internal regions. Moreover, some of these possibilities may occur simultaneously or also they could change during drying. In addition, if water evaporates from the surface or inside vegetables, the heating method should be taken into account because it has a great influence on the mass transfer mechanisms inside vegetables. Considering some of these key factors, Okos et al (2007) classified the most important internal water transfer mechanisms reported in scientific literature (table 2).

Vapor	**Liquid**
Mutual diffusion	Diffusion
Knudsen diffusion	Capillary flow
Effusion	Surface diffusion
Slip flow	Hydrodynamic mechanisms
Hydrodynamic flow	
Stepan diffusion	
Poiseuille flow	
Evaporation/condensation	

Table 2. Proposed internal mass transfer mechanisms during drying process (from Okos et al., 2007)

Although classical literature recognized that these internal transport mechanisms have a great importance during drying processes (Craspite & Rotstein, 1997; Genkoplis, 2003; Okos et al., 2007), their knowledge and their use in the planning of dehydration processes is very limited. The difficulty to theoretically study these mechanisms, to measures the microstructure properties of food and to obtain easy mathematical model, lead to assume, in practical application, liquid diffusion as the only molecular motion during drying of fruits and vegetables. Nevertheless, in the last years some pioneering researches focused their aims on the study of mass transfer in food, taking into account their nature of porous media. So, below the most important internal mass transfer mechanisms are discussed with particular attention on diffusion and capillary flow.

3.2.1 Water diffusion

Water diffusion is probably the most studied transport mechanism during drying of vegetables. Diffusion is the process by which molecules are transferred from a region to another on the basis of random motions in which no molecules have a preferred direction. Moreover, during diffusion the molecules move from the region of high concentration to that lower. Fick (1855) was the first scientist that translate diffusion in mathematical language stating that the *diffusion in a isotropic substance is based on the hypothesis that the rate transfer of diffusing substance through unit area of a section is proportional to the concentration gradient measured normal to the section* (Crank, 1975). So, often it is generally assumed that during drying water diffuses from internal regions (with a high moisture content) toward its surface (with low moisture content) where it evaporates if sufficient heat is supplied. This mechanism is described by the second Fick's law which may be expressed as:

$$\frac{\partial m}{\partial t} = D_{eff} \frac{\partial^2 m}{\partial x^2} \tag{5}$$

where m is the moisture content, t is time, x is the spatial coordination and D_{eff} is the effective diffusion coefficient. If the diffusion occurs in three dimension, eq. 5 becomes:

$$\frac{\partial m}{\partial t} = D_{eff} \left(\frac{\partial^2 m}{\partial x^2} + \frac{\partial^2 m}{\partial y^2} + \frac{\partial^2 m}{\partial z^2} \right) \tag{6}$$

The solutions of equation 6 are different depending on the geometry of samples. Solutions for simple geometries such as finite and infinite slabs, infinite cylinders, finite cylinders, spheres, rectangular parallelepipeds, were developed from Crank (1975). For instance, equations 8 and 9 are the solution of Fick's law for infinite slabs and spheres.

$$MR = \frac{M - M_e}{M_0 - M_e} = \frac{8}{\pi^2} \sum_{n=0}^{\infty} \frac{1}{(2n+1)^2} \exp\left[(2n+1)^2 \frac{\pi^2}{4} \frac{D_{eff}\, t}{L^2} \right] \tag{7}$$

$$MR = \frac{M - M_e}{M_0 - M_e} = \frac{6}{\pi^2} \sum_{n=1}^{\infty} \frac{1}{n^2} \exp\left[n^2 \frac{D_{eff}\, t}{r^2} \right] \tag{8}$$

Where MR is the moisture ratio, M is the moisture content at time t, M_o and M_e are the moisture content respectively at time zero and at equilibrium; D_{eff} is the effective diffusion coefficient (m^2/s), L is the half thickness of slab, r is the radius (m) of sphere and t is time (s). Diffusion is strictly related to the random motion of molecules, hence, with their kinetic energy, the effective diffusion coefficient may be increased by increasing temperature. Arrhenius type equation is the most used model to represent the dependence from temperature and D_{eff} (Craspite et al., 1997; Okos et al., 2007; Orikasa et al., 2008):

$$D_{eff} = D_0 \exp\left(-\frac{Ea}{RT}\right) \tag{9}$$

where D_0 is a constant (m^2/s), Ea is activation energy (KJ/mol), R is the gas constant (8.134 J/mol/K) and T is temperature (K). A wide list of diffusion coefficient for several foods and at different temperature may be look up in Okos et al. (2007). An enormous number of scientific papers used the Fick's law to study the kinetic of drying processes of fruits and vegetables and on its capacity to model the moisture content as a function of time no doubts exist (Ponciano et al., 1996; Sarvacos & Maroulis, 2001; Rastogi et al., 2002; Orikasa et al., 2008; Margaris & Ghiaus, 2007; Giner, 2009). Nevertheless, as reported from Saguy et al. (2005) Fick's laws contain several assumptions that are often unrealistic for food: fruits and vegetables are considered to have simple geometries; they are considered homogeneous and isotropic media; the heat transfer during the motion of water is completely neglected; the collapse, which refers to dramatic changes in shape and dimension during drying, is completely dropped. Moreover, as it is possible to observe from the Eq. (7) and (8) only the shape and the dimension of samples are taken into account as internal variables. For these reasons, the use of Fick's law on the basis of the idea that water transfer inside vegetable is driven only by concentration gradient shows to have several limits from the theoretical point of view. However, it allows us to estimate with good approximation the effective average of water diffusion coefficients during drying.

3.2.2 Capillary flow

On the basis of above consideration and taking into account the mass transfer mechanisms reported in table 2, some researchers begun to consider food as porous media rather than homogeneous materials and studying mass transfer on the basis of a different approach. Porous media were defined those having a clearly recognizable pores space (Vanbrakel, 1975). Moreover, by using a definition of Khaled & Vafai (2003), food may be defined as biological material volume consisting of solid matrix with interconnected void. These definitions recognizes the importance of the three dimensional microstructure of food, stating that the mass transport of water is a more complex phenomena than in a non-porous material (Datta, 2007a). Starting from this idea, capillary forces flow must be considered as one of the most important mass transfer mechanism during drying (Datta, 2007a), rehydration (Saguy et al., 2005) as well as during frying, and fat migration in chocolate (Aguilera et al., 2004).

As known, capillary forces are responsible to the attraction among liquid molecules and between them and the solid matrix. Moreover, capillary rise into a pore space is a consequence of an interfacial pressure difference (Hamraoui & Nylander, 2002). These force are very important in food science; for instance, food cannot be completely drained by gravity because capillary forces held water inside capillaries. Moreover, as a consequence of different intensity of capillary forces, water is hardly held in the regions in which solid matrix has low water content and it is less held in the regions highly moist. So, capillary force are among the reasons of: a. the water transport from a region with more water to a region with less one due to the differences in capillary force; b. the difficulty to remove water from vegetable structure. Historically, Lucas-Washburn equation it is recognized as best equation to model capillary rise into a small pores (Aguilera et al., 2004). The equation shows that the pressure inside a cylindrical capillary is balanced by viscous drag and gravity (Lucas, 1912; Washburn, 1921; Krotov and Rusanov, 1999). From this it is possible to observe that the equilibrium height within a capillary (when the hydrostatic pressure balances the interfacial pressure differences) may be expressed as:

$$h_e = \frac{2\gamma \cos(\theta_0)}{r\rho g} \tag{10}$$

where h_e is the height of capillary, γ is the surface tension of liquid, r is the radius of capillaryθ_0 is the equilibrium contact angle, ρ is the density of liquid and g is the gravitational acceleration. So, as the radius of pores reduces as the height of capillary increases. For instance, Hamraoui & Nylander (2002) showed that for glass capillaries with different radius the equilibrium height may change such as those reported in figure 3.

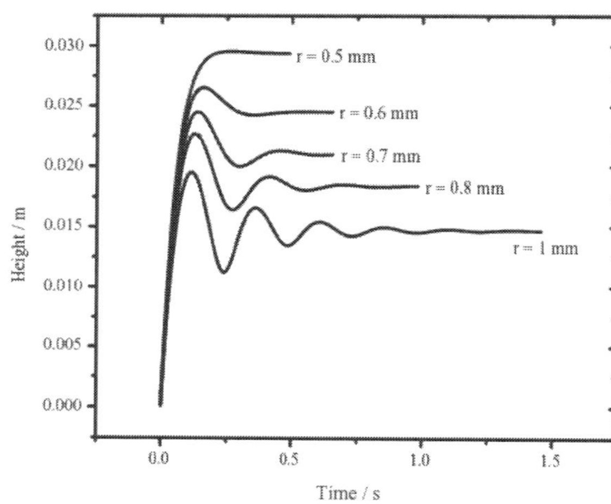

Fig. 3. Equilibrium height of water inside glass capillaries with different radius as a function of time (From Hamraoui & Nylander, 2002)

However, on the basis of a porous media approach, capillary flow in food may be expressed by Darcy's law (Khaled and Vafai, 2003; Saguy et al., 2005; Datta, 2007a):

$$u = -\frac{k_l}{\mu}\frac{\partial P}{\partial x} \tag{11}$$

where u, P, μ and k_l are the Darcy velocity (the average of the fluid velocity over a cross section), fluid pressure, dynamic viscosity of the fluid and the permeability of the porous medium, respectively. In the case of liquid transport and taking into account that $u = n^{press}/\rho_l$, where n^{press} is the mass flux of liquid and ρ_l is the density of liquid, it is possible to define the hydraulic conductivity (Saguy et al., 2005; Datta, 2007a):

$$K = \rho_l k_l / \mu_l \tag{12}$$

where k_l is the permeability of the medium (m^2) given by $k_l = k^* k_{lr}$, where k is the intrinsic permeability and k_{lr} is the relative permeability, in the liquid phase, ρ is fluid density (kg/m^3), (Datta, 2007a; Weerts et al. 2003). By substituting Eq (12) in Eq. (11) it is possible to observe that hydraulic conductivity, that is the coefficient determining the velocity of flow in the Eq (11), is affected from both liquid and solid matrix properties. The formers are expressed by density and viscosity of liquid (water in the case of vegetable dehydration); instead the latter are characterized by the three dimensional structure of vegetable tissues such as size distribution and shape of pores, porosity and tortuosity. For instance, particles with small size show a high surface area that increases the drag of water molecules that through the porous medium. The result is a reduced intrinsic permeability, hence a reduced hydraulic conductivity and capillary flow (Saguy et al., 2005). Moreover, with the aim to better express in mathematical language the importance of solid matrix on capillary flow, Datta (2007a) reported the hydraulic conductivity in the following form:

$$K = \rho_l \frac{k_l}{\mu_l} = \frac{\rho_l}{\mu_l} \frac{1}{8\tau} \sum_i \beta_i r_i^2 \tag{13}$$

where ρ_l and μ_l are respectively the density and viscosity of gas, k_l is the permeability in the liquid phase and $\Delta\beta_i$ is the volume fraction of pores the *i-th* class having radius r_i.. Again, in the Eq. (13) hydraulic conductivity is affected by two factors: 1. fluid properties by density, ρ_l, and viscosity, μ_l; 2. matrix properties. In particular, matrix properties were included into a parameter called intrinsic permeability (Datta, 2007a):

$$k = \frac{1}{8\tau} \sum_i \beta_i r_i^2 \tag{14}$$

Starting from these basic equations and with the aim to study the capillary flow inside vegetable tissues during drying, it is necessary to consider some aspects. The negative pressure of Eq. (11) (opposite with gravity) due to capillary forces is a function of water content and temperature. The effect of water content is specific for each food (see below) but in general two main cases are reported: porous medium close to saturation (food in which the pores are filled with water); porous medium unsaturated (food in which air is trapped within the structure). Datta (2007a) with the purpose to highlight the effect of water content and temperature on capillary flow, reported Darcy's law in the following form:

$$n_1^{press,cap} = -\rho_1 \frac{k_1}{\mu_1} \frac{\partial P}{\partial s} + \rho_1 \frac{k_1}{\mu_1} \frac{\partial P_c}{\partial c_1} \frac{\partial c_1}{\partial s} + \rho_1 \frac{k_1}{\mu_1} \frac{\partial P_c}{\partial T} \frac{\partial T}{\partial s} \tag{15}$$

where the first, second and third terms on the right hand are the mass flux due to gas pressure, the capillary flux due to concentration gradient (i.e. the gradient of water content) and the capillary flux due to temperature gradient, respectively. Also Datta (2007a) reported that in the case of food close to saturation, only the first term may be considered because the capillary pressure of water (P_c) is very small. Instead, for an unsaturated food (as in the case of drying process) into the Eq. (15) may be included only the second and third terms because the pressure of gas phase (P) is negligible. Nevertheless, as above reported, the effect of water content on capillary force is hardly to obtain and little (almost none) data are available in food science. In particular, the relation between moisture content and capillary pressure head and/or hydraulic conductivity are commonly available for soil science (retention curves) but very hard to find in literature concerning food. In general, capillary pressure head (h) is inversely related to moisture content; instead hydraulic conductivity (k) shows a direct correlation (figure 4). Retention curves are difficult to obtain in food but, as reported from Saguy et al. (2005), a possible approach is to convert moisture content into a volumetric water content (θ, m^3/m^3) and the a_w values in a capillary pressure head (m); briefly, the approach is to convert isotherm into a water retention curve. Again this is experimentally possible measuring h by common techniques used in soil science (Klute, 1986) or by using the Kelvin equation:

$$h = \frac{RT}{\rho_w g V_m} ln(a_w) \qquad (16)$$

where h is the capillary pressure head (m), R is the gas constant (m^3 Pa/mol K), T the temperature (K), ρ_w is the density of water (kg/m^3), g is the acceleration due to gravity and V_m is the molar volume of water (0.018 m^3/mol). In figure 5 the adsorption isotherms and the water retention curves obtained from Eq. (16) for tea (type I), wheat (type II) and apricot (type III) are reported. Another example of water retention curve was reported from Weerts et al. (2003) which studied the rehydration of tea leaf. Now, the concept of hysteresis of isotherms shown in figure 3 may be explained on the basis of different phenomena: the ink bottle effect due to the non uniformity of shape and size of interconnected pores; different liquid-solid contact angle during dehydration or rehydration process; the entrapped air in newly wetted porous media; swelling and shrinking during dehydration or rehydration (Saguy et al., 2005). At last, since water hardly interacts with biological tissues of food it is important to consider that the parameters such as porosity, size and shape of pores and tortuosity may significantly change during dehydration due to collapse, leading to a change of intrinsic permeability, hence, the capillary flow.

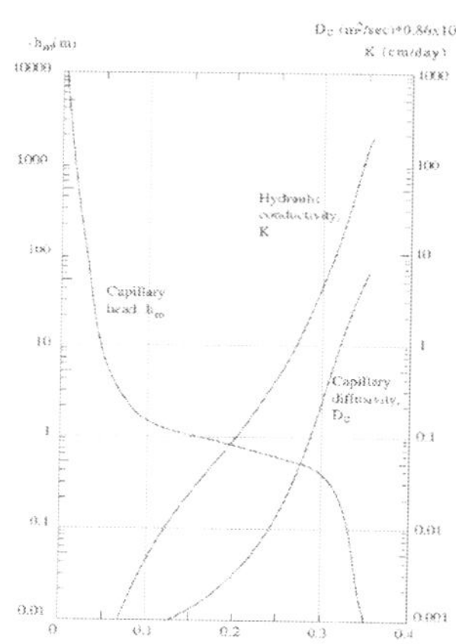

Fig. 4. Relation between volumetric moisture content and capillary head (h), hydraulic conductivity (k) and capillary diffusivity (Dc) for porous soil structure (from Datta, 2007b)

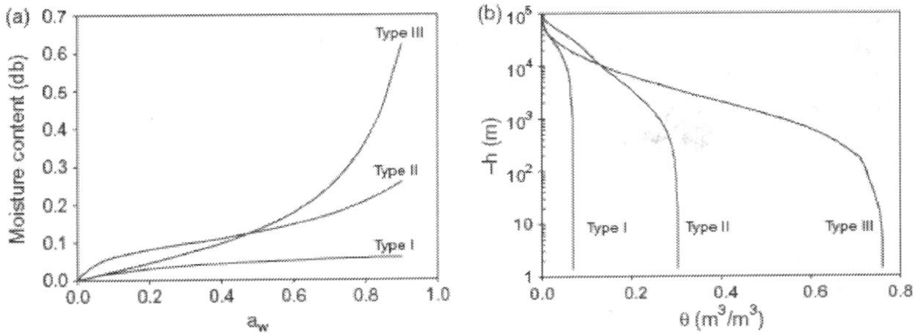

Fig. 5. Adsorption isotherm (left hand) and water retention curve (right hand) for tea (type I), wheat (type II) and apricot (type III) in which capillary pressure head values were obtained by using Kelvin equation. (from Rahman, 1995)

3.2.3 Gas flow due to pressure gradient

Gas flows such as water vapor and air flow due to differences in pressure gradient inside pores may be expressed by Darcy's law with the same equation previously discussed. We reported only the analogue of Eq (11) for gases:

$$n_g^{press} = -\rho_g \frac{k_g}{\mu_g} \frac{\partial P}{\partial s} \tag{17}$$

where n_g^{press} is the mass flux of gas, ρ_g is the density, μ_g is the viscosity, P is the total pressure in the gas phase.

3.2.4 Other mechanisms

Others mass transport mechanisms have been hypothesized in literature although the difficult to modeling vapor, gases, and liquid motion, which are also greatly affected by heat transfer, makes the literature very poor. Stephan diffusion refers to motion of vapor across a layer of stagnant air that is the case of convective drying. Knudsen diffusion occurs when the mean-free path of the molecules is long in comparison with the pore diameter. A combined condensation and evaporation phenomena may promote water flow. In fact, in a closed pore, water may be vaporized by heating at its end and it may condense at the opposite end. So, liquid is transport in the opposite direction of vapor flow along the wall of the pore. Furthermore, the importance of the cross influence of mass and heat transfer should be considered in food science but, in general, it is completely drop. Irreversible thermodynamic theory studies these cross influence. In particular, when heat and mass transfers occur simultaneously, the temperature gradient may influence mass transfer (Soret effect) and the concentration gradient may influence heat transfer (Dufour effect) (Hallstrom et al., 2007).

3.3 Air drying

During air dehydration heat is transferred from surrounding air to the surface of vegetables by convection and inside it by conduction as predominant mechanisms (due to internal thermal gradient) and by convection (due to moisture migration) with less extent. At the same time, water evaporates from the surface of vegetable toward the surrounding hot air (which has a low humidity) and it moves inside vegetables toward its surface by liquid diffusion, vapor diffusion, capillary flow and viscous flow. A schematic representation of air drying is shown in figure 6. Even though it is recognized that all the above water transfer mechanisms may occurs simultaneously, the trend of drying processes are usually represented by a drying curve obtained plotting moisture content as a function of time (figures 7a) and the kinetics by plotting rate constant as a function of moisture content (Figure 7b). Traditional literature divides the drying curve into three regions. When the drying is at time zero the moisture concentration may be at the points A or A' respectively in the case in which food is at cold or hot temperature. As time as process proceeds, a region called "constant rate period" is met. Here, during heating water evaporates from the surface of food and the vapor moves away by convective air. At the same time, water molecules move from the core of vegetable to the surface replacing the water just evaporated. This phenomenon, usually considered as diffusive, maintains the water concentration at the surface and drying rate at constant levels. At point C, named "critical moisture content", the time necessary at the water molecules to reach the surface becomes significantly high due to the increment of the path length. In this condition the rate of diffusing water is lower than the evaporation rate leading to a progressive reduction of the overall drying rate. This behavior is represented from the region C-D and it is called as "falling rate period". Moreover due to this condition, the surface of food progressively dried until it becomes completely dehydrated (point D). Once this condition is reached the evaporation will continue from the internal regions of vegetable pieces.

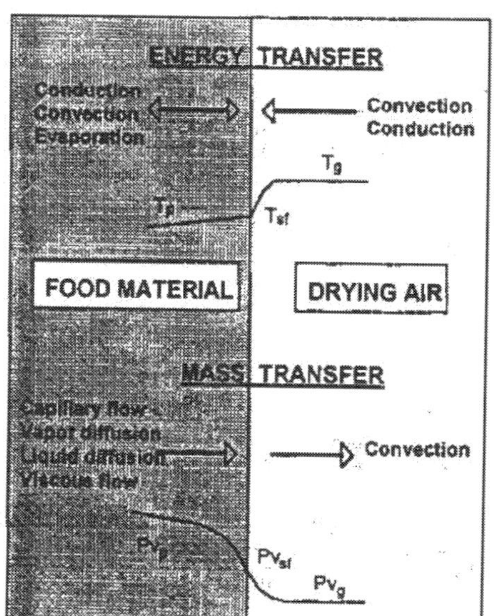

Fig. 6. Schematic representation of air drying technology (from Craspite et al., 2007)

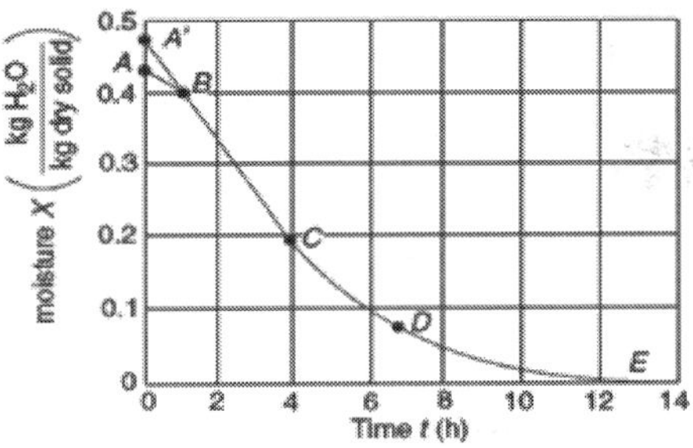

A

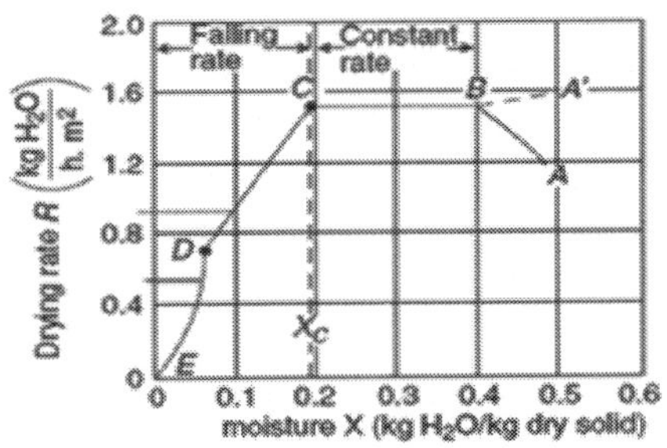

B

Fig. 7a, 7b. Typical trend of convective drying. A) moisture content as a function of time. B) drying rate as a function of moisture content. (Adapted from Okos et al., 2007)

Datta (2007) reported the temperature, moisture content and total pressure profiles across general biological food systems with high moisture content as in the case of several fresh vegetables during air drying (figure 8a, 8b and 8c). It is possible to observe that temperature increases slowly reaching a values of 55°C after 60 minute; moreover, it remains almost constant across the sample. Moisture content profile shows a maximum value (close to saturation) at time zero which begins to reduce after 20 minutes of heating. The trend across the sample did not show difference in the first 40 minutes of heating. Nevertheless, as heating proceeds a slightly decrease of the moisture content at the surface occurs. This is because the rate of internal diffusion of water is less and it becomes unable to replace the evaporated water. Pressure profile shows values close to atmospheric pressure leading to the absence of intense vapor formation.

From an engineering point of view the constant rate period is recognized as externally controlled and the falling rate period as internally controlled. External control means that the process is controlled by variables which are independent from the properties of fresh vegetables (external variables) such as air temperature, relative humidity, air flow. Instead, internal control means that the process is controlled from vegetable characteristics (internal variables) among which the size and shape, the collapse phenomenon during drying, chemical composition of vegetables and its three dimensional microstructure are the most important. So, it is worth noting that the falling rate period is specific for each systems and it

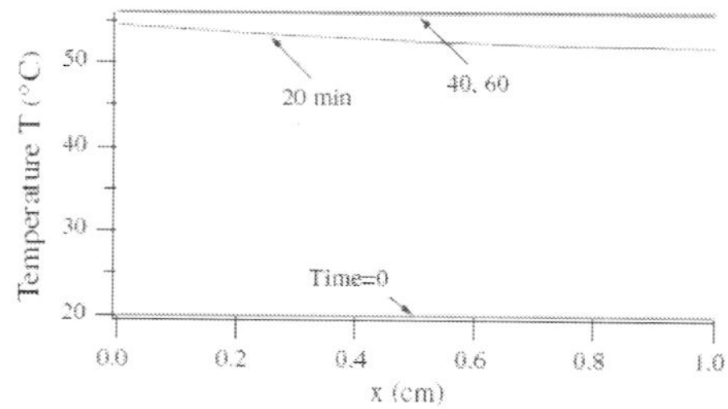

A

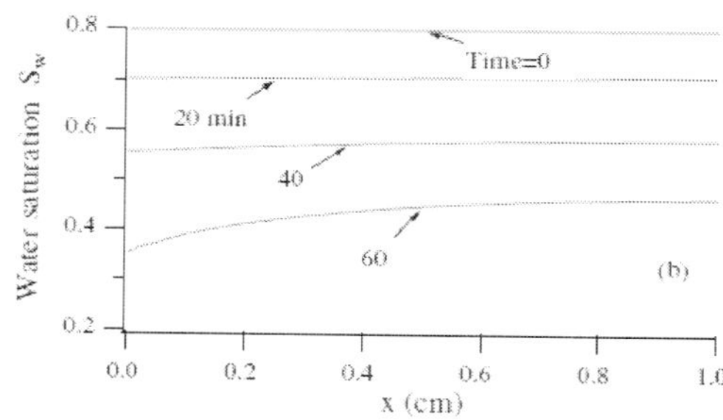

B

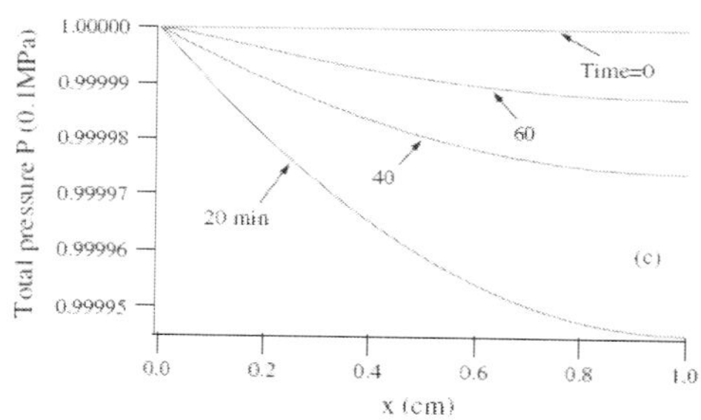

C

Fig. 8a, 8b, 8c. General temperature, water content and total pressure profiles of high moisture material submitted to convective drying (From Datta, 2007)

could significantly change among different vegetables. As reported from Okos et al. (2007) the importance of the external and internal mass transfer may be highlighted by using the concept of overall mass transfer coefficient:

$$1/K = 1/K_c + L/D_{eff} \tag{18}$$

where K is the overall mass transfer coefficient (m^2/s), Kc is the external controlled mass transfer (m^2/s), L is the characteristic dimension of the sample and D_{eff} is the effective diffusion coefficient (m^2/s).

The effects of external variables have been extensively studied (Lee et al., 1983; Mulet et al., 1989; Karathanos & Belessiotis, 1997; Krokida et al., 2003). For instance, Krokida et al. (2003) studied the effect of air temperature, air humidity and air velocity on the drying kinetics of several vegetables such as green, yellow and red peppers, pumpkin, green peas, carrots, tomatoes, corn, garlic, mushrooms, spinaches, onions, celery and leek. As expected, the authors reported that air temperature is directly correlated with drying rate and that is the most important external variables. This result is in accordance with several researchers (Senadeera et al., 2003; Leiva Diaz et al., 2007; Orikasa et al., 2008). For example, Senadeera et al. (2003) showed an increase of rate constant as temperature increased from 30°C to 40°C and 50°C during air dehydration of green beans, potatoes and peas. Orikasa et al (2008) studying the air drying of kiwifruit slices showed a significantly increase of rate constants as temperature increased from 40°C to 70°C. Krokida et al. (2003) studied the effects of both the flow and the humidity of air on drying rate of several vegetables showing that these variables were less important in comparison with temperature. In fact, air velocity (in a range of 1.5 and 2.6 m/s), which is considered important to limit the external resistance to the drying, was shown to be almost negligible. Thus the authors stated that the water diffusion toward surface was high and that the external resistance was not very important. Furthermore, they showed that the effect of air humidity was significant only when its value increased from 20% to 40%. Senadeera et al. (2003) studied the effect of shape of samples on drying rate of some vegetables. In particular, green beans with a length to diameter ratio (L:D) of 1:1, 2:1 and 3:1 and potatoes with an aspect ratio (A:R) of 1:1, 2:1 and 3:1 were used during drying in fluidized bed. Results showed that in both cases the rate constants decreased increasing L:D and A:R values as a consequence of an improvement of surface area per unit of volume.

The importance of the externally and internally controlled period during air drying is historically recognized but in the last years some new finding need to be considered. In general assuming that constant rate period is only externally controlled none water diffusion due to moisture gradient inside food should be detected. Analyzing the drying curve of onions, carrots, mushrooms and garlic Pabis (1999) found that the initial linear segment could be significantly extended; the author ascribed this behavior to an external control of the process. Fiorentini et al. (2008) studying the drying curve of a tomato pectic gels, showed that the falling rate period may be divided in two sub-regions among which the first (at high moisture content) appeared to be both internally and externally controlled and the second one was strictly controlled by internal diffusion. Giner (2009) studied the drying curves of a sucrose-added apple pectic gels in the first 90 minutes of the process, showing, in accordance with Pabis (1999), a linear trend. Nevertheless, the author showed that the experimental data were well fitted from a form of Fick's law valid for both internal and external controlled drying process. An accurate fitting was obtained by using a Biot number of 2 which states that the external resistance is twice of internal one. Furthermore, the author estimated the local moisture content along the thickness of the samples showing that a moisture gradient exists hence, this constant rate period cannot be exclusively governed by external variables.

3.4 Microwave drying

Microwaves (MW) are electromagnetic waves whit a frequency in the range of 300 MHz and 300 GHz. Among these only two are permitted for food application: 915 MHz for industrial application and 2450 MHz for microwave ovens. The use of microwaves is regulated by the maximum exposure or absorption of human working in a microwave environment and by the maximum leakage of the microwave equipment (Reiger & Schubert, 2005). MW dehydration of fruits and vegetables is based on the absorption of the electromagnetic waves from biological tissues and their conversion in thermal energy. As known, these factors are included in the relative permittivity of food (ε^*):

$$\varepsilon^* = \varepsilon' - j\varepsilon'' \tag{19}$$

where ε' is the dielectric constant which refers the ability of food to absorb microwave and ε'' is the dielectric loss factor which is a measure of the conversion of electromagnetic energy into a thermal one. These two constants are affected by frequency, temperature, and composition of food materials with particular reference to salt, fat and water content. The influence of these variables have been studied in details and several information may be find in the book of Schubert & Reiger (2005). Briefly, since microwave are absorbed from polar molecules of water, due to its high content and its homogeneous distribution, that is the most important internal variable affecting microwave processing of food. For these reasons, microwaves promote a volumetric heating that proceeds from the core of vegetables foods (with higher moisture content) to the surface (the region with less moisture content). In this way, when microwaves are used as drying method the mass transfer mechanism is extremely different from the convective dehydration. At first, it is during the falling rate period that microwave drying exhibit the most important advantage. As previously reported this phase of the process is usually considered as internally controlled by diffusion mechanism; so, the volumetric heating promotes a fast water diffusion inside the core of food (Erle, 2005). However, although the internal diffusion is very fast, a reduction of drying rate as a function of time, is observed. This is because ε' and ε'' values decrease during the process due to the reduction of water content. However, the main peculiar characteristic associated with MW drying is the formation of a high vapor pressure into the internal region of sample that push water toward the surface significantly increasing water transport. Ni et al. (1999) and Datta (2007) by using a multiphase porous media model, showed the characteristic profiles of temperature, water content and total pressure across a generalized food system during microwave heating (figures 9a, 9b and 9c). In the case of fruits and vegetables (high moisture porous material), the increase of temperature across all layer of samples is very fast reaching the boiling temperature (with a maximum of 107°C) in only three minutes. In particular, an heating rate of 0.69°C/s was estimated. After, temperature begins to reduce due to the decreasing of moisture content that lead a drop of dielectric properties of vegetable tissues; nevertheless, temperature on the surface is always lower than the internal part. Pressure profile is extremely different in comparison with convective drying. After three minutes, the intensive heating in the internal regions leads to the formation of a high vapor pressure gradient across the sample with its maximum value of 32 kPa (0.32 atm). Again, this is the cause of a fast water transport from the core of samples toward its surface. Obviously, total pressure at the surface remains at its initial value (atmospheric pressure) because there is a open structure in contact with surrounding air. Moisture content at time zero is homogeneous across the sample but after three minutes the generated vapor pressure push water promoting its accumulation at the surface of sample which becomes fully saturated (Figure 9c). As time proceeds (after 6 minutes), moisture content inside the sample reduces but at the surface it remain approximately constant. Different behavior is observed in the case of microwave heating of low moisture materials (Ni et al., 1999). In this case, the temperature across the sample increases slowly as a function of time due to the lower initial moisture content. Also, the heating rate was estimated to be 0.27C/s. Pressure gradient shows the same trend of temperature and its maximum values (about 1.20 kPa) is less than those estimated for high moisture materials. Again, the moisture profile shows a fairly reduction at the surface in the first 6 minutes but after at 7 and 8 minutes the prolonged heating produce an internal vapor pressure sufficient to promote the water motion toward the surface that consequently becomes more moist. On the other hand, the core of sample dehydrated progressively (Ni et al., 1999).

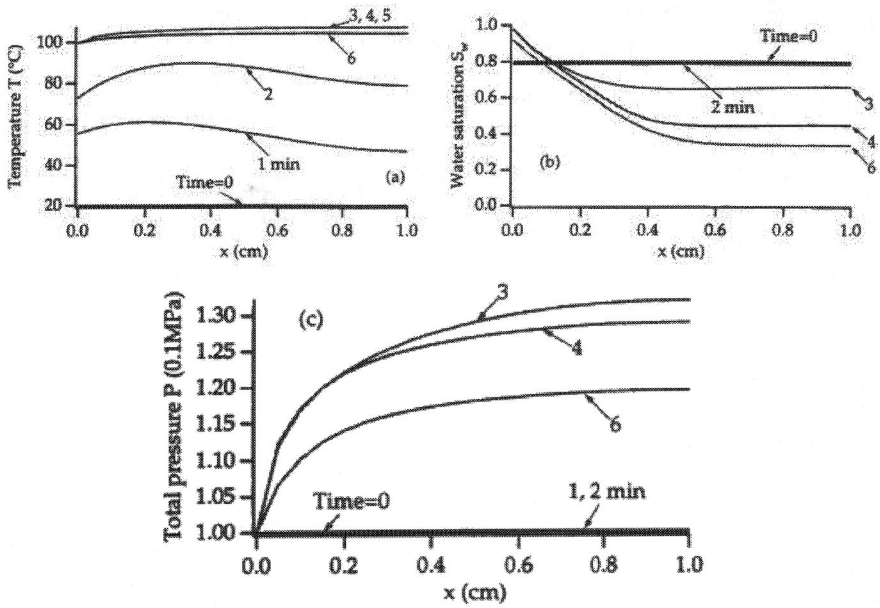

Fig. 9a, 9b and 9c. General profile of temperature, moisture content and total pressure across sample submitted to microwave drying (from Ni et al., 1999)

Microwave drying has been used to remove water from several vegetable such as herbs (Giese, 1992), potato (Bouraout et al., 1994), carrot (Prabhanjan et al., 1995), banana (Maskan, 2000), kiwifruit (Maskan, 2001). Since microwave heating is volumetric the number of process variables are reduced and the predominant are the microwave power, the dielectric properties of food, size and shape of samples. For instance Maskan et al. (2000) studied the effect of microwave power and the thickness of banana slices on drying rate. As expected, the author reported a significantly increase of drying rate as a function of energy power in a range of 350 W and 790 W. Also, the rate constants were 10-fold greater that the estimated value during air convective drying. However, when banana slices with a thickness of 4.3 mm, 7.3 mm and 14 mm were respectively used, the rate of the process directly increased. This result seems to be in disagreement with the increase of path length of water; nevertheless it was explained by taking into account that the volumetric heating may produce a vapor pressure gradient as greater is the size of the sample (Maskan et al., 2000).

3.5 Osmotic dehydration

Osmotic dehydration (OD) is widely applied to partial removal of water from fruits and vegetables. OD occurs when vegetables are immersed into a hypertonic solution leading to the formation of an osmotic pressure gradient able to remove water from vegetable tissues. However, since cell membranes are not completely semi permeable a countercurrent flow occurs: 1. water moves from vegetable tissues toward hypertonic solution; 2. osmotic agent flows inside vegetable. Also, some chemical compounds of vegetables such as pigments, vitamins, salts, etc., may be leak into osmotic solution promoting changes of nutritional and sensorial properties of food products. The scientific and industrial interest for osmotic dehydration is related to the possibility to drying fruits and vegetables at room temperature or by a low heating minimizing the heat damages on food and significantly reducing energy cost of the treatment. Moreover, OD may improve texture, pigments stability and color of dried products (Rault-Wack, 1994; Krokida et al., 2000). In this way, OD has been used as pretreatment for air drying, vacuum drying, freeze-drying, freezing, microwave drying, etc., with the aim to increase nutritional, sensorial and functional properties by maximizing the integrity of vegetable tissues (Torreggiani, 1993). In terms of water and solutes transport, osmotic dehydration is usually considered as a diffusive phenomenon in which the driving force is the osmotic pressure gradient between vegetables and osmotic solution. The predominant resistance for water and solutes flow is characterized from cell membranes, their mechanical properties and their changes during the process. Unfortunately, these changes are not commonly considered in literature and several papers are still based on a macroscopic point of view which implies that the dehydration occurs under a uniform moisture gradient and that diffusion coefficient is constant through the food (Figure 10a). Instead, this is not realistic assumption because several changes on cell structure have been reported in literature (Marcotte et al., 1991; Alzamora et al., 1997; Ferrando and Spies, 2001; Lewicki et al., 2005). For instance, Ferrando and Spiess (2001) observed a degradation of middle lamella for onion tissue and a reduction of cross section area of 40%-60%; moreover, in the case of strawberries the authors showed a typical plasmolysis with the detachment of plasmalemma from the cell wall. Lewicki et al., (2005) showed that osmotic dehydration causes changes in size and shape of apple cells. More generally, Barat et al., (2001) reported that cells surrounding intercellular spaces shrunk, the solid matrix is deformed and the porosity increases. On the basis of these results, it must be assumed that the resistance of cell membranes significantly change during drying, hence different constant diffusion coefficients across the vegetables should be expected. Rastogi et al. (2000) proposed a mechanism to describe the behavior of biological tissue submitted to osmotic dehydration taking into account the structure changes of cell membranes (figure 10b). The mechanism is based on the existence of a dehydration front (Δx) that during drying moves toward the centre of the sample. Dehydration front is considered a region in which the cell membranes is damaged and shrunk as a results of a critical value of osmotic pressure gradient. So, diffusion coefficient value at dehydration front (D_2) is the greater across the sample. In the region close to the surface (in contact with osmotic solution) cell membranes are damaged and shrunk and water molecules flow toward hypertonic solution with a diffusion coefficient $D_3<D_2$. This is because the osmotic pressure gradient is less than in dehydration front. Moreover, the internal region shows cells at their natural turgor pressure hence, with a diffusion coefficient (D_1) much lower than D_2 and D_3. Moreover,

Rastogi et al (2000) proposed to use of a cell disintegration index (*Zp*) (also called permeabilization index) to measure the relative reduction of intact cells across the sample. Figure 10b shows the profiles of moisture content (M/M_0) and *Zp* as a function of the relative distance from the hypertonic solution. It is possible to observe that in the dehydration front cell disintegration index sharply increased and moisture content reduces significantly because of the cells are dehydrating.
However, some researchers showed that this model does not completely explain the water mass transfer mechanism. For instance, Derossi et al. (2008), studying the osmotic dehydration of apple cylinders by MRI techniques showed a reduction of spin-spin relaxation time (T2) in all layers of samples leading to the idea that drying occurs in all layer of samples (figure 11).

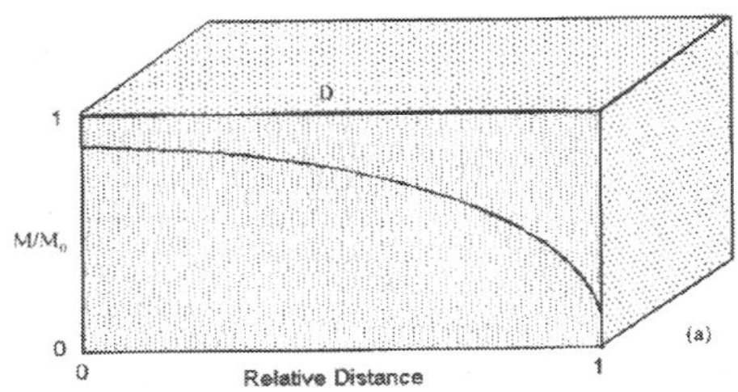

A)

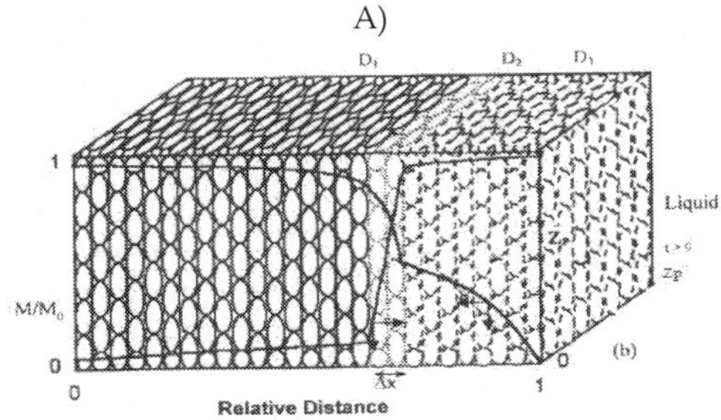

B)

Fig. 10a, 10b. Schematical representation of osmotic dehydration in homogeneous (non biological) material (a) and for biological system (b). (from Rastogi et al., 2000)

It should be noted that several factors may affect the rate of the osmosis among which external variables such as concentration of osmotic solution, type of osmotic agents, temperature, agitation rate, product solution mass ratio as well as the internal variables such as dimension and shape of vegetables, structure properties of vegetables (porosity, permeability of cell membrane, ripening index, etc) are the most important. A wide number of scientific papers described in considerable details the influence of these variables on the kinetics of OD (Lerici et al., 1985; Torreggiani, 1993; Raoult-Wack, 1994; Rastogi and Raghavarao, 1994; Rastogi and Raghavarao, 1995; Rastogi et al., 1997; Mavroudis et al., 1998; Spiazzi and Mascheroni, 1997; Chiralt and Talens, 2005; Sacchetti et al., 2001; Rastogi et al.,

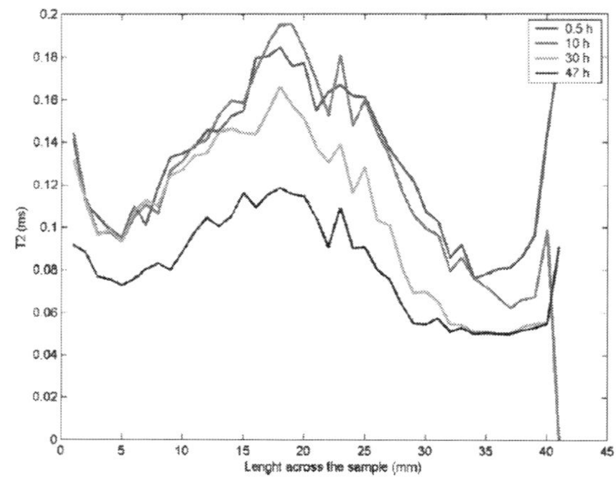

Fig. 11. T2 profiles along the cross of apple cylinders submitted to osmotic dehydration in sucrose solution (from Derossi et al., 2008)

2002; Azoubel et al., 2004; Tsamo et al., 2005). Obviously, temperature and concentration of osmotic solution are directly correlated with osmotic dehydration rate. However, by using a solution with high concentration, water loss is favored in comparison with solid gain; instead, the use of osmotic solution at low concentration favors the impregnation at the expense of dehydration. Due to the dilution effect that proceed during the process, osmotic solution with a product:solution mass ratio of 1:20 is considered sufficient to maintain a uniform driving force during drying. Nevertheless, it should be considered that this condition is unrealistic in practical application due to the lower production output and high costs. With an industrial point of view a product:solution mass ration of 1:4 or 1:5 may be used. The type of osmotic agents is an important process variables. Traditionally, hypertonic solutions are prepared by dissolving different type of sugars in tap water. In these cases the use of compounds with high molecular weight promotes the dehydration rather than the impregnation; on the other hand, sugars with low molecular weights are used from candy industries (Rastogi et al., 2002). However, in the last years the use of complex solutions prepared with water, sugar, sodium chloride, etc., received much attention but it appeared to be more complex in terms of dehydration kinetics (Sacchetti et al., 2001; Tsamo et al., 2005; Derossi et al., 2010). For instance, in disagreement with the higher impregnation effect caused by solutes with low molecular weight, Tsamo et al. (2005) reported that water loss of tomato and onion samples submitted to OD in salt solution was higher than in sucrose solution. This behavior was explained taking into account that the small molecules of NaCl may through cell membranes producing a double source of pressure gradient: at cytoplasm and vacuole level. In this way, more water could be removed from cells. Moreover, the authors showed that the maximum dehydration was obtained when a mixed solution was used probably because the increase of concentration gradient. Nevertheless, when the rate of OD in sucrose, sodium chloride and mixed solution were studied, different behaviors were observed (Sacchetti et al., 2001). The authors showed that at low concentration of sucrose, the initial rate constant of OD reduced as NaCl concentration increased. This antagonist effect was explained by a reduction of the cell membrane permeability. Also, in accordance with other researchers (Collignan and Raoult-Wack, 1994) it was hypothesized that sodium chloride due to its small molecular dimension may penetrate into the cells leading to a reduction of the osmotic pressure gradient. The agitation rate of the solution is an important variable able to modify the mass exchanges at the surface of vegetable tissues. Mavroudis et al. (1998) studying the effect of agitation level on osmotic dehydration rate of apple cylinders showed that water loss increased rising, the Reynolds number; in fact a higher dehydration in the turbulent flow region rather than in the laminar flow region, was observed. So, an external obstacle to the water transfer was confirmed. Instead, solid gain was not influenced from the agitation rate.

As previously reported, diffusion coefficients of water and solutes are greatly hindered from the resistance of cell membranes which is related to the Zp value. Under these consideration, a technological approach to improve mass transfer is the imposed increase of cell's permeability. This is possible by performing some treatments such as high hydrostatic pressure (Rastogi et al., 1999), pulsed electric field (Omowayne et al., 2003; Amami et al., 2007), ultrasound (Simal et al., 1998; Fernandes et al., 2008), centrifugal force before osmotic dehydration. As known, high hydrostatic pressure damage the cell walls increasing their permeability hence, the mass transfer. It was shown that water diffusion coefficients for pineapple samples submitted to osmotic dehydration increased of 4-fold by using a pressure range between 100 and 800 Mpa (Rastogi et al., 1999). Also, Rastogi et al. (2000) showed that hydrostatic pressure and osmotic dehydration had a synergistic effect on the enhancement of water removal from potato samples. Pulsed electric field has been shown to have a permeabilization effect on cells due to local structure damage and the breakdown of cell membranes. In general, it is widely accepted that if an electrical potential is applied between the both side on cell membrane the related motion of the charges along the electric field lines produces a deformation of cell membranes. If the electric potential (E) is greater than a critical value (E_{crit}), which depends from the properties of membranes, rupture and breakdown occurs (Lebovka et al., 2001; Amami et al., 2007). In general an electric potential of 1 V is sufficient to produce the permeabilization phenomenon. Amami et al. (2007) studying the effects of PEF and centrifugal force on osmotic dehydration, showed that the application of PEF improved water loss and solid gain. The same results were reported from Ade-Omowaye et al., (2003). Ultrasounds are characterized from a series of compression and rarefaction waves. The most important effect of these waves is the cavitation whit a formation of bubbles of vapor or gases. Under certain conditions this bubbles becomes unstable and collapse, generating mechanical forces that damage the its increasing cell permeability. Simal et al (1998) applying ultrasound of 40 KHz showed an increase of water loss of 14%-17%. Other results on the use of these pretreatments to increase mass transfer of fruits and vegetables during OD are available on Bohuon et al. (1998), Carlcel et al., (2002); Taiwo et al., (2003) Fernandes et al., (2008), Stojanovic & Silva (2007).

In the first section of this paragraph, although the importance of internal factor such as permeability of cell membranes variables are reported, OD has been exclusively considered as diffusive phenomenon and none different mass transfer mechanisms was taken into account. Nevertheless studies at microscopic level detected an effective diffusion coefficient of 2-fold greater than the D_e value obtained or estimated taking into account the only diffusion (by experiment at macroscopic level). These results stated the existence of an additional mass transport mechanism. This is the hydrodynamic mechanism (HDM) which is a result of a capillary flow caused by a pressure gradient (externally imposed or generated) as a consequence of the changes in intercellular spaces during OD (Fito, 1994;

Fito & Pastor, 1994; Chiralt & Fito, 2003). In figure 12, HDM is schematically subdivided in five steps. Before the immersion of vegetables inside osmotic solution, the pressure inside pores is equal to atmospheric pressure (step 1, t_0). After the immersion, osmotic solution partially penetrates inside capillaries and it compresses internal gas with a result of an increase of internal pressure (step 2, t_1). Nevertheless, drying process begins due to osmotic pressure gradient, cells leak water and shrink (step 3, t_2). In this condition, the volume of capillary pores increases leading to a decrease of internal pressure that becomes less that external one. In this way a pressure gradient is created and the suction of osmotic solution is promoted (step 4, t_3). This phenomenon will continue until the complete impregnation occurs or the equilibrium between external and internal pressure is reached (step 5, t_n) (Chiralt & Fito, 2003). This mechanism based on the involving of the internal volume of vegetable structure was exploited to increase the rate of osmotic dehydration by externally imposing an high pressure gradient. Vacuum osmotic dehydration (VOD) is a treatment based on the filling of capillary pores of vegetables that increases the liquid-solid surface contact area improving mass transfer. VOD is performed by applying two steps: the application of a vacuum pressure for a short period called vacuum period (t_1); the restoration of atmospheric pressure and its maintaining for a relaxation period (t_2). In the first period, by the application of a pressure usually in a range of 50 mbar and 400 mbar, the removal of native liquids and gases from capillary pores as well as gas expansion, are promoted. In the second step the restoration of atmospheric pressure produces the suction of the osmotic solution inside the pores and the deformation-relaxation phenomenon which is able to modify the dimension of pores (Fito, 1994; Zhao & Xie, 2004; Atares et al., 2008). A wide bibliography concerning the main process variables of VOD, the mathematical modeling and the effect on quality of vegetables is available (Fito, 1994; Fito & Chiralt, 1995; Salvatori et al., 1998; Fito et al., 2001; Gras et al., 2002; Mujica-Paz et al, 2003; Zhao & Xie, 2004; Atares et al., 2008). Moreover, if the relaxation period is extended like those of a traditional osmotic dehydration, the process is defined pulsed vacuum osmotic dehydration (PVOD). In this case a short vacuum pulse is applied at the initial stage of osmotic treatment. It is worth nothing that the interest in the exploiting of capillary pores to impregnate food by different substances such as ingredients, antioxidants, pigments, organic acids, nutritional or functional compounds etc. has exponentially increased in the last 10 years (Zhao & Xie, 2004; Betoret et al., 2003; Derossi et al., 2010)

3.6 Infrared drying

In the last years, the interest on the use of infrared as method to dehydrate thin food is greatly increased. In general the supply of heat by infrared radiation has given good results for thin vegetables with a high surface area exposed at radiation. However, the transmissivity characteristics of food depends on their chemical composition: FAR-infrared radiation seems to be more effective for thin layer and NEAR-infrared (NIR) should give better results for thick material. The use of infrared has several advantages in comparison with traditional convective methods. Novak and Lewicki (2004) summarized them as follow: heat efficiency, high diffusion coefficient, low drying time, the process is performed at room temperature because of air is transparent to infrared; the equipment may be compact. In terms of mass transfer mechanism during infrared drying of food, few information are available. Infrared are absorbed on the surface of a moist food and the hot point is at a distance from the surface depending from the extinction coefficient. As small is

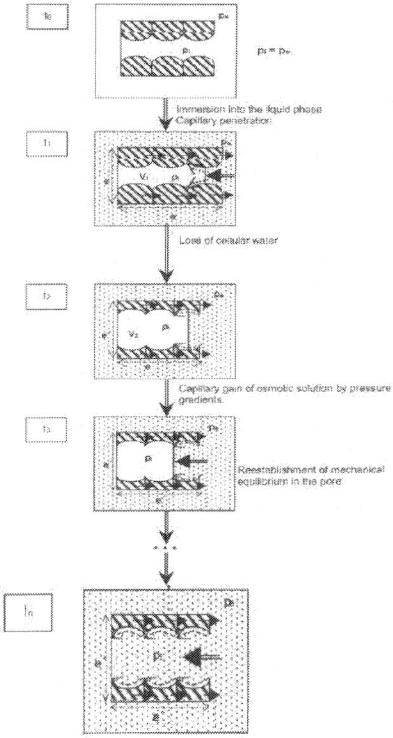

Fig. 12. Schematic representation of hydrodynamic mechanisms (HDM) (adapted from Chiralt & Fito, 2003)

this coefficient as more is the distance between surface and the layer with the maximum temperature. From this region, heat moves simultaneously toward surface and the core of sample by conduction promoting the evaporation of water (at the surface) and increasing diffusion coefficient of water (in internal regions). So, water and heat move in the same direction above the absorbent layer and in countercurrent below it. Nowak and Lewicki (2004) studied the effect of some process variables such as air velocity and the distance of the emitters from the samples on the kinetics of infrared drying of apples slices. The results showed that emitter distance is directly related to drying rate. The flux of water was 1.25 (g/m^2s) , 1.10 (g/m^2s) and 1.0 (g/m^2s) when a distance of 10 cm, 20 cm and 30 cm were used with an air velocity of 0.5 m/s. Moreover, it is worth noting that drying rate resulted inversely correlated with air velocity. This behavior was explained taking into account that air is transparent at IR but, in turn, it may be heated from the surface of food. During drying the surrounding air is heated by the surface of food leading to its cooling. Obviously as greater is air velocity as faster is the cooling of food surface hence, water evaporation is reduced. Other examples on the use of infrared drying are available by Lampinen et al (1991); Sakai & Hanzawa (1994), Ratti & Mujumdar (1995), Nowak & Lewicki (1998).

3.7 Ultrasonic drying

As previously reported ultrasonic waves are characterized from a subsequent series of compression and expansion. This forces, crossing vegetable tissue, may increase their permeability creating microscopic channels in which water may easily flow. Moreover, the cavitation promotes the formation of unstable gas bubbles which may collapse reducing the mass transfer resistance of cell membranes (Mason et al., 2005). Even if wide evidence on the mechanism involved during ultrasonic wave not exists and although this is an emerging technology not still used at industrial scale, some researchers reported interesting results. Fluente-Blanco et al (2006) studied the ultrasonic dehydration applied to carrot cylinder by using a laboratory system equipped with a trasductor, a vacuum chamber and a forced air-generator. The vacuum chamber performs the suction of water from the samples and the fan was used to increase the removal of internal moisture. The authors showed a significantly reduction of moisture content in comparison with traditional convective drying. Also, it was shown a direct influence of ultrasonic power on the weight reduction of carrots.

3.8 Hybrid technologies for vegetable dehydration

The analysis of the above dehydration methods allows to highlight their extremely different characteristics such as mass transfer mechanism involved, heating rate, pressure gradient inside vegetable tissues, effect of cell membranes, etc. So, each of these shows to have some advantages and some disadvantages. For instance, the most important problem related to the use of hot air dehydration is the low water diffusivity during the falling rate period. Osmotic dehydration may be performed at room temperature drastically reducing heat damage but it is characterized from a low drying rate. So, when the use of a single drying techniques is not the best way to obtain vegetable food with high quality, the possibility to apply hybrid technologies is very interesting. Microwave-air assisted drying is probably the most studied hybrid dehydration techniques. Since air drying is low efficient during the falling rate period due to internal resistance, the use of microwave heating may significantly increase the rate of process. Microwave-vacuum dehydration is used to obtain a fast removal of water from the food structure. This is the result of an external negative pressure which is added at the internal pressure gradient promoted from the vapor generation. Osmotic dehydration may be used as pre-treatment to air drying, microwave drying, microwave-vacuum drying, etc. For instance, Erle & Schubert (2001) studied the application of a combined osmotic and microwave-vacuum dehydration of apples and strawberries. The use of osmosis during the first stage of dehydration allowed to remove water reducing the energy consumption of the treatment. In the last stage of drying curve (when osmosis is less effective) microwave-vacuum allowed to remove water minimizing heat damage. In fact the authors showed a vitamin C retention of 60% and the volume of fresh sample was preserved of 60% for apple and of 50% for strawberries. Other example of hybrid technologies are: ultrasonic-air drying, radio-frequency, infrared-microwave drying; microwave-vacuum drying, infrared-heat pump drying, infrared convective drying. A more deep description of these technologies was reported from Chua & Chou (2005).

4. Mathematical modeling of water and solute transfer during vegetable dehydration

Mathematical modeling is a fundamental research field in food science. The term model is difficult to precisely define. An intuitive definition may be expressed as follows: *a model is a structure similar to the original* (Pabis, 2007). A model is an abstract system that may be obtained also from the results of experimental process performed on the existing original. If the model is an empirical equation, it may give only information on the existing original under the condition in which the experiments have been carried out. Empirical model is obtained using as starting point a phenomenon experimentally observed. From it, the model is extracted and proposed with the aim to obtain information on the principal characteristics of the existing original. In other word, empirical model states the existence of a new empirical regularity between independent and dependent variables but it cannot provides any scientific explanation. Instead, if the model is theoretical it may provide information on both the action of the original and its structure which promoted that action (Pabis, 2007). Theoretical model is obtained deductively from the true statements of the theory; it results logically from the theory. Often, empirical models are also called *phenomenological* and theoretical models as named *mechanistic*.

In terms of vegetable dehydration a wide number of equations have been proposed and used to model mass transfer during several dehydration techniques. Instead, less are the theoretical models available. Here, the most important are going to discuss taking into account both traditional and innovative equations.

4.1 Empirical models

As known, the most important empirical models for mass transfer during drying processes of food are the Fick's laws. As previously reported, it must be emphasized that Fick's laws are diffusion equations. So, when these models are used, the main assumption is that water and solutes diffuse inside vegetable tissues on the basis of the following equation:

$$F = -D\frac{\partial C}{\partial X} \tag{20}$$

where F is the rate of transfer per unit area of section, C the concentration of diffusing substance, x the space coordinate, D is the diffusion coefficient. From the equation (20) it was obtained the fundamental differential equation for diffusion:

$$\frac{\partial C}{\partial t} = D\frac{\partial^2 C}{\partial x^2} \tag{21}$$

The equations (20) and (21) are usually referred as the first and second Fick's laws. Solutions of the second Fick's law are reported for simple geometries such as infinite flat plate, rectangular parallelepiped, infinite cylinder, finite cylinder and sphere (Crank, 1975). The solutions for infinite flat plates and spheres were previously reported in Eq (7) and (8). Instead, Eq. (22), (23) and (24) report the solution of second Fick's law for rectangular parallelepiped, infinite and finite cylinder, respectively (Crank, 1975; Rastogi et al., 2002; Azoubel et al., 2004)

$$MR = \frac{M - M_e}{M_0 - M_e} = \sum_{n=1}^{\infty} C_n \exp\left[-D_{ew} t q_n \left(\frac{1}{a^2} + \frac{1}{b^2} + \frac{1}{c^2}\right)\right] \tag{22}$$

$$MR = \frac{M_t}{M_e} = 1 - \sum_{n=1}^{\infty} \frac{4}{(a\alpha_n)^2} \exp\left[-F_0 (a\alpha_n)^2\right] \tag{23}$$

$$MR = \frac{M - M_e}{M_0 - M_e} = \sum_{n=1}^{\infty} C_{pn} C_{cn} \exp\left[- D_{ew} t \left(\frac{q_{cn}^2}{A^2} \right) \right] \tag{24}$$

where for Eq (22): C_n is equal to $2\alpha(1+\alpha)/(1+\alpha+\alpha^2 q_n^2)$, where q_n values are the non zero positive roots of the equation $\tan q_n = -\alpha q_n$; a, b and c are the sides of the parallelepiped.
For Eq (23): $a\alpha_n$'s are the roots of the equations $J_0(a\alpha_n) = 0$; F_0 is the Fourier number that is given by $D_e t/a^2$, where D_e is diffusion coefficient.
For Eq (24): C_{pn} is equal to C_n; C_{cn} is equal to $4\alpha\ (1+\alpha)/(4+4\alpha+\alpha^2 q_{cn}^2)$, where q_{pn} are the roots of the equation $\alpha q_{cn} J_0(q_{cn}) + 2J_1(q_{cn}) = 0$. The $J_0(q_{cn})$ and $J_1(q_{cn})$ are given by the roots of Bessel function of zero and first order respectively. Furthermore, $1/A^2 = 1/r^2\ [1+(r/l)^2(q_{pn}/q_{cn})^2]$.
In general when the Fourier number is greater than 0.1 only the first term of the equation is significant and other terms may be neglected (Rastogi et al., 2002). Moreover, some author simplified the second Fick's law by using the following (Torregiani et al., 1986; Giangiacomo et al., 1987; Mujaffar & Sankat, 2005):

$$\frac{X_t}{X_\infty} = K\sqrt{t} \tag{25}$$

Second Fick's law has been widely used to study the kinetics of several drying technologies also applied to different fruits and vegetables Rastogi & Niranjan (1998), Rastogi et al. (1999), Senadeera et al., (2003) Azoubel & Murr (2004), Babalis & Belessiotis (2004); Margaris & Chiaus (2007); Orikasa et al. (2008), Giner (2009). In all cases a good accuracy between experimental data and the fits was obtained. However, as previously reported it should be considered that several assumptions, often unrealistic for vegetable dehydration, are assumed into Fick's laws. Some of these were previously discussed in this chapter (paragraph 3.2.1) but others may be find from Ertekin & Sutannoglu (2000) and Saguy et al (2005). Nevertheless, it is worth nothing that no doubt exist of the capacity of Fick's laws to well fit experimental drying curves hence, they are very useful if our aim is to study dehydration kinetics in different operative conditions or to predict the time length of the process necessary to reach a desired moisture content into the product. Maybe the most simple empirical model was born on the observation that drying curves has, in general, an exponential trend. In this way, some researchers (Orikasa et al., 2008; Senadeera et al. 2003) modeled experimental data of moisture content as a function of time by a first order kinetics:

$$MR = \frac{M - M_e}{M_0 - M_e} = \exp(-Kt) \tag{26}$$

where MR is the moisture ratio, K is kinetic constant and t is time. Nevertheless, as reported from Senadeera et al. (2003), Eq (26) tends to over-predict the first region and under-predict the last region of drying curve. To avoid this problem an exponent n was introduce in the equation leading to the Page's model which is commonly used for vegetable dehydration:

$$MR = \frac{M - M_e}{M_0 - M_e} = \exp\left(-Kt^n\right) \tag{27}$$

where MR is the moisture ratio, K is the drying constant (min^{-1}), t (min) is time, n is a dimensionless exponent and the subscript 0 and e represent the time zero and the equilibrium. Another empirical equation, widely used to model vegetable dehydration process, is the Peleg's model:

$$MC_t = MC_0 \pm \frac{t}{k_1 + k_2 t} \quad (28)$$

where *MC* is the moisture content, *t* is time, k_1 and k_2 are kinetic constants. Moreover, $1/k_1$ and $1/k_2$ represent respectively the initial rate mass transfer (at t=0) and the value at equilibrium (at t=∞).

Page's model has been used to fit experimental results of several dehydration methods such as microwave drying (Ozkan et al., 2007), hot air dehydration (Sharma & Prasad, 2001; Soysal, 2004; Margaris & Ghiaus, 2007), microwave/air dehydration (Maskan, 2000), solar drying of apricots (Togrul & Pehlivan, 2002) etc., showing, in all cases, a high accuracy. In the same way, Peleg's model well fitted dehydration data of several vegetables such as papaya (Palou et al., 1994), apples (Sacchetti et al., 2001), cherry tomato (Azoubdel & Murr, 2004), chesnut (Moreira et al., 2007), etc.

Again, Azuara et al. (1992) starting from second Fick's law obtained the following equation to model water loss and solid gain during osmotic dehydration:

$$ML_\theta = \frac{S_1\theta(ML_\infty)}{1+S_1\theta} = \frac{\theta(ML_\infty)}{\frac{1}{S_1}+\theta} \quad (29)$$

where ML_θ is the moisture loss fraction at any time, S_1 is constant related to the rate of water diffusion, θ is the time and ML_∞ is the moisture loss fraction at equilibrium. Similar equation may be written to model solid gain:

$$SG_\theta = \frac{S_2\theta(SG_\infty)}{1+S_2\theta} = \frac{\theta(SG_\infty)}{\frac{1}{S_2}+\theta} \quad (30)$$

where SGL_θ is the solid gain fraction at any time, S_2 is constant related to the rate of solid diffusion, θ is the time and SG_∞ is the solid gain fraction at equilibrium. This model was shown to be accurate in the modeling of osmotic dehydration of several vegetables such as apples (Erketein & Sultanoglu, 2000) and onion slices (Sutar & Gupta. 2007). Since the lack of space in this chapter a list of others empirical models commonly used to model vegetable drying curves is reported in table 3.

4.2 Theoretical models

Mechanistic models for drying process are difficult to find in literature. In general, some theoretical models have been developed using as starting point the biological theory related to the cell membrane changes as in the case of osmotic dehydration, the volume changes during drying or on the basis of porous media approach.

Pabis (2007) developed a non linear theoretical model of the kinetics of convective drying of cut vegetables that takes into account the shrinkage of vegetable tissue:

$$M_t = M_0\left[\frac{1}{1-b}\left(1-\frac{1-b}{NM_0}k_0 t\right)^N - \frac{b}{1-b}\right] \quad (31)$$

where M_t and M_0 are the moisture content at each time and at time zero, N is equal to 3n/3n-2, *b* is the shrinkage coefficient. This model was used in different operative conditions showing a good capability to fit experimental moisture content values in the first region of drying curves (Pabis, 1999; Sojak, 1999; Pabis, 2007).

Empirical equation	Name	Field of application	References
$MR = a\exp(-K_t)$	Handerson and Pabis	General application	Chhinman, 1984
$t = a\ln(MR) + (\ln(MR))^2$	Thomson	General application	Paulsen & Thomson (1973)
$MR = \exp(-(kt)^{n)}$	Modified Page's model	General application	White et al. (1981)
$MR = \exp\left[-k\left(\frac{t}{L^2}\right)^n\right]$	Modified page equation II	Air drying, general application	Diamante & Murno (1991)
$\frac{WL_\infty - WL}{WL_\infty} = C_{ce}\exp^{-k_{ce}t} + C_{de}\exp^{-k_{de}t}$	Amami	Osmotic dehydration	Amami et al (2007)
$MR = a\exp(-kt) + c$	Logaritmic	Air drying, solar drying	Rahman et al, (1998)
$WL = [WL]_\downarrow e(1 - \exp(-K_\downarrow WLt)$	Panagiotou	Osmotic dehydration	Panagiotou et al. (1999)
$M_t = M_e + (M_o - M_e)\exp(-kt)$	Mulet	General application	Pabis (2007)

Table 3. Empirical equation generally used to model vegetable drying processes

Based on the biology, Toupin et al (1989) developed a mathematical model able to describe intercellular transport and the thermodynamic irreversible processes for transmembrane transport. Although the model is recognized as valid, it requires a wide number of physical and biological parameters that, often, are difficult to find in literature. Starting to this model, Spiazzi & Mascheroni (1997) developed a model composed by a system of ordinary differential equations that take into account the main mass transfer ways in cellular tissues, product shape, size and variation during drying as well as the chemical composition of the product, of the osmotic solution and their mass ratio.

In the last years, some theoretical models have been studied on the basis of the increasing interest of porous media approach. Maybe, the most deep work on mass transfer is by Whitaker (1997) who, starting from consideration at microscopic level as well as conservation equation for heat and mass for each phase (liquid, solid and gas plus phases), obtained representative information at macroscopic level. However, since the complexity of the theoretical modeling of fruits and vegetables submitted to different drying technologies and also the difficulty to experimentally validate these models, these equations are still far to be commonly used. Details of two mechanistic models such as distributed evaporation formulation and sharp interface formulation may be found from Ni et al. (1999), Farid (2002), Datta (2007).

5. Diffusion in inhomogeneous structures

The classical mathematical models of diffusion, usually applied to food systems, strongly depend on the hypotheses of homogeneity of the medium where the diffusion process occurs. However, food matter is far from being homogenous: at every scale at which it is observed, it presents relevant irregularities, making the description in terms of continuous space models essentially unrealistic.

The mathematical study of the most typical inhomogeneous structures occurring in nature is rather recent. Even if the deviation of the geometry of nature from analytical models was already pointed out by Perrin at the beginning of 20th century (Perrin, 1913), only in 1975 Mandelbrot introduced the concept of fractal to define a general class of geometrical patterns suitable to describe natural shapes (Mandelbrot, 1975).

"Clouds are not spheres, - Mandelbrot writes - mountains are not cones, coastlines are not circles, and bark is not smooth, nor does lightning travel in a straight line."(Mandelbrot, 1982).

According to Mandelbrot, a fractal is "a rough or fragmented geometric shape that can be split into parts, each of which is (at least approximately) a reduced-size copy of the whole" (Mandelbrot, 1982). This defining property, called self-similarity, is often summarized saying that in a fractal the part is similar to the whole.

Some examples of fractal shapes in vegetables are shown in Figure 13, 14, 15 an 16.

Fig. 13. Romanesco Broccoli (Brassica Oleracea Botrytis)

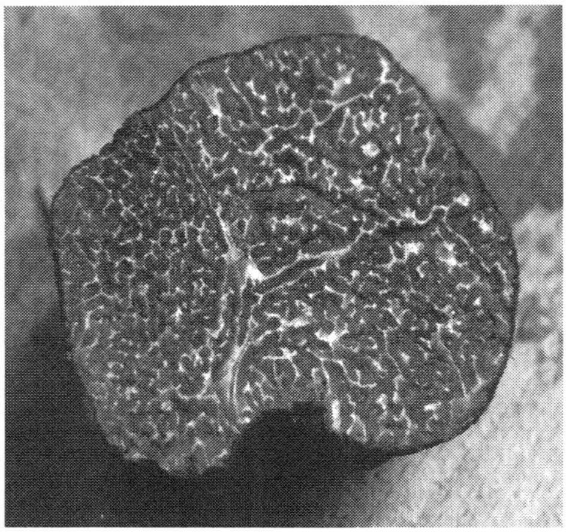

Fig. 14. Section of a truffle

An important consequence of self-similarity is the absence, in a fractal, of a characteristic length scale and the presence of details of any size. However, this is true only for ideal mathematical fractals. Real objects have a finite size and are usually constituted of elementary parts having a characteristic size. Therefore, they exhibit fractal behavior only in a limited range of length, called the *fractal range* and defined by a minimal and a maximal length, l_{min} and l_{max}.

Obviously, every fractal structure has its specific features and we need to classify all possible fractals in order to study their properties. The first classification, introduced by Mandelbrot, is based on a parameter called the *fractal dimension*, which is a generalization of the usual concept of dimensionality of a system.

There exist several possible definitions of fractal dimension, differing only in some specific case of purely theoretical interest. The most useful definition for practical application is the following:

Fig. 15. Mushrooms (Gyromitra Esculenta and Morhcella esculenta)

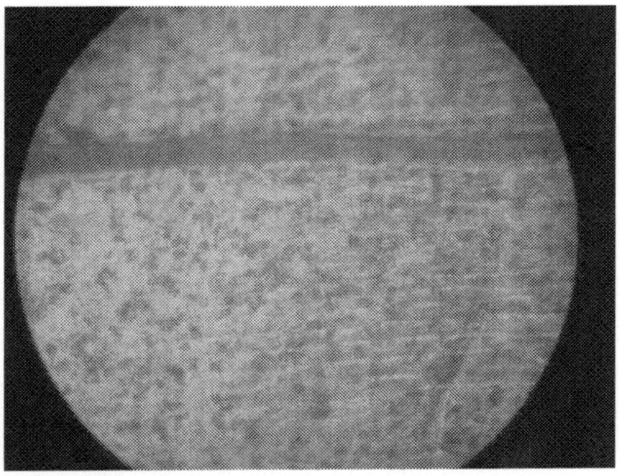

Fig. 16. Microscope image of a leaf

Take a point O on the fractal and a series of spheres with varying radius r and with centre in O: on a fractal, number of elementary building blocks (or, equivalently in most cases, the mass) of the part of the fractal contained in each sphere depends on r according to the law:

$$N(r) \propto r^{d_f} \tag{32}$$

where d_f is a real number less or equal to the dimension of the space where the fractal is embedded. Such a number is defined to be the *fractal dimension* of the fractal, and it does not depend on the choice of O. Notice that, for regular (non fractal) shapes, it coincides with the usual space dimension (e.g. it is 1 for a straight line, 2 for a plane figure, 3 for a solid figure). In Figures 17, 18, 19 and 20 we show some examples of mathematically generated fractals, together with their fractal dimension.

Fig. 17. Sierpinski Gasket (d_f=1.5850, $\tilde{d}$=1.36521)

Fig. 18. DLA cluster (d_f=1.71, $\tilde{d}$ =1.2)

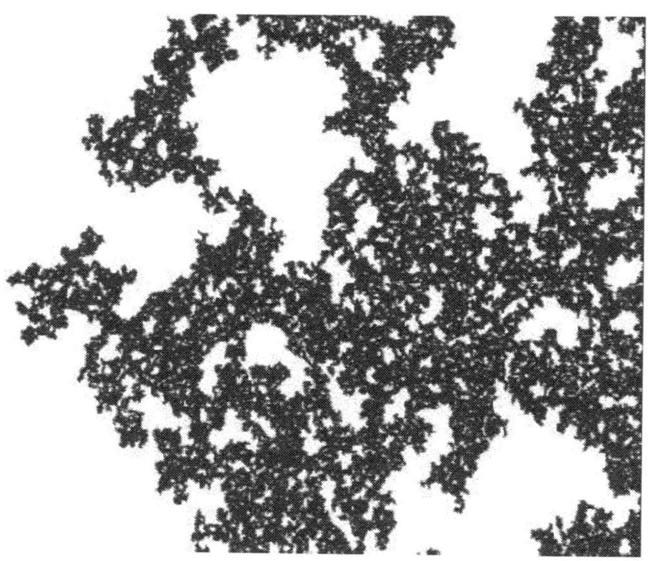

Fig. 19. Percolation cluster (d_f=1.89, $\tilde{d}$ =1.3)

Fig. 20. Menger sponge (d_f=2.727)

A further step in the understanding of the geometry of natural objects is the extension of the concept of fractal dimension to structures which are not self similar. Indeed, while self-similarity implies Eq. (32), the inverse is not true. Therefore, whenever Eq.(32) holds, even in absence of self-similarity, we can define a fractal dimension and we can introduce the concept of *fractal behavior* of a particular structure, replacing and generalizing the concept of *fractal geometry*. Two mathematical examples of non self-similar structures for which fractal dimension can be defined are shown in Figure 21 and 22.

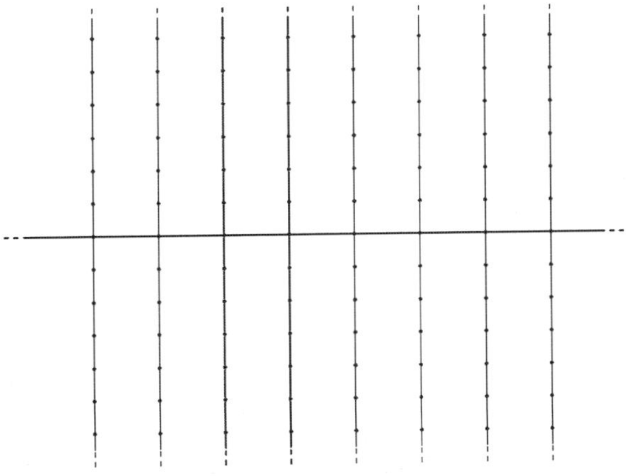

Fig. 21. Comb lattice (d_f=2, $\tilde{d}$ =1.5)

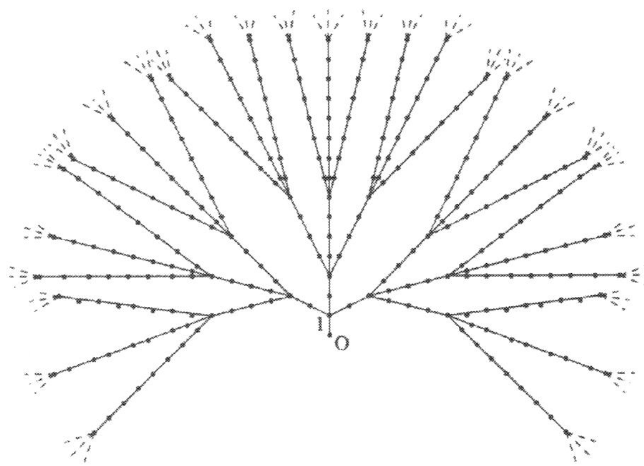

Fig. 22. "Nice Tree" (d_f=2.58=$\tilde{d}$)

According to its definition given in Eq. (32), fractal dimension can be experimentally determined using several techniques. The simplest ones are imaging techniques combined with image analysis, but also elastic scattering of photons or neutrons can be very useful.
In the years following the discovery of fractals, many studies proved the presence of fractal structures in several food systems (Peleg & Barbosa, 1993; Barrett & Peleg, 1995) and many authors began to apply the new ideas to describe physical and chemical processes occurring in inhomogeneous structures.
In particular, fractal dimension plays a fundamental role in the understanding of hydration and dehydration processes. Indeed, the water and vapor exchange rate between an object and its environment is proportional to the interface between them, i.e. to the object surface. In regular three dimensional objects, the surface area scales as the square of the object linear size, but in the case of fractal surfaces, the area scales with an exponent greater than 2, making the hydration-dehydration mechanism more rapid and effective (De Marchi and Cassi, 1993). A simplified dimensional analysis can shed more light on the role of fractality in this kind of processes. Assume that the total amount of water to be removed from a fractal object is proportional to its volume V, and that the dehydration rate is proportional to its surface S. Let us call d_V the fractal dimension of this object and d_S the fractal dimension of its surface. Now, the total time t required to remove all the water is given by the ratio between the amount of water and the dehydration rate, i.e., calling l the linear size of the object:

$$t \propto \frac{V}{S} = \frac{l^{d_V}}{l^{d_S}} = l^{d_V - d_S} \qquad (33)$$

For regular, homogeneous solid shapes, d_V=3 and d_S=2, therefore t is proportional to l. For fractal objects, usually $d_V \leq 3$ and $d_S \geq 2$, therefore t is proportional to l^a with $a < 1$. This would imply that dehydration processes in fractals are more rapid and that their effectiveness increases with the fractals linear size.
However, in this simplified analysis, we have completely neglected the water diffusion mechanism inside the fractal, which turns out to have a competing role with the surface dehydration enhancement due to fractal geometry.

In general, geometrical or topological inhomogeneities produce a dramatic slowing down of diffusion processes. This fact was first pointed out by De Gennes in a seminal paper where he proposed to study the random motion of "the ant in a labyrinth", (De Gennes, 1976). The problem was studied in the following years in the framework of fractal geometry, but the first attempts to describe such anomalous diffusion in terms of the fractal dimension failed. Only in 1982 Alexander and Orbach (Alexander & Orbach, 1982) succeeded in finding a suitable diffusion law, by introducing a further generalized dimension, called the *spectral dimension.* In order to understand the meaning of such a new parameter, we first notice that, to mathematically investigate diffusion on fractals and inhomogeneous structures, we cannot apply the usual analytical formalism based on continuous space and differential operators (Perrin, 1913; Mandelbrot, 1975), but we have to introduce discrete models. Diffusion on inhomogeneous structures is usually described by random walks on graphs, i.e. by stochastic hopping of a particle (walker) over e set of sites joined pairwise by links (Burioni & Cassi, 2005). The sites represents the elementary building blocks of the structure, and the links the connections between them, making it possible the passage of the diffusing particle. According to Eq. (32), the fractal dimension describes the spatial distributions of the sites. However, diffusion is affected not only by the sites position, but also by the positions of the links between the sites. This is a crucial point, since the same sites can be connected in very different ways by links: for example, the set of sites of the comb lattice of Figure 21, and the set of sites of a square lattice are the same, but, due to the different link distribution, diffusion processes on these two structures are very different.

Spectral dimension takes into account the effect of links distribution on diffusion. It is denoted by $\tilde{d}$ and it can be defined by the long time behavior of the random walk autocorrelation function $P_{ii}(t)$, giving the probability that a walker, starting from site i, returns to i after time t:

$$P_{ii}(t) \underset{t\to\infty}{\sim} t^{-\frac{\tilde{d}}{2}} \tag{34}$$

Notice that it can be shown that $\tilde{d}$ is independent of the choice of the starting site i and that, for regular structures, it coincides with the usual spatial dimension d.

Spectral dimension is independent of fractal dimension. The only constraint relating them is the inequality

$$\tilde{d} \le d_f \tag{35}$$

The spectral dimension is the relevant parameter to describe also a variety of physical phenomena different from diffusion, ranging from vibrational dynamics to phase transitions, and including electrical conductivity and anomalous specific heath at low temperature; therefore it can be measured in real systems using very different techniques: inelastic scattering , calorimetry, time resolved spectroscopy, and many others. It is rather interesting to notice that, in all experimental measure of spectral dimension of fractals, its value always turned out to be less than 2.

Now, the main result concerning diffusion on fractals and inhomogeneous structures involves both d_f and $\tilde{d}$: the mean square displacement of a diffusing particle after time t, at long times is given by

$$\left\langle r^2(t) \right\rangle \sim t^{\frac{\tilde{d}}{d_f}} \tag{36}$$

Notice that, for regular structures, $\tilde{d} \le d_f$, and therefore Eq. (36) reduces to the classical diffusion law

$$\left\langle r^2(t) \right\rangle \sim t \tag{37}$$

On the other hand, on fractals, due to Eq. (35), the exponent of t is usually smaller than 1, therefore the diffusion is slower with a respect to the classical case. Due to this property, such anomalous diffusion is often called *subdiffusion.*

However, anomalous diffusion can occur even in absence of fractal geometry as well as a non integer value of fractal dimension does not imply anomalous diffusion: in the case of the comb lattice (Figure 21) we have subdiffusion without fractality while for the Nice Tree (Figure 22) diffusion is not anomalous.

The effect of subdiffusion on dehydration processes is a significant slowing down, which is most relevant for large objects with smooth surfaces. The presence of rough, fractal surfaces, as we discussed before, tends to accelerate water removal. Therefore, the dehydration kinetics has to be considered as the results of two competing mechanism, both due to inhomogeneous and fractal geometry. From a theoretical point of view, Monte-Carlo simulations and analytical trapping-diffusion models are the most suitable tools to investigate them and we expect that in the next years new relevant results will shed more light on this class of processes.

6. Conclusion

The water removal from fruits and vegetables is the older method to increase their shelf life. In terms of mass transfer, drying processes could be summarized the motion of water molecules from the core of vegetables to their surface and from it to the external region (i.e. surrounding air). The control of these motion is essential to obtain vegetables with a high quality also maintaining a high output production as well as minimum energy costs. The deep knowledge of mass transfer mechanisms is the only way to reach the desired moisture content minimizing the microbial growth and the degradation reactions of vegetable products. In general, the internal mass transfer during vegetable drying is still considered as diffusive phenomena on the basis of which water molecules randomly move without a preferred direction under the control of a moisture gradient. On this basis, several empirical mathematical models have been proposed and are commonly used to fit the macroscopic features (i.e. moisture content) of food during drying. Among these Fick's laws are the most used mathematical equation to model drying curves of all traditional and innovative dehydration techniques.

However, molecular motion during drying is a much more complex phenomenon. Water may flow inside vegetables and may leave their surface in liquid and/or vapor state. Moreover, each of the drying technologies promotes unique (usually combined) mass transfer mechanisms rather than the only diffusion. Also, it is well known that due to the high number of assumptions, some of which are unrealistic for food, diffusive phenomenon should be used with the utmost care. Nowadays, several mechanisms such as capillary flow, surface diffusion, vapor diffusion in the pores due to pressure gradient, water vaporization-condensation, etc., have been recognized as participating to the overall mass transfer. For instance, capillary forces are responsible of a negative pressure inside capillary pores able to maintain water inside vegetable structure at atmospheric pressure and to hinder water removal during drying. Moreover, due to the dependence of capillary pressure from moisture content, water moves from the region at low moisture concentration toward the highly moist regions. So, the study of water retention curve of vegetable tissues under different operative conditions and/or different dehydration methods rather than the only isotherm should be recommended.

Nowadays, it is clear the great importance of three dimensional structure properties of biological tissues such as pore size distribution, shape of pores, permeability, tortuosity, etc., on mass transfer of food. Food should be considered as inhomogeneous structures inside which water and solutes move inside unique motion pathways before to reach their surface and leave food. Moreover, these pathways are subject to significant changes during drying due to the collapse of vegetable tissues. Under this approach, geometrical or topological inhomogeneities produce a dramatic slowing down of diffusion processes. Unfortunately, apart few exceptions, these aspects are rarely considered in food science.

The future challenge in the field of mass transfer mechanisms of drying food is to obtain microstructure information of biological tissue and their changes during dehydration process. Also, starting from these information, it will be necessary to develop new theoretical models that include the different concentration and pressure gradients involved during each dehydration process as well as the three dimensional pathway inside which the vapor, the liquid and/or the gas phase may move. This approach will be the only way to precisely control fruits and vegetable drying, to obtain high quality of the products.

7. References

Aguilera, J.M, Michel, M., & Mayor, G. (2004). Fat migration in chocolate: Diffusion or capillary Flow in Particulate Solid? A Hypothesis paper. *Journal of Food Science*, 69 (7), 167-174.

Alexander, S., & Orbach, J. (1982). Density of states on fractals: "fractons"*Journal de Physique Lettres* 43, 17 (1982) 625-631

Alzamora, S.M., Gerschenson, L.N., Vidales, S.L., & Nieto, A. (1997). Strcutural changes in the minimal processino of fruits: some effects of blanching and sugar impregnation. In: *Food engineering 2000,* P. Fito, E. Ortega Rodriguez, & G.V.

Barbosa-canovas (Eds.), New York: Chapman & Hall. Pp 117-139.

Amami, E., Fersi, A., Vorobiev, E., & Kechaou, N. (2007). Osmotic dehydration of carrot tissue enhanced by pulsed electric field, salt, and centrifugal force. *Journal of Food Engineering*, 83, 605-613.

Amezaga-Johnston M.R., McLaggan D., & Booth, I.R. (1999). *Surviving* Osmotic Stress: The Role of Natural Food Components in Limiting Preservation Action. In: Y.H. Roos, R.B. Leislie, P.J. Lilliford. *Water Management in the Design and Distribution of Quality Foods.* (pp. 325-352). ISOPOW 7. Pennsylvania: Echnomic Pubblishing Company.

Atares, L., Chiralt, A., & Gonzales-Martinez, C. (2008). Effect of solute on osmotic dehydration and rehydration of vacuum impregnated apple cylinders (cv. Granny Smith). *Journal of Food Engineering*, 89, 49-56.

Azoubel, P.M., & Murr, F.E.X. (2004). Mass transfer of osmotic dehydration of cherry tomato. *Journal of Food Engineering*, 61, 291-295.

Azuara, E., Cortes, R., Garcia, H.S., & Beristain, C.I. (1992). Kinetic model for osmotic dehydration and its relationship with Fick's second law. *International journal of Food Science and technology*, 27, 239-242.

Babalis, J.S., & Belessiotis, V.G. (2004). Influence of the drying conditions on the drying constants and moisture diffusivity during the thin-layer drying of figs. *Journal of Food Engineering*, 65, 449-458.

Baltes, W. (1982). Chemical changes in food by the Maillard reaction. *Food Chemistry*, 9: 59.

Barat, J.M., Fito, P., & Chiralt, A. (2001). Modeling of simultaneous mass transfer and structural changes in fruits tissues. *Journal of Food Engineering*, 49, 77-85.

Barrett, A.H., & Peleg, M. (1995). Applications of fractal analysis to food structure. *Lebensmittel-Wissenschaft und-Technologie* Volume 28, Issue 6, July 1995, Pages 553-563

Baucour, P., & Daudin, J.D. (2000). Development of a new method for fast measurement of water sorption isotherm in the high humidity rage validation on gelatine gel. *Journal of Food Engineering*, 44, 97 - 107.

Betoret, N., Puente, L., Diaz, M.J., Pagan, M.J., Garcia, M.J., Gras, M.L., Marto, J., & Fito, P. (2003). Development of prebiotic-enrichment dried fruits by vacuum impregnation. *Journal of Food Engineering*, 56 (2-3), 273-277.

Beuchat L.R. (1985). *Influence* of Water Activity on Sporulation, Germination, Outgrowth, and Toxin production. In Rockland, L.B. & Beuchat L.R. (Eds.). *Water Activity: Theory and Applications to Food.* (pp. 101-107). Basic Symposium Series.Chicago, Illinois: Marcel Dekker Inc. pp 137 - 151.

Bohuon, P., Collignan, A., Rios, G.M., & Raoult-Wack, A.L. (1998). Soaking process in ternary liquids: Experimental study of mass transport under natural and forced convection. *Journal of Food engineering*, 37, 451-469.

Bouraout, M., Richard, P., & Durance, T. (1994). Microwave and convective drying of potato slices. *Journal of Food Process Engineering*, 17, 353-363.

Brands, M.J., Alink, G.M., van Boekel, M.A.J.S., & Jongen, W.M.F. (2000). Mutagenicity of Heat sugar-casein systems: effect of the Maillard reaction. *J. Agric. Food. Sci.* 48, 2271-2275.

Buchanan, R.L., & Bagi, L.K. (1997). Effect of water activity and humectant identity on the growth of Escherichia coli O157:H7. *Food Microbiology*, 14, 413-423.

Burioni, R., & Cassi, D. (2005). Random walks on graphs: ideas, techniques and results. *J. Phys. A: Math. Gen.* 38 R45 doi: 10.1088/0305-4470/38/8/R01

Carlcel, J.A., Golas, Y., Benedito, J., & Mulet, A. (2002). *Influence of power ultrasound in osmotic dehydration of apple slices*. Proceedings of the International drying Symposium (IDS 2002). Beijingm China, pp. 1281-1286.

Chhninman, M.S. (1984). *Evaluation of selected mathematical models for describing thin layer drying in-shell pecans*. Transaction of the ASAE, 27, 610-615.

Chiralt, A., & Fito, P. (2003). Transport mechanism in osmotic dehydration: the role of the structure. *Food Sci Tech Int.*, 9 (3), 179-186.

Chiralt, A., & Talens, P. (2005). Physical and chemical changes induced by osmotic dehydration in plant tissues. *Journal of Food Engineering*, 67, 167-177.

Choen, E., & Saguy, I. (1983). Effect of water activity and moisture content on stability of beet powder pigments. *J. Food Science*, 48: 703.

Christian, J.H.B. (1981). Specific solute effects on microbial water relations. In Water Activity: Influence of Food Quality. (Eds. Rockland L.B. and Steward, G.F.) pp. 825-854. New York, Academic Press.

Chua, K.J., & Chou, S.K. (2005). New Hybrid Technologies. In: *Emerging technologies for food processing*., 535 - 551.

Colatoni, A. & Magri, M. (1997). Il genere Clostridium e Clostridium perfringens. In: *Metodi Microbiologici per lo studio delle matrici alimentari*. Pp. 37-44. CDS, Azienda USL Bologna (Italy).

Collingnan, A., & Raoult-Wack, A.L. (1994). Dewatering and salting of cod by immersion in concentrated sugar/salt solutions. *Lebensm Wiss.-u. Technologie*, 27, 259-264.

Crank, J. (1975). *The mathematics of Diffusion*. Second Edition. Oxford Science Pubblication.

Craspite, G.H., & Rotstein, E. (1997). Design and Performance Evaluation of Dryers. In: Handobook *of Food Engineering Practice*, Ed. Valentas, K.J., Rotstein, E., Singh R.P. CRC Press, 132 - 173.

Datta, A.K. (2007). Porous media approaches to studying simultaneous heat and mass transfer in food processes. II: Property data and representative results. *Journal of Food Engineering*, 80, 96-110.

Datta, A.K. (2007a). Porous media approaches to studying simultaneous heat and mass transfer in food processes. I: Problem formulations. *Journal of Food Engineering*, 80, 80-95.

Datta, A.K. (2007b). Porous media approaches to studying simultaneous heat and mass transfer in food processes. II: Property data and representative results. *Journal of Food engineering*, 80, 96-110.

De Gennes, P.G. (1976). La fourmi dans un labyrinthe. *La recherche Volume 7, Pag. 919*

De Marchi, A., & Cassi, D. (1993). Fractal geometry and ecology of lichens. *Fractals* 1:346-353.

Derossi A., De Pilli, T., Giuliani, R., Orlando, I., Palmieri, L., & Severini, C. (2010). Study on prestabilization of pumpkin (Cucurbita Moschata) by osmotic dehydration in quaternary complex solution. *Journal of Food Process Engineering*. In press.

Derossi, A., De Pilli, T., Severini, C. (2010). Reduction in the pH of vegetables by vacuum impregnation: A study on pepper. *Journal of Food Engineering*, 99, 9-15.

Derossi, A., De Pilli, T., Severini, C., & McCarhty, M.J. (2008). Mass transfer during osmotic dehydration of apples. *Journal of Food Engineering*, 86, 519-528.

Diamante, L.M., & Murno, P.A. (1991). Mathematical modelling of hot air drying of sweet potato slices. *International Journal of Food Science and Technology*, 26, 99.

Drapon, R. (1985). Enzyme activity as a function of water activity. In Simato, D. and Multon, J.L. (Ed.) *Properties of Water in Foods*. (p. 26). Dordrecht, The Netherlands. Martinus Nijhoff Pubblishers.

Dukworth, R.B. (1975). *Water Relations of Foods*. Academic Press, New York.

Eichner, K. (1975). The influence of water content on non-enzymatic browning reactions in dehydrated foods and model systems and the inhibition of fat oxidation by browning intermediates. In *Water Relations of Foods*, Duckworth, R.B. (Eds.). (p. 417). Academic Press, New York.

Erle, U. (2005). Drying using microwave processing. In: *The microwave processing of foods*. Ed. Schubert, H., Reiger, M. CRC press, pp 143 - 152.

Farid, M. (2002). The moving boundary problems from melting and freezing to drying and frying of food. *Chemical Engineering and Processing*, 41 (1), 1-10.

Fennema O. (1999). Water: the star of biomanipulators obscured in a cloud of superficial familiarity. In: *Water Management in the Design and Distribution of Quality Foods*. (pp.

3-22). Y.H. Roos, R.B. Leislie & P.J. Lilliford (Eds.) ISOPOW 7. Pennsylvania: Echnomic Pubblishing Company.

Fernandes, F.A.N., Gallao, M.I., & Rodrigues, S. (2008). Effect of osmotic dehydration and ultrasound pre-treatment on cell structure: Melon dehydration. *Lebensm. Wiss.-u. Technologie*, 41, 604-610.

Ferrando, M., & Spiess, W.E.L. (2001). Cellular response of plant tissue during the osmotic treatment with sucrose, maltose and trehalose solutions. *Journal of Food Engineering*, 49, 115-127.

Fiorentini, C., Leiva Diaz, E., & Giner, S.A. (2008). A mass transfer model for the drying o fan innovative tomato gel. *Food Science and Technology International*, 14 (1), 39-46.

Fito, P. & Chiralt, A. (1995). An update on vacuum osmotic dehydration. In: *Food Preservation by moisture control: Fundamentals and Applications*, Barbosa-Canovas, G.V., & Welti-Chanes, J. (Eds.),. Technomic Publishig Company, Lancaster, pp 351-372.

Fito, P. (1994). Modelling of vacuum osmotic dehydration of foods. *Journal of Food Engineering*, 22, 313-318.

Fito, P., & Pastor, R. (1994). Non-diffusional mechanism occurring during vacuum osmotic dehydration (VOD). *Journal of Food Engineering*, 21, 513-519.

Fito, P., Chiralt, A., Barat, J.M., Martinez-Monzo, J., & Martinez-Navarrate, N. (2001). Vacuum impregnation for development of new dehydrated products. Journal of Food Engineering, 49, 297-302.

Fluente-Blanco, S., de Sarabia, E.R.F., Acosta-Aparico, V., Blanco-Blanco, A., & Gallego-Juarez, J.A. (2006). Food drygin process by power ultrasound. *Ultrasonics*, 44, 523-527.

Fors, S. (1983). Sensory properties of volatile Maillard reaction products and related compounds. A literature review. In the Maillard Reaction in Food Nutrition. *ACS symp. Series*, 215, 185-286.

Frank, H.S. & Wen, W.Y. (1957). *Discuss. Faraday Soc. 24*,133.

Geankopolis, C.J. (2003). Transport Processes and Separation Process Principles. 4th edn Prentice Hall, New Jersey.

Giangiacomo, R., Torreggiani, D., & Abbo, E. (1987). Osmotic dehydration of fruit. Part 1. Sugar exchanges between fruit and extracting syrups. *Journal Food Process Preservation*, 11, 183-195.

Giese, J. (1992). Advances in microwave food processing. *Food Technology*, 46, 35-46.

Giner, S.A. (2009). Influence of internal andexternal reistance to mass transfer on the constant drying rate period in high-moisture foods. *Biosystems Engineering*, 102, 90-94.

Gould, G.W. (1985). *Present state of knowledge of aw effects on microorganisms. In properties of Water in Foods.* Simatos, D., and

Multon J.L. (Ed.). Dordrecht, Netherlands.

Gould, G.W., & Measures, J.C. (1977). Water relations in single cells. *Phil. Trans. R. Soc. Lond. B.*, 278: 151.

Gras, M., Vidal-Brotons, D., Betoret, N., Chiralt, A., & Fito, P. (2002). The response of some vegetables to vacuum impregnation. *Innovative food Science & Emerging Technologies*, 3, 263-269.

Griffith, T., & Johnson, J.A. (1957). Relation of the browning reaction to storage of sugar cookies. *Cereal Chem.* 34, 159-169.

Hallstrom, B., Vassilis, G., Sjoholm, I., & Romulus, A.M. (2007). Mass transfer in Foods. In: *Handobook of Food Engineering, Second Edition*. Edited by Heldman, D.R., and Lund, D.B. CRC press. pp 471 - 492.

Hamraoui, A., & Nylander, T. (2002). Analytical Approach for the Lucas-Washburn Equation. *Journal of Colloid and Interface Science*, 250, 415 - 421.

Haralampu, S.G., & Karel, M. (1983). Kinetc models for moisture dependence of ascorbic acid and b-carotene degradation in dehydrated sweet potato. *Journal of Food. Science*, 48: 1872

Ibarz, A., & Barbosa-Canovas, G.V. (2003). Dehydration. *In Unit Operations in Food Engineering*, Edited by A. Ibarz and G.V. Barbosa-Canovas, CRC press, 573 - 623.

Kang, C.K., Woodburn, M., Pagenkopf, A., & Chency, R. (1969). Growth, sporulation and germination of Clostridium perfringes in media of controlled water activity. *Appl. Microbiol.*, 18: 798.

Karathanos, V.T., & Belessiotis, V.G. (1997). Sun and artificial air drying kinetics of some agricultural products. *Journal of Food engineering*, 31 (1), 35-46.

Karel, M., & Lund, D.B. (2003). Physical Principles of Food Preservation. Second Edition. Marcel Dekker Inc., New York.

Kaymak-Ertekin, F., & Sultanoglu, M. (2000). Modelling of mass transfer during osmotic dehydration of apples. *Journal of Food Engineering*, 46, 243-250.

Khaled, A.R.A., & Vafai, K. (2003). The role of porous media in modeling flow and heat transfer in biological tissues. *International Journal of Heat and Mass Transfer*, 46, 4989-5003.

Kinsella, J.E., & Fox, P.F. (1986). Water sorption y proteins: Milk and Whey Proteins. *CRC Crit. Rev. Food SCi. Nut.*, 24:91

Klute, A. (1986). *Methods of soil analysis. Part I.* Madison, WI: ASA and SSSA.

Krokida, M.K., Karathanos, V.T., Maroulis, Z.B., & Marinos-Kouris, D. (2003). Drying kinetics of some vegetables. *Journal of Food Engineering*, 59, 391-403.

Krotov, V.V., & Rusanov, A.I. (1999). *Physicochemical hydrodynamics of capillary systems.* London: Imperial College Press.

Labuza, T.P. (1970). Properties of water as related to the keeping quality of Foods. *Proceedings of the third International Congress of Food Science Technology*. p. 618. Washington, DC.

Labuza, T.P. (1980). The effect of water activity on reaction kinetics of food deterioration. *Food Technology. 34 (4)*. p. 36.

Labuza, T.P., & Saltmarch, M. (1981). The nonenzymatic browning reaction affected by water in foods. In *Water Activity: Influences on Food Quality*. (p. 605).Rockland, L.B. and Stewart, G.F. (Eds.). Academic Press, New York.

Lahsasni, S., Kouhila, M., Mahrouz, M., & Kechaou, N. (2002). Experimental study and modelling of adsorption and desorption isotherms of prickly pear peel (Opuntia ficus indica). *Journal of Food Engineering*, 55, 201-207.

Lajollo, F., & Marquez, U.L.M. (1982). Chlorophyll degradation in a spinach system at low intermediate water activities. *Journal of Food Science*, 47: 1995.

Lampinen, M.J., Ojala, K.T., & Koski, E. (1991). Modeling and measurement of dryers fro coated paper. *Drying Technology*, 9 (4), 973-1017.

Lebvoca, N.I., Bazhal, M.I., & Vorobiev, E. (2001). Pulsed electric field breakage of cellular tissues: Visualization per locative properties. *Innovative of Food Science and Emerging Technologies*, 2, 113-125.

Lee, S., Voung, R.E., Schiffman, P.M., & Coggins, C.W.Jr. (1983). Maturity studies of arocade fruit based on picking dates and dry weight. *Journal of American Society of Agricultural Science*, 180, 390-394.

Leiva Diaz, E., Giannuzzi, L., & Giner, S.A. (2007). Apple pectic gel produced by dehydration. *Food ad Bioprocess Technology*.

Lerici, C.L., Pinnavaia, G., Dalla Rosa, M., & Bartolomucci, L. (1985). Osmotic ehydration of fruit: influence of osmotic agents on drying behavior and product quality. *Journal of Food Science*, 50, 1217-1219.

Lewicki, P.P., & Przeka-pawlak, R. (2005). Effect of osmotic dewatering on apple tissue structure. Journal of Food Engineering, 66, 43-50.

Lucas, R. (1918). Ueber das Zeitgesets des kapillar Aufstieg von Flussigkeiten. Kolloid-Zeitschrift, 23: 12 - 22.

Maillard, L.C. (1912). Action Des Acides Amines sur les sucres. Formation Des Melanoidins Par Voie Methodique.

Compt. Rend. 154, 66-68.

Maltini, E., Torreggiani, D., Venir, E., & Bertolo, G. (2003). Water activity and the preservation of plant foods. *Food Chemistry*. 82, pp. 79-86.

Mandelbrot, B.B. (1975). *Les objets fractals: forme, hasard et dimension. Flammarion, Paris*

Mandelbrot, B.B. (1982). The Fractal Geometry of Nature. W.H. Freeman and Company.. ISBN 0-7167-1186-9.

Marcote, M, Toupin, C.J., & Le Mauger, M. (1991). Mass transfer in cellular tissues. Part I: the mathematical model. *Journal of Food Engineering*, 13, 199-200.

Margaris, D.P., & Ghiaus, A.G. (2007). Experimental study of hot air dehydration of Sultana grapes. *Journal of Food Engineering*, 79, 1115-1121.

Margaris, D.P., & Ghiaus, A.G. (2007). Experimental study of hot air dehydration of Sultana grapes. *Journal of Food Engineering*, 79, 1115-1121.

Martins, S.I.F.S., Jongen, W.M.F., & van Bokel, M.A.J.S. (2001). A review of Maillard reaction in food implication to kinetic modelling. *Trends in Food Science & Technology*, 11, 364 - 373.

Maskan, M. (2000). Microwave/air and microwave finish drying of banana. Journal of Food Engineering, 44, 71-78.

Maskan, M. (2001). Drying, shrinkage and rehydration characteristics of kiwifruits during hot air and microwave drying. *Journal of Food Engineering*, 48, 177-182.

Mason, T.J., Riera, E., Vercet, A., & Lopez-Buesa, P. (2005). Application of Ultrasound. In: *Emerging tehnologies for food processing*, 323 - 351.

Mauron, J. (1981). The Maillard reaction in food: A critical review from the nutritional standpoint. *Progr. Fd. Nutr. Sci.*, 5:5.

Mavroudis, N.E., Gekas, V., & Sjoholm, I. (1998). Osmotic dehydration of apples - Effects of agitation and raw material characteristics. *Journal of Food Engineering*, 35, 191-209.

Mujaffar, S., & Sankat, C.K. (2005). The mathematical modeling of the osmotic dehydration of shark fillets at different brine temperatures. *International Journal of Food Science and Technology*, 40, 1-12.

Mujica-Paz, H., Fragos-Valdez, A., Lopez-Malo, A., Palou, E., & Welti-Chanes, J. (2003). Impregnation properties of some fruits at vacuum pressure. *Journal of Food Engineering*, 56, 307-314.

Mulet, A., Berna, A., & Rossello, C. (1989). Drying of carrots. I. Drying Models. *Drying Technology*, 7 (3), 537-557.

Ni, H., Datta, A.K., & Torrance, K.E. (1999). Moisture transport in intensive microwave heating of biomaterials: a multiphase porous media model. *International journal of Heat and Mass Transfer*, 42, 1501-1512.

Nowak, D., & Lewicki, P.P. (1998). *Characteristics of infrared-convective dryer*. Proceedings of the XVIth National Conference of Chemical and Process Engineering. Krakow; Zaklad Graficzny Politechniki Krakiwskiej. In Polish, 194-199.

Nowak, D., & Liwicki, P.P. (2004). Infrared drying of apple slices. *Innovative Food Science and Emerging Technologies*, 5, 353-360.

Okos, M.R, Campanella, O., Narsimhan, G., Singh, R.K., & Weitnauer, A.C. (2007). Food Dehydration. In: *Handobook of Food Engineering, Second Edition*. Edited by Heldman, D.R., and Lund, D.B. CRC press, 601-744.

Omowaye, A.B.I.O, Taiwo, K.A., Eshtiaghi, N.M., Angersbach, A., & Knorr, D. (2003). Comparative evaluation of the effects of pulsed electric field and freezing on cell membrane permeabilisation and mass transfer during dehydration of red bell peppers. *Innovative Food Science and Emerging Technologies*, 4, 177-188.

Orikasa, T., Wu, L., Shiina, T., & Tagawa, A. (2008). Drying characteristics of kiwifruit during hot air drying. *Journal of Food Engineering*, 85, 303-308.

Ozkan, I.A., Akbudak, B., & Akbudak, N. (2007). Microwave drying characteristics of spinach. *Journal of Food Engineering*, 78, 577-583.

Pabis, S. (1999). The initial phase of convection drying of vegetables and mushrooms and the effect of shrinkage. *Journal of Agricultural Engineering Research*, 72, 187-195.

Pabis, S. (2007). Theoretical models of vegetable drying by convection. *Transp. Porous Med*, 66, 77-87.

Palou, E., Lopez-Malo, A., Argaiz, A., & Welti, J. (1994). The use of the Peleg's equativo to model osmotic concnetration of papaya. *Drying Technology*, 12, 965-978.

Panagiotou, N.M., Karathanos, V.T., & Maroulis, Z.B. (1999). Effect of osmotic agent on osmotic dehydration of fruits. *Drying Technology*, 17, 175-189.

Panchariya, P.C., Popovic, D., & Sharma, A.L. (2001). Modeling of desorption isotherm of black tea. *Drying Tech.*, 19, 1177-88.

Paulsen, M.R., & Thomson, T.L. (1973). Drying endysus of grain sorghum. *Transaction of the ASAE*, 16, 537-540.

Peleg, M., & Barbosa, G. V. (1993). Fractals and foods. *Critical Reviews in Food Science and Nutrition*, *33*(2), 149-165. doi:10.1080/10408399309527617

Perrin, J. (1913). *Les atomes.* Félix Alcan, Paris

Ponciano, S.M., Robert, H.D., & Ken, A.B. (1996). The thin-layer drying characteristics of garlic slices. *Journal of Food Engineering*, 29, 75-97.

Prabhanjan, D.G., Ramaswamy, H.S., & Raghavan, G.S.V. (1995). Microwave-assisted convective air drying of thin layer carrots. *Journal of Food Engineering*, 25, 283-293.

Rahman, M.S., Perera, C.O., & Theband, C. (1998). Desorption isotherm and heat pump drying kinetics of peas. *Food Research International*, 30, 485-491.

Rahman, S. (1995). *Food Properties handbook.* Boca Raton: CRC Press.

Raoult-Wack, A.L. (1994). Advances in osmotic dehydration. *Trends in Food Science and Technology*, 5, 255-260.

Rastogi, N.K., & Niranjan, K (1998). Enhanced mass trasfer during osmotic dehydration of high pressure treeated pineapple. *Journal of Food Science*, 63, 508-511.

Rastogi, N.K., & Raghavarao, K.S.M.S. (1994). Effect of temperature and concentration of osmotic dehydration of coconut. *Lebensm. Wiss.u-. Technologie*, 27, 264-267.

Rastogi, N.K., & Raghavarao, K.S.M.S. (1995). Kinetics of osmotic dehydration of coconut. *Journal of Food Process Engineering*, 18, 187-197.

Rastogi, N.K., & Raghavarao, K.S.M.S. (1997). Water and solute diffusion coefficients of carrot as a function of temperature and concentration. *Journal of Food Engineering*, 34, 429-440.

Rastogi, N.K., Eshtiaghi, M.N., & Knorr, D. (1999). Accellerated mass transfer during osmotic dehydration of high intensity electrical field pulse pretreated carrots. *Journal of Food Science*, 64, 1020 - 1023.

Rastogi, N.K., Angersbach, A., & Knorr, D. (2000). Evaluation of Mass Transfer mechanisms during osmotic treatment of plant materials. Journal of Food Science, 65 (6), 1016-1019.

Rastogi, N.K., Angersbach, A., & Knorr, D. (2000). Synergistic effect of high hydrostatic pressure pretreatment and osmotic stress on mass transfer during osmotic dehydration. *Journal of Food Engineering*, 45, 25-31.

Rastogi, N.K., Raghavarao, K.S.M.S., Niranjan, K., & Knorr, D. (2001). Recent developments in osmotic dehydration: methods to enhance mass transfer. *Trends in Food Science & Technologies*, 13, 48-59.

Ratti, C., & Mujumdar, A.S. (1995). Infrared drying. In: *Handbook of industrial drying*, Mujumdar, A.S. (ed.), vol. 1. New York: Marcel Dekker, pp. 567-588.

Reiger, M., & Schubert, H. (2005). Introducing microwave processing processing of food: principles and technologies. In: *The microwave processing of foods*. Ed. Schubert, H., Reiger, M. CRC press.

Rockland, L.B., & Beuchat, L.R. (1981). *Water activity: theory and application to food*. Basic Symposium series (IFT).

Rockland, L.B., & Nishi, S.K. (1980). Influence of water activity on food product quality and stability. *Food Technology. 34 (4)*. p 42.

Rockland, L.B., & Stewart, G.F. (1981). *Water activity: Influences on Food Quality*. Academic Press, New York.

Roos, Y.H., Leslie, R.B., & Lillford, P.J. (1999). *Water management in the design and distribution of quality foods*. ISOPOW 7

Ruan, R.R., & Chen, P.L. (1998). *Water in Foods and Biological Materials. A Nuclear Magnetic Resonance approach*. Technomic, Publ. Inc.

Sacchetti, G., Gianotti, A., & Dalla Rosa, M. (2001). Sucrose-salt combined effects on mass transfer kinetics and product acceptability. Study on apple osmotic treatments. *Journal of Food Engineering*, 49, 163-173.

Saguy, I., Goldman, M., Bord, A., & Cohen, E. (1984). Effect of oxygen on beet powder on the stability of betanine and vulgaxanthin-I. *Journal of Food Science*, 49:99.

Saguy, I., Kopelman, J., & Mizrahi, S. (1980). Computer-aided prediction of beet pigment (betanine and vulgaxanthin-I) retention during air-drying. *Journal of Food Science*. 45: 230.

Saguy, I.S., Marabi, A., & Wallach, R. (2005). New approach to model rehydration of dry food particulates utilizing principles of liquid transport in porous media. *Trends in Food Science & Technology*, 16, 495 - 506.

Sakai, N, & Hanzawa, T. (1994). Application and advances in far-infrared heating in Japan. *Trends in Food Science and Technology*, 5 (11), 357-362

Salvatori, D., Andres, A., Chiralt, A., & Fito, P. (1998). The response of some properties of fruits to vacuum impregnation. *Journal of Food Process Engineering*, 21, 59-73.

Saravacos, G.D., & Maroulis, Z.B. (2001). *Transport properties of foods*. New York: Marcel Dekker.

Schmidt, S.J. (2004). Water and solids mobility in foods. *Adv. Food Nutr. Res*. 48, 1 - 101.

Scott, W.J. (1957). Water Relations of food spoilage microorganisms. *Adv. Food Res*. 7:83-127.

Senadeera, W., Bhandari, B.R., Young, G., & Wijesinghe, B. (2003). Influence of shapes of selected vegetable materials on drying kinetics during fluidized drying. *Journal of Food Engineering*, 58, 277-283.

Sharma, G.P., & Prasad, S. (2001). Drying of garlic (Allium sativum) cloves by microwave-hot air combination. Journal *Journal of Food Engineering*, 50, 99-105.

Simal, S., Benedito, J., Sanchez, E.S., & Rosello, C. (1998). Use of ultrasound to increase mass tran sport rates during osmotic dehydration. *Journal of Food Engineering*, 36, 323-336.

Simato, D., & Multon, J.L. (1985). *Properties of Water in Foods*. Martinus Nijhoff Pubblishersm Dordrecht, The Netherlands.

Slade, L., & Levine, H (1988a). Water as a plasticizer: physico-chemical aspects of low-moisture polymeric systems. In *Water Science Reviews*, Vol 3, Franks, F., Ed., Cambridge University Press, Cambridge, 79.

Slade, L., & Levine, H. (1985) Intermediate moisture systems; concentrated and supersatured solutions; pastes and dispersions; water as plasticizer; the mystique of "bound" water; Thermodynamics versus kinetic. In: *Water Activity: A credible Measure of Technological Performance and Physiological Viability* Presented at Faraday Division, Royal Society of Chemistry Discussion Conference –? Cambridge, July 1-3, 16.

Slade, L., & Levine, H. (1988b). Structural stability of intermediate moisture foods – a new understanding? In *Food Structure – Its Creation and Evaluation*, Mitchell, J.R. and Blanshard, J.M.V., Eds., Butterworths, London, 115.

Sojak, M. (1999). Modeling of the kinetics of pumpkin dehydration (in Polish). *Inzynieria Rolnicza*, 2(8), 87-94.

Soysal, Y. (2004). Microwave drying characteristics of parsley. *Biosystems engineering*, 89, 167-173.

Sperber, W.H. (1983). *Influence of water activity on food-borne bacteria – A review*. J. Food Prot., 46, p. 142.

Spiazzi, E., & Mascheroni, R. (1997). Mass Transfer for osmotic dehydration of fruits and vegetables-I. Development of the simulation model. *Journal of Food Engineering*, 34, 387-410.

Stephanovic, A.F., & Karel, M. (1982). Kinetics of beta-carotene degradation at temperatures typical of air drying of foods. *Journal Food Proc. Preserv.*, 6: 227-242.

Stojanovic, J., & Silva, J.L. (2007). Influence of osmotic concentration, continuous high frequency ultrasound and dehydration on antioxidants, colour and chemical properties of rabbiteye blueberries. *Food Chemistry*, 101, 898-906.

Sutar, P.P., & Gupta, D.K. (2007). Mathematical modeling of mass transfer in osmotic dehydration of onion slices. *Journal of Food Engineering*, 78, 90-97.

Taiwo, K.A., Angersbach, A., & Knorr, D. (2003). Effects of pulsed electric field on quality factors and mass transfer during osmotic dehydration of apples. *Journal of Food Process Engineering*, 26 (1), 31-48.

Togrul, I.T., & Pehlivan, D. (2002). Mathematical modeling of solar drying of apricots in thin layers. *Journal of Food Engineering*, 55, 209-216.

Torreggiani, D. (1993). Osmotic dehydration in fruits and vegetable processing. Food research International, 26, 59-68.

Torregiani, D., Giangiacomo, R., Bertolo, G., & Abbo, E. (1986). Ricerche sulla disidratazione osmotica della frutta I. idoneità varietale delle ciliegie. *Industria Conserve*, 61, 101-107.

Toupin, C.J., Marcotte, M., & Le Maguer, M. (1989). Osmotically induced mass transfer in plant storage tissues: a mathematical model – Part 1. *Journal of Food Engineering*, 10, 13-38.

Troller, J.A. (1985). *Adaptation of Microorganisms in Environmentals with Reduced Water Activity*. In Rockland, L.B. & Beuchat L.R. (Eds.). *Water Activity: Theory and Applications to Food*. (pp. 101-107). Basic Symposium Series. Chicago, Illinois: Marcel Dekker Inc.

Tsamo, C.V.P., Bilame, A.F., Ndjouenkeu, R., & Nono, Y.J. (2005). Study of material transfer during osmotic dehydration of onion slices (Allium cepa) and tomato fruits (Lycopersicon esculentum). *Lebesm. Wiss-u. Technologie*, 38, 495-500.

Van Den Berg, C. & Bruin, S. (1981). Water activity and its estimation in food systems: theoretical aspects. In: L. Rockland & G.F. Stewart, (Eds.). *Water Activity: Influence in Food Quality*. (pp. 1-64). Academic Press: New York.

Van den Berg, C. (1985). On the significance of water activity in low moisture systems; water vapor sorption equilibrium and hysteresis; the starch/water system as a model. In: *Water Activity: A credible Measure of Technological Performance and Physiological Viability* Presented at Faraday Division, Royal Society of Chemistry Discussion Conference -? Cambridge, July 1-3, 16.

Van den Berg, C. (1986). *Water activity. Concentration and Drying of Foods*. MacCarthy, D., Ed., Elsevier Applied Science, London, 11.

Vanbrakel, J. (1975). Pore space models for transport phenomena in porous media review and evaluation with special emphasis on capillary liquid transport. *Powder Technology*, 11 (3), 205-236.

Washbur, E.W. (1921). The dynamics of capillary rise. *Physic Rev*, 17 (3), 273-83.

Weerts, A.H., Lian, G., & Martin, D. (2003). Modeling Rehydration of Porous Biomaterials: Anisotropy Effects. *Journal of Food Science*, 68, 3, 937 - 942.

White, G.M., Roos, I.J., & Ponelert, R. (1981). Fully exposed drying of popcorn. *Transactions of the ASAE*, 24, 466-468.

Wolf, M., Walker, J.E., Jr., & Kapsalis, J.G. (1972). Water vapor sorption hysteresis in dehydrated foods. *J. Agric. Food Chem*, 20: 1073.

Yaylayan, V.A. (1997). Classification of the Maillard reaction: A conceptual approach. *Trends in Food Science & Technology*, 8, 13 - 18.

Zhao, Y., & Xie, J. (2004). Practical applications of vacuum impregnation in fruit and vegetable processing. *Trends in Food Science & Techology*, 15, 434-451.

16

Mass Transfer in Cheese

Jorge F. Vélez-Ruiz
Departamento de Ingeniería Química, Alimentos y Ambiental
Universidad de las Américas-Puebla.
Sta. Catarina Mártir, Cholula, Puebla.
MËXICO

1. Introduction

1.1 Introduction to food preservation

Food transformation of raw items, preservation of food characteristics, and supplying of food products are the main goals for the food industry. The transformation and preservation of food materials involve a diversity of processes oriented to produce intermediate or final items, maintaining their nutritional, physical, and other desired properties. In order to transform foods and preserve their properties at the best conditions, the operational variables of the food processes should be correctly identified to get maximum benefits.

Among the many transformation processes used for foods making, there are some frequently applied in the milk industry, such as refrigeration, standardization, homogenization, centrifugation, pasteurization, evaporation, salting, and dehydration, among others, depending of the dairy product to be manufactured. Each one of them offers characteristics and advantages that may be combined to get a better dairy product.

Furthermore, the combination of preservation treatments has been and should be oriented to influence positively the properties of the final products. In general, the preservation processes may be divided in conventional and minimal, the first group produces foods that are shelf stable, they can be stored by several months at normal environmental. Whereas the minimally processed foods have a shorter shelf life, they retain their freshness, and nowadays, they are more acceptable by consumers. Most of the studies about the conventional, minimal, and combination of methods, also known as hurdle technology, have covered microbiological, chemical, and quality aspects, and only a minor quantity of works have focused on engineering and physical aspects. Therefore, more efforts and studies should be dedicated to cover the engineering approach, such is the case of mass transfer in food systems, cheeses among them.

1.2 System equilibrium

From the engineering point of view, a food item is a system strongly influenced by the surroundings (Figure 1), or it influences importantly the immediate environment. When the food material is not suffering any change, the system is balanced or in equilibrium, but if some of the food properties change due to a physicochemical driven force, the specific food will develop a transport phenomenon approaching to the equilibrium. Thus, a dairy product may be in chemical, mass, mechanical, phase and/or thermal equilibrium (Cengel and

Boles, 2006), depending of the particular conditions, otherwise the food system will present changes as a result of natural or artificial processes in which a physicochemical potential exists. The physical processes developed in a food system are normally an expression of one of the transport phenomena, momentum, heat or mass transport, even as a single or simultaneous change, in which the processes are also identifies as unit operations or food process operations.

Fig. 1.

When thermodynamic aspects are considered for the state of a food system, there is a Gibbs free energy that determines the equilibrium. A null free energy implies an equilibrium state, while a free energy different to zero is for food systems with a changing nature or exposing to a given process. Gibbs free energy includes enthalpy, temperature and entropy properties (Karel and Lund, 2003).

The lack of equilibrium of any food system requires specific considerations of the involved phases in the mass transfer phenomenon; thus, vapor (or gas)-liquid equilibrium is implied in dehydration and distillation, whereas the liquid-liquid equilibrium is involved in extraction, and solid-liquid equilibrium is considered in lixiviation. Further the gas-solid or vapor-solid equilibrium is too much transcendental in food systems transformations.

1.3 Transport phenomena

A transport phenomenon is the evolution of a system toward equilibrium; that is to say, it is a change of the food system, some or several of the food properties are modified due to the given change and those transformations are mathematically modeled by the so named equations of change, in which the quantity or volume of the dairy product will affect the rate of transport, whereas the geometry of the changing system will affect the direction. If a momentum gradient is present between the food system and the surroundings a transport of momentum will happen. When a difference of temperatures exists between them, a heat transfer will occur. And finally, if a chemical potential or a concentration driven force among the milk components is observed, then a mass transfer will be experienced (Vélez-Ruiz, 2009).

Any change developed within process equipment, also identified as food process operation may be analyzed from a basic principle in which one, two or three transport phenomena are taking place. As examples of food process operations in the milk industry, in which a transport of momentum is present, are: milk pumping and transportation through pipes, homogenization of fat globules in milk, and separation of fat from skim milk by centrifugation. Cooling, heating, pasteurization, and evaporation are unit operations in which thermal treatments or heat transfer are mainly involved; whereas salting, drying, and volatiles loss/gain or cheese components migration through of packaging films, are processes involving mass transfer, just to mention a few.

The work of many food engineers or scientists in the industrial or manufacturing role involves the development or selection of processes, the design or evaluation of the required equipments, and the successful operation of food plants, that are based on their fundamental concepts.

1.4 Water activity

The adsorption and desorption of water vapor by foods, is highly related to their stability and perishability. And although the water content is a control factor, several food items with the same moisture concentration exhibit different stability or perishability; thus the term of water activity (A_w) expressing the water associated to nonaqueous constituents, has became the physicochemical or thermodynamic concept more related to microbial, biochemical and physical stability. Water activity as an objective concept, that has been defined from the activity or fugacity relationship between the solvent and the pure solvent; it is expressed by the equation 1, a practical expression of it, in which the assumptions of solution ideality and the existence of thermodynamic equilibrium are been considered (Saravacos, 1986; Fennema, 1996; Vélez Ruiz, 2001; Toledo, 2007):

$$A_w = \frac{p_w}{p_w^0} = \frac{\%RH}{100} = \frac{\%ERH}{100} \tag{1}$$

Where: A_w is the water activity (dimensionless), p_w is the partial pressure of water in the food (Pa or mm Hg), $p_w{}^0$ is the partial pressure of pure water (Pa or mm Hg), $\%RH$ is the percentage of relative humidity, and $\%ERH$ is the percent of equilibrium relative humidity.

As it is known and expected, water activity (0 - 1.0) has been associated with stability problems and several reactions developed during the storage, such as microbiological growth, kinetics of nutrients loss, browning reactions, and also with physical changes, like dehydration or rehydration and textural modifications. Particularly, the A_w is different for each cheese type, due to variability in composition and moisture gradients, as well as salt content. For this reason, several authors have proposed to evaluate the A_w for cheeses, by utilizing the chemical composition through of empirical relationships (Saurel et al., 2004). Some examples of cheeses in which empirical equations have been obtained for water activity evaluation, are the following: European varieties (Marcos et al., 1981), Emmental (Saurel et al., 2004), and Manchego type (Illescas-Chávez and Vélez-Ruiz, 2009). A couple of examples for evaluation of A_w are presented next:

i. Saurel et al. (2004) obtained a practical relationship for French Emmental cheese as a function of three variables, water, salt and free NH_2 concentrations ($R^2 = 0.92$):

$$A_w = 1.07 - 0.19X_{water} - 3.49X_{NaCl} - 0.33X_{NH_2} + 6.51X_{water}X_{NaCl} + 0.57X_{water}X_{NH_2} \tag{2}$$

X is the component content (mass fraction) of water, salt and free NH_2.

ii. Illescas-Chávez and Vélez-Ruiz (2009) used an empirical correlation between salt content and water activity (R^2 = 0.996) for Manchego type cheese, showed by a quadratic expression:

$$264A_w - 175.9A_w^2 = 89.77 + X_{NaCl} \quad (3)$$

A_w is the water activity in cheese, and X_{NaCl} is the salt concentration (g/100g).

2. Cheese as a system

Cheese as a biological system and as a dairy product, is one of the first, most popular and universal elaborated food item. Cheese represents a product in which the milk components are preserved. This food item, is known as cheese (in English), "fromaggio" (in Italian), "fromage" (in French), "kase" (in German), and "queso" (in Spanish). Thus, a cheese is a food system in which due to many components, it is exposed to many changes, either biochemical and/or physical. Thus, a cheese is a dairy product made to preserve most of the milk components, including fat, protein and minor constituents from the milk, eliminating water and/or serum and adding salt and other ingredients, with a special flavor and with a solid or semisolid consistency (Vélez-Ruiz, 2010).

2.1 Cheese manufacturing

Though there are a lot of cheese types, the elaboration process involves common stages in which the variations in some of the steps contribute to generate a diversity of cheese products. These treatments, food process operations or unit operations may be summarized in a number of six, in which some specific equipments and process conditions may vary (Vélez-Ruiz, 2010).

i. Milk recollection. Milk is recollected, clarified and cooled down, to ensure a hygienic raw material.
ii. Milk preparation. Basic processes such as, standardization, mixing, homogenization, heating and/or addition of microorganisms may be carried out in this part. The fat-protein ratio is frequently standardized, $CaCl_2$ is normally added, and pH is sometimes controlled to a needed value. On the other hand, pasteurization destroys pathogenic microorganisms and most of enzymes.
iii. Milk coagulation. Addition of rennet, coagulant or acid is completed in order to transform milk into a coagulum. The enzyme acts on a specific amino acid of the casein, whereas the acid generates precipitation of proteins.
iv. Whey elimination. The formed coagulum contracts and expel part of the entrapped serum, constituting the syneresis phenomenon. Whey elimination from the cheese is favored by cutting, scalding, and/or stirring, and lately by salting.
v. Curd brining/salting. Salt is added to the curd, as a solid material or as a solution to favor elimination of whey, to develop desired flavor, and to preserve cheese.
vi. Final treatments. Agitation, milling, heating, pressing, casing, turning, packing, waxing, wrapping, ripening and/or other treatments, are some of the final operations than may be utilized as part of the cheese making, to reach those specific characteristics of each type. Of all these possible treatments, ripening is the most important due to the biochemical, microbial and physical modifications occurring during this period.

Each operation contributes to the milk/curd/cheese transformation in which the biochemical (enzymatic, acidification, hydrolysis, lipolysis, proteolysis, etc), microbial (bacteria or molds) and physical changes (homogenization, shearing, mixing, gelation, syneresis, curd fusion, solids diffusion, etc) are important parts of this food system. In summary, through the manufacturing process, there are three stages affecting most importantly the cheese characteristics a) the milk formulation, with a huge number of ingredients such as calcium chloride, cream, lactoperoxidase, ropy microorganisms, milk powder, just to mention a few; b) the used operational variables, rennet, salt, forces or stresses (centrifuge, pressure, shear), temperatures of cooling and heating, treatment times, and shear rates, among others; and c) the biochemical and/or physicochemical transformations developed during the elaboration and maturing stages.

2.2 Classification of cheeses

Grouping of cheese types is extremely complicated due to the enormous variety of them, in addition to the aforementioned factors, there are variants due to size, shape, as well as culture of the region of manufacture. A number of efforts have been realized to classify cheese, taking different points of view in order to meaningfully group them. Some classifications are based on the cheese origin (animal, country), involved coagulation process, applied manufacturing operations, presence of microorganism, rheological parameters, moisture content, or other considerations.

A simple and practical classification of cheeses, that may be very useful, is based on the existence of a ripening process stage, grouping them in fresh and ripened cheeses. This classification ignore if cheese ripening is completed by bacteria or molds, neither includes size and external appearance.

Fresh cheeses have a shorter shelf life, they are high in moisture content; and if a package is used, a null or insignificant mass transfer through the film may be considered, being the salting the main treatment in which a mass transfer phenomenon is developed. In contrary, a ripened cheese will have a larger shelf life, normally they are more dried and packaged with different types of films; and in these cheeses three mass transport changes can occur: salting in the manufacturing process, drying during maturation in the cave of ripening, and migration of volatiles and components through the package.

2.3 Mechanisms of mass transfer

A good number of food process operations are based on the mass transfer phenomenon involving changes in concentrations of foods and cheese components, depending of the phases and particular components in the food or cheese item, considered as a multi-component mixture, or as a binary one to simplified the physical analysis.

Mass transfer is the result of a concentration difference or driven force of a specific component, the component moves out from a portion of the food item or cheese with a portion or phase of high concentration to one of low, without to forget the influence of the surroundings.

Mass transfer is analogous to heat transfer and depends upon the dynamics of the food systems in which it occurs. It is known that there are two mechanisms of mass transfer, the diffusion and convection phenomena; in the first one, the mass may be transferred by a random molecular movement in quiescent food fluids or static solid items; and in the second one, the mass is transferred from the food surface to a moving fluid. And such it happens in many food processes, both mechanisms are developed simultaneously. Mass diffusion and convection may be more or less important depending of the specific operation. In salting and constituents migration of cheese, the diffusion is by far the most important; whereas in dehydration of cheese by exposing to a dry atmosphere, both mechanisms are very important.

Diffusion

The basic relation for molecular diffusion for a food system defines the molar flux related to the component concentration, for steady processes it is modeled by the Fick´s first law (Bird et al., 1960; Welty et al., 1976; Crank, 1983; Welti-Chanes et al., 2003; Vélez-Ruiz, 2009):

$$J_{iz} = -D_{im}\frac{dC_i}{dz} \tag{4}$$

Where: J_{iz} is the molar or mass flux of the i component in the z direction (mol/m²s or mg/m²s), D_{im} is the mass diffusivity or diffusion constant (m²/s), being specific for the i component in a given medium, dC_i/dz is the concentration gradient of the i component in the z direction (mol/m⁴ or mg/m⁴), dC_i is the concentration difference or driven force (mol/m³ or mg/ m³), and dz is the interface separation or separation distance between two points or portions with different concentration of the i component (m).
The molar flux of the involved component, in equation 4, may be converted to mass units of kilogram by considering the molar weight. Some diffusion constants have been evaluated for particular systems, few data are included in Table 1 (Welty et al., 1976; Okos et al., 1992). As it may be observed, gas diffusion is easier than liquid and solid diffusion, as well as liquid diffusion is easier than solid diffusion.

System	T (°C)	Dim (m^2/s) at 1 atm	Reference
Air in Water	25	6.37×10^{-2}	Welty et al., 1976
Carbon dioxide in water	25	5.90×10^{-2}	"
Hydrogen in water	20	3.06×10^{-1}	"
Sodium chloride in water	18	4.36×10^{-6} at 0.2 kg mole/m^3	"
Sodium chloride in water	18	4.46×10^{-6} at 1.0 kg mole/m^3	"
Sodium chloride in water	18	4.90×10^{-6} at 3.0 kg mole/m^3	"
Acetic acid in water	12.5	3.28×10^{-6} at 0.10 kg mole/m^3	"
Acetic acid in water	12.5	3.46×10^{-6} at 1.0 kg mole/m^3	"
Water in whole milk foam	35	8.50×10^{-10}	Okos et al., 1992
Water in whole milk foam	40	1.40×10^{-9}	"
Water in whole milk foam	50	2.00×10^{-10}	"
Water in nonfat milk	25	2.13×10^{-11}	"

Table 1. Diffusion Constants or Effective Diffusion of Some Particular Systems

Convection

Convective mass transport occurs in fluids as a result from the bulk flow, natural and forced motion is involved. It is very similar to heat convection, therefore the properties of the two

interacting phases, in which any of them may be a cheese or food item are very important. The supplying medium and the flowing phase, as well as some physical parameters of the system, are also involved through of dimensionless groups for the evaluation of the convective mass transfer coefficient.
The molar flux of a given component may be computed from the equation 5 (Bird et al., 1960; Welty et al., 1976; Welti-Chanes et al., 2003; Vélez-Ruiz, 2009), and as in the case of diffusion, it occurs in the decreasing concentration direction:

$$N_i = k_m \Delta C_i = k_m \left(C_{is} - C_{if} \right) \tag{5}$$

Where: N_i is the molar or mass flux of the i component in the flow stream direction (mol/m^{2}s or mg/m^{2}s), k_m is the convective mass transfer coefficient (m/s), ΔC_i is the concentration difference or driving force (mol/m^3 or mg/m^3), involving a concentration difference between the boundary surface concentration (C_{is}) and the average concentration of the fluid stream (C_{if}).
Mass transfer coefficients are expected to vary as a function of the dynamic conditions, geometrical aspects of the involved system, and physical properties of the fluid and solid phases. Although there are a good number of equations for the evaluation of the convective mass transfer coefficient, food systems and processes particularities are demanding for more specific correlations.

2.4 Salting, drying and migration through packege

Three are three mass transfer phenomena related to cheese manufacturing and storing, that are briefly commented next.

Cheese salting

Salting process during cheese manufacturing favors the development of well accepted quality attributes, both organoleptic and textural, it also suppresses unwanted microorganisms, affects acceptability favorably, causes volume reduction, and determines ripening in some degree. And although salt concentration and distribution play an important role on the aforementioned aspects, there is a limited knowledge about engineering principles of the salting phenomena in cheese, related with the mass transfer.

Cheese drying

Cheese dehydration as a mass transfer phenomenon involves the removal of moisture from the food material, the dehydration or drying process in a cheese reduces its moisture content. This process is not intentionally favored in cheese manufacturing, except during the coagulation part by mechanical means. It is developed as a consequence of the moisture difference between the cheese type and the surroundings (atmosphere, refrigerator, and maturation cave, for instance). Thus the control of relative humidity of the surroundings is needed to avoid undesirable and excessive dehydration; as an udesired phenomenon it is identified as weight loss.
A model of the mass loss of Camembert type cheese was established experimentally during ripening by Hélias et al. (2009), in which the O_2 and CO_2 mass concentrations, A_w, vapor pressure, and convective coefficients for mass transport phenomena were considered (weight loss as the most important).

Migration through a package

Migration of cheese components through a package may become other mass transfer phenomenon, commonly found in these dairy systems. Of those cheese components (volatiles and water vapor), moisture loss or gain is the most important that influences the shelf life of cheese. A cheese system has a micro-climate within a package, determined by the vapor/gas pressure of cheese moisture at the temperature of storage and the permeability of the specific package; in the case of cheeses with appreciable quantity of fat or other oxygen-sensitive components, the uptake of oxygen is also important. Therefore the control of vapor and gases exchange is needed to avoid undesirable spoilage, dehydration, condensing, texture changes, and oxidation, among others. Oxygen and off-odors scavengers may be utilized when the correspondent damages are serious problems. Some interchange of gases is also involved in modified atmospheres in order to preserve cheese characteristics.

Most of the studies of mass transfer in cheese have been focused on salting to favor it, and properly, the other two mass transfer phenomena (drying and migration through a package, without consideration of modified atmospheres as preservation method) are undesirable for most of cheese varieties.

3. Salting of cheeses

Cheese is a matrix of protein, fat and aqueous phase (with salt and minerals), that is subjected to salting as a very important stage. From the engineering viewpoint, salting as a mass transfer process involving salt uptake and water loss at the same time, that are the main studied mass transport phenomena.

3.1 Mass transfer characteristics

In cheese mass transfer, generally it has been recognized that the weight of salt taken up is smaller than the quantity of water expelled from the cheese, giving a loss of weight as consequence of the difference in mass balance. Salt travels from the external medium to the center of a piece within the liquid phase of the cheese, whereas in a contrary direction and mayor flow, there is a movement of water out from the cheese interior into the salt solution or to the atmosphere.

Some factors involved in the mass transfer through of cheese salting are cited next. These factors and their effects have been studied by different researchers, porosity (in Gouda cheese by Payne and Morison, 1999; in Manchego type by González-Martínez et al., 2002; Illescas-Chavez and Vélez-Ruiz, 2009; in Ragusano cheese by Mellili et al., 2005) and tortuosity (in experimental Gouda by Geurts et al., 1974) within the structure of the cheese, geometry and shape of cheese samples (in spherical geometry of experimental Gouda cheese with different weights by Geurts et al., 1974, 1980; in wheel shaped Romano type by Guinee and Fox, 1983; in finite slabs of Cuartirolo cheese by Luna and Bressan, 1986, 1987; in small cubes of Cuartirolo cheese by De Piante el al., 1989; in cylinders of Fynbo cheese by Zorrilla and Rubiolo, 1991, 1994; in cylinders and parallelepipeds of fresh cheese by Sánchez et al., 1999; in blocks of Ragusano cheese by Mellili et al., 2003a; in rectangular samples of white cheese by Izady et al., 2009), relation in which water is bound in cheese, viscosity of the free water portion, volume ratios of brine and solid (in Fynbo cheese by Zorrilla and Rubiolo, 1991), as well as the interaction of salt with protein matrix as the main; presalting and brine concentration (in experimental Gouda by Geurts et al., 1974, 1980; in white cheese by Turhan and Kaletunc, 1992; in Cheddar cheese by Wiles and Baldwin, 1996a, b; in Gouda cheese by Payne and Morison, 1999; in Emmental cheese by Pajonk et al., 2003; in Ragusano cheese by Mellili et al., 2003a; in Pategras cheese by Gerla and Rubiolo., 2003), brine temperature (in experimental Gouda by Geurts et al., 1974; in white cheese by Turhan and Kaletunc, 1992; in Ragusano cheese by Mellili et al., 2003b; in Emmental cheese by Pajonk et al., 2003; in white cheese by Izady et al., 2009). Internal pressure (in Manchego type by González-Martínez et al., 2002; Illescas-Chavez and Vélez-Ruiz, 2009), and ultrasound (in fresh cheese by Sánchez et al., 1999) have been also considered.

The water loss of cheese causes some shrinkage of the structure and decrease in porosity, limiting both mass transfer phenomena, moisture flow out of the item and salt movement into the cheese matrix. In general terms, water diffusivity has been related with temperature and moisture contents, it increases as a function of temperature and salt content in cheese aqueous phase.

3.2 Modeling of the salting process

Diffusion phenomenon is pretty much the main approach used to fit the mass transfer of components through a cheese system. Diffusion rates are expressed using effective coefficients of solutes in the solid; solutes such as sodium chloride, potassium chloride, and lactic acid have been modeled, as well as the water diffusion. The unstable equation or second Fick´s law (Eqn. 6) has been used for modeling of this diffusion process, in which different mathematical solutions have been applied depending of the particular cheese characteristics and process conditions. With the same meaning for the included variables (diffusivity of salt in water) and taking just one dimension for the mass transport, taking the external mass transfer as negligible.

$$\frac{\partial C}{\partial t} = D_{NaCl/water}\left(\frac{\partial^2 C}{\partial x^2}\right) \tag{6}$$

When more than one direction is considered in the mass transfer phenomenon, the corresponding dimensions should be incorporated (y, z, and r for radial effects). Most of the applied mathematical solutions have been based on Crank (1983) considerations. Table 2, includes reported data for mass diffusivity, obtained for salting of different types of cheese in a variety of process conditions.

If more than one component is considered in the diffusion process, the following relation (Eqn. 7), as a variation of equation 4, expresses the mass flux of n-1 solutes and the solvent, in a solid in contact with a homogeneous solution, without chemical reaction and insignificant convective mass transfer (Gerka and Rubiolo, 2003):

$$J_i = \sum_{j=1}^{n-1} D_{ij} \nabla x_j \tag{7}$$

Where: J_{iz} is the mass flux of the i solute or component (g/cm^{2}s), D_{ij} is the diffusion coefficient (cm^2/s), of the i component in a multicomponent system, ∇ is the gradient operator, and x is the local concentration of the j component (g/cm^3). Other mathematical approaches include empirical fittings, analytical solutions different to Fick´s law, such as the Boltzmann equation, hydrodynamic mechanisms and numerical solutions, among others.

Cheese type	Experimental conditions	Dx10^6 (m^2/h)	Comments	Author(s)
Experimental Gouda	Pseudo diffusion of salt at 12.6, 18 and 20.1°C at different days	0.73 - 1.17	Brine concentrations of 19.6-20.0 g NaCl/ 100g H_2O	Geurts et al., 1974
Romano	Apparent salt diffusion at 20°C	0.96	Cylinders of 8 cm height and 20-21 cm diameter	Guinee and Fox, 1983
Cheddar	Salt diffusion in 20 kg blocks	0.63	Periods of 24 and 48 h	Morris et al., 1985
White	Salt diffusion at 4, 12.5 and 20°C with 15 and 20% w/w	0.76 - 1.40	As a function of temperature and salt concentration	Turhan and Kaletunc, 1992
Fynbo	KCl and Na Cl diffusion at 12°C	1.41 & 1.49	Cylinders of 6 cm height and 12 cm diameter	Zorrilla and Rubiolo, 1994
Cheddar	Salt diffusivity in 20 kg blocks	4.17	Stored at 10°C	Wiles and Baldwin, 1996a
Fresco	Water diffusivity at two temperatures	1.73 & 4.68	Acoustic brining at 5 & 20°C	Sánchez et al., 1999
	Na Cl diffusivity at two temperatures	2.84 & 4.32	Acoustic brining at 5 & 20°C	
Manchego	Na Cl pseudodiffusion	1.58 & 1.87	Upper & lower parts, brine immersion	González et al., 2002
	Na Cl pseudodiffusion	2.20 & 3.02	Upper & lower parts, pulse vacuum impregnation	
	Na Cl average pseudodiffusion	2.54 & 3.60	Upper & lower parts, vacuum impregnation	
Pategrass	NaCl and lactic acid diffusion by two approaches	1.15 or 1.26	Series or short solution/ternary	Gerka and Rubiolo, 2003
	and lactic acid by two approaches	0.34 or 0.36	Series or short solution/ternary	
Emmental	Na Cl effective diffusion as a function of temperature and time	0.22 & 0.27	At 4°C, 24 and 48 h	Pajonk et al., 2003
		0.44	At 8°C and 48 h	
		0.35 & 0.68	At 13°C, 24 and 48 h	
		0.80	At 18°C and 48 h	
Camembert	NaCl and KCl diffusion. Agitation	1.01 (NaCl)	Reduction in NaCl	Bona et al., 2007
	Numerical solution to Fick´s eqn.	1.06 (KCl)	Finite element method	

Table 2. Effective Diffusivity Coefficients for Cheese Salting

3.3 Integral approach

An average velocity factor (AVF) was proposed as an integral mathematical relationship, that considers the cumulative mass transfer of salt in cheese through a selected period of time. It is obtained from those kinetic parameters evaluated from the Peleg's equation (Illescas-Chavez and Vélez-Ruiz, 2009), and is defined as:

$$\frac{dNaCl_t}{dt} = \frac{1}{k_1 + k_2 t} - k_2 \frac{t}{(k_1 + k_2 t)^2}$$

$$AVF = \frac{\int_0^{t_p} \frac{dNaCl_t}{dt}}{24} \qquad (8)$$

Where: $dNaCl/dt$ is the sodium chloride flux or mass transfer (= J_{NaCl}, g/h); k_1 (h/g) and k_2 (1/g) are constants of the Peleg's model, $NaCl_t$ is the sodium chloride concentration at any time t (g), and AVF is the thus defined, average velocity factor (g/h) as an integral value for a given process time (t_p in h).

Peleg's (1988) equation has been applied to many sorption/desorption processes as an empirical non-exponential model with the two aforementioned parameters, in which the $NaCl_0$ is the sodium chloride concentration at the beginning of the process:

$$\frac{t}{NaCl_t - NaCl_0} = k_1 + k_2 t \qquad (9)$$

The averaged velocity factor pretend to be a most representative value of the overall salting process, in which certainly the computed values are based on Peleg's constants. Therefore, if this approach is selected, the two constants of the Peleg model should be previously evaluated. Illescas-Chavez and Vélez-Ruiz (2009), applied this AVF approach to three different salting treatments of Manchego type cheese. The sample cheese was divided in twelve zones, 3 vertical, of 1.1 cm each (1 for the upside, 2 for the center portion, and 3 for the down part), and 4 radial divisions, of 2.6 cm each (A for the center, D for the external ring, B and C for the intermediate rings). For the correspondent calculations (differential and integral equations), a proper software was utilized (Maple V, Maplesoft, Ontario, Canada), some results are commented next. Table 3 shows the corresponding Peleg's constants for the three salting processes (conventional by immersion, pulsed vacuum with immersion, and vacuum with immersion) applied in Manchego cheese manufacturing, after manipulation of salt concentrations determinations.

The AVF calculations for three zones with different salting method are presented as examples of this approach:

i. zone C1 (third ring, upside) by conventional immersion (CI):

$$NaCl_{C1} = 0.131 + \frac{t}{1.489 + 0.188t}$$

$$\frac{dNaCl_t}{dt} = \frac{1}{1.489 + 0.188t} - 0.188 \frac{t}{(1.489 + 0.188t)^2}$$

$$\int_0^{24} \left(\frac{dNaCl}{dt} \right) = 3.984, \quad AVF_{CI} = \frac{3.984g}{24h} = 0.166\, g/h$$

Process	Zone	k_1 (h/g)	k_2 (g^{-1})	R^2
Conventional Immersion (CI)	A1	1.41	0.22	0.99
	B1	1.33	0.20	0.99
	C1	1.49	0.19	0.99
	D1	1.10	0.12	0.99
	A2	23.8	0.42	0.62
	B2	14.5	0.56	0.97
	C2	13.5	0.62	0.86
	D2	1.67	0.16	0.97
	A3	2.05	0.17	0.99
	B3	1.54	0.20	0.99
	C3	1.61	0.18	0.98
	D3	1.02	0.13	0.99
Pulsed Vacuum Immersion (PI)	A1	1.61	0.19	0.99
	B1	1.70	0.19	0.99
	C1	1.84	0.18	0.96
	D1	0.87	0.14	0.99
	A2	12.7	1.12	0.98
	B2	14.6	0.99	0.88
	C2	16.7	0.55	0.92
	D2	1.82	0.15	0.95
	A3	1.83	0.19	0.98
	B3	1.63	0.19	0.99
	C3	1.85	0.17	0.99
	D3	0.90	0.14	0.99
Vacuum Immersion (VI)	A1	1.82	0.18	0.94
	B1	1.37	0.19	0.93
	C1	0.96	0.20	0.99
	D1	0.86	0.13	0.99
	A2	47.1	0.48	0.99
	B2	45.5	0.49	0.83
	C2	17.1	0.09	0.98
	D2	1.81	0.15	0.95
	A3	2.27	0.17	0.92
	B3	1.02	0.26	0.97
	C3	1.66	0.17	0.92
	D3	0.60	0.14	0.99

Table 3. Peleg´s Kinetics Contants (k_1 and k_2) for Salting of Manchego Cheese

ii. zone B2 (second ring, center) by pulsed vacuum immersion (PI):

$$NaCl_{B2} = 0.131 + \frac{t}{14.610 + 0.992t}$$

$$\frac{dNaCl_t}{dt} = \frac{1}{14.610 + 0.992t} - 0.992\frac{t}{(14.610 + 0.992t)^2}$$

$$\int_0^{24}\left(\frac{dNaCl}{dt}\right) = 0.624, \quad AVF_{PI} = \frac{0.624g}{24h} = 0.026\, g/h$$

iii. zone D3 (external ring, down zone) by pulsed vacuum immersion (VI):

$$NaCl_{D3} = 0.131 + \frac{t}{0.601 + 0.142t}$$

$$\frac{dNaCl_t}{dt} = \frac{1}{0.601 + 0.142t} - 0.142\frac{t}{(0.601 + 0.142t)^2}$$

$$\int_0^{24}\left(\frac{dNaCl}{dt}\right) = 5.976, \quad AVF_{VI} = \frac{5.976g}{24h} = 0.249\, g/h$$

In according with the AVF values, lower salt rates were developed in cheese zones A2, B2 and C2; similar rates (with a mean of 0.163 g/h) were obtained for A1, B1, C1, A3, B3, C3, and even the D2 zone with a small increasing (0.184 g/h vs. 0.163 g/h). Whereas D1 and D3 exhibited the highest salting velocities (with a mean of 0.246 g/h), for the three salting treatments.
From the AVF, the salt uptake ranged from 0.029 to 0.245 g/h (with a mean value of 0.145 for the twelve zones) for conventional immersion, 0.025 to 0.241 (with a mean value of 0.143 for the twelve portions) for pulsed vacuum immersion, and 0.018 to 0.256 g/h (with a mean value of 0.148 for all the zones) for vacuum immersion. The comparison of mean value for the three salting processes, did not show a significant difference ($p > 0.05$) utilizing this approach, that was attributed to the influence of cheese porosity, therefore additional studies are recommended.
A graphic expression of the AVF values is presented by Figure 2, showing a similar trend of the three salting treatments.

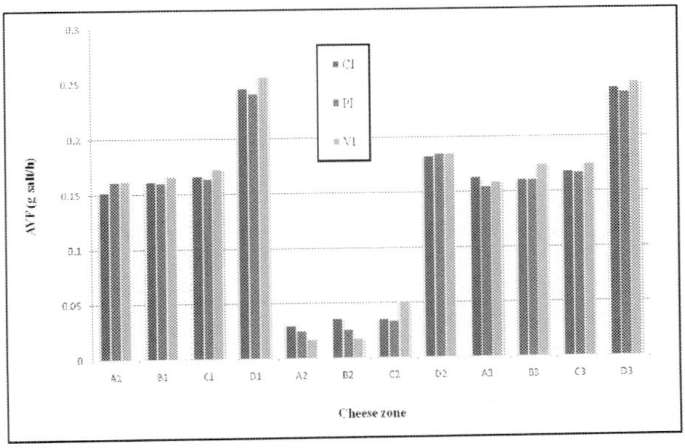

Fig. 2.

Thus, to model the salt or other component diffusion, there are several mathematical approaches, that imply limitations, advantages and disadvantages as well. To select the proper modeling will be function of the focus of the particular study.

4. Final remarks

The mass transfer phenomenon is very important through food transformation, manufacturing and preservation. Cheese as a biological system is characterized by a complex matrix in which all its components are exposed to mass transfer, either by diffusion as the most common or by convection. Although there are works related to mass transfer in cheese, mainly covering diffusion aspects, still there is a necessity of additional studies in order to achieve a more complete knowledge.

Salting as the most transcendental and analyzed mass transport process of cheese manufacturing has been satisfactory characterized, being the Fick´s mathematical approach the most utilized. Diffusion coefficients for various solutes involved in the brining or salting stage, have exhibited values in a range of 0.22 - 4.17 x 10^{-6} m^2/s for NaCl, obtained for different cheese types in an enormous variety of process and experimental conditions; other solute diffusivities have been scarcely quantified.

Furthermore to the diffusion approach, other mathematical solutions have been applied, such the average velocity factor, finite element, hydrodynamic mechanism and numerical approaches, offering advantages and limitations to each salt transport in cheese. The average velocity factor as an integral approach used to model the salting process, imply disadvantages as the rest of the analytical alternatives. More experimental studies are recommended in order to complete a clear scope and to model accurately this outstanding mass transport process of cheese salting.

5. References

Bird R.B., Stewart W.E. and Lightfoot E.N. 1960. Transport Phenomena. John Wiley and Sons, Inc. NY, USA

Bona E., Carneiro R.L., Borsato D., Silva R.S.S.F., Fidelis D.A.S. and Monkey e Silva L.H. 2007. Simulation of NaCl and KCl mass transfer during salting of Prato cheese in brine with agitation: A numerical solution. *Brazilian Journal of Chemical Engineering*, 24 (3): 337-349.

Cengel Y.A. and Boles M.A. 2006. Thermodynamics. An Engineering Approach. Mc Graw-Hill. Singapure.

Crank J. 1983. The Mathematics of Diffusion. Clarendon Press. Oxford, England

De Piante D., Castelao E. and Rubiolo A. 1989. Diffusion coefficient of salt in salting and ripening soft cheese. *Proceedings of ICEF 5*. Vol 1: 493-502.

Fennema O.W. 1996. Food Chemistry. Marcel Dekker, Inc. N.Y., USA.

Gerla P.E. and Rubiolo A.C. 2003. A model for determination of multicomponent diffusion coefficients in foods. *Journal of Food Engineering*, 56, 401–410.

Geurts T.J., Walstra P. and Mulder H. 1974. Transport of salt and water during salting of cheese. 1. Analysis of the processes involved. *Netherland Milk Dairy Journal*, 28: 102-129.

Geurts T.J., Walstra P. and Mulder H. 1980. Transport of salt and water during salting of cheese. 2. Quantities of salt taken up and moisture lost. *Netherland Milk Dairy Journal*, 34: 229-254.

González-Martínez C., Chafer M., Fito P. and Chiralt A. 2002. Development of salt profiles on Manchego type cheese during brining. Influence of vacuum pressure. *Journal of Food Engineering*, 53: 67-73.

Guinee T. and Fox P.F. 1983. Sodium chloride and moisture changes in Romano-type cheese during salting. *Journal of Dairy Research*, 50: 511–518.

Hélias A., Mirade P.S. and Corrieu G. 2009. Modeling of Camembert-type cheese mass loss in a ripening chamber: Main biological and physical phenomena. *Journal of Dairy Science*, 90: 5324–5333.

Illescas-Chávez E. and Vélez-Ruiz J.F. 2009. Effect of the salting process on the mass transfer kinetics of manchego-type cheese. *International Journal of Food Properties*. 12 (4): 791-807.

Izady L., Hamdami N. and Mohammadi S.A. 2009. Determination of moisture diffusivity in Iranian white cheese during brining. *Journal of Science and Technology. Agriculture and Natural Resources*. 13 (47A): 25-33.

Karel M. and Lund D.B. 2003. Physical Principles of Food Preservation. Marcel Dekker, Inc. N.Y., USA.

Luna J.A. and Bressan J.A. 1986. Mass transfer during brining of cuartirolo Argentino cheese. *Journal of Food Science*, 51 (3): 829-831.

Luna J.A. and Bressan J.A. 1987. . Mass transfer during ripening of cuartirolo Argentino cheese. *Journal of Food Science*, 52 (2): 308-311.

Marcos A., Alcalá M., León F., Fernández-Salguero, J. and Esteban, M.A. 1981. Water activity and chemical composition of cheese. *Journal of Dairy Science*, 64: 622–626.

Melilli, C. Barbano, D.M. Licitra, G. Tumino, G. Farina, G. and Carpino, S. 2003a. Influence of presalting and brine concentration on salt uptake by Ragusano cheese. *Journal of Dairy Science*, 86 (4): 1083–1100.

Melilli C., Barbano D.M., Licitra G., Portelli G., Di Rosa, G. and Carpino, S. 2003b. Influence of the temperature of salt brine on salt uptake by Ragusano cheese. *Journal of Dairy Science*, 86 (9): 2799–2812.

Mellili C., Carcó D., Barbano D.M., Tumino G., Carpino, S. and Licitra G. 2005. Composition, microstructure and surface barrier layer development during brine salting. *Journal of Dairy Science*, 68: 2329–2340.

Morris H.A., Guinee T.P. and Fox P.F. 1985. Salt diffusion in Cheddar cheese. *Journal of Dairy Science*, 88: 1851–1858.

Okos M.R., Narsimhan G., Singh R.K. and Weitnauer A.C. 1992. Food dehydration. In *Handbook of Food Engineering* (D.R Heldman. and D.B. Lund; Eds.). Marcel Dekker, Inc. NY, USA. 437- 562.

Pajonk A.S., Saurel R. and Andrieu J. 2003. Experimental study and modeling of effective NaCl diffusion coefficient values during Emmental cheese brining. *Journal of Food Engineering*, 60, 307–313.

Payne M.R. and Morison K.R. 1999. A multi-component approach to salt and water diffusion in cheese. *International Dairy Journal*, 9: 887-894.

Peleg, M. An empirical model for the description of moisture sorption curves. 1988. *Journal of Food Science*, 53 (4): 1216–1219.

Sánchez E. S., Simal S., Femenia A., Benedito J. and Roselló C. 1999. Influence of ultrasound on mass transport during cheese brining. *European Food Research and Technology*, 209: 215-219.

Saravacos G. 1986. Mass Transfer Properties of Foods. In *Engineering Properties of Foods* (M.A. Rao and S.S.H. Rizvi; Eds.). Marcel Dekker, Inc. N.Y., USA.

Saurel R., Pajonk A. and Andrieu J. 2004. Modeling of French Emmental cheese water activity during salting and ripening periods. *Journal of Food Engineering*, 63: 163-170.

Toledo R. 2007. Fundamentals of Food Process Engineering. Springer Science+Business Media, LLC. N.Y., USA.

Turhan M. and Kaletunc G. 1992. Modeling of salt diffusion in white cheese during long-term brining. *Journal of Food Science, 57* (5), 1082–1085.

Vélez-Ruiz J. F. 2001. Water Activity. In *Methods to Measure Physical Properties in Food Industries*. (J.D. Alvarado and J.M. Aguilera; Eds.) Ed. Acribia, S.A. Zaragoza, Spain. (In Spanish).

Vélez-Ruiz J.F. 2009. Notes of Transport Phenomena for Food Engineering. Universidad de las Américas, Puebla. Unpublished document. Puebla, Mexico.

Vélez-Ruiz J.F. 2010. Notes of Milk Science and Processing. Universidad de las Américas, Puebla. Unpublished document. Puebla, Mexico.

Welti-Chanes J., Mújica Paz H., Valdez Fragoso A. and León Cruz R. 2003. Fundamentals of Mass Transport. In *Transport Phenomena in Food Processing*. (J Welti-Chanes, J. Vélez-Ruiz and G. Barbosa-Cánovas; Eds.). CRC Press. Boca Raton, USA. 3-23.

Welty J.R., Wicks C.E. and Wilson R.E. 1976. Fundamentals of Momentum Heat and Mass Transfer. John Wiley and Sons, New York, NY, USA.

Wiles P.G. and Baldwin A.J. 1996a. Dry salting of cheese. Part I: Diffusion. *Transactions Institution of Chemical Engineers,* 74 (C): 127-132.

Wiles P.G. and Baldwin A.J. 1996b. Dry salting of cheese. Part II: Variability. *Transactions Institution of Chemical Engineers*, 74 (C): 133-139.

Zorrilla S.E. and Rubiolo A.C. 1991. Average NaCl concentration in cheese for different volume ratios of brine and solid during salting. *Journal of Food Science*, 56 (6): 1548- 1551.

Zorrilla S.E. and Rubiolo A.C. 1994. Modeling NaCl and KCl movement in Fynbo cheese during salting. *Journal of Food Science*, 59 (5): 976–980.

Part 4

Mass Transfer in Large-Scale Applications

17

Comprehensive Survey of Multi-Elements in Coastal Sea and Stream Sediments in the Island Arc Region of Japan: Mass Transfer from Terrestrial to Marine Environments

Atsuyuki Ohta and Noboru Imai
Geological Survey of Japan, AIST
Japan

1. Introduction

The dispersion of sediments across terrestrial areas, estuaries, shelves, slopes, and basins in island arc regions is extremely active. Because Japan is surrounded by vast oceans, it is fundamentally important to know sediment characteristics and dispersal systems upon the shelf and slope in terms of the erosion, transport, and deposition of sediments. Coastal seas, especially inner bays, often sustain severe damage from anthropogenic activities. Geochemical characteristics of marine sediments provide valuable clues regarding the dispersion of sediments. So far, the diffusion and fixation process of elements in the vertical direction and the past sedimentary environment found in core samples have attracted considerable scientific interest. However, regional spatial distributions of the chemical compositions of sediments in land, estuaries, and coastal areas are less understood (e.g., Balls *et al.*, 1997; Degens *et al.*, 1991; Ibbeken & Schleyer, 1991; Irion *et al.*, 1995; Karageorgis *et al.*, 2005; Voutsinou-Taliadouri & Varnavas, 1995; Wang *et al.*, 2008).

To better understand the spatial distribution of the elements in the earth's surface, a geochemical map is used. The purpose of the map is to obtain natural geochemical baselines on the earth's surface because the geochemical history of the earth's surface is a fundamental part of geo-information. Many countries have produced national geochemical atlases to explore mineral recourses and address growing concerns about environmental problems (Fauth *et al.*, 1985; Gustavsson *et al.*, 2001; Lis & Pasieczna, 1995; Shacklette & Boerngen, 1984; Thalmann *et al.*, 1988; Weaver *et al.*, 1983; Webb *et al.*, 1978; Xie *et al.*, 1997; Zheng, 1994). Simultaneously, cross-boundary and sub-continental geochemical mapping projects have actively been carried out (e.g., Bølviken *et al.*, 1986; Reimann *et al.*, 1998; Salminen *et al.*, 2005). Furthermore, the International Geochemical Mapping Project (IGCP 259) was initiated in 1988 under the auspices of the UNESCO to compile data on the global geochemical composition of the earth's surface (Darnley *et al.*, 1995).

Thus, geochemical mapping has been undertaken globally, but it is restricted to terrestrial areas because surveys of soils, stream sediments, and river water samples are easy. However, 70% of the earth's surface is covered by water. Surveying marine sediments is important to elucidate the mass transfer from land to sea. The Geological Survey of Japan, National Institute of Advanced Industrial Science and Technology, has conducted research on the fundamental geochemical properties of earth surface materials (Imai *et al.*, 2004, 2010). For a geochemical mapping project, 3,024 river sediments and 4,905 surface marine sediments around Japan were collected and were subsequently analyzed for 53 elements including toxic elements (e.g., As, Cd, and Hg). We have taken the initiative in determining the regional elemental distribution for both the land and sea. We intend to elucidate the geochemical differences of terrestrial surface sediments and coastal and open sea sediments comprehensively.

Geochemical variation occurring among terrestrial areas has been explained conclusively in relation to the lithology and distribution of mineral deposits. Japanese geology has particularly complicated distributions, and many mineral deposits occur on a smaller scale than those of Europe, the United States, or China, which are situated on older continental terrains. Basically, regional geochemical maps are intended to be utilized for environmental assessments. To achieve a more objective and quantitative interpretation of geochemical maps, we carried out a watershed hydrological estimation of geology using the Geographic Information System and statistical methods such as the analysis of variance and multiple comparison tests to analyze geochemical datasets (Ohta *et al.*, 2004a, 2004b, 2005a, 2005b).

In contrast, large geochemical mapping datasets in the ocean present difficulties in determining the dominant geochemical factors controlling geochemical distributions in coastal areas (Imai *et al.*, 1997; Ohta *et al.*, 2004b, 2007, 2010). This is because of the following effects: (1) mass transport from the land to coastal seas, (2) particle size effects, (3) transportation of coastal sediments by a gravity flow or by an oceanic current, (4) mixing of volcanic materials such as tephra and pumice, (5) early diagenetic processes, (6) denudation or resedimentation of basement rocks, (7) biogenic products such as shell fragments, foraminifera tests, and fecal pellets, and (8) contamination caused by human activity. In this paper, we introduce comprehensive land-marine geochemical maps that can be used by persons unfamiliar with geochemical maps. Furthermore, we discuss the mass transfer process on the earth's surface using these geochemical maps.

2. Study area

2.1 Terrestrial areas

Japan consists of over 6,800 islands of various sizes. In this study, we focus on four major islands: the Hokkaido, Honshu, Shikoku, and Kyushu Islands. Japanese geography is characterized by variable topography (Fig. 1a). The central region contains the high mountains. From this region, the elevation decreases steeply from 1,000–3,000 m to sea level within a distance of only 20–30 km. The main streams flowing from these mountains formed large coastal plains during the Pliocene-Pleistocene. Rice fields, residential areas, and industrial areas extend throughout these flood plains.

Figure 1b shows a geological map of the studied area, simplified from the Geological Map of Japan (scale: 1:1,000,000) (Geological Survey of Japan, 1992). Japan originated at the east end of the Eurasian plate and formed as an island arc during the Neogene period. The geology of Japan is quite different from that of Europe, America, and China, which are old continental terrains. Around 40% of Japan is covered by poorly lithified sediments formed in the Quaternary-Neogene age and sedimentary rocks deposited before the Paleogene era. Accretionary complexes (accretionary prisms or accretionary wedge) consist mainly of melange, mudstone, and sandstone and are associated with small amounts of chert, limestone, greenstone, gabbro, and ultramafic rocks. They are subducted sediments and were formed by scraped trench-fill deposits and pelagic sediments on the oceanic plate during the subduction process. The sedimentary rocks and igneous rocks of accretionary complexes reside in 15% and 2% of the area, respectively.

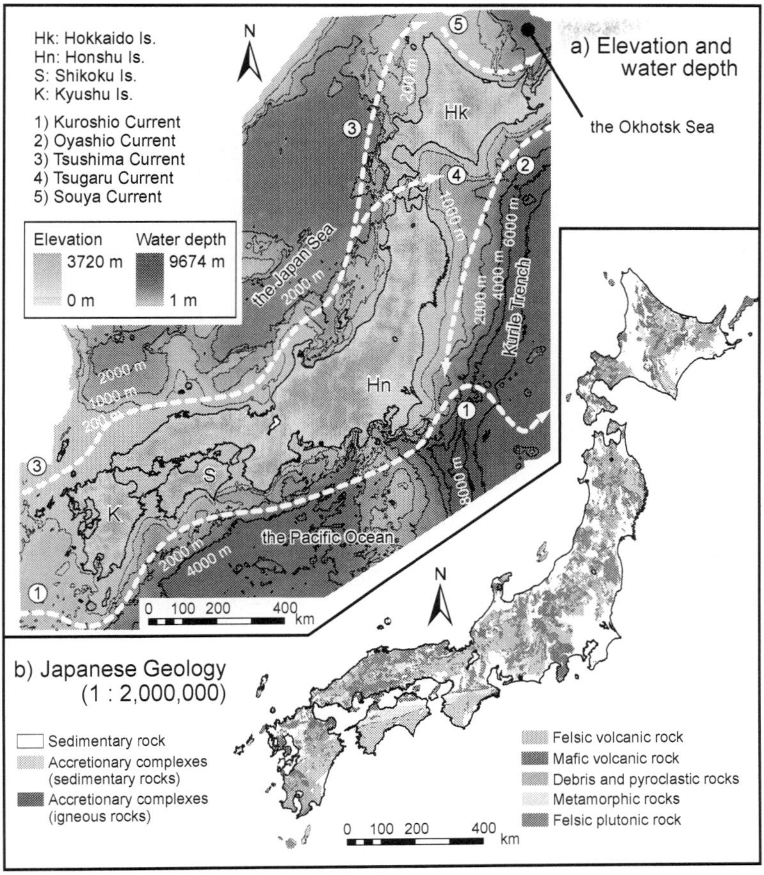

Fig. 1. a) Japanese topography and bathymetric features. b) Japanese geological map

Japan is a very active volcanic island. Non-alkaline volcanic rocks (mainly dacitic-andesitic lava, tuff, and pyroclastic flow materials) were formed in the Neogene-Quaternary and crop out in 23% of the area. Granitic rocks mostly intruded during the Cretaceous age and compose 10% of the area. Metamorphic rocks are scarce in the area (less than 4%). Many mineral deposits of hydrothermal vein-, skarn-, Kuroko-, and Kieslager-type bearing Au, Ag, As, Cu, Zn, Cd, Sb, Hg, and Pb are distributed in Japan.

2.2 Coastal sea area

Japan is surrounded by three oceans: the Pacific Ocean, the Japan Sea, and the Okhotsk Sea (Fig. 1a). Japan has a narrow continental shelf except in the west and north. In the Pacific

Ocean, the marine environment has a steep slope from the coast: the water depth reaches 1,000 m just 20–40 km offshore. The Kuril Trench is a deep valley where the Pacific Plate is subducting and the water depth is greater than 7,000 m. The west part of the Pacific Ocean has some deep valleys where the Philippine Plate is subducting and the water depth is greater than 2,000 m. Gravity flow deposits are found extensively on the slope and in these submarine canyons. The Oyashio Current flows in the counter clockwise direction and the Kuroshio Current flows in the SW-NE direction. The Japan Sea has an average depth of 1,750 m. The wide continental shelf spread is in the western and northern part of the Japan Sea. The Tsushima Current, the branch current of the Kuroshio Current, flows from SW to NE. The current conveys a warm water mass to the surface of the sea. In the deep regions of the sea, cold and oxygen-rich water exists, which is formed in the winter season. The wide continental shelf is also present in the Okhotsk Sea. The Soya Current, a branch current of the Tsushima Current, flows on the shelf in the counter clockwise direction. Drift ice covers the area from February to April, thereby transporting allochthonous debris. Taken as a whole, coarse and fine sands are distributed on the shelf. Very fine sand, silt, and clay are distributed on inner bays, slopes, and basins. Coarse sediments distributing in the shelf generally contain a large amount of shell fragments.

3. Materials and methods

The 3,024 stream sediments used for the combined geochemical mapping were collected from adjacent terrestrial regions covering an area of 378,000 km^2 during 1999–2003 (Fig. 2). The sampling density is one sample per 80–120 km^2 (Imai *et al.*, 2004). The stream sediments were dried in air and sieved with an 83-mesh (180 µm) screen. Magnetic minerals were removed using a magnet, thereby minimizing the effect of the accumulation of magnetic minerals. The samples were not milled. Sieving might change elemental concentrations from the bulk stream sediments. However, Imai (1987) suggested that sieving with an 83-mesh (180 µm) screen hardly changes elemental concentrations from bulk stream sediments, although enrichment of Cu, Ni, Zn, and Pb appears in the finer fractions passing through a 390-mesh (38 µm) screen.

In all, 4,905 marine sediments were collected using a grab sampler in 1980–2007 (Fig. 2). These samples represent an area of c.a. 460,000 km^2. The mean sampling density of the coastal sea sediments is one sample per 90 km^2. The uppermost area of 0–3 cm of the sediments collected with the grab sampler was separated, dried in air, ground with an agate mortar and pestle, and retained for chemical analyses. Rocky, cobble, pebble, and gravel sediments were not used for geochemical mapping. Alternatively, matrix sediments were collected from these samples.

First, 0.2 g of each sample was digested using 5 mL HF, 4 mL HNO_3, and 3 mL $HClO_4$ solutions at 120°C for 2 h (Imai, 1990). The degraded product was evaporated to dryness under 200°C, and the residue was dissolved with 100 mL of 0.35 M HNO_3 solution. The concentrations of the 51 elements were determined using ICP-AES (for Na, Mg, Al, P, K, Ca, Ti, Mn, Fe, V, Sr, and Ba) and ICP-MS (for Li, Be, Sc, Cr, Co, Ni, Cu, Zn, Ga, Rb, Zr, Nb, Y, Mo, Cd, Sn, Sb, Cs, La, Ce, Pr, Nd, Sm, Eu, Gd, Tb, Dy, Ho, Er, Tm, Yb, Lu, Hf, Ta, Tl, Pb, Bi, Th, and U). The As and Hg analyses were subcontracted to ALS Chemex (Vancouver, BC).

The geochemical reference samples JB-1 and JB-3 (Imai *et al.*, 1995) were inserted at the rate of 10 and 20 samples, respectively, for quality control. The methodology was examined by Imai (1990) and Ujiie-Mikoshiba *et al.* (2006). They concluded that the precision and accuracy

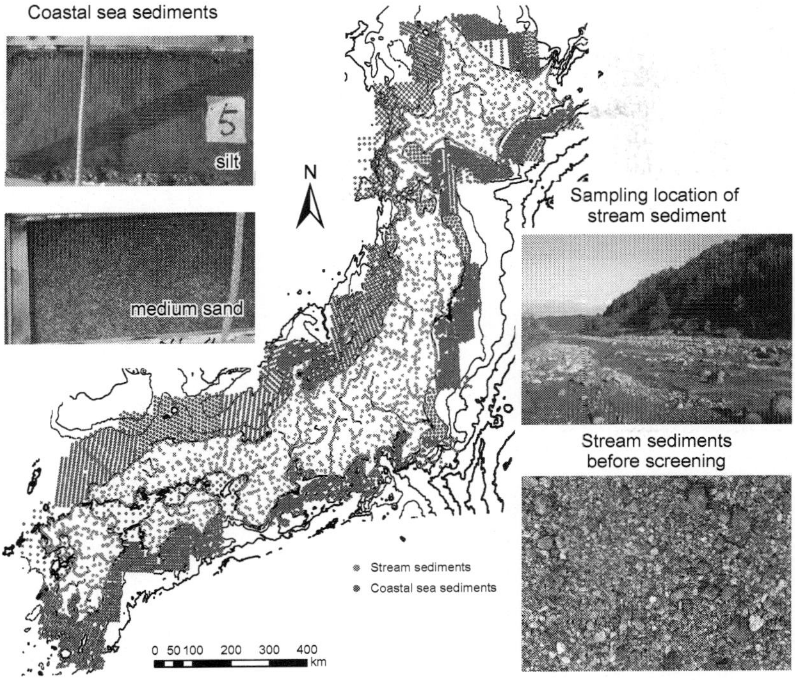

Fig. 2. Sampling locations. Bathymetric features are the same as Fig. 1

of the analytical data were satisfactory, although the method gave higher concentrations of Mg and lower concentrations of Al and heavy rare earth elements for some samples. A part of the heavy mineral fraction, especially zircon, is not digested satisfactorily by HF, HNO_3, and $HClO_4$. Incomplete decomposition of these minerals seriously affects a few elements, such as Zr and Hf (Imai, 1987, 1990). Therefore, the data regarding Zr and Hf in the sediments should be used only as a guide. In addition, the marine sediments were not desalinated. The Na concentrations in marine sediments are also used as a guide. Table 1 presents the median value of the analytical results obtained for the coastal sea and river sediments. Major elements (Na, Mg, Al, P, K, Ca, Ti, Mn, and Fe) in the sediments are expressed as oxides because they are most abundant in the Earth's crust and their abundance is frequently expressed in terms of weight percent oxide. Although Mn and Fe exhibit three and two valences in the environment, respectively, their oxides are expressed as MnO and Fe_2O_3 in this paper.

4. Geochemical map preparation

Geographic Information System software (ArcGIS 9.3; Environmental Systems Research Institute, ESRI) was used to prepare geochemical maps. Geochemical maps were created as mesh maps. The elemental concentration of each cell was an interpolation of scatter points based on the inverse distance weighted method (Watson and Philip, 1985). This method

Elements	Units	River Sed.	Marine Sed.	Elements	Units	River Sed.	Marine Sed.
Li	mg/kg	28	31	Sn	mg/kg	2.1	1.5
Be	mg/kg	1.32	1.17	Sb	mg/kg	0.60	0.52
Na_2O	wt. %	2.16	3.08	Cs	mg/kg	3.6	3.0
MgO	wt. %	2.56	2.81	Ba	mg/kg	410	320
Al_2O_3	wt. %	10.87	9.07	La	mg/kg	16	13
P_2O_5	wt. %	0.12	0.12	Ce	mg/kg	30	26
K_2O	wt. %	1.74	1.71	Pr	mg/kg	3.9	3.2
CaO	wt. %	2.34	4.01	Nd	mg/kg	16	13
Sc	mg/kg	13	10	Sm	mg/kg	3.3	2.7
TiO_2	wt. %	0.72	0.46	Eu	mg/kg	0.82	0.68
V	mg/kg	120	86	Gd	mg/kg	3.1	2.5
Cr	mg/kg	53	47	Tb	mg/kg	0.55	0.44
MnO	wt. %	0.12	0.06	Dy	mg/kg	2.9	2.2
Fe_2O_3	wt. %	5.74	4.34	Ho	mg/kg	0.56	0.43
Co	mg/kg	14	9	Er	mg/kg	1.7	1.3
Ni	mg/kg	21	20	Tm	mg/kg	0.27	0.20
Cu	mg/kg	27	16	Yb	mg/kg	1.7	1.2
Zn	mg/kg	107	75	Lu	mg/kg	0.25	0.18
Ga	mg/kg	16	13	Hf	mg/kg	1.5	1.3
As	mg/kg	8	7	Ta	mg/kg	0.57	0.46
Rb	mg/kg	66	54	Hg	mg/kg	0.04	0.05
Sr	mg/kg	150	220	Tl	mg/kg	0.47	0.40
Y	mg/kg	17	13	Pb	mg/kg	21	19
Zr	mg/kg	54	47	Bi	mg/kg	0.23	0.25
Nb	mg/kg	7.2	5.3	Th	mg/kg	5.4	4.2
Mo	mg/kg	1.2	0.9	U	mg/kg	1.3	1.1
Cd	mg/kg	0.13	0.08				

Table 1. Median values of 53 elemental concentrations of river (N = 3,024) and coastal sea sediments (N = 4,905) in Japan

assumes that interpolating surface data is influenced predominately by the nearby points and less by distant points. In this study, the concentration data for a 2 km × 2 km cell was calculated from the nearest 12 points. A power value of 2 was chosen to decrease the weight of distant values. The stream sediment is considered to consist of soil and rocks in the catchment area upstream from the sampling site (Howarth and Thornton, 1983). A sampling site is presumed to express the average chemical concentrations in a drainage basin. Therefore, several additional data points were generated within each river basin in addition to the original survey points (3,024 points); that is, the catchment area was expressed by a set of points. This method facilitates the creation of mesh maps considering the watershed and the exploration of the data (Ohta *et al.*, 2004b). In contrast, the marine geochemical maps were simply interpolated as a mesh map with a 2 km × 2 km cell, because it was not necessary to consider the watershed. The resultant marine geochemical maps were combined with the existing land geochemical maps, which were depicted by considering the watershed.

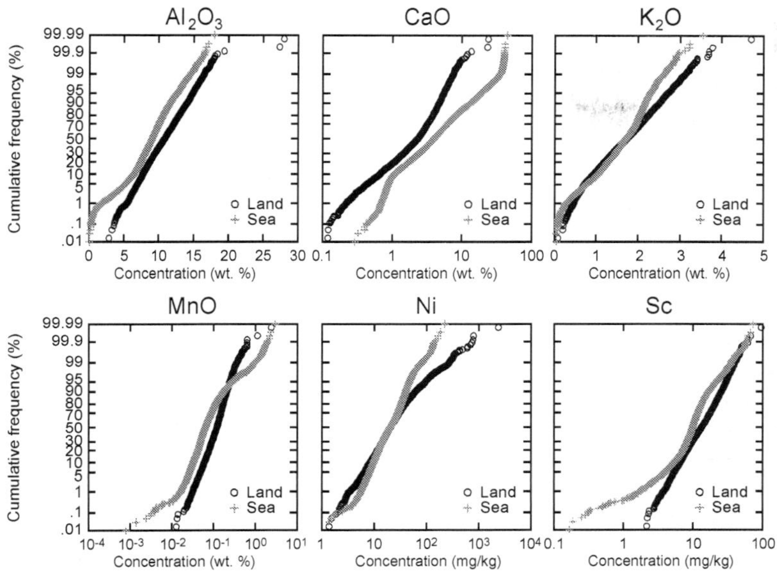

Fig. 3. The cumulative probability plots of six elements

A geochemical map is expressed as a color map, where each cell is color-coded according to a particular classification. In this paper, we applied percentiles ($0 \le x \le 5$, $5 < x \le 10$, $10 < x \le 25$, $25 < x \le 50$, $50 < x \le 75$, $75 < x \le 90$, $90 < x \le 95$, and $95 < x \le 100\%$, where x represents the elemental concentration) as a classification according to Reimann (2005). Furthermore, the classification of elemental concentrations was calculated separately in terrestrial and coastal areas because marine sediments and stream sediments differ in their chemical and mineralogical compositions, particle size, and origin (Ohta *et al.*, 2004b, 2007, 2010). Figure 3 portrays the cumulative probability plots of six elements. The Al (Al_2O_3) contents in land are systematically higher than those in the coastal sea; the Ca (CaO) contents are in stark contrast to Al (Al_2O_3). The ranges of K (K_2O), Mn (MnO), and Sc concentrations in stream sediments are smaller than those in coastal sea sediments; the Ni contents show an opposite result. Thus, the cumulative probability trends of most elements in coastal sea sediments are different from those of stream sediments. Therefore, the regional geochemical differences (color variation) of most elements in the sea or the land will be obscured if a common class selection is applied to geochemical maps combining the two sample materials. The geochemical patterns displayed in different sample materials using percentiles can be compared directly because the same range (e.g. 90–95%) implies the same statistical weight, even on different scales (Reimann, 2005).

5. Comparison of elemental abundance in coastal sea sediment, stream sediment, and Japanese upper crusts

Figure 4 shows the median concentrations of 41 elements for the stream and coastal sea sediments. These concentrations have been normalized to the continental upper crust data

(Taylor & McLennan, 1995). Coastal sea and stream sediments contain water and organic matters that are different from rock samples. For example, Ujiie-Mikoshiba *et al.* (2006) reported that the lost of ignition at 1000 °C for 1 h of 16 stream sediments in northern Japan was 8.1% average. Their elemental concentrations became systematically lower than those of rock samples. Therefore, each elemental concentration is normalized by Al_2O_3 content.

Overall, the coastal sea sediments show a pattern similar to that of the stream sediments. In particular, the relative abundance of Be, K, Rb, Nb, Cs, light rare earth elements (light REEs), and Ta, which are lithophile elements abundant in felsic rocks, are consistent to one another. This result supports the theory that coastal sea sediments were originally adjacent terrestrial materials. The coastal sea sediments have higher concentrations of Na (Na_2O), Mg (MgO), Ca (CaO), and Sr than stream sediments. This enrichment is caused by calcareous shells and sea salt. Coastal sea sediments have systematically lower concentrations of Ti (TiO_2), V, Mn (MnO), Fe (Fe_2O_3), Co, Cu, and Zn, which are lithophile and siderophile elements that are abundant in mafic rocks. Cu, Zn, As, Mo, Cd, Sn, Sb, Hg, Pb, and Bi are chalcophile elements. The abundance of As, Mo, Sb, and Pb in coastal sea sediments is comparable to those of stream sediments. The coastal sea sediments have lower Cu, Zn, Cd, and Sn abundance and higher Hg and Bi abundance than stream sediments.

For comparison, the geochemical data of the Japanese upper crust is also portrayed in Fig. 4. This is compiled data from analytical values of 166 rock types distributed in Japan (Togashi *et al.*, 2000). The most important characteristic exhibited is that the elemental abundance of coastal sea and stream sediments are comparable to that of the Japanese upper crust. The Japanese upper crust is enriched in Li, Sc, V, Cr, Mn (MnO), Co, Ni, As, Sb, and Cs and poor in Na (Na_2O), P (P_2O_5), K (K_2O), Sr, Nb, La, Ce, and Ta as compared to the upper continental crust. These chemical compositions reflect the geology of Japan. Sedimentary rocks in the accretionary complexes are abundant in Li, As, Sb, Cs, Tl, and Bi. Mafic volcanic rocks with high concentrations of 3d transition metals and unmetamorphosed sedimentary rocks enriched in As and Sb are distributed more widely in Japan as compared to the continental crust (Togashi *et al.*, 2000; Ujiie-Mikoshiba *et al.*, 2006). Rocks associated with mineral deposits are not compiled in the Japanese upper crust. The abundance of chalcophile elements such as Cu, Zn, Cd, and Pb is poorer than that in stream and coastal sea sediments. Except for these elements, stream and coastal sea sediments are useful to assess the chemical composition of the earth's surface instead of using rock samples.

6. Particle size dependence of elemental concentrations in coastal sea sediments

As we described above, stream sediments retained for geochemical mapping were sieved with 180-μm mesh and composed of fine sand and very fine sand. In contrast, coastal sea sediments used here vary in size. In some places, the spatial distribution of elemental concentrations in the marine environment is predominantly determined by the variation of particle sizes (Ohta *et al.*, 2004b). To validate this point, the effect of particle size on elemental concentrations in coastal sea sediments is evaluated.

Figure 5 shows the median elemental concentrations plotted against nine types of sediment particle sizes. It can be observed that most of the elemental concentrations increase with a decrease in the particle size, and become constant through sandy silt to silty clay. However, there are some exceptions to this rule. Calcium (CaO) and Sr are abundant in very coarse and coarse sand because they contain a large amount of calcareous shells, shell fragments,

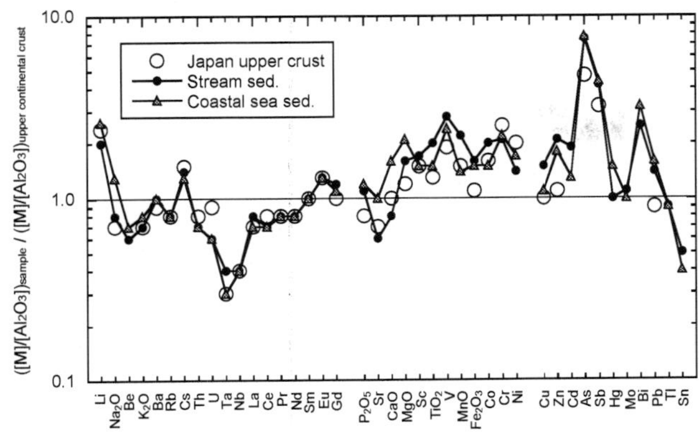

Fig. 4. Chemical compositions of coastal sea sediments, stream sediments, and the Japanese upper crust normalized to the continental upper crust data. The [M] means elemental concentration

and foraminiferas. The concentrations of Mn (MnO), Fe (Fe_2O_3), and Co are high in coarse and medium sands, which indicate the presence of Fe-Mn oxides. The Mn^{2+}, Fe^{2+}, and Co^{2+} are easily oxidized in water and precipitate on coarse grains. Little systematic differences are found in Al (Al_2O_3), Ti (TiO_2), Y, and heavy REEs. The Mg (MgO), V, Mn (MnO), and As are enriched both in coarse sand and very fine sediments (silt and clay). Except for these elements, most elemental concentrations increase with decreasing particle size. This trend is attributed mainly to the dilution effect by quartz and calcareous materials. Finer sediments are distributed in the outer sea. We must pay attention to the fact that elemental concentrations become gradually higher with increasing distance out to sea because particle size distribution produces geochemical graduation in marine environments irrespective of mass transfer from land to sea.

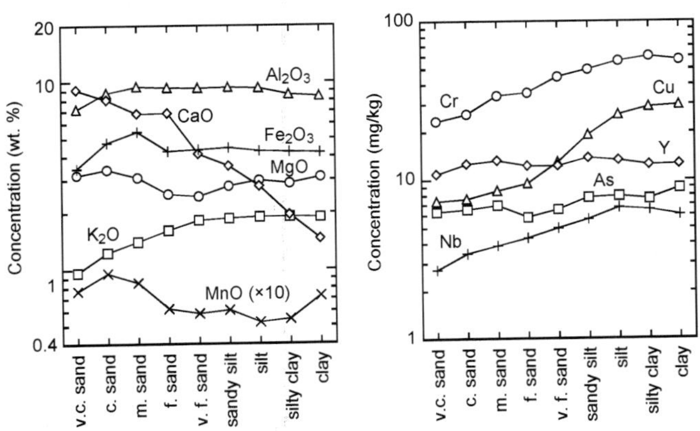

Fig. 5. Median of elemental concentrations of coastal sea sediments according to the particle size classification. The abbreviations v. c. sand, c. sand, m. sand, f. sand, v. f. sand refer to vary coarse sand, coarse sand, medium sand, fine sand, and very fine sand, respectively

7. Spatial distribution patterns of elemental concentrations in coastal seas and adjoining terrestrial areas

The spatial distribution of elements in the terrestrial area is predominantly controlled by parent lithology. The influence of mineral deposits and anthropogenic activity are restricted in small areas, but they cause extreme enrichments of elements. The analysis of controlling factors using the geographic information system and statistical analysis demonstrate that the following seven factors are the most dominant cause for the spatial distribution of elemental concentrations (Ohta *et al.*, 2004a, b, 2005a, b):

a. Accretionary complexes (sedimentary rocks): Li, Be, K (K_2O), Rb, Cs, and Tl
b. Felsic volcanic rocks: Be, K (K_2O), Rb, Cs, Ba, and Tl
c. Granitic rocks: Be, Na (Na_2O), K (K_2O), Nb, Sn, REEs, Ta, Th, and U
d. Mafic volcanic rocks: Mg (MgO), P (P_2O_5), Ca (CaO), Sc, Ti (TiO_2), V, Cr, Mn (MnO), Fe (Fe_2O_3), Co, Ni, Cu, and Zn
e. Ultramafic rocks: Mg (MgO), Cr, Co, and Ni
f. Mineral deposits: Cu, Zn, As, Mo, Cd, Sn, Sb, Hg, Pb, and Bi
g. Anthropogenic activities: P (P_2O_5), Cr, Co, Ni, Cu, Zn, As, Mo, Cd, Sn, Sb, Hg, and Pb

Figures 6 and 7 show geochemical maps of K (K_2O), Ca, (CaO), Sc, and Cu. Overall, elemental concentrations are continuous between the terrestrial area and the coastal sea area. K (K_2O) is abundant in the eastern Honshu Island and Sc has a reverse distribution as compared to K (K_2O). Their spatial distributions in the marine environment correspond well to the adjoining terrestrial area, where K (K_2O) is enriched in the area covered by accretionary complexes, felsic volcanic, and granitic rocks and Sc is enriched in the area covered by mafic volcanic rocks (Fig. 1b). Figure 4 suggests that K (K_2O) and Sc abundances in coastal seas match that of stream sediments and the Japanese upper crust materials. It is obvious that marine sediments originated from stream sediments on the adjacent terrestrial area.

In contrast, the spatial distribution of Ca (CaO) and Cu concentrations in coastal seas is not continuous to that in terrestrial areas, which is different from the cases of K (K_2O) and Sc. The Ca (CaO) and Cu are enriched in stream sediments derived from mafic volcanic rocks, which are similar to Sc. In the coastal sea, especially in shallow water depth, marine sediments are composed of shell fragments and foraminifera tests. The extremely high Ca (CaO) contents in such calcareous materials hide the contribution of Ca (CaO) in sediments derived from the adjacent terrestrial area. In contrast, the Cu concentration is low in shallow water depths but increases with water depth, except at the inner basin. The increasing trend of Cu concentrations in marine sediments also corresponds to systematic changes in particle size (Fig. 5). Although we will discuss the details later, early diagenetic processes and organic matters elevate Mn (MnO), Cu, Mo, Cd, Sb, Hg, Pb, and U concentrations in silty and clayey sediments. The abundance of Ca (CaO) and Cu in coastal sea sediments is different from that in stream sediments because of the input of biogenic materials and early diagenetic process. Therefore, these geochemical maps do not reflect the horizontal mass transfer process.

8. Mass transport processes

8.1 Mass transport processes: contribution of terrestrial materials and transportation of coastal sediments by a gravity flow or by an oceanic current

The elemental abundance pattern and regional geochemical maps support mass transfer from land to sea. Next, we discuss the details of the horizontal mass transfer process using

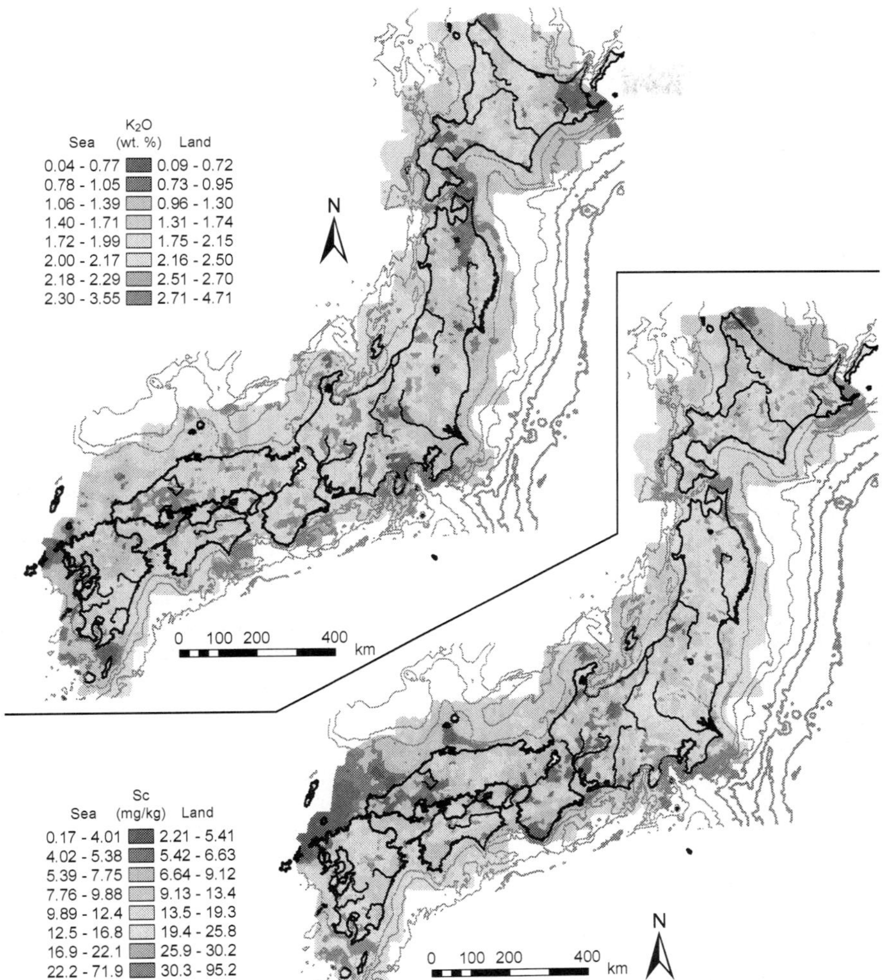

Fig. 6. Geochemical maps of K (K_2O) and Sc in stream and coastal sea sediments. Small islands are expressed as slightly darkened gray areas

local geochemical maps. Figure 8 shows some examples of visible mass transfers from land to sea. The area is characterized by the narrow continental shelf and the Toyama submarine valley, whose water depth is over 1,000 m, and is 500 km long. In the terrestrial area, granitic outcrops influence the distribution of Be, Na (Na_2O), K (K_2O), Rb, Y, rare earth elements (REEs), Tl, Th, and U (see K_2O in Fig. 8). Ultramafic rocks (mainly serpentinite) associated with accretionary complexes occupy only a small area. However, these rocks are

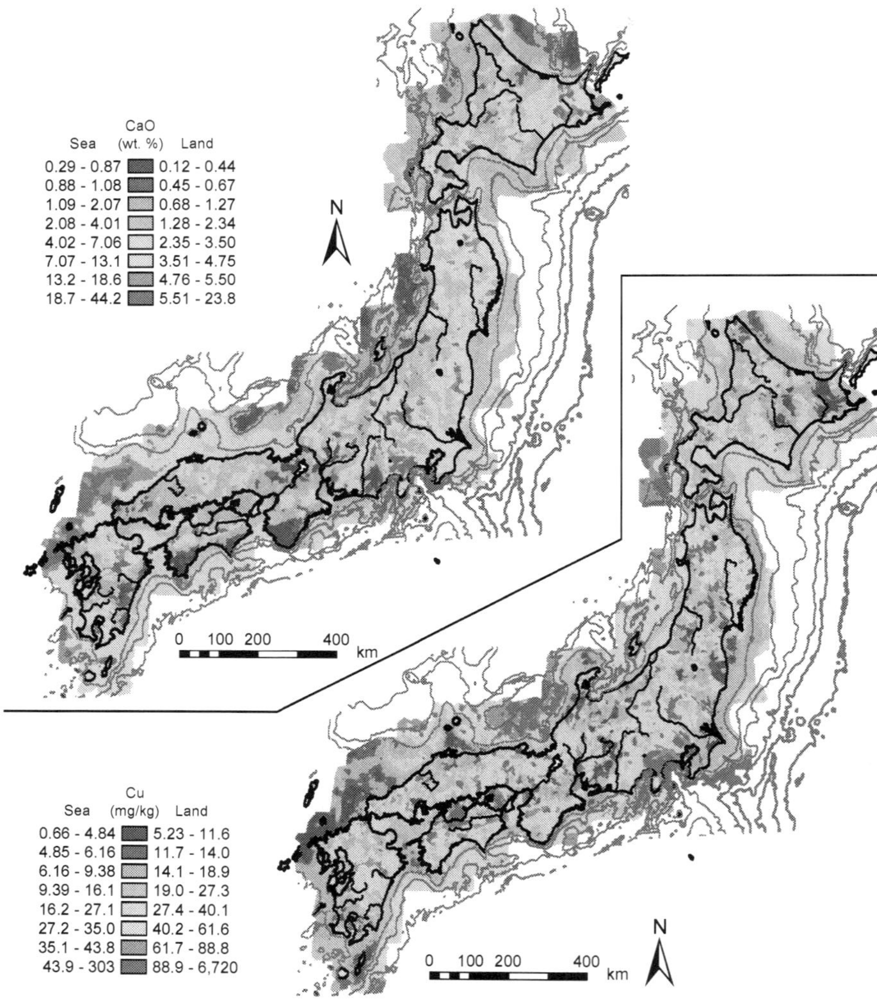

Fig. 7. Geochemical maps of Ca (CaO) and Cu in stream and coastal sea sediments. Small islands are expressed as slightly darkened gray areas

the source of the extremely high concentrations of Mg (MgO), Cr, Co, and Ni (see Cr in Fig. 8). It can be observed that the distribution of high K (K_2O) and Cr concentrations continues from land to sea. The distribution of K (K_2O) in coastal seas suggests that sediments supplied through rivers initially spread out in a fan from the continental shelf and slope (~20 km). In contrast, the spatial distribution of high levels of Mg (MgO), Cr, Co, and Ni contents extends to the outer sea along the deep valley (Fig. 8). Sediments deposited on an oceanic floor with depths of several hundred meters are rarely moved by a tidal wave and an oceanic current.

They are conveyed by the gravity flow that occurs over 100 km through the slope and the Toyama submarine canyon. The high concentration of K (K_2O) along the submarine canyons does not extend as far as the high concentration of Cr area. K (K_2O) is possibly associated with coarse particles such as K-feldspar, which is not transported far. In contrast, Cr is associated with fine particles such as serpentine, chromite, and chromium spinel.

In the northern part of Japan, fine sediments supplied from rivers are conveyed by an oceanic current up to a distance of 100–200 km along the coast. The Hokkaido Island, which is located in the northernmost region of Japan, is characterized by a wide continental shelf. The distribution of sediments is influenced strongly by a coastal current. Distribution of high Cr and Ni concentrations extends over 200 km on the shelf (see dotted arrows numbered 1 for Cr in Fig. 9). Underlying ultramafic rocks (mainly serpentinite) in the watershed upstream of rivers are the origin of the Cr and Ni enrichments on the shelf. However, the coastal current direction changes with the seasons. The extension of high Cr and Ni concentration to the north and northwest direction is explained by two reasons. Large amounts of sediments are supplied from rivers especially during spring and summer because of snowmelt (Katayama, 2008). The northward coastal current prevails during those seasons and transports large amounts of sediments associated with Cr and Ni northward (Katayama, 2008). Serpentine, chromite, and chromium spinel, including Cr and Ni, have very fine minerals (clay size) that can be transported long distances.

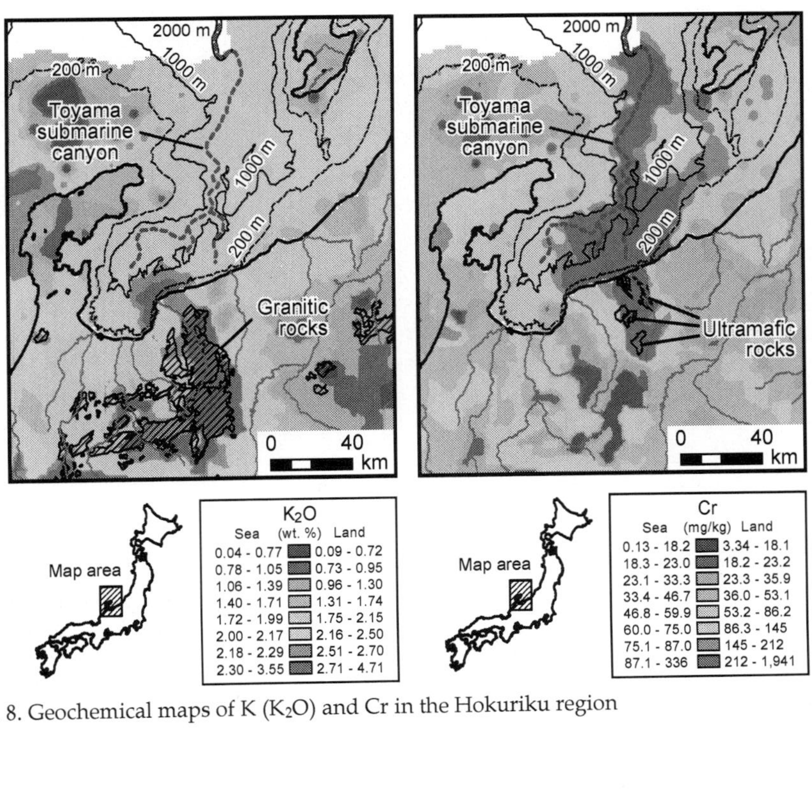

K2O Sea (wt. %)	Land
0.04 - 0.77	0.09 - 0.72
0.78 - 1.05	0.73 - 0.95
1.06 - 1.39	0.96 - 1.30
1.40 - 1.71	1.31 - 1.74
1.72 - 1.99	1.75 - 2.15
2.00 - 2.17	2.16 - 2.50
2.18 - 2.29	2.51 - 2.70
2.30 - 3.55	2.71 - 4.71

Cr Sea (mg/kg)	Land
0.13 - 18.2	3.34 - 18.1
18.3 - 23.0	18.2 - 23.2
23.1 - 33.3	23.3 - 35.9
33.4 - 46.7	36.0 - 53.1
46.8 - 59.9	53.2 - 86.2
60.0 - 75.0	86.3 - 145
75.1 - 87.0	145 - 212
87.1 - 336	212 - 1,941

Fig. 8. Geochemical maps of K (K_2O) and Cr in the Hokuriku region

In the southern part of the Hokkaido Island, high-concentration areas of Li, Be, Mg (MgO), Al (Al_2O_3), Ti (TiO_2), V, Cr, Fe (T-Fe_2O_3), Co, Ni, Nb, light REEs, and Ta exist in the shelf and adjoining terrestrial area (see a dotted arrow numbered 2 for Cr and a circle numbered 1 for Nb in Fig. 9). These spatial distribution patterns suggest that marine sediments are transported northwestward. The coastal current direction in this region also changes with the seasons (Kuroda *et al.*, 2006). Sediments are supplied in large amounts from rivers during spring-summer because of snowmelt. At that time, the northwestward coastal current prevails (Kuroda *et al.*, 2006), so that the sediments are transported northwestward. Sediments deposited on the shelf are further moved by gravity flows to the deep sea (the Hidaka Trough). In contrast, the areas of Nb, REEs, and Ta enrichment are limited to the near shore. Granitic rocks host Li, Be, Nb, REEs, and Ta, but they are scarce on the Hokkaido Island where mafic volcanic rocks and andesitic-dacitic pyroclastic rocks dominant the lithology. It is possible that sediments enriched in these elements supplied to the coastal sea are not abundant. Otherwise, Nb, REEs, and Ta are associated with coarse sediments, which are not transported far by the coastal current. In contrast, Mg (MgO), Cr, Co, and Ni are possibly included in fine minerals (such as serpentine), which are transported more easily by the coastal current.

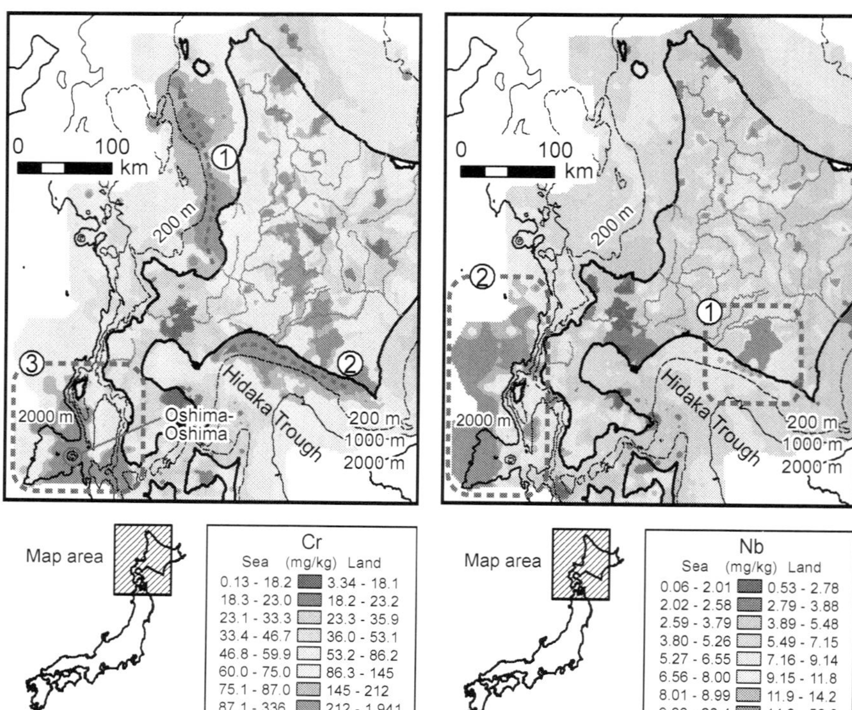

Fig. 9. Geochemical maps of Cr and Nb in the Hokkaido region

The mass transfer process of marine sediments is also influenced by a current rip and a periodical current. Figure 10 portrays the geochemical map of Cr around the Shikoku Island. Greenstones and ultramafic rocks (mainly serpentinite) in accretionary complexes host these elements in the adjacent terrestrial area. It can be observed that the silty and clayey sediments are distributing on the shelf and basin. These fine sediments have high Mg (MgO), Cr, Co, and Ni concentrations. The distribution pattern of high Cr concentrations follows a reverse J-shape (see a circle numbered 1 in Fig. 10). That characteristic distribution is explained by the sedimentation process occurring between two water masses and a gravity flow. Ikehara (1988) explained that the silty and clayey sediments supplied from rivers deposit on the shelf where the boundary of superjacent waters and oceanic water resides. Oceanic water is possibly a counter current of the Kuroshio Current. These fine sediments deposited on the shelf are further conveyed to the deep-sea basin by a gravity flow. In contrast, the NNW-SSE trend of high Mg (MgO), Cr, Co, and Ni concentrations is found in the straight region (see a circle numbered 2 in Fig. 10). The feature clearly indicates that the fine sediments abundant in Mg (MgO), Cr, Co, and Ni are supplied mainly from the Shikoku Island and conveyed by a periodical current.

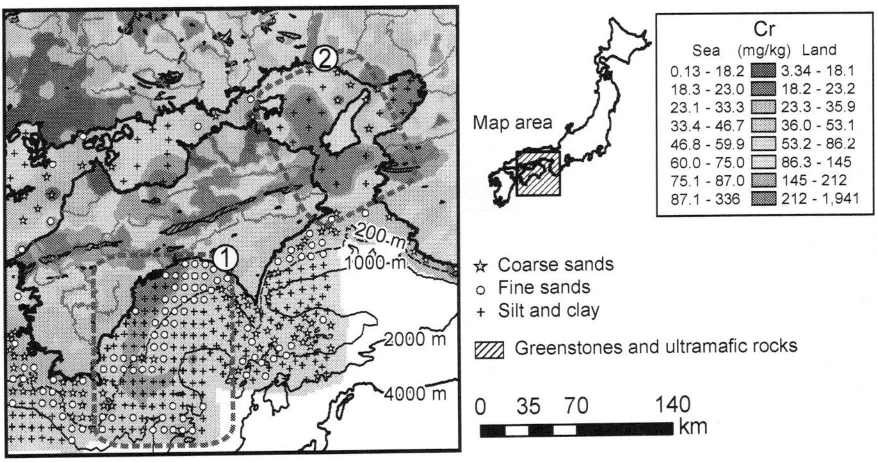

Fig. 10. Geochemical map of Cr around the Shikoku Island

The mass movement in the sea reflects the sea topography and is followed by a gravity flow and an oceanic current. However, the continuous distribution of elements between the land and sea is not always found everywhere. Relict sediments distributing on the shelf and slope might conceal the modern mass transfer process. Their distribution is indicative of the past mass transfer process because they were deposited during the last glacial age and the early transgression age. Ohta *et al.* (2007; 2010) found, however, that the influence on elemental distribution is restricted in small areas. Instead, they suggest that continuous distribution is found in the case where parent lithology covers widely adjacent terrestrial areas. Otherwise a rock type having extremely high concentrations of elements existed on land. Figures 8, 9, and 10 represent the latter case. In addition, the mass transfer process in the marine environment is traceable up to 30–50 km for all geochemical maps, except for Cr and Ni maps. According to Ohta *et al.* (2002; 2007), variations of chemical compositions in samples become small by the following order,

Parent lithology > stream sediment > sandy marine sediment > clayey marine sediment

Finer particles are mixed well and become homogeneous as compared to coarser particles because they take a longer time to deposit as compared to coarse particles according to Stokes' law. It becomes hard to find the systematic differences between elemental concentrations in finer sediments during the mass transfer process. For these reasons, a geochemical map is effective to elucidate short-range transfer processes.

8.2 Mass transport processes: volcanic materials transported through the atmosphere

Japan possesses many active volcanoes and both terrestrial and marine areas are strongly influenced by volcanic eruptions. Volcanic materials, mainly volcanic ash, are transported extensively through the atmosphere. A volcanic eruption is episodic so that the occurrence is concealed by active erosion or active sedimentation processes. Conversely, the distribution of volcanic materials found in geochemical maps may be used to determine where modern sediments are supplied. Furthermore, they provide a sedimentation rate of modern sediments and the presence of a strong bottom sea current that sweeps away fine recent sediments. In this chapter, we discuss the mass transport process of recent volcanic materials erupted in the Holocene.

The distribution pattern of high Al (Al_2O_3), Ti (TiO_2), Fe (T-Fe_2O_3), Sc, Cr, Co, Ni, Sr, and Eu contents follow an L shape around the southeast part of Hokkaido Island (see the circle numbered in 3 for Cr in Fig. 9). The feature suggests the contribution of basaltic-andesitic materials, including ultramafic inclusions erupted in 1741–1742 from a volcanic island (Oshima-Oshima island) (Inouchi *et al.*, 1995). The lower concentration of Cr in the north part of the circle numbered 3 in Fig. 9 might be caused by modern turbidity sediments supplied from adjacent terrestrial areas and steep slopes in the coastal sea.

The deep basin located in the western part of the Hokkaido Island is widely influenced by distant volcanic activity, i.e. regional tephra (Machida, 1999). Core profiles suggest that many tephra layers are included in sediment core samples. The Baegdusan-Tomakomai tephra (B-Tm) is supplied from the Baegdusan Volcano of North Korea 1,000 years ago, which is 900 km away from Hokkaido Island (Machida, 1999). It is characterized by having extremely higher Zn, Zr, Nb, REEs (except Eu), Hf, Ta, Th, and U concentrations and lower Sc, Ti (TiO_2), V, Cr, Ni, Sr, and Ba concentrations than the other tephras found in northern Japan (Fukuoka, 1991; Kikkawa *et al.*, 1989). Actually, there is an extreme enrichment of Zr, Nb, REEs, Hf, Ta, and Th in the western part of the Hokkaido Island (see a circle numbered in 2 for Nb in Fig. 9). Although the B-Tm tephra widely covers northern Japan, the enrichments of Zr, Nb, REEs, Hf, Ta, and Th are found only in deep basins located in the west. This fact is explained by the increase of the sedimentation rate from west to east. The B-Tm tephra layer is found in sediment cores at a depth of 0–25 cm in the western part and at a depth of 128–176 cm in the eastern part (Noda *et al.*, 2007). Because the sedimentation rate is very slow in the deep basins, we can determine the past volcanic activity from geochemical maps.

Regional tephras supplied from Japanese giant volcanoes are found all over the place as well as in the B-Tm tephra (Machida, 1999). The Kikai Caldera, which was formed by large-scale violent eruptions 7,300 years ago, is located in the southern part of the Kyushu Island (Fig.

11). The pyroclastic flows from the Kikai Caldera extended over 100 km, with some of them reaching the Kyushu Island. Its volcanic ash covered all of Japan (Machida, 1999). The coarse sediments distributed in the southern part of the Kyushu Island contained volcanic materials including pumice. Strong bottom currents flowing from west to east sweep away fine sediments there. On account of this, the past volcanic materials supplied from the Kikai Caldera eruptions remain in the region. It is possible that a part of them have been supplied from the southern Kyushu Island, where pyroclastic flow deposits erupted from giant volcanoes during the Pleistocene era (26,000–110,000 years ago). The distribution of high Mg (MgO), Ti (TiO_2), Mn (MnO), Fe (Fe_2O_3), Sc, V, Co, and Zn concentrations correspond fairly well with that of the coarse sediments (Fig. 11). These elements are not highly abundant in the fine sediments located in the northern and southeastern regions of the Kikai Caldera, which suggests that terrestrial materials free of volcanic materials were deposited here.

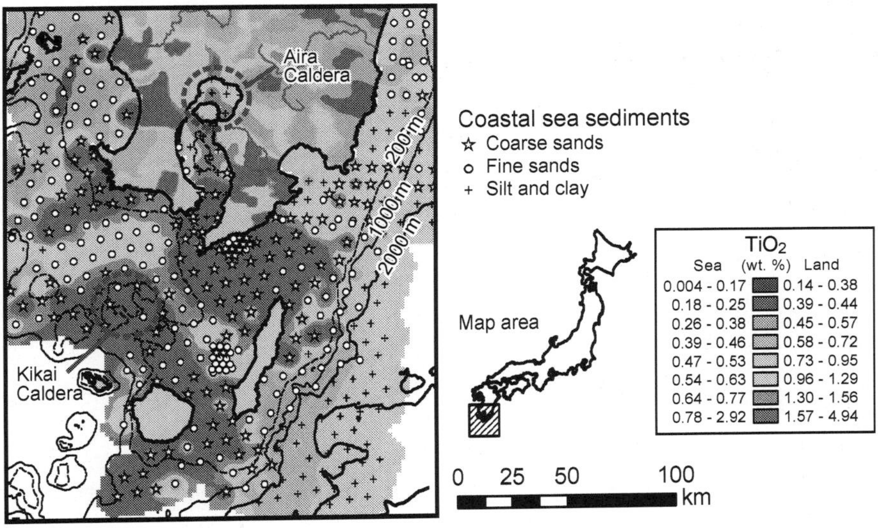

Fig. 11. Geochemical map of Ti (TiO_2) in the southern part of the Kyushu Island

8.3 Mass transport processes: contribution of metalliferous deposits and anthropogenic activity

Cu, Zn, As, Mo, Cd, Sn, Sb, Hg, Pb, and Bi are chalcophile elements that are highly enriched in stream sediments associated with metalliferous deposits and anthropogenic activity. The spatial distribution of these elements is of environmental interest. Figure 12 shows the spatial distribution of Cd in the northern and the southeastern parts of the Honshu Island. Numerous Kuroko-style deposits occur in the northern part of the Honshu Island. High elevated Cu, Zn, As, Cd, Sb, Pb, Hg, and Bi concentrations of stream sediments are present in the region (Ujiie-Mikoshiba *et al.*, 2006). However, no enrichment of these elements, except for Zn and Pb, is found in the adjoining coastal seas (see the circle numbered in 1 for Cd in Fig. 12a). The same trend is found around the Shimokita Peninsula where the active hydrothermal field is located and a metalliferous deposit is currently forming. We see no enrichment of Cu, As, Cd, Sb, Hg, and Bi in the coastal seas and bays around the Shimokita Peninsular (see the circle numbered in 2 for Cd in Fig. 12a). These facts suggest that sulfide minerals are not supplied directly to coastal seas because the sulfide ores are oxidized, consequently releasing Cu, Zn, As, Cd, Sb, Pb, and Bi during transport from terrestrial areas to coastal waters (Hudson-Edwards *et al.*, 1996). It is also possible that Zn and Pb sulfides are extremely resistant to weathering or that their mass concentrations are the highest among these metals. Alternatively, aqueous Zn and Pb are easily sorbed on the sediment surface in coastal seas.

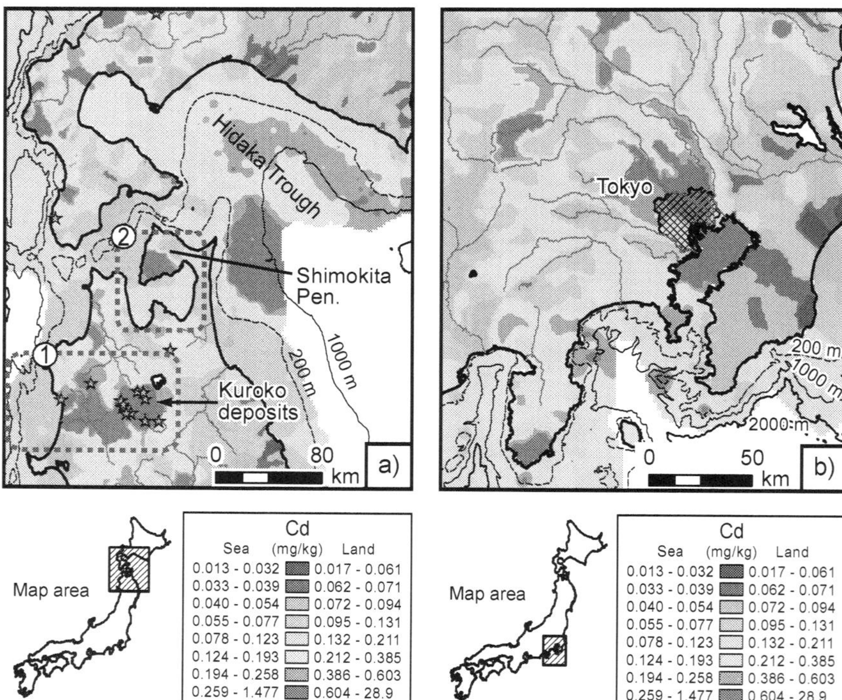

Fig. 12. Geochemical maps of Cd

In contrast, the influence of anthropogenic activity on geochemical maps is somewhat different from that of metalliferous deposits. The P (P_2O_5), Cr, Ni, Cu, Zn, Mo, Cd, Sn, Sb, Hg, Pb, and Bi concentrations are elevated in both the metropolitan area and adjacent inner bay. Figure 12b shows that the spatial distribution of Cd in the southeast part of the Honshu Island where the Tokyo metropolitan area exists. The high concentrations of chalcophile elements such as Zn and Cd are found in both the terrestrial area and inner bay. Their spatial distribution patterns suggest that the contaminated materials remain in the bay without extending to the outer sea. This is because of the distribution of sandy sediments, which have a low content of heavy metals, around the entrance of the bay. Another possible explanation is the influence of water circulation in the bay. A strong bottom current (estuarine circulation) might prevent fine particles with heavy metals from reaching the outer sea because it flows from the outer sea to the bay.

9. Vertically varying element transport

In deep water (over 1,000 m), Mn (MnO), Cu, Zn, Mo, Cd, Sn, Sb, Pb, Hg, and Bi are particularly concentrated. The presence of high concentration areas of these elements found far from the adjacent terrestrial area are not explained only by materials from rivers, gravity flows, volcanic materials, metalliferous deposits, and anthropogenic acidities. For example, Figure 13 shows the geochemical maps of Mn (MnO) and Cu in the central part of Japan. Both elements are highly enriched in deep water, but the spatial distributions differ from one another. The enrichments of these elements in surface marine sediments are caused by early diagenetic processes, the supply of organic remains, and reductive-oxidative conditions.

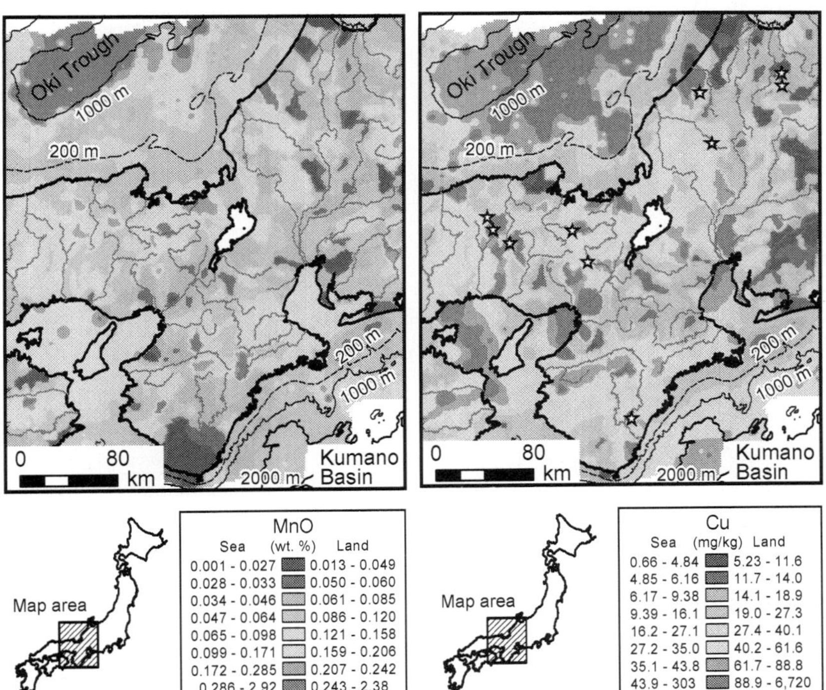

Fig. 13. Geochemical maps of Mn (MnO) and Cu in the central part of Japan

Mn (MnO), Cu, Mo, Sb, Pb, and Bi are dissolved at greater depths in sediments under reducing conditions. They diffuse upward and precipitate with Mn oxides or on the sediment surface under oxic conditions. This enrichment is caused by early diagenetic processes (e.g., Macdonald *et al.*, 1991; Shaw *et al.*, 1990). These processes are found in pelagic areas where the sedimentation rate is very slow. The organic remains are also an important source of elements in deep seas. Cu, Zn, Cd, Mo, Sn, Sb, Hg, Pb, and Bi are removed from surface seawater by organic matter. After they sink into deep basins, they are released into porewater during the organic matter's decomposition. We assumed that these elements are ultimately precipitated as diagenetic sulfide (authigenic precipitation) or associated with residual organic matter in marine environments (Chaillou *et al.*, 2008; Rosenthal *et al.*, 1995; Zheng *et al.*, 2000). Mn (MnO), Cu, Mo, Sb, Pb, and Bi are dissolved in anoxic conditions and are immobile in oxic conditions, but the geochemistries of Cd and U are opposite to these elements during the early diagenetic process (Rosenthal *et al.*, 1995). Hg is released from surface sediments to seawater during decomposition of organic matters (Bothner *et al.*, 1980; Mason *et al.*, 1994). Thus, various controlling factors affect the elemental concentrations of surface sediments in deep seas.

Figure 13 shows that Mn (MnO), Cu, Mo, Sb, Pb, and Bi are particularly concentrated in fine sediments of the Oki Trough below 1,000 m. The ocean floor in the deep sea (the Japan Sea Proper Water) is covered by a thick layer of cold and oxygen-rich water, and the surface sediments are under oxidative conditions (Katayama *et al.*, 1993). Their enrichments are possibly caused by early diagenetic processes. In contrast, high Cu, Cd, Hg, and U concentrations and the low concentration of Mn (MnO) are found in fine sediments of the Kumano Basin (<2,000 m). It is possible that the input of a large amount of organic matter engenders reductive conditions in surface sediments and causes high Cu, Cd, Hg, and U concentrations. Thus, the enrichment of elements differs among deep basins.

Although Mn (MnO) enrichment occurs in the Oki Trough, the spatial distribution of high Cu concentrations is present even in the marginal terrace (200–1,000 m). Its distribution corresponds to distribution of silty and clayey sediments. The spatial distribution of Cr, Ni, Zn, Cd, Sn, Sb, Pb, Hg, Bi, and U are also similar to that of Cu. These results are consistent with the result that Cu concentration increases with decreasing particle size (Fig. 5). Ikehara (1991) suggests that muddy sediments deposit around current rips and between surface water and deep water in the Japan Sea. The results suggest that muddy sediments deposit under 200 m, where the boundary of water mass is located between the surface water (the Tsushima Current) and deep water (the Japan Sea Proper Water). The organic remains might cause the enrichments of Cr, Ni, Cu, Zn, Cd, Sn, Sb, Pb, Hg, Bi, and U in the marginal terrace. In the Japan sea, however, the sedimentation process of silty and clayey particles at the boundary of water mass predominantly determine the spatial distribution of Cr, Ni, Cu, Zn, Cd, Sn, Sb, Pb, Hg, Bi, and U concentrations. Early diagenetic processes influence the enrichments of Mn (MnO), Cu, Mo, Sb, Pb, and Bi in water at a depth of >1,000 m.

10. Conclusion

The spatial distribution patterns of the elemental concentrations found in geochemical maps in coastal seas floor along with terrestrial areas are useful to define the natural geochemical background variation, mass transport, and contamination processes. We intend to elucidate geochemical differences between terrestrial surface sediments and coastal and open sea sediments comprehensively. The elemental abundance patterns of coastal sea sediments are consistent with those of stream sediments and the Japanese upper crust materials. This fact suggests that coastal sea sediments are originally adjacent terrestrial materials. However, the mineralogical compositions of coastal sea sediments change with particle size, resulting in a change in the chemical compositions. Coarse sediments in the marine environment contain quartz and calcareous shells, which enhance Si, Ca, and Sr concentrations and deplete the other elements. Consequently, the concentrations of most elements increase with decreasing particle size. The particle size effect often conceals the horizontal mass transfer process.

Because Japan is located in the subducting zone, the Japanese marine environment has a narrow continental shelf and a steep slope from the coast. The horizontal mass movement in the sea reflects the sea topography and is followed by a gravity flow and an oceanic current. Terrestrial materials supplied through rivers initially fan out on the shore (~20 km); subsequently, they are gradually transported off shore (over 100 km) by gravity over a long period of time. An oceanic current conveys fine sediments up to a distance of 100–200 km from the coast, along the coast. Heavy metals and toxic elements such as Zn, Cd, and Hg are present in high concentrations in urban areas and are exposed to an adjacent inner bay. However, their high concentration area is found only in the bay: the contaminated materials remain in the bays without extending to the outer sea. These elements are also abundant in terrestrial areas having metalliferous deposits. However, the adjoining coastal seas are only enriched in Zn and Pb. The mass transfer process of these elements from sediments associated with metalliferous deposits to sea is different from that of anthropogenic disposed elements. We can see some extensive distributions of volcanic materials in marine environment. The distribution of volcanic materials such as pyroclastic materials, pumice, and ash is indicative of mass transfer through atmosphere, although that is not the direct mass transfer from land to sea. Thus, we can see various kinds of horizontal mass transfer processes from these comprehensive geochemical maps. In contrast, the spatial distribution of Cu, Zn, Cd, Mo, Sn, Sb, Hg, Pb, Bi, and U in the deep-sea basins is determined by early diagenetic processes in sediments, oxidation-reduction potentials in surface sediments, and the input of organic remains from surface water. These processes represent vertical element transport processes. The enrichments of these elements are not continuous between land and sea. That is, the vertical element transport process conceals the horizontal mass transfer process.

11. Acknowledgements

The authors express their special appreciation to Ken Ikehara, Takeshi Nakajima, Hajime Katayama, and Atsushi Noda for offering marine sediment samples stored in a sample chamber and collecting new samples; Shigeru Terashima and Yoshiko Tachibana for their technical assistance in preparing samples and analyzing elemental concentrations of stream and coastal sea sediment samples; and Daisaku Kawabata for assisting in GIS analyses. We are also grateful to Takumi Tsujino, Masumi Ujiie-Mikoshiba, and Takashi Okai for their useful suggestions, which helped to improve an earlier version of the manuscript. The authors are grateful to the Japan Oceanographic Data Centre (JODC) for offering data files.

12. References

Balls, P.W., Hull, S., Miller, B.S., Pirie, J.M. & Proctor, W. (1997) Trace metal in Scottish estuarine and coastal sediments. *Marine Pollution Bulletin*, 34(1), 42-50, ISSN: 0025-326X.

Bølviken, B., Bergstrøm, J., Bjørklund, A., Kontio, M., Lehmuspelto, P., Lindholm, T., Magnusson, J., Ottesen, R.T., Steenfelt, A. & Volden, T. (1986) Geochemical Atlas of Northern Fennoscandia. Geological Surveys of Finland, Norway and Sweden, ISBN: 91-7158-383-1, Helsinki, Trondheim and Stockholm.

Bothner, M.H., Jahnke, R.A., Peterson, M.L. & Carpenter, R. (1980) Rate of mercury loss from contaminated estuarine sediments. *Geochimica et Cosmochimica Acta*, 44(2), 273-285, ISSN: 0016-7037.

Chaillou, G., Schäfer, J., Blanc, G. & Anschutz, P. (2008) Mobility of Mo, U, As, and Sb within modern turbidites. *Marine Geology*, 254(3-4), 171-179, ISSN: 0025-3227.

Darnley, A.G., Björklund, A., Bølviken, B., Gustavsson, N., Koval, P.V., Plant, J.A., Steenfelt, A., Tauchid, M., Xie, X., Garrett, R.G. & Hall, G.E.M. (1995) *A global geochemical database for environmental and resource management: recommendations for international geochemical mapping. Science Report 19.* UNESCO Publishing, ISBN: 92-3-103085-X, Paris.

Degens, E., Kempe, S. & Richey, J. (1991) *Biogeochemistry of major world rivers.* John Wiley & Sons, ISBN: 0-471-92676-0, Chichester.

Fauth, H., Hindel, R., Siewers, U. & Zinner, J. (1985) *Geochemischer Atlas Bundesrepublik Deutschland.* Bundesanstatt für Geowissenschaften and Rohstoffe (BGR), ISBN: 3-9801097-0-4, Hannover.

Fukuoka, T. (1991) Identification of particular tephras by trace element compositions of volcanic glass. *The Earth Montly*, 13(3), 186-192, ISSN: 0387-3498 (in Japanese).

Geological Survey of Japan (1992). *Geological Map of Japan 1:1,000,000, third ed.* Geological Survey of Japan, Tsukuba.

Gustavsson, N., BøLviken, B., Smith, D.B. & Severson, R.C. (2001) Geochemical landscapes of the conterminous United States: New map presentations for 22 elements. *US Geological Survey professional paper*, 1648, 1-38, ISSN: 1044-9612.

Howarth, R.J. & Thornton, I. (1983) Regional Geochemical Mapping and its Application to Environmental Studies. In *Applied Environmental Geochemistry*, Thornton, I. (Ed.) p. 41-73. Academic Press, ISBN: 0-12-690640-8, London.

Hudson-Edwards, K.A., Macklin, M.G., Curtis, C.D. & Vaughan, D.J. (1996) Processes of formation and distribution of Pb-, Zn-, Cd-, and Cu-bearing minerals in the Tyne Basin, northeast England: Implications for metal-contaminated river systems. *Environmental Science & Technology*, 30(1), 72-80, ISSN: 0013-936X.

Ibbeken, H. & Schleyer, R. (1991) *Source and sediment: a case study of provenance and mass balance at an active plate margin (Calabria, Southern Italy).* Springer-Verlag, ISBN: 0-387-53282-X, Berlin.

Ikehara, K. (1988). *Sedimentological Map of Tosa Wan.* 1:200,000 Marine Geology Map Series 34.

Ikehara, K. (1991). Modern sedimentation off San'in district in the southern Japan Sea. In: *Oceanography of Asian Marginal Seas*, Takano, K. (Ed.), 143-161, Elsevier, ISBN: 0-444-88805-5, Amsterdam.

Imai, N. (1987) Multielement analysis of stream sediment by ICP-AES. *Bunseki Kagaku*, 36(4), T41-T45, ISSN: 0525-1931 (in Japanese).

Imai, N. (1990) Multielement analysis of rocks with the use of geological certified reference material by inductively coupled plasma mass spectrometry. *Analytical Sciences*, 6(3), 389-395, ISSN: 0910-6340.

Imai, N., Terashima, S., Itoh, S. & Ando, A. (1995) 1994 compilation of analytical data for minor and trace-elements in 17 GSJ geochemical reference samples, igneous rock series. *Geostandards Newsletter*, 19(2), 135-213, ISSN: 0150-5505.

Imai, N., Terashima, S., Katayama, H., Nakajima, T., Ikehara, K. & Taniguchi, M. (1997) Geochemical behavior of heavy metals in coastal marine sediments from the eastern margin of the Japan Sea. *Bulletin of the Geological Survey of Japan*, 48(9), 511-530, ISSN: 0016-7665 (in Japanese, with English abstr.).

Imai, N., Terashima, S., Ohta, A., Mikoshiba, M., Okai, T., Tachibana, Y., Togashi, S., Matsuhisa, Y., Kanai, Y. , Kamioka, H. & Taniguchi, M. (2004) *Geochemical map of Japan*. Geological Survey of Japan, AIST (in Japanese, with English abstr.).

Imai, N., Terashima, S., Ohta, A., Mikoshiba, M., Okai, T., Tachibana, Y., Ikehara, K., Katayama, H., Noda, A., Togashi, S., Matsuhisa, Y., Kanai, Y. & Kamioka, H. (2010) Elemental distribution in Japan -Geochemical map of Japan- (in Japanese, with English abstract). Geological Survey of Japan, AIST, Tsukuba. Database of elemental distribution in Japan <http://riodb02.ibase.aist.go.jp/geochemmap/>.

Inouchi, Y., Ohtsuka, M., Kumon, F., Motoyama, I. & Katayama, H. (1995) Surface sediment characteristics around Ishikari Bay. In: *Comprehensive Study on Environmental Changes in the Southwestern Hokkaido Area*, Okamura, Y. & Inouchi Y. (Eds.), 63-88, Geological Survey of Japan (in Japanese).

Irion, G., Müller, J., de Mello, J.N. & Junk, W.J. (1995) Quaternary geology of the Amazonian lowland. *Geo-Marine Letters*, 15(3-4), 172-178, ISSN: 0276-0460.

Karageorgis, A.P., Anagnostou, C.L. & Kaberi, H. (2005) Geochemistry and mineralogy of the NW Aegean Sea surface sediments: implications for river runoff and anthropogenic impact. *Applied Geochemistry*, 20(1), 69-88, ISSN: 0883-2927.

Katayama, H. (2008). Sedimentological Map of Ishikari Bay. 1:200,000 Marine Geology Map Series 66 (CD-ROM version).

Katayama, H., Sato, M. & Ikehara, K. (1993). Sedimentological Map offshore of Kyo–ga Misaki. 1:200,000 Marine Geology Map Series 38.

Kikkawa, K., Imai, N., Okumura, K. & Mizuno, K. (1989). Identification of tephra layers by chemical analysis of volcanic glass using inductively coupled plasma emission spectrometry (ICP). *Bulletin of the Geological Survey of Japan*, 40(1), 1-18, ISSN: 0016-7665 (in Japanese, with English abstr.).

Kuroda, H., Isoda, Y., Takeoka, H. & Honda, S. (2006) Coastal current on the eastern shelf of Hidaka Bay. *Journal of Oceanography*, 62(5), 731-744, ISSN: 0916-8370.

Lis, J. & Pasieczna, A. (1995) *Geochemical atlas of Poland 1: 2500000*. Polish Geological Institute, ISBN: 83-903705-9-X, Warszawa.

Macdonald, R.W., Macdonald, D.M., O'Brien, M.C. & Gobeil, C. (1991) Accumulation of heavy-metals (Pb, Zn, Cu, Cd), carbon and nitrogen in sediments from strait of Georgia, B.C., Canada. *Marine Chemistry*, 34(1-2), 109-135, ISSN: 0304-4203.

Machida, H. (1999) The stratigraphy, chronology and distribution of distal marker-tephras in and around Japan. *Global and Planetary Change*, 21(1-3), 71-94, ISSN: 0921-8181.

Mason, R.P., Fitzgerald, W.F. & Morel, F.M.M. (1994) The biogeochemical cycling of elemental mercury: Anthropogenic influences. *Geochimica et Cosmochimica Acta*, 58(15), 3191-3198, ISBN: 0016-7037.

Noda, A., Katayama, H., Arai, K. & Furukawa, R. (2007). Surface sediment characteristics off Hidaka region. In: *Cruise Report: Marine geological and Geophysical Studies. On the Collision Zone of Kurile and Northeast Japan Arcs - off Hidaka Area -*, Katayama, H. (Ed.), 76-96. Geological Survey of Japan, AIST 39 (in Japanese).

Ohta, A., Imai, N., Okai, T., Endo, H., Kawanabe, S., Ishii, T., Taguchi, Y. & Kamioka, H. (2002) The characteristics of chemical distribution patterns in and around Yamagata city -Geochemical map in the southern area of Yamagata Basin-. *Chikyukagaku (Geochemistry)*, 36(3), 109-125, ISSN: 0386-4073 (in Japanese, with English abstr.).

Ohta, A., Imai, N., Terashima, S. & Tachibana, Y. (2004a) Investigation of elemental behaviors in Chugoku region of Japan based on geochemical map utilizing stream sediments. *Chikyukagaku (Geochemistry)*, 38(4), 203-222, ISSN: 0386-4073 (in Japanese, with English abstr.).

Ohta, A., Imai, N., Terashima, S., Tachibana, Y., Ikehara, K. & Nakajima, T. (2004b) Geochemical mapping in Hokuriku, Japan: influence of surface geology, mineral occurrences and mass movement from terrestrial to marine environments. *Applied Geochemistry*, 19(9), 1453-1469, ISSN: 0883-2927.

Ohta, A., Imai, N., Terashima, S. & Tachibana, Y. (2005a) Influence of surface geology and mineral deposits on the spatial distributions of elemental concentrations in the stream sediments of Hokkaido, Japan. *Journal of Geochemical Exploration*, 86(2), 86-103, ISSN: 0375-6742.

Ohta, A., Imai, N., Terashima, S. & Tachibana, Y. (2005b) Application of multi-element statistical analysis for regional geochemical mapping in Central Japan. *Applied Geochemistry*, 20(5), 1017-1037, ISSN: 0883-2927.

Ohta, A., Imai, N., Terashima, S., Tachibana, Y., Ikehara, K., Okai, T., Ujiie-Mikoshiba, M. & Kubota, R. (2007) Elemental distribution of coastal sea and stream sediments in the island-arc region of Japan and mass transfer processes from terrestrial to marine environments. *Applied Geochemistry*, 22(12), 2872-2891, ISSN: 0883-2927.

Ohta, A., Imai, N., Terashima, S., Tachibana, Y., Ikehara, K., Katayama, H. & Noda, A. (2010) Factors controlling regional spatial distribution of 53 elements in coastal sea sediments in northern Japan: Comparison of geochemical data derived from stream and marine sediments. *Applied Geochemistry*, 25(3), 357-376, ISSN: 0883-2927.

Reimann, C., Äyräs, M., Chekushin, V., Bogatyrev, I., Boyd, R., Caritat, P.d., Dutter, R., Finne, T.E., Halleraker, J.H., Jæger, Ø., Kashulina, G., Lehto, O., Niskavaara, H., Pavlov, V., Räisänen, M.L., Strand, T. & Volden, T. (1998) *Environmental Geochemical Atlas of the Central Barents Region*. Geological Survey of Norway, ISBN: 82-7385-176-1, Trondheim.

Reimann, C. (2005) Geochemical mapping: technique or art? *Geochemistry: Exploration, Environment, Analysis*, 5(4), 359-370, ISSN: 1467-7873.

Rosenthal, Y., Lam, P., Boyle, E.A. & Thomson, J. (1995) Authigenic cadmium enrichments in suboxic sediments: Precipitation and postdepositional mobility. *Earth and Planetary Science Letters*, 132(1-4), 99-111, ISSN: 0012-821X.

Salminen, R., Batista, M.J., Bidovec, M., Demetriades, A., B., D.V., De Vos, W., Duris, M., Gilucis, A., Gregorauskiene, V., Halamic, J., Heitzmann, P., Lima, A., Jordan, G., Klaver, G., Klein, P., Lis, J., Locutura, J., Marsina, K., Mazreku, A., O'Connor, P.J., Olsson, S.Å., Ottesen, R.-T., Petersell, V., Plant, J.A., Reeder, S., Salpeteur, I., Sandström, H., Siewers, U., Steenfelt, A. & Tarvainen, T. (2005) *Geochemical atlas of Europe. Part 1 - Background Information, Methodology and Maps*. Geological Survey of Finland, ISBN: 951-690-921-3, Espoo.

Shacklette, H.T. & Boerngen, J.G. (1984) *Element concentrations in soils and other surficial materials of the conterminous United States. USGS Geological Survey Professional paper.* United States Government Printing Office, ISSN: 0096-0446.

Shaw, T.J., Gieskes, J.M. & Jahnke, R.A. (1990) Early diagenesis in differing depositional environments: The response of transition metals in pore water. *Geochimica et Cosmochimica Acta*, 54(5), 1233-1246, ISSN: 0016-7037.

Taylor, S.R. & McLennan, S.M. (1995) The geochemical evolution of the continental crust. *Reviews of Geophysics*, 33(2), 241-265, ISSN: 8755-1209.

Thalmann, F., Schermann, O., Schroll, E. & Hausberger, G. (1988) *Geochemical Atlas of the Republic of Austria 1:1,000,000*. Geological Survey of Austria.

Togashi, S., Imai, N., Okuyama-Kusunose, Y., Tanaka, T., Okai, T., Koma, T. & Murata, Y. (2000) Young Upper Crustal Chemical Composition of the Orogenic Japan Arc. *Geochemistry Geophysics Geosystems*, 1(11), 2000GC000083, ISSN: 1525-2027.

Ujiie-Mikoshiba, M., Imai, N., Terashima, S., Tachibana, Y. & Okai, T. (2006) Geochemical mapping in northern Honshu, Japan. *Applied Geochemistry*, 21(3), 492-514, ISSN: 0883-2927.

Voutsinou-Taliadouri, F. & Varnavas, S.P. (1995) Geochemical and Sedimentological Patterns in the Thermaikos Gulf, North-West Aegean Sea, Formed from a Multisource of Elements. *Estuarine Coastal and Shelf Science*, 40(3), 295-320, ISSN: 0272-7714.

Wang, S.S., Cao, Z.M., Lan, D.Z., Zheng, Z.C. & Li, G.H. (2008) Concentration distribution and assessment of several heavy metals in sediments of west-four Pearl River Estuary. *Environmental Geology*,.55(5), 963-975, ISSN: 0943-0105.

Watson, D.F. & Philip, G.M. (1985) A refinement of inverse distance weighted interpolation. Geo-Processing, 2(4), 315-327, ISSN: 0165-2273.

Weaver, T.A., Broxton, D.E., Bolivar, S.L. & Freeman, S.H. (1983) *The Geochemical Atlas of Alaska: Compiled by the Geochemistry Group, Earth Sciences Division, Los Alamos National Laboratory*. GJBX-32(83), Los Alamos.

Webb, J.S., Thornton, I., Thompson, M., Howarth, R.J. & Lowenstein, P.L. (1978) *The Wolfson Geochemical Atlas of England and Wales*. Clarendon Press, ISBN: 0-19-891113-0, Oxford.

Xie, X.J., Mu, X.Z. & Ren, T.X. (1997) Geochemical mapping in China. *Journal of Geochemical Exploration*, 60(1), 99-113, ISSN: 0375-6742.

Zheng, C. (1994). *Atlas of Soil Environmental Background Value in the People's Republic of China*. China Environmental Science Press, ISBN: 7800931374, Beijing.

Zheng, Y., Anderson, R.F., van Geen, A. & Kuwabara, J. (2000) Authigenic molybdenum formation in marine sediments: A link to pore water sulfide in the Santa Barbara Basin. *Geochimica et Cosmochimica Acta*, 64(24), 4165-4178, ISSN: 0016-7037.

18

Mass Transfers and Sedimentary Budgets in Geomorphologic Drainage Basin Studies

Achim A. Beylich
Geological Survey of Norway (NGU), Quaternary Geology & Climate group and Norwegian University of Science and Technology (NTNU), Department of Geography Trondheim, Norway

1. Introduction

Geomorphologic processes, operating at the Earth surface and being responsible for transferring sediments and effecting landform change, are highly dependent on climate, and it is anticipated that climate change will have a major impact on the behaviour of Earth surface systems. Accordingly is geomorphologic research on mass transfers in a variety of different climatic environments represented by a substantial body of literature.

Studies on mass transfers and sedimentary source-to-sink fluxes generally refer to the development of sedimentary budgets. A sediment budget is an accounting of the sources and disposition of sediment as it travels from its point of origin to its eventual exit from a defined landscape unit like a drainage basin (e.g. Reid & Dunne, 1996). Accordingly, the development of a sediment budget necessitates the identification of processes of weathering, erosion, transport and storage / deposition within a defined area, and their rates and controls (Reid & Dunne, 1996; Slaymaker, 2000; Beylich & Warburton, 2007). A thorough understanding of the current sediment production and flux regime within a system is fundamental to predict likely effects of changes to the system, whether climatic induced or human-influenced. Source-to-sink sedimentary flux and sediment budget research therefore enables the prediction of changes to erosion and sedimentation rates, knowledge of where sediment will be deposited, how long it will be stored and how much sediment will be remobilised (Gurnell & Clark, 1987; Reid & Dunne, 1996; Beylich & Warburton, 2007).

Sedimentary mass transfers move eroded sediments from their source area to an area of temporal storage or long-term deposition in sinks. Rates of sediment transfer are not only conditioned by competence of geomorphic processes but also by the availability of sediment for transport. Accordingly, in assessing sediment transfer we need to quantify the forces, which drive transport processes but equally account for the factors, which control sediment supply (e.g. Ballantyne, 2002; Warburton, 2007). Small-scale geomorphologic process and sediment budget studies focus on sedimentary fluxes from areas of weathering and erosion to areas of storage within defined landscape units like drainage basins (Beylich & Warburton, 2007; Beylich & Kneisel, 2009), whereas large-scale sediment systems couple headwaters to oceanic sinks.

The identification of storage elements and sinks is critical to the effective study and understanding of source-to-sink sedimentary fluxes (Reid & Dunne, 1996). The setting of a particular drainage basin defines the boundary conditions for storage within that landscape unit. Within a defined landscape unit like a drainage basin, the slope and valley infill elements constitute the key storage units and storage volumes are important for addressing time-dependent sediment budget dynamics. Dating of storage in sedimentary source-to-sink flux studies is applied to determine or estimate the ages and chronology of the storage components within the system. An understanding of the nature of primary stores, secondary stores and the potential storage capacities of different types of drainage basins is important along with knowledge of sediment residence times. Of growing importance within geomorphologic drainage basin research is the development of innovative field methods, such as modern surface process monitoring techniques (Beylich & Warburton, 2007) and geophysical techniques for estimating sediment storage volumes (Schrott et al., 2003; Sass, 2005; Hansen et al., 2009). Within large-scale sediment systems oceanic sinks are most important and provide the opportunity to estimate rates of sediment production and delivery at long-term temporal as well as continental spatial scales (e.g. Rise et al., 2005; Dowdeswell et al., 2006).

In this chapter on mass transfers and sedimentary budgets in geomorphologic drainage basin studies results on mass transfers and sedimentary budgets from small and non-glaciated drainage basin geo-systems in Iceland (Hrafndalur and Austdalur) and Sweden (Latnjavagge) are presented and discussed as selected examples for field-based and process oriented geomorphologic drainage basin research. The presented material is a summary of key results from longer-term geomorphic studies (starting in 1996 in Iceland and in 1999 in Sweden) and relates to a number of publications where more details on methodology, drainage basin instrumentation and the spatio-temporal variability of geomorphic process rates within the drainage basins can be found.

2. Mass transfers, sediment budgets and relief development in small rainage basin geo-systems

Until today, there has only been a very limited number of truly integrated-quantitative studies of geomorphologic mass transfers, sediment budgets and relief development in drainage basins (e.g. Jäckli, 1957; Rapp, 1960; Caine, 1974; 2004; Caine & Swanson, 1989; Barsch, 1981; Barsch et al., 1994; Warburton, 1993; 2006; Becht, 1995; Beylich, 2000; 2008; Beylich & Warburton, 2007; Beylich & Kneisel, 2009; Beylich et al., 2005; in press; Schrott et al., 2002; 2003; Otto & Dikau, 2004; Slaymaker, 2008; Burki et al., 2009; Hansen et al., 2009). There is especially a significant lack of longer-term (about ten or more years of continuous field research and process monitoring in the defined drainage basin) quantitative process studies despite the fact that longer-term monitoring programmes are necessary for the calculation of reliable contemporary process rates, mass transfers and sediment budgets (e.g. Beylich & Warburton, 2007).

Geomorphic processes, operating within drainage basins, transferring sediments and changing landforms are highly dependent on climate, vegetation cover and human impact and will be significantly affected by climate change (e.g. Rapp, 1985; Barsch, 1993; Evans & Clague, 1994; Haeberli & Beniston, 1998; Lamoureux, 1999; Lamoureux et al., 2007; Slaymaker et al., 2003; Orwin & Smart, 2004; Beylich et al., 2006b; 2008; Beylich & Warburton, 2007; Beylich & Kneisel, 2009; Cockburn & Lamoureux, 2007; Slaymaker, 2008). An improved quantitative knowledge of mass transfers by sedimentary transfer processes operating in present-day climates is needed to model and determine the possible consequences of predicted climate change. It is therefore necessary to collect and compare extended data on both contemporary sedimentary fluxes and on storage elements (for the calculation of long-term process rates, e.g. on the Holocene timescale) from a wide range of different global environments and to apply more standardised methods for research on sediment fluxes and relationships between climate and sedimentary transfer processes (Beylich et al., 2006a; 2008; Beylich & Warburton, 2007; Lamoureux et al., 2007). Comparable datasets on process rates and mass transfers collected in drainage basins from different environments can then be used to model possible effects of predicted climate change as well as trends of relief development by applying the Ergodic principle of space-for-time substitution (e.g. Beylich et al., 2006a; 2008; Beylich & Kneisel, 2009).

2.1 Study areas

Results of field-based geomorphologic research on mass transfers, sediment budgets and relief development from three small and non-glaciated selected drainage basin geo-systems in eastern Iceland (Hrafndalur and Austdalur) and northern Swedish Lapland (Latnjavagge) are presented.

The Hrafndalur (7 km², 6 - 731 m a.s.l.; 65°28`N, 13°42`W) and the Austdalur drainage basin (23 km², 0 - 1028 m a.s.l.; 65°16`N, 13°48`W) are situated in the Easter Fjords region (Austfirðir) of eastern Iceland (Figs 1 - 3).

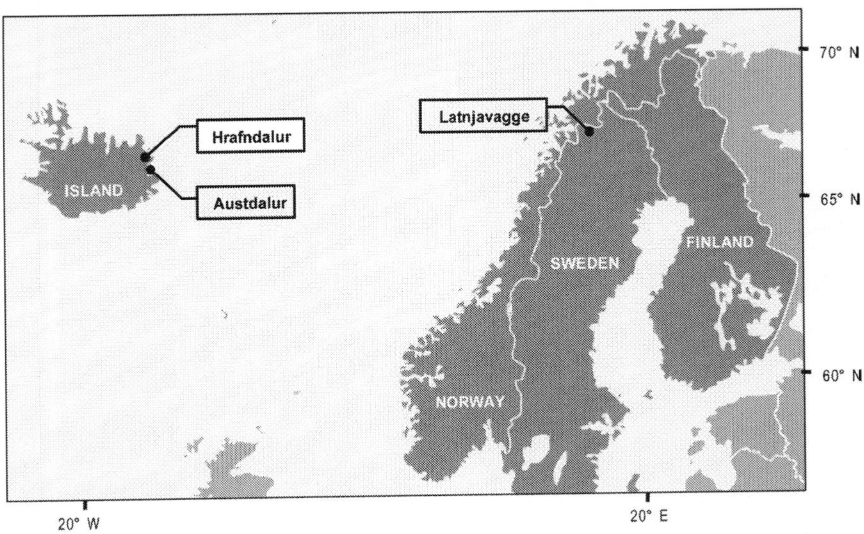

Fig. 1. Locations of the Hrafndalur, Austdalur and Latnjavagge drainage basins

The climate of the Eastern Fjords region is sub-Arctic oceanic, with a mean annual precipitation of 1719 mm yr^{-1} in Hrafndalur and 1431 mm yr^{-1} in Austdalur, and a mean annual air temperature of 3.6°C in both drainage basins. Runoff occurs year-round with the highest channel discharges happening during spring snow melt (normally April - June), wintry thaw events and especially during extreme rainfall events, which are normally most frequent in fall (September - November) (Beylich, 1999; 2003; 2009). During dry spells in

Fig. 2. The Hrafndalur drainage basin in eastern Iceland

Fig. 3. The Austdalur drainage basin in eastern Iceland

summer and frost spells in winter runoff can be very low and the Hrafndalur drainage basin can be even without any runoff (Beylich, 2003; 2009; Beylich & Kneisel, 2009). The steep and glacially sculptured relief of both drainage basins is Alpine, with slopes being composed of rock faces and talus cones, and main channels changing between resistance-limited bedrock channels and channel stretches with temporary storage of bed load material (Figs 2 and 3). Regional deglaciation occurred about 8000 yr BP. The lithology in Hrafndalur is clearly dominated by rhyolites and basalt occurs only in some smaller intrusions and dykes. Compared to that, Austdalur is clearly dominated by basalt. Vegetation in both drainage basins includes lichens, mosses, meadows, bogs and dwarf shrubs. Relevant denudative processes are rock and boulder falls, avalanches, debris flows and slides, creep processes, slope wash, chemical denudation, fluvial transport of solutes, suspended sediments and bed load, and deflation. The main storage elements in the valleys are extended talus cones, which are partly inter-fingering with till deposits. In addition, Holocene valley infills are found in the lower parts of both valleys. Dominant soils are regosols and lithosols. There is no permafrost within both Hrafndalur and Austdalur. Direct human impact exists in form of grazing which has caused a significant disturbance of the vegetation cover in larger parts of the drainage basins (Figs 2 and 3) (Beylich, 2000; 2007; Beylich & Kneisel, 2009).

Fig. 4. The Latnjavagge drainage basin in Swedish Lapland

The Latnjavagge drainage basin (68°20`N, 18°30`E; 9 km^2; 950 - 1440 m a.s.l.) is situated in the Abisko Mountain Area in northernmost Swedish Lapland (Figs 1 and 4). The Arctic-oceanic climate of the area (Beylich, 2003) is characterized by a mean annual temperature of - 2.0°C and a mean annual precipitation of 852 mm yr^{-1}. July is the warmest month (mean 8.6 °C). The coldest month is February (mean - 9.4 °C). About 2/3 of the annual precipitation is temporarily stored as snow during the winter. Snowmelt normally starts at the end of

May/beginning of June. Stable freezing temperatures with little daily fluctuation at 10 cm above ground and autumn snow accumulation usually occur from September/October onwards. Regarding the summer months June - August, August shows the highest mean precipitation (82 mm) and also the highest frequency of extreme rainfall events (Beylich, 2003; Beylich & Gintz, 2004). Precipitation from June to August accounts for about one quarter of the mean annual precipitation. The hydrological regime is nival, with runoff being limited to the period from end of May until October / November (Beylich, 2003). The bedrock is mainly composed of Cambro-Silurian mica-garnet schists and inclusions of marble (Beylich et al., 2004a). Intrusions of acidic granites can be found in the northern part of the valley. Regional deglaciation occurred about 8000 - 10000 yr BP (André, 1995). The drainage basin is dominated by large and flat plateau areas at 1300 m a.s.l., steep slopes which bound the glacially sculptured valley, and a flat valley floor situated between 950 and 1200 m a.s.l. (Fig. 4). The plateaux are best described as bare bedrock and boulder fields. The transition between slopes and plateaux is generally abrupt and, on the very steep, east-facing slope, covered by perennial snow and ice patches. The lower part of the valley floor is dominated by a lake, Latnjajaure, and a series of moraine ridges. As based on extended geophysical mapping, regolith thicknesses are shallow and reach locally only a few meters (Beylich et al., 2004b). Main present soils are regosols and lithosols. The drainage basin area belongs to the mid-Alpine zone with a continuous and closed vegetation cover up to 1300 m a.s.l. comprising dwarf shrub heaths and Alpine meadows and bogs. The exact distribution of permafrost is not directly known but drilling outside the drainage basin at 1200 m a.s.l. indicates at least sporadic permafrost down to 80 m below the surface (see also Beylich et al., 2004b). There seems to be no ice-rich permafrost on the valley floor around 1000 m a.s.l. and on the lower parts of the gentle, west-facing valley slope (Beylich et al., 2004b). Denudative slope processes include chemical weathering and denudation, mechanical weathering, rock falls, boulder falls, ground avalanches, debris flows, translation slides, creep processes, solifluction, ploughing boulders, and slope wash. Slush flows occur in certain areas of the valley and deflation is active where the vegetation cover is disturbed or lacking. In the channels dissolved, suspended and bed load is transported. Direct human impact on the natural system is presently small and is limited to reindeer husbandry (extensive grazing), some hiking tourism and field research at the Latnjajaure Field Station (LFS) (Beylich et al., 2006b; Beylich, 2008).

2.2 Aims of this geomorphologic research

The major goals of this longer-term geomorphologic research are to: (1) analyse the rates and the spatio-temporal variability of denudative processes and sedimentary transfers within the three different small drainage basin geo-systems; (2) investigate the absolute and the relative importance of these different geomorphic processes; (3) quantify the mass transfers and sediment budgets for the entire three drainage basins; (4) analyse trends of relief development in the three different study areas.

2.3 Approach and methods

This field-based geomorphologic research focuses on quantifying rates of denudative processes, mass transfers and the sediment budgets in three small drainage basins in sub-Arctic-oceanic Eastern Iceland and Arctic-oceanic Swedish Lapland. The collected data can be used for direct comparisons with other drainage basins worldwide (Beylich, 1999; 2000; 2002; Beylich & Warburton, 2007; Beylich et al., 2007b; 2008; Lamoureux et al., 2007).

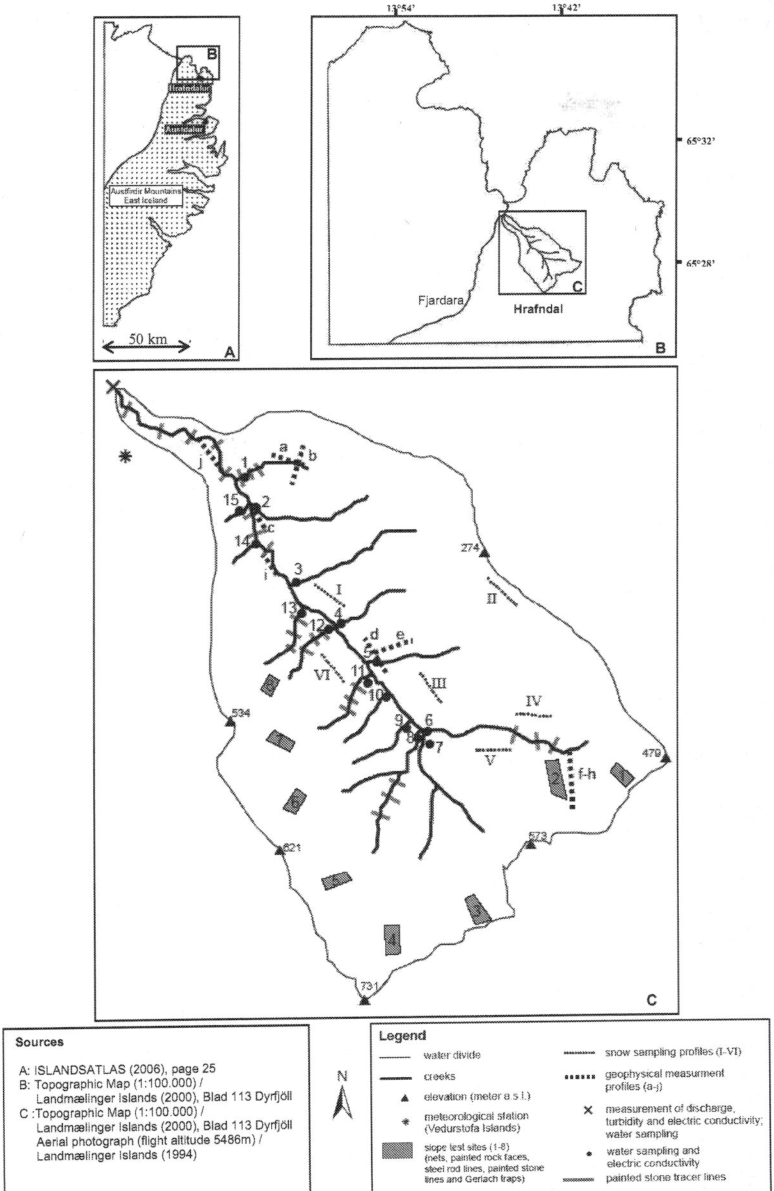

Fig. 5. Slope test sites, measuring points and instrumentation in the Hrafndalur drainage basin (eastern Iceland)

By a combined, quantitative recording of the relevant denudative slope processes and the stream work information on the absolute and relative importance of the different denudative processes is collected. This kind of drainage basin - based quantitative study, applying unified and simple (i.e. reliable and low costs) geomorphic field methods and techniques in combination with selected advanced techniques (techniques for continuous surface processes monitoring and geophysical techniques), and being carried out in a larger number of different environments having different climatic, vegetational, human impact topographic, lithological / geological and tectonic features shall contribute to gain better understanding of the internal differentiation of different global environments (e.g. Barsch, 1984; 1986; Beylich; 2000; 2002; Beylich & Warburton, 2007; Beylich et al., 2008; Beylich & Kneisel, 2009). Furthermore, information on the controls of geomorphic processes, the quantitative role of extreme events for longer-term mass transfer rates and sediment budgets, the general intensity of geomorphic processes, and the relative importance of different geomorphic processes for slope and valley formation and recent relief and landform development in different environments shall be improved.

Measurement of relevant denudative processes:

A combination of surface processes monitoring, geomorphologic and geophysical mapping as well as further field observations and detailed photo documentation were used to analyse relevant denudative processes and sedimentary fluxes in the three drainage basins (Beylich, 2008; Beylich & Kneisel, 2009). An adequate number of defined slope test sites within the three different drainage basins were selected after studying aerial photographs and after field investigations. The slope test sites differ with respect to slope form, aspect, elevation and local geographical setting and were selected to cover the complete range of different settings, which can be found within the three drainage basin systems. Figure 5 shows the location of slope test sites, measuring points and the instrumentation in Hrafndalur as an example. On the basis of detailed geomorphologic mapping (including mapping of areas being affected by certain processes) and process rates measured at the defined slope test sites and at defined measuring points (e.g. outlet of the drainage basin, etc.) process rates for the entire drainage basins were computed (inter- and extrapolations) using aerial photographs, DEM and GIS techniques in combination with fieldwork (Beylich, 2008; Beylich & Kneisel, 2009).

Rock falls and boulder falls:

Rock falls and boulder falls were investigated by applying a combination of process monitoring and detailed photo documentation. At each investigated slope test site a net was installed with its longer side placed along the rock wall / face foot on the talus cone developed below clearly defined vertical rock walls/rock faces. The nets were efficient in collecting debris produced by mechanical weathering at the rock walls and transferred to the nets by primary and secondary rock falls. The collected debris was repeatedly quantified by weighing it with a portable field balance ([kg m^{-2}], [kg $m^{-2}yr^{-1}$]). Rock wall retreat rate [mm yr^{-1}] was calculated by estimating the surface area of the defined and debris supplying rock face and relating it to the mass of debris accumulated below the rock face (Beylich et al., 2007a). The mass of accumulated debris smaller than ca. 1.0 cm in diameter was quantified with help of painted rock faces. At each slope test site with debris-supplying rock face two squares of 1 m^2 were painted at the beginning of the investigations and repainted in each following year. Fine debris accumulated below the painted squares could be identified by colour on the debris and the total mass of fine debris was quantified by

weighing the debris with a portable field balance. The total mass of fine debris was then related to the clearly defined source area of 1 m^2 rock surface.
Boulder falls were investigated by detecting, mapping, counting and measuring fresh boulders accumulated below the boulder supplying rock walls. For detecting and mapping of fresh boulder falls a detailed photo documentation was carried out each summer taking photos from ground and from helicopter. The detailed boulder size measurements (a-, b- and c-axis) were carried out in field (Beylich, 2008; Beylich & Kneisel, 2009).

Avalanches:

Total annual accumulations [t yr^{-1}] of inorganic material (including fine material, debris and boulders) by avalanches were quantified by combining a detailed sampling, measuring and weighing of newly deposited material with an estimation of the entire affected deposition area at the valley slope and a detailed photo documentation of the entire valley slope systems. Newly accumulated and dried debris was weighted at defined 1-m^2 plots within the accumulation areas of avalanches and fine material was sampled for the quantification of the inorganic mass (burning of the material over 12h in 550°C in the laboratory). The mapping of the entire deposition area as well as the detection and mapping of fresh boulder falls were carried out each summer during the investigation periods. Boulders were measured in field (a-, b-, c-axis) (Beylich, 2008; Beylich & Kneisel, 2009).

Debris flows and slides:

Debris flows and debris slides were investigated by detailed and yearly repeated photo documentation of slopes both from ground and from helicopter. Both new and old traces of debris flows and debris slides were mapped. The volumes of transferred material as well as the transport distances were measured in field. Debris flows can be significant for transferring material from slope to main channel systems (Beylich & Kneisel, 2009).

Creep processes:

Creep processes were analysed at the different slope test sites by detailed monitoring of movements of painted stone tracer lines and steel rod lines as well as depth-integrating peg lines (Beylich, 2008). At each slope test site lines with a number of painted stones were installed. Down-slope movements of all painted stones were measured each year in field. At each second location where a painted stone was placed a steel rod (1.0 cm diameter) was installed vertically 10 cm down into the ground. Down-slope movements of all steel rods as well as of depth-integrating peg lines (one per stone and steel rod line) were measured every year together with the measurements of movements of painted stones (Beylich, 2008).

Chemical slope denudation:

Chemical slope denudation was investigated by analysing water samples collected from small creeks and pipes on the slopes. Solute yields and chemical denudation rates were calculated based on measurements of atmospheric solute inputs to the drainage basin, runoff and solute concentrations in the main creeks (see below) (Beylich, 2008; Beylich & Kneisel, 2009).

Slope wash:

Slope wash was studied by using Gerlach traps, which were installed at selected slope test sites. In addition, suspended sediment concentrations in small creeks draining the slope systems were analysed (Beylich et al., 2006b; Beylich, 2008; Beylich & Kneisel, 2009).

Estimating the importance of deflation and aeolian deposition:

The importance of deflation and aeolian deposition were estimated by using sediment traps and by analysing sediment concentrations in snow cores collected along defined profiles within the drainage basins (Beylich, 2008; Beylich et al., 2006b; Beylich & Kneisel, 2009).

Runoff and fluvial transport:

Channel discharge was measured by continuous and year-round monitoring of water level using a pressure sensor (GLOBAL WATER) and collecting data every hour, in combination with propeller measurements at different selected water level stages using an Ott-propeller (model C2) during the field campaigns. Daily specific runoff [mm d-1] was calculated by dividing calculated daily discharge by the contributing drainage basin area (Beylich, 1999; Beylich & Kneisel, 2009).

Fluvial suspended sediment and solute transport were analysed by combining continuous and year-round monitoring of turbidity and electric conductivity (GLOBAL WATER) with hourly readings with discrete water sampling (1 and 5 l samples) during the field campaigns (Beylich & Kneisel, 2009). Vertically integrated water samples were taken with 1000 ml wide-necked polyethylene bottles. In addition, 1 l water samples were also collected at different high-resolution time-intervals by automatic water samplers (ISCO). The samples were filtered at the field bases with a pressure filter and ash-free filter papers (Munktell quantitative filter papers). After the field campaigns the filter papers were burned (550 °C) to analyse the concentrations of mineralogenic suspended solids [mg l^{-1}]. The estimation of annual solute yields was based on the relationship between electric conductivity and concentration of total dissolved solids (Beylich, 2008; Beylich & Kneisel, 2009). The stability of creeks and channel stone pavements as well as the range of bed load transport was estimated by using painted stone tracer lines at selected creeks and channel stretches. In addition, fresh accumulations of debris/bed load were analysed by weighing of debris (portable field balance) and by a detailed measuring of the volumes of fresh deposits. The estimation of annual bedload transport rates presented in this chapter might include errors. Anyway, the repeated detailed mapping and analysis (two to three times per year) of selected channel stretches using photo documentation and the careful measuring of the volumes of fresh bedload deposits allows at least rough estimates (Beylich, 2008; Beylich & Kneisel, 2009).

3. Results and discussion

On the basis of the process rates which were calculated for the Hrafndalur, Austdalur and Latnjavagge drainage basins after longer-term field studies (several years of process monitoring, mapping and observation) (Beylich, 2008; Beylich & Kneisel, 2009) the absolute and the relative importance of present-day denudative processes in the entire catchments was estimated by the quantification of the mass transfers caused by the different processes. To allow direct comparison of the different processes, all mass transfers are shown as tonnes multiplied by meter per year [t m yr^{-1}], i.e. as the product of the annually transferred mass and the corresponding transport distance (see Jäckli, 1957; Rapp, 1960; Barsch, 1981; Beylich, 2000; 2008; Beylich & Kneisel 2009). The mass transfers calculated for the Hrafndalur drainage basin are shown in Table 1.

It is stressed that these mass transfers are based on detailed process studies, extended mapping and process monitoring carried out over an eight-years period (2001 - 2009) (Beylich & Kneisel, 2009). In computing the mass transfer caused by rock falls and boulder

	Process	Volume [m^3 yr^{-1}]	Mass [t yr^{-1}]	Area [km^2]	Mass/Area [t km^{-2} yr^{-1}]	Average movement [m]	Mass transfer [t m yr^{-1}]
Slope denudation	Rockfalls and boulder falls	3196	7031	0.90	7812	50	351550
	Avalanches	23	46	0.02	2300	235	10810
	Debris flows	15	30	0.000062	483871	110	3300
	Translational slides	5	10	0.000043	232558	26	263
	Creep	450000	900000	1.8	500000	0.022	19800
	Chemical denudation	92	203	7	29	400	*81200*
	Mechanical fluvial denudation	67	133	7	19	400	53200
	Deflation	0.02	0.04	5	0.008	400	16
Stream work	Solute transport	92	203	7	29	1750	355250
	Suspended sediment transport	67	133	7	19	1750	*232750*
	Bedload transport	81	178	0.18	989	875	155750

Table 1. Annual mass transfers by different denudative processes in Hrafndalur (eastern Iceland), as based on investigations carried out from 2001 until 2009

falls, the following assumptions were made: a total debris-supplying rock wall surface of 0.9 km^2, a mean annual rock wall retreat rate of 3.55 mm yr^{-1}, an average transport distance of 50 m and an average rock density of 2.2 (Beylich et al., 2007a; Beylich & Kneisel, 2009). The material transported and deposited by avalanches was presumed to have an average density of 2.0 (Beylich & Kneisel, 2009). Calculating the mass transfer caused by creep was led by the following assumptions: an affected surface of 1.8 km^2, an average movement rate of 0.022 m yr^{-1} in a 0.25 m thick layer, and a medium density of 2.0 (Beylich & Kneisel, 2009). In contrast to the other denudative processes, chemical denudation affects the total surface of the slope systems. The transport distance of 400 m is about half the medium distance between water divide and main channel. The transport distance of 1750 m for fluvial solute and suspended sediment transport is about half the distance between the water divide opposite to the drainage basin outlet and the outlet. The presented estimation of annual

mass transfers might include some errors. Anyway, the scale of the calculated mass transfers and the relative importance of the different processes should be correct.
Ranking the different processes according to their annual mass transfers shows that stream work dominates over slope denudation, with fluvial solid transport, including both suspended sediment transport plus bed load transport, being more important than fluvial solute transport. Rock falls plus boulder falls are clearly the most important slope process, followed by chemical denudation. As a result, according to their relative importance, the different processes can be ranked as follows (see Table 1):

1. Fluvial suspended sediment plus bed load transport
2. Fluvial solute transport
3. Rock falls plus boulder falls
4. Chemical slope denudation
5. Mechanical fluvial slope denudation (slope wash)
6. Creep processes
7. Avalanches
8. Debris flows
9. Translation slides
10. Deflation

With respect to the temporal variability of process intensities and/or process frequencies the main snow melt period (April - June), autumn showing heavy rainfalls (especially September - November) and intensive thaw periods in winter can be pointed out as periods with comparably high activity of denudative processes.
The combined quantitative analysis of the slope and channel systems, or of slope and stream work respectively, allows statements on trends of relief development in Hrafndalur. By the retreat of rock faces and the connected formation of talus cones located below these rock faces, slope processes cause a gradually valley widening in the Hrafndalur drainage basin (Fig. 2). The valley floor is characterised by a fluvial throughput of material delivered from the slope systems. Coupling between slope and fluvial systems exists especially in the upper parts of the drainage basin where talus cones reach down into the channels. Altogether material transfer from slopes to channels via (i) avalanches, (ii) debris flows, (iii) creep processes and (iv) fluvial transfers is intensive, with material being fluvially exported from the drainage basin without longer-term storage within the main channels.
Geoelectrical surveys in the upper parts of Hrafndalur (Beylich & Kneisel, 2009) documented larger volumes of debris with thickness of talus cones of about 15-20 m. The geophysical surveys were also aimed to infer Holocene valley filling and regolith thickness in the lower parts of the drainage basin using geoelectrical techiquess. A detected generally high conductivity within the subsurface can be related to a high porosity and high water storage capacity of the weathered rhyolites. Due to indistinct resistivity contrasts the differentiation between regolith and bedrock is made difficult. It seems that the boundary between bedrock and overlying regolith / valley filling is not reached with the applied survey length, which was limited to 175 m due to 5 m spacing of the multi-electrode cable. This indicates that the depth of the boundary between bedrock and regolith / valley filling is larger than 25 m. The high porosity of the rhyolite and the large regolith thicknesses found in lower parts of Hrafndalur explain the comparably high rates of chemical slope denudation, with the values in this rhyolite area clearly exceeding the rates of chemical slope denudation in basalt areas of Austfirđir (see Tables 1 and 2) (Beylich, 2000; 2003; 2007; Beylich & Kneisel, 2009). The larger volumes of Postglacial valley fillings in the lower part of Hrafndalur indicate high rates of fluvial solid transport during the Holocene.

Due to the comparably high mechanical weathering and sedimentary transfer rates, Postglacial modification of the Pleistocene glacially formed landscape is clearly more advanced than in the extended basalt areas of Austfirðir. Anyway, due to the short time since the deglaciation (about 8000 yr) there has been no adjustment of the glacially formed landforms to the geomorphic processes, which have been operating under Holocene climates until today.

Comparing the results from Hrafndalur with data collected in the Austdalur drainage basin, located south from Hrafndalur in the basalt area of Austfirðir (Fig. 1, Table 2) (see also Beylich, 1999; 2000; 2003; 2007) reveals significant differences between the two drainage basins. Both areas are very similar with respect to climate, vegetation, human impact and topography, but significantly different with respect to lithology.

	Process	Volume [m^3 yr^{-1}]	Mass [t yr^{-1}]	Area [km^2]	Mass/Area [t km^{-2} yr^{-1}]	Average movement [m]	Mass transfer [t m yr^{-1}]
Slope denudation	Rockfalls and boulder falls	149	373	1.50	249	100	37300
	Avalanches	73	146	0.04	3650	290	42340
	Debris flows	12	24	0.000051	470588	80	1920
	Translational slides	1.5	3	0.000032	93750	14	42
	Creep	1080000	1944000	5.4	360000	0.01	19440
	Chemical denudation	74	184	23	8	750	*138000*
	Mechanical fluvial denudation	439	966	23	42	750	724500
	Deflation	0.06	0.12	17	0.007	750	90
Stream work	Solute transport	74	184	23	8	2800	*515200*
	Suspended sediment transport	439	966	23	42	2800	2704800
	Bedload transport	19	48	0.4	120	500	24000

Table 2. Annual mass transfers by different denudative processes in Austdalur (eastern Iceland), as based on investigations carried out from 1996 until 2009

The rhyolites in Hrafndalur are clearly less resistant than the basalts found in Austdalur (Beylich et al., 2007a), and Postglacial modification of the Pleistocene glacial landforms is only very little in Austdalur but significant in Hrafndalur (Beylich, 2000; 2007; Beylich & Kneisel, 2009). The high level of coupling between slope and channel systems given in Hrafndalur does not exist to that extent in Austdalur, resulting in a clearly lower relevance of bed load transport in Austdalur as compared to Hrafndalur. The intensive coupling between slope and fluvial systems in Hrafndalur, with significant transfers of material from the slopes into main channels, is strongly connected to the Holocene development of extended talus cones, which partly reach and directly feed into the main channels. Compared to that, slope systems in Austdalur are only connected to the main channels via small creeks draining the slope systems (Beylich, 1999; 2000), and slope and channel systems are largely decoupled in this drainage basin.

Suspended sediment yields in Hrafndalur are clearly lower than in Austdalur and chemical denudation dominates over suspended sediment yields in the Hrafndalur drainage basin whereas suspended sediment yields clearly dominate over chemical denudation rates in Austdalur (Tables 1 and 2). These differences between the two areas can be explained by the significantly higher porosity and connected higher infiltration and water storage capacity of the rhyolites in Hrafndalur as compared to the basalts in Austdalur (see above, Beylich et al., 2007a; Beylich & Kneisel, 2009).

According to the mass transfers, which were computed after thirteen years (1996 - 2009) of geomorphic process studies in Austdalur, the different denudative processes in the Austdalur drainage basin can be ranked as follows according to their relative importance (see Table 2):

1. Fluvial suspended sediment plus bedload transport
2. Fluvial solute transport
3. Mechanical fluvial slope denudation (slope wash)
4. Chemical denudation
5. Avalanches
6. Rock falls plus boulder falls
7. Creep processes
8. Debris flows
9. Deflation
10. Translation slides

The mass transfers calculated for the Latnjavagge drainage basin in northern Swedish Lapland are provided in Table 3.

It has to be stressed that these mass transfers are based on detailed geomorphologic process studies, extended mapping and process monitoring which were carried out over a ten-years period (1999 - 2009), and that the values shown in Table 3 are excluding the geomorphic effects of a rare rainfall event that happened on July 20-21, 2004 (Beylich & Sandberg, 2005) and of a mega slush flow event that occurred in May 1999. The role of these extreme events is further discussed below. In computing the mass transfers caused by rock- and boulder falls, the following assumptions were made: a total debris-supplying rock wall and rock ledge surface of 3.15 km^2, a mean annual rock wall retreat rate of 0.44 mm yr^{-1}, an average transport distance of 7.5 m and an average rock density of 2.5. Also the material transported and deposited by ground avalanches was presumed to have an average density of 2.5 (Beylich, 2008). Calculating the mass transfer caused by creep and solifluction was led by the following assumptions: an affected surface of 2.8 km^2, an average movement rate of 0.003 m

yr^{-1} in a 0.3 m thick layer, and a medium material density of 2.0 (Beylich, 2008). In contrast to the other denudative processes, chemical denudation affects the total surface of the slope systems (Beylich et al., 2004a; 2004b). The transport distance of 500 m is about half the medium distance between water divide and main channel. The transport distance of 2300 m for fluvial solute transport is about half the medium distance between the water divide opposite to the outlet and the drainage basin outlet. Mass transfers by fluvial suspended sediment transport and fluvial bedload transport are based on an average movement of 1400 m, which is about half the medium distance between water divide and inlet Latnjajaure (lake), for suspended sediments and 250 m, based on monitoring of tracer movements, for bed load (Beylich et al., 2006b; Beylich, 2008).

	Process	Volume [m^3 yr^{-1}]	Mass [t yr^{-1}]	Area [km^2]	Mass/Area [t km^{-2} yr^{-1}]	Average movement [m]	Mass transfer [t m yr^{-1}]
Slope denudation	Rockfalls and boulder falls	1377	3441	3.15	1092	7.5	25808
	Avalanches	27	68	0.02	3400	150	10200
	Debris flows	3	6	0.000016	375000	47	279
	Translational slides	1.5	3	0.000022	136364	3.7	11
	Creep	840000	1680000	2.8	600000	0.003	5040
	Chemical denudation	22	44	9	4.9	500	22000
	Mechanical fluvial denudation	10.5	21	9	2.4	500	10500
	Deflation	0.005	0.01	2.5	0.004	500	5
	Slush flows	13	32	0.000097	329897	110	3520
Stream work	Solute transport	22	44	9	4.9	2300	101200
	Suspended sediment transport	10.5	21	9	2.4	1400	29400
	Bedload transport	3	8	0.04	200	250	2000

Table 3. Annual mass transfers by different denudative processes in Latnjavagge (Swedish Lapland), as based on investigations carried out from 1999 until 2009

Fluvial solute transport clearly dominates over fluvial sediment transport. Ranking the different processes according to their annual mass transfers shows that stream work dominates over slope denudation, with fluvial solute transport being clearly the most important process. Rock falls plus boulder falls is the most important factor in slope denudation, followed by chemical denudation. As a result, according to their relative importance, the different processes can be ranked as follows (see Table 3):

1. Fluvial solute transport
2. Fluvial suspended sediment transport plus fluvial bedload transport
3. Rock falls plus boulder falls
4. Chemical slope denudation
5. Mechanical fluvial slope denudation (slope wash)
6. Avalanches
7. Creep processes and solifluction
8. Slush flows
9. Debris flows
10. Translation slides
11. Deflation

With respect to the temporal variability of process intensities and/or process frequencies the main snow melt period (late May - June/July) which is characterised by high runoff, increased fluvial transport, slush flow activity, ground avalanches and rock and boulder falls as well as August which is the month showing the highest frequency of extreme rainfall events (Beylich, 2003; Beylich & Gintz, 2004) triggering debris flows, translation slides, secondary rock falls, boulder falls, peak runoff and increased fluvial transport, can be pointed out as periods with comparably high activity. The period from late October until early May is compared to that characterised by only very little (single avalanches) or no activity (Beylich et al., 2006b; Beylich, 2008).

On July 20th -21st, 2004, an extreme rainfall event, with 58.4 mm of rain within 24 h and 71.7 mm of rain within 48 h as measured at the weather station of the Latnjajaure Field Station (LFS), occurred (Beylich & Sandberg, 2005). This extreme rainfall event is considered a "rare event" (Beylich, 2003), which caused extreme peak runoff in the fluvial systems of the Abisko mountain area and significant damage to the road between Abisko and Kiruna (Beylich & Sandberg, 2005). The rare rainfall event triggered high sediment transfers on the slopes and in the fluvial systems of the Latnjavagge drainage basin, and the quantified mass transfers at only one day during this rare rainfall event exceeded the mean annual mass transfers, calculated for the nine-years period 1999 - 2008 (and excluding the rare rainfall event of July 2004), by several times (see Beylich & Sandberg, 2005; Beylich et al., 2006b). The rare rainfall event caused several debris flows, translation slides, boulder falls, secondary rock falls, locally very high slope wash, bank erosion in creeks above the Latnjajaure delta and high suspended sediment and bed load transport in creeks. Table 2 in Beylich & Sandberg (2005, page 415) provides mass transfers [t m] by debris flows, translation slides and fluvial debris transport in creeks and channels, which were caused in different parts of the Latnjavagge drainage basin by the July 20th -21st, 2004 rainfall event. The mass transfers triggered by the rare event can be discussed in direct comparison with the mean annual mass transfers calculated for the period 1999 - 2009 (and excluding the rare rainfall event) (Table 3). The mass transfer by debris flows, which were triggered during the rare rainfall event was 2030 t m in the entire Latnjavagge drainage basin which is 7.3 times higher than the mean annual mass transfer by debris flows (279 t m). It has to be noted that the annual mass transfers by debris flows have clearly increased after the July 2004 event which is also reflected in significantly higher annual mean mass transfer for the ten-years period 1999 - 2009 (279 t m) as compared to the annual mean mass transfer for the five-years period 1999 - 2004 (212 t m). The mass transfer by translation slides which were triggered during the rare event reached 2446 t m in the entire drainage basin which is 222.4 times higher than the mean annual mass transfer by slides (11 t m). It is again necessary to stress that also the annual mass transfers caused by translation slides have clearly increased after the July 2004 rare rainfall event, which is documented by a significantly higher mean annual mass transfer for the ten-years period 1999 - 2009 (11 t m) as compared to the five-years period 1999 - 2004 (2.5 t m). The total mass transfer by fluvial debris transport in creeks and channels during the rare rainfall event was 7480 t m which is 16.9 times higher than the corresponding mean annual mass transfer within the entire Latnjavagge drainage basin quantified to be 444 t m for the period 1999 - 2004 (Beylich & Sandberg, 2005).

During the rare event the mass transfer by fluvial debris transport in creeks and channels (7480 t m) had clearly the highest relative importance, followed by mass transfers by slides (2446 t m) and debris flows (2030 t m). Compared to that, the mean annual mass transfers (excluding the July 20-21, 2004 event) show again a relative dominance of fluvial debris transport in creeks, but are followed by debris flows whereas translation slides are of only minor importance.

During the rare rainfall event the mass transfers caused by debris flows, translation slides and fluvial debris transport as well as suspended sediment transport were several times higher than the mean annual mass transfers.

The mega slush flow event of May 1999 caused a mass transfer, which was about 107 times higher than the mean annual total mass transfer caused by several small slush flows between 1999 and 2009.

Both the rare rainfall event of July 20th -21st, 2004 and the mega slush flow event of May 1999 underline the high importance of rare events for the calculation of longer-term sedimentary budgets. Both rare events also reveal the difficulties of defining an adequate length for monitoring programmes (Beylich & Sandberg, 2005; Beylich & Warburton, 2007).

Comparing the results from Latnjavagge with data collected in other cold environment drainage basins in northern Sweden shows some similarities but also some significant differences. Rapp (1960) stressed in his seminal study of denudation and slope development in Kärkevagge, northernmost Swedish Lapland, the high importance of chemical weathering and denudation and his work stands out as the study which opened awareness of chemical denudation in cold environments. His findings for Kärkevagge and a high relative importance of chemical weathering and denudation were confirmed by later investigations carried out in this valley (e.g. Dixon et al., 2008) and have also been postulated for other cold climate environments (see e.g. French, 1996). Compared to that, and in accordance with findings in Latnjavagge, slope wash processes are of only little importance in Kärkevagge (Rapp, 1960; Strömquist & Rehn, 1981). A significant difference between the mass transfers calculated for Latnjavagge and the mass transfers published by Rapp (1960) is the high importance of slides and debris flows in Kärkevagge. The high importance of slides and debris flows found by Rapp (1960) is mainly based on mass transfers which were triggered by a rare rainfall event with 107 mm in 24h and 175 mm in 72h in October 1959. This rare event occurred at the end of Rapp`s 9-years investigation period (1952 - 1960) and the mean annual mass transfers caused by slides and debris flows as calculated by Rapp for his 9-years investigation period might actually be too high, given

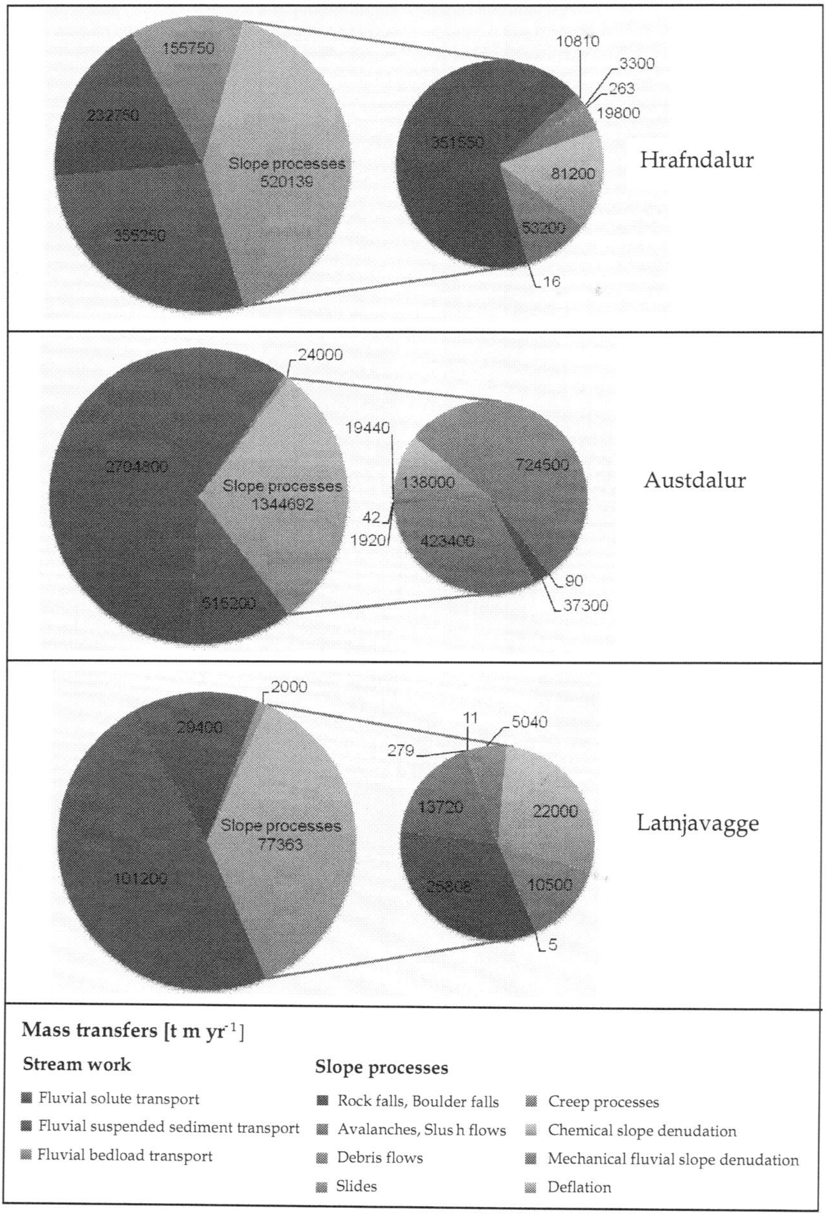

Fig. 6. Annual mass transfers in Hrafndalur (eastern Iceland), Austdalur (eastern Iceland) and Latnjavagge (Swedish Lapland)

the long recurrence intervals of such rare rainfall events in northernmost Swedish Lapland (Beylich, 1999; 2003; see also French, 1996). The problem of defining an adequate length of monitoring programmes becomes, as in Latnjavagge with respect to the rare rainfall event of July 2004 and the mega slush flow event of May 1999, again obvious (Beylich & Sandberg, 2005).

As in eastern Iceland, fluvial transport dominates over slope processes in Swedish Lapland but, differently from eastern Iceland, in Swedish Lapland fluvial solute transport dominates over fluvial sediment transport (Beylich, 2009; Beylich & Kneisel, 2009). This difference is partly due to human impact and the connected disturbance and partly removal of the vegetation cover in eastern Iceland.

The conducted combined quantitative analysis of the slope systems and channel systems, or of slope denudation and stream work, respectively, allows also statements on current trends of relief development in Latnjavagge. By the retreat of rock walls and rock ledges and the continued formation of talus cones located below the rock walls and rock ledges, slope processes cause also in the Latnjavagge drainage basin a general valley widening (Fig. 4). The valley floor is characterised by a fluvial throughput of material above the delta of Latnjajaure, net accumulation of material in the delta of Latnjajaure and in the lake and a moderate fluvial down cutting between the outlet of Lake Latnjajaure and the outlet of the entire Latnjavagge drainage basin. Coupling between slope systems and fluvial systems exists only to a limited extend (i) via slush flows in defined areas within the valley as well as (ii) via ground avalanches at a short main channel stretch directly above the delta of Latnjajaure. Altogether, Postglacial modification of the glacial relief is little (Fig. 4). Due to the also in Swedish Lapland short time since the deglaciation (8000 - 10000 yr) and the altogether low intensity of the active denudative surface processes, until today, there has been no adjustment of the Pleistocene glacial landforms to the geomorphic processes which have been operating until present under the Holocene climates.

As based on the presented quantitative investigations, in all selected study areas in sub-Arctic oceanic eastern Iceland and Arctic oceanic Swedish Lapland the intensity of contemporary denudative processes and mass transfers is altogether rather low which is in opposition to the earlier postulated (e.g. Büdel, 1981; Tricart, 1970) high intensity of geomorphic processes in cold climate environments, with these earlier statements have only being based on purely qualitative descriptions.

A direct comparison of the mass transfers within the three investigated drainage basins (Fig. 6., see also Tables 1 - 3) summarises that there are differences between process intensities and the relative importance of different denudative processes within the study areas in eastern Iceland and Swedish Lapland.

The major controls of these detected differences are

i. Climate:

   The higher annual precipitation along with the larger number of extreme rainfall events and the higher frequency of snowmelt and rainfall generated peak runoff events in eastern Iceland as compared to Swedish Lapland leads to higher mass transfers. All three study areas are located in cold regions and projected climate change is expected to alter melt season duration and intensity, along with an increased number of extreme rainfall events, total annual precipitation and the balance between snowfall and rainfall. In addition, changes in the thermal balance are expected to reduce the extent of permafrost and seasonal ground frost and increase active layer depths (ACIA, 2004). Looking at the existing differences between Hrafndalur / Austdalur (eastern Iceland)

and Latnjavagge (Swedish Lapland) it seems obvious that the projected changes in climate will cause significant changes of mass transfers,

ii. Lithology:
The low resistance of the rhyolites in Hrafndalur causes especially high weathering rates and connected mass transfers in this drainage basin. Due to the lower resistance of the rhyolites as compared to the basalts found in Austdalur Postglacial modification of the glacially formed relief is clearly further advanced in Hrafndalur as compared to Austdalur,

iii. Relief:
The greater steepness of the Icelandic drainage basin leads to higher mass transfers here as compared to Latnjavagge,

iv. Vegetation cover:
The significant disturbance of the vegetation cover by direct human impacts in Hrafndalur / Austdalur (eastern Iceland) causes higher mass transfers by slope wash here whereas restricted sediment availability is a major reason for lower mass transfers in Latnjavagge (Swedish Lapland).

4. Conclusion

By the combined, quantitative recording of the relevant denudative slope processes and stream work in three selected drainage basins, information on the absolute and relative importance of the different denudative processes has been collected. The selected drainage basin - based approach (Beylich, 1999; 2000; 2002; 2008; Beylich & Kneisel, 2009) seems to be effective for analysing the sediment budget and trends of Postglacial relief development in selected study areas with given environmental settings. More studies to the present one, carried out with unified geomorphologic field methods (Beylich & Warburton, 2007; Beylich et al., 2006a; 2008; Lamoureux et al., 2007) in environments having different climatic, vegetational, human impact, topographic, lithological / geological and tectonic features will help to gain better understanding of the internal differentiation of different global environments (see e.g. Barsch, 1984; 1986; Beylich et al., 2006a; 2008). Furthermore, additional information on the control mechanisms of processes, the role of extreme geomorphic events for longer-term mass transfers and sediment budgets, the general intensity of geomorphic processes and mass transfers, and the relative importance of different processes for slope and valley formation and relief development under different environmental conditions can be collected. Direct comparisons of drainage basins and the application of the Ergodic principle of space-for-time substitution will improve the possibilities to model relief development as well as possible effects of projected climate change.

5. References

ACIA (2004). Impacts of a warming Arctic: Arctic Climate Impact Assessment. ACIA Overview Report. Arctic Climate Impact Assessment. Cambridge University Press.

André, M.-F. (1995). Postglacial microweathering of granitic roches mountonnees in Northern Scandinavia. In: Slaymaker, O. (Ed.), Steepland Geomorphology, pp. 103-127, Wiley, London.

Ballantyne, C.K. (2002). Paraglacial Geomorphology. *Quaternary Science Reviews*, 21, pp. 1935-2017.

Barsch, D. (1981). Studien zur gegenwartigen Geomorphodynamik im Bereich der Oobloyah Bay, N-Ellesmere Island, N.W.T., Kanada. *Heidelberger Geographische Arbeiten*, 69, pp. 123-161.

Barsch, D. (1984). Geomorphologische Untersuchungen zum periglazialen Milieu polarer Geosysteme. *Zeitschrift für Geomorpologie. N.F., Suppl.-Bd.*, 50, pp. 107-116.

Barsch, D. (1986). Forschungen in Polargebieten. *Heidelberger Geowissenschaftliche Abhandlungen*, 6, pp. 33-50.

Barsch, D. (1993). Periglacial geomorphology in the 21st century. *Geomorphology*, 7, pp. 141-163.

Barsch, D.; Gude, M.; Mäusbacher, R.; Schukraft, G. & Schulte, A. (1994). Recent fluvial sediment budgets in glacial and periglacial environments, NW Spitsbergen. *Zeitschrift für Geomorphologie, Suppl.*, 97, pp. 111-122.

Becht, M. (1995). Untersuchungen zur aktuellen Reliefentwicklung in alpinen Einzugsgebieten. *Münchener Geographische Abhandlungen*, A47. München.

Beylich, A.A. (1999). Hangdenudation und fluviale Prozesse in einem subarktisch-ozeanisch geprägten, permafrostfreien Periglazialgebiet mit pleistozäner Vergletscherung - Prozessgeomorphologische Untersuchungen im Bergland der Austfirđir (Austdalur, Ost-Island). *Berichte aus der Geowissenschaft*, Aachen. 130pp.

Beylich, A.A. (2000). Geomorphology, sediment budget, and relief development in Austdalur, Austfirdir, East Iceland. *Arctic, Antarctic, and Alpine Research*, 32, 4, pp. 466-477.

Beylich, A.A. (2002). Sediment budgets and relief development in present periglacial environments - a morphosystem analytical approach. *Hallesches Jahrbuch für Geowissenschaften*, A24, pp. 111-126.

Beylich, A.A. (2003). Present morphoclimates and morphodynamics in Latnjavagge, the northern Swedish Lapland and Austdalur, East Iceland. *Jökull*, 52, pp. 33-54.

Beylich, A.A. (2007). Quantitative studies on sediment fluxes and sediment budgets in changing cold environments - potential and expected benefit of coordinated data exchange and the unification of methods. *Landform Analysis*, 5: pp. 9-10.

Beylich, A.A. (2008). Mass transfers, sediment budget and relief development in the Latnjavagge catchment, Arctic-oceanic Swedish Lapland. *Zeitschrift für Geomorphologie. N.F.*, 52, 1, pp. 149-197.

Beylich, A.A. (2009). Chemical and mechanical fluvial denudation rates in cold environments - Comparison of denudation rates from three catchments in sub-Arctic Eastern Iceland, sub-Arctic Finnish Lapland and Arctic Swedish Lapland. *Jökul*, 59, pp. 19-32.

Beylich, A.A.; Etienne, S.; Etzelmüller, B.; Gordeev, V.V.; Käyhkö, J.; Rachold, V.; Russell, A.J.; Schmidt, K.-H.; Sæmundsson, Þ.; Tweed, F.S. & Warburton, J. (2006a). The European Science Foundation (ESF) Network SEDIFLUX - An introduction and overview. *Geomorphology*, 80, pp. 3-7.

Beylich, A.A. & Gintz, D. (2004). Effects of high-magnitude/low-frequency fluvial events generated by intense snowmelt or heavy rainfall in Arctic periglacial environments in northern Swedish Lapland and northern Siberia. *Geografiska Annaler*, 86A, pp. 11-29.

Beylich, A.A.; Gustavsson, M. & Kolstrup, E. (2007a). Experimental weathering of selected non-calcareous rock types under wet/moist conditions. *Zeitschrift für Geomorphologie N.F.*, 51, 1, pp. 1-26.

Beylich, A.A. & Kneisel, Ch. (2009). Sediment budget and relief development in Hrafndalur, subarctic oceanic eastern Iceland. *Arctic, Antarctic and Alpine Research*, 41, 1, pp. 3-17.

Beylich, A.A.; Kolstrup, E.; Thyrsted, T. & Gintz, D. (2004a). Water chemistry and its diversity in relation to local factors in the Latnjavagge drainage basin, Arctic-oceanic Swedish Lapland. *Geomorphology*, 58, pp. 125-143.

Beylich, A.A.; Kolstrup, E.; Thyrsted, T.; Linde, N.; Pedersen, L.B. & Dynesius, L. (2004b). Chemical denudation in Arctic-alpine Latnjavagge (Swedish Lapland) in relation to regolith as assessed by radio magnetotelluric-geophysical profiles. *Geomorphology*, 57, pp. 303-319.

Beylich, A.A.; Lamoureux, S.F. & Decaulne, A. (2007b). Coordinated quantitative studies on sediment fluxes and sediment budgets in changing cold environments - examples from three SEDIBUD key test areas in Canada, Iceland and Norway. *Landform Analysis*, 5, pp. 11-12.

Beylich, A.A.; Lamoureux, S.F. & Decaulne, A. (2008). The global I.A.G./A.I.G. SEDIBUD (Sediment Budgets in Cold Environments) programme: Introduction and overview. *Norsk Geografisk Tidsskrift – Norwegian Journal of Geography*, 62, 2, pp. 50-51.

Beylich, A.A.; Lamoureux, S.F. & Decaulne, A. (in press). Developing quantitative frameworks for studies on sedimentary fluxes and budgets in changing cold environments. *Polish Polar Research.*

Beylich, A.A.; Molau, U.; Luthbom, K. & Gintz, D. (2005). Rates of chemical and mechanical fluvial denudation in an Arctic-oceanic periglacial environment, Latnjavagge drainage basin, northernmost Swedish Lapland. *Arctic, Antarctic, and Alpine Research*, 37, 1, pp. 75-87.

Beylich, A.A. & Sandberg, O. (2005). Geomorphic effects of the extreme rainfall event of 20-21 July, 2004 in the Latnjavagge catchment, northern Swedish Lapland. *Geografiska Annaler*, 87A, pp. 409-419.

Beylich, A.A.; Sandberg, O.; Molau, U. & Wache, S. (2006b). Intensity and spatio-temporal variability of fluvial sediment transfers in an Arctic-oceanic periglacial environment in northernmost Swedish Lapland (Latnjavagge catchment). *Geomorphology*, 80, pp. 114-130.

Beylich, A.A. & Warburton, J. (Eds) (2007). Analysis of Source-to-Sink-Fluxes and Sediment Budgets in Changing High-Latitude and High-Altitude Cold Environments. SEDIFLUX Manual. *NGU Report*, 2007.053. 158pp.

Büdel, J. (1981). Klima-Geomorphologie. 2., veränderte Auflage, Berlin, Stuttgart.

Burki, V.; Hansen, L.; Fredin, O.; Andersen, T.A.; Beylich, A.A.; Jaboyedoff, M.; Larsen, E. & Tønnesen, J.-F. (2009). Little Ice Age advance and retreat sediment budgets for an outlet glacier in western Norway. *Boreas*, 10.1111/j.1502-3885.2009.00133.x.ISSN 0300-9483

Caine, N. (1974). The geomorphic processes of the alpine environment. In: Ives, J.D. & Barry, R.G. (Eds). *Arctic and Alpine Environments*. Methuen, pp. 721-748.

Caine, N. (2004). Mechanical and chemical denudation in mountain systems. In: Owens, P.N. & Slaymaker, O. (Eds.). *Mountain Geomorphology*. Arnold, London.

Caine, N. & Swanson, F.J. (1989). Geomorphic coupling of hillslope and channel systems in two small mountain basins. *Zeitschrift für Geomorphologie*, 33, pp. 189-203.

Cockburn, J.M.H. & Lamoureux, S.F. (2007). Hydroclimate controls over seasonal sediment yield in two adjacent high arctic watersheds. *Hydrological Processes*, doi: 10.1002/hyp.6798

Dixon, J.C.; Thorn, C.E. & Darmody, R.G. (2008). Spatial scale and chemical weathering in Kärkevagge: Influences on landscape evolution. *Zeitschrift für Geomorphologie N.F.*, 52, 1, pp. 27-49.

Dowdeswell, J.A.; Ottesen, D. & Rise, L. (2006). Flow switching and large-scale deposition by ice streams draining former ice sheets. *Geology*, 34, pp. 313-316.

Evans, S.G. & Clague, J.J. (1994). Recent climate change and catastrophic geomorphic processes in mountain environments. *Geomorphology*, 10, pp. 107-128.

French, H.M. (1996). The Periglacial Environment. 2nd ed., London.

Gurnell, A.M. & Clark, M.J. (Eds) (1987). Glacio-fluvial sediment transfer: An Alpine Perspective. Chichester, Wiley.

Haeberli, W. & Beniston, M. (1998). Climate change and its impacts on glaciers and permafrost in the Alps. *Ambio*, 27, pp. 258-265.

Hansen, L.; Beylich, A.A.; Burki, V.; Eilertsen, R.; Fredin, O.; Larsen, E.; Lyså, A.; Nesje, A.; Stalsberg, K. & Tønnesen, J.-F. (2009). Stratigraphic architecture and infill history of a deglaciated bedrock valley based on georadar, seismic profiling and drilling. *Sedimentology*, 56, pp. 1751-1773.

Jäckli, H. (1957). Gegenwartsgeologie des Bündnerischen Rheingebietes. Beitrag zur Geologischen Karte der Schweiz, *Geotechn. Serie*, 36.

Lamoureux, S.F. (1999). Spatial and inter-annual variations in sedimentation patterns recorded in nonglacial varved sediments from the Canadian high Arctic. *Journal of Paleolimnology*, 21, pp. 73-84.

Lamoureux, S.F.; Beylich, A.A. & Decaulne, A. (2007): Sediment Fluxes and Budgets in High-Latitude and High-Altitude Cold Environments. Sediment Budgets in Cold Environments (SEDIBUD) Second Workshop; Abisko, Sweden, 15-19 September 2007. *EOS*, Vol. 88, 52, 25 December 2007: 580.

Orwin, J.F. & Smart, C.C. (2004). Short-term spatial and temporal patterns of suspended sediment transfer in proglacial channels, Small River Glacier, Canada. *Hydrological Processes*, 18, pp. 1521-1542.

Otto, J.-C. & Dikau, R. (2004). Geomorphologic system analysis of a high mountain valley in the Swiss Alps. *Zeitschrift für Geomorphologie*, 48, pp. 323-341.

Rapp, A. (1960). Recent development of mountain slopes in Kärkevagge and surroundings, Northern Scandinavia. *Geografiska Annaler*, 42, pp. 71-200.

Rapp, A. (1985). Extreme rainfall and rapid snowmelt as causes of mass movements in high latitude mountains. In: Church, M. & Slaymaker, O. (Eds). Field and theory: lectures in geocryology. *University of British Columbia Press*, Vancouver, pp. 36-56.

Reid, L.M. & Dunne, T. (1996). Rapid evaluation of sediment budgets. Cremlingen, Catena.

Reynolds, J.M. (1997). An introduction to applied and environmental geophysics. Wiley, Chichester.

Rise, L.; Ottesen, D.; Berg, K. & Lundin, E. (2005). Large-scale development of the mid-morwegian margin during the last 3 million years. *Marine and Petroleum Geology*, 22, pp. 33-44.

Sass, O. (2005). Spatial patterns of rockfall intensity in the northern Alps. *Zeitschrift für Geomorphologie N.F.*, Suppl.-Bd. 138, pp. 51-65.

Schrott, L.; Hufschmidt, G.; Hankammer, M.; Hoffmann, T. & Dikau, R. (2003). Spatial distribution of sediment storage types and quantification of valley fill deposits in an alpine basin, Reintal, Bavarian Alps, Germany. *Geomorphology*, 55, pp. 45-63.

Schrott, L.; Niederheide, A.; Hankammer, M.; Hufschmidt, G. & Dikau, R. (2002). Sediment storage in a mountain catchment: geomorphic coupling and temporal variability (Reintal, Bavarian Alps, Germany). *Zeitschrift für Geomorphologie N.F.*, Suppl. 127, pp. 175-196.

Slaymaker, O. (2000). Research developments in the hydrological sciences in Canada (1995-1998): Surface water - quantity, quality and ecology. *Hyrological Prccesses*, 14, pp. 1539-1550.

Slaymaker, O., (2008). Sediment budget and sediment flux studies under accelerating global change in cold environments. *Zeitschrift für Geomorphologie N.F.*, 52, 1, pp. 123-148.

Slaymaker, O.; Souch, C.; Menounos, B. & Filippelli, G. (2003). Advances in Holocene mountain geomorphology inspired by sediment budget methodology. *Geomorphology*, 55, pp. 305-316.

Strömquist, L. & Rehn, J. (1981). Rainfall measurements, runoff and sediment transport in Kärkevagge, Swedish Lapland. *Transactions, Japanese Geomorphological Union*, 2, pp. 211-222.

Tricart, J. (1970). Geomorphology of cold environments. London.

Warburton, J. (1993). Energetics of Alpine proglacial geomorphic processes. *Transactions of the Institute of British Geographers*, 18, pp. 197 - 206.

Warburton, J. (2006). Mountain Environments. In: Perry, C. & Taylor, K. (Eds). *Environmental Sedimentology*. Blackwell, Oxford.

Warburton, J. (2007): Sediment budgets and rates of sediment transfers across cold environments in Europe: a commentary. *Geografiska Annaler*, 89A, 1, pp. 95-100.

Part 5

Combined Heat and Mass Transfer

19

Successive Linearisation Solution of Free Convection Non-Darcy Flow with Heat and Mass Transfer

Sandile Motsa[1] and Stanford Shateyi[2]
[1]*University of Swaziland*
[2]*University of Venda*
[1]*Swaziland*
[2]*South Africa*

1. Introduction

Coupled convective heat and mass transfer processes in porous media has received a great deal of attention amongst researchers in recent years because of its numerous applications. The heat and mass tranfer phenomena in porous media is often encountered in industrial and engineering applications including transport processes through packed bed reactors in the chemical industry, ground water flows, oil industry, nuclear reactors, ceramic industry and in many other applications where heat and mass transfer with convection in porous media can be observed. Examples of recent studies where heat and mass transfer in porous media has been analysed include Amin (2004); Murthty & Singh (1999); Narayana & Murthy (2006); Partha (2008).

The boundary layer equations governing the non-Darcy flow with heat and mass transfer are highly non-linear and do not usually possess exact analytical solutions. In recent years, a great deal of attention has been focussed towards developing new methods of finding solutions to non-linear boundary value problems (BVPs) in general. Traditional numerical techniques for solving non-linear BVPs arising in fluid mechanics and other engineering applications include finite differences, finite element methods, Keller-Box method, Runge-Kutta with shooting methods, and spectral collocation methods. Recent advances in computational fluid dynamics and the advent of inexpensive high-speed computers have resulted in the creation of powerful easy-to-use software tools which make use of some of the traditional numerical techniques. Solving complex non-linear BVPs can now be done in a matter of seconds using functions or routines which are built into scientific computation software packages such as Maple, Matlab, Mathematica, Scilab and many others. Despite their apparent simplicity and power, these in-built functions along with the numerical methods have their own limitations such as dealing with problems with singularities, problems with multiple solutions, problems requiring stringent accuracies and for problems with solutions that have very sharp changes. Besides, from numerical results, it is also difficult to have a clear understanding and give essential interpretation of the numerically generated solutions of the underlying non-linear problem.

Apart from the traditional numerical methods for solving boundary layer equations recent purely analytic or semi-analytic approaches have been developed. These include the Adomian decomposition method Adomian (1976; 1991), homotopy perturbation method He (1999; 2000; 2006), homotopy analysis method Liao (1992; 1999; 2003; 2009), differential transform methods Zhou (1986); Chen & Ho (1999); Ayaz (2004), variational iteration method He (1999) . Apart from the homotopy analysis method (HAM), these methods converge very slowly or fail to converge at all in highly non-linear problems with very large parameters. The advantage of the HAM is that it uses a convergence controlling parameter which may accelerate convergence to the solution of the non-linear problem. But, in some cases, even with the use of this convergence controlling parameter, convergence of the HAM may be very slow in problems with very large parameters and in problems defined on unbounded domains such as the boundary layer problems. The mathematical difficulties associated with non-linear boundary layer equations have prompted several researchers to continually seek ways of improving on these numerical and analytic solution methods. A strategy that is commonly used to address the problems of slow convergence of analytical solutions is to use the Padé technique such as in the Hermite Padé and Homotopy Padé approach. A novel idea of blending spectral collocation methods and the homotopy analysis method to obtain the spectral-homotopy analysis method (SHAM) has recently been proposed in Motsa et al. (2010a;b). The advantage of the SHAM is that it uses the best ideas from both the analytic and numerical approaches such as the use of convergence controlling parameters (from the HAM) and the promise of spectral accuracy and the ability to deal with coupled equations (from the spectral collocation methods).

The aim of this work is to present a new numerical perturbation scheme for solving complex nonlinear boundary value problems arising in problems of heat and mass transfer. The proposed method of solution, hereinafter referred to as the successive linearisation method (SLM), is based on a novel idea of iteratively linearising the underlying governing non-linear boundary layer equations , which are written in similarity form, and solving the resulting equations using spectral methods. The SLM approach has been successfully applied to different fluid flow problems (see for example Motsa et al. (2010c); Makukula et al. (2010a;b); Shateyi & Motsa (2010)). The flow velocity, temperature and concentration profiles, skin friction and rate of surface heat and transfer are computed using the proposed method. The influence of the governing parameters on these flow characteristics is illustrated graphically and using tables. The results of the proposed method are validated by comparing them with some available results published in literature and against results obtained using other numerical solutions. The comparison indicates that the proposed SLM approximation converges very rapidly to the true solutions and is at least as good as or more accurate than the most efficient methods of solutions currently being used. Because of its simplicity, rapid convergence and accuracy, we conclude that the proposed method of successive linearisation has great potential of being used in other related studies in non-linear science and engineering. The SLM method can therefore be used in place of traditional methods such as finite differences, Runge-Kutta shooting methods, finite elements, etc, in solving non-linear boundary value problems.

2. Mathematical formulation

We consider the problem of free convection heat and mass transfer from a vertical surface embedded in a fluid saturated non-Darcy porous medium. The $x-$ axis is along the wall surface and the $y-$ axis is normal to it. The wall is maintained at a constant temperature T_w

and concentration C_w which are greater than the ambient temperature T_∞ and concentration C_∞, respectively. The governing equations for the steady non-Darcy flow can be written as

$$\frac{\partial u}{\partial x} + \frac{\partial v}{\partial y} = 0 \tag{1}$$

$$\frac{\partial u}{\partial y} + \frac{c\sqrt{K}}{\nu}\frac{\partial (u^2)}{\partial y} = \frac{Kg}{\nu}\left[\beta_T \frac{\partial T}{\partial y} + \beta_C \frac{\partial C}{\partial y}\right] \tag{2}$$

$$u\frac{\partial T}{\partial x} + v\frac{\partial T}{\partial y} = \alpha \frac{\partial^2 T}{\partial y^2} + \frac{D_m K_T}{c_s c_p}\frac{\partial^2 C}{\partial y^2} \tag{3}$$

$$u\frac{\partial C}{\partial x} + v\frac{\partial C}{\partial y} = D_m \frac{\partial^2 C}{\partial y^2} + \frac{D_m K_T}{T_m}\frac{\partial^2 T}{\partial y^2} \tag{4}$$

The boundary conditions are defined as follows;

$$v = 0, \quad T = T_w, \quad C = C_w, \quad \text{at} \quad y = 0, \tag{5}$$

$$u = 0, \quad T = T_\infty, \quad C = C_\infty, \quad \text{as} \quad y \Rightarrow \infty, \tag{6}$$

where u and v denote velocity components in the $x-$ and $y-$ directions, respectively, T and C are the temperature and concentration, respectively, K is the permeability parameter, β_T and β_C are the coefficients of thermal and solutal expansions, respectively, c is the Forchheimer constant, ν is the kinematic viscosity, g is the acceleration due to gravity, α and D_m are the thermal and mass diffusivity, respectively, c_p is the specific heat capacity, c_s is the concentration susceptibility, T_m is the mean fluid temperature and K_T is the thermal diffusion ratio.

By making use of the following similarity variable transformations,

$$\eta = \frac{y}{x}\sqrt{Ra_x}, \quad u = \frac{\alpha}{x}Ra_x f'(\eta), \quad v = -\frac{\alpha}{2x}\sqrt{Ra_x}\left[f(\eta) - \eta f'(\eta)\right], \tag{7}$$

$$\theta(\eta) = \frac{T - T_\infty}{T_w - T_\infty}, \quad \phi(\eta) = \frac{C - C_\infty}{C_w - C_\infty}, \tag{8}$$

the governing equations (1 - 6) are reduced to the following system of non-linear ordinary differential equations,

$$f'' + 2Gr^* f' f'' = \theta' + N_1 \phi', \tag{9}$$

$$\theta'' + \frac{1}{2} f\theta' + D_f \phi'' = 0, \tag{10}$$

$$\phi'' + \frac{1}{2} Lef\phi' + LeS_r \theta'' = 0. \tag{11}$$

The transformed dimensionless boundary conditions are:

$$f = 0, \quad \theta = 1, \quad \phi = 1, \quad \text{at} \quad \eta = 0, \tag{12}$$

$$f' = 0, \quad \theta = 0, \quad \phi = 0, \quad \text{as} \quad \eta \to \infty, \tag{13}$$

where the primes denote differentiation with respect to η. In equations (9 - 11) $Gr^* = c\sqrt{K}Kg\beta_T(T_w - T_\infty)/\nu^2$ represents the inertial parameter, $Ra_x = Kg\beta_T(T_w - T_\infty)x/\alpha\nu$ is the Darcy-Rayleigh number, $N_1 = \dfrac{\beta_c(C_w - C_\infty)}{\beta_T(T_w - T_\infty)}$ is the buoyancy ratio, $Le = \alpha/D_m$ is the Lewis number, $D_f = \dfrac{DmK_T(C_w - C_\infty)}{\alpha c_s c_p}$ is the Dufour number and $S_r = \dfrac{K_T D_m(T_w - T_\infty)}{\alpha T_m(C_w - C_\infty)}$ is the Soret number. Other parameters of engineering interest for the present problem are the local Nusselt number and the local Sherwood number, which are given in dimensionless terms by the expressions

$$Nu_x/Ra_x^{1/2} = -\theta'(0), \quad Sh_x/Ra_x^{1/2} = -\phi'(0) \tag{14}$$

Equations (9 - 11) are fully coupled and highly nonlinear. In the next section we propose a simple new method for solving the nonlinear system of equations.

3. Successive linearisation method (SLM) solution

The SLM algorithm starts with the assumption that the independent variables $f(\eta), \theta(\eta)$ and $\phi(\eta)$ can be expressed as

$$f(\eta) = f_i(\eta) + \sum_{n=0}^{i-1} f_n(\eta), \quad \theta(\eta) = \theta_i(\eta) + \sum_{n=0}^{i-1} \theta_n(\eta), \quad \phi(\eta) = \phi_i(\eta) + \sum_{n=0}^{i-1} \phi_n(\eta) \tag{15}$$

where f_i, θ_i, ϕ_i $(i = 1,2,3,\ldots)$ are unknown functions and f_n, θ_n and ϕ_n $(n \geq 1)$ are approximations which are obtained by recursively solving the linear part of the equation system that results from substituting (15) in the governing equations. The main assumption of the SLM is that f_i, θ_i and ϕ_i become increasingly smaller when i becomes large. that is

$$\lim_{i\to\infty} f_i = \lim_{i\to\infty} \theta_i = \lim_{i\to\infty} \phi_i = 0. \tag{16}$$

Thus, starting from the initial guesses $f_0(\eta)$, $\phi_0(\eta)$, $\theta_0(\eta)$, which are chosen to satisfy boundary conditions, the subsequent solutions for f_i, ϕ_i, θ_i $i \geq 1$ are obtained by successively solving the linearised form of the equations which are obtained by substituting equation (15) in the governing equations and considering only the linear terms. The linearised equations to be solved are given as

$$a_{1,i-1}f_i'' + a_{2,i-1}f_i' - \theta_i' - N_1\phi_i' = r_{1,i-1}, \tag{17}$$

$$\theta_i'' + b_{1,i-1}\theta_i' + b_{2,i-1}f_i + D_f\phi_i'' = r_{2,i-1}, \tag{18}$$

$$\phi_i'' + c_{1,i-1}\phi_i' + c_{2,i-1}f_i + LeSr\theta_i'' = r_{3,i-1}, \tag{19}$$

subject to the boundary conditions

$$f_i(0) = f_i'(\infty) = 0, \ \theta_i(0) = \theta_i(\infty) = 0, \ \phi_i(0) = \phi_i(\infty) = 0. \tag{20}$$

where the coefficient parameters $a_{k,i-1}, b_{k,i-1}, c_{k,i-1}$ $(k = 1,2)$, $r_{j,i-1}$ $(j = 1,2,3)$ are defined as,

$$a_{1,i-1} = 1+2Gr^*\sum_{n=0}^{i-1} f'_n,\quad a_{2,i-1} = 2Gr^*\sum_{n=0}^{i-1} f''_n, \tag{21}$$

$$b_{1,i-1} = \frac{1}{2}\sum_{n=0}^{i-1} f_n,\quad b_{2,i-1} = \frac{1}{2}\sum_{n=0}^{i-1} \theta'_n, \tag{22}$$

$$c_{1,i-1} = \frac{1}{2}\sum_{n=0}^{i-1} f_n,\quad c_{2,i-1} = \frac{Le}{2}\sum_{n=0}^{i-1} \phi'_n, \tag{23}$$

$$r_{1,i-1} = \sum_{n=0}^{i-1}\theta'_n + N_1\sum_{n=0}^{i-1}\phi'_n - \sum_{n=0}^{i-1} f''_n - 2Gr^*\sum_{n=0}^{i-1} f'_n\sum_{n=0}^{i-1} f''_n, \tag{24}$$

$$r_{2,i-1} = -\sum_{n=0}^{i-1}\theta''_n - \frac{1}{2}\sum_{n=0}^{i-1} f_n\sum_{n=0}^{i-1}\theta'_n - D_f\sum_{n=0}^{i-1}\phi''_n, \tag{25}$$

$$r_{3,i-1} = -\sum_{n=0}^{i-1}\phi''_n - \frac{Le}{2}\sum_{n=0}^{i-1} f_n\sum_{n=0}^{i-1}\phi'_n - LeS_r\sum_{n=0}^{i-1}\theta''_n. \tag{26}$$

Once each solution for f_i, θ_i, ϕ_i ($i \geq 1$) has been found from iteratively solving equations (17 - 20), the approximate solutions for $f(\eta), \theta(\eta)$ and $\phi(\eta)$ are obtained as

$$f(\eta) \approx \sum_{m=0}^{M} f_m(\eta),\quad \theta(\eta) \approx \sum_{m=0}^{M}\theta_m(\eta),\quad \phi(\eta) \approx \sum_{m=0}^{M}\phi_m(\eta), \tag{27}$$

where M is the order of SLM approximation. Since the coefficient parameters and the right hand side of equations (17 - 19), for $i = 1,2,3,\ldots$, are known (from previous iterations). The equation system can easily be solved analytically (wherever possible) or using any numerical method such as finite differences, finite elements, Runge-Kutta based shooting methods or collocation methods. In this work, equations (17 - 20) are solved using the Chebyshev spectral collocation method. This method is based on approximating the unknown functions by the Chebyshev interpolating polynomials in such a way that they are collocated at the Gauss-Lobatto points defined as

$$\xi_j = \cos\frac{\pi j}{N},\quad j = 0,1,\ldots,N. \tag{28}$$

where $N+1$ is the number of collocation points used (see for example Canuto et al. (1988); Don & Solomonoff (1995); Trefethen (2000)). In order to implement the method, the physical region $[0,\infty)$ is transformed into the region $[-1,1]$ using the domain truncation technique in which the problem is solved on the interval $[0,L]$ instead of $[0,\infty)$. This leads to the mapping

$$\frac{\eta}{L} = \frac{\xi+1}{2},\quad -1 \leq \xi \leq 1 \tag{29}$$

where L is the scaling parameter used to invoke the boundary condition at infinity. The unknown functions f_i, θ_i and ϕ_i are approximated at the collocation points by

$$f_i(\xi) \approx \sum_{k=0}^{N} f_i(\xi_k)T_k(\xi_j),\ \theta_i(\xi) \approx \sum_{k=0}^{N}\theta_i(\xi_k)T_k(\xi_j),\ \phi_i(\xi) \approx \sum_{k=0}^{N}\phi_i(\xi_k)T_k(\xi_j),\ j = 0,1,\ldots,N \tag{30}$$

where T_k is the kth Chebyshev polynomial defined as

$$T_k(\xi) = \cos[k\cos^{-1}(\xi)]. \tag{31}$$

The derivatives of the variables at the collocation points are represented as

$$\frac{d^a f_i}{d\eta^a} = \sum_{k=0}^{N} \mathbf{D}^a_{kj} f_i(\xi_k), \quad \frac{d^a \theta_i}{d\eta^a} = \sum_{k=0}^{N} \mathbf{D}^a_{kj} \theta_i(\xi_k), \quad \frac{d^a \phi_i}{d\eta^a} = \sum_{k=0}^{N} \mathbf{D}^a_{kj} \phi_i(\xi_k), \; j = 0,1,\ldots,N \tag{32}$$

where a is the order of differentiation and $\mathbf{D} = \frac{2}{L}\mathcal{D}$ with $\mathcal{D}$ being the Chebyshev spectral differentiation matrix (see for example, Canuto et al. (1988); Trefethen (2000)). Substituting equations (29 - 32) in (17) - (20) leads to the matrix equation given as

$$\mathbf{A}_{i-1}\mathbf{X}_i = \mathbf{R}_{i-1}, \tag{33}$$

subject to the boundary conditions

$$f_i(\xi_N) = \sum_{k=0}^{N} \mathbf{D}_{0k} f_i(\xi_k) = \theta_i(\xi_N) = \theta_i(\xi_0) = \phi_i(\xi_N) = \phi_i(\xi_0) = 0 \tag{34}$$

in which $\mathbf{A}_{i-1}$ is a $(3N+3) \times (3N+3)$ square matrix and $\mathbf{X}_i$ and $\mathbf{R}_{i-1}$ are $(3N+1) \times 1$ column vectors defined by

$$\mathbf{A}_{i-1} = \begin{bmatrix} A_{11} & A_{12} & A_{13} \\ A_{21} & A_{22} & A_{23} \\ A_{31} & A_{32} & A_{33} \end{bmatrix}, \quad \mathbf{X}_i = \begin{bmatrix} \mathbf{F}_i \\ \mathbf{\Theta}_i \\ \mathbf{\Phi}_i \end{bmatrix}, \quad \mathbf{R}_{i-1} = \begin{bmatrix} \mathbf{r}_{1,i-1} \\ \mathbf{r}_{2,i-1} \\ \mathbf{r}_{3,i-1} \end{bmatrix}, \tag{35}$$

where

$$\mathbf{F}_i = [f_i(\xi_0), f_i(\xi_1), \ldots, f_i(\xi_{N-1}), f_i(\xi_N)]^T, \tag{36}$$

$$\mathbf{\Theta}_i = [\theta_i(\xi_0), \theta_i(\xi_1), \ldots, \theta_i(\xi_{N-1}), \theta_i(\xi_N)]^T, \tag{37}$$

$$\mathbf{\Phi}_i = [\phi_i(\xi_0), \phi_i(\xi_1), \ldots, \phi_i(\xi_{N-1}), \phi_i(\xi_N)]^T, \tag{38}$$

$$\mathbf{r}_{1,i-1} = [r_{1,i-1}(\xi_0), r_{1,i-1}(\xi_1), \ldots, r_{1,i-1}(\xi_{N-1}), r_{1,i-1}(\xi_N)]^T, \tag{39}$$

$$\mathbf{r}_{2,i-1} = [r_{2,i-1}(\xi_0), r_{2,i-1}(\xi_1), \ldots, r_{2,i-1}(\xi_{N-1}), r_{2,i-1}(\xi_N)]^T, \tag{40}$$

$$\mathbf{r}_{3,i-1} = [r_{3,i-1}(\xi_0), r_{3,i-1}(\xi_1), \ldots, r_{3,i-1}(\xi_{N-1}), r_{3,i-1}(\xi_N)]^T. \tag{41}$$

$$A_{11} = \mathbf{a}_{1,i-1}\mathbf{D}^2 + \mathbf{a}_{2,i-1}\mathbf{D}, \quad A_{12} = -\mathbf{D}, \quad A_{13} = -N_1\mathbf{D} \tag{42}$$

$$A_{21} = \mathbf{b}_{2,i-1}, \quad A_{22} = \mathbf{D}^2 + \mathbf{b}_{1,i-1}\mathbf{D}, \quad A_{23} = D_f\mathbf{D}^2, \tag{43}$$

$$A_{31} = \mathbf{c}_{2,i-1}, \quad A_{32} = LeS_r\mathbf{D}^2, \quad A_{33} = \mathbf{D}^2 + \mathbf{c}_{1,i-1}\mathbf{D}. \tag{44}$$

In the above definitions, $\mathbf{a}_{k,i-1}, \mathbf{b}_{k,i-1}, \mathbf{c}_{k,i-1}$ $(k = 1,2)$ are diagonal matrices of size $(N+1) \times (N+1)$ and the superscript T is the transpose.

The boundary conditions (34) are imposed on equation (33) by modifying the first and last rows of $\mathbf{A}_{mn}$ $(m,n = 1,2,3)$ and $\mathbf{r}_{m,i-1}$ in such a way that the modified matrices $\mathbf{A}_{i-1}$ and $\mathbf{R}_{i-1}$ take the form;

$$\mathbf{A}_{i-1} = \left(\begin{array}{cccc|cccc|cccc}
\mathbf{D}_{0,0} & \mathbf{D}_{0,1} & \cdots & \mathbf{D}_{0,N-1} & \mathbf{D}_{0,N} & 0 & 0 & \cdots & 0 & 0 & 0 & 0 & \cdots & 0 & 0 \\
 & & \mathbf{A}_{11} & & & & & \mathbf{A}_{12} & & & & & \mathbf{A}_{13} & & \\
0 & 0 & \cdots & 0 & 1 & 0 & 0 & \cdots & 0 & 0 & 0 & 0 & \cdots & 0 & 0 \\
\hline
0 & 0 & \cdots & 0 & 0 & 1 & 0 & \cdots & 0 & 0 & 0 & 0 & \cdots & 0 & 0 \\
 & & \mathbf{A}_{21} & & & & & \mathbf{A}_{22} & & & & & \mathbf{A}_{23} & & \\
0 & 0 & \cdots & 0 & 0 & 0 & 0 & \cdots & 0 & 1 & 0 & 0 & \cdots & 0 & 0 \\
\hline
0 & 0 & \cdots & 0 & 0 & 0 & 0 & \cdots & 0 & 0 & 1 & 0 & \cdots & 0 & 0 \\
 & & \mathbf{A}_{31} & & & & & \mathbf{A}_{32} & & & & & \mathbf{A}_{33} & & \\
0 & 0 & \cdots & 0 & 0 & 0 & 0 & \cdots & 0 & 0 & 0 & 0 & \cdots & 0 & 1
\end{array}\right), \tag{45}$$

$$\mathbf{R}_{i-1} = \left(\begin{array}{c}
0 \\ r_{1,i-1}(\xi_1) \\ \vdots \\ r_{1,i-1}(\xi_{N-2}) \\ r_{1,i-1}(\xi_{N-1}) \\ 0 \\
\hline
0 \\ r_{2,i-1}(\xi_1) \\ \vdots \\ r_{2,i-1}(\xi_{N-2}) \\ r_{2,i-1}(\xi_{N-1}) \\ 0 \\
\hline
0 \\ r_{3,i-1}(\xi_1) \\ \vdots \\ r_{3,i-1}(\xi_{N-2}) \\ r_{3,i-1}(\xi_{N-1}) \\ 0
\end{array}\right) \tag{46}$$

After modifying the matrix system (33) to incorporate boundary conditions, the solution is obtained as

$$\mathbf{X}_i = \mathbf{A}_{i-1}^{-1}\mathbf{R}_{i-1}. \tag{47}$$

4. Results and discussion

In this section we give the successive linearization method results for the main parameters affecting the flow. To check the accuracy of the proposed successive linearisation method (SLM), comparison was made with numerical solutions obtained using the MATLAB routine `bvp4c`, which is an adaptive Lobatto quadrature scheme. The graphs and tables presented in this work, unless otherwise specified, were generated using $N = 150$, $L = 30$, $N_1 = 1$, $Le = 1$, $D_f = 1$, $Gr = 1$ and $S_r = 0.5$

Tables 1 - 8 give results for the Nusselt and Sherwood numbers at different orders of approximation when varying the values of the main parameters. Table 1 and 2 depict the numerical values of the local Nusselt Number and the Sherwood number, respectively, for various modified Grashof numbers. In this chapter, the modified Grashof numbers are used to evaluate the relative importance of inertial effects and viscous effects. It is clearly observed that the local Nusselt number and the local Sherwood number tend to decrease as the modified Grashof number Gr^* increases. Increasing Gr^* values retards the flow, thereby thickening the thermal and concentration boundary layers and thus reducing the heat and mass transfer rates between the fluid and the wall. We also observe in both of these tables that the successive linearisation method rapidly converges to a fixed value.

Gr^*	2nd order	3rd order	4th order	6th order	8th order	10th order
0.5	0.25449779	0.25459016	0.25459014	0.25459014	0.25459014	0.25459014
1.0	0.23356479	0.23357092	0.23357092	0.23357092	0.23357092	0.23357092
1.5	0.21998679	0.21998820	0.21998820	0.21998820	0.21998820	0.21998820
2.0	0.21001909	0.21001959	0.21001959	0.21001959	0.21001959	0.21001959
2.5	0.20218859	0.20218881	0.20218881	0.20218881	0.20218881	0.20218881
3.0	0.19576817	0.19576829	0.19576829	0.19576829	0.19576829	0.19576829

Table 1. Values of the Nusselt Number, $-\theta'(0)$ for different values of Gr^* at different orders of the SLM approximation using $L = 30$, $N = 150$ when $Le = 1$, $N_1 = 1$, $D_f = 1$, $S_r = 0.5$

Table 3 and 4 represent the numerical values of the local Nusselt number and Sherwood number, respectively, for various buoyancy ratios (N_1). We observe that the local Nusselt number and Sherwood number tend to increase as the buoyancy ratio N_1 increases. Increasing the buoyancy ratio accelerates the flow, decreasing the thermal and concentration boundary layer thickness and thus increasing the heat and mass transfer rates between the fluid and the wall.

Gr^*	2nd order	3rd order	4th order	6th order	8th order	10th order
0.5	0.49979076	0.49970498	0.49970494	0.49970494	0.49970494	0.49970494
1.0	0.45388857	0.45388333	0.45388333	0.45388333	0.45388333	0.45388333
1.5	0.42518839	0.42518709	0.42518709	0.42518709	0.42518709	0.42518709
2.0	0.40449034	0.40448985	0.40448985	0.40448985	0.40448985	0.40448985
2.5	0.38841782	0.38841759	0.38841759	0.38841759	0.38841759	0.38841759
3.0	0.37534971	0.37534959	0.37534959	0.37534959	0.37534959	0.37534959

Table 2. Values of the Sherwood Number, $-\phi'(0)$ for different values of Gr^* at different orders of the SLM approximation using $L = 30$, $N = 150$ when $Le = 1$, $N_1 = 1$, $D_f = 1$, $S_r = 0.5$

Table 5 and 6 show the values of the local Nusselt number and local Sherwood number, respectively for various values of the Soret number Sr. It is noticed that the magnitude of the local Nusselt number increases for Sr values less than a unit. However it decreases for

N_1	2nd order	3rd order	4th order	6th order	8th order	10th order
0	0.10955207	0.13941909	0.17140588	0.18806653	0.18822338	0.18822338
1	0.23329934	0.23357050	0.23357092	0.23357092	0.23357092	0.23357092
2	0.26550968	0.26554932	0.26554933	0.26554933	0.26554933	0.26554933
3	0.29071828	0.29072940	0.29072940	0.29072940	0.29072940	0.29072940
4	0.31170490	0.31170917	0.31170917	0.31170917	0.31170917	0.31170917
5	0.32980517	0.32980715	0.32980715	0.32980715	0.32980715	0.32980715
10	0.39608249	0.39638607	0.39638637	0.39638637	0.39638637	0.39638637

Table 3. Values of the Nusselt Number, $-\theta'(0)$ for different values of N_1 at different orders of the SLM approximation using $L = 30$, $N = 150$ when $Gr^* = 1$, $Le = 1$, $D_f = 1$, $S_r = 0.5$

N_1	2nd order	3rd order	4th order	6th order	8th order	10th order
0	0.43723544	0.39927412	0.37520193	0.36285489	0.36273086	0.36273086
1	0.45378321	0.45388312	0.45388333	0.45388333	0.45388333	0.45388333
2	0.51638584	0.51639273	0.51639273	0.51639273	0.51639273	0.51639273
3	0.56500365	0.56500444	0.56500444	0.56500444	0.56500444	0.56500444
4	0.60522641	0.60522641	0.60522641	0.60522641	0.60522641	0.60522641
5	0.63977204	0.63977193	0.63977193	0.63977193	0.63977193	0.63977193
10	0.76607162	0.76609949	0.76609963	0.76609963	0.76609963	0.76609963

Table 4. Values of the Sherwood number, $-\phi'(0)$ for different values of N_1 at different orders of the SLM approximation using $L = 30$, $N = 150$ when $Gr^* = 1$, $Le = 1$, $D_f = 1$, $S_r = 0.5$

larger values of Sr. The magnitude of the local Sherwood number decreases for Soret number values less than unit but increases for large values of Sr.

As the Dufour effect Df increases (Table 7 and 8) heat transfer decreases and mass transfer increases.

Figure 1 illustrates the temperature and concentration profiles as a function of the similarity variable η for various values of the modified Grashof number. It is observed from these figures that an increase in the modified Grashof number leads to increases in the temperature and the concentration distributions in the boundary layer and as a result both the thermal and solutal boundary layers become thicker.

The effect of the Lewis number Le on the temperature and concentration distributions are shown in Figure 2. We observe here that the temperature increases with increases values of Le for small values of the similarity variable $\eta(< 4)$. Thereafter, the temperature decreases with increasing values of Le. We observe further that the species concentration distributions

S_r	2nd order	3rd order	4th order	6th order	8th order	10th order
0.0	0.11828553	0.15103149	0.18809470	0.20356483	0.20360093	0.20360093
0.5	0.23358338	0.23357092	0.23357092	0.23357092	0.23357092	0.23357092
1.5	0.28601208	0.28601478	0.28601478	0.28601478	0.28601478	0.28601478
2.0	0.25626918	0.25626899	0.25626899	0.25626899	0.25626899	0.25626899
3.0	0.21828033	0.21825939	0.21825951	0.21825951	0.21825951	0.21825951
4.0	0.19258954	0.19260872	0.19260881	0.19260881	0.19260881	0.19260881
5.0	0.17369571	0.17369607	0.17369607	0.17369607	0.17369607	0.17369607

Table 5. Values of the Nusselt Number, $-\theta'(0)$ for different values of S_r at different orders of the SLM approximation using $L = 30$, $N = 150$ when $Gr^* = 1$, $Le = 1$, $D_f = 1$, $N_1 = 1$

S_r	2nd order	3rd order	4th order	6th order	8th order	10th order
0.0	0.57981668	0.51916309	0.50219360	0.49634082	0.49632751	0.49632751
0.5	0.45386831	0.45388333	0.45388333	0.45388333	0.45388333	0.45388333
1.5	0.35029249	0.35029514	0.35029514	0.35029514	0.35029514	0.35029514
2.0	0.36241935	0.36241908	0.36241908	0.36241908	0.36241908	0.36241908
3.0	0.37807254	0.37803636	0.37803657	0.37803657	0.37803657	0.37803657
4.0	0.38517909	0.38521742	0.38521761	0.38521761	0.38521761	0.38521761
5.0	0.38839482	0.38839561	0.38839562	0.38839562	0.38839562	0.38839562

Table 6. Values of the Sherwood Number, $-\phi'(0)$ for different values of S_r at different orders of the SLM approximation using $L = 30$, $N = 150$ when $Gr^* = 1$, $Le = 1$, $D_f = 1$, $N_1 = 1$

D_f	2nd order	3rd order	4th order	6th order	8th order	10th order
0.0	0.53407939	0.49152600	0.48562621	0.48464464	0.48464458	0.48464458
0.5	0.38477915	0.38479579	0.38479579	0.38479579	0.38479579	0.38479579
0.8	0.30341973	0.30342078	0.30342078	0.30342078	0.30342078	0.30342078
1.2	0.14317488	0.14317766	0.14317766	0.14317766	0.14317766	0.14317766
1.4	0.01721331	0.01721348	0.01721348	0.01721348	0.01721348	0.01721348
1.8	-0.60409800	-0.60409008	-0.60409008	-0.60409008	-0.60409008	-0.60409008

Table 7. Values of the Nusselt Number, $-\theta'(0)$ for different values of D_f at different orders of the SLM approximation using $L = 30$, $N = 150$ when $Gr^* = 1$, $Le = 1$, $Sr = 0.5$, $N_1 = 1$

D_f	2nd order	3rd order	4th order	6th order	8th order	10th order
0.0	0.32610052	0.32342809	0.33553652	0.33829291	0.33829308	0.33829308
0.5	0.38480751	0.38479579	0.38479579	0.38479579	0.38479579	0.38479579
0.8	0.42201071	0.42200998	0.42200998	0.42200998	0.42200998	0.42200998
1.2	0.49527611	0.49527429	0.49527429	0.49527429	0.49527429	0.49527429
1.4	0.55343701	0.55343690	0.55343690	0.55343690	0.55343690	0.55343690
1.8	0.84855160	0.84854722	0.84854722	0.84854722	0.84854722	0.84854722

Table 8. Values of the Sherwood Number, $-\phi'(0)$ for different values of D_f at different orders of the SLM approximation using $L = 30$, $N = 150$ when $Gr^* = 1$, $Le = 1$, $S_r = 0.5$, $N_1 = 1$

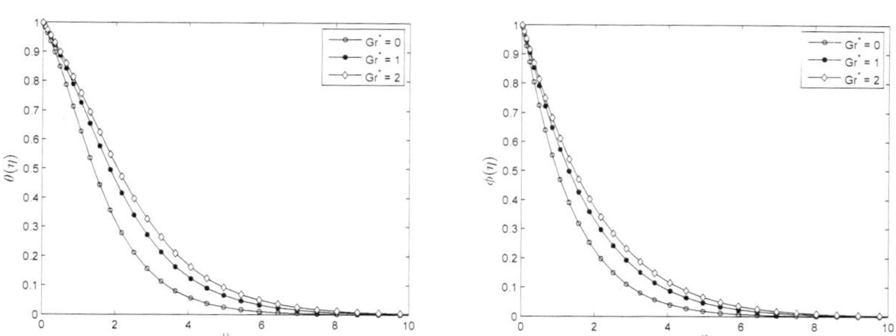

Fig. 1. Effect of Gr^* on the temperature and concentration profiles

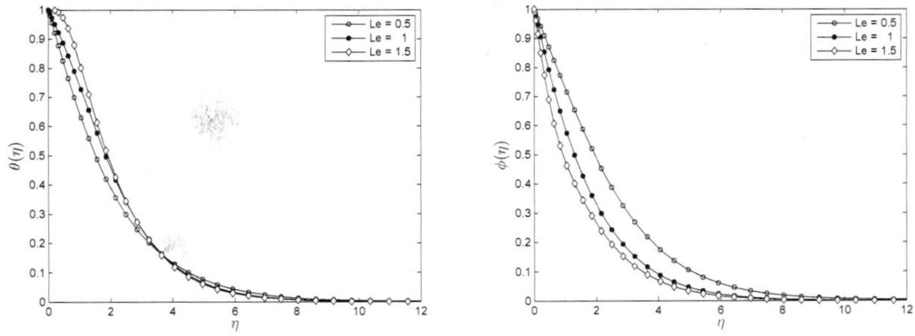

Fig. 2. Effect of *Le* on the temperature and concentration profiles

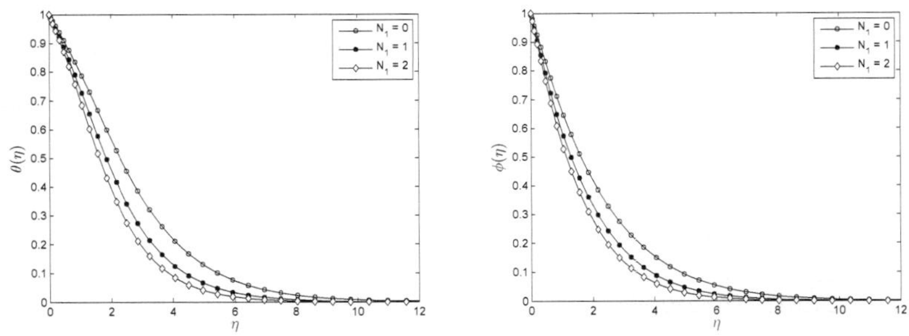

Fig. 3. Effect of N_1 on the temperature and concentration profiles

decrease due to an increase in the value of the Lewis number. Increasing *Le* leads to the thickening of the temperature boundary layer and to thin the concentration boundary layer. The temperature profiles and concentration profiles for aiding buoyancy are presented in Figure 3. It is seen in these figures that as the buoyancy parameter N_1 increases the temperature and concentration decrease. This is because the effect of the buoyancy ratio is to increase the surface heat and mass transfer rates. Therefore, the temperature and concentration gradients are increased and hence, so are the heat and mass transfer rates.

Figure 4 illustrates the effect of the Dufour parameter on the dimensionless temperature and concentration. It is observed that the temperature of fluid increases with an increase of Dufour number while the concentration of the fluid decreases with increases of the value of the Dufour number.

Figure 5 depict the effects of the Soret parameter on the dimensionless temperature and concentration distributions. It is clear from these figures that as the Soret parameters increases concentration profiles increase significantly while the temperature profiles decrease.

5. Conclusion

In the present chapter, a new numerical perturbation scheme for solving complex nonlinear boundary value problems arising in problems of heat and mass transfer. This numerical

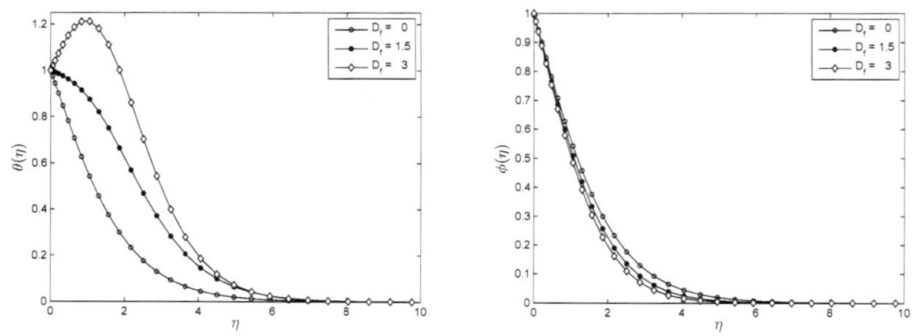

Fig. 4. Effect of D_f on the temperature and concentration profiles

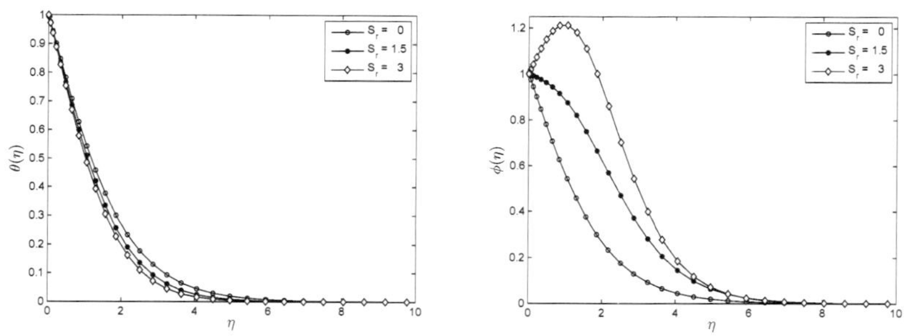

Fig. 5. Effect of S_r on the temperature and concentration profiles

method is based on a novel idea of iteratively linearising the underlying governing non-linear boundary equations, which are written in similarity form, and then solving the resultant equations using spectral methods. Extensive numerical integrations were carried out, to investigate the non-Darcy natural convection heat and mass transfer from a vertical surface with heat and mass flux. The effects with the modified Grashof number, the buoyancy ratio, the Soret and Dufour numbers on the Sherwood and Nusselt numbers have been studied. From the present analysis, we conclude that (1) both the local Nusselt number, Nu_x, and local Sherwood number, Sh_x, decrease due to increase in the value of the inertial parameter (modified Grashof number, Gr^*); (2) An increase in the buoyancy ratio tends to increase both the local Nusselt number and the Sherwood number; (3) The Lewis number has a more pronounced effect on the local mass transfer rate than it does on the local heat transfer rate; (4) Increases in Soret number tends to decrease the local heat transfer rate and the Dufour effects greatly affect the mass and heat transfer rates. Numerical results for the temperature and concentration were presented graphically. These results might find wide applications in engineering, such as geothermal system, heat exchangers, fibre and granular insulation, solar energy collectors and nuclear waste depositors.

6. References

El-Amin, M.F. (2004) Double dispersion effects on natural convection heat and mass transfer in non-Darcy porous medium, *Applied Mathematics and Computation* Vol.156, 1–17

Adomian, G. (1976) Nonlinear stochastic differential equations. *J Math. Anal. Appl.* Vol.55, 441 – 52

Adomian, G. (1991) A review of the decomposition method and some recent results for nonlinear equations. *Comp. and Math. Appl.* Vol.21, 101 – 27

Ayaz,F. (2004) Solutions of the systems of differential equations by differential transform method, *Applied Mathematics and Computation*, Vol.147, 547-567

Canuto,C., Hussaini, M. Y., Quarteroni,A. and Zang, T. A. (1988) *Spectral Methods in Fluid Dynamics*, Springer-Verlag, Berlin

Chen,C.K., Ho,S.H. (1999) Solving partial differential equations by two dimensional differential transform method, Applied Mathematics and Computation Vol.106, 171-179.

Don, W. S., Solomonoff, A. (1995) Accuracy and speed in computing the Chebyshev Collocation Derivative. *SIAM J. Sci. Comput*, Vol.16, No.6, 1253–1268.

He,J.H, (1999) Homotopy perturbation technique, *Comput. Methods Appl. Mech. Eng.* Vol.178, 257 – 262.

He,J.H. (2000) A new perturbation technique which is also valid for large parameters, *J. Sound and Vibration*, Vol.229, 1257 – 1263.

He, J.H. (1999) Variational iteration method a kind of nonlinear analytical technique:some examples, *Int. J. Nonlinear Mech.* Vol.34, 699–708

He, J.H. (2006) New interpretation of homotopy perturbation method, *Int. J. Modern Phys.B* vol.20, 2561 – 2568.

Liao,S.J. (1992) The proposed homotopy analysis technique for the solution of nonlinear problems, PhD thesis, Shanghai Jiao Tong University, 1992.

Liao,S.J. (1999) A uniformly valid analytic solution of 2D viscous flow past a semi-infinite flat plate. *J. Fluid Mech.* Vol.385, 101 – 128.

Liao,S.J. (2003) Beyond perturbation: *Introduction to homotopy analysis method.* Chapman & Hall/CRC Press.

Liao,S. J.(2009) Notes on the homotopy analysis method: Some definitions and theories, *Commun. Nonlinear Sci. Numer. Simul.* Vol.14, 983–997.

Motsa,S.S.,Sibanda, P., Shateyi,S. (2010) A new spectral-homotopy analysis method for solving a nonlinear second order BVP, *Commun. Nonlinear Sci. Numer. Simul.* Vol.15 2293-2302.

Motsa, S.S., Sibanda, P., Awad, F.G., Shateyi,S. (2010) A new spectral-homotopy analysis method for the MHD Jeffery-Hamel problem, *Computer & Fluids* Vol.39, 1219–1225.

Motsa,S.S., Shateyi,S., (2010) A New Approach for the Solution of Three-Dimensional Magnetohydrodynamic Rotating Flow over a Shrinking Sheet, *Mathematical Problems in Engineering*, vol. 2010, Article ID 586340, 15 pages, 2010. doi:10.1155/2010/586340

Makukula,Z., Sibanda,P., Motsa,S.S. (2010) A Note on the Solution of the Von Kármán Equations Using Series and Chebyshev Spectral Methods, *Boundary Value Problems*, Volume 2010 (2010), Article ID 471793, 17 pages doi:10.1155/2010/471793

Makukula,Z., Sibanda,P., Motsa,S.S. (2010) , A Novel Numerical Technique for Two-dimensional Laminar Flow Between Two Moving Porous Walls, *Mathematical Problems in Engineering*, Vol. 2010, Article ID 528956, 15 pages, 2010. doi:10.1155/2010/528956

Shateyi,S., Motsa,S.S.,(2010) Variable viscosity on magnetohydrodynamic fluid flow and heat transfer over an unsteady stretching surface with Hall effect, *Boundary Value Problems*, Vol. 2010, Article ID 257568, 20 pages, doi:10.1155/2010/257568

Murthy, P.V.S.N, Singh, P. (1999) Heat and mass transfer by natural convection in a non-Darcy porous medium, *Acta Mechanica*, Vol.138, 243–254

Lakhshmi Narayana, P.A., Murthy, P.V.S.N, (2006) Free convective heat and mass transfer in a doubly stratified non-Darcy porous medium, *Journal of Heat Transfer*, Vol.128, 1204–1212

Partha, M.K. (2008) Thermophoresis particle deposition in a non-Darcy porous medium under the influence of Soret, Dufour effects, *Heat Mass Transfer* Vol.44, 969–977

Trefethen, L.N. (2000) Spectral Methods in MATLAB, SIAM

Zhou, J.K. (1986) Differential Transformation and Its Applications for Electrical Circuits, *Huazhong University Press*, Wuhan, China (in Chinese)

20

Explicit and Approximated Solutions for Heat and Mass Transfer Problems with a Moving Interface

Domingo Alberto Tarzia
CONICET and Universidad Austral
Argentina

1. Introduction

The goal of this chapter is firstly to give a survey of some explicit and approximated solutions for heat and mass transfer problems in which a free or moving interface is involved. Secondly, we show simultaneously some new recent problems for heat and mass transfer, in which a free or moving interface is also involved. We will consider the following problems:

1. Phase-change process (Lamé-Clapeyron-Stefan problem) for a semi-infinite material:

i. The Lamé-Clapeyron solution for the one-phase solidification problem (modeling the solidification of the Earth with a square root law of time);
ii. The pseudo-steady-state approximation for the one-phase problem;
iii. The heat balance integral method (Goodman method) and the approximate solution for the one-phase problem;
iv. The Stefan solution for the planar phase-change surface moving with constant speed;
v. The Solomon-Wilson-Alexiades model for the phase-change process with a mushy region and its similarity solution for the one-phase case;
vi. The Cho-Sunderland solution for the one-phase problem with temperature-dependent thermal conductivity;
vii. The Neumann solution for the two-phase problem for prescribed surface temperature at the fixed face;
viii. The Neumann-type solution for the two-phase problem for a particular prescribed heat flux at the fixed face, and the necessary and sufficient condition to have an instantaneous phase-change process;
ix. The Neumann-type solution for the two-phase problem for a particular prescribed convective condition (Newton law) at the fixed face, and the necessary and sufficient condition to have an instantaneous phase-change process;
x. The similarity solution for the two-phase Lamé-Clapeyron-Stefan problem with a mushy region.
xi. The similarity solution for the phase-change problem by considering a density jump;
xii. The determination of one or two unknown thermal coefficients through an over-specified condition at the fixed face for one or two-phase cases.
xiii. A similarity solution for the thawing in a saturated porous medium by considering a density jump and the influence of the pressure on the melting temperature.

2. Free boundary problems for the diffusion equation:
   i. The oxygen diffusion-consumption problem and its relationship with the phase-change problem;
   ii. The Rubinstein solution for the binary alloy solidification problem;
   iii. The Zel'dovich-Kompaneets-Barenblatt solution for the gas flow through a porous medium;
   iv. Luikov coupled heat and mass transfer for a phase-change process;
   v. A mixed saturated-unsaturated flow problem representing absorption of water by a soil with a constant pond depth at the surface and an explicit solution for a particular diffusivity;
   vi. Estimation of the diffusion coefficient in a gas-solid system;
   vii. The coupled heat and mass transfer during the freezing of the high-water content materials with two free boundaries: the freezing and sublimation fronts.

2. Explicit solutions for phase-change process (Lamé-Clapeyron-Stefan problem) for a semi-infinite material

Heat transfer problems with a phase-change such as melting and freezing have been studied in the last century due to their wide scientific and technological applications. A review of a long bibliography on moving and free boundary problems for phase-change materials (PCM) for the heat equation is shown in (Tarzia, 2000a). Some previous reviews on explicit or approximated solutions were presented in (Garguichevich & Sanziel, 1984; Howison, 1988; Tarzia, 1991b & 1993). Some reviews, books or booklets in the subject are (Alexiades & Solomon, 1993; Bankoff, 1964; Brillouin, 1930; Cannon, 1984; Carslaw & Jaeger, 1959; Crank, 1984; Duvaut, 1976; Elliott & Ockendon, 1982; Fasano, 1987 & 2005; Friedman, 1964; Gupta, 2003; Hill, 1987; Luikov, 1968; Lunardini, 1981 & 1991; Muehlbauer & Sunderland, 1965; Primicerio, 1981; Rubinstein, 1971; Tarzia, 1984b & 2000b; Tayler, 1986).

2.1 The Lamé-Clapeyron solution for the one-phase solidification problem (modeling the solidification of the Earth with a square root law of time)

We consider the solidification of semi-infinite material, represented by $x>0$. We will find the interface solid-liquid $x=s(t)$ and the temperature $T=T(x,t)$ of the solid phase defined by

$$T(x,t)=\begin{cases} T(x,t) & if \quad 0<x<s(t), \quad t>0 \\ T_f & if \quad s(t)\le x, \quad t>0 \end{cases} \tag{1}$$

which satisfy the following free boundary problem:

$$\rho c T_t - kT_{xx}=0\,, \quad 0<x<s(t)\,, \quad t>0 \tag{2}$$

$$T(0,t)=T_0<T_f\,, \quad t>0 \tag{3}$$

$$T(s(t),t)=T_f\,, \quad t>0 \tag{4}$$

$$kT_x(s(t),t)=\rho\ell\dot{s}(t)\,, \quad t>0 \tag{5}$$

$$s(0) = 0 \tag{6}$$

Eq. (2) represents the heat equation for the solid phase, k is the thermal conductivity, ρ is the mass density, c is the heat capacity, ℓ is the latent heat of fusion by unit of mass, T_0 is the imposed temperature at the fixed face $x = 0$, and the material is initially at the melting temperature T_f. The problem (2)-(6) is known in literature as the one-phase Stefan problem (Lamé-Clapeyron-Stefan problem) and the condition (5) as the Stefan condition. Free boundary problems of this type were presented by the first time in (Lamé & Clapeyron, 1831) in order to study the solidification of the Earth and was continued independently by (Stefan, 1891a, b & 1990) in order to study the thickness of polar ice. We remark here that Lamé & Clapeyron found the important law for the phase-change interface with a square root of time.

Theorem 1. (Lamé-Clapeyron solution).

The explicit solution to the free boundary problem (2)-(6) is given by

$$T(x,t) = T_0 + \frac{T_f - T_0}{f(\xi)} erf(\frac{x}{2a\sqrt{t}}), \quad s(t) = 2a\xi\sqrt{t} \tag{7}$$

where $a^2 = \alpha = \frac{k}{\rho c}$ is the diffusion coefficient and $\xi > 0$ is the unique solution to the equation

$$E(x) = \frac{Ste}{\sqrt{\pi}}, \quad x > 0 \tag{8}$$

with

$$erf(x) = \frac{2}{\sqrt{\pi}} \int_0^x \exp(-u^2)du, \quad E(x) = x\, erf(x) \exp(x^2), \tag{9}$$

$$Ste = \frac{c(T_f - T_0)}{\ell} : \text{Stefan number}, \tag{10}$$

and the total heat flux at the fixed face $x = 0$ is given by

$$Q(t) = \int_0^t kT_x(0,\tau)d\tau = \rho\ell s(t)\exp(\xi^2). \tag{11}$$

Proof.

We have the following properties:

$$E(0) = 0, \quad E(+\infty) = +\infty, \quad E'(x) > 0, \quad \forall x > 0. \tag{12}$$

Remark 1.

From (4) we have

$$T_x(s(t),t)\dot{s}(t) + T_t(s(t),t) = 0, \quad t > 0 \tag{13}$$

and therefore the Stefan condition (5) is transformed in

$$kT_x^2(s(t),t) = -\rho\ell T_t(s(t),t) = -\frac{\ell k}{c}T_{xx}(s(t),t), \quad t>0 \tag{14}$$

which implies that the problem (2)-(6) is always a nonlinear problem (Pekeris & Slichter, 1939).

Remark 2.

A generalization of the Lamé-Clapeyron solution is given in (Menaldi & Tarzia, 2003) for a particular source in the heat equation. A study of the behaviour of the Lamé-Clapeyron solution when the latent heat goes to zero is given in (Guzman, 1982; Sherman, 1971).

2.2 The pseudo-steady-state approximation for the one-phase problem

An approximated solution to problem (2)-(6) is given by the pseudo-steady-state approximation which must satisfy the following conditions: (3)-(6) and the steady-state equation

$$T_{xx} = 0, \quad 0 < x < s(t), \quad t > 0. \tag{15}$$

Theorem 2 (Stefan, 1989a)

The solution to the problem (15), (3)-(6) is given by

$$T(x,t) = T_0 + \frac{T_f - T_0}{s(t)}x, \qquad 0 < x < s(t), \quad t > 0 \tag{16}$$

$$s(t) = 2a\xi_{ap}\sqrt{t}, \qquad \xi_{ap} = \sqrt{\frac{Ste}{2}} \tag{17}$$

Proof.

The solution to (15), (3) and (4) is given by (16). Therefore the condition (5) is transformed in the ordinary differential equation

$$k(T_f - T_0)/s(t) = \rho\lambda\dot{s}(t) \tag{18}$$

with the initial condition (6), whose solution is given by

$$s^2(t) = 2k(T_f - T_0)t/(\rho\ell) = 4a^2\frac{Ste}{2}t \tag{19}$$

that is

$$s(t) = \sqrt{\frac{2k(T_f - T_0)}{\rho\ell}t} \tag{20}$$

Remark 3.

If the Stefan number is very small, i.e.

$$Ste = \frac{c(T_f - T_0)}{\ell} << 1 \tag{21}$$

then the solution ξ to the equation (8) for the Lamé-Clapeyron solution can be taken as ξ_{ap}, given in (17). This can be obtained by using the following first approximation:

$$\exp(x^2) \approx 1, \qquad f(x) \approx \frac{2}{\sqrt{\pi}} x, \quad 0 < x << 1 . \tag{22}$$

Remark 4.
A study of sufficient conditions on data to estimate the occurrence of a phase-change process is given in (Solomon et al., 1983; Tarzia & Turner, 1992 & 1999).

2.3 The heat balance integral method (Goodman method) and the approximate solution for the one-phase problem

An approximated solution for the following fusion problem (similar to the solidification problem (2)-(6))

$$\rho c T_t - k T_{xx} = 0 , \quad 0 < x < s(t) , \quad t > 0 \tag{23}$$

$$T(0,t) = T_0 > 0 , \quad t > 0 \tag{24}$$

$$T(s(t),t) = 0 , \quad t > 0 \tag{25}$$

$$k T_x(s(t),t) = -\rho \ell \dot{s}(t) , \quad t > 0 \tag{26}$$

$$s(0) = 0 \tag{27}$$

is given by the heat balance integral method, known by the Goodman method (Goodman,1958). This method consists of replacing the Stefan condition (26) by

$$T_x^2(s(t),t) = \frac{\ell}{c} T_{xx}(s(t),t), \quad t > 0 \tag{28}$$

and the heat equation (23) by its integral on the domain $(0,s(t))$ given by

$$\begin{aligned} \frac{d}{dt}\int_0^{s(t)} T(x,t)dx &= \int_0^{s(t)} T_t(x,t)dx + T(s(t),t)\dot{s}(t) = \int_0^{s(t)} T_t(x,t)dx = \frac{k}{\rho c}\int_0^{s(t)} T_{xx}(x,t)dx \\ &= \frac{k}{\rho c}[T_x(s(t),t) - T_x(0,t)] = -\frac{k}{\rho c}[\frac{\rho\lambda}{k}\dot{s}(t) + T_x(0,t)] \end{aligned} \tag{29}$$

that is

$$\frac{d}{dt}\int_0^{s(t)} T(x,t)dx = -\frac{k}{\rho c}[\frac{\rho\ell}{k}\dot{s}(t) + T_x(0,t)] . \tag{30}$$

In order to solve (30), (28), (24), (25) and (27), we propose an approximated temperature profile

$$T(x,t) = \alpha(t)(s(t)-x) + \beta(t)(s(t)-x)^2, \quad 0 < x < s(t), \ t > 0 \tag{31}$$

where $\alpha = \alpha(t), \beta = \beta(t),$ and $s = s(t)$ are real functions to be determined. Firstly, we can obtain α and β as a function of s and, therefore, we solve the corresponding ordinary differential equation for $s = s(t)$.

Theorem 3.

The Goodman approximated solution is given by:

$$\alpha(t) = \frac{\ell}{c}\frac{\sqrt{1+2Ste}-1}{s(t)}, \quad \beta(t) = \frac{\alpha(t)s(t)+T_0}{s^2(t)} \tag{32}$$

$$s(t) = 2a\xi_g\sqrt{t}, \quad \xi_g = \sqrt{3}\frac{1+2Ste-\sqrt{1+2Ste}}{5+Ste+\sqrt{1+2Ste}}, \quad Ste = \frac{cT_0}{\ell} \tag{33}$$

Remark 5.

Other refinements of the Goodman method are given in (Bell, 1978; Lunardini, 1981; Lunardini 1991). In (Reginato & Tarzia, 1993; Reginato et al, 1993; Reginato et al., 2000) the heat balance method was applied to root growth of crops and the modelling nutrient uptake. In (Tarzia, 1990a) the heat balance method was applied to obtain the exponentially fast asympotic behaviour of the solutions in heat conduction problems with absorption.

2.4 The Stefan solution for the planar phase-change surface moving with constant speed

When the phase-change interface is moving with constant speed we can consider the following inverse Stefan problem: find the temperature $T = T(x,t)$ and $f(t) = T(0,t)$ such that:

$$\alpha T_{xx} = T_t, \qquad 0 < x < s(t), \quad t > 0 \quad (\alpha = k/\rho c) \tag{34}$$

$$T(s(t),t) = 0, \qquad t > 0 \tag{35}$$

$$kT_x(s(t),t) = \rho\ell\dot{s}(t), \qquad t > 0 \tag{36}$$

$$\dot{s}(t) = m > 0, \quad s(0) = 0 \quad (s(t) = mt) \tag{37}$$

Theorem 4. (Stefan, 1989b & 1991)

The solution to (34)-(37) is given by

$$T(x,t) = \frac{\ell}{c}[1-\exp(\frac{m}{\alpha}(mt-x))] \tag{38}$$

and the temperature at the fixed face is variable in time given by the expression:

$$f(t) = T(0,t) = -\frac{\ell}{c}[\exp(\frac{m^2 t}{\alpha})-1] < 0 = T_f, \quad t > 0 . \tag{39}$$

Remark 6.

More details with respect to the inverse Stefan problem can be found in (Quilghini, 1967).

2.5 The Solomon-Wilson-Alexiades model for the phase-change process with a mushy region and its similarity solution for the one-phase case

We consider a semi-infinite material in the liquid phase at the melting temperature T_f. We impose a temperature $T_0 < T_f$ at the fixed face $x = 0$, and the solidification process begins, and three regions can be distinguished, as follows (Solomon et al., 1982):

i. the liquid phase, at temperature $T = T_f$, occupying the region $x > r(t),\ t > 0$;
ii. the solid phase, at temperature $T(x,t) < T_f$, occupying the region $0 < x < s(t),\ t > 0$;
iii. the mushy zone, at temperature T_f, occupying the region $s(t) < x < r(t),\ t > 0$. We make the following two assumptions on its structure:

a. the material in the mushy zone contains a fixed fraction $\varepsilon\ell$ (with constant $0 < \varepsilon < 1$) of the total latent heat ℓ.
b. the width of the mushy zone is inversely proportional (with constant $\gamma > 0$) to the temperature gradient at $s(t)$.

Therefore the problem consists of finding the free boundaries $x = s(t)$ and $x = r(t)$, and the temperature $T = T(x,t)$ such that the following conditions are satisfied:

$$\rho c T_t - k T_{xx} = 0, \quad 0 < x < s(t), \quad t > 0 \tag{40}$$

$$T(0,t) = T_0 < T_f, \quad t > 0;\ s(0) = r(0) = 0 \tag{41}$$

$$T(s(t),t) = T_f, \quad t > 0 \tag{42}$$

$$kT_x(s(t),t) = \rho\ell[\varepsilon\dot{s}(t) + (1-\varepsilon)\dot{r}(t)], \quad t > 0 \tag{43}$$

$$T_x(s(t),t)(r(t) - s(t)) = \gamma, \quad t > 0. \tag{44}$$

<u>Theorem 5.</u> (Solomon et al., 1982):
The explicit solution to problem (40)-(44) is given by:

$$T(x,t) = T_0 + \frac{(T_f - T_0)}{erf(\xi)} erf\left(\frac{x}{2a\sqrt{t}}\right), \quad s(t) = 2\xi a\sqrt{t}, \quad r(t) = 2\mu a\sqrt{t} \tag{45}$$

where

$$\mu = \xi + \frac{\gamma\sqrt{\pi}}{2(T_f - T_0)} erf(\xi)\exp(\xi^2), \quad a = \sqrt{k/\rho c} \tag{46}$$

and $\xi > 0$ is the unique solution to the equation

$$D(x) = \frac{Ste}{\sqrt{\pi}}, \quad x > 0 \quad \left(Ste = c(T_f - T_0)/\ell \right) \tag{47}$$

with

$$D(x) = x\,erf(x)\exp(x^2) + \frac{\gamma(1-\varepsilon)\sqrt{\pi}}{2(T_f - T_0)}[\exp(x^2)erf(x)]^2. \tag{48}$$

Remark 7.
The classical Lamé-Clapeyron solution can be obtained for the particular case $\varepsilon = 1, \quad \gamma = 0$. If the Stefan number is small, then an approximated solution for ξ and μ is given by:

$$\xi = \left[\frac{Ste}{2[1+\gamma(1-\varepsilon)/(T_f - T_0)]} \right]^{\frac{1}{2}}, \quad \mu = \xi[1+\gamma/(T_f - T_0)] \quad . \tag{49}$$

2.6 The Cho-Sunderland solution for the one-phase problem with temperature-dependent thermal conductivity

We consider the following solidification problem for a semi-infinite material

$$\rho\, c\, T_t(x,t) = (k(T) T_x(x,t))_x, \quad 0 < x < s(t), t > 0 \tag{50}$$

$$T(0,t) = T_0 < T_f, \quad t > 0 \tag{51}$$

$$T(s(t),t) = T_f, \quad t > 0 \tag{52}$$

$$k(T_f)\; T_x(s(t),t) = \rho\, \ell\, \dot{s}(t), \quad t > 0 \tag{53}$$

where $T(x,t)$ is the temperature of the solid phase, $\rho > 0$ is the density of mass, $\ell > 0$ is the latent heat of fusion by unity of mass, $c > 0$ is the specific heat, $x = s(t)$ is the phase-change interface, T_f is the phase-change temperature, T_o is the temperature at the fixed face $x=0$. We suppose that the thermal conductivity has the following expression:

$$k = k(T) = k_0[1 + \beta(T - T_0)/(T_f - T_0)] \quad , \; \beta \in \mathbb{R}. \tag{54}$$

Let $a_o = k_o/\rho c$ be the diffusion coefficient at the temperature T_o. We observe that if $\beta = 0$, the problem (50)-(53) becomes the classical one-phase Lamé-Clapeyron-Stefan problem.
Theorem 6. (Cho & Sunderland, 1974)
The solution to problem (50)-(54) is given by:

$$T(x,t) = T_0 + \frac{(T_f - T_0)}{\Phi(\lambda)} \Phi(\eta) \quad , \quad \eta = \frac{x}{2\sqrt{\alpha_0 t}} \quad , \quad 0 < \eta < \lambda \tag{55}$$

$$s(t) = 2\lambda\sqrt{\alpha_0 t} \tag{56}$$

where $\Phi = \Phi(x) = \Phi_\delta(x)$ is the modified error function, for $\delta > -1$, the unique solution to the following boundary value problem in variable x, i.e:

$$\begin{cases} i) \;\; [(1+\delta\, \Phi'(x))\Phi'(x)]' + 2x\Phi'(x) = 0, \quad x > 0, \\ ii) \;\; \Phi(0^+) = 0 \quad , \quad \Phi(+\infty) = 1 \end{cases} \tag{57}$$

and the unknown thermal coefficients λ and δ must satisfy the following system of equations:

$$\beta - \delta\, \Phi(\lambda) = 0 \tag{58}$$

$$[1 + \delta\, \Phi(\lambda)] \frac{\Phi'(\lambda)}{\lambda\, \Phi(\lambda)} - \frac{2\ell}{c(T_f - T_0)} = 0 \ . \tag{59}$$

Remark 8.
Explicit solutions are given in (Briozzo et al., 2007 & 2010; Briozzo & Tarzia, 2002; Natale & Tarzia, 2006; Rogers & Broadbridge, 1988; Tirskii, 1959; Tritscher & Broadbridge, 1994) where nonlinear thermal coefficients are considered and in (Natale & Tarzia, 2000; Rogers, 1986) for Storm's materials.

2.7 The Neumann solution for the two-phase problem for prescribed surface temperature at the fixed face

We consider a semi-infinite material with null melting temperature $T_f = 0$, with an initial temperature $-C < 0$ and having a temperature boundary condition $B > 0$ at the fixed face $x = 0$. The model for the two-phase Lamé-Clapeyron-Stefan problem is given by: find the free boundary $x = s(t)$, defined for $t > 0$, and the temperature $T = T(x,t)$ defined by

$$T(x,t) = \begin{cases} T_2(x,t) > T_f & \text{if } 0 < x < s(t), \ t > 0 \\ T_f & \text{if } x = s(t), \ t > 0 \\ T_1(x,t) < T_f & \text{if } s(t) < x, \ t > 0 \end{cases} \tag{60}$$

for $x > 0$ and $t > 0$, such that (i=1: solid phase; i=2: liquid phase):

$$\rho c_2 T_{2_t} - k_2 T_{2_{xx}} = 0, \qquad 0 < x < s(t), \quad t > 0\,, \tag{61}$$

$$\rho c_1 T_{1_t} - k_1 T_{1_{xx}} = 0, \qquad x > s(t), \quad t > 0\,, \tag{62}$$

$$T_1(x,0) = -C < 0, \qquad x > 0, \tag{63}$$

$$T_2(0,t) = B > 0, \qquad t > 0, \tag{64}$$

$$T_1(s(t),t) = T_f = 0, \qquad t > 0\,, \tag{65}$$

$$T_2(s(t),t) = T_f = 0, \qquad t > 0\,, \tag{66}$$

$$k_1 T_{1_x}(s(t),t) - k_2 T_{2_x}(s(t),t) = \rho \ell \dot{s}(t), \quad t > 0\,, \tag{67}$$

$$s(0) = 0\,. \tag{68}$$

Theorem 7. (Neumann solution (Webber, 1901))
The explicit solution to problem (61)-(68) is given by:

$$T_2(x,t) = B - \frac{B}{erf(\sigma / a_2)} erf(\frac{x}{2a_2\sqrt{t}}), \quad 0 \le x \le s(t), t > 0 \tag{69}$$

$$T_1(x,t) = -C + \frac{B}{erfc(\sigma / a_1)} erfc(\frac{x}{2a_1\sqrt{t}}), \quad s(t) \le x, \quad t > 0 \tag{70}$$

$$s(t) = 2\sigma\sqrt{t} \qquad (a_2^2 = \frac{k_2}{\rho c_2}, \quad a_1^2 = \frac{k_1}{\rho c_1}) \tag{71}$$

where $\sigma > 0$ is the unique solution to the following equation:

$$F(x) = x, \quad x > 0 \tag{72}$$

where

$$F(x) = \frac{Bk_2}{\rho \ell a_2 \sqrt{\pi}} F_2(\frac{x}{a_2}) - \frac{Ck_1}{\rho \ell a_1 \sqrt{\pi}} F_1(\frac{x}{a_1}) \tag{73}$$

$$F_1(x) = \frac{exp(-x^2)}{erfc(x)}, \quad F_2(x) = \frac{exp(-x^2)}{erf(x)}, \quad erfc(x) = 1 - erf(x). \tag{74}$$

Remark 9.
It is very interesting to answer the following question: When is the Neumann solution for a semi-infinite material applicable to a finite material $(0, x_0)$? (Solomon, 1979).

Taking into account that $erf(x) \cong 1 \ for \ 2 \le x$, we have an affirmative answer for a short period of time because $T_1(x_0, t) \cong -C$ is equivalent to

$$erf(\frac{x_0}{2a_1\sqrt{t}}) \approx 1 \tag{75}$$

that is

$$t \le \frac{x_0^2}{16a_1^2}. \tag{76}$$

Remark 10.
A generalization of Neumann solution is given in (Briozzo et al, 2004 & 2007b) for particular sources in the heat equations for both phases. A study of the behaviour of the Neumann solution when the latent heat goes to zero is given in (Tarzia & Villa, 1991). A generalization of Neumann solution in multi-phase media is given in (Sanziel & Tarzia, 1989; Weiner, 1955; Wilson, 1978 & 1982), and when we have shrinkage or expansion (Fi & Han, 2007; Natale et al., 2010; Wilson & Solomon, 1986).

2.8 The Neumann-type solution for the two-phase problem for a particular prescribed heat flux at the fixed face, and the necessary and sufficient condition to have an instantaneous phase-change process

If we consider the problem (61)-(68) by changing the boundary condition (64) at $x=0$ by a heat flux condition of the type

$$k_2 T_{2_x}(0,t) = -\frac{q_0}{\sqrt{t}} \tag{77}$$

then we can obtain the following result:

Theorem 8. (Tarzia, 1981)

i. If q_0 verifies the inequality

$$q_0 > \frac{Ck_1}{a_1\sqrt{\pi}} \tag{78}$$

then we have an instantaneous change of phase and the corresponding explicit solution is given by:

$$T_2(x,t) = A_2 + B_2 erf(\frac{x}{2a_2\sqrt{t}}), \quad 0 \le x \le s(t), \ t > 0 \tag{79}$$

$$T_1(x,t) = A_1 + B_1 erf(\frac{x}{2a_1\sqrt{t}}), \quad s(t) \le x, \ t > 0 \tag{80}$$

$$s(t) = 2w\sqrt{t} \qquad \left(a_2^2 = \frac{k_2}{\rho c_2}, a_1^2 = \frac{k_1}{\rho c_1} \right) \tag{81}$$

where

$$A_1(w) = C\frac{erf(w/a_1)}{erfc(w/a_1)}, \quad B_1(w) = \frac{-C}{erfc(w/a_1)} \tag{82}$$

$$A_2(w) = \frac{a_2 q_0\sqrt{\pi}}{k_2} erf(w/a_2), \quad B_2(w) = -\frac{a_2 q_0\sqrt{\pi}}{k_2} \tag{83}$$

and $w > 0$ is the unique solution to the equation

$$F_0(x) = x, \quad x > 0, \tag{84}$$

where

$$F_0(x) = \frac{q_0}{\rho\lambda}\exp(-x^2/a_2^2) - \frac{Ck_1}{\rho\lambda a_1\sqrt{\pi}} F_1(x/a_1). \tag{85}$$

ii. If $q_0 \le Ck_1/a_1\sqrt{\pi}$ the corresponding problem represents only a heat conduction problem for the initial solid phase, and the temperature is given by

$$T(x,t) = T_1(x,t) = -C + \frac{q_0 a_1 \sqrt{\alpha\pi}}{k_1} erfc(\frac{x}{2a_1\sqrt{\alpha t}}), \quad x>0, t>0 . \tag{86}$$

Corollary 9 (Tarzia, 1981)
The coefficient σ that characterizes the free boundary $s(t) = 2\sigma\sqrt{t}$ of Neumann solution (69)-(74) must satisfy the following inequality:

$$erf(\frac{\sigma}{a_2}) < \frac{B}{C}\sqrt{\frac{k_2 c_2}{k_1 c_1}} . \tag{87}$$

2.9 The Neumann-type solution for the two-phase problem for a particular prescribed convective condition (Newton law) at the fixed face, and the necessary and sufficient condition to have an instantaneous phase-change process

We consider the following free boundary problem: find the solid-liquid interface $x = s(t)$ and the temperature $T(x,t)$ defined by

$$T(x,t) = \begin{cases} T_s(x,t) & if \;\; 0 < x < s(t), \;\; t>0, \\ T_f & if \;\; x = s(t), \;\; t>0, \\ T_l(x,t) & if \;\; x > s(t), \;\; t>0, \end{cases} \tag{88}$$

which satisfy the following equations and boundary conditions

$$T_{s_t} = \alpha_s T_{s_{xx}}, \quad 0 < x < s(t), \; t>0 \tag{89}$$

$$T_{l_t} = \alpha_l T_{l_{xx}}, \quad x > s(t), \; t>0 \tag{90}$$

$$T_s(s(t),t) = T_l(s(t),t) = T_f, \quad x = s(t), \quad t>0 \tag{91}$$

$$T_l(x,0) = T_l(+\infty,t) = T_i, \quad x>0, \quad t>0 \tag{92}$$

$$k_s T_{s_x}(0,t) = \frac{h_0}{\sqrt{t}}(T_s(0,t) - T_\infty), \quad t>0 \tag{93}$$

$$k_s T_{s_x}(s(t),t) - k_l T_{l_x}(s(t),t) = \rho\ell\dot{s}(t), \quad t>0 \tag{94}$$

$$s(0) = 0 \tag{95}$$

where the subscripts s and l represent the solid and liquid phases respectively, ρ is the common density of mass and ℓ is the latent heat of fusion, and $T_\infty < T_f < T_i$. We have the following results:
Theorem 10. (Tarzia, 2004)
If the coefficient h_0 verifies the inequality

$$h_0 > \frac{k_l}{\sqrt{\pi\alpha_l}} \frac{T_i - T_f}{T_i - T_\infty} \tag{96}$$

there exists an instantaneous solidification process and then the free boundary problem (89)-(95) has the explicit solution to a similarity type given by

$$s(t) = 2\lambda\sqrt{\alpha_l t} \tag{97}$$

$$T_s(x,t) = T_\infty + \frac{(T_f - T_\infty)[1 + \frac{h_0\sqrt{\pi\alpha_s}}{k_s} erf(\frac{x}{2\sqrt{\alpha_s t}})]}{1 + \frac{h_0\sqrt{\pi\alpha_s}}{k_s} erf(\lambda\sqrt{\frac{\alpha_l}{\alpha_s}})} \tag{98}$$

$$T_l(x,t) = T_i - (T_i - T_f)\frac{erfc(\frac{x}{2\sqrt{\alpha_l t}})}{erfc(\lambda)} \tag{99}$$

and the dimensionless parameter $\lambda > 0$ satisfies the following equation

$$F(x) = x, \quad x > 0 \tag{100}$$

where function F and the $b's$ coefficients are given by

$$F(x) = b_1\frac{\exp(-bx^2)}{1 + b_2 erf(x\sqrt{b})} - b_3\frac{\exp(-x^2)}{erfc(x)} \tag{101}$$

$$b = \frac{\alpha_l}{\alpha_s} > 0; \quad b_1 = \frac{h_0(T_f - T_\infty)}{\rho\ell\sqrt{\alpha_l}} > 0 \tag{102}$$

$$b_2 = \frac{h_0}{h_s}\sqrt{\pi\alpha_s} > 0; \quad b_3 = \frac{c_l(T_i - T_f)}{\ell\sqrt{\pi}} > 0 \tag{103}$$

Proof.
Function F has the following properties:

$$F(0^+) = b_1 - b_3 = \frac{h_0(T_f - T_\infty)}{\rho\ell\sqrt{\alpha_l}} - \frac{c_l(T_i - T_f)}{\ell\sqrt{\pi}} \tag{104}$$

$$F(+\infty) = -\infty, \quad F'(x) < 0, \quad \forall x > 0 \tag{105}$$

Therefore, there exists a unique solution λ >0 of the Eq. (100) if and only if $F(0^+) > 0$, that is inequality (96) holds.

2.10 The similarity solution for the two-phase Lamé-Clapeyron-Stefan problem with a mushy region

We consider a semi-infinite material initially in the solid phase at the temperature $-C < T_f = 0$. We impose a temperature $B > T_f = 0$ at the fixed face $x = 0$, and the fusion process begins, and three regions can be distinguished, as follows: (Tarzia, 1990b):

i. the liquid phase, at temperature $T_2 = T_2(x,t) > 0$, occupying the region $0 < x \leq s(t),\ t > 0$;

ii. the solid phase, at temperature $T_1 = T_1(x,t) < 0$, occupying the region $x > r(t),\quad t > 0$;

iii. the mushy zone, at temperature $T_f = 0$, occupying the region $s(t) < x < r(t),\quad t > 0$. We make the following two assumptions on its structure:

a. the material in the mushy zone contains a fixed fraction $\varepsilon\ell$ (with constant $0 < \varepsilon < 1$) of the total latent heat ℓ;

b. the width of the mushy zone is inversely proportional (with constant $\gamma > 0$) to the temperature gradient at $s(t)$.

Therefore, the problem consists of finding the free boundaries $x = s(t), x = r(t)$, and the temperature:

$$T(x,t) = \begin{cases} T_2(x,t) > 0 & \textit{if } 0 < x < s(t), t > 0 \\ 0 & \textit{if } s(t) \leq x \leq r(t), t > 0 \\ T_1(x,t) < 0 & \textit{if } r(t) < x, t > 0 \end{cases} \tag{106}$$

defined for $x > 0$ and $t > 0$, such that the following conditions are satisfied:

$$\alpha_2 T_{2_{xx}} = T_{2_t}, \qquad 0 < x < s(t), t > 0 \tag{107}$$

$$\alpha_1 T_{1_{xx}} = T_{1_t}, \qquad r(t) < x, t > 0 \tag{108}$$

$$s(0) = r(0) = 0, \tag{109}$$

$$T_2(s(t),t) = T_1(r(t),t) = 0, \quad t > 0 \tag{110}$$

$$k_1 T_{1_x}(r(t),t) - k_2 T_{2_x}(s(t),t) = \rho\ell[(1-\varepsilon)\dot{s}(t) + \varepsilon\dot{r}(t)], \tag{111}$$

$$-T_{2_x}(s(t),t)\,(r(t) - s(t)) = \gamma, \quad t > 0 \tag{112}$$

$$T_1(x,0) = T_1(+\infty,t) = -C, \quad x > 0, t > 0 \tag{113}$$

$$T_2(0,t) = B > 0, \quad t > 0 \tag{114}$$

Theorem 11. (Tarzia, 1990b)

i. The explicit solution to the problem (107)-(114) is given by

$$T_1(x,t) = A_1 + B_1 erf(\frac{x}{2a_1\sqrt{t}}), \quad T_2(x,t) = A_2 + B_2 erf(\frac{x}{2a_2\sqrt{t}}) \tag{115}$$

$$s(t) = 2\sigma\sqrt{t}, \quad r(t) = 2\omega\sqrt{t} \quad (a_2^2 = \frac{k_2}{\rho c_2}, \quad a_1^2 = \frac{k_1}{\rho c_1}) \tag{116}$$

where

$$A_2 = B, \quad B_2 = -\frac{B}{erf(\frac{\sigma}{a_2})}, \quad A_1 = \frac{C\,erf(\frac{\omega}{a_1})}{erfc(\frac{\omega}{a_1})}, \quad B_1 = -\frac{C}{erfc(\frac{\omega}{a_1})} \tag{117}$$

$$\omega = \omega(\sigma) = \sigma + \frac{\gamma a_2 \sqrt{\pi}}{2B} \exp(\frac{\sigma^2}{a_2^2}) erf(\frac{\sigma}{a_2}) \tag{118}$$

where $\sigma > 0$ is the unique solution to the equation

$$K_1(x) = K_2(x), \quad x > 0 \tag{119}$$

with

$$K_1(x) = \frac{k_2 B}{a_2\sqrt{\pi}} F_2(\frac{x}{a_2}) - \frac{k_1 B}{a_1\sqrt{\pi}} F_1(\frac{\omega(x)}{a_1}), \quad F_1(x) = \frac{\exp(-x^2)}{erfc(x)}$$
$$K_2(x) = \rho\ell[x + \frac{\varepsilon\gamma a_2\sqrt{\pi}}{2B}\exp(\frac{x^2}{a_2^2}) erf(\frac{x}{a_2})], \quad F_2(x) = \frac{\exp(-x^2)}{erf(x)} \tag{120}$$

Proof.
We have the following properties

$$K_1(0^+) = +\infty, \quad K_1(+\infty) = -\infty, \quad K_1' < 0, \forall x > 0, \tag{121}$$

$$K_2(0^+) = 0, \quad K_2(+\infty) = +\infty, \quad K_2' < 0, \forall x > 0, \tag{122}$$

and the thesis holds.
Remark 11
If the boundary condition (114) is replaced by a heat flux condition of the type (77) then we will have an instantaneous change of phase if and only if the coefficient q_0 that characterizes the heat flux (77) verifies an inequality (Tarzia, 1990b).

2.11 The similarity solution for the phase-change problem by considering a density jump

We will consider the two-phase Lamé-Clapeyron-Stefan problem for a semi-infinite material taking into account the density jump under the change of phase. We will find the interface $s = s(t) > 0$ (free boundary), defined for $t > 0$, and the temperature

$$\theta(x,t) = \begin{cases} \theta_1(x,t) < 0 & if \ \ 0 < x < s(t), \ \ t > 0, \\ 0 & if \ \ x = s(t), \ \ t > 0, \\ \theta_2(x,t) > 0 & if \ \ x > s(t), \ \ t > 0, \end{cases} \tag{123}$$

defined for $x > 0$ and $t > 0$, such that they satisfy the following conditions:

$$\alpha_1 \theta_{1_{xx}} = \theta_{1_t}, \qquad 0 < x < s(t), t > 0 \tag{124}$$

$$\alpha_2\theta_{2_{xx}} + \frac{\rho_1 - \rho_2}{\rho_2}\dot{s}(t)\theta_{2_x} = \theta_{2_t}, \quad x > s(t), \quad t > 0 \tag{125}$$

$$\begin{aligned} \theta_1(s(t),t) &= 0, \quad t > 0 \\ \theta_2(s(t),t) &= 0, \quad t > 0 \end{aligned} \tag{126}$$

$$k_1\theta_{1_x}(s(t),t) - k_2\theta_{2_x}(s(t),t) = \rho_1\ell\dot{s}(t), \quad t > 0 \tag{127}$$

$$\theta_2(x,0) = \theta_0 > 0, \quad x > 0 \tag{128}$$

$$s(0) = 0 \tag{129}$$

$$\theta_1(0,t) = -d < 0, \quad t > 0\,. \tag{130}$$

Theorem 12 (Carslaw & Jaeger, 1959; Rubinstein, 1971)
The explicit solution to the free boundary problem (124)-(130) is given by

$$\begin{cases} \theta_1(x,t) = A_1 + B_1 erf(\dfrac{x}{2a_1\sqrt{t}}) \\ \theta_2(x,t) = A_2 + B_2 erf(\delta_1 + \dfrac{x}{2a_2\sqrt{t}}) \\ s(t) = 2\gamma\sqrt{t}, \qquad \gamma > 0 \end{cases} \tag{131}$$

where

$$A_1(\gamma) = -d, \qquad B_1(\gamma) = \frac{d}{erf(\gamma/a_1)}, \tag{132}$$

$$A_2(\gamma) = \frac{-\theta_0\, erf(\gamma/a_0)}{erfc(\gamma/a_0)}, \qquad B_2(\gamma) = \frac{\theta_0}{erfc(\gamma/a_0)}, \tag{133}$$

$$\varepsilon = \frac{\rho_1 - \rho_2}{\rho_2}, \quad \delta_1 = \frac{\gamma}{a_2}|\varepsilon|, \quad a_0 = \frac{a_2}{1+|\varepsilon|}, \tag{134}$$

and γ is the unique solution to the following equation:

$$F(x) = x, \qquad x > 0\,, \tag{135}$$

with

$$F(x) = \frac{k_1}{\ell\rho_1 a_1\sqrt{\pi}} B_1(x)\exp\left(\frac{-x^2}{a_1^2}\right) - \frac{k_2}{\ell\rho_1 a_2\sqrt{\pi}} B_2(x)\exp\left(\frac{-x^2}{a_0^2}\right). \tag{136}$$

Proof.
We have the following properties:

$$F(0^+) = +\infty, \qquad F(+\infty) = -\infty, \qquad F'(x) < 0, \forall x > 0. \tag{137}$$

Theorem 13 (Bancora & Tarzia, 1985)

i. If we replace the boundary condition (130) by the following one given by:

$$k_1 \theta_{1_x}(0,t) = \frac{q_0}{\sqrt{t}}, \quad t > 0, \tag{138}$$

then there exists an explicit solution corresponding to the free boundary problem (124)-(130) and (138) if and only if the coefficient q_0 satisfies the inequality

$$q_0 > \theta_0 \sqrt{\frac{k_2 \rho_2 c_2}{\pi}}. \tag{139}$$

In this case the explicit solution is given by:

$$\begin{cases} \theta_1(x,t) = C_1 + D_1 erf\left(\frac{x}{2a_1\sqrt{t}}\right) \\ \theta_2(x,t) = C_2 + D_2 erf\left(\delta_2 + \frac{x}{2a_2\sqrt{t}}\right) \\ s(t) = 2w\sqrt{t}, \qquad w > 0 \end{cases} \tag{140}$$

where

$$C_1(w) = -\frac{a_1 q_0 \sqrt{\pi}}{k_1} erf\left(\frac{w}{a_1}\right), \qquad D_1(w) = \frac{a_1 q_0 \sqrt{\pi}}{k_1}, \tag{141}$$

$$C_2(w) = \frac{-\theta_0\, erf(w/a_0)}{erfc(w/a_0)}, \qquad D_2(\gamma) = \frac{\theta_0}{erfc(w/a_0)}, \tag{142}$$

$$\varepsilon = \frac{\rho_1 - \rho_2}{\rho_2}, \quad \delta_2 = \frac{\gamma}{a_2}|\varepsilon|, \quad a_0 = \frac{a_2}{1+|\varepsilon|}, \tag{143}$$

and w is the unique solution to the following equation:

$$F_0(x) = x, \qquad x > 0, \tag{144}$$

with

$$F_0(x) = \frac{q_0}{\ell \rho_1} \exp\left(\frac{-x^2}{a_1^2}\right) - \frac{k_2}{\ell \rho_1 a_2 \sqrt{\pi}} B_2(x) \exp\left(\frac{-x^2}{a_0^2}\right). \tag{145}$$

Proof.

We have the following properties:

$$F_0(0^+) = \frac{1}{\ell \rho_1}\left(q_0 - \frac{k_2 \theta_0}{a_2 \sqrt{\pi}}\right), \qquad F(+\infty) = -\infty, \qquad F_0'(x) < 0, \forall x > 0. \tag{146}$$

Remark 12 .
When the boundary condition at the fixed face $x = 0$ is given by (93) the explicit solution was given in (Tarzia, 2007).

2.12 The determination of one or two unknown thermal coefficient through an over-specified condition at the fixed face for one or two-phase cases

We consider the one-phase Lamé-Clapeyron-Stefan problem with unknown thermal coefficients. If we give an overspecified boundary condition at the fixed face $x = 0$ we can determine one or two unknown coefficients following (Arderius et al., 1996; Cannon, 1963 & 1964; Garguichevich et al., 1985; Jones, 1962 & 1963; Tarzia, 1982,1983 & 1984)).

2.12.1 Determination of one unknown thermal coefficient through a one-phase case

The problem consists of finding the free boundary $x = s(t)$, the temperature $T = T(x,t)$, and one unknown thermal coefficient chosen among $\{k, \rho, c, \ell\}$ such that they must satisfy the following conditions (we have a free boundary problem):

$$T_t = \alpha T_{xx}, \qquad 0 < x < s(t), \quad t > 0 \quad (a^2 = \alpha = k/\rho c) \tag{147}$$

$$T(s(t),t) = 0, \qquad t > 0 \tag{148}$$

$$kT_x(s(t),t) = \rho \ell \dot{s}(t), \qquad t > 0, \tag{149}$$

$$T(0,t) = T_0 > 0 \tag{150}$$

$$kT_x(0,t) = -\frac{q_0}{\sqrt{t}}, \quad t > 0, \tag{151}$$

$$s(0) = 0, \tag{152}$$

where q_0 is the coefficient that characterizes the heat flux at the fixed face $x = 0$ and it must be obtained experimentally.
Theorem 14 (Tarzia, 1982)
Let T_0 and q_0 be determinated experimentally. The solution for the determination of one thermal coefficient is given by:

$$T(x,t) = T_0 - \frac{T_0}{erf(\frac{\sigma}{a})} erf(\frac{x}{2a\sqrt{t}}), \quad 0 < x < s(t), \ t > 0 \quad , \tag{153}$$

$$s(t) = 2\sigma\sqrt{t}, \quad t > 0 \quad , \tag{154}$$

where σ and the unknown thermal coefficient are computed in the summarized way in the following Table 1:

Case #	Formulae for unknown coefficients	Parameter ξ is the unique solution to the equation	Restrictions on data
1	$\sigma=\xi\sqrt{\frac{k}{\rho c}}$ $\ell=\frac{q_0}{\rho}\sqrt{\frac{\rho c}{k}}\frac{\exp(-\xi^2)}{\xi}$	$erf(x)=\frac{T_0}{q_0}\sqrt{\frac{k\rho c}{\pi}}$ $x>0$	$\frac{T_0}{q_0}\sqrt{\frac{k\rho c}{\pi}}<1$
2	$\sigma=\frac{kT_0}{q_0\sqrt{\pi}}\frac{\xi}{erf(\xi)}$ $\rho=\frac{\pi q_0^2}{kcT_0^2}erf^2(\xi)$	$E(x)=\frac{cT_0}{\ell\sqrt{\pi}}$ $x>0$	-------
3	$\sigma=\frac{q_0}{\rho\ell}\exp(-\xi^2)$ $k=\frac{\pi q_0^2}{\rho cT_0^2}erf^2(\xi)$	As in Case 2	-------
4	$\sigma=\frac{q_0}{\rho\ell}\exp(-\xi^2)$ $c=\frac{\pi q_0^2}{\rho kT_0^2}erf^2(\xi)$	$\frac{erf(x)}{x}=\frac{k\rho\ell T_0}{q_0^2\sqrt{\pi}}\exp(x^2)$ $x>0$	$\frac{k\rho\ell T_0}{2q_0^2}<1$

Table 1. Summary of the determination of one thermal coefficient through a one-phase Lamé-Clapeyron-Stefan problem (4 cases)

Remark 13.
The determination of one unknown thermal coefficient for phase-change problems with temperature-dependent thermal conductivity of the type (54) was given in (Tarzia, 1998).

2.12.2 Determination of two unknown thermal coefficients for the one-phase case

If the interface solid-liquid is given by the law:

$$s(t)=2\sigma\sqrt{t},\quad t>0 \tag{155}$$

where the coefficient $\sigma>0$ was determined experimentally then the problem consists of finding the temperature $T=T(x,t)$ and two unknown thermal coefficients chosen among $\{k,\rho,c,\ell\}$ such that they satisfy the conditions (147)-(152) (we have a moving boundary problem).

Theorem 15 (Tarzia, 1983)
Let T_0, σ and q_0 be determined experimentally. The solution for the determination of two thermal coefficients is given by

$$T(x,t)=T_0-\frac{T_0}{erf(\frac{\sigma}{a})}erf(\frac{x}{2a\sqrt{t}}),\quad 0<x<s(t),\ t>0, \tag{156}$$

and the two coefficients are computed in the summarized way in the following Table 2.

Case #	Formulae for unknown coefficients	Parameter ξ is the unique solution to the equation	Restrictions on data
1	$c = \frac{q_0\sqrt{\pi}}{\rho\sigma T_0}\xi\, erf(\xi)$ $k = \frac{\sigma q_0\sqrt{\pi}}{T_0}\frac{erf(\xi)}{\xi}$	$\xi = \sqrt{\log(\frac{q_0}{\rho\ell\sigma})}$	$\frac{q_0}{\rho\ell\sigma} > 1$
2	$\ell = \frac{q_0}{\rho\sigma}\exp(-\xi^2)$ $k = \frac{\rho c\sigma^2}{\xi^2}$	$x erf(x) = \frac{\rho c\sigma T_0}{q_0\sqrt{\pi}}$ $x > 0$	---------
3	$\rho = \frac{q_0}{\ell\sigma}\exp(-\xi^2)$ $k = \frac{q_0 c\sigma}{\ell}\frac{\exp(-\xi^2)}{\xi^2}$	$E(x) = \frac{cT_0}{\ell\sqrt{\pi}}$ $x > 0$	---------
4	$c = \frac{k}{\rho\sigma^2}\xi^2$ $\ell = \frac{q_0}{\rho\sigma}\exp(-\xi^2)$	$\frac{erf(x)}{x} = \frac{kT_0}{\sigma q_0\sqrt{\pi}}$ $x > 0$	$\frac{kT_0}{2\sigma q_0} < 1$
5	$\rho = \frac{k}{c\sigma^2}\xi^2$ $\ell = \frac{c\sigma q_0}{k}\frac{\exp(-\xi^2)}{\xi^2}$	As in case 4	As in case 4
6	$\rho = \frac{q_0}{\ell\sigma}\exp(-\xi^2)$ $c = \frac{k\ell}{\sigma q_0}\xi^2\exp(\xi^2)$	As in case 4	As in case 4

Table 2. Summary of the simultaneous determination of two thermal coefficients through a one-phase Lamé-Clapeyron-Stefan problem (6 cases)

Remark 14.

The determination of thermal coefficients for the Solomon-Wilson-Alexiades mushy region was obtained in (Tarzia, 1987). The simultaneous determination of two unknown thermal coefficients for phase-change problems with temperature-dependent thermal conductivity of the type (54) was given in (Salva & Tarzia, 2010) with a sensitivity analysis.

2.12.3 Determination of one or two unknown thermal coefficients through a two-phase Lamé-Clapeyron-Stefan problem

The determination of one or two unknown thermal coefficients for a two-phase solidification or fusion problem was obtained in (Stampella & Tarzia, 1989). The determination of one or two unknown thermal coefficients for a mushy region was obtained in (González & Tarzia, 1996). Another variant for the simultaneous determination of the thermal coefficients is given in (Tarzia, 1991a).

Remark 15.

Explicit solutions for the determination of unknown coefficients are given in (Briozzo et al., 1999) for Storm's type materials.

2.13 A similarity solution for the thawing in a saturated porous medium by considering a density jump and the influence of the pressure on the melting temperature

We consider the problem of thawing of a partially frozen porous medium, saturated with an incompressible liquid. For a detailed exposition of the physical background we refer to (Charach & Rubinstein, 1992; Fasano et al. 1993; Fasano & Primicerio, 1984; Nakano, 1990; O'Neill & Miller, 1985; Talamucci, 1997 & 1998). More specifically, we deal with the following situations:

i. a sharp interface between the frozen part and the unfrozen part of the domain exists (sharp, in the macroscopic sense);
ii. the frozen phase is at rest with respect to the porous skeleton, which will be considered to be undeformable;
iii. due to the density jump between the liquid and solid phases, thawing can induce either desaturation or water movement in the melting regíon. We will consider the latter situation, assuming that liquid is continuously supplied to keep the medium saturated.

The unknowns of the problem are the function $x=s(t)$, representing the free boundary, and the two functions $u(x, t)$ and $v(x, t)$ representing the temperature of the unfrozen and of the frozen zone respectively which must satisfy the following conditions:

$$u_t = a_1 u_{xx} - b\rho\dot{s}(t)u_x, \qquad 0 < x < s(t), \quad t > 0 \tag{157}$$

$$v_t = a_2 v_{xx}, \qquad x > s(t), \quad t > 0 \tag{158}$$

$$u(s(t),t) = v(s(t),t) = d\rho s(t)\dot{s}(t), \quad t > 0 \tag{159}$$

$$k_F v_x(s(t),t) - k_U u_x(s(t),t) = \alpha\dot{s}(t) + \beta\rho s(t)\,(\dot{s}(t))^2, \quad t > 0 \tag{160}$$

$$u(0,t) = B > 0, \quad t > 0. \tag{161}$$

$$v(x,0) = v(+\infty,t) = -A < 0, \quad x > 0, t > 0 \tag{162}$$

$$s(0) = 0 \tag{163}$$

with

$$a_1 = \alpha_1^2 = \frac{k_U}{\rho_U c_U}, \quad a_2 = \alpha_2^2 = \frac{k_F}{\rho_F c_F}, \quad b = \frac{\varepsilon\rho_W c_W}{\rho_U c_U}, \quad d = \frac{\varepsilon\gamma\mu}{K}$$
$$\rho = \frac{\rho_W - \rho_I}{\rho_W}, \quad \alpha = \varepsilon\rho_I \ell, \quad \beta = \frac{\varepsilon^2\rho_I\gamma\mu(c_W - c_I)}{K} = \varepsilon d\rho_I(c_W - c_I) \tag{164}$$

where
ε : porosity,
ρ_W and ρ_I : density of water and ice,
c: specific heat at constant density,
k_U and k_F : conductivity of the unfrozen and frozen zones,
$u=v=0$: the melting point at atmospheric pressure,
ℓ : latent heat at $u=0$,
γ : coefficient in the Clausius-Clapeyron law,
$\mu > 0$: viscosity of liquid,
$K>0$: hydraulic permeability,
$B>0$: boundary temperature at the fixed face $x=0$,
$-A<0$: initial temperature.

Theorem 16 (Fasano et al., 1999)
The free boundary problem (157) - (163) has the similarity solution

$$s(t) = 2\xi\alpha_1\sqrt{t}, \tag{165}$$

$$u(x,t) = B + \frac{m\xi^2 - B}{g(p,\xi)} \int_0^{x/(2\alpha_1\sqrt{t})} \exp(-r^2 + pr\xi)\, dr \tag{166}$$

$$v(x,t) = \frac{m\xi^2 erfc\left(\dfrac{x}{2\alpha_2\sqrt{t}}\right) + A\left(erf(\gamma_0\xi) - erf\left(\dfrac{x}{2\alpha_2\sqrt{t}}\right)\right)}{erfc(\gamma_0\xi)} \tag{167}$$

if and only if the coefficient $\xi > 0$ satisfies the following equation:

$$K_1(B - my^2)H(p,y) - K_2F(m,y) = \delta y + \upsilon y^3, \quad y > 0, \tag{168}$$

where

$$g(p,y) = \int_0^y \exp(-r^2 + pyr)\, dr, \qquad H(p,y) = \frac{\exp((p-1)y^2)}{2\sqrt{\pi}\, g(p,y)} \tag{169}$$

$$F(m,y) = (A + my^2)\frac{\exp(-\gamma_0^2 y^2)}{erfc(\gamma_0 y)} \tag{170}$$

and the constants $K_1, K_2, \gamma_0, \delta, p, m \text{ and } \upsilon$ are defined as follows:

$$K_1 = \frac{k_U}{\alpha_1\sqrt{\pi}} > 0, \quad K_2 = \frac{k_F}{\alpha_2\sqrt{\pi}} > 0, \quad \gamma_0 = \frac{\alpha_1}{\alpha_2} > 0, \quad \delta = \alpha\alpha_1 > 0 \tag{171}$$

$$p = 2b\rho, \quad m = 2d\rho\alpha_1^2 > 0, \quad \upsilon = 2\beta\rho\alpha_1^3 \tag{172}$$

Moreover, the existence and uniqueness of the unknown coefficient $\xi > 0$ depends on the sign of the three parameters p, m and υ of the problem.

If we replace the boundary condition (161) by the following one:

$$k_U u_x(0,t) = -\frac{q_0}{\sqrt{t}}, \quad t>0, \tag{173}$$

then we can consider the free boundary problem (157) - (160), (173), (162)-(163) and we can obtain the following result.

Theorem 17 (Lombardi & Tarzia, 2001)

The free boundary problem (157) - (160), (173), (162)-(163) has the following similarity solution:

$$s(t) = 2\xi^* \alpha_1 \sqrt{t}, \tag{174}$$

$$u(x,t) = m(\xi^*)^2 + \frac{2q_0\alpha_1}{K_U} g(p,\xi^*) - \frac{2q_0\alpha_1}{K_U} \int_0^{x/(2\alpha_1\sqrt{t})} \exp(-r^2 + pr\xi^*)\, dr \tag{175}$$

$$v(x,t) = \frac{m(\xi^*)^2 + A erf(\gamma_0\xi^*)}{erfc(\gamma_0\xi^*)} - \frac{m(\xi^*)^2}{erfc(\gamma_0\xi^*)} erf\left(\frac{x}{2\alpha_1\sqrt{t}}\right) \tag{176}$$

if and only if the coefficient $\xi^* > 0$ satisfies the following equation:

$$q_0 \exp((p-1)y^2) - K_2 F(m,y) = \delta y + \upsilon y^3, \quad y>0, \tag{177}$$

or its equivalent

$$Q_0(y) = q_0, \quad y>0, \tag{178}$$

where

$$Q_0(y) = \frac{K_2 F(m,y) + \delta y + \upsilon y^3}{\exp((p-1)y^2)}, \quad y>0. \tag{179}$$

Moreover, the existence and uniqueness of the unknown coefficient $\xi^* > 0$ depends on the sign of the three parameters p, m and υ of the problem.

3. Explicit solutions for free boundary problems for the diffusion equation

Heat and mass transfer with phase change problems, taking place in a porous medium, such as evaporation, condensation, freezing, melting, sublimation and desublimation, have wide application in separation processes, food technology, heat and mixture migration in soils and grounds, etc. Due to the non-linearity of the problem, solutions usually involve mathematical difficulties and only a few exact solutions have been found. Mathematical formulation of the heat and mass transfer in capillary porous bodies has been established by (Luikov, 1964, 1966, 1975 & 1978). Some books or booklets on the subject for the diffusion equation are (Crank, 1975; Duvaut, 1976; Fasano, 2005; Fasano & Primicerio, 1986; Froment & Bischoff, 1979; Levenspiel, 1962; Primicerio & Gianni, 1989; Szekely et al., 1976).

3.1 The oxygen diffusion-consumption problem and its relationship with the phase-change problem

The diffusion-consumption of oxygen in absorbing tissue consists in finding the free boundary $x=s(t)$ and the concentration $C(x,t)$ such that they satisfy the following conditions (Crank & Gupta, 1972; Crank, 1984; Liapis et al., 1982):

$$C_t - C_{xx} = -1, \quad 0 < x < s(t), \quad 0 < t < T_0 \tag{180}$$

$$C(x,0) = H(x), \quad 0 \leq x \leq 1, \tag{181}$$

$$C_x(0,t) = -G(t) \qquad (C(0,t) = F(t)), \quad 0 < t < T_0, \tag{182}$$

$$C(s(t),t) = 0, \qquad 0 < t < T_0, \tag{183}$$

$$C_x(s(t),t) = 0, \qquad 0 < t < T_0, \tag{184}$$

$$s(0) = 1, \qquad 0 < t < T_0. \tag{185}$$

Remark 16.

We remark that $\dot{s}(t)$ does not appear in both conditions (183) and (184) on the free boundary $x=s(t)$ and for this reason the free boundary problem is of an implicit type; the Lamé-Clapeyron-Stefan problem is one of an explicit type.

Theorem 18. (Fasano, 1974; Schatz, 1966)

The free boundary problem of the diffusion-consumption of oxygen in absorbing tissue is equivalent to the following Lamé-Clapeyron-Stefan problem:

$$z_t - z_{xx} = 0, \quad 0 < x < s(t), \quad 0 < t < T_0 \tag{186}$$

$$z(x,0) = h(x), \quad 0 \leq x \leq 1 \tag{187}$$

$$z_x(0,t) = -g(t) \quad (z(0,t) = f(t)), \quad 0 < t < T_0 \tag{188}$$

$$z(s(t),t) = 0, \qquad 0 < t < T_0 \tag{189}$$

$$z_x(s(t),t) = -\dot{s}(t), \quad 0 < t < T_0 \tag{190}$$

$$s(0) = 1, \qquad 0 < t < T_0. \tag{191}$$

i. If (s,z,T_0) is a solution to the problem (186) - (191) then (s,C,T_0) is solution to the problem (180) - (185) where we define:

$$C(x,t) = \int_x^{s(t)} d\xi \int_\xi^{s(t)} [1 + z(y,t)]dy, \tag{192}$$

$$H(x) = \int_x^1 d\xi \int_\xi^1 [1 + h(y)]dy, \quad F(t) = H(0) + \int_0^t f(\tau)d\tau \tag{193}$$

$$G(t) = 1 + \int_0^1 h(x)dx + \int_0^t g(\tau)d\tau, \qquad (194)$$

ii. If (s, C, T_0) is a solution to the problem (180) - (185) then (s, z, T_0) is solution to the problem (186) - (191) where we define:

$$z(x,t) = C_t(x,t) \qquad (195)$$

$$h(x) = H''(x) - 1, \quad g(t) = G'(t), \quad f(t) = F'(t) \qquad (196)$$

Remark 17.
The oxygen diffusion-consumption free boundary problem was applied to the anaerobiosis in saturated soil aggregates in (González et al., 2008).

3.2 The Rubinstein solution for the binary alloy solidification problem

We consider a semi-infinite slab of a binary alloy consisting of two components *A, B*. Let *C* be the concentration of "*A*". We suppose that solidification of the alloy is governed by an equilibrium phase diagram consisting of liquidus curve $T = f_L(C)$, and a solidus curve $T = f_S(C)$, $0 < C < 1$ and we assume f_L, f_S to be monotonically increasing, $f_L(C) > f_S(C)$ and $f_L(0) = f_S(0) = T_{cr}^A$, $f_L(1) = f_S(1) = T_{cr}^B$. Material is in its solid state if $T \le f_S(C)$ and liquid if $f_L(C) \le T$. If $f_S(C) < T < f_L(C)$ then the material state is not well defined (it is known as mushy region).
We consider that the semi-infinite alloy is initially liquid at constant temperature T_{in} and concentration C_{in}, for which $f_L(C_{in}) \le T_{in}$. Beginning at time $t = 0$, a cold temperature $T_B < T_{cr}^A$ is imposed at $x = 0$. Freezing occurs with, in principle, a sharp phase change front $s = s(t)$ separating solid alloy $(x < s(t))$ from liquid alloy $(x > s(t))$. The mathematical formulation of the solidification process is given in (Rubinstein, 1971) as follows:
Find temperature $T(x,t)$, concentration $C(x,t)$ and phase-change front $x = s(t)$, such that the following conditions must be satisfied:

$$C_{S_t} = D_S C_{S_{xx}}, \quad 0 < x < s(t), \quad t > 0 \qquad (197)$$

$$C_{\ell_t} = D_\ell C_{\ell_{xx}}, \quad x > s(t), \quad t > 0 \qquad (198)$$

$$T_{S_t} = \alpha_S T_{S_{xx}}, \quad 0 < x < s(t), \quad t > 0 \qquad (199)$$

$$T_{\ell_t} = \alpha_\ell T_{\ell_{xx}}, \quad x > s(t), \quad t > 0 \qquad (200)$$

$$T_{cr} = T_S(s(t),t) = f_S[C_S(s(t),t)], \quad t > 0 \qquad (201)$$

$$T_{cr} = T_\ell(s(t),t) = f_L[C_\ell(s(t),t)], \quad t > 0 \qquad (202)$$

$$D_\ell C_{\ell_x}(s(t),t) - D_S C_{S_x}(s(t),t) = [C_S(s(t),t) - C_\ell(s(t),t)]\dot{s}(t), \quad t > 0 \qquad (203)$$

$$k_S T_{S_x}(s(t),t) - k_\ell T_{\ell_x}(s(t),t) = \rho \ell \dot{s}(t), \quad t>0 \tag{204}$$

$$T_S(0,t) = T_B < T_{cr}^A, \quad t>0 \tag{205}$$

$$T_\ell(x,0) = T_{in} > f_L(C_{in}), \quad t>0 \tag{206}$$

$$C_\ell(x,0) = C_{in}, \quad x>0 \tag{207}$$

$$C_{S_x}(0,t) = 0, \quad t>0 \tag{208}$$

$$s(0) = 0 \tag{209}$$

where ρ, k, α, D, ℓ represent the mass density, the thermal conductivity, the thermal diffusivity, the mass diffusion and the latent heat of fusion, being S and ℓ the subscripts that denote the solid and liquid phase respectively.

Theorem 19 (Rubinstein, 1971; Solomon et al., 1983)

There exists a unique solution to the coupled free boundary problem (197)-(209), moreover, the solidus $C_1(s(t),t)$ and liquidus $C_2(s(t),t)$ concentrations as well as the phase-change temperature $T_{Cr} = T_1(s(t),t) = T_2(s(t),t)$ are constants in time. The explicit solution is given by the following expressions:

$$s(t) = 2\lambda\sqrt{\alpha_1 t} \tag{210}$$

$$C_S(x,t) = C_S, \quad T_{cr} = f_S(C_S) = f_L(C_L), \tag{211}$$

$$C_\ell(x,t) = C_{in} + (C_L - C_{in})\frac{erfc(x/2\sqrt{D_\ell t})}{erfc(\lambda\sqrt{\alpha_S/D_\ell})} \tag{212}$$

$$T_S(x,t) = T_B + (T_{cr} - T_B)\frac{erf(x/2\sqrt{\alpha_S t})}{erf(\lambda)} \tag{213}$$

$$T_\ell(x,t) = T_{in} + (T_{cr} - T_{in})\frac{erf(x/2\sqrt{\alpha_\ell t})}{erf(\lambda\sqrt{\alpha_S/\alpha_\ell})} \tag{214}$$

where λ and T_{cr} (or equivalently C_S and C_ℓ) must satisfy the following conditions:

$$\frac{C_{in} - C_L}{C_S - C_L} = Q(\lambda\sqrt{\alpha_s/D_\ell}), \quad T_{cr} = W_1(\lambda) \tag{215}$$

where

$$Q(x) = \sqrt{\pi}\, x \exp(x^2) erfc(x), \quad A = \rho \ell \alpha_\ell / k_\ell \tag{216}$$

$$W_1(x) = T_B + [(T_{in} - T_B) + AQ(x\sqrt{\alpha_S / \alpha_\ell})]\frac{\Psi(x)}{1+\Psi(x)} \tag{217}$$

$$\Psi(x) = \frac{k_\ell \alpha_S \sqrt{\pi}}{k_S \alpha_\ell} \frac{E(x)}{Q(x\sqrt{\alpha_S / \alpha_\ell})} \tag{218}$$

Proof.
The system of equations (215) has a unique solution because of the following properties

$$Q(0^+) = 0, \quad Q(+\infty) = 1, \quad Q' > 0 \tag{219}$$

$$W_1(0^+) = T_B, \quad W_1(+\infty) = T_{in} + \frac{\ell}{c_\ell}, \quad W_1' > 0 \,. \tag{220}$$

Remark 18.
Some other references on the binary alloy solidification problems are (Alexandrov & Malygin, 2006; Gupta et al., 1997; Tien & Geigen, 1967; Tien & Koump, 1970; Tsubaki & Boley, 1977; Voller, 2006, 2008 a&b; White, 1985; Wilson et al., 1982). In (Cirelli & Tarzia, 2010) the binary alloy solidification problem (197)-(209) is solved by changing the boundary condition (205) at the fixed face by a heat flux boundary condition of the type (77) or a convective boundary condition of the type (93).

3.3 The Zel'dovich-Kompaneets-Barenblatt solution for the gas flow through a porous medium

The porous medium equation for a unidimensional material is given by

$$u_t = (u^m)_{xx}, \quad m > 1 \tag{221}$$

which appears in a natural way, mainly to describe processes involving fluid flow, heat transfer or diffusion, the flow of a gas through a porous medium and groundwater infiltration (Vázquez, 2007). The diffusion coefficient of the equation (221) is

$$D(u) = m u^{m-1} \tag{222}$$

assuming $0 \le u$. Equation (221) is parabolic only at those points where $u \neq 0$ and it is in general a degenerate parabolic equation because degenerates wherever $u = 0$.
Its main qualitative property with respect to the classical heat or diffusion equation is the finite propagation which implies the appearance of a free boundary that separates the regions where the solution $u > 0$ (where there is gas, according to the standard interpretation of u as a gas density), from the empty region where $u = 0$.
There exists an explicit solution to the Eq. (221) with a free boundary given by:
Theorem 20 (Zel'dovich & Kompaneets, 1950; Barenblatt, 1952; Pattle, 1959)
The function

$$u(x,t)=\begin{cases}\frac{1}{s(t)}\left[1-\left(\frac{x}{s(t)}\right)^2\right]^{\frac{1}{m-1}} & for \quad |x|\le s(t),\, 0\le t\\ 0 \quad for \quad |x|>s(t),\, 0\le t\end{cases} \tag{223}$$

with

$$s(t)=\left[\frac{2m(m+1)}{m-1}(t+1)\right]^{\frac{1}{m+1}} \tag{224}$$

is a solution to the Cauchy problem for the equation (221) with initial data $u(x,0)=u_0(x),\, x\in\mathbb{R}$ given by

$$u_0(x)=\begin{cases}\frac{1}{s(0)}\left[1-\left(\frac{x}{s(0)}\right)^2\right]^{\frac{1}{m-1}} & for \quad |x|\le s(0)\\ 0 \quad for \quad |x|>s(0)\end{cases}. \tag{225}$$

3.4 The Luikov solution for the coupled heat and mass transfer for a phase-change process

3.4.1 Drying with coupled phase-change in a porous medium

A semi-infinite porous medium is dried by maintaining a heat flux condition at $x=0$ of the type $-q_0/\sqrt{t}$, with $q_0>0$. Initially, the whole body is at uniform temperature t_0 and uniform moisture potential u_0. The moisture is assumed to evaporate completely at a constant temperature, evaporation point t_v. It is also assumed that the moisture potential in the first region $0<x<s(\tau)$, is constant at u_v, where $x=s(\tau)$ locates the evaporation front at time $\tau>0$. It is further assumed that the moisture in vapour form does not take away any appreciable amount of heat from the system. Neglecting mass diffusion due to temperature variation, the problem can be expressed as (Cho, 1975; Gupta, 1974; Luikov, 1978; Santillan Marcus & Tarzia, 2003):

$$\frac{\partial t_1}{\partial \tau}(x,\tau)=\alpha_1\frac{\partial^2 t_1}{\partial x^2}(x,\tau), \qquad 0<x<s(\tau),\, \tau>0 \tag{226}$$

$$u_1=u_v, \quad 0<x<s(\tau),\, \tau>0 \tag{227}$$

$$\frac{\partial t_2}{\partial \tau}(x,\tau)=\alpha_2\frac{\partial^2 t_2}{\partial x^2}+\frac{\varepsilon L c_m}{c_2}\frac{\partial u_2}{\partial \tau}, \qquad x>s(\tau),\, \tau>0 \tag{228}$$

$$\frac{\partial u_2}{\partial \tau}(x,\tau)=\alpha_m\frac{\partial^2 u_2}{\partial x^2}(x,\tau), \qquad x>s(\tau),\, \tau>0 \tag{229}$$

$$k_1 \frac{\partial t_1}{\partial x} = -\frac{q_0}{\sqrt{\tau}}, \quad \text{at} \quad x = 0, \tau > 0 \tag{230}$$

$$t_2 = t_0 \quad \text{in} \quad x > 0, \tau = 0 \tag{231}$$

$$u_2 = u_0 \quad \text{in} \quad x > 0, \tau = 0 \tag{232}$$

$$t_1(s(\tau),\tau) = t_2(s(\tau),\tau) = t_v > t_0 \quad \text{on} \quad x = s(\tau) \tag{233}$$

$$u_1(s(\tau),\tau) = u_2(s(\tau),\tau) = u_v < u_0 \quad \text{on} \quad x = s(\tau) \tag{234}$$

$$-k_1 \frac{\partial t_1}{\partial x}(s(\tau),\tau) + k_2 \frac{\partial t_2}{\partial x}(s(\tau),\tau) = (1-\varepsilon)\rho_m L \frac{ds}{dt} \quad \text{on} \quad x = s(\tau) \tag{235}$$

where t_1 : temperature of the dried porous medium; t_2 : temperature of the humid porous medium; u_2 :mass-transfer potential of the humid porous medium; α_i $(i=1,2)$: thermal diffusivity of the phase i; α_{12} : ratio of thermal diffusivities from phase 1 to phase 2; α_m : moisture diffusivity; c_m : specific mass capacity; c_2 : specific heat capacity; k_i $(i=1,2)$: thermal conductivity of the phase i; $k_{21} = \frac{k_2}{k_1}$; $K_0 = Lc_m \frac{(u_0 - u_v)}{c_2(t_v - t_0)}$: Kossovitch number; L : latent heat evaporation of liquid per unit mass-transfer potential; ε : coefficient of internal evaporation; ρ_m : density of moisture; $L_u = \alpha_m / \alpha_1$: Luikov number, and $\nu = (1-\varepsilon)L\rho_m \alpha_1 / k_1(t_v - t_0) > 0$.

<u>Theorem 21</u> (Santillan Marcus & Tarzia, 2003)

i. If the Luikov number is equal to one, and the coefficient q_0 verifies the condition

$$q_0 > \frac{k_2(t_v - t_0)}{2\sqrt{\pi\alpha_1}}(\varepsilon K_0 + 2) \quad , \tag{236}$$

then there exists one and only one solution $\lambda > 0$ to the following equation:

$$\frac{k_{21}}{\sqrt{\pi}} F_1(x)[-\frac{2\varepsilon K_0}{\sqrt{\pi}} x F_1(x) + 2\varepsilon K_0 x^2 - \varepsilon K_0 - 2] + \frac{2\sqrt{\alpha_1} q_0}{k_1(t_v - t_0)} \exp(-x^2) = 2\nu x, \quad x > 0 \tag{237}$$

Furthermore, the solution to the problem (226)-(235) is given by:

$$u_1(x,\tau) = u_v, \quad 0 < x < s(\tau), \tau > 0, \quad \eta = \frac{x}{2\sqrt{\alpha_1 \tau}} \tag{238}$$

$$t_1(x,\tau) = 1 + \frac{q_0\sqrt{\pi\alpha_1}}{k_1(t_v - t_0)}(erf(\lambda) - erf(\eta)), \quad 0 < x < s(\tau), \quad \tau > 0 \tag{239}$$

$$u_2(x,\tau)=\frac{erfc\left(\frac{\eta}{\sqrt{L_u}}\right)}{erfc(\lambda)},\qquad x>s(\tau),\quad \tau>0 \tag{240}$$

$$t_2(\eta)=\frac{\varepsilon K_0}{\sqrt{\pi}erfc(\lambda)}[\lambda\exp(-\lambda^2)\frac{erfc(\eta)}{erfc(\lambda)}-\eta\exp(-\eta^2)]+\frac{erfc(\eta)}{erfc(\lambda)},\quad x>s(\tau),\tau>0 \tag{241}$$

$$s(\tau)=2\lambda\sqrt{\alpha_1\tau}\,. \tag{242}$$

ii. If the Luikov number is different than one, (that is $\alpha_m\neq\alpha_1$) and the coefficient q_0 verifies the condition

$$q_0>k_2\left(1+\frac{\sqrt{L_u\varepsilon K_0}}{1+\sqrt{L_u}}\right)\frac{t_v-t_0}{\sqrt{\pi\alpha_1}} \tag{243}$$

then there exist one and only one solution $\lambda>0$ to the equation:

$$\phi(x)=\varphi(x),\quad x>0\,, \tag{244}$$

where

$$\phi(x)=\frac{\sqrt{\pi\alpha_1}q_0}{(t_v-t_0)}\exp(-x^2)+P(x) \tag{245}$$

$$\varphi(x)=k_2F_1(x)+\sqrt{\pi}k_1\nu x \tag{246}$$

$$P(x)=\frac{L_u\varepsilon K_0}{L_u-1}k_2\left(\frac{1}{\sqrt{L_u}}F_1(\frac{x}{\sqrt{L_u}})-F_1(x)\right). \tag{247}$$

Furthermore, the solution to the problem (226)-(235) is given by (238)-(240) , (242) and $\left(\eta=\frac{x}{2\sqrt{\alpha_1\tau}}\right)$

$$t_2(\eta)=\frac{\varepsilon K_0L_u}{L_u-1}\left[-\frac{erfc\left(\frac{x}{\sqrt{L_u}}\right)}{erfc\left(\frac{\lambda}{\sqrt{L_u}}\right)}+\frac{erfc(\eta)}{erfc(\lambda)}\right]+\frac{erfc(\eta)}{erfc(\lambda)},\quad x>s(\tau),\quad \tau>0. \tag{248}$$

iii. If the Luikov number L_u verifies the condition

$$L_u>\frac{1}{\varepsilon K_0+1} \tag{249}$$

then the temperature distribution t_2 reaches to a minimum value which is smaller than the initial temperature or its limit value at $+\infty$. The minimum value is attained when the dimensionless variable $\eta = x / 2\sqrt{\alpha_1 \tau}$ takes the value

$$\eta = \sqrt{\left(\frac{L_u}{L_u - 1}\right) \log\left(\frac{((\varepsilon K_0 + 1)L_u - 1)}{\varepsilon K_0 \sqrt{L_u}} \frac{erfc\left(\frac{\lambda}{\sqrt{L_u}}\right)}{erfc(\lambda)}\right)} . \tag{250}$$

3.4.2 Other free boundary problem in a porous medium

There are some explicit solutions for the following free boundary problems for the diffusion equation corresponding to evaporation, freezing, sublimation or desublimation processes in (Lin, 1981, 1982a & 1982b; Mikhailov, 1975 & 1976; Santillan Marcus & Tarzia, 2000a & b). The simultaneous determination of one or two unknown thermal coefficients of a semi-infinite material through a desublimation process with coupled heat and moisture flows is given in (Santillan Marcus & Tarzia, 2007; Santillan Marcus, et al., 2008).

3.5 A mixed saturated-unsaturated flow problem representing absorption of water by a soil with a constant pond depth at the surface and an explicit solution for a particular diffusivity

In wet soils, zones of saturation develop naturally in the vicinity of impermeable strata, surface ponds and subterranean cavities. Hydrology must be concerned with transient flow through coexisting unsaturated and saturated zones. Models of advancing saturated zones necessarily involve a nonlinear free boundary problem (Broadbridge & White, 1990; Knigh & Philip, 1974; Philip, 1957 & 1958; Warrick & Broadbridge, 1992).

We consider a homogeneous soil which initially has some uniform volumetric water content θ_n. At times $t > 0$, water is supllied at the surface $x = 0$ under pressure head ψ_0. Then, a mixed saturated-unsaturated flow problem representing absorption of water by a soil with constant pond depth at the surface is presented. At any time t the zone of saturation extends from $x = 0$ to $x = s(t)$. Assuming Darcy's law and neglecting gravity, the water flux is given by

$$v = -K(\psi)\frac{\partial \psi}{\partial x} \tag{251}$$

where ψ is the soil water matric potential and K is the hydraulic conductivity.
In the saturated zone we have

$$\psi(x,t) = \psi_0 - \frac{\psi_0 - \psi_s}{s(t)}, \quad 0 < x < s(t) \tag{252}$$

and we obtain a free boundary problem for the unsaturated zone:

$$\theta(s(t)^+, t) = \theta_s, \quad t > 0 \tag{253}$$

$$\frac{\partial \theta}{\partial t} = \frac{\partial}{\partial x}\left[D(\theta)\frac{\partial \theta}{\partial x}\right], \quad x > s(t), \quad t > 0 \tag{254}$$

$$-D(\theta)\frac{\partial \theta}{\partial x}(s(t)^+, t) = K_s \frac{\psi_0 - \psi_s}{s(t)}, \quad t > 0 \tag{255}$$

$$\theta(x,0) = \theta(+\infty, t) = \theta_n, \quad x > s(t), \quad t > 0 \tag{256}$$

$$s(0) = 0 \tag{257}$$

where
x : spatial coordinate,
t : time,
θ : volumetric water content,
θ_n : initial volumetric water content,
θ_s : volumetric water content at saturation,
ψ : soil water matric potential,
ψ_0 : pond depth,
ψ_s : soil water potential at $x = s(t)$, $\psi_s < \psi < \psi_0$
K : hydraulic conductivity,
K_s : hydraulic conductivity at saturation,
D : soil water diffusivity ($D = K\frac{d\psi}{d\theta}$).

We consider the free boundary (253)-(257) where the position $s(t)$ of the free boundary and the water content field $\theta(x,t)$ must be determined; and we restrict our attention to the special functional form of the soil water diffusivity

$$D(\theta) = \frac{a}{(b-\theta)^2} \tag{258}$$

where a, and b are positive constants. With this form of diffusivity the nonlinear diffusion equation (254) may be transformed to a linear diffusion equation. We consider the following parameter:

$$C = \frac{b-\theta_n}{\theta_s - \theta_n} > 1\,. \tag{259}$$

Remark 19.
In (Briozzo & Tarzia, 1998) a closed-form analytic solution can be obtained for a nonlinear diffusion model under conditions of ponding surface. The explicit solution depends on a parameter C (determined by the data of the problem), according to two cases: $1 < C < C_1$ or $C_1 \le C$, where C_1 is a constant which is obtained as the unique solution to an equation. This results complements the study given in (Broadbridge, 1990) in order to established when the explicit solution is available. The behaviour of the bifurcaton parameter C_1 as a function of the driving potential is studied with the corresponding limits for small and large

values. We also prove that the sorptivity is continuously differentiable as a function of variable C.

3.6 Estimation of the diffusion coefficient in a gas-solid system

Looking for a competitive separation process like as the permeation, the development and optimal choice of membrane materials become necessary. On this subject, equations modelling the permeation process are required. The parameters contained in such a model must be obtained from simple experiments. The knowledge of solubility and diffusivity are very important to solve the separation problem.

We consider a polymeric membrane swelling for a hydrocarbon solution. The following assumptions are considered: Once the gaseous component reaches a threshold concentration on the gas-polymer interface, it diffuses through the membrane in the x direction being immobilized by a quickly and irreversible transformation. Then a swelling front is generated whose position is given by the free boundary $x = s(t)$, $t > 0$ with the initial condition $s(0)=0$.

Moreover, the hydrocarbon diffusion coefficient D in the saturated o swollen region of the polymer is considered a constant for each experimental condition. A free boundary model (Castro et al., 1987; Crank, 1975; Villa, 1987) with an overspecified condition for the one-dimensional diffusion equation under the preceding assumptions is given:

$$c_t = Dc_{xx}, \qquad 0 < x < s(t), \quad t > 0 \tag{260}$$

$$c(s(t),t) = 0, \qquad t > 0 \tag{261}$$

$$Dc_x(s(t),t) = -\beta\dot{s}(t), \qquad t > 0, \tag{262}$$

$$c(0,t) = C_0 > 0, \quad t > 0 \tag{263}$$

$$A\int_0^t Dc_x(0,\tau)d\tau = -\alpha\sqrt{t}, \quad t > 0, \tag{264}$$

$$s(0) = 0, \tag{265}$$

where $c=c(x,t)$ denotes the concentration profile of the hydrocarbon in the swollen area, $s(t)$ gives the position at time t of the free interface and separates the two regions in the membrane, the saturated and unsaturated, D is the unknown diffusion coefficient in the system, and C_0, α and β are positive parameters and A is a positive constant which must be obtained experimentally.

Theorem 22. (Destefanis et al., 1993)

The concentration profile and the free boundary position are given by:

$$c(x,t) = C_0 - \frac{C_0}{erf(\frac{\sigma}{\sqrt{D}})} erf(\frac{x}{2\sqrt{Dt}}), \quad 0 < x < s(t), \ t > 0, \tag{266}$$

$$s(t) = 2\sigma\sqrt{t}, \quad t > 0, \tag{267}$$

and the unknown coefficients D and σ are obtained by the following expressions:

$$\begin{aligned} D &= \frac{\pi\alpha^2}{4A^2C_0} erf(\xi) = \frac{\alpha^2}{4A^2\beta^2}\frac{\exp(-2\xi^2)}{\xi^2} \\ \sigma &= \frac{\alpha\sqrt{\pi}}{2AC_0}\xi erf(\xi) = \frac{\alpha}{2A\beta}\exp(-\xi^2) \end{aligned} \tag{268}$$

where ξ is the unique solution to the equation:

$$E(x) = \frac{C_0}{\beta\sqrt{\pi}}, \quad x > 0 . \tag{269}$$

Remark 20.

The methodology used in this determination of the unknown diffusion coefficient is a variant of those developed in (Tarzia, 1982 & 1983) for the determination of thermal coefficients for a semi-infinite material through a phase-change process.

3.7 The coupled heat and mass transfer during the freezing of high-water content materials with two free boundaries: the freezing and sublimation fronts

Ice sublimation takes place from the surface of high water-content systems like moist soils, aqueous solutions, vegetable or animal tissues and foods that freeze uncovered or without an impervious and tight packaging material. The rate of both phenomena (solidification and sublimation) is determined both by material characteristics (mainly composition, structure, shape and size) and cooling conditions (temperature, humidity and rate of the media that surrounds the phase change material). The sublimation process, in spite of its magnitude being much less than that of freezing process, determines fundamental features of the final quality for foods and influences on the structure and utility of frozen tissues. Modelling of these simultaneous processes is very difficult owed to the coupling of the heat and mass transfer balances, the existence of two moving phase change fronts that advance with very different rates and to the involved physical properties which are, in most cases, variable with temperature and water content.

When high water-content materials like foods, tissues, gels, soils or water solutions of inorganic or organic substances, held in open, permeable or untightly-sealed containers are refrigerated to below their initial solidification temperature, two simultaneous physical phenomena take place:

- Liquid water solidifies (freeze), and
- Surface ice sublimates.

For the description of the freezing process, the material can be divided into three zones: unfrozen, frozen and dehydrated. Freezing begins from the refrigerated surface/s, at a temperature (T_{if}) lower than that of pure water, due to the presence of dissolved materials, and continues along an equilibrium line. Simultaneously, ice sublimation begins at the frozen surface and a dehydration front penetrates the material, whose rate of advance is again determined by all the abovementioned characteristics of the material and environmental conditions. Normally this rate is much lower than that of the freezing front. A complete mathematical model has to solve both, the heat transfer (freezing) and the mass transfer (weight loss) simultaneously (Campañone et al., 2005a & b).

Phase change is accounted for in the following way:

- Solidification (freezing) as a freezing front ($x = s_f(t)$) located in the point where material temperature reaches the initial freezing temperature (T_{if}), determined by material composition. For temperatures lower than T_{if} (the zone nearer to the refrigerated surface) the amount of ice formed is determined by an equilibrium line (ice content vs temperature and water content) specific to the material.
- On the dehydration front ($x = s_d(t)$) we impose Stefan-like conditions for temperature distribution and vapor concentration.

We consider a semi-infinite material with characteristics similar to a very dilute gel (whose properties can be supposed equal to those of pure water). The system has initial uniform temperature equal to T_{if} and uncovered flat surface which at time t=0 is exposed to the surrounding medium (with constant temperature T_s (lower than T_{if}) and heat and mass transfer coefficients h and K_m). We assume that $T_s < T_0(t) < T_{if}$, $t > 0$ where $T_0(t)$ is the unknown sublimation temperature.

To calculate the evolution of temperature and water content in time, we will consider the following free boundary problem: Find the temperatures $T_d = T_d(x,t)$ and $T_f = T_f(x,t)$, the concentrations $C_{va} = C_{va}(x,t)$, the free boundaries $s_d = s_d(t)$ and $s_f = s_f(t)$ and the temperature $T_0 = T_0(t)$ at the sublimation front $x = s_d(t)$ which must satisfy the following:

- Differential equations at the dehydrated region:

$$\rho_d C_{p_d} \frac{\partial T_d}{\partial t} = k_d \frac{\partial^2 T_d}{\partial x^2}, \quad 0 < x < s_d(t), \quad t > 0 \tag{270}$$

$$\varepsilon \frac{\partial C_{va}}{\partial t} = D_{ef} \frac{\partial^2 C_{va}}{\partial x^2}, \quad 0 < x < s_d(t), \quad t > 0 \tag{271}$$

- Differential equations at the frozen region:

$$\rho_f C_{p_f} \frac{\partial T_f}{\partial t} = k_f \frac{\partial^2 T_f}{\partial x^2}, \quad s_d(t) < x < s_f(t), \quad t > 0 \tag{272}$$

Free boundary conditions at the sublimation front $x = s_d(t)$:

$$T_d(s_d(t),t) = T_f(s_d(t),t) = T_0(t), \quad t > 0 \tag{273}$$

$$k_f \frac{\partial T_f(s_d(t),t)}{\partial x} - k_d \frac{\partial T_d(s_d(t),t)}{\partial x} = L_s m_s \dot{s}_d(t), \quad t > 0 \tag{274}$$

$$D_{ef} \frac{\partial C_{va}(s_d(t),t)}{\partial x} = m_s \dot{s}_d(t) \tag{275}$$

$$C_{va}(s_d(t),t) = \frac{M P_{sat}(T)}{R_g T_0(t)} = M a \frac{\exp\left(b - \dfrac{c}{T_0(t)}\right)}{R_g T_0(t)} \tag{276}$$

where $C_{va}(s_d(t),t)$ is the equilibrium vapor concentration at $T_0(t)$ and the saturation pressure $P_{sat}(T)$ is evaluated according to (Fennema & Berny, 1974).
Free boundary conditions at the freezing front $x = s_f(t)$:

$$T_f(s_f(t),t) = T_{if}, \quad t>0 \tag{277}$$

$$k_f \frac{\partial T_f(s_f(t),t)}{\partial x} = m_f L_f \dot{s}_f(t), \quad t>0 \tag{278}$$

- The convective boundary conditions at the fixed interphase $x = 0$:

$$k_d \frac{\partial T_d(0,t)}{\partial x} = h(T_d(0,t) - T_s), \quad t>0 \tag{279}$$

$$D_{ef} \frac{\partial C_{va}(0,t)}{\partial x} = K_m(C_{va}(0,t) - C_a), \quad t>0 \tag{280}$$

- The initial conditions at $t = 0$:

$$s_f(0) = s_d(0) = 0 \tag{281}$$

$$T = T_{if} \text{ for } x \geq 0. \tag{282}$$

We will solve the system (270) - (282) by using the quasi-steady method. In general, it is a good approximation when the Stefan number tends to zero, i.e. when the latent heat of the material is high with respect to the heat capacity of the solid material. This approximation has often been used when modelling the freezing of high-water content materials.
Theorem 23. (Olguin et al., 2008)
The temperatures T_f, T_d and the concentration C_{va} are given by the following expressions:

$$T_d(x,t) = A(t) + B(t)x, \quad 0<x<s_d(t), \quad t>0 \tag{283}$$

$$C_{v_a}(x,t) = D(t) + E(t)x, \quad 0<x<s_d(t), \quad t>0 \tag{284}$$

$$T_f(x,t) = F(t) + G(t)x, \quad s_d(t)<x<s_f(t), \quad t>0 \tag{285}$$

where $A(t), B(t), D(t)$ and $E(t)$ as a function of $T_0(t)$ and $s_d(t)$, as well as $F(t)$ and $G(t)$ as a function of $T_0(t)$, $s_d(t)$ and $s_f(t)$, given by the following expressions:

$$A(t) = \frac{T_0(t) + T_s \frac{h}{k_d} s_d(t)}{1 + \frac{h}{k_d} s_d(t)}, \quad B(t) = \frac{h}{k_d} \frac{T_0(t) - T_s}{1 + \frac{h}{k_d} s_d(t)} \tag{286}$$

$$D(t)=\frac{\frac{K_m}{D_{ef}}C_a s_d(t)+M\,a\frac{\exp\left(b-\frac{c}{T_0(t)}\right)}{R_g\,T_0(t)}}{1+\frac{K_m}{D_{ef}}s_d(t)},\quad E(t)=\frac{K_m}{D_{ef}}\frac{M\,a\frac{\exp\left(b-\frac{c}{T_0(t)}\right)}{R_g\,T_0(t)}-C_a}{1+\frac{K_m}{D_{ef}}s_d(t)} \tag{287}$$

$$F(t)=\frac{T_0(t)s_f(t)-T_{if}s_d(t)}{s_f(t)-s_d(t)},\quad G(t)=\frac{T_{if}-T_0(t)}{s_f(t)-s_d(t)} \tag{288}$$

and we obtain the following system of two ordinary differential equations and one algebraic equation for $s_d(t)$, $s_f(t)$ and $T_0(t)$ given by:

$$1+\frac{h}{k_f}\frac{T_s}{T_{if}}s_f(t)+hs_d(t)\left(\frac{1}{k_d}-\frac{T_s}{T_{if}k_f}\right)-\frac{T_0(t)}{T_{if}}\left[1+\frac{h}{k_f}s_f(t)+hs_d(t)\left(\frac{1}{k_d}-\frac{1}{k_f}\right)\right]=$$
$$=\frac{L_sK_m}{T_{if}k_f}\frac{\left(1+\frac{h}{k_f}s_f(t)\right)\left(s_f(t)-s_d(t)\right)}{1+\frac{K_m}{D_{ef}}s_d(t)}\left[M\,a\frac{\exp\left(b-\frac{c}{T_0(t)}\right)}{R_gT_0(t)}-C_a\right] \tag{289}$$

$$\dot{s}_d(t)=\frac{k_fT_{if}}{m_sL_s}\frac{1+\frac{h}{k_f}\frac{T_s}{T_{if}}s_f(t)+hs_d(t)\left(\frac{1}{k_d}-\frac{T_s}{k_fT_{if}}\right)-\frac{T_0(t)}{T_{if}}\left[1+\frac{h}{k_f}s_f(t)+hs_d(t)\left(\frac{1}{k_d}-\frac{1}{k_f}\right)\right]}{\left(1+\frac{h}{k_d}s_f(t)\right)\left(s_f(t)-s_d(t)\right)} \tag{290}$$

$$\dot{s}_f(t)=\frac{k_fT_{if}}{m_fL_f}\frac{1-\frac{T_0(t)}{T_{if}}}{s_f(t)-s_d(t)} \tag{291}$$

$$s_f(0)=s_d(0)=0. \tag{292}$$

Remark 21.

There exist some approximate or explicit solutions for some other free boundary problems for the heat-diffusion equation, e.g.: model for a single nutrient uptake by a growing root system by using a moving boundary approach; explicit estimate for the asymptotic behavior of the solution of the porous media equation with absorption (reaction-diffusion processes of a gas inside a chemical reactor); penetration of solvents in polymers; filtration of water through oil in a porous medium; the Wen model for an isothermal mono-catalytic diffusion-reaction process of a gas with a solid. The solid is chemically attacked from its surface with a quick and irreversible reaction and, at the same time, a free boundary begins, etc.

4. Conclusion

We have given a review on explicit and approximated solutions for heat and mass transfer problems in which a free or moving interface is involved. We have also showed some new recent problems for heat and mass transfer in which a free or moving interface is also involved.

5. Acknowledgements

This paper was partially sponsored by the project PIP No. 0460 of CONICET - UA (Rosario, Argentina), and Grant FA9550-10-1-0023.

6. References

Alexandrov, D.V. & Malygin, A.P. (2006). Self-similar solidification of an alloy from a cooled boundary. *Int. J. Heat Mass Transfer*, 49, 763-769.

Alexiades, V. & Solomon, A. D. (1993). *Mathematical modelling of melting and freezing processes*, Hemisphere - Taylor & Francis, Washington.

Arderius, J.C.; Lara, M. & Tarzia, D.A. (1996). Experimental-numerical determination of thermal coefficients through a phase-change process. *Int. Comm. Heat Mass Transfer*, 23, 745-754.

Bancora, A.B. & Tarzia, D.A. (1985). On the Neumann solution for the two-phase Stefan problem including the density jump at the free boundary. *Latin Amer. J. Heat Mass Transfer*, 9, 215-222.

Bankoff, S.G. (1964). Heat conduction of diffusion with change of phase. *Advances in Chemical Engineering*, 5, 75-150.

Barenblatt, G.I. (1952). On some unsteady motions of a liquid or a gas in a porous medium. *Prink. Mat. Mekh*, 16, 67-78 (in Russian).

Bell, G.E. (1978). A refinement of the heat balance integral method applied to a melting problem. *Int. J. Heat Mass Transfer*, 21, 1357-1361.

Brillouin, M. (1930/31). Sur quelques problèmes non résolus de physique mathématique classique : Propagation de la fusion. *Annales de lInst. H. Poincaré*, 1, 285-308.

Briozzo, A.C. & Tarzia, D.A. (1998). The explicit solution of a free boundary problem for a nonlinear absorption model of mixed saturated-unsaturated flow. *Advances in Water Resources*, 21, 713-721.

Briozzo, A.C. ; Natale, M.F. & Tarzia, D.A. (1999). Determination of unknown thermal coefficients for Storm's type materials through a phase-change process. *International Journal of Non-Linear Mechanics*, 34, 329-340.

Briozzo, A.C.; Natale, M.F. & Tarzia, D.A. (2004). An explicit solution for a two-phase Stefan problem with a similarity exponential heat sources. *MAT - Serie A*, 8, 11-19.

Briozzo, A.C.; Natale, M.F. & Tarzia, D.A. (2007a). Explicit solutions for a two-phase unidimensional Lamé-Clapeyron-Stefan problem with source terms in both phases. *J. Math. Anal. Appl.*, 329, 145-162.

Briozzo, A.C.; Natale, M.F. & Tarzia, D.A. (2007b). Existence of an exact solution for a one-phase Stefan problem with nonlinear thermal coefficients from Tirskii's method. *Nonlinear Analysis*, 67, 1989-1998.

Briozzo, A.C.; Natale, M.F. & Tarzia, D.A. (2010). The Stefan problem with temperature-dependent thermal conductivity and a convective term with a convective condition at the fixed face. *Comm. Pure and Applied Analysis*, 9, 1209-1220.

Briozzo, A.C. & Tarzia, D.A. (2002). An explicit solution for the instantaneous two-phase Stefan problem with nonlinear thermal coefficients. *IMA J. Applied Math*, 67, 249-261.

Broadbridge, P. (1990). Solution of a nonlinear absorption model of mixed saturated-unsaturated flow. *Water Resour. Research*, 26, 2435-2443.

Broadbridge, P. & White, I. (1988). Constant rate rainfall infiltration: A versatile nonlinear model, 1, Analytic solutions. *Water Resources Research*, 24, 145-154.

Campañone, L.A., Salvadori, V.O. & Mascheroni, R.H. (2005a). Food freezing with simultaneous surface dehydration. Approximate prediction of freezing time, *Int. J. Heat Mass Transfer*, 48 (6), 1205-1213.

Campañone, L.A., Salvadori, V.O. & Mascheroni, R.H. (2005b). Food freezing with simultaneous surface dehydration. Approximate prediction of weight loss during freezing and storage, *Int. J. Heat Mass Transfer*, 48 (6), 1195-1204.

Cannon, J.R. (1963). Determination of an unknown coefficient in a parabolic differential equation. *Duke Math. J.*, 30, 313-323.

Cannon, J.R. (1964). Determination of certain parameters in heat conduction problems. *J. Math. Anal. Appl.*, 8, 188-201.

Cannon, J.R. (1984). *The one-dimensional heat equation*, Addison-Wesley, Menlo Park, California.

Charach, Ch. & Rubisnstein, I. (1992). Pressure-temperature effects in planar Stefan problems with density change, *J. Appl. Phys.*, 71, 1128-1137.

Carslaw, H.S. & Jaeger, J.C. (1959). *Conduction of heat in solids*, Clarendon Press, Oxford.

Castro, E.F., Gonzo, E.E. & Gottifredi, J.C. Gottifredi (1987). Thermodynamic of the absorption of hydrocarbon vapor in poliethylene films, *J. Mem. Sci.*, 31, 235-248.

Cho, S.H. (1975). An exact solution of the coupled phase change problem in a porous medium. *Int. J. Heat Mass Transfer*, 18, 1139-1142.

Cho, S.H. & Sunderland, J.E. (1974). Phase change problems with temperature-dependent thermal conductivity. *J. Heat Transfer*, 96C, 214-217.

Cirelli, M. B. & Tarzia, D.A. (2010), "Explicit solutions for the Rubinstein binary-alloy solidification problem with a heat flux or a convective condition at the fixed face", In preparation.

Crank, J. (1975). *The mathematics of diffusion*, Clarendon Press, Oxford.

Crank, J. (1984). *Free and moving boundary problems*, Clarendon Press, Oxford.

Crank, J. & Gupta, R.S. (1972). A moving boundary problem arising from the diffusion of oxygen in absorbing tissue.*J. Inst. Math. Appl.*, 10, 19-33.

Destefanis, H.A.; Erdmann, E.; Tarzia, D.A. & Villa, L.T. (1993). A free boundary model applied to the estimation of the diffusion coefficient in a gas-solid system. *International Communications Heat Mass Transfer*, 20, 103-110.

Duvaut, G. (1976). *Problèmes à frontière libre en théorie des milieux continus*, Rapport de Recherche No. 185, LABORIA IRIA, Rocquencourt.

Elliott, C.M. & Ockendon, J.R. (1982). *Weak and variational methods for moving boundary problems, Research Notes in Math., No. 59*, Pitman, London.

Fasano, A. (1974). Alcune osservazioni su una classe di problemi a contorno libero per l'equazione del calore, *Le Matematiche*, 29, 397-411.

Fasano, A. (1987). *Las zonas pastosas en el problema de Stefan*, Cuadern. Inst. Mat. Beppo Levi, 13, 1-60.

Fasano, A. (2005). *Mathematical models of some diffusive processes with free boundaries*, MAT - Serie A, 11, 1-128.

Fasano, A.; Guan, Y.; Primicerio, M. & Rubinstein, I. (1993). *Thawing in saturated porous media*, Meccanica, 28, 103-109.

Fasano, A. & Primicerio, M. (1984). Freezing in porous media: a review of mathematical model, In: *Applications of Mathematics in Technology*, V. Boffi & H. Neunzert (Eds.), 288-311, Teubner.

Fasano, A. & Primicerio, M. (1986). *Nonlinear diffusion problems, Lecture Notes in Mathematics No. 1224*, Springer Verlag, Berlin.

Fasano, A.; Primicerio, M. & Tarzia, D.A. (1999). Similarity solutions in Stefan-like problems. *Math Models and Methods in Appl Sci*, 9, 1-10.

Fennema, O, Berny, L.A. (1974). Equilibrium vapour pressure and water activity of food at subfreezing temperature. In *Proceedings of IV International Congress of Food Science and Technology*, 2, 27-35.

Fi, F. & Han, X. (2007). An one -dimensional two-phase free boundary problem in an angular domain. *Nonlinear Anal. RWA*, 8, 959-979

Friedman, A. (1964). *Partial differential equations of parabolic type*, Prentice Hall, Englewood Cliffs, N. J.

Froment, G.F. & Bischoff, K.B. (1979). *Chemical reactor analysis and design*, J. Wiley, New York.

Garguichevich, G.G. & Sanziel, M.C. (1984). Una introducción general a la resolución aproximada del problema de Stefan unidimensional. *Cuadern. Inst. Mat Beppo Levi*, 11, 167-177.

Garguichevich, G.G.; Sanziel, M.C. & Tarzia, D.A. (1985). Comparison of approximate methods for the determination of thermal coefficients through a phase-change problems. *Int. Comm. Heat Mass Transfer*, 12, 451-464.

González, A.M.; Reginato, J.C. & Tarzia, D.A. (2008). A free-boundary model for anaerobiosis in saturated soil aggregates. *Soil Science*, 173, 758-767.

González, A.M. & Tarzia, D.A. (1996). Determination of unknown coefficient of a semi-infinite material through a simple mushy zone model for the two-phase Stefan problem. *International Journal of Engineering Science*, 34, 799-817.

Goodman, T.R. (1958). The heat balance integral and its application to problems involving a change of phase. *Trans. of the ASME*, 80, 335-342.

Gupta, L.N. (1974). *An approximate solution to the generalized Stefan's problem in a porous medium*, Int. J. Heat Mass Transfer, 17, 313–321.

Gupta, S.C. (2003). *The classical Stefan problem. Basic concepts, modelling and analysis*, Elsevier, Amsterdam.

Gupta, S.C.; Sanziel, M.C. & Tarzia, D.A. (1997). A similarity solution for the binary alloy solidification problem with a simple mushy zone model. *Anales de la Academia Nacional de Ciencias Exactas, Físicas y Naturales*, 49, 75-82.

Guzman, J.A. (1982). Application of the heat transfer solution for the case of one-dimensional solidification with negligible latent heat. *Lat. Amer. J. Heat Mass Transfer*, 6, 41-51.

Hill, J.M. (1987). *One-dimensional Stefan Problems : an introduction*, Longman, Essex .

Howison, S.D. (1988). Similarity solutions to the Stefan problem and the binary alloy problem. *IMA J. Appl. Math.*, 40, 147-161.

Jones, Jr. B. F. (1962). The determination of a coefficient in a parabolic differential equation, Part I : Existence and uniqueness. *J. Math. Mech.*, 11, 907-918.

Jones, Jr. B. F. (1963). Various methods for finding unknown coefficients in parabolic differential equation. *Comm. Pure Appl. Math.*, 16, 33-44.

Knight, J. & Philip, J.R. (1974). Exact solutions in nonlinear diffusion. *J. Eng. Math*, 8, 219-227.

Lamé, G. & Clapeyron, B.P. (1831). Memoire sur la solidification par refroidissement dun globe liquide. *Annales Chimie Physique*, 47, 250-256.

Levenspiel, O. (1962). *Chemical reaction engineering*, J. Wiley, New York.

Liapis, A. I.; Lipscomb, G.G.; Crosser, O.K. & Liapis-Tsiroyianni, E. (1982). A model of oxygen diffusion in absorbing tissue. *Mathematical Modelling*, 3, 83-92.

Lin, S. (1981). An exact solution of the sublimation problem in a porous medium. *J. Heat Transfer*, 103, 165-168.

Lin, S. (1982a). An exact solution of the sublimation problem in a porous medium, Part II With an unknow temperature and vapor concentration at the moving sublimation front. *J. Heat Transfer*, 104, 808-811.

Lin, S. (1982b). An exact solution of the desublimation problem in a porous medium. *Int. J. Heat Mass Transfer*, 25, 625-630.

Lombardi, A. & Tarzia, D.A. (2001). Similarity solutions for thawing processes with a heat flux on the fixed boundary. *Meccanica*, 36, 251-264.

Luikov, A.V. (1964), *Heat and mass transfer in capillary-porous bodies*, Adv. HeatTransfer, 1, 123–184.

Luikov, A.V. (1966). *Heat and mass transfer in capillary-porous bodies*, Pergamon Press, Oxford.

Luikov, A. V. (1968). *Analytical heat diffusion theory*, Academic Press, New York.

Luikov, A.V. (1975). *Systems of differential equations of heat and mass transfer in capillary porous bodies*, Int. J. Heat Mass Transfer, 18, 1–14.

Luikov, A.V. (1978). *Heat and mass transfer*, MIR Publishers, Moscow.

Lunardini, V.J. (1981). *Heat transfer in cold climates*, Van Nostrand, New York.

Lunardini, V.J. (1991). *Heat transfer with freezing and thawing*, Elsevier, Amsterdam.

Menaldi, J.L. & Tarzia, D.A. (1993). Generalized Lamé-Clapeyron solution for a one-phase source Stefan problem. *Comput. Appl. Math.*, 12, 123-142.

Mikhailov, M.D. (1975). Exact solution of temperature and mosture distributions in a porous half-space with moving evaporation front. *Int. J. Heat Mass Transfer*, 18, 797-804.

Mikhailov, M.D. (1976). Exact solution for freezing of humid porous half-space. *Int. J. Heat Mass Transfer*, 19, 651-655.

Muehlbauer, J.C. & Sunderland, J.E. (1965). Heat conduction with freezing or melting. *Appl. Mech. Reviews*, 18, 951-959.

Nakano, Y. (1990). Quasi-steady problems in freezing soils, *Cold Reg. Sci. Techn.*, 17, 207-226.

Natale, M.F.; Santillan Marcus, E.A. & Tarzia, D.A. (2010). Explicit solutions for one dimensional two-phase free boundary problems with either shrinkage or expansion. *Nonlinear Analysis Series B: Real World Applications*, 11, 1946-1952.

Natale, M.F. & Tarzia, D.A. (2000). Explicit solutions to the two-phase Stefan problem for Storm's type materials. *Journal of Physics A: Mathematical and General*, 33, 395-404.

Natale, M.F. & Tarzia, D.A. (2006). Explicit solutions for a one-phase Stefan problem with temperature-dependent thermal conductivity. *Bollettino dell'Unione Matematica Italiana*, 9B, 79-99.

Olguin, M.C.; Salvadori, V.O.; Mascheroni, R.H. & Tarzia, D.A. (2008). An analytical solution for the coupled heat and mass transfer during the freezing of high-water content materials. *International Journal of Heat and Mass Transfer*, 51, 4379-4391.

O'Neill, K. & Miller, R.D. (1985). Explanation of a rigid ice model of frost heave, *Water Resour. Res.*, 21, 281-296.

Pattle, R.E. (1959). Diffusion from an instantaneous point source with concentration dependent coefficient. *Quart. J. Mech. Appl. Math*, 12, 407-409.

Pekeris, C.L. & Slichter, L.B. (1939). Problem of ice formation. *J. Appl. Physics*, 10, 135-137.

Philip, J.R. (1957). The theory of infiltration, 4, Sorptivity and algebraic infiltration equations. *Soil Sci.*, 84, 257-264.

Philip, J.R. (1958). The theory of infiltration, 6, Effect of water depth over soil. *Soil Sci.*, 85, 278-286.

Primicerio, M. (1981). Problemi di diffusione a frontiera libera. *Bollettino Un. Mat. Italiana*, 18A, 11-68.

Primicerio, M. & Gianni, R. (1989). La filtración en medios porosos. *Cuadern. Inst. Mat. Beppo Levi*, 18 1-92.

Quilghini, D. (1967). Sul problema inverso di quello di Stefan. *Riv. Mat. Univ. Parma*, 8, 131-142.

Reginato, J.C.; Palumbo, M.C.; Moreno, I.S.; Bernardo, I.Ch. & Tarzia, D.A. (2000). Modeling nutrient uptake using a moving boundary approach. Comparison with the Barber-Cushman model. *Soil Science Society of America Journal*, 64, 1363-1367.

Reginato, J.C. & Tarzia, D.A. (1993). The balance integral method applied to root growth of crops. *International Journal of Engineering Science*, 31, 61-70.

Reginato, J.C.; Tarzia, D.A. & Dzioba, M.A. (1993). A model for root growth with root competition. *Plant and Soil*, 15, 185-196.

Rogers, C. & Broadbridge, P. (1988). On a nonlinear moving boundary problem with heterogeneity: application of a reciprocal transformation. *ZAMP*, 39, 122-128.

Rogers, C. (1986). On a class of moving boundary problems in non-linear heat conduction: Application of a Backlund transformation. *Int. J. Nonlinear Mech.*, 21, 249-256.

Rubinstein, L.I. (1971). *The Stefan problem*, Translations of Mathematical Monographs, 27, Amer. Math. Soc., Providence, R.I.

Salva, N.N. & Tarzia, D.A. (2010). Simultaneous determination of unknown coefficients through a phase-change process with temperature-dependent thermal conductivity, Submitted.

Santillan Marcus, E.A.; Natale, M.F. & Tarzia, D.A. (2008). Simultaneous determination of two unknown thermal coefficients of a semi-infinite porous material through a desublimation moving boundary problem with coupled heat and moisture flows. *JP Journal of Heat and Mass Transfer*, 2, 73-116

Santillan Marcus, E.A. & Tarzia, D.A. (2000a). Soluciones explícitas para el problema de desublimación en un semi-espacio húmedo poroso con una condición de flujo de calor. *Información Tecnológica*, 11, 19-24.

Santillan Marcus, E.A. & Tarzia, D.A. (2000b). Explicit solution for freezing of humid porous half-space with heat flux condition. *International Journal of Engineering Science*, 38, 1651-1665.

Santillan Marcus, E.A. & Tarzia, D.A. (2003). Exact solutions for drying with coupled phase change in a porous medium with a heat flux condition on the surface. *Computational and Applied Mathematics*, 22, 293-311.

Santillan Marcus, E.A. & Tarzia, D.A. (2007). Determination of one unknown thermal coefficient of a semi-infinite porous material through a desublimation problem with coupled heat and moisture flows. *JP Journal of Heat and Mass Transfer*, 1, 251-270.

Sanziel, M.C. & Tarzia, D.A. (1989). Necessary and sufficient condition to obtain n phases in a one-dimensional medium with a flux condition on the fixed face. *Math. Notae*, 33, 25-32.

Schatz, A. (1969). Free boundary problems of Stephan type with prescribed flux. *J. Math. Anal. Appl.* 28, 569-580.

Sherman, B. (1971). Limiting behaviour in some Stefan problems as the latent goes to zero. *SIAM J. Appl. Math.*, 20, 319-327.

Solomon, A.D. (1979). An easily computable solution to a two-phase Stefan problem. *Solar Energy*, 23, 525-528.

Solomon, A.D.; Wilson, D.G. & Alexiades, V. (1982). A mushy zone model with an exact solution. *Letters Heat Mass Transfer*, 9, 319-324.

Solomon, A.D.; Wilson, D.G. & Alexiades, V. (1983). Explicit solutions to change problems. *Quart. Appl. Math.*, 41, 237-243.

Stampella, M.B. & Tarzia, D.A. (1989). Determination of one or two unknown thermal coefficients of a semi-infinite material through a two-phase Stefan problem. *Int. J. Eng. Science*, 27, 1407-1419.

Stefan, J. (1889a). Ueber einige Probleme der Theorie der Waermeleitung. *Zitzungsberichte der Kaiserlichen Akademie der Wissenschaften, Mathematisch-naturwissenschaftliche classe*, 98, 473-484.

Stefan, J. (1889b). Ueber die Theorie der Eisbildung, insbesondere ueber die Eisbildung im Polarmeere. *Zitzungsberichte der Kaiserlichen Akademie der Wissenschaften, Mathematisch-naturwissenschaftliche Classe*, 98, 965-983.

Stefan, J. (1890). Ueber die Theorie der Eisbildung. *Monatshefte fur Mathematik und Physik*, 1, 1-6.

Szekely, J.; Evans, J.W. & Sohn, H.Y. (1976). *Gas-solid reactions*, Academic Press, New York.

Talamucci, F. (1997). Analysis of the coupled heat-mass transport in freezing porous media, *Survey in Math. in Industry*, 7, 93-139.

Talamucci, F. (1998). Freezing processes in saturated soils, *Math. Models Meth. Appl. Sci.*, 8, 107-138.

Tarzia, D.A. (1981). An inequality for the coefficient ↦ of the free boundary $s(t) = 2\sigma\sqrt{t}$ of the Neumann solution for the two-phase Stefan problem. *Quart. Appl. Math.*, 39, 491-497.

Tarzia, D.A. (1982). Determination of the unknown coefficients in the Lame´-Clapeyron problem (or one-phase Stefan problem). *Adv. Appl. Math.*, 3, 74-82.

Tarzia, D.A. (1983). Simultaneous determination of two unknown thermal coefficients through an inverse one-phase Lamé-Clapeyron (Stefan) problem with an overspecified condition on the fixed face. *Int. J. Heat Mass Transfer*, 26, 1151-1158.

Tarzia, D.A. (1984a). A new variant for the simultaneous calculation of some thermal coefficients of a semi-infinite material through a phase-change problem with an over-condition on the fixed face. *Latin Amer. J. Heat Mass Transfer*, 8, 227-235.

Tarzia, D.A. (1984b). Soluciones exactas del problema de Stefan unidimensional. *Cuadern. Inst. Mat. Beppo Levi*, 12, 5-36.

Tarzia, D.A. (1987). Determination of unknown thermal coefficients of a semi-infinite material for the one-phase Lamé-Clapeyron (Stefan) problem through the Solomon-Wilson-Alexiades mushy zone model. *Int. Comm. Heat. Mass Transfer*, 14, 219-228.

Tarzia, D.A. (1990a). A variant of the heat balance integral method and a new proof of the exponentially fast asymptotic behaviour of the solutions in heat conduction problems with absorption. *Int. J. Engineering Science*, 28, 1253-1259.

Tarzia, D.A. (1990b). Neumann-like solution for the two-phase Stefan problem with a simple mushy zone model. *Computational and Applied Mathematics (ex Matemática Aplicada Computacional)*, 9, 201-211.

Tarzia, D.A. (1991a). Sobre una nueva variante para el cálculo simultáneo de coeficientes térmicos de un material semi-infinito a través de un problema directo o inverso de Stefan a dos fases. *Math. Notae*, 35, 25-41.

Tarzia, D.A. (1991b). Six free boundary problems for the heat-diffusion equation. *Revista Un. Mat. Argentina*, 37, 294-309.

Tarzia, D.A. (1993). Approximate and analytic methods to solve some parabolic free boundary problems, In *Free Boundary Problems Involving Solids (FBP'90)*, J.M. Chadam - H. Rasmussen (Eds.), Pitman Research Notes in Mathematics, Vol. 281, pp. 190-196, Longman, Essex.

Tarzia, D.A. (1998). The determination of unknown thermal coefficients through phase change process with temperature-dependent thermal conductivity. *International Communications in Heat and Mass Transfer*, 25, 139-147.

Tarzia, D.A. (2000a). A bibliography on moving-free boundary problems for the heat-diffusion equation. The Stefan and related problems, *MAT - Serie A*, 2, 1-297. See: http://web.austral.edu.ar/descargas/facultad-cienciasEmpresariales/mat/Tarzia-MAT-SerieA-2(2000).pdf

Tarzia, D.A. (2000b). Explicit solution to some free boundary problems for the heat equation, In *International Conference on Free Boundary Problems (FBP'99)*, N. Kenmochi (Ed.), GAKUTO International Series, Mathematical Sciences and Applications, Vol. 13, pp. 385-398, Gottotosho, Tokyo.

Tarzia, D.A. (2004). An explicit solution for a two-phase unidimensional Stefan problem with a convective boundary condition at the fixed face. *MAT – Serie A*, 8, 21-27.

Tarzia, D.A. (2007). Exact solution for a Stefan problem with convective boundary condition and density jump. *PAMM – Proceedings on Applied Mathematics and Mechanics*, 7, 1040307-1040308.

Tarzia, D.A. & Turner, C.V. (1992). A note on the existence of a waiting time for a two-phase Stefan problem. *Quart. Appl. Math.*, 50, 1-10.

Tarzia, D.A. & Turner, C.V. (1999). Estimation of the occurance of the phase-change process in spherical coordinates. *International Communications in Heat and Mass Transfer*, 26, 559-568.

Tarzia, D.A. & Villa, L.T. (1991). Mathematical considerations on the heat transfer with phase change with negligible latent heat. *Latin American Applied Research*, 21, 1-6.

Tayler, A.B. (1986). *Mathematical models in applied mechanics*, Clarendon Press, Oxford .

Tien, R.H. & Geiger, G.E. (1967). A heat-transfer analysis of the solidification of a binary eutectic system. *J. Heat Transfer*, 89C, 230-234.

Tien, R.H. & Koump, V. (1970). Effect of density change on the solidification of alloys. *J. Heat Transfer*, 92C, 11-16.

Tirskii, G.A. (1959). Two exact solutions of Stefans nonlinear problems. *Soviet Physics Dokl.*, 4, 288-292.

Tritscher, P. & Broadbridge, P. (1994). A similarity solution of a multiphase Stefan problem incorporating general non-linear heat conduction. *Int. J. Heat Mass Transfer*, 37, 2113-2121.

Tsubaki, T. & Boley, B.A. (1977). One-dimensional solidification of binary mixtures. *Mech. Res. Comm.*, 4, 115-122.

Vázquez, J.L. (2007). *The porous medium equation. Mathematical theory*, Clarendon Press, Oxford.

Villa, L.T. (1987). La ecuación de la difusión y su aplicación a problemas de frontera libre. *Cuadern. Inst. Mat. Beppo Levi*, 14, 99-116.

Voller, V.R. (2006). A similarity solution for solidification of an under-cooled binary alloy. *Int. J. Heat Mass Transfer*, 49, 1981-1985.

Voller, V.R. (2008a). A numerical method for the Rubinstein binary- alloy problem in the presence of an under-cooled liquid. *Int. J. Heat Mass Transfer*, 51, 696-706.

Voller, V.R. (2008b). An enthalpy method for modeling dendritic growth in a binary alloy. *Int. J. Heat Mass Transfer*, 51, 823-834.

Warrick, A.W. & Broadbridge, P. (1992). Sorptivity and macroscopic capillary length relationships. *Water. Resour. Researh*, 28, 427-431.

Weber, H. (1901). *Die partiellen Differential-Gleinchungen der Mathematischen Physik, nach Riemann's Vorlesungen, t. II*, Braunwschweig.

Weiner, J.H. (1955). Transient heat conduction in multi-phase media. *British J. Appl. Physics*, 6, 361-363.

White, R.E. (1985). The binary alloy problem: existence, uniqueness, and numerical approximation. *SIAM J. Numer. Anal.*, 22, 205-244.

Wilson, D.G. (1978). Existence and uniqueness for similarity solutions of one dimensional multi-phase Stefan problems. *SIAM J. Appl. Math.*, 35, 135-147.

Wilson, D.G. (1982). Lagrangian coordinates for moving boundary problems. *SIAM J. Appl. Math*, 42, 1195-1201.

Wilson, D.G. & Solomon, A.D. (1986). A Stefan-type problem with void formation and its explicit solution. *IMA J. Appl. Math.*, 37, 67-76.

Wilson, D.G. & Solomon, A.D & Alexiades, V. (1982). A shortcoming of the explicit solution for the binary alloy solidification problem. *Letters Heat Mass Transfer*, 9, 421-428.

Zel'dovich, Y.B. & Kompaneets, A.S. (1950). *Towards a theory of heat conduction with thermal conductivity depending on the temperature.* In collection of papers dedicated to 70th anniversary of A.F. Ioffe, Izd. Akad. Nauk. SSSR, Moscow, 61-72 (in Russian).

21

Boundary Control Problems for Oberbeck–Boussinesq Model of Heat and Mass Transfer

Gennady Alekseev[1], Dmitry Tereshko[1] and Vladislav Pukhnachev[2]
[1]*Institute of Applied Mathematics of Far Eastern Branch RAS*
[2]*Lavrentyev Institute of Hydrodynamics of Siberian Branch RAS*
Russia

1. Introduction

Much attention has recently been given to statement and investigation of new problems for the models of heat and mass transfer. The control problems for Navier-Stokes and Oberbeck–Boussinesq equations are examples of such kind problems. This interest to control problems is connected with a variety of technical applications in science and engineering such as the crystal growth process, the aerodynamic drag reduction, the suppression of a turbulence and mass flow separation. In control problems the unknown densities of boundary or distributed sources, the coefficients of model differential equations or boundary conditions are recovered from minimum of certain cost functionals depending on controls and state (the solution to the original boundary value problem). The number of papers is devoted to study of control problems for models of heat and mass transfer. We mention in particular papers (Gunzburger et al., 1991; 1993; Ito & Ravindran, 1998; Alekseev, 1998a;b; Alekseev & Tereshko, 1998a;b; Capatina & Stavre, 1998; Lee & Imanuvilov, 2000a;b; Lee, 2003) devoted to theoretical study of control problems for stationary Oberbeck–Boussinesq equations.

Along with control problems, the inverse problems for models of heat and mass transfer play an important role. Importantly, these inverse problems can be reduced to corresponding control problems by choosing a suitable minimized cost functional that adequately describes the inverse problem in question (Alekseev, 2000; Alekseev & Adomavichus, 2001; Alekseev, 2001; 2002; 2006; 2007a;b; Alekseev et al., 2008; Alekseev & Soboleva, 2009; Alekseev & Tereshko, 2010a). As a result, both control and inverse problems can be analyzed by applying a unified approach based on the constrained optimization theory in Hilbert or Banach spaces. These theoretical results, the increasing power of computers and the development of numerical methods for the flow simulation itself motivate the numerical study of the optimal flow control problems under consideration.

Oberbeck–Boussinesq equations are the most often used model of convection. Rigorous derivation of these equations from full model of viscous compressible heat conducting fluid can be found in (Gershuni & Zhukhovitskii, 1976; Joseph, 1976). Limits of applicability of this model are indicated in (Pukhnachev, 1992) for heat convection description and in (Perera & Sekerka, 1997) for concentration convection. In these papers new models of convection with non-solenoidal velocity field were proposed. In (Pukhnachev, 2004) the hierarchy of thermal gravitational convection models in closed domains, including the traditional Oberbeck–Boussinesq equations, was constructed using the asymptotic expansions of the original equations with respect to the parameters of weak compressibility and microconvection. This theory was further developed in (Andreev et al., 2008). It is shown there that the non-solenoidal effect results in a minor correction to the velocity field in the case of steady flows. At the same time, there are significant differences in predictions of classical and new models for dramatically unsteady flows, caused by large gradients of the initial temperature field or by long-term time-periodic changes in the boundary temperature regime. These differences are most noticeable in the case when the microconvection parameter $\eta = gl^3/(\nu\lambda_1)$ has the order one. Here g is the acceleration of gravity, l is the characteristic linear scale, λ_1 is the thermal diffusivity.

Let Ω be a bounded domain in the space $\mathbb{R}^d$ $(d = 2,3)$ with a Lipschitz boundary Γ consisting of two parts Γ_D and Γ_N. In this chapter we consider the model of heat and mass transfer in a viscous incompressible heat conducting fluid. The model consists of the Navier-Stokes equations and the convection-diffusion equations for the substance concentration and the temperature that are nonlinearly related via buoyancy in the Oberbeck–Boussinesq approximation and via convective mass and heat transfer. It is described by the equations

$$-\nu\Delta\mathbf{u} + (\mathbf{u}\cdot\nabla)\mathbf{u} + \nabla p = \mathbf{f} + (\beta_C C - \beta_T T)\mathbf{G}, \ \operatorname{div}\mathbf{u} = 0 \text{ in } \Omega, \ \mathbf{u} = \mathbf{g} \text{ on } \Gamma, \tag{1}$$

$$-\lambda\Delta C + \mathbf{u}\cdot\nabla C + kC = f \text{ in } \Omega, \ C = \psi \text{ on } \Gamma_D, \ \lambda\partial C/\partial n = \chi \text{ on } \Gamma_N, \tag{2}$$

$$-\lambda_1\Delta T + \mathbf{u}\cdot\nabla T = f_1 \text{ in } \Omega, \ T = \varphi \text{ on } \Gamma_D, \ \lambda_1(\partial T/\partial n + \alpha T) = \eta \text{ on } \Gamma_N. \tag{3}$$

Here $\mathbf{u}$, T and C denote the velocity, temperature and substance concentration fields respectively, $p = P/\rho$ where P is a pressure and $\rho =$ const is the density of medium, ν is the kinematic viscosity coefficient, $\mathbf{f}$ is the mass density of body forces, $\mathbf{G}$ is the gravitational acceleration vector, k is the coefficient of decomposition of the substance due to chemical reactions, f and f_1 are volume densities of the substance and heat sources, β_C is the solutal expansion coefficient, β_T is the volumetric thermal expansion coefficient, λ and λ_1 are the diffusivity and thermal diffusivity coefficients, $\mathbf{g}$ is a given vector-function on Γ, ψ and φ are given functions on a part Γ_D of Γ, χ and α, η are functions given on another part $\Gamma_N = \Gamma \setminus \Gamma_D$ of Γ, $\mathbf{n}$ is the unit outer normal. We note that all the quantities in (1)–(3) are dimensional and their dimensions are defined in terms of SI units.

Hereinafter it is assumed that the parameters $\nu, \lambda, \lambda_1, \beta_C, \beta_T, k$ are constants. This assumption is physically justified if the temperature T varies in a small range (Batchelor, 2000). The temperature change has the greatest influence on the viscosity. For example the viscosity of water decreases by more than twice during heating from 1°C to 99°C. As to the solvability of the boundary value problems for thermal convection model with temperature-dependent transport coefficients see (Goncharova, 2002) and references therein. Group-theoretical properties of the equations of this model and their exact solutions are considered in (Andreev et al., 1998). We note the anomalous property of water concerning with the nonmonotonic dependence of density on temperature near $T = 4$°C. The dependence of parameters in equations (1)–(3) on the substance concentration C is less significant. It should be noted also that the applicability of our concentration convection model is limited to small values of concentration C.

In the special case where $\beta_T = 0$ and $k = 0$ problem (1)–(3) splits into two: problem (1), (2) for the mass transfer model in the Oberbeck–Boussinesq approximation (see (Joseph, 1976)), and the linear boundary value problem (3) (for given $\mathbf{u}$) for temperature T, which describes

the convection-diffusion of a heat. We shall refer to problem (1), (2) at $\beta_T = 0$ as Model 1. In another special case where $\beta_C = 0$ problem (1)–(3) splits into problem (1), (3) for a heat transfer model in the Oberbeck–Boussinesq approximation (see (Joseph, 1976)), and the linear boundary value problem (2) for substance concentration C. We shall refer to problem (1), (3) at $\beta_C = 0$ as Model 2. Finally problem (1)–(3) itself is referred to as Model 3.
Our goal is the study of the boundary control problems for the models under consideration. The problems consist in minimization of certain cost functionals depending on the state and controls. In order to formulate a boundary control problem for Model 1 we divide the set of all data of Model 1 into two groups: the group of controls containing the functions $\mathbf{g} \in \mathbf{H}^{1/2}(\Gamma)$ and $\chi \in L^2(\Gamma_N)$, which play the role of controls, and the group of fixed data comprising the invariable functions $\mathbf{f}, f, k, \beta_C$ and ψ. We assume that the controls $\mathbf{g}$ and χ vary in some closed convex sets $K_1 \subset \mathbf{H}^{1/2}(\Gamma)$ and $K_2 \subset L^2(\Gamma_N)$. The mathematical statement of the boundary control problem for Model 1 is as follows: to find a pair $(\mathbf{x}, u)$, where $u = (\mathbf{g}, \chi) \in K = K_1 \times K_2$, $\mathbf{x} = (\mathbf{u}, p, C) \in X = \mathbf{H}^1(\Omega) \times L_0^2(\Omega) \times H^1(\Omega)$ such that $F(\mathbf{x}, u) \equiv 0$ and

$$J(\mathbf{x}, \mathbf{g}, \chi) \equiv \frac{\mu_0}{2} I(\mathbf{x}) + \frac{\mu_1}{2} \|\mathbf{g}\|^2_{\mathbf{H}^{1/2}(\Gamma)} + \frac{\mu_2}{2} \|\chi\|^2_{L^2(\Gamma_N)} \to \inf.$$

Here $F(\mathbf{x}, u) = 0$ is the operator constraint in the form of a weak formulation of Model 1; μ_0, μ_1, μ_2 are nonnegative constants, $I(\mathbf{x})$ is a cost functional.
Similar boundary control problems can be formulated for Models 2 and 3. We divide the set of all data of Model 2 into two groups: the group of controls containing the functions $\mathbf{g} \in \mathbf{H}^{1/2}(\Gamma)$ and $\eta \in L^2(\Gamma_N)$ and the group of fixed data comprising the invariable functions $\mathbf{f}, f_1, \beta_T, \alpha$ and φ. It is assumed that the controls $\mathbf{g}$ and η vary in some closed convex sets $K_1 \subset \mathbf{H}^{1/2}(\Gamma)$ and $K_3 \subset L^2(\Gamma_N)$. A boundary control problem for Model 2 consists of finding a pair $(\mathbf{x}, u)$, where $u = (\mathbf{g}, \eta) \in K = K_1 \times K_3$, $\mathbf{x} = (\mathbf{u}, p, T) \in X = \mathbf{H}^1(\Omega) \times L_0^2(\Omega) \times H^1(\Omega)$ such that $G(\mathbf{x}, u) = 0$ and

$$J(\mathbf{x}, \mathbf{g}, \eta) \equiv \frac{\mu_0}{2} I(\mathbf{x}) + \frac{\mu_1}{2} \|\mathbf{g}\|^2_{\mathbf{H}^{1/2}(\Gamma)} + \frac{\mu_3}{2} \|\eta\|^2_{L^2(\Gamma_N)} \to \inf. \quad (4)$$

Here $G(\mathbf{x}, u) = 0$ is the operator constraint in the form of a weak formulation of Model 2, μ_0, μ_1 and μ_3 are nonnegative constants. Finally, a boundary control problem for Model 3 consists of finding a pair $(\mathbf{x}, u)$, where $u = (\mathbf{g}, \chi, \eta) \in K = K_1 \times K_2 \times K_3$, $\mathbf{x} = (\mathbf{u}, p, C, T) \in X = \mathbf{H}^1(\Omega) \times L_0^2(\Omega) \times H^1(\Omega) \times H^1(\Omega)$ such that $H(\mathbf{x}, u) = 0$ and

$$J(\mathbf{x}, \mathbf{g}, \chi, \eta) \equiv \frac{\mu_0}{2} I(\mathbf{x}) + \frac{\mu_1}{2} \|\mathbf{g}\|^2_{\mathbf{H}^{1/2}(\Gamma)} + \frac{\mu_2}{2} \|\chi\|^2_{L^2(\Gamma_N)} + \frac{\mu_3}{2} \|\eta\|^2_{L^2(\Gamma_N)} \to \inf.$$

Here $H(\mathbf{x}, u) = 0$ is the operator constraint in the form of a weak formulation of Model 3.
The work consists of two parts. In the first part the solvability theorems for boundary control problems under study are formulated and proved. Optimality systems describing the first-order necessary optimality conditions are derived and analyzed. Sufficient conditions to the data ensuring the local uniqueness and stability of optimal solutions for concrete tracking-type functionals are established. The local uniqueness conditions are rather cumbersome. To simplify them, we introduce analogues of dimensionless parameters widely used in fluid dynamics, namely, the Reynolds number and the diffusion or temperature Rayleigh and Prandtl numbers. In terms of these parameters, the uniqueness conditions can be written in a relatively simple form and are similar to those for the coefficient inverse problems for the stationary linear convection-diffusion-reaction equation (see e.g. (Alekseev & Tereshko, 2008)).

In the second part a numerical algorithm based on Newton's method for the optimality system and finite element method for linearized boundary value problems is formulated and analyzed. Some computational results connected with the vortex reduction in the steady 2D viscous fluid flow around a cylinder in a channel by means of the temperature and hydrodynamic controls on some parts of the boundary are given and discussed. The details of theoretical and numerical studies can be found in (Alekseev & Tereshko, 2008).

2. Statement of boundary control problem for mass transfer model

We begin our study with consideration of Model 1 having in denotions of Sect. 1 the form

$$-\nu\Delta\mathbf{u}+(\mathbf{u}\cdot\nabla)\mathbf{u}+\nabla p=\mathbf{f}+\beta_C C\mathbf{G},\ \mathrm{div}\,\mathbf{u}=0 \text{ in } \Omega,\ \mathbf{u}=\mathbf{g} \text{ on } \Gamma, \tag{5}$$

$$-\lambda\Delta C+\mathbf{u}\cdot\nabla C+kC=f \text{ in } \Omega,\ C=\psi \text{ on } \Gamma_D,\ \lambda\partial C/\partial n=\chi \text{ on } \Gamma_N. \tag{6}$$

Under theoretical study of control problems for Model 1 we shall use the Sobolev spaces $H^s(D)$ with $s\in\mathbb{R}$ and the spaces $L^r(D)$ with $r\geq 2$ where D denotes Ω, its subset Q, Γ or its part Γ_0 with positive measure. The corresponding spaces of vector functions are denoted by $\mathbf{H}^s(D)$ and $\mathbf{L}^r(D)$. The norms and inner products in $H^s(Q)$, $H^s(\Gamma)$ and in their vector analogies are denoted by $\|\cdot\|_{s,Q}$, $\|\cdot\|_{s,\Gamma}$ and $(\cdot,\cdot)_{s,Q}$, $(\cdot,\cdot)_{s,\Gamma}$. The inner products and norms in $L^2(Q)$ or in $\mathbf{L}^2(Q)$ are denoted by $(\cdot,\cdot)_Q$ and $\|\cdot\|_Q$. If $Q=\Omega$ then we set $\|\cdot\|_\Omega=\|\cdot\|$, $(\cdot,\cdot)_\Omega=(\cdot,\cdot)$. The inner product and norm in $L^2(\Gamma_N)$ are denoted by $(\cdot,\cdot)_{\Gamma_N}$ and $\|\cdot\|_{\Gamma_N}$. The norm and seminorm in $H^1(\Omega)$ or in $\mathbf{H}^1(\Omega)$ are denoted by $\|\cdot\|_1$ and $|\cdot|_1$. The duality relation for the pair of dual spaces X and X^* is denoted by $\langle\cdot,\cdot\rangle_{X^*\times X}$ or simply $\langle\cdot,\cdot\rangle$. Let the following assumptions hold:

(i) Ω is a bounded domain in $\mathbb{R}^d$, $d=2,3$ with a boundary $\Gamma\in C^{0,1}$ consisting of N connected components $\Gamma_i, i=1,2,...,N$. The open segments Γ_D and Γ_N of Γ obey the conditions $\Gamma_D\in C^{0,1}$, $\Gamma_N\in C^{0,1}$, $\Gamma_D\neq\emptyset$, $\Gamma_D\cap\Gamma_N=\emptyset$, $\Gamma=\overline{\Gamma}_D\cup\overline{\Gamma}_N$.

Let $\mathcal{D}(\Omega)$ be the space of infinitely differentiable finite in Ω functions, $H_0^1(\Omega)$ be a closure of $\mathcal{D}(\Omega)$ in $H^1(\Omega)$, $\mathbf{H}_0^1(\Omega)=H_0^1(\Omega)^d$, $\mathbf{V}=\{\mathbf{v}\in\mathbf{H}_0^1(\Omega):\mathrm{div}\,\mathbf{v}=0\}$, $\mathbf{H}^{-1}(\Omega)=\mathbf{H}_0^1(\Omega)^*$, $L_0^2(\Omega)=\{p\in L^2(\Omega):(p,1)=0\}$, $\mathcal{T}=H^1(\Omega,\Gamma_D)\equiv\{S\in H^1(\Omega):S|_{\Gamma_D}=0\}$, $L_+^2(D)=\{u\in L^2(D):u\geq 0\}$. We shall use the following inequalities which are implied by the embedding theorems and the continuity of the trace operator:

$$\|\mathbf{u}\|_Q\leq c_Q\|\mathbf{u}\|_1,\ \ \|\mathrm{rot}\,\mathbf{u}\|\leq c_r\|\mathbf{u}\|_1,\ \ \|\mathbf{u}\|_{1/2,\Gamma}\leq c_\Gamma\|\mathbf{u}\|_1\ \ \forall\mathbf{u}\in\mathbf{H}^1(\Omega). \tag{7}$$

Here c_Q, c_r, c_Γ are constants depending on Ω.

Together with $\mathbf{H}^1(\Omega)$ and $\mathbf{H}^{1/2}(\Gamma)$ we shall consider their closed subspaces $\tilde{\mathbf{H}}^1(\Omega)=\{\mathbf{u}\in\mathbf{H}^1(\Omega):\mathbf{u}\cdot\mathbf{n}|_{\Gamma_N}=0,(\mathbf{u},\mathbf{n})_{\Gamma_i}=0,\ i=\overline{1,N}\}$, $\mathbf{H}_{\mathrm{div}}^1(\Omega)=\{\mathbf{v}\in\mathbf{H}^1(\Omega):\mathrm{div}\,\mathbf{v}=0\}$, $\tilde{\mathbf{H}}_{\mathrm{div}}^1(\Omega)=\{\mathbf{v}\in\tilde{\mathbf{H}}^1(\Omega):\mathrm{div}\,\mathbf{v}=0\}$, $\tilde{\mathbf{H}}^{1/2}(\Gamma)=\{\mathbf{u}|_\Gamma:\mathbf{u}\in\tilde{\mathbf{H}}^1(\Omega)\}$, and also duals $\tilde{\mathbf{H}}^1(\Omega)^*$, $\tilde{\mathbf{H}}^{1/2}(\Gamma)^*$ of the spaces $\tilde{\mathbf{H}}^1(\Omega)$, $\tilde{\mathbf{H}}^{1/2}(\Gamma)$. Let us introduce the following bilinear and trilinear forms: $a_0:\mathbf{H}^1(\Omega)^2\to\mathbb{R}$, $b:\mathbf{H}^1(\Omega)\times L_0^2(\Omega)\to\mathbb{R}$, $a_1:H^1(\Omega)^2\to\mathbb{R}$, $b_1:H^1(\Omega)\times\mathbf{H}_0^1(\Omega)\to\mathbb{R}$, $c:\mathbf{H}^1(\Omega)^3\to\mathbb{R}$, $c_1:\mathbf{H}^1(\Omega)\times H^1(\Omega)\times H^1(\Omega)\to\mathbb{R}$ by

$$a_0(\mathbf{u},\mathbf{v})=(\nabla\mathbf{u},\nabla\mathbf{v}),\ b(\mathbf{v},q)=-(\mathrm{div}\mathbf{v},q),\ c(\mathbf{u},\mathbf{v},\mathbf{w})=((\mathbf{u}\cdot\mathrm{grad})\mathbf{v},\mathbf{w}),$$

$$a_1(C,S)=(\nabla C,\nabla S),\ b_1(S,\mathbf{v})=(\mathbf{b}S,\mathbf{v}),\ c_1(\mathbf{u},C,S)=(\mathbf{u}\cdot\nabla C,S),\ \ \mathbf{b}\equiv\beta_C\mathbf{G}.$$

We note that forms c and c_1 possess the next properties (Alekseev & Tereshko, 2008; Girault & Raviart, 1986):

$$c(\mathbf{u},\mathbf{v},\mathbf{w}) = -c(\mathbf{u},\mathbf{w},\mathbf{v}),\ c(\mathbf{u},\mathbf{v},\mathbf{v}) = 0\ \forall \mathbf{u} \in \mathbf{H}^1_{\rm div}(\Omega),\ (\mathbf{v},\mathbf{w}) \in \mathbf{H}^1_0(\Omega) \times \mathbf{H}^1(\Omega), \tag{8}$$

$$c_1(\mathbf{u},C,S) = -c_1(\mathbf{u},S,C),\ c_1(\mathbf{u},C,C) = 0\ \forall \mathbf{u} \in \tilde{\mathbf{H}}^1_{\rm div}(\Omega), (C,S) \in \mathcal{T} \times H^1(\Omega). \tag{9}$$

Besides all the forms are continuous and the following technical lemma holds (Alekseev & Tereshko, 2008).

Lemma 1. *Under conditions (i) there exist constants δ_0, δ_1, γ_0, γ_1, γ_2 and β_1 depending on Ω such that*

$$|a_0(\mathbf{u},\mathbf{v})| \le \|\mathbf{u}\|_1\|\mathbf{v}\|_1\ \forall(\mathbf{u},\mathbf{v}) \in \mathbf{H}^1(\Omega)^2,\ a_0(\mathbf{v},\mathbf{v}) \ge \delta_0\|\mathbf{v}\|_1^2\ \forall \mathbf{v} \in \mathbf{H}^1_0(\Omega), \tag{10}$$

$$|a_1(C,S)| \le \|C\|_1\|S\|_1\ \forall(C,S) \in H^1(\Omega)\times \in H^1(\Omega),\ a_1(C,C) \ge \delta_1\|C\|_1^2\ \forall C \in \mathcal{T}, \tag{11}$$

$$|c(\mathbf{u},\mathbf{v},\mathbf{w})| \le \gamma_0\|\mathbf{u}\|_1\|\mathbf{v}\|_1\|\mathbf{w}\|_1\ \forall(\mathbf{u},\mathbf{v},\mathbf{w}) \in \mathbf{H}^1(\Omega)^3, \tag{12}$$

$$|c_1(\mathbf{u},C,S)| \le \gamma_1\|\mathbf{u}\|_1\|C\|_1\|S\|_1\ \forall \mathbf{u} \in \mathbf{H}^1(\Omega),\ (C,S) \in H^1(\Omega)\times \in H^1(\Omega), \tag{13}$$

$$|b_1(C,\mathbf{v})| \le \beta_1\|C\|_1\|\mathbf{v}\|_1\ \forall C \in H^1(\Omega), \mathbf{v} \in \mathbf{H}^1_0(\Omega), \tag{14}$$

$$|(\chi,C)_{\Gamma_N}| \le \|\chi\|_{\Gamma_N}\|C\|_{\Gamma_N} \le \gamma_2\|\chi\|_{\Gamma_N}\|C\|_1\ \forall C \in H^1(\Omega). \tag{15}$$

Bilinear form $b(\cdot,\cdot)$ satisfies inf-sup condition

$$\inf_{q\in L^2_0(\Omega), q\neq 0}\ \sup_{\mathbf{v}\in \mathbf{H}^1_0(\Omega), \mathbf{v}\neq 0} \frac{b(\mathbf{v},q)}{\|\mathbf{v}\|_1\|q\|} \ge \beta = \text{const} > 0. \tag{16}$$

Let in addition to (i) the following conditions take place:
(ii) $\mathbf{f} \in \mathbf{H}^{-1}(\Omega)$, $\mathbf{b} \in \mathbf{L}^2(\Omega)$, $f \in L^2(\Omega)$, $k \in L^2_+(\Omega)$, $\psi \in H^{1/2}(\Gamma_D)$;
(iii) $\mathbf{g} \in \tilde{\mathbf{H}}^{1/2}(\Gamma)$, $\chi \in L^2(\Gamma_N)$.
In order to formulate an extremum problem for Model 1 we divide the set of all input data in problem (5), (6) into two groups. One consists of control functions $\mathbf{g}$ and χ, and the other consists of fixed data, namely $\mathbf{f}, \mathbf{b}, f, k$ and ψ. Assume that controls $\mathbf{g}$, χ vary over some sets K_1 and K_2 such that
(j) $K_1 \subset \tilde{\mathbf{H}}^{1/2}(\Gamma)$, $K_2 \subset L^2(\Gamma_N)$ are nonempty convex closed subsets.
Let $X = \tilde{\mathbf{H}}^1(\Omega) \times L^2_0(\Omega) \times H^1(\Omega)$, $Y = \mathbf{H}^{-1}(\Omega) \times L^2_0(\Omega) \times \tilde{\mathbf{H}}^{1/2}(\Gamma) \times \mathcal{T}^* \times H^{1/2}(\Gamma_D)$, $\mathbf{x} = (\mathbf{u}, p, C) \in X$. Introduce an operator $F \equiv (F_1, F_2, F_3, F_4, F_5) : X \times K_1 \times K_2 \to Y$, defined by

$$\langle F_1(\mathbf{x},u),\mathbf{v}\rangle = \nu a_0(\mathbf{u},\mathbf{v}) + c(\mathbf{u},\mathbf{u},\mathbf{v}) + b(\mathbf{v},p) - b_1(C,\mathbf{v}) - \langle \mathbf{f},\mathbf{v}\rangle,$$

$$\langle F_2(\mathbf{x},u),r\rangle = b(\mathbf{u},r) \equiv -(\operatorname{div}\mathbf{u},r),\ F_3(\mathbf{x},u) = \mathbf{u}|_\Gamma - \mathbf{g},\ F_5(\mathbf{x},u) = C|_{\Gamma_D} - \psi,$$
$$\langle F_4(\mathbf{x},u),S\rangle = \lambda a_1(C,S) + (kC,S) + c_1(\mathbf{u},C,S) - (f,S) - (\chi,S)_{\Gamma_N}.$$

We multiply the first equation in (5) by $\mathbf{v} \in \mathbf{H}^1_0(\Omega)$, the equation in (6) by $S \in \mathcal{T}$, integrate the results over Ω with use of Green formulas, and use boundary conditions in (5), (6) to obtain a weak formulation of problem 1. It consists of finding a triple $\mathbf{x} = (\mathbf{u}, p, C) \in X$ satisfying the relations

$$\nu a_0(\mathbf{u},\mathbf{v}) + c(\mathbf{u},\mathbf{u},\mathbf{v}) + b(\mathbf{v},p) - b_1(C,\mathbf{v}) = \langle \mathbf{f},\mathbf{v}\rangle \;\; \forall \mathbf{v} \in \mathbf{H}_0^1(\Omega), \tag{17}$$

$$\lambda a_1(C,S) + (kC,S) + c_1(\mathbf{u},C,S) = \langle l,S\rangle \equiv (f,S) + (\chi,S)_{\Gamma_N} \;\forall S \in \mathcal{T}, \tag{18}$$

$$\operatorname{div}\mathbf{u} = 0 \text{ in } \Omega,\; \mathbf{u}|_\Gamma = \mathbf{g},\; C|_{\Gamma_D} = \psi, \tag{19}$$

which one can rewrite in an equivalent form of the operator equation

$$F(\mathbf{x},u) \equiv F(\mathbf{u},p,C,\mathbf{g},\chi) = 0. \tag{20}$$

This triple $(\mathbf{u},p,C) \in X$ will be called the weak solution to problem (5), (6).
Let $I : X \to \mathbb{R}$ be a weakly lower semicontinuous cost functional. Setting $K = K_1 \times K_2$, $u = (\mathbf{g},\chi)$, $u_0 = (\mathbf{f},\mathbf{b},f,k,\psi)$ we formulate the following constrained minimization problem

$$J(\mathbf{x},u) = (\mu_0/2)I(\mathbf{x}) + (\mu_1/2)\|\mathbf{g}\|_{1/2,\Gamma}^2 + (\mu_2/2)\|\chi\|_{\Gamma_N}^2 \to \inf,\; F(\mathbf{x},u) = 0,\; (\mathbf{x},u) \in X \times K. \tag{21}$$

Here $\mu_0 > 0$ and $\mu_1 \geq 0, \mu_2 \geq 0$ are positive dimensional parameters which serve to regulate the relative importance of each of the terms in (21). Another purpose of introducing μ_l is to ensure the uniqueness and stability of solutions to control problems under consideration (see below). The possible cost functionals are defined as

$$I_1(\mathbf{x}) = \|\mathbf{v} - \mathbf{v}_d\|_Q^2,\; I_2(\mathbf{x}) = \|\mathbf{v} - \mathbf{v}_d\|_{1,Q}^2,\; I_3(\mathbf{x}) = \|\operatorname{rot}\mathbf{v} - \eta_d\|_Q^2,\; I_4(x) = \|p - p_d\|_Q^2. \tag{22}$$

Here Q is a subset of Ω, $\mathbf{v}_d \in \mathbf{L}^2(Q)$ (or $\mathbf{v}_d \in \mathbf{H}^1(Q)$), $\eta_d \in \mathbf{L}^2(Q)$ and $p_d \in L^2(Q)$ are functions which are interpreted as measured velocity, vorticity or pressure fields. Define $Z_{ad} = \{(\mathbf{x},u) \in X \times K : F(\mathbf{x},u) = 0, J(\mathbf{x},u) < \infty\}$. Let us assume in addition to (j) that
(jj) $\mu_0 > 0$, $\mu_1 \geq 0$, $\mu_2 \geq 0$ and K is a bounded subset or $\mu_0 > 0$, $\mu_1 > 0$, $\mu_2 > 0$ and functional I is bounded from below.
According to general theory of extremum problems (see (Ioffe & Tikhomirov, 1979)) we introduce an element $\mathbf{y}^* = (\xi,\sigma,\zeta,\theta,\zeta^c) \in Y^* = \mathbf{H}_0^1(\Omega) \times L_0^2(\Omega) \times \tilde{\mathbf{H}}^{1/2}(\Gamma)^* \times \mathcal{T} \times H^{1/2}(\Gamma_D)^*$ which is reffered to as the adjoint state and define the Lagrangian $\mathcal{L} : X \times K \times \mathbb{R}^+ \times Y^* \to \mathbb{R}$, where $\mathbb{R}^+ = \{\lambda \in \mathbb{R} : \lambda \geq 0\}$, by the formula

$$\mathcal{L}(\mathbf{x},u,\lambda_0,\mathbf{y}^*) = \lambda_0 J(\mathbf{x},u) + \langle F_1(\mathbf{x},u),\xi\rangle + (F_2(\mathbf{x},u),q) +$$
$$\langle \zeta, F_3(\mathbf{x},u)\rangle_\Gamma + \varkappa\langle F_4(\mathbf{x},u),\theta\rangle + \varkappa\langle \zeta^c, F_5(\mathbf{x},u)\rangle_{\Gamma_D}.$$

Here $\langle \zeta,\mathbf{g}\rangle_\Gamma = \langle \zeta,\mathbf{g}\rangle_{\tilde{\mathbf{H}}^{1/2}(\Gamma)^* \times \tilde{\mathbf{H}}^{1/2}(\Gamma)}$ for $\zeta \in \tilde{\mathbf{H}}^{1/2}(\Gamma)^*$, $\langle \zeta^c,\psi\rangle_{\Gamma_D} = \langle \zeta^c,\psi\rangle_{H^{1/2}(\Gamma_D)^* \times H^{1/2}(\Gamma_D)}$ for $\zeta^c \in H^{1/2}(\Gamma_D)^*$, $\varkappa$ is a dimensional parameter. Let the dimension $[\varkappa]$ be chosen so that the dimensions of ξ, s, θ at the adjoint state $\mathbf{y}^*$ coincide with those at the basic state $\mathbf{x} = (\mathbf{u},p,C)$ i.e.

$$[\xi] = [\mathbf{u}] = L_0 T_0^{-1},\; [\theta] = [C] = M_0 L_0^{-3},\; [s] = [p] = L_0^2 T_0^{-2}. \tag{23}$$

Here L_0, T_0, M_0 denote the SI dimensions of the length, time and mass units expressed in meters, seconds and kilograms respectively. A simple analysis of (23) shows that necessary condition for fulfillment of (23) is $[\varkappa] = L_0^8 T_0^{-2} M_0^{-2}$ (see (Alekseev & Tereshko, 2008)).
Below we shall use some results concerning problem (5), (6) and extremum problem (21). The proofs of the theorems are simular to those in (Alekseev & Tereshko, 2008).

Theorem 1. *Let conditions (i), (ii) be satisfied. Then for any $u \in K$ problem (5), (6) has a weak solution $(\mathbf{u},p,C) \in X$ that satisfies the estimates $\|\mathbf{u}\|_1 \le M_{\mathbf{u}}(u_0,u)$, $\|p\| \le M_p(u_0,u)$, $\|C\|_1 \le M_C(u_0,u)$. Here $M_{\mathbf{u}}(u_0,u)$, $M_p(u_0,u)$ and $M_C(u_0,u)$ are nondecreasing continuous functions of the norms $\|\mathbf{f}\|_{-1}$, $\|\mathbf{b}\|$, $\|f\|$, $\|k\|$, $\|\psi\|_{1/2,\Gamma_D}$, $\|\mathbf{g}\|_{1/2,\Gamma}$, $\|\chi\|_{\Gamma_N}$. If the functions $\mathbf{f},\mathbf{b},f,k$, ψ, $\mathbf{g}$, χ are small (or the viscosity ν is high) in the sense that*

$$\frac{\gamma_0 M_{\mathbf{u}}(u_0,u)}{\delta_0 \nu} + \frac{1}{\delta_0 \nu}\frac{\beta_1\gamma_1 M_C^0(u_0,u)}{\delta_1 \lambda} < 1, \tag{24}$$

then the weak solution to problem 1 is unique. Here $\delta_0,\delta_1,\gamma_0,\gamma_1,\beta_1$ are the constants from (10)–(14).

Theorem 2. *Under conditions (i), (ii), (j) and (jj) let $I : X \to \mathbb{R}$ be a weakly lower semicontinuous functional and $Z_{ad} \neq \emptyset$. Then control problem (21) has at least one solution.*

Theorem 3. *Under conditions (i), (ii), (j) let $\mu_0 > 0$, $\mu_l > 0$ or $\mu_0 > 0$, $\mu_l \ge 0$ and K_l be the bounded sets, $l = 1,2$. Then control problem (21) has at least one solution for $I = I_k$, $k = 1,2,3,4$.*

Theorem 4. *Under conditions (i), (ii), (j) and (jj) let $(\hat{\mathbf{x}},\hat{u}) \equiv (\hat{\mathbf{u}},\hat{p},\hat{C},\hat{\mathbf{g}},\hat{\chi}) \in X \times K$ be a local minimizer in problem (21) and let the functional I be continuously differentiable at the point $\hat{\mathbf{x}}$. Then, there exists a nonzero Lagrange multiplier $(\lambda_0,\mathbf{y}^*) = (\lambda_0,\xi,\sigma,\zeta,\theta,\zeta^c) \in \mathbb{R}^+ \times \mathbf{H}_0^1(\Omega) \times L_0^2(\Omega) \times \tilde{\mathbf{H}}^{1/2}(\Gamma)^* \times \mathcal{T} \times H^{1/2}(\Gamma_D)^*$ that satisfies the Euler-Lagrange equation $F'_{\mathbf{x}}(\hat{\mathbf{x}},\hat{u})^*\mathbf{y}^* = -\lambda_0 J'_{\mathbf{x}}(\hat{\mathbf{x}},\hat{u})$, which is equivalent to the identities*

$$\nu a_0(\mathbf{w},\xi) + c(\hat{\mathbf{u}},\mathbf{w},\xi) + c(\mathbf{w},\hat{\mathbf{u}},\xi) + \varkappa c_1(\mathbf{u},\hat{C},\theta) + b(\mathbf{w},\sigma) + \langle\zeta,\mathbf{w}\rangle_\Gamma =$$

$$-\lambda_0(\mu_0/2)\langle I'_{\mathbf{u}}(\hat{\mathbf{x}}),\mathbf{w}\rangle \ \forall \mathbf{w} \in \tilde{\mathbf{H}}^1(\Omega),\ b(\xi,r) \equiv -(\operatorname{div}\xi,r) = -\lambda_0(\mu_0/2)(I'_p(\hat{\mathbf{x}}),r) \ \forall r \in L_0^2(\Omega), \tag{25}$$

$$\varkappa[\lambda a_1(\tau,\theta) + (k\tau,\theta) + c_1(\hat{\mathbf{u}},\tau,\theta) + \langle\zeta^c,\tau\rangle_{\Gamma_D}] - b_1(\tau,\xi) =$$

$$-\lambda_0(\mu_0/2)\langle I'_C(\hat{\mathbf{x}}),\tau\rangle \ \forall \tau \in H^1(\Omega), \tag{26}$$

and satisfies the minimum principle $\mathcal{L}(\hat{\mathbf{x}},\hat{u},\lambda_0,\mathbf{y}^*) \le \mathcal{L}(\hat{\mathbf{x}},u,\lambda_0,\mathbf{y}^*)$ for all $u \in K$ and the variational inequality

$$\lambda_0\mu_1(\hat{\mathbf{g}},\mathbf{g}-\hat{\mathbf{g}})_{1/2,\Gamma} - \langle\zeta,\mathbf{g}-\hat{\mathbf{g}}\rangle_\Gamma + \lambda_0\mu_2(\hat{\chi},\chi-\hat{\chi})_{\Gamma_N} - \varkappa\langle\theta,\chi-\hat{\chi}\rangle_{\Gamma_N} \ge 0 \ \forall u = (\mathbf{g},\chi). \tag{27}$$

Theorem 5. *Let the assumptions of Theorem 4 be satisfied and inequality (24) hold for all $u \in K$. Then: 1) homogeneous problem (25), (26) (if $\lambda_0 = 0$) has only trivial solution $\mathbf{y}^* \equiv (\xi,\sigma,\zeta,\theta,\zeta^c) = 0$; 2) any nontrivial Lagrange multiplier satisfying (25), (26) is regular, i.e. it has the form $(1,\mathbf{y}^*)$.*

Relations (25), (26), together with variational inequality (27) and operator constraint (20) constitute an optimality system. It consists of three parts. The first part has the form of a weak formulation (17)–(19) of problem (5), (6), which is equivalent to operator equation (20). The second part consists of identities (25), (26) for the Lagrange multipliers ξ, σ, ζ, θ and ζ^c. Finally, the last part of the optimality system is the variational inequality (27) with respect to controls $\mathbf{g}$ and χ which is the consequence of the minimum principle.

Remark 1. We emphasize that the multiplier (adjoint velocity) ξ is in a general case a nonsolenoidal vector-function except the situation when the cost functional I independent of pressure p. Only in this case it follows from (25) that $\operatorname{div}\xi = 0$ and moreover $\xi \in \mathbf{V}$.

Remark 2. Denote by $\mathbf{g}_i = \mathbf{g}|_{\Gamma_i}$ the restriction of the boundary vector $\mathbf{g}$ to the component Γ_i of Γ and introduce values (flows) q_i of the vector $\mathbf{g}_i$ through Γ_i by $q_i = (\mathbf{g}_i, \mathbf{n})_{\Gamma_i} \equiv \int_{\Gamma_i} \mathbf{g} \cdot \mathbf{n} d\sigma$. We note that the incompressibility condition $\operatorname{div} \mathbf{u} = 0$ in (5) results in the following necessary condition for q_i:

$$(\mathbf{g}, \mathbf{n})_\Gamma = q_1 + q_2 + \ldots + q_N = 0.$$

At the same time Theorem 1 is proved under more strict condition $\mathbf{g} \in \tilde{\mathbf{H}}^{1/2}(\Gamma)$ on $\mathbf{g}$ equivalent to N conditions $q_1 = 0, q_2 = 0, \ldots, q_N = 0$ for the vector $\mathbf{g}$. The latter is connected with the fact that the proof of Theorem 1 is based on the generalization of the Hopf's lemma see (Hopf, 1941). According to this generalized Hopf's lemma for any vector $\mathbf{g} \in \tilde{\mathbf{H}}^{1/2}(\Gamma)$ and any $\varepsilon > 0$ there exists such a solenoidal expansion $\mathbf{u}_0 \in \mathbf{H}^1_{\text{div}}(\Omega)$ into Ω for which

$$|((\mathbf{v} \cdot \nabla)\mathbf{u}_0, \mathbf{v})| \leq \varepsilon \|\mathbf{g}\|_{1/2,\Gamma} \|\mathbf{v}\|_1^2 \ \forall \mathbf{v} \in \mathbf{V}.$$

Using this lemma one can look for a weak solution of problem (5) at $\beta_C = 0$ in the form $\mathbf{u} = \mathbf{u}_0 + \tilde{\mathbf{u}}$, where $\tilde{\mathbf{u}} \in \mathbf{V}$ is a new unknown function and to obtain a "coercitive" nonlinear operator equation for function $\tilde{\mathbf{u}}$. The existence of the solution $\tilde{\mathbf{u}}$ of the latter equation can be proved using Schauder theorem (see e.g. (Alekseev & Tereshko, 2008)). It should be noted that the question of Hopf's lemma validity and proof of the existence theorem for problem (5) at $\beta_C = 0$ under fulfillment only condition $(\mathbf{g}, \mathbf{n})_\Gamma = 0$ to the vector $\mathbf{g} \in \mathbf{H}^{1/2}(\Gamma)$ is till open. One can read about this problem (so called Leray problem) in more details in (Alekseev & Tereshko, 2008, Appendix 5) and in (Pukhnachev, 2009; 2010).

The sufficient conditions of solvability of the stationary boundary value problem for the Navier-Stokes equations without the assumption $q_i = 0$, $i = 1, \ldots, N$ are stated in mentioned papers. Besides a detailed bibliography on Leray problem is provided. As to the general boundary value problem (1)–(3) for stationary heat and mass transfer equations, its uniqueness, even when the equalities $q_i = 0$ hold, can be proved only in the case when values of thermal and diffusion Rayleigh numbers are small. Moreover the branching of stationary solutions is possible in the case of large values of Rayleigh numbers (Gershuni & Zhukhovitskii, 1976; Joseph, 1976).

3. General property of the optimality system solutions

Let us consider control problem (21). Denote by $(\mathbf{x}_1, u_1) \equiv (\mathbf{u}_1, p_1, C_1, \mathbf{g}_1, \chi_1) \in X \times K$ its solution. By $(\mathbf{x}_2, u_2) \equiv (\mathbf{u}_2, p_2, C_2, \mathbf{g}_2, \chi_2) \in X \times K$ we denote a solution of problem

$$\tilde{J}(\mathbf{x}, u) = \frac{\mu_0}{2}\tilde{I}(\mathbf{x}) + \frac{\mu_1}{2}\|\mathbf{g}\|^2_{1/2,\Gamma} + \frac{\mu_2}{2}\|\chi\|^2_{\Gamma_N} \to \inf,\ F(\mathbf{x}, u) = 0,\ (\mathbf{x}, u) \in X \times K, \tag{28}$$

which is obtained from (21) by replacing a functional I in (21) with another one $\tilde{I}$. In view of Theorem 1 the following estimates for pairs $(\mathbf{u}_i, p_i)$ hold

$$\|\mathbf{u}_i\|_1 \leq M^0_{\mathbf{u}} = \sup_{u \in K} M_{\mathbf{u}}(u_0, u),\ \|p_i\| \leq M^0_p = \sup_{u \in K} M_p(u_0, u),\ \|C_i\|_1 \leq M^0_C = \sup_{u \in K} M_C(u_0, u). \tag{29}$$

We introduce "model" Reynolds number $\mathcal{R}e$, Rayleigh number $\mathcal{R}a$ and Prandtl number $\mathcal{P}$ by

$$\mathcal{R}e = \frac{\gamma_0 M^0_{\mathbf{u}}}{\delta_0 \nu},\ \mathcal{R}a = \frac{\gamma_1}{\delta_0 \nu}\frac{\beta_1 M^0_C}{\delta_1 \lambda},\ \mathcal{P} = \frac{\delta_0 \nu}{\delta_1 \lambda} \tag{30}$$

and assume that

$$\mathcal{R}e + \mathcal{R}a \equiv \frac{\gamma_0 M^0_{\mathbf{u}}}{\delta_0 \nu} + \frac{\gamma_1}{\delta_0 \nu}\frac{\beta_1 M^0_C}{\delta_1 \lambda} < 1/2. \tag{31}$$

Denote by $(1,\mathbf{y}_i^*) \equiv (1,\xi_i,\sigma_i,\zeta_i,\theta_i,\zeta_i^c)$, $i = 1,2$, the Lagrange multipliers corresponding to the solutions $(\mathbf{x}_i,u_i)$ (these multipliers are uniquely determined under condition (31)). By definition elements $(\xi_i,\sigma_i,\zeta_i,\theta_i,\zeta_i^c)$ satisfy relations

$$\nu a_0(\mathbf{w},\xi_i) + c(\mathbf{u}_i,\mathbf{w},\xi_i) + c(\mathbf{w},\mathbf{u}_i,\xi_i) + \varkappa c_1(\mathbf{w},C_i,\theta_i) + b(\mathbf{w},\sigma_i) + \langle\zeta_i,\mathbf{w}\rangle_\Gamma =$$

$$= -(\mu_0/2)\langle(I^i)'_{\mathbf{u}}(\mathbf{x}_i),\mathbf{w}\rangle \ \forall\mathbf{w}\in\tilde{\mathbf{H}}^1(\Omega),\ b(\xi_i,r) = -(\mu_0/2)((I^i)'_p(\mathbf{x}_i),r)\ \forall r\in L_0^2(\Omega), \quad (32)$$

$$\varkappa[\lambda a_1(\tau,\theta_i) + (k\tau,\theta_i) + c_1(\mathbf{u}_i,\tau,\theta_i) + \langle\zeta_i,\tau\rangle_{\Gamma_D}] - b_1(\tau,\xi_i) =$$

$$-(\mu_0/2)\langle(I^i)'_C(\mathbf{x}_i),\tau\rangle\ \forall\tau\in H^1(\Omega). \quad (33)$$

Here we renamed $I = I^1$, $\tilde{I} = I^2$. Let $\mathbf{g} = \mathbf{g}_1 - \mathbf{g}_2$, $\chi = \chi_1 - \chi_2$,

$$\mathbf{u}=\mathbf{u}_1-\mathbf{u}_2,\ p=p_1-p_2,\ C=C_1-C_2,\ \xi=\xi_1-\xi_2,\ \sigma=\sigma_1-\sigma_2,\ \zeta=\zeta_1-\zeta_2,\ \theta=\theta_1-\theta_2,\ \zeta^c=\zeta_2^c-\zeta_1^c. \quad (34)$$

Subtracting equations (17)–(19), written for $\mathbf{u}_2,p_2,C_2,u_2$, from corresponding equations (17)–(19) for $\mathbf{u}_1,p_1,C_1,u_1$ gives

$$\nu a_0(\mathbf{u},\mathbf{v}) + c(\mathbf{u},\mathbf{u}_1,\mathbf{v}) + c(\mathbf{u}_2,\mathbf{u},\mathbf{v}) + b(\mathbf{v},p) - b_1(C,\mathbf{v}) = 0\ \forall\mathbf{v}\in\mathbf{H}_0^1(\Omega), \quad (35)$$

$$\lambda a_1(C,S) + (kC,S) + c_1(\mathbf{u},C_1,S) + c_1(\mathbf{u}_2,C,S) = (\chi,S)_{\Gamma_N}\ \forall S\in\mathcal{T}, \quad (36)$$

$$\operatorname{div}\mathbf{u} = 0 \text{ in } \Omega,\ \mathbf{u}|_\Gamma = \mathbf{g},\ C|_{\Gamma_D} = 0. \quad (37)$$

Setting $\mathbf{g} = \mathbf{g}_1$, $\chi = \chi_1$ in (27) under $\lambda_0 = 1$, written for $\hat{\mathbf{g}} = \mathbf{g}_2$, $\hat{\chi} = \chi_2$, $\zeta = \zeta_2$ and setting $\mathbf{g} = \mathbf{g}_2$, $\chi = \chi_2$ in (27) written for $\hat{\mathbf{g}} = \mathbf{g}_1$, $\hat{\chi} = \chi_1$, $\zeta = \zeta_1$, we obtain $\mu_1(\mathbf{g}_2,\mathbf{g})_{1/2,\Gamma} - \langle\zeta_2,\mathbf{g}\rangle_\Gamma + \mu_2(\chi_2,\chi)_{\Gamma_N} - \varkappa(\theta_2,\chi)_{\Gamma_N} \geq 0$, $-\mu_1(\mathbf{g}_1,\mathbf{g})_{1/2,\Gamma} + \langle\zeta_1,\mathbf{g}\rangle_\Gamma - \mu_2(\chi_1,\chi)_{\Gamma_N} + \varkappa(\theta_1,\chi)_{\Gamma_N} \geq 0$. Adding up these inequalities yields the relation

$$-\langle\zeta,\mathbf{g}\rangle_\Gamma - \varkappa(\theta,\chi)_{\Gamma_N} \leq -\mu_1\|\mathbf{g}\|^2_{1/2,\Gamma} - \mu_2\|\chi\|^2_{\Gamma_N}. \quad (38)$$

Subtract equations (32), (33) written for $(\mathbf{x}_2,u_2,\mathbf{y}_2^*)$ from corresponding equations for $(\mathbf{x}_1,u_1,\mathbf{y}_1^*)$. We obtain

$$\nu a_0(\mathbf{w},\xi) + c(\mathbf{u}_1,\mathbf{w},\xi) + c(\mathbf{u},\mathbf{w},\xi_2) + c(\mathbf{w},\mathbf{u}_1,\xi) + c(\mathbf{w},\mathbf{u},\xi_2) + \varkappa c_1(\mathbf{w},C_1,\theta) + \varkappa c_1(\mathbf{w},C,\theta_2) +$$

$$b(\mathbf{w},\sigma) = -\langle\zeta,\mathbf{w}\rangle_\Gamma - (\mu_0/2)\langle I'_{\mathbf{u}}(\mathbf{x}_1) - \tilde{I}'_{\mathbf{u}}(\mathbf{x}_2),\mathbf{w}\rangle\ \forall\mathbf{w}\in\tilde{\mathbf{H}}^1(\Omega), \quad (39)$$

$$\varkappa[\lambda a_1(\tau,\theta) + (k\tau,\theta) + c_1(\mathbf{u}_1,\tau,\theta) + c_1(\mathbf{u},\tau,\theta_2) + \langle\zeta,\tau\rangle_{\Gamma_D}] - b_1(\tau,\xi) =$$

$$-(\mu_0/2)\langle I'_C(\mathbf{x}_1) - \tilde{I}'_C(\mathbf{x}_2),\tau\rangle\ \forall\tau\in H^1(\Omega), \quad (40)$$

$$b(\xi,r) = -(\mu_0/2)(I'_p(\mathbf{x}_1) - \tilde{I}'_p(\mathbf{x}_2),r)\ \forall r\in L_0^2(\Omega). \quad (41)$$

Set $\mathbf{w} = \mathbf{u}$, $\tau = C$, $r = p$ in (39), (40), (41) and add up the results. Taking into account (37) we obtain

$$\nu a_0(\mathbf{u},\xi) + c(\mathbf{u}_1,\mathbf{u},\xi) + 2c(\mathbf{u},\mathbf{u},\xi_2) + c(\mathbf{u},\mathbf{u}_1,\xi) + \varkappa c_1(\mathbf{u},C_1,\theta) + \varkappa c_1(\mathbf{u},C,\theta_2) + \langle \zeta,\mathbf{g}\rangle_\Gamma +$$

$$\varkappa[\lambda a_1(C,\theta) + (kC,\theta) + c_1(\mathbf{u}_1,C,\theta) + c_1(\mathbf{u},C,\theta_2)] -$$

$$b_1(C,\xi) + b(\xi,p) = -(\mu_0/2)\langle I'_{\mathbf{u}}(\mathbf{x}_1) - \tilde{I}'_{\mathbf{u}}(\mathbf{x}_2),\mathbf{u}\rangle - (\mu_0/2)\langle I'_C(\mathbf{x}_1) - -\tilde{I}'_C(\mathbf{x}_2),C\rangle -$$

$$-(\mu_0/2)(I'_p(\mathbf{x}_1) - \tilde{I}'_p(\mathbf{x}_2),p). \quad (42)$$

Set $\mathbf{v} = \xi$ in (35), $S = \varkappa\theta$ in (36) and subtract the results from (42). Using (38) and identities

$$2c(\mathbf{u},\mathbf{u},\xi_2) + c(\mathbf{u}_1,\mathbf{u},\xi) - c(\mathbf{u}_2,\mathbf{u},\xi) = 2c(\mathbf{u},\mathbf{u},\xi_2) + c(\mathbf{u},\mathbf{u},\xi) = c(\mathbf{u},\mathbf{u},\xi_1 + \xi_2),$$

$$2c_1(\mathbf{u},C,\theta_2) + c_1(\mathbf{u},C_1,\theta) - c_1(\mathbf{u},C_2,\theta) = c_1(\mathbf{u},C,\theta_1 + \theta_2),$$

we obtain

$$c(\mathbf{u},\mathbf{u},\xi_1 + \xi_2) + \varkappa c_1(\mathbf{u},C,\theta_1 + \theta_2) + (\mu_0/2)\langle I'_{\mathbf{u}}(\mathbf{x}_1) - \tilde{I}'_{\mathbf{u}}(\mathbf{x}_2),\mathbf{u}\rangle +$$

$$(\mu_0/2)\langle I'_C(\mathbf{x}_1) - \tilde{I}'_C(\mathbf{x}_2),C\rangle + (\mu_0/2)(I'_p(\mathbf{x}_1) - \tilde{I}'_p(\mathbf{x}_2),p) \le -\mu_1\|\mathbf{g}\|^2_{1/2,\Gamma} - \mu_2\|\chi\|^2_{\Gamma_N}. \quad (43)$$

Thus the following result holds.

Theorem 6. *Let under conditions of Theorem 4 for $I_1 = I$ and $I_2 = \tilde{I}$ pairs $(\mathbf{x}_1,u_1) = (\mathbf{u}_1,p_1,C_1,\mathbf{g}_1,\chi_1) \in X \times K$ and $(\mathbf{x}_2,u_2) = (\mathbf{u}_2,p_2,C_2,\mathbf{g}_2,\chi_2) \in X \times K$ are solutions to problems (21) and (28) respectively, $\mathbf{y}_i^* = (\xi_i,\sigma_i,\zeta_i,\theta_i,\zeta_i^c)$, $i = 1,2$, are the Lagrange multipliers corresponding to these solutions $(\mathbf{x}_i,u_i)$. Then the relation (43) for differences $\mathbf{u} = \mathbf{u}_1 - \mathbf{u}_2$, $p = p_1 - p_2$, $C = C_1 - C_2$, $\mathbf{g} = \mathbf{g}_1 - \mathbf{g}_2$, $\chi = \chi_1 - \chi_2$ holds.*

Below we shall need some estimates of differences $\mathbf{u} = \mathbf{u}_1 - \mathbf{u}_2$ and $p = p_1 - p_2$ via differences $C = C_1 - C_2$ and $\mathbf{g} = \mathbf{g}_1 - \mathbf{g}_2$. In order to deduce them denote by $\mathbf{u}_0 \in \mathbf{H}^1(\Omega)$ a function such that $\operatorname{div}\mathbf{u}_0 = 0$, $\mathbf{u}_0|_\Gamma = \mathbf{g}$, $\|\mathbf{u}_0\|_1 \le c_0\|\mathbf{g}\|_{1/2,\Gamma}$. Here a constant c_0 depends on Ω. The existence of $\mathbf{u}_0$ follows from (Girault & Raviart, 1986, p. 24). Set $\mathbf{u} \equiv \mathbf{u}_1 - \mathbf{u}_2 = \mathbf{u}_0 + \tilde{\mathbf{u}}$ where $\tilde{\mathbf{u}} \in \mathbf{V}$ is a certain function. Set $\mathbf{u} = \mathbf{u}_0 + \tilde{\mathbf{u}}$, $\mathbf{v} = \tilde{\mathbf{u}}$ in (35). Taking into account (8) we obtain

$$\nu a_0(\tilde{\mathbf{u}},\tilde{\mathbf{u}}) = -\nu a_0(\mathbf{u}_0,\tilde{\mathbf{u}}) - c(\mathbf{u}_0,\mathbf{u}_1,\tilde{\mathbf{u}}) - c(\tilde{\mathbf{u}},\mathbf{u}_1,\tilde{\mathbf{u}}) - c(\mathbf{u}_2,\mathbf{u}_0,\tilde{\mathbf{u}}) + b_1(C,\tilde{\mathbf{u}}).$$

Using estimates (10), (12), (14), (29) and this relation we deduce, that

$$\delta_0\nu\|\tilde{\mathbf{u}}\|_1^2 \le \nu\|\mathbf{u}_0\|_1\|\tilde{\mathbf{u}}\|_1 + \gamma_0 M^0_{\mathbf{u}}\|\tilde{\mathbf{u}}\|_1^2 + 2\gamma_0 M^0_{\mathbf{u}}\|\mathbf{u}_0\|_1\|\tilde{\mathbf{u}}\|_1 + \beta_1\|C\|_1\|\tilde{\mathbf{u}}\|_1. \quad (44)$$

It follows from (31) that

$$(\delta_0\nu/2) < \delta_0\nu - \gamma_0 M^0_{\mathbf{u}} - \frac{\beta_1\gamma_1}{\delta_1\lambda}M^0_C \le \delta_0\nu - \gamma_0 M^0_{\mathbf{u}}. \quad (45)$$

Rewriting (44) in view of (45) as

$$(\delta_0\nu/2)\|\tilde{\mathbf{u}}\|_1^2 \le (\delta_0\nu - \gamma_0 M^0_{\mathbf{u}})\|\tilde{\mathbf{u}}\|_1^2 \le (\nu + 2\gamma_0 M^0_{\mathbf{u}})\|\mathbf{u}_0\|_1\|\tilde{\mathbf{u}}\|_1 + \beta_1\|C\|_1\|\tilde{\mathbf{u}}\|_1,$$

we obtain that

$$\|\tilde{\mathbf{u}}\|_1 \le (2/\delta_0\nu)(\nu + 2\gamma_0 M_{\mathbf{u}}^0)\|\mathbf{u}_0\|_1 + (2\beta_1/\delta_0\nu)\|C\|_1 \le (2\delta_0^{-1} + 4\mathcal{R}e)\|\mathbf{u}_0\|_1 +$$

$$+(2\beta_1/\delta_0\nu)\|C\|_1 \le 2\mathcal{R}\|\mathbf{u}_0\|_1 + (2\beta_1/\delta_0\nu)\|C\|_1,\ \ \mathcal{R} \equiv \delta_0^{-1} + 2\mathcal{R}e.$$

Taking into account that $\mathbf{u} = \mathbf{u}_0 + \tilde{\mathbf{u}}$, we deduce the following estimate:

$$\|\mathbf{u}\|_1 \le \|\mathbf{u}_0\|_1 + \|\tilde{\mathbf{u}}\|_1 \le c_0(2\mathcal{R}+1)\|\mathbf{g}\|_{1/2,\Gamma} + (2\beta_1/\delta_0\nu)\|C\|_1. \tag{46}$$

Analogous estimate holds and for pressure difference $p = p_1 - p_2$. We make use inf-sup condition (16) for obtaining this estimate. By (16) for the function $p = p_1 - p_2$ and any (small) number $\delta > 0$ there exists a function $\mathbf{v}_0 \in \mathbf{H}_0^1(\Omega)$, $\mathbf{v}_0 \neq 0$ such that $b(\mathbf{v}_0, p) \ge \beta_0\|\mathbf{v}_0\|_1\|p\|$, $\beta_0 = (\beta - \delta) > 0$. Set $\mathbf{v} = \mathbf{v}_0$ in (35). Using this estimate and (10), (12), (14) we have

$$\beta_0\|\mathbf{v}_0\|_1\|p\| \le b(\mathbf{v}_0, p) \le (\nu + 2\gamma_0 M_{\mathbf{u}}^0)\|\mathbf{v}_0\|_1\|\mathbf{u}\|_1 + \beta_1\|C\|_1\|\mathbf{v}_0\|_1.$$

As $\|\mathbf{v}_0\|_1 \neq 0$ we deduce from this relation that

$$\|p\| \le \frac{\nu + 2\gamma_0 M_{\mathbf{u}}^0}{\beta_0}\|\mathbf{u}\|_1 + \frac{\beta_1}{\beta_0}\|C\|_1 = \frac{\delta_0\nu}{\beta_0}\mathcal{R}\|\mathbf{u}\|_1 + \frac{\beta_1}{\beta_0}\|C\|_1. \tag{47}$$

Using (46) we obtain the following estimate for $\|p\|$:

$$\|p\| \le \frac{\delta_0\nu}{\beta_0}c_0\mathcal{R}(2\mathcal{R}+1)\|\mathbf{g}\|_{1/2,\Gamma} + \frac{\beta_1}{\beta_0}(2\mathcal{R}+1)\|C\|_1. \tag{48}$$

Based on Theorem 6 and estimates (46)–(48) we establish in the next section sufficient conditions to input data which provide uniqueness and stability of the solution $(\hat{\mathbf{x}}, \hat{u})$ to problem (21) for a number of concrete cost functionals and controls.

Remark 3. Let us note that if $u = \mathbf{g}$ (or $u = \chi$) then this boundary control problem can be considered as a particular case of the general boundary control problem (21) corresponding to situation when K_2 (or K_1) is singleton: $K_2 = \{\chi\}$ (or $K_1 = \{\mathbf{g}\}$).

4. Uniqueness and stability of solutions of boundary control problems

In this section we firstly consider the problem (21) in the case where $I = I_1$ and $u = \mathbf{g} \in K_1$, i.e. we consider one-parameter boundary control problem

$$J(\mathbf{x}, \mathbf{g}) \equiv \frac{\mu_0}{2}\|\mathbf{v} - \mathbf{v}_d\|_Q^2 + \frac{\mu_1}{2}\|\mathbf{g}\|_{1/2,\Gamma}^2 \to \inf,\ F(\mathbf{x}, \mathbf{g}) = 0,\ \mathbf{x} = (\mathbf{v}, q, S) \in X, \mathbf{g} \in K_1. \tag{49}$$

Let $(\mathbf{x}_1, u_1) \equiv (\mathbf{u}_1, p_1, C_1, \mathbf{g}_1)$ be a solution to problem (49) which corresponds to a function $\mathbf{v}_d \equiv \mathbf{u}_d^{(1)} \in \mathbf{L}^2(Q)$, $(\mathbf{x}_2, u_2) \equiv (\mathbf{u}_2, p_2, C_2, \mathbf{g}_2)$ be a solution to problem (49) which corresponds to another function $\tilde{\mathbf{v}}_d \equiv \mathbf{u}_d^{(2)} \in \mathbf{L}^2(Q)$. Setting $\mathbf{g} = \mathbf{g}_1 - \mathbf{g}_2$, $\mathbf{u}_d = \mathbf{u}_d^{(1)} - \mathbf{u}_d^{(2)}$ in addition to (34), we note that

$$\langle (I_1)'_{\mathbf{u}}(\mathbf{x}_i), \mathbf{w}\rangle = 2(\mathbf{u}_i - \mathbf{u}_d, \mathbf{w})_Q,\ \langle (I_1)'_{\mathbf{u}}(\mathbf{x}_1) - (I_1)'_{\mathbf{u}}(\mathbf{x}_2), \mathbf{u}\rangle = 2(\mathbf{u} - \mathbf{u}_d, \mathbf{u})_Q =$$

$$2(\|\mathbf{u}\|_Q^2 - (\mathbf{u}, \mathbf{u}_d)_Q),\ (I_1)'_C = 0. \tag{50}$$

The relations (35), (37) for problem (49) do not change while identities (36), (32), (33) and the main inequality (43) take by (50) and remark 1 a form

$$\lambda a_1(C,S)+(kC,S)+c_1(\mathbf{u},C_1,S)+c_1(\mathbf{u}_2,C,S)=0\ \forall S\in\mathcal{T},\tag{51}$$

$$\nu a_0(\mathbf{w},\xi_i)+c(\mathbf{u}_i,\mathbf{w},\xi_i)+c(\mathbf{w},\mathbf{u}_i,\xi_i)+\varkappa c_1(\mathbf{w},C_i,\theta_i)+b(\mathbf{w},\sigma_i)+\langle\zeta_i,\mathbf{w}\rangle_\Gamma=$$

$$-\mu_0(\mathbf{u}_i-\mathbf{u}_d^{(i)},\mathbf{w})_Q\ \forall\mathbf{w}\in\tilde{\mathbf{H}}^1(\Omega),\ \xi_i\in\mathbf{V},\ i=1,2,\tag{52}$$

$$\varkappa[\lambda\tilde{a}(\tau,\theta_i)+(k\tau,\theta_i)+c_1(\mathbf{u}_i,\tau,\theta_i)+\langle\zeta_i,\tau\rangle_{\Gamma_D}]-b_1(\tau,\xi_i)=0\ \forall\tau\in H^1(\Omega),\tag{53}$$

$$c(\mathbf{u},\mathbf{u},\xi_1+\xi_2)+\varkappa c_1(\mathbf{u},C,\theta_1+\theta_2)+\mu_0(\|\mathbf{u}\|_Q^2-(\mathbf{u},\mathbf{u}_d)_Q)\le-\mu_1\|\mathbf{g}\|_{1/2,\Gamma}^2.\tag{54}$$

It follows from (37) that $C\in\mathcal{T}$. Set $S=C$ in (51). Using (9) we obtain that

$$\lambda a_1(C,C)+(kC,C)=-c_1(\mathbf{u},C_1,C).\tag{55}$$

It follows from (13), (14), (29) that

$$|c_1(\mathbf{u},C_1,C)|\le\gamma_1M_C^0\|\mathbf{u}\|_1\|C\|_1,\ \ |b_1(C,\mathbf{u})|\le\beta_1\|\mathbf{u}\|_1\|C\|_1.\tag{56}$$

Taking into account (11), (56), we obtain from (55) that $\delta_1\lambda\|C\|^2\le\gamma_1M_C^0\|\mathbf{u}\|_1\|C\|_1$. From this inequality we deduce the following estimate for $\|C\|_1$:

$$\|C\|_1\le\frac{\gamma_1M_C^0}{\delta_1\lambda}\|\mathbf{u}\|_1.\tag{57}$$

Using (46), (57) and (30) we have

$$\|\mathbf{u}\|_1\le c_0(2\mathcal{R}+1)\|\mathbf{g}\|_{1/2,\Gamma}+\frac{2\beta_1}{\delta_0\nu}\frac{\gamma_1M_C^0}{\delta_1\lambda}\|\mathbf{u}\|_1=c_0(2\mathcal{R}+1)\|\mathbf{g}\|_{1/2,\Gamma}+2\mathcal{R}a\|\mathbf{u}\|_1.\tag{58}$$

It follows from (58) and (30) that $(1-2\mathcal{R}a)\|\mathbf{u}\|_1\le c_0(2\mathcal{R}+1)\|\mathbf{g}\|_{1/2,\Gamma}$. Taking into account that $2\mathcal{R}a<1$ by (31) we obtain from this estimate, (57) and (48) that

$$\|\mathbf{u}\|_1\le\frac{c_0(2\mathcal{R}+1)}{1-2\mathcal{R}a}\|\mathbf{g}\|_{1/2,\Gamma},\tag{59}$$

$$\|C\|_1\le\frac{c_0(2\mathcal{R}+1)}{(1-2\mathcal{R}a)}\frac{\gamma_1M_C^0}{\delta_1\lambda}\|\mathbf{g}\|_{1/2,\Gamma},\ \|p\|\le\frac{\delta_0\nu c_0(2\mathcal{R}+1)(\mathcal{R}+\mathcal{R}a)}{\beta_0(1-2\mathcal{R}a)}\|\mathbf{g}\|_{1/2,\Gamma}.\tag{60}$$

Set $\mathbf{w}=\xi_i$, $\tau=\theta_i$ in (52), (53). Using (8), (9) and conditions $\xi_i\in\mathbf{V}$, $\theta_i\in\mathcal{T}$ we deduce that

$$\nu a_0(\xi_i,\xi_i)=-c(\xi_i,\mathbf{u}_i,\xi_i)-\varkappa c_1(\xi_i,C_i,\theta_i)-\mu_0(\mathbf{u}_i-\mathbf{u}_d^{(i)},\xi_i)_Q,\tag{61}$$

$$\varkappa[\lambda a_1(\theta_i,\theta_i)+(k\theta_i,\theta_i)]=b_1(\theta_i,\xi_i),\ i=1,2.\tag{62}$$

It follows from (10)–(14), (7), (29) that

$$a_0(\xi_i,\xi_i)\ge\delta_0\|\xi_i\|_1^2,\ |c(\xi_i,\mathbf{u}_i,\xi_i)|\le\gamma_0\|\mathbf{u}_i\|_1\|\xi_i\|_1^2\le\gamma_0M_{\mathbf{u}}^0\|\xi_i\|_1^2,\tag{63}$$

$$a_1(\theta_i,\theta_i) \geq \delta_1\|\theta_i\|_1^2,\ |b_1(\theta_i,\xi_i)| \leq \beta_1\|\theta_i\|_1\|\xi_i\|_1,\ |c_1(\xi_i,C_i,\theta_i)| \leq \gamma_1 M_C^0\|\xi_i\|_1\|\theta_i\|_1, \quad (64)$$

$$|(\mathbf{u}_i-\mathbf{u}_d^{(i)},\xi_i)_Q| \leq \|\mathbf{u}_i-\mathbf{u}_d^{(i)}\|_Q\|\xi_i\|_Q \leq c_Q(c_Q M_{\mathbf{u}}^0+\|\mathbf{u}_d^{(i)}\|_Q)\|\xi_i\|_1. \quad (65)$$

Taking into account (63)–(65) we deduce from (61) and (62) that

$$\|\theta_i\|_1 \leq \frac{\beta_1}{\delta_1\lambda\varkappa}\|\xi_i\|_1,\ \left(\delta_0\nu-\gamma_0 M_{\mathbf{u}}^0-\frac{\beta_1\gamma_1}{\delta_1\lambda}M_C^0\right)\|\xi_i\|_1^2 \leq \mu_0 c_Q(c_Q M_{\mathbf{u}}^0+\|\mathbf{u}_d^{(i)}\|_Q)\|\xi_i\|_1.$$

Combining these inequalities with (45) and (30) gives

$$\|\xi_i\|_1 \leq \frac{2\mu_0\gamma_3}{\gamma_0}(\mathcal{R}e+\mathcal{R}e^0),\ \ \|\theta_i\|_1 \leq \frac{\beta_1}{\delta_1\lambda\varkappa}\frac{2\mu_0\gamma_3}{\gamma_0}(\mathcal{R}e+\mathcal{R}e^0), \quad (66)$$

where

$$\gamma_3=c_Q^2,\ \mathcal{R}e^0=\frac{\gamma_0}{\delta_0\nu c_Q}\max(\|\mathbf{u}_d^{(1)}\|_Q,\|\mathbf{u}_d^{(2)}\|_Q). \quad (67)$$

Taking into account (12), (13), (57), (66) and (30) we have

$$|c(\mathbf{u},\mathbf{u},\xi_1+\xi_2)| \leq \gamma_0\|\mathbf{u}\|_1^2(\|\xi_1\|_1+\|\xi_2\|_1) \leq 4\mu_0\gamma_3(\mathcal{R}e+\mathcal{R}e^0)\|\mathbf{u}\|_1^2,$$

$$\varkappa|c_1(\mathbf{u},C,\theta_1+\theta_2)| \leq \varkappa\gamma_1\|\mathbf{u}\|_1\|C\|_1(\|\theta_1\|_1+\|\theta_2\|_1) \leq$$

$$\frac{\gamma_1 M_C^0}{\delta_1\lambda}\frac{\gamma_1\beta_1}{\delta_1\lambda}\frac{4\mu_0\gamma_3(\mathcal{R}e+\mathcal{R}e^0)}{\gamma_0}\|\mathbf{u}\|_1^2=4\mu_0\gamma_3(\mathcal{R}e+\mathcal{R}e^0)\frac{\gamma_1}{\gamma_0}\mathcal{PR}a\|\mathbf{u}\|_1^2$$

which yields

$$|c(\mathbf{u},\mathbf{u},\xi_1+\xi_2)+\varkappa c_1(\mathbf{u},C,\theta_1+\theta_2)| \leq 4\mu_0\gamma_3(\mathcal{R}e+\mathcal{R}e^0)\,[1+(\gamma_1/\gamma_0)\mathcal{PR}a]\,\|\mathbf{u}\|_1^2. \quad (68)$$

Using (59) we deduce from (68) that

$$|c(\mathbf{u},\mathbf{u},\xi_1+\xi_2)+\varkappa c_1(\mathbf{u},C,\theta_1+\theta_2)| \leq$$

$$\leq \frac{4\mu_0\gamma_3 c_0^2(2\mathcal{R}+1)^2(\mathcal{R}e+\mathcal{R}e^0)[1+(\gamma_1/\gamma_0)\mathcal{PR}a]}{(1-2\mathcal{R}a)^2}\|\mathbf{g}\|_{1/2,\Gamma}^2. \quad (69)$$

Let input data for problem (49) and parameters μ_0,μ_1 be such that

$$(1-\varepsilon)\mu_1 \geq \frac{4\mu_0\gamma_3 c_0^2(2\mathcal{R}+1)^2(\mathcal{R}e+\mathcal{R}e^0)[1+(\gamma_1/\gamma_0)\mathcal{PR}a]}{(1-2\mathcal{R}a)^2},\ \varepsilon=\text{const}>0. \quad (70)$$

Here and further $\varepsilon>0$ is a (small) constant. In view of (70) we find from (69) that

$$|c(\mathbf{u},\mathbf{u},\xi_1+\xi_2)+\varkappa c_1(\mathbf{u},C,\theta_1+\theta_2)| \leq (1-\varepsilon)\mu_1\|\mathbf{g}\|_{1/2,\Gamma}^2. \quad (71)$$

Taking into account (71) we come from (54) to the inequality

$$\mu_0(\|\mathbf{u}\|_Q^2-(\mathbf{u},\mathbf{u}_d)_Q) \leq -c(\mathbf{u},\mathbf{u},\xi_1+\xi_2)-\varkappa c_1(\mathbf{u},C,\theta_1+\theta_2)-\mu_1\|\mathbf{g}\|_{1/2,\Gamma}^2 \leq -\varepsilon\mu_1\|\mathbf{g}\|_{1/2,\Gamma}^2. \quad (72)$$

It follows from (72) that $\|\mathbf{u}\|_Q^2 \le (\mathbf{u},\mathbf{u}_d)_Q \le \|\mathbf{u}\|_Q\|\mathbf{u}_d\|_Q$, which yields $\|\mathbf{u}\|_Q \le \|\mathbf{u}_d\|_Q$. As $\mathbf{u} = \mathbf{u}_1 - \mathbf{u}_2$, $\mathbf{u}_d = \mathbf{u}_d^{(1)} - \mathbf{u}_d^{(2)}$ we deduce the following estimate:

$$\|\mathbf{u}_1 - \mathbf{u}_2\|_Q \le \|\mathbf{u}_d^{(1)} - \mathbf{u}_d^{(2)}\|_Q. \tag{73}$$

The estimate (73) in the case where $Q = \Omega$ has the sense of the stability estimate of the component $\hat{\mathbf{u}}$ of the solution $(\hat{\mathbf{u}},\hat{p},\hat{C},\hat{\mathbf{g}})$ to problem (49) with respect to small disturbances in the $\mathbf{L}^2(\Omega)$-norm of the function $\mathbf{v}_d \in \mathbf{L}^2(\Omega)$ which enters into the expression for the functional I_1 in (22). In the case where $\mathbf{u}_d^{(1)} = \mathbf{u}_d^{(2)}$ it follows from (73) that $\mathbf{u}_1 = \mathbf{u}_2$. This yields together with (57), (47) and condition $\mathbf{u}|_\Gamma = \mathbf{g} \equiv \mathbf{g}_1 - \mathbf{g}_2$ in (35) that $C_1 = C_2$, $p_1 = p_2$ and $\mathbf{g}_1 = \mathbf{g}_2$. The latter means the uniqueness of the solution to problem (49) when $Q = \Omega$ and (70) holds.
We note that the uniqueness and stability of the solution to problem (49) under condition (70) take place and in the case where $Q \subset \Omega$, i.e. Q is a part of Ω. In order to prove this fact let us consider the inequality (72). Using (73) rewrite it in the form

$$\varepsilon\mu_1\|\mathbf{g}\|_{1/2,\Gamma}^2 \le c(\mathbf{u},\mathbf{u},\xi_1+\xi_2) + \varkappa c_1(\mathbf{u},C,\theta_1+\theta_2) + \mu_1\|\mathbf{g}\|_{1/2,\Gamma}^2 \le$$

$$\le -\mu_0\|\mathbf{u}\|_Q^2 + \mu_0\|\mathbf{u}\|_Q\|\mathbf{u}_d\|_Q \le \mu_0\|\mathbf{u}_d\|_Q^2.$$

From this relation, (59) and (60) we deduce the following stability estimates:

$$\|\mathbf{g}_1 - \mathbf{g}_2\|_{1/2,\Gamma} \le \sqrt{\frac{\mu_0}{\varepsilon\mu_1}}\|\mathbf{u}_d^{(1)} - \mathbf{u}_d^{(2)}\|_{\mathbf{H}^s(Q)},\ \|\mathbf{u}_1 - \mathbf{u}_2\|_1 \le \frac{c_0(2\mathcal{R}+1)}{1-2\mathcal{R}a}\sqrt{\frac{\mu_0}{\varepsilon\mu_1}}\|\mathbf{u}_d^{(1)} - \mathbf{u}_d^{(2)}\|_{\mathbf{H}^s(Q)},$$

$$\|C_1 - C_2\|_1 \le \frac{\gamma_1 M_C^0 c_0(2\mathcal{R}+1)}{\delta_1\lambda(1-2\mathcal{R}a)}\sqrt{\frac{\mu_0}{\varepsilon\mu_1}}\|\mathbf{u}_d^{(1)} - \mathbf{u}_d^{(2)}\|_{\mathbf{H}^s(Q)},$$

$$\|p_1 - p_2\| \le \frac{\delta_0\nu c_0(2\mathcal{R}+1)(\mathcal{R}+\mathcal{R}a)}{\beta_0(1-2\mathcal{R}a)}\sqrt{\frac{\mu_0}{\varepsilon\mu_1}}\|\mathbf{u}_d^{(1)} - \mathbf{u}_d^{(2)}\|_{\mathbf{H}^s(Q)}\ (\mathcal{R} \equiv \delta_0^{-1} + 2\mathcal{R}e) \tag{74}$$

where $s = 0$. Thus we have proved the following theorem.
Theorem 7. *Let under conditions (i), (ii), (j) and (31) the quadruple $(\mathbf{u}_i,p_i,C_i,\mathbf{g}_i)$ be the solution to problem (49) corresponding to a given function $\mathbf{u}_d^{(i)} \in \mathbf{L}^2(Q)$, $i = 1,2$, and the condition (70) holds where γ_3 and $\mathcal{R}e^0$ are defined in (67). Then stability estimates (73) and (74) under $s = 0$ hold true.*
We emphasize that the uniqueness and stability of the solution to problem (49) both under $Q = \Omega$, and under $Q \subset \Omega$ is proved only if parameter μ_1 in (49) is positive and satisfies (70). This means that term $(\mu_1/2)\|\mathbf{g}\|_{1/2,\Gamma}^2$ in the expression for the minimized functional J in (49) has a regularizing effect on control problem (49).
In the same manner one can study uniqueness and stability of solutions to boundary control problems for another cost functionals depending on the velocity $\mathbf{u}$. Let us consider for example the control problem

$$J(\mathbf{x},\mathbf{g}) \equiv \frac{\mu_0}{2}\|\mathbf{v} - \mathbf{v}_d\|_{1,Q}^2 + \frac{\mu_1}{2}\|\mathbf{g}\|_{1/2,\Gamma}^2 \to \inf,\ F(\mathbf{x},\mathbf{g}) = 0,\ \mathbf{x} = (\mathbf{v},q,S) \in X,\ \mathbf{g} \in K_1, \tag{75}$$

which corresponds to the cost functional $I_2(\mathbf{v}) = \|\mathbf{v} - \mathbf{v}_d\|_{1,Q}^2$. Denoting by $(\mathbf{x}_i,u_i) \equiv (\mathbf{u}_i,p_i,C_i,\mathbf{g}_i)$ the solution to problem (75) corresponding to a function $\mathbf{v}_d \equiv \mathbf{u}_d^{(i)} \in \mathbf{H}^1(Q)$, $i = 1,2$, and setting $\mathbf{g} = \mathbf{g}_1 - \mathbf{g}_2$, $\mathbf{u}_d = \mathbf{u}_d^{(1)} - \mathbf{u}_d^{(2)}$ in addition to (34) we note that

$$\langle (I_2)'_{\mathbf{u}}(\mathbf{x}_i),\mathbf{w}\rangle = 2(\mathbf{u}_i - \mathbf{u}_d^{(i)},\mathbf{w})_{1,Q},\ \langle (I_2)'_{\mathbf{u}}(\mathbf{x}_1) - (I_2)'_{\mathbf{u}}(\mathbf{x}_2),\mathbf{u}\rangle = 2(\mathbf{u} - \mathbf{u}_d,\mathbf{u})_{1,Q}$$

$$= 2(\|\mathbf{u}\|_{1,Q}^2 - (\mathbf{u},\mathbf{u}_d)_{1,Q}),\ (I_2)'_C = 0. \tag{76}$$

In view of (76) relations (35), (37), (51), (53), (62) and estimates (57), (59), (60) do not change while (43) and (32) under $\mathbf{w} = \xi_i$ take instead of (54), (61) the form

$$c(\mathbf{u},\mathbf{u},\xi_1+\xi_2) + \varkappa c_1(\mathbf{u},C,\theta_1+\theta_2) + \mu_0(\|\mathbf{u}\|_{1,Q} - (\mathbf{u},\mathbf{u}_d)_{1,Q}) \le -\mu_1\|\mathbf{g}\|_{1/2,\Gamma}^2, \tag{77}$$

$$\nu a(\xi_i,\xi_i) = -c(\xi_i,\mathbf{u}_i,\xi_i) - \varkappa c_1(\xi_i,C_i,\theta_i) - \mu_0(\mathbf{u}_i - \mathbf{u}_d^{(i)},\xi_i)_{1,Q},\ \xi_i \in \mathbf{V}. \tag{78}$$

Using the estimates (29) we deduce (instead of (65)) that

$$|(\mathbf{u}_i - \mathbf{u}_d^{(i)},\xi_i)_{1,Q}| \le \|\mathbf{u}_i - \mathbf{u}_d^{(i)}\|_{1,Q}\|\xi_i\|_{1,Q} \le (M_{\mathbf{u}}^0 + \|\mathbf{u}_d^{(i)}\|_{1,Q})\|\xi_i\|_1.$$

Proceeding as above we obtain estimates (66) for $\|\xi_i\|_1$, $\|\theta_i\|_1$ and inequality (69) where

$$\gamma_3 = 1,\ \mathcal{R}e^0 = (\gamma_0/\delta_0\nu)\max(\|\mathbf{u}_d^{(1)}\|_{1,Q}, \|\mathbf{u}_d^{(2)}\|_{1,Q}). \tag{79}$$

Let us assume that condition (70) takes place where γ_3 and $\mathcal{R}e^0$ are defined in (79). Using (70) we deduce (71). Taking into account (71) we obtain from (77) that

$$\mu_0(\|\mathbf{u}\|_{1,Q}^2 - (\mathbf{u},\mathbf{u}_d)_{1,Q}) \le -c(\mathbf{u},\mathbf{u},\xi_1+\xi_2) - \varkappa c_1(\mathbf{u},C,\theta_1+\theta_2) - \mu_1\|\mathbf{g}\|_{1/2,\Gamma}^2 \le -\varepsilon\mu_1\|\mathbf{g}\|_{1/2,\Gamma}^2. \tag{80}$$

It follows from (80) that $\|\mathbf{u}\|_{1,Q}^2 \le (\mathbf{u},\mathbf{u}_d)_{1,Q}$ which yields $\|\mathbf{u}\|_{1,Q} \le \|\mathbf{u}_d\|_{1,Q}$ or

$$\|\mathbf{u}_1 - \mathbf{u}_2\|_{1,Q} \le \|\mathbf{u}_d^{(1)} - \mathbf{u}_d^{(2)}\|_{1,Q}. \tag{81}$$

In the case where $Q = \Omega$ we deduce from (81), relation $\mathbf{u}|_\Gamma = \mathbf{g} = \mathbf{g}_1 - \mathbf{g}_2$, (7), (57) and (47) the following estimates:

$$\|\mathbf{u}_1 - \mathbf{u}_2\|_1 \le \|\mathbf{u}_d^{(1)} - \mathbf{u}_d^{(2)}\|_1,\ \|\mathbf{g}_1 - \mathbf{g}_2\|_{1/2,\Gamma} \le c_\Gamma\|\mathbf{u}_d^{(1)} - \mathbf{u}_d^{(2)}\|_1,$$

$$\|C_1 - C_2\|_1 \le \frac{\gamma_1 M_C^0}{\delta_1\lambda}\|\mathbf{u}_d^{(1)} - \mathbf{u}_d^{(2)}\|_1,\ \|p_1 - p_2\| \le \frac{\delta_0\nu(\mathcal{R} + \mathcal{R}a)}{\beta_0}\|\mathbf{u}_d^{(1)} - \mathbf{u}_d^{(2)}\|_1. \tag{82}$$

The estimates (82) have the sense of stability estimates for the solution $(\hat{\mathbf{u}},\hat{p},\hat{C},\hat{\mathbf{g}})$ to problem (75) under $Q = \Omega$ with respect to small disturbances in $\mathbf{H}^1(\Omega)$-norm of the function $\mathbf{v}_d$ which enters into the expression for the functional I_2. In the case where $\mathbf{u}_d^{(1)} = \mathbf{u}_d^{(2)}$ we deduce from (82) that $\mathbf{u}_1 = \mathbf{u}_2$, $\mathbf{g}_1 = \mathbf{g}_2$, $C_1 = C_2$, $p_1 = p_2$ which means the uniqueness of solution to control problem (75) under $Q = \Omega$. If $Q \subset \Omega$ the estimates (82) do not hold true but using (80) one can obtain more rough estimates of kind (74) instead of them. In fact rewriting (80) in view of (81) in the form

$$\varepsilon\mu_1\|\mathbf{g}\|_{1/2,\Gamma}^2 \le -\mu_0\|\mathbf{u}\|_{1,Q}^2 + \mu_0\|\mathbf{u}\|_{1,Q}\|\mathbf{u}_d\|_{1,Q} \le \mu_0\|\mathbf{u}_d\|_{1,Q}^2$$

and using (59), (60) we come to to the estimates (74) under $s = 1$. Thus we have proved the following result.

Theorem 8. *Let under conditions (i), (ii), (j) and (31) the quadruple $(\mathbf{u}_i, p_i, C_i, \mathbf{g}_i)$ be a solution to problem (75) corresponding to a given function $\mathbf{u}_d^{(i)} \in \mathbf{H}^1(Q)$, $i = 1,2$, and the condition (70) holds* *where* γ_3 *and* $\mathcal{R}e^0$ *are defined in (79). Then stability estimates (81) and (74) under* $s = 1$ *hold true. Furthermore estimates (82) hold if* $Q = \Omega$.

We again note that the uniqueness and stability of the solution to problem (75) both under $Q = \Omega$ and under $Q \subset \Omega$ is proved above under condition that the parameter μ_1 in (75) satisfies (70). We can not prove the stability of the solution to problem (75) as well as to problem (49) in the case where $\mu_1 = 0$. But we can establish the local uniqueness of the solution to problem (75) under $\mu_1 = 0$ in the case where $Q = \Omega$. In fact setting $\mu_1 = 0$, $Q = \Omega$, $\mathbf{u}_d^{(1)} = \mathbf{u}_d^{(2)}$ in (77) we obtain the inequality

$$c(\mathbf{u},\mathbf{u},\xi_1+\xi_2) + \varkappa c_1(\mathbf{u},C,\theta_1+\theta_2) \le -\mu_0\|\mathbf{u}\|_1^2. \tag{83}$$

Let input data for problem (75) be such that

$$4(\mathcal{R}e + \mathcal{R}e^0)[1 + (\gamma_1/\gamma_0)\mathcal{P}\mathcal{R}a] < 1. \tag{84}$$

It follows from (68) under $\gamma_3 = 1$ and (83) that $\mathbf{u} = \mathbf{0}$, and from (57), (47) and relation $\mathbf{u}|_\Gamma = \mathbf{g}$ we deduce that $C_1 = C_2$, $p_1 = p_2$, $\mathbf{g}_1 = \mathbf{g}_2$. So the next theorem holds.

Theorem 9.*Let under conditions (i), (ii), (j) and (31)* $\mathbf{v}_d \in \mathbf{H}^1(\Omega)$ *be a given function,* $\mu_0 > 0$, $\mu_1 \ge 0$ *and the condition (84) takes place where* $\mathcal{R}e^0 = (\gamma_0/\delta_0\nu)\|\mathbf{v}_d\|_1$. *Then the solution* $(\hat{\mathbf{u}},\hat{p},\hat{C},\hat{\mathbf{g}})$ *to problem (75) under* $Q = \Omega$ *is unique.*

Let us consider the one-parameter control problem

$$J(\mathbf{x},\mathbf{g}) \equiv \frac{\mu_0}{2}\|\mathrm{rot}\,\mathbf{v} - \eta_d\|_Q^2 + \frac{\mu_1}{2}\|\mathbf{g}\|_{1/2,\Gamma}^2 \to \inf,\ F(\mathbf{x},\mathbf{g}) = 0,\ \mathbf{x} = (\mathbf{v},g,S) \in X,\ \mathbf{g} \in K_1, \tag{85}$$

corresponding to the cost functional $I_3(\mathbf{v}) = \|\mathrm{rot}\,\mathbf{v} - \eta_d\|_Q^2$. Denoting by $(\mathbf{x}_i,u_i) = (\mathbf{u}_i,p_i,C_i,\mathbf{g}_i)$, $i = 1,2$, the solution to problem (85) corresponding to the function $\eta_d = \zeta_d^{(i)} \in \mathbf{L}^2(Q)$, $i = 1,2$, and setting $\mathbf{g} = \mathbf{g}_1 - \mathbf{g}_2$, $\zeta_d = \zeta_d^{(1)} - \zeta_d^{(2)}$ in addition to (34) we note that

$$\langle (I_3)'_{\mathbf{u}}(\mathbf{x}_i),\mathbf{w}\rangle = 2(\mathrm{rot}\,\mathbf{u}_i - \zeta_d^{(i)},\mathrm{rot}\,\mathbf{w})_Q,\ \langle (I_3)'_{\mathbf{u}}(\mathbf{x}_1) - (I_3)'_{\mathbf{u}}(\mathbf{x}_2),\mathbf{u}\rangle =$$

$$= 2(\mathrm{rot}\,\mathbf{u} - \zeta_d,\mathrm{rot}\,\mathbf{u})_Q,\ (I_3)'_C = 0. \tag{86}$$

In view of (86) relations (35), (37), (51), (53), (62) and estimates (57), (59), (60) do not change while (43) and (32) under $\mathbf{w} = \xi_i$ transform to

$$c(\mathbf{u},\mathbf{u},\xi_1+\xi_2) + \varkappa c_1(\mathbf{u},C,\theta_1+\theta_2) + \mu_0(\|\mathrm{rot}\,\mathbf{u}\|_Q^2 - (\zeta_d,\mathrm{rot}\,\mathbf{u})_Q) \le -\mu_1\|\mathbf{g}\|_{1/2,\Gamma}^2, \tag{87}$$

$$\nu a_0(\xi_i,\xi_i) = -c(\xi_i,\mathbf{u}_i,\xi_i) - \varkappa c_1(\xi_i,C_i,\theta_i) - \mu_0(\mathrm{rot}\,\mathbf{u}_i - \zeta_d^{(i)},\mathrm{rot}\,\xi_i)_Q,\ \xi_i \in \mathbf{V}. \tag{88}$$

Using (7) and (29) we have

$$|(\mathrm{rot}\,\mathbf{u}_i - \zeta_d^{(i)},\mathrm{rot}\,\xi_i)_Q| \le (\|\mathrm{rot}\,\mathbf{u}_i\|_Q + \|\zeta_d^{(i)}\|_Q)\|\mathrm{rot}\,\xi_i\|_Q \le c_r(c_r M_{\mathbf{u}}^0 + \|\zeta_d^{(i)}\|_Q)\|\xi_i\|_1. \tag{89}$$

Taking into account (63), (64), (89) we deduce from (88) and (62) that

$$\|\theta_i\|_1 \le \frac{\beta_1\|\xi_i\|_1}{\delta_1\lambda\varkappa},\ (\delta_0\nu - \gamma_0 M_{\mathbf{u}}^0 - \frac{\beta_1\gamma_1}{\delta_1\lambda}M_C^0)\|\xi_i\|_1^2 \le \mu_0 c_r(c_r M_{\mathbf{u}}^0 + \|\zeta_d^{(i)}\|_Q)\|\xi_i\|_1.$$

In view of (45) we obtain from this inequality that $\|\xi_i\|_1 \le (2\mu_0/\delta_0\nu)c_r^2(M_{\mathbf{u}}^0 + c_r^{-1}\|\zeta_d^{(i)}\|_Q)$ which yields (66), (68) and (69) where

$$\gamma_3 = c_r^2, \ \mathcal{R}e^0 = \frac{\gamma_0}{\delta_0 \nu c_r}\max(\|\zeta_d^{(1)}\|_Q, \|\zeta_d^{(2)}\|_Q). \tag{90}$$

Let us assume that the condition (70) takes place where γ_3 and $\mathcal{R}e^0$ are defined in (90). Using (70) we deduce (71). Taking into account (71) we obtain from (87) that

$$\mu_0(\|\mathrm{rot}\,\mathbf{u}\|_Q^2 - (\mathrm{rot}\,\mathbf{u}, \zeta_d)_Q) \le -\varepsilon\mu_1\|\mathbf{g}\|_{1/2,\Gamma}^2. \tag{91}$$

It follows from (91) that $\|\mathrm{rot}\,\mathbf{u}\|_Q^2 \le (\mathrm{rot}\,\mathbf{u}, \zeta_d)_Q$ which yields $\|\mathrm{rot}\,\mathbf{u}\|_Q \le \|\zeta_d\|_Q$ or

$$\|\mathrm{rot}\,\mathbf{u}_1 - \mathrm{rot}\,\mathbf{u}_2\|_Q \le \|\zeta_d^{(1)} - \zeta_d^{(2)}\|_Q. \tag{92}$$

Rewriting (91) by (92) in the form $\varepsilon\mu_1\|\mathbf{g}\|_{1/2,\Gamma}^2 \le \mu_0\|\zeta_d\|_Q^2$ and using (59), (60) we obtain the following stability estimates:

$$\|\mathbf{g}_1 - \mathbf{g}_2\|_{1/2,\Gamma} \le \sqrt{\frac{\mu_0}{\varepsilon\mu_1}}\|\zeta_d^{(1)} - \zeta_d^{(2)}\|_Q, \ \|\mathbf{u}_1 - \mathbf{u}_2\|_1 \le \frac{c_0(2\mathcal{R}+1)}{1-2\mathcal{R}a}\sqrt{\frac{\mu_0}{\varepsilon\mu_1}}\|\zeta_d^{(1)} - \zeta_d^{(2)}\|_Q,$$

$$\|C_1 - C_2\|_1 \le \frac{\gamma_1 M_C^0}{\delta_1\lambda}\frac{c_0(2\mathcal{R}+1)}{1-2\mathcal{R}a}\sqrt{\frac{\mu_0}{\varepsilon\mu_1}}\|\zeta_d^{(1)} - \zeta_d^{(2)}\|_Q,$$

$$\|p_1 - p_2\| \le \frac{\delta_0\nu c_0(2\mathcal{R}+1)(\mathcal{R}+\mathcal{R}a)}{\beta_0(1-2\mathcal{R}a)}\sqrt{\frac{\mu_0}{\varepsilon\mu_1}}\|\zeta_d^{(1)} - \zeta_d^{(2)}\|_Q. \tag{93}$$

Thus we have proved the following theorem.

Theorem 10. *Let under conditions (i), (ii), (j) and (31) the quadruple $(\mathbf{u}_i, p_i, C_i, \mathbf{g}_i)$ be a solution to problem (85) corresponding to a given function $\zeta_d^{(i)} \in \mathbf{L}^2(Q)$, $i = 1,2$, and condition (70) holds where γ_3 and $\mathcal{R}e^0$ are defined in (90). Then stability estimates (92) and (93) hold true.*

We can not prove the stability of the solution to problem (85) in the case where $\mu_1 = 0$. But we can establish the local uniqueness of the solution to problem (85) under more strict conditions on Ω and boundary vector $\mathbf{g}$ if we replace condition (j) by the next condition:

(j′) Ω is a simply connected domain with the boundary $\Gamma \in C^{1,1}$; $K_1 \subset \tilde{\mathbf{H}}^{1/2}(\Gamma)$ is a convex closed set consisting of functions $\mathbf{g}$ which satisfy the condition $\mathbf{g}\cdot\mathbf{n}|_\Gamma = q$ where $q \in H^{1/2}(\Gamma)$ is a given function.

Indeed, let us note that (87) takes under $\mu_1 = 0$, $\zeta_d^{(1)} = \zeta_d^{(2)}$, $Q = \Omega$ a form

$$c(\mathbf{u}, \mathbf{u}, \xi_1 + \xi_2) + \varkappa c_2(\mathbf{u}, C, \theta_1 + \theta_2) \le -\mu_0\|\mathrm{rot}\,\mathbf{u}\|^2. \tag{94}$$

Under the first condition in (j′) the difference $\mathbf{g} = \mathbf{g}_1 - \mathbf{g}_2$ has the zero normal component on Γ. Therefore taking into account the simple connectedness of the domain Ω we have the estimate $\|\mathbf{u}\|_1 \le c_3\|\mathrm{rot}\,\mathbf{u}\|$ with the constant c_3 depending on Ω (Girault & Raviart, 1986). Using this estimate we deduce from (68) where γ_3 and $\mathcal{R}e^0$ are given in (90) that

$$|c(\mathbf{u}, \mathbf{u}, \xi_1 + \xi_2) + \varkappa c_1(\mathbf{u}, C, \theta_1 + \theta_2)| \le 4\mu_0 c_r^2 c_3^2(\mathcal{R}e + \mathcal{R}e^0)[1 + (\gamma_1/\gamma_0)\mathcal{P}\mathcal{R}a]\|\mathrm{rot}\,\mathbf{u}\|^2. \tag{95}$$

Let input data for problem (85) be such that

$$4c_r^2 c_3^2(\mathcal{R}e + \mathcal{R}e^0)[1 + (\gamma_1/\gamma_0)\mathcal{P}\mathcal{R}a] < 1. \tag{96}$$

It follows from (94) and (95) that $\mathrm{rot}\,\mathbf{u} = \mathbf{0}$ which yields $\mathbf{u} = \mathbf{0}$ or $\mathbf{u}_1 = \mathbf{u}_2$. From (57), (47) and the condition $\mathbf{u}|_\Gamma = \mathbf{g}$ we deduce that $C_1 = C_2$, $p_1 = p_2$, $\mathbf{g}_1 = \mathbf{g}_2$. Thus we have proved the following theorem.

Theorem 11. *Let under conditions (i), (ii), (j') and (31) $\eta_d \in \mathbf{L}^2(\Omega)$ be a given function, $\mu_0 > 0$, $\mu_1 \geq 0$ and the condition (96) takes place where $\mathcal{R}e^0 = (\gamma_0/\delta_0 \nu c_r)\|\eta_d\|$. Then the solution $(\hat{\mathbf{u}}, \hat{p}, \hat{C}, \hat{\mathbf{g}})$ to problem (85) under $Q = \Omega$ is unique.*

In conclusion we consider two-parameter control problem corresponding to the cost functional I_1, i.e. we consider the control problem

$$J(\mathbf{x},u) \equiv \frac{\mu_0}{2}\|\mathbf{v}-\mathbf{v}_d\|_Q^2 + \frac{\mu_1}{2}\|\mathbf{g}\|_{1/2,\Gamma}^2 + \frac{\mu_2}{2}\|\chi\|_{\Gamma_N}^2 \to \inf,\ F(\mathbf{x},u) = 0,$$

$$\mathbf{x} = (\mathbf{v}, q, S) \in X,\ u = (\mathbf{g}, \chi) \in K_1 \times K_2. \tag{97}$$

Let $(\mathbf{x}_1, u_1) \equiv (\mathbf{u}_1, p_1, C_1, \mathbf{g}_1, \chi_1)$ be a solution to problem (97) which corresponds to a function $\mathbf{v}_d \equiv \mathbf{u}_d^{(1)} \in \mathbf{L}^2(Q)$, $(\mathbf{x}_2, u_2) \equiv (\mathbf{u}_2, p_2, C_2, \mathbf{g}_2, \chi_2)$ be a solution to problem (97) which corresponds to another function $\tilde{\mathbf{v}}_d \equiv \mathbf{u}_d^{(2)} \in \mathbf{L}^2(Q)$. Setting $\mathbf{g} = \mathbf{g}_1 - \mathbf{g}_2$, $\mathbf{u}_d = \mathbf{u}_d^{(1)} - \mathbf{u}_d^{(2)}$ in addition to (34), we note that the relations (50) hold true for problem (97). In view of (50) relations (52), (53), (61), (62) and estimates (66), (67) do not change while (43) takes the form

$$c(\mathbf{u},\mathbf{u},\xi_1+\xi_2) + \varkappa c_1(\mathbf{u},C,\eta_1+\eta_2) + \mu_0(\|\mathbf{u}\|_Q^2 - (\mathbf{u},\mathbf{u}_d)_Q) \leq -\mu_1\|\mathbf{g}\|_{1/2,\Gamma}^2 - \mu_2\|\chi\|_{\Gamma_N}^2. \tag{98}$$

Moreover instead of (51) we have to use the original identity (36). It follows from (37) that $C \in \mathcal{T}$. Setting $S = C$ in (36) and using (9) we obtain that

$$\lambda a_1(C,C) + (kC,C) = -c_1(\mathbf{u},C_1,C) + (\chi,C)_{\Gamma_N}. \tag{99}$$

Using (11), (15) and the first estimate in (56) we deduce from (99) that $\delta_1\lambda\|C\|^2 \leq \gamma_1 M_C^0\|\mathbf{u}\|_1\|C\|_1 + \gamma_2\|\chi\|_{\Gamma_N}\|C\|_1$. This yields the following estimate for $\|C\|_1$:

$$\|C\|_1 \leq \frac{\gamma_1 M_C^0}{\delta_1\lambda}\|\mathbf{u}\|_1 + \frac{\gamma_2}{\delta_1\lambda}\|\chi\|_{\Gamma_N}. \tag{100}$$

Taking into account (46) we obtain from (100) that

$$\|C\|_1 \leq \frac{2\beta_1}{\delta_0\nu}\frac{\gamma_1 M_C}{\delta_1\lambda}\|C\|_1 + c_0(2\mathcal{R}+1)\frac{\gamma_1 M_C^0}{\delta_1\lambda}\|\mathbf{g}\|_{1/2,\Gamma} + \frac{\gamma_2}{\delta_1\lambda}\|\chi\|_{\Gamma_N}. \tag{101}$$

Using (30) and (46), (47) we deduce the following estimates for C, $\mathbf{u}$ and p:

$$\|C\|_1 \leq \frac{c_0(2\mathcal{R}+1)}{(1-2\mathcal{R}a)}\frac{\gamma_1 M_C^0\|\mathbf{g}\|_{1/2,\Gamma}}{\delta_1\lambda} + \frac{\gamma_2\|\chi\|_{\Gamma_N}}{\delta_1\lambda(1-2\mathcal{R}a)},$$

$$\|\mathbf{u}\|_1 \leq \frac{c_0(2\mathcal{R}+1)\|\mathbf{g}\|_{1/2,\Gamma}}{1-2\mathcal{R}a} + \frac{2\beta_1\gamma_2\|\chi\|_{\Gamma_N}}{\delta_0\nu\delta_1\lambda(1-2\mathcal{R}a)},$$

$$\|p\| \leq \frac{\delta_0\nu c_0(2\mathcal{R}+1)(\mathcal{R}+\mathcal{R}a)\|\mathbf{g}\|_{1/2,\Gamma}}{\beta_0(1-2\mathcal{R}a)} + \frac{\beta_1(2\mathcal{R}+1)\gamma_2\|\chi\|_{\Gamma_N}}{\beta_0\delta_1\lambda(1-2\mathcal{R}a)}. \tag{102}$$

It follows from (12), (13), (66) and (102) that

$$|c(\mathbf{u},\mathbf{u},\xi_1+\xi_2)| \leq \gamma_0\|\mathbf{u}\|_1^2(\|\xi_1\|_1+\|\xi_2\|_1) \leq 4\mu_0\gamma_3\frac{(\mathcal{R}e+\mathcal{R}e^0)}{(1-2\mathcal{R}a)^2}\left[c_0(2\mathcal{R}+1)\|\mathbf{g}\|_{1/2,\Gamma}+\right.$$

$$\left.\frac{2\beta_1}{\delta_0\nu}\frac{\gamma_2}{\delta_1\lambda}\|\chi\|_{\Gamma_N}\right],\ \varkappa|c_1(\mathbf{u},C,\theta_1+\theta_2)| \leq \varkappa\gamma_1\|\mathbf{u}\|_1\|C\|_1(\|\theta_1\|_1+\|\theta_2\|_1) \leq 4\mu_0\gamma_3\frac{\gamma_1}{\gamma_0}\frac{\beta_1}{\delta_1\lambda}\times$$

$$\frac{(\mathcal{R}e+\mathcal{R}e^0)}{\gamma_0(1-2\mathcal{R}a)^2}\left[c_0(2\mathcal{R}+1)\|\mathbf{g}\|_{1/2,\Gamma}+\frac{2\beta_1}{\delta_0\nu}\frac{\gamma_2\|\chi\|_{\Gamma_N}}{\delta_1\lambda}\right]\left[c_0(2\mathcal{R}+1)\frac{\gamma_1 M_C^0\|\mathbf{g}\|_{1/2,\Gamma}}{\delta_1\lambda}+\frac{\gamma_2\|\chi\|_{\Gamma_N}}{\delta_1\lambda}\right].$$

Here γ_3 and $\mathcal{R}e^0$ are defined in (67). From these inequalities and (30) we deduce that

$$|c(\mathbf{u},\mathbf{u},\xi_1+\xi_2)+\varkappa c_1(\mathbf{u},C,\theta_1+\theta_2)| \leq \mu_0(a_1\|\mathbf{g}\|_{1/2,\Gamma}^2+a_2\|\chi\|_{\Gamma_N}^2). \tag{103}$$

Here constants a_1 and a_2 are given by

$$a_1=\frac{2\mu_0\gamma_0\gamma(\mathcal{R}e+\mathcal{R}e^0)}{(1-2\mathcal{R}a)^2}C_0^2(2\mathcal{R}+1)^2\left[6+\frac{\gamma_1}{\gamma_0}(5\mathcal{R}a+1)\mathcal{P}^2\right],$$

$$a_2=\frac{4\mu_0\gamma_0\gamma(\mathcal{R}e+\mathcal{R}e^0)}{(\delta_0\nu)^2(1-2\mathcal{R}a)^2}\left(\frac{2\beta_1 r\gamma_2}{\delta_1\lambda}\right)^2\left[3+\frac{\gamma_1}{\gamma_0}(\mathcal{R}a+2)\mathcal{P}^2\right]. \tag{104}$$

Let input data for problem (97) and parameters μ_0, μ_1, μ_2 be such that

$$(1-\varepsilon)\mu_1 \geq \mu_0 a_1,\ (1-\varepsilon)\mu_2 \geq \mu_0 a_2,\ \varepsilon=\text{const}>0. \tag{105}$$

In view of (105) we deduce from (103) that

$$|c(\mathbf{u},\mathbf{u},\xi_1+\xi_2)+\varkappa c_1(\mathbf{u},C,\theta_1+\theta_2)| \leq (1-\varepsilon)\mu_1\|\mathbf{g}\|_{1/2,\Gamma}^2+(1-\varepsilon)\mu_2\|\chi\|_{\Gamma_N}^2. \tag{106}$$

Combining (98) and (106) we obtain the inequality

$$\mu_0(\|\mathbf{u}\|_Q^2-(\mathbf{u},\mathbf{u}_d)_Q) \leq -c(\mathbf{u},\mathbf{u},\xi_1+\xi_2)-\varkappa c_1(\mathbf{u},C,\theta_1+\theta_2)-\mu_1\|\mathbf{g}\|_{1/2,\Gamma}^2-\mu_2\|\chi\|_{\Gamma_N} \leq$$

$$-\varepsilon\mu_1\|\mathbf{g}\|_{1/2,\Gamma}^2-\varepsilon\mu_2\|\chi\|_{\Gamma_N}. \tag{107}$$

Using (107) we deduce (73) which takes place under conditions (105). The uniqueness of the solution to problem (97) follows from this estimate and (107), (102) when $Q=\Omega$.
Rewriting (107) in view of (73) in the form

$$\varepsilon\mu_1\|\mathbf{g}\|_{1/2,\Gamma}^2+\varepsilon\mu_2\|\chi\|_{\Gamma_N}^2 \leq c(\mathbf{u},\mathbf{u},\xi_1+\xi_2)+\varkappa c_1(\mathbf{u},C,\theta_1+\theta_2)+\mu_1\|\mathbf{g}\|_{1/2,\Gamma}^2+\mu_2\|\chi\|_{\Gamma_N}^2 \leq$$

$$\leq -\mu_0\|\mathbf{u}\|_Q^2+\mu_0\|\mathbf{u}\|_Q\|\mathbf{u}_d\|_Q \leq \mu_0\|\mathbf{u}_d\|_Q^2,$$

and using (102) we obtain the following stability estimates:

$$\|\mathbf{g}_1-\mathbf{g}_2\|_{\Gamma_N} \leq \sqrt{\frac{\mu_0}{\varepsilon\mu_1}}\|\mathbf{u}_d^{(1)}-\mathbf{u}_d^{(2)}\|_Q,\ \|\chi_1-\chi_2\|_{\Gamma_N} \leq \sqrt{\frac{\mu_0}{\varepsilon\mu_2}}\|\mathbf{u}_d^{(1)}-\mathbf{u}_d^{(2)}\|_Q,$$

$$\|\mathbf{u}_1-\mathbf{u}_2\|_1 \le \left[\frac{c_0(2\mathcal{R}+1)}{(1-2\mathcal{R}a)\sqrt{\mu_1}}+\frac{2\beta_1\gamma_2}{\delta_0\nu\delta_1\lambda(1-2\mathcal{R}a)\sqrt{\mu_2}}\right]\sqrt{\frac{\mu_0}{\varepsilon}}\|\mathbf{u}_d^{(1)}-\mathbf{u}_d^{(2)}\|_Q,$$

$$\|C_1-C_2\|_1 \le \left[\frac{c_0(2\mathcal{R}+1)}{(1-2\mathcal{R}a)}\frac{\gamma_1 M_C^0}{\delta_1\lambda\sqrt{\mu_1}}+\frac{\gamma_2}{\delta_1\lambda(1-2\mathcal{R}a)\sqrt{\mu_2}}\right]\sqrt{\frac{\mu_0}{\varepsilon}}\|\mathbf{u}_d^{(1)}-\mathbf{u}_d^{(2)}\|_Q,$$

$$\|p_1-p_2\| \le \frac{(2\mathcal{R}+1)}{\beta_0(1-2\mathcal{R}a)}\left[\frac{c_0\delta_0\nu(\mathcal{R}+\mathcal{R}a)}{\sqrt{\mu_1}}+\frac{\beta_1\gamma_2}{\delta_1\lambda\sqrt{\mu_2}}\right]\sqrt{\frac{\mu_0}{\varepsilon}}\|\mathbf{u}_d^{(1)}-\mathbf{u}_d^{(2)}\|_Q. \qquad (108)$$

So the next theorem holds.

Theorem 12.*Let under conditions (i), (ii), (j) and (31) the quintuple $(\mathbf{u}_i, p_i, C_i, \mathbf{g}_i, \chi_i)$ be a solution to problem (97) corresponding to a given function $\mathbf{u}_d^{(i)} \in \mathbf{L}^2(Q)$, $i = 1,2$, and conditions (105) hold where parameters a_1 and a_3 are given by relations (104) in which γ_3 and $\mathcal{R}e^0$ are defined in (67). Then stability estimates (73) and (108) hold true.*

In the same manner one can study two-parameter control problems for another cost functionals entering into (22). Consider for example the following control problem:

$$J(\mathbf{x},\mathbf{g},\chi) \equiv \frac{\mu_0}{2}\|p-p_d\|_Q^2+\frac{\mu_1}{2}\|\mathbf{g}\|_{1/2,\Gamma}^2+\frac{\mu_2}{2}\|\chi\|_{\Gamma_N}^2 \to \inf,$$

$$F(\mathbf{x},u)=0,\ \mathbf{x}=(\mathbf{v},q,S)\in X,\ u=(\mathbf{g},\chi)\in K_1\times K_2, \qquad (109)$$

corresponding to the functional I_4. The following theorem holds.

Theorem 13.*Let under conditions (i), (ii), (j) and (31) the quintuple $(\mathbf{u}_i, p_i, C_i, \mathbf{g}_i, \chi_i)$ be a solution to problem (109) corresponding to a given function $p_d^{(i)} \in L^2(Q)$, $i = 1,2$, and the conditions (105) hold where parameters a_1 and a_3 are given by relations*

$$a_1 = \frac{2\mu_0\gamma_0(\mathcal{R}+\mathcal{R}a)\tilde{M}_p}{(1-2\mathcal{R}a)^2}C_0^2(2\mathcal{R}+1)^2\left[6+\frac{\gamma_1}{\gamma_0}(5\mathcal{R}a+1)\mathcal{P}^2\right],$$

$$a_2 = \frac{4\mu_0\gamma_0(\mathcal{R}+\mathcal{R}a)\tilde{M}_p}{(\delta_0\nu)^2(1-2\mathcal{R}a)^2}\left(\frac{2\beta_1 r\gamma_2}{\delta_1\lambda}\right)^2\left[3+\frac{\gamma_1}{\gamma_0}(\mathcal{R}a+2)\mathcal{P}^2\right]$$

in which

$$\tilde{M}_p = \beta_0^{-1}\left[M_p^0+\max(\|p_d^{(1)}\|_Q,\|p_d^{(2)}\|_Q)\right].$$

Then the estimate

$$\|p_1-p_2\|_Q \le \|p_d^{(1)}-p_d^{(2)}\|_Q$$

and stability estimates

$$\|\mathbf{g}_1-\mathbf{g}_2\|_{1/2,\Gamma} \le \sqrt{\frac{\mu_0}{\varepsilon\mu_1}}\|p_d^{(1)}-p_d^{(2)}\|_Q,\ \|\chi_1-\chi_2\|_{\Gamma_N} \le \sqrt{\frac{\mu_0}{\varepsilon\mu_2}}\|p_d^{(1)}-p_d^{(2)}\|_Q,$$

$$\|\mathbf{u}_1-\mathbf{u}_2\|_1 \le \left[\frac{c_0(2\mathcal{R}+1)}{(1-2\mathcal{R}a)\sqrt{\mu_1}}+\frac{2\beta_1\gamma_2}{\delta_0\nu\delta_1\lambda(1-2\mathcal{R}a)\sqrt{\mu_2}}\right]\sqrt{\frac{\mu_0}{\varepsilon}}\|p_d^{(1)}-p_d^{(2)}\|_Q,$$

$$\|C_1-C_2\|_1 \le \left[\frac{c_0(2\mathcal{R}+1)}{(1-2\mathcal{R}a)}\frac{\gamma_1 M_T^0}{\delta_1\lambda\sqrt{\mu_1}}+\frac{\gamma_2}{\delta_1\lambda(1-2\mathcal{R}a)\sqrt{\mu_2}}\right]\sqrt{\frac{\mu_0}{\varepsilon}}\|p_d^{(1)}-p_d^{(2)}\|_Q,$$

$$\|p_1 - p_2\| \leq \frac{(2\mathcal{R}+1)}{\beta_0(1-2\mathcal{R}a)}\left[\frac{c_0\delta_0\nu(\mathcal{R}+\mathcal{R}a)}{\sqrt{\mu_1}} + \frac{\beta_1\gamma_2}{\delta_1\lambda\sqrt{\mu_2}}\right]\sqrt{\frac{\mu_0}{\varepsilon}}\|p_d^{(1)} - p_d^{(2)}\|_Q$$

hold true.

Similar theorems can be formulated and proved for Models 2 and 3. Details can be found in (Alekseev, 2006; 2007a; Alekseev & Soboleva, 2009; Alekseev & Tereshko, 2010a; Alekseev & Khludnev, 2010).

5. Numerical algorithm. Results of numerical experiments for Model 2

In this section we discuss results of computational experiments related to the numerical solution of control problem (4) for Model 2. Our numerical algorithm will be based on optimality system for control problem (4) which is the analogue of relations (25), (26) and (27) for control problem (21). For the sake of simplicity we consider the case where control sets K_1 and K_3 coincide with spaces $\tilde{\mathbf{H}}^1(\Omega)$ and $L^2(\Gamma_N)$ respectively. In this case the minimum is reached in an internal point of set K and it is possible to express optimal controls $\mathbf{g}$ and η via adjoint state by explicit formulas (see (Alekseev & Tereshko, 2010c))

$$\mathbf{g} = (\sigma\mathbf{n} - \nu\partial\xi/\partial n)/\mu_1, \ \eta = \theta/\mu_2.$$

Using these expressions we can rewrite the optimality system as a nonlinear operator equation

$$\Phi(\mathbf{u}, p, T, \xi, \sigma, \theta) = 0.$$

For its numerical solution the iterative algorithm based on Newton's method is proposed. This algorithm consists of following steps:

1. For given $(\mathbf{u}_0, p_0, T_0, \xi_0, \sigma_0, \theta_0)$ and supposing $\mathbf{u}_n$, p_n, T_n, ξ_n, σ_n, θ_n are known, we define $\tilde{\mathbf{u}}$, $\tilde{p}$, $\tilde{T}$, $\tilde{\xi}$, $\tilde{\sigma}$, $\tilde{\theta}$ by solving the following linear problem:

$$\Phi'(\mathbf{u}_n, p_n, T_n, \xi_n, \sigma_n, \theta_n)(\tilde{\mathbf{u}}, \tilde{p}, \tilde{T}, \tilde{\xi}, \tilde{\sigma}, \tilde{\theta}) = -\Phi(\mathbf{u}_n, p_n, T_n, \xi_n, \sigma_n, \theta_n).$$

2. Then we calculate new approximations $\mathbf{u}_{n+1}$, p_{n+1}, T_{n+1}, ξ_{n+1}, σ_{n+1}, θ_{n+1} for $\mathbf{u}$, p, T, ξ, σ, θ as

$$\mathbf{u}_{n+1} = \mathbf{u}_n + \tilde{\mathbf{u}}, \ p_{n+1} = p_n + \tilde{p}, \ T_{n+1} = T_n + \tilde{T},$$

$$\xi_{n+1} = \xi_n + \tilde{\xi}, \ \sigma_{n+1} = \sigma_n + \tilde{\sigma}, \ \theta_{n+1} = \theta_n + \tilde{\theta}.$$

3. If the condition $\|T_{n+1} - T_n\|/\|T_n\| < \varepsilon$ for some sufficiently small number ε is not satisfied, then we go to step 1.

Below we shall present some numerical tests which illustrate an application of the proposed algorithm to the control problem (4). The first example is connected with the vortex reduction in the steady 2D viscous fluid flow around a cylinder in a channel. As a control we use heat flux η on some parts of the boundary. The initial uncontrolled flow is the solution of the boundary-value problem for the dimensionless stationary Navier-Stokes equations

$$-\frac{1}{\mathrm{Re}}\Delta\mathbf{u} + (\mathbf{u}\cdot\mathrm{grad})\mathbf{u} + \mathrm{grad}\, p = \mathbf{0}, \ \mathrm{div}\,\mathbf{u} = 0 \text{ in } \Omega = \Omega_2 \setminus \Omega_1,$$

$$\mathbf{u}|_{\Gamma_0} = \mathbf{0}, \ \mathbf{u}|_{\Gamma_1} = \mathbf{g}_1(y), \ \left(\frac{1}{\mathrm{Re}}\frac{\partial\mathbf{u}}{\partial n} - p\mathbf{n}\right)\bigg|_{\Gamma_2} = \mathbf{0}. \qquad (110)$$

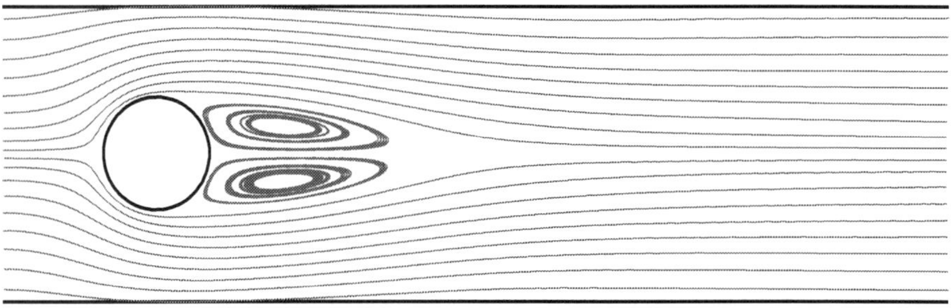

Fig. 1. Streamlines for uncontrolled flow (Re=100)

Here $\mathrm{Re} = UL/\nu$ is the Reynolds number, U and L are characteristic velocity and length for a flow. We prescribe the no-slip condition on the solid boundaries Γ_0 (the surface of the cylinder and channel walls), a parabolic inflow profile for the velocity on the inlet segment Γ_1 and "do-nothing" boundary conditions on the outlet Γ_2. The streamlines for uncontrolled flow are shown in Fig. 1. They were obtained by solving the boundary-value problem (110) using Newton's method. The open source software freeFEM++ (www.freefem.org) with an adaptive triangular mesh (about 3000 elements) is used for numerical solution of a linearized boundary-value problem. Usually 4 iterations were required for convergence of Newton's method when the solution to the corresponding Stokes problem plays the role of an initial guess.

The flow separation past the body can be clearly identified in Fig. 1. In order to reduce this recirculation the vorticity functional I_3 with $\zeta_d = 0$ in (22) is minimized using the proposed algorithm for the dimensionless analogue of Model 2. In this case we need to use additional dimensionless parameter - the Rayleigh number $\mathrm{Ra} = \tilde{\beta} G L^3 \Delta T/(\nu\lambda)$, where ΔT is a characteristic temperature difference. Firstly we choose the heat flux η only on the cylinder surface as control. The streamlines of the controlled flow for this case are shown in Fig. 2. One can see that flow separation past the body is eliminated.

In order to obtain a more laminarized flow we expand the area of action of temperature control. In Fig. 3 the controlled flow is presented for the case when the temperature control acts both on the cylinder surface and on the nearest parts of the channel walls (marked with dashed lines in Fig. 3). The vorticity of this flow is small and there is no recirculation zone past the cylinder. An analysis of the temperature field shows that the heat fluxes on additional parts of the boundary compensate a high temperature action on the cylinder surface.

		Re=20	Re=40	Re=60	Re=80	Re=100
Fig. 1	$\|\mathrm{rot}\,\mathbf{u}\|$	7.68	8.47	9.14	9.71	10.22
	C_D	4.25	2.77	2.26	2.01	1.85
Fig. 2	$\|\mathrm{rot}\,\mathbf{u}\|$	7.27	7.56	7.84	8.10	8.35
	C_D	4.06	2.60	2.16	1.95	1.83
Fig. 3	$\|\mathrm{rot}\,\mathbf{u}\|$	7.81	8.00	8.11	8.22	8.33
	C_D	5.42	3.03	2.12	1.64	1.34

Table 1. Norm of the vorticity and drag coefficient values

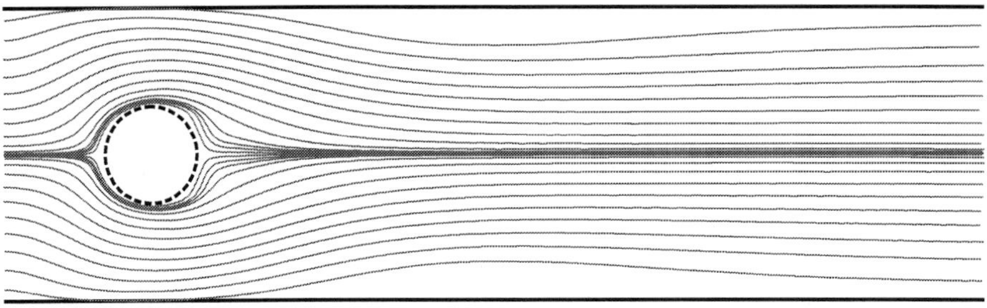

Fig. 2. Streamlines for the heat flux control on the cylinder (Re=100)

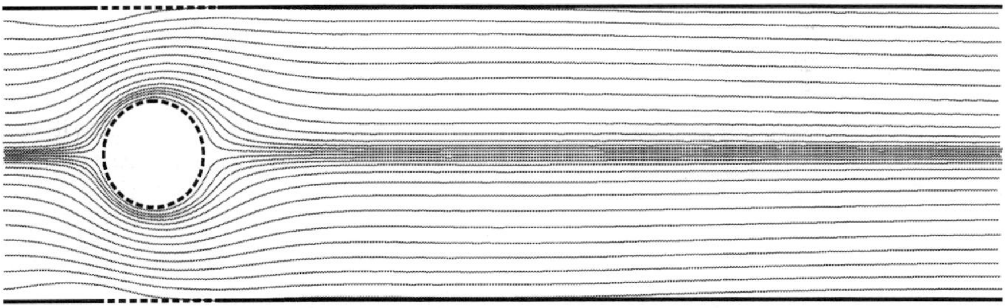

Fig. 3. Streamlines for the heat flux control on the cylinder and on the channel walls (Re=100)

In order to analyze the efficiency of different types of boundary control we calculate the norm of the vorticity $\|\mathrm{rot}\,\mathbf{u}\|$ and the drag coefficient C_D for different values of Reynolds number Re. These values for uncontrolled and controlled flows are shown in Table 1. The smallest values of these parameters correspond to the last case of boundary control.
Analogous results were obtained for the steady 2D viscous fluid flow around a cylinder in a channel with forward-facing step. The streamlines for uncontrolled flow are shown in Fig. 4. One can see a flow separation past the body and vortex at the corner. In order to reduce these recirculations the vorticity functional I_3 with $\zeta_d = 0$ in (22) is minimized. In Fig. 5 the controlled flow is presented for the case when the heat flux control acts both on the cylinder surface and on the nearest parts of the channel walls.
The regularization parameter μ_3 plays an important role in computations. If μ_3 is small then we usually obtain a flow with high temperature gradients on the cylinder surface because a substantial change of the velocity field requires powerful temperature action. This is the reason for the simultaneous use of two controls: $\mathbf{g}$ and η. If main change in the velocity field is achieved due to hydrodynamic control $\mathbf{g}$ then temperature control η will play only auxiliary role and temperature gradients will be small.
In the next example we consider two types of control on different parts of the boundary. As in the first example the heat flux on the cylinder surface and on the nearest parts of the channel walls is used to avoid the flow separation past the body. Additionally the Dirichlet control $\mathbf{g}$ on a part Γ_c of the boundary Γ (marked with dotted lines) is used for corner vortex suppression. Streamlines of corresponding controlled flow are shown in Fig. 6.

Fig. 4. Streamlines for uncontrolled flow (Re=50)

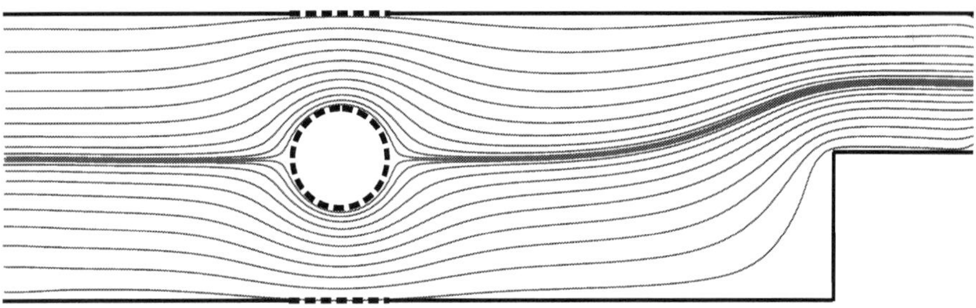

Fig. 5. Streamlines for the heat flux control on the cylinder and on the channel walls (Re=50)

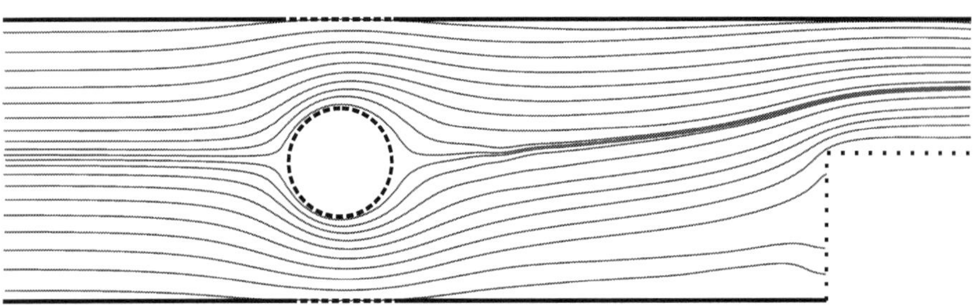

Fig. 6. Streamlines for the heat flux control on the cylinder and on the channel walls and velocity control on the step (Re=50)

		Re=10	Re=20	Re=30	Re=40	Re=50
Fig. 4	$\|\mathrm{rot}\,\mathbf{u}\|$	9.04	9.46	9.89	10.29	10.64
	C_D	7.16	4.09	3.11	2.63	2.34
Fig. 5	$\|\mathrm{rot}\,\mathbf{u}\|$	8.88	9.04	9.22	9.41	9.60
	C_D	6.21	2.98	2.02	1.59	1.34
Fig. 6	$\|\mathrm{rot}\,\mathbf{u}\|$	8.38	8.52	8.76	8.84	9.00
	C_D	6.36	3.00	2.00	1.54	1.28

Table 2. Norm of the vorticity and drag coefficient values

The flow separation past the body is eliminated and the corner vortex is absent in this case because of fluid suction at the corner. Calculated values of the norm of the vorticity and of the drag coefficient C_D are shown in Table 2. It can be observed that a simultaneous application of temperature and hydrodynamic controls results in the smallest values of $\|\mathrm{rot}\,\mathbf{u}\|$. Moreover in this case we obtain the smallest values of the temperature gradients.

The computational experiments showed that if the initial guess is selected sufficiently close to the exact solution, then the algorithm converges for several iterations. The regularization parameter μ_3 plays an important role. If its values are relatively large then we can not obtain small values of the functional I_3. But, on the other hand, the very small values of the regularization parameter can lead to the instability and oscillations in the numerical solution. An analysis of computational results shows that proposed method can be effectively used to produce flows in channels with desired dynamical properties. Some results of other numerical experiments can be found in (Tereshko, 2009; Alekseev & Tereshko, 2010b;c).

5.1 Acknowledgments

This work was supported by the Russian Foundation for Basic Research (projects no. 09-01-98518-r-vostok-a and 10-01-00219-a), Siberian Branch of the Russian Academy of Sciences (project no. 116-2009) and Far Eastern Branch of the Russian Academy of Sciences (projects no. 09-I-P29-01 and 09-I-OMN-03).

6. References

Alekseev, G.V. (1998a). Solvability of stationary boundary control problems for heat convection equations. *Siberian Mathematical Journal*, Vol. 39, No. 5, Sep 1998, 844–858. ISSN 0037-4466

Alekseev, G.V. (1998b). Stationary problems of boundary control for the heat-convection equations. *Doklady Akademii Nauk*, Vol. 362, No. 2, Sep 1998, 174–177. ISSN 1064-5624

Alekseev, G.V. & Tereshko, D.A. (1998a). On solvability of inverse extremal problem for stationary equations of viscous heat conducting fluid. *Journal of Inverse and Ill-Posed Problems*, Vol. 6, No. 6, Jan 1998, 521–562. ISSN 0928-0219

Alekseev, G.V. & Tereshko, D.A. (1998b). Stationary problems of optimal control for equations of the hydrodynamics of a viscous heat-conducting fluid. *Sibirskii Zhurnal Industrial'noi Matematiki*, Vol. 1, No. 2, May 1998, 24–44. ISSN 1560-7518 (In Russian)

Alekseev, G.V. (2000). Inverse extremum problems for stationary equations of heat and mass transfer. *Doklady Mathematics*, Vol. 62, No. 3, Dec 2000, 420–424. ISSN 1064-5624

Alekseev, G.V. & Adomavichus, E.A. (2001). Theoretical analysis of inverse extremum problems of admixture diffusion in viscous fluids. *Journal of Inverse and Ill-Posed Problems*, Vol. 9, No. 5, Jan 2001, 435–468. ISSN 0928-0219

Alekseev, G.V. (2001). Solvability of inverse extremum problems for stationary equations of heat and mass transfer. *Siberian Mathematical Journal*, Vol. 42, No. 5, Sep 2001, 811–827. ISSN 0037-4466

Alekseev, G.V. (2002). Inverse extremal problems for stationary equations in mass transfer theory. *Computational Mathematics and Mathematical Physics*, Vol. 42, No. 3, Sep 2002, 363–376. ISSN 0965-5425

Alekseev, G.V. (2006). Inverse extremum problems for stationary equations of heat convection. *Vestnik NGU*, Vol. 6, No. 2, Jul 2006, 6–32. ISSN 1818-7994 (In Russian)

Alekseev, G.V. (2007a). Coefficient inverse extremum problems for stationary heat and mass transfer equations. *Computational Mathematics and Mathematical Physics*, Vol. 47, Feb 2007, 1055–1076. ISSN 0965-5425

Alekseev, G.V. (2007b). Uniqueness and stability in coefficient identification problems for a stationary model of mass transfer. *Doklady Mathematics*, Vol. 76, No. 2, Feb 2007, 797–800. ISSN 1064-5624

Alekseev, G.V. & Kalinina, E.A. (2007). Coefficient identification problem for stationary convection-reaction-diffusion equation *Sibirskii Zhurnal Industrialnoi Matematiki*, Vol. 10, No. 1, Jan 2007, 3-16. ISSN: 1560-7518 (In Russian)

Alekseev, G.V.; Soboleva, O.V. & Tereshko, D.A. (2008). Identification problems for stationary model of mass transfer. *Journal of Applied Mechanics and Technical Physics*, Vol. 49, No. 4, Apr 2008, 24-35. ISSN 0021-8944

Alekseev, G.V. & Tereshko, D.A. (2008). *Analysis and optimization in viscous fluid dynamics*, Dalnauka, ISBN 978-5-8044-1045-3, Vladivostok (In Russian)

Alekseev, G.V. & Soboleva, O.V. (2009). On stability of solutions of extremum problems for stationary equations of mass transfer. *Dal'nevostochnyi matematicheskii zhurnal*, Vol. 9, No. 1-2, Sep 2009, 5–14, ISSN 1608-845X (In Russian)

Alekseev, G.V. & Khludnev, A.M. (2010). The stability of solutions to extremal problems of boundary control for stationary heat convection equations. *Sibirskii Zhurnal Industrial'noi Matematiki*, Vol. 13, No. 2, May 2010, 5–18. ISSN 1560-7518 (In Russian)

Alekseev, G.V. & Tereshko D.A. (2010a). Boundary control problems for stationary equations of heat transfer, In: *New Directions in Mathematical Fluid Mechanics*, Fursikov, A.V.; Galdi, G.P.; Pukhnachev, V.V. (Eds.), 1–21, Birkhauser Verlag, ISBN 978-3-0346-0151-1, Basel

Alekseev, G.V. & Tereshko, D.A. (2010b). Extremum problems of boundary control for a stationary thermal convection model. *Doklady Mathematics*, Vol. 81, No. 1, Feb 2010, 151–155. ISSN 1064-5624

Alekseev, G.V. & Tereshko, D.A. (2010c). Extremum problems of boundary control for the stationary model of heat convection. *Journal of Applied Mechanics and Technical Physics*, Vol. 51, No. 4, Jul 2010, 453–463. ISSN 0021-8944

Andreev, V.K.; Kaptsov, O.V.; Pukhnachov, V.V. & Rodionov, A.A. (1998). *Applications of group-theoretical methods in hydrodynamics*. Kluwer Academic Publishers, ISBN 978-0-7923-5215-0, Dordrecht

Andreev, V.K.; Gaponenko, Yu.A.; Goncharova, O.A. & Pukhnachev, V.V. (2008). *Modern mathematical models of convection*. Fizmatlit, ISBN 978-5-9221-0905-5, Moscow (in Russian)

Batchelor, G.K. (2000). *An introduction to fluid dynamics*. Cambridge University Press, ISBN

978-0-5216-6396-0, Cambridge

Capatina, A. & Stavre, R. (1998). A control problem in bioconvective flow. *Journal of Mathematics of Kyoto University*, Vol. 37, No. 4, Oct 1998, 585–595. ISSN 0023-608X

Gershuni, G.Z. & Zhukhovitskii, E.M. (1976). *Convective stability of incompressible fluids*. Keter, ISBN 978-0-7065-1562-6, Jerusalem

Girault, V., Raviart, P.A. (1986). *Finite element methods for Navier-Stokes equations. Theory and algorithms*. Springer-Verlag, ISBN 978-0-3871-5796-2, Berlin, New York

Goncharova, O.N. (2002). Unique solvability of a two-dimensional non-stationary problem for the convection equations with temperature depending viscosity. *Differential Equations*, Vol. 38, No. 2, Feb 2002, 249-258. ISSN 0012-2661

Gunzburger, M.D.; Hou, L. & Svobodny, T.P. (1991). Analysis and finite element approximation of optimal control problems for the stationary Navier-Stokes equations with distributed and Neumann controls. *Mathematics of Computation*, Vol. 57, Jul 1991, 123–151. ISSN 0025-5718

Gunzburger, M.D.; Hou, L. & Svobodny, T.P. (1993). The approximation of boundary control problems for fluid flows with an application to control by heating and cooling. *Computers & Fluids*, Vol. 22, Mar 1993, 239–251. ISSN 0045-7930

Hopf, E. (1941). Ein allgemeiner Endlichkeitssatz der Hydrodynamik. *Mathematische Annalen*, Vol. 117, 1940–1941, 764–775. ISSN 0025-5831

Ito, K. & Ravindran, S.S. (1998). Optimal control of thermally convected fluid flows. *SIAM Journal on Scientific Computing*, Vol. 19, No. 6, Nov 1998, 1847–1869. ISSN 1064-8275

Ioffe, A.D. & Tikhomirov, V.M. (1979). *Theory of extremal problems*. North Holland, ISBN 978-0-4448-5167-3, Amsterdam

Joseph, D.D. (1976). *Stability of fluid motions*. Springer-Verlag, ISBN 978-0-4711-1621-9, New York

Lee, H.C. & Imanuvilov, O.Yu. (2000a). Analysis of optimal control problems for the 2-D stationary Boussinesq equations. *Journal of Mathematical Analysis and Applications*, Vol. 242, No. 2, Feb 2000, 191–211. ISSN 0022-247X

Lee, H.C. & Imanuvilov, O.Yu. (2000b). Analysis of Neumann boundary optimal control problems for the stationary Boussinesq equations including solid media. *SIAM Journal on Control and Optimization*, Vol. 39, No. 2, Sep 2000, 457–477. ISSN 0363-0129

Lee, H.C. (2003). Analysis and computational methods of Dirichlet boundary control problems for 2D Boussinesq equations. *Advances in Computational Mathematics*, Vol. 19, No. 1-3, Jul 2003, 255–275. ISSN 1019-7168

Perera, P.S. & Sekerka, R.F. (1997). Nonsolenoidal flow in a liquid diffusion couple. *Physics of Fluids*. Vol. 9, No. 2, Feb 1997, 376-391. ISSN 1070-6631

Pukhnachov, V.V. (1992). Model of convective flow under low gravity. *Microgravity Quarterly*, Vol. 2, 1992, 251-252. ISSN 0958-5036

Pukhnachov, V.V. (2004). Hierarchy of models in the theory of convection. *Journal of Mathematical Sciences*, Vol. 123, No. 6, Oct 2004, 4607-4620. ISSN 1072-3374

Pukhnachev, V.V. (2009). J. Leray problem and V.I. Yudovich conjecture. *Proceedings of High Schools. Northern-Caucasian Region. Actual Problems of Mathematical Hydrodynamics*, Special Issue, 2009, 185–194 (in Russian)

Pukhnachev, V.V. (2010). Viscous flows in domains with multiply connected boundary, In: *Advances in Mathematical Fluid Mechanics*, In: *New Directions in Mathematical Fluid Mechanics*, Fursikov, A.V.; Galdi, G.P.; Pukhnachev, V.V. (Eds.), 333–348, Birkhauser Verlag, ISBN 978-3-0346-0151-1, Basel

Tereshko, D.A. (2009). Numerical solution of control problems for stationary model of heat convection. *Dal'nevostochnyi matematicheskii zhurnal*, Vol. 9, No. 1-2, Sep 2009, 168–175, ISSN 1608-845X (In Russian)

22

Heat and Mass Transfer in Desiccant Wheels

Celestino Ruivo[1,2], José Costa[2] and António Rui Figueiredo[2]
[1]University of Algarve,
[2]ADAI-University of Coimbra
Portugal

1. Introduction

1.1 Background

Nowadays the interest in heating, ventilation, air-conditioning and refrigerating systems (HVAC&R) based on desiccant wheels is increasing due to the possibility of using renewable energy sources, making them an attractive alternative or complement to conventional systems. The thermally driven desiccant systems can potentially reduce the peak electricity demand and associated electricity infrastructure costs. They generally incur in higher initial cost compared with equivalent conventional systems, but cost reduction can be achieved at the design stage through careful cycle selection, flow optimisation and size reduction.

The performance of these systems can be evaluated by experimental or numerical approaches. To date there still exists a lack of data of real manufactured wheels enabling to perform a dynamic energy analysis of such alternative systems with reasonable accuracy at design stage.

The data given by the manufacturers of desiccant wheels are usually restricted to particular sets of operating conditions. Besides, the available software for sizing is usually appropriate to run only stationary operating conditions. For these reasons, it is recognized the importance of the use of a simple predicting method to perform the dynamic simulation of air handling units equipped with desiccant wheels.

In this chapter, the results of a detailed numerical model are used to determine the effectiveness parameters for the coupled heat and mass transfer processes in desiccant wheels, allowing the use of the effectiveness method as an easy prediction tool for designers.

1.2 General characterization and modelling aspects

Desiccant wheels are air-to-air heat and mass exchangers used to promote the dehumidification of the process airflow. The rotor matrix, as illustrated in Fig. 1, is compact and mechanically resistant, and consists of a high number of channels with porous desiccant walls. The rotation speed of the wheel is relatively low. The hygroscopic matrix is submitted to a cyclic sequence of adsorption and desorption of water molecules. The regeneration process of the matrix (desorption) is imposed by a hot airflow. In each channel of the matrix, a set of physical phenomena occurs: heat and mass convection on the gas side as well as heat and mass diffusion and water sorption in the desiccant wall. The regeneration airflow should be heated by recovering energy from the system and using renewable energy sources whenever possible.

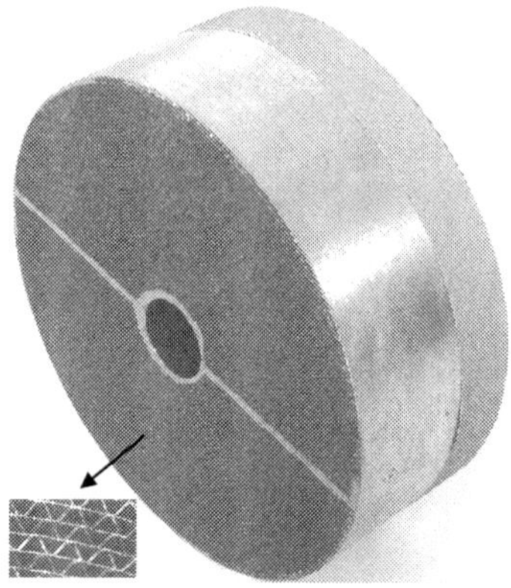

Fig. 1. Desiccant wheel and detail of the porous structure of the matrix

In the schematic representation of a desiccant wheel in Fig. 2, airflow 1 (process air) and airflow 2 (regeneration air) cross the matrix in a counter-current configuration, with equal or different mass flow rates. The desorption zone is generally equal or smaller than the adsorption zone.

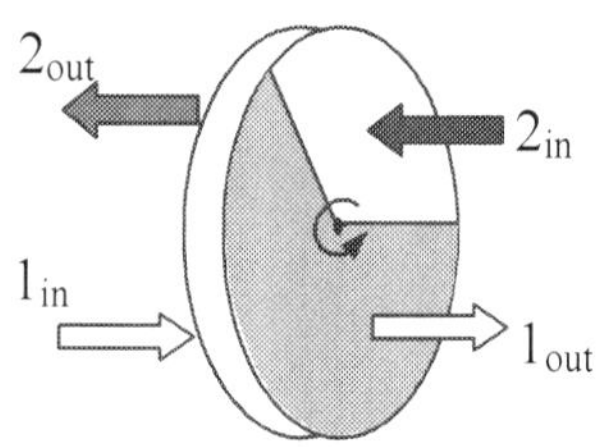

Fig. 2. Desiccant wheel (▭ - Adsorption zone; ▭ - Desorption zone)

The approaching airflows in each zone can present instabilities and heterogeneities and are generally turbulent. However, the relatively low values of hydraulic diameter of the channels (frequently less than 5 mm) together with moderate values of the frontal velocity (usually between 1 and 3 m s^{-1}) impose laminar airflows. Besides, in very short matrixes, the entrance effects can be relevant, particularly for larger hydraulic diameters of the channels.

During the adsorption/desorption cycle, the matrix exhibits non uniform distributions of adsorbed water content and temperature, and the angular gradients depend on the constitution of the wall matrix and also on the rotation speed.

The desiccant wheels are mainly used in dehumidification systems to control the humidity of airflows or the indoor air conditions in process rooms of some industries. Fig. 3.a schematically represents a system with a heating coil, operating by Joule effect, or actuating as a heat exchanger, to heat the regeneration airflow. In Fig. 3.b, a desiccant hybrid system with two stages of air dehumidification is shown. The first stage occurs in a cooling coil of the compression vapour system and the second corresponds to the adsorption in the desiccant wheel. The heat released by the condenser is recovered to heat the regeneration airflow, improving the global efficiency of the system.

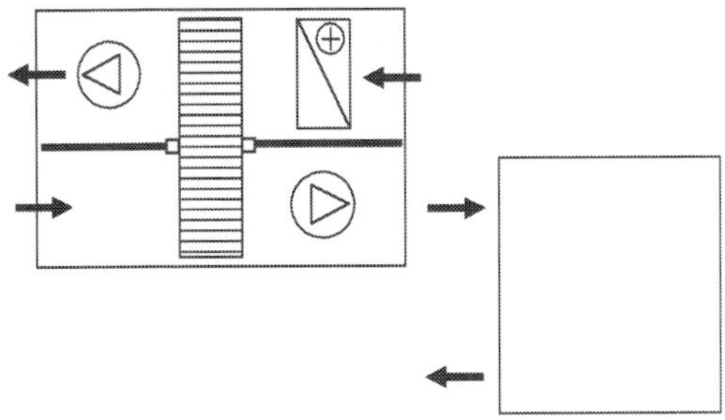

a)

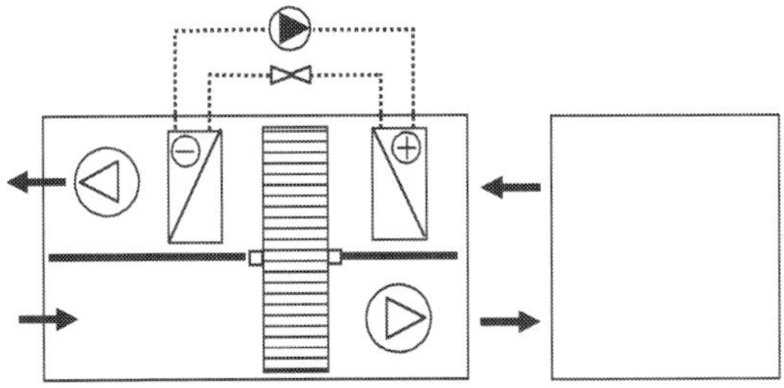

b)

Fig. 3. Dehumidification systems based on desiccant wheels: a) simple dehumidification system and b) hybrid dehumidification system

Another possible interesting application, although less common, is for air cooling operations, combining the evaporative cooling with the solid adsorption dehumidification, as schematically represented in Fig. 4.

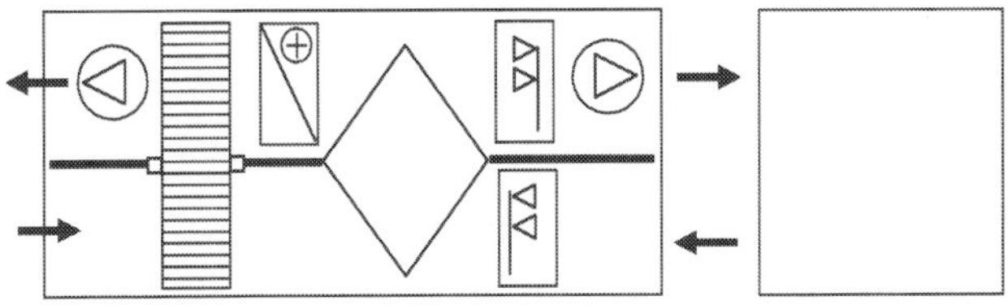

Fig. 4. Desiccant evaporative cooling system

The moisture removal capacity of the desiccant wheel can exhibit significant time variations according to the load profile and weather conditions, a fact that must be taken into account at design stage. On the other hand, the operational costs depend on the control strategy chosen for the system. The capacity control alternatives can be based on: *a)* fan modulation, *b)* by-pass of the process airflow or of the regeneration airflow, *c)* modulation of the heating device for regeneration or *d)* modulation of the rotation speed of the wheel. The strategies based on variable airflow by fan modulation are generally more efficient, presenting higher potential to reduce the running costs.

Different numerical modelling methods of solution supported by different simplified treatments of the flow and the solid domains have been used. Several numerical difficulties are related with the coupling between the different phenomena and the computational time consumption, mainly in detailed numerical models. One crucial aspect is the characterization of the matrix material of the desiccant wheel, namely the knowledge of its thermal properties, diffusion coefficients, phase equilibrium laws, hysteresis effects, etc. In Pesaran (1983), the study of water adsorption in silica gel particles is focussed on the importance of the internal resistances to mass transfer. The investigation of Kodama (1996) deals with the experimental characterization of the matrix of a desiccant rotor made of a composite desiccant medium, a fibrous material impregnated with silica gel.

It is recognized the importance of validating the numerical models by comparison with experimental data, necessarily covering a wide range of conditions, but the published data on this matter are scarce. In some cases, the degree of accuracy of the measured results is not indicated and, in other works, a poor degree of accuracy is reported. Moreover, some examples of exhaustive experimental research on the behaviour of a desiccant wheel (Cejudo et al., 2006) show significant mass and energy imbalances between the regeneration and the process air streams.

1.3 Real and ideal psychrometric evolutions

An example of the psychrometric evolutions in both air flows is schematically represented in Fig. 5. A decrease of the water vapour content and a temperature increase of the process air are observed and opposite changes are observed in the regeneration air.

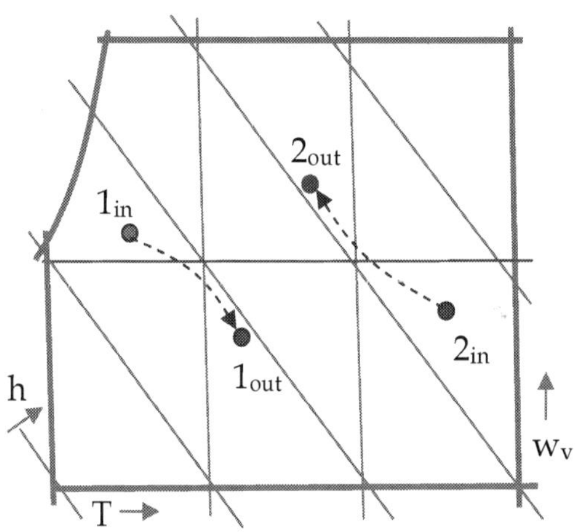

Fig. 5. Psychrometric evolutions of the airflows in a desiccant wheel

The outlet states of both airflows are influenced by the rotation speed, the airflow rates, the transfer area in the adsorption and the desorption zones of the wheel, the thickness of channel wall and its properties. The expected influence of the channel length and of the adsorption/desorption cycle duration on the outlet states of both airflows is schematically represented in Fig. 6, for the particular case of equal mass flow rates. The outlet state of each airflow is defined by the interception of the isolines of the channel length L_c and of the cycle duration τ_{cyc}. The solid curves L_{c1}, L_{c2} and L_{c3} correspond to rotor matrix with short, medium-length and long channels, respectively. The solid curves τ_{cyc1}, τ_{cyc2} and τ_{cyc3} correspond to low, medium and high cycle durations, respectively. For each channel length an optimum value of the cycle duration exists, i.e. the optimum rotation speed that maximizes the dehumidification rate. This optimal rotation diminishes with the channel length.

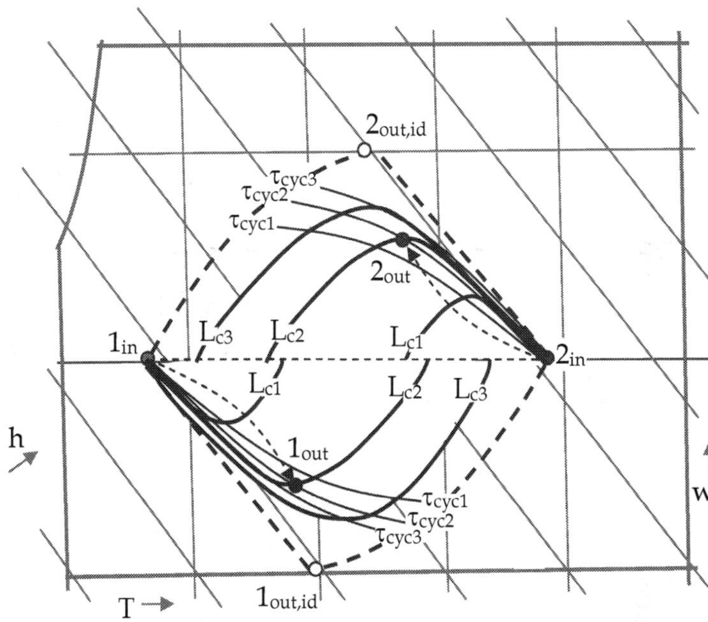

Fig. 6. Influence of the channel length and of the cycle duration on the psychrometric evolutions

The ideal behaviour of a desiccant wheel corresponds to cases with infinite transfer area of the channel. It is common to take the maximum ideal dehumidification rate as a reference, the corresponding outlet states being represented in Fig. 6 by $1_{out,id}$ and $2_{out,id}$. The identification of the ideal outlet states requires the knowledge of the equilibrium curves of the hygroscopic matrix, i.e. the sorption isotherms. Such information is schematically represented in Fig. 7 by the adsorbed water content X_ℓ as a function of the water vapour content w_v and of the temperature T.

The adsorbed water content in the hygroscopic matrix at the equilibrium condition imposed by the inlet state of the process airflow corresponds to the ideal maximum value. The minimum value of the adsorbed water content that can be achieved in ideal operating conditions is dictated by the inlet conditions of the regeneration airflow. The horizontal lines c′1out and c′2out in Fig. 7 represent those minimum and maximum values, respectively, and correspond to the dashed curves c′1out and c′2out in Fig. 8. In most hygroscopic matrices, those curves correspond to constant or quite constant values of the ratio of the water vapour partial to saturation pressure (p_v/p_{vs}). This ratio corresponds strictly to the relative humidity concept of the moist air only in the cases where the temperature of the moisture air is lower than the water saturation temperature at local atmospheric pressure (ASHRAE, 1989).

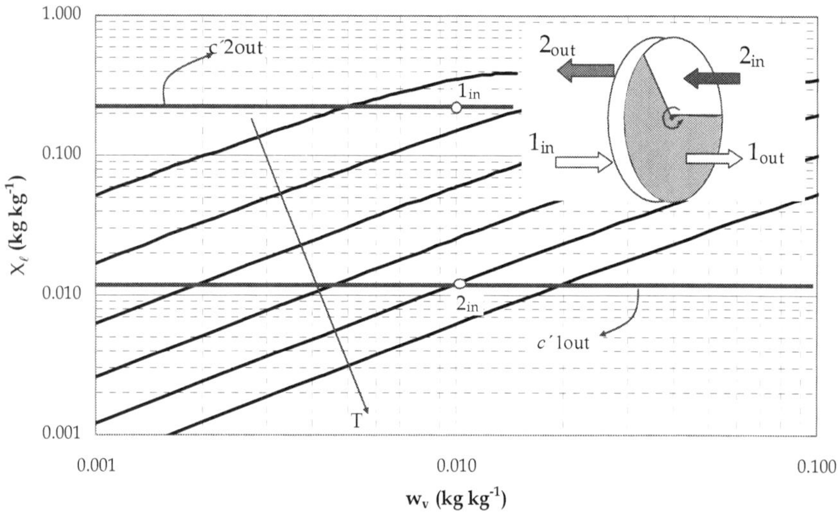

Fig. 7. Representation of the equilibrium curves between the desiccant and the moist air

The ideal outlet state of the process air ($1_{out,id'}$) is defined by the interception of the curve c'1out with the line of constant specific enthalpy h_{1in}. In a similar way, the ideal outlet state of the regeneration air ($2_{out,id'}$) is defined by the interception of the curve c'2out with the line of constant specific enthalpy h_{2in}. Consequently, the ideal (maximum) mass transfer rates are in a first step estimated as:

$$\dot{m}_{w1,id'} = \dot{m}_1 \left(w_{1in} - w_{1out,id'} \right) \tag{1.a}$$

and

$$\dot{m}_{w2,id'} = \dot{m}_2 \left(w_{2out,id'} - w_{2in} \right), \tag{1.b}$$

which can most probably present different values, the lower value indicating the limiting airflow (hereafter called *critical airflow*). The equality between the mass transfer rates in both airflows is imposed by the principle of mass conservation, which implies the redefinition of the outlet ideal sate of the non-critical airflow ($1_{out,id}$ or $2_{out,id}$). This rationale is illustrated in Fig. 8, a case where the critical airflow is the process air (airflow 1).

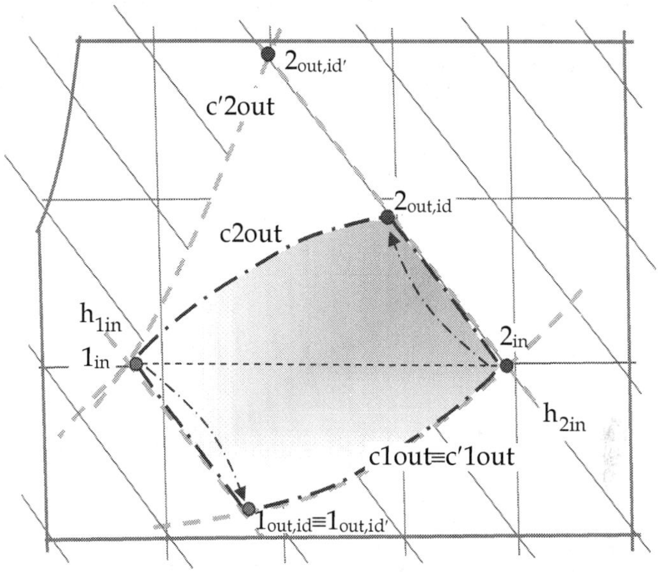

Fig. 8. Ideal air evolutions in a desiccant wheel, where the critical airflow is the process air

1.4 Pair of effectiveness parameters

Following the classical analysis of the behaviour of heat exchangers, the concept of effectiveness results from the comparison between a real heat exchanger and an ideal one adopted as a reference. The application of the so-called *effectiveness method* to a desiccant wheel requires the use of two effectiveness independent parameters due to the existence of the simultaneous and coupled processes of heat and mass transfer. Furthermore, those parameters should be quite independent of the inlet states of both airflows or, at least, easily correlated with them.

The use of the effectiveness method has practical interest, mainly to perform quick simulations of desiccant wheels, but it needs the prior knowledge of the ideal outlet conditions or of the ideal transfer rates, as described in the previous section.

The deviation of the outlet states of both airflows relatively to the ideal ones, as illustrated in Fig. 6, is an indicator of the effectiveness of the heat and mass transfer phenomena in the desiccant wheel. So, the state changes registered in both airflows in a real application should be compared with those of the ideal operation. Taking into account the analogy with the classical analysis of heat exchangers, the following generic definition for the effectiveness is purposed:

$$\eta_\phi = \frac{\phi_{1in} - \phi_{1out}}{\phi_{1in} - \phi_{1out,id}} = \frac{\phi_{2in} - \phi_{2out}}{\phi_{2in} - \phi_{2out,id}}, \qquad (2)$$

where the generic variable ϕ can assume different meanings such as the adsorbed water content at equilibrium between the moist air and the desiccant.

According to preliminary investigation, the recommended independent parameters for a desiccant wheel are those based on the changes of adsorbed water content X_ℓ (kg of adsorbed water/kg of dry desiccant,) and of the specific enthalpy h (J/kg of dry air), respectively, $\eta_{X\ell}$ and η_h. Taking into account, for example, the changes occurring in the process airflow, $\eta_{X\ell}$ can be calculated by:

$$\eta_{X\ell} = \frac{X_{\ell,1in} - X_{\ell,1out}}{X_{\ell,1in} - X_{\ell,1out,id}}, \tag{3}$$

where $X_{\ell,1out,id} = X_{\ell,2in}$. Concerning the evaluation of η_h, it is not possible to consider that $h_{1out,id} = h_{1in}$. So the following definition is proposed, by convenience:

$$\eta_h = \frac{h_{1in} - h_{1out}}{h_{1in} - h_{2in}}. \tag{4}$$

At real conditions, it is expected that both effectiveness parameters exhibit a dependence on the airflow rates, channel length and rotation speed, as well on the inlet states of both airflows. In an optimized case, operating near the ideal conditions, the effectiveness parameter η_h should be low, near zero, while $\eta_{X\ell}$ should be as high as possible, near the unity, the dependence on the operating parameters and conditions being quite negligible. The application of the effectiveness method is highly helpful in perform quick energy dynamic simulations of HVAC&R systems integrating desiccant wheels at the design stage, thus promoting the use of more efficient systems that allow the incorporation of renewable energy or waste energy recovery.

2. Modelling of desiccant wheels

2.1 Objectives and outline

The aim of this chapter consists mainly of the use of a detailed numerical model to study the behaviour of desiccant wheels. Focused on a representative channel of a compact matrix, which is hypothetically treated as a parallel-plate channel, the detailed mathematical formulation takes into account the important changes of properties in both the porous solid and airflow domains that generally occur in transient sorption processes.

Although the detailed model is not an appropriate tool to perform the dynamic simulation of a real desiccant wheel, due to its complexity and the required computational effort, it is an interesting complementary tool to be used in the product optimization by the manufacturer, in the investigation of the validity of the assumptions supporting simplified models (e.g., the lumped capacitance method) and also to evaluate the dependence of the effectiveness parameters on the operating parameters and conditions.

2.2 Detailed numerical modelling of a representative channel

The physical domain of the hygroscopic wheel can be considered as a set of small angular slices, the channels in each slice having the same behaviour. The transient three-dimensional problem is too complex to be solved in a very detailed way and, therefore, it is necessary to adopt a set of simplifications. The most common simplification about the physical domain is the consideration of two-dimensional airflow between desiccant parallel plates. The

hypothesis of two-dimensionality, together with the consideration of cyclic inlet conditions, real wall thickness and ratio of airflow rate to wetted perimeter is frequently adopted when modelling the behaviour of desiccant wheels (Dai et. al, 2001 and Zhang et al., 2003).
The wall domain of a channel of the hygroscopic matrix is modelled in a detailed way, by taking into account the simultaneous heat and mass transfer together with the adsorption/desorption process. Fig. 9 illustrates the physical domain of the channel to be modelled. Two phases co-exist in equilibrium inside the desiccant porous medium, the equilibrium being characterized by sorption isotherms. The ordinary diffusion of vapour is neglected due to the small dimension of the pores (Pesaran, 1983). Therefore, only two mechanisms of mass transport are considered: surface diffusion of adsorbed water and Knudsen diffusion of water vapour. For simplification purposes, the wall is considered to be a homogeneous desiccant porous material. The upper boundary of the domain is considered impermeable and adiabatic. The treatment of the airflow as a bulk flow and the use of suitable convective heat and mass transfer coefficients are considered to evaluate the exchanges occurring at the interface between the airflow and the desiccant wall surface.

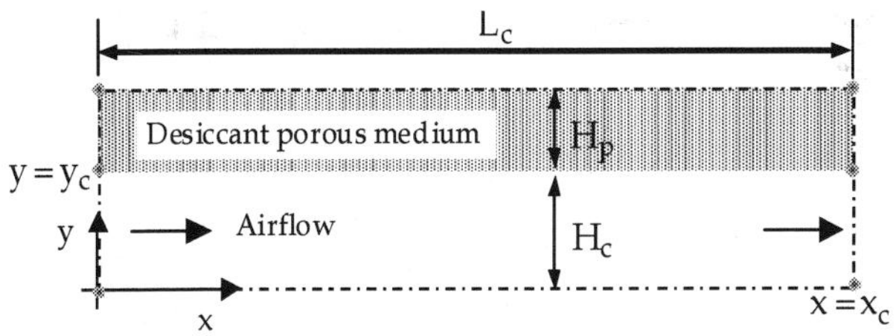

Fig. 9. Physical domain of the modelled channel

For the wall domain, the complete set of conservation equations to be solved by the model can be reduced to the general form:

$$\frac{\partial}{\partial t}(\rho_\phi \phi)+\frac{\partial}{\partial x_j}\left(-\Gamma_\phi \frac{\partial \phi}{\partial x_j}\right)-S_\phi=0, \qquad (4)$$

where the density ρ_ϕ, the diffusion coefficient Γ_ϕ and the source-term S_ϕ assume different meanings depending on the nature of the generic variable ϕ considered ($\phi = X_\ell$ - mass conservation equation of adsorbed water, $\phi = T$ - energy conservation equation, $\phi = \varphi_v$ - mass conservation equation of water vapour). According to the local equilibrium condition assumption, only one of the two differential mass conservation equations is solved, the mass conservation equation for water adsorbed water. The mass fraction of water vapour inside the porous medium φ_v is calculated through the knowledge of the sorption isotherm.
For the airflow domain, the simplified one-dimensional conservation equation is considered:

$$\frac{\partial}{\partial t}(\rho_f \phi)+\frac{\partial}{\partial x}(\rho_f u_f \phi)-S_\phi=0 \qquad (5)$$

where the source-term S_ϕ assumes also different meanings depending on the nature of the generic variable ϕ considered ($\phi = 1$ - global mass conservation equation, $\phi = \varphi_v$ - mass conservation equation of water vapour, $\phi = T$ - energy conservation equation, $\phi = u_f$ -

momentum conservation equation). At the interface ($y = y_c$), the mass and heat convection transfers are modelled assuming that the low mass transfer rate theory is valid (Bird, 1960 and Mills, 1994). The heat convection coefficient h_h is estimated after the Nusselt number Nu for developed laminar channel flow. As for the mass convection coefficient h_m, the Sherwood number Sh is related to Nu according to the Chilton-Colburn analogy. The convective fluxes at the interface are calculated as:

$$j_{v,gs} = h_m \rho_f \frac{\varphi_{v,f} - \varphi_{v,i}}{1 - \varphi_{v,i}} \tag{6}$$

$$j_{h,gs} = h_h \left(T_f - T_i \right) \tag{7}$$

where $\varphi_{v,i}$ and T_i are values at the interface, respectively, for the vapour mass fraction and the temperature.

The water vapour content in the airflow or inside the pores of the desiccant medium is related with the mass fraction of water vapour by $w_v = \varphi_v / (1 - \varphi_v)$.

The modelling of a channel requires the definition of the initial conditions and of the conditions of the airflow entering the channel. The initial conditions are imposed by specifying uniform distributions of T and X_ℓ in the desiccant wall. The airflow domain is assumed to be initially in thermodynamic equilibrium with the desiccant wall. The condition of the airflow entering the channel is imposed by specifying the inlet velocity of the airflow $u = u_{in}$ (or the corresponding mass inlet velocity, $F_m = F_{m,in}$), as well the inlet temperature T_{in} and the water vapour fraction $\varphi_{v,in}$. The total pressure is assumed to be constant and its value is imposed.

The numerical solution procedure is based on the solution of the discretized partial differential equations using the finite volume method. The values of the diffusion coefficients at the control-volume interfaces are estimated by the harmonic mean, thus allowing the conjugate and simultaneous solution in both gas and solid domains (Patankar, 1980). The energy and the vapour mass transport equations are solved in a conjugate procedure that covers simultaneously both sub-domains. Within the desiccant wall sub-domain, the equilibrium value of the vapour mass fraction is locally specified.

Additional data and the complete description of the formulation of different versions of the model can be found in previous works (Ruivo et al, 2006; Ruivo et al, 2007[a,b]; Ruivo et al, 2008[a,b] and Ruivo et al. 2009). The numerical model has been used in simulating the cyclic behaviour of a typical channel of desiccant wheels and also the behaviour of a wall element of the channel, namely to inspect the validity of some assumptions that support simplified numerical methods.

2.3 Prediction of the behaviour of desiccant wheels

The behaviour of the modelled channel enables the prediction of the global behaviour of the desiccant wheel crossed by two airflows at steady state conditions. The adsorption mode corresponds to the adsorption zone of the rotor matrix, where the dehumidification of the process airflow occurs, while the desorption mode corresponds to the desorption zone, where the rotor matrix is regenerated. The cyclic process with a duration τ_{cyc} is divided into the adsorption and the desorption modes, with durations τ_{ads} and τ_{des}. From the point of

view of the modelled channel, the desorption and the adsorption processes occur, respectively, when $0< t\leq \tau_{des}$ and $\tau_{des} < t\leq \tau_{cyc}$. The modelled channel that is representative of the matrix is submitted to an initial transient process that must be started at a certain condition. The transition of mode, from desorption to adsorption, or vice-versa, is done by suddenly changing the inlet airflow conditions and reversing the airflow direction in the channel. After a certain number of desorption/adsorption cycles, the differences between two consecutive cycles are negligible, meaning that the stationary cyclic regime was achieved.
The initial condition for the sequence of the cycles corresponds to the beginning of one of the modes of the cycle (desorption or adsorption), imposing uniform distributions for temperature and adsorbed water content in the desiccant and assuming that the airflows are initially in thermodynamic equilibrium with the desiccant medium.
At steady state conditions, the mass transfer rate occurring in the desorption zone is equal to that occurring in the adsorption zone. Therefore, considering the desorption mode, the following expressions can be deduced, respectively, for the mass and heat transfer rates between both airflows, per unit of transfer area of the matrix:

$$\dot{j}_{v,gs} = \frac{1}{x_c}\int_0^{x_c} j_{v,gs}\,dx \tag{8}$$

$$\dot{j}_{h,gs} = \frac{1}{x_c}\int_0^{x_c} j_{h,gs}\,dx \tag{9}$$

The global mass and heat transfer rates in the desiccant wheel at steady state operating conditions, per unit of transfer area of the matrix, are:

$$\dot{j}_m = \int_0^{\tau_{des}} \frac{\dot{j}_{v,gs}}{\tau_{cyc}}dt \tag{10}$$

$$\dot{j}_h = \int_0^{\tau_{des}} \frac{\dot{j}_{h,gs}}{\tau_{cyc}}dt \tag{11}$$

At the outlet of each zone, the air state exhibits a non uniform angular distribution. The downstream average of temperatures and of water vapour contents at the outlet of the channel in each operation mode are evaluated, the achieved values representing the outlet states of the regeneration and process airflows crossing the desiccant wheel at steady state condition (Ruivo, 2007b).

2.4 Properties and coefficients

The numerical model takes into account the changes occurring in the airflow properties and in the convection and diffusion coefficients. The major part of the relations for the dry air, water vapour and liquid water was derived from thermodynamics tables (Çengel, 1998) in the form of polynomial expressions (Ruivo, 2005). The properties of the air-mixture such as the specific heat and the thermal conductivity are weighted averages based on the dry air and water vapour mass fractions. Similarly, the specific heat and the thermal conductivity of

the wet desiccant medium are weighted averages based on the mass fraction of each component (dry-air, water vapour, adsorbed water and dry desiccant).
The properties of silica gel RD, the relations for the equilibrium condition, the heat of wetting, the adsorbed water enthalpy and the adsorption heat are indicated in Ruivo et al. (2007[a]). The dependences of the mass diffusion coefficients on the temperature and on the adsorbed water content are also presented in Ruivo et al. (2007[a]), and were derived after the expressions in Pesaran (1983). The equilibrium curve for the pair silica gel-moist air is represented in Fig. 10.

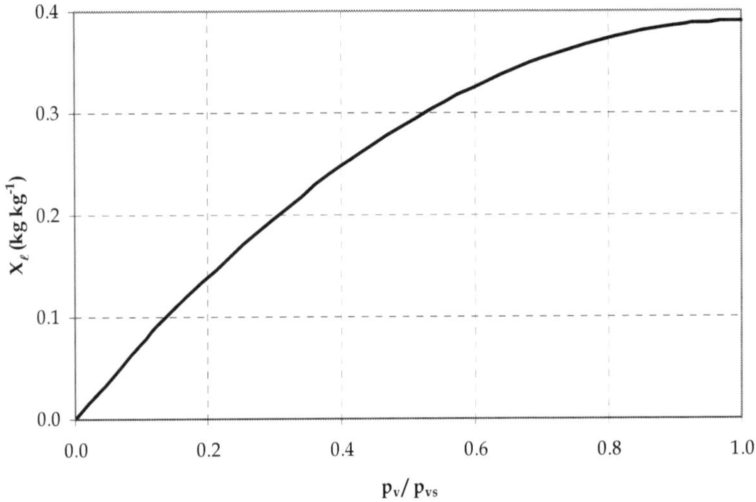

Fig. 10. Equilibrium curve for the pair silica gel-moist air

3. Study cases and results

3.1 Prediction of the performance of desiccant wheels

One of the potentialities of the present numerical model is the calculation of the transient evolutions of the internal fields of temperature and of water vapour content, both in the airflow and in the channel wall domains. Different parametric studies have been conducted using the numerical model to investigate the influence of a set of parameters, namely the rotation speed, the cell dimensions, the wall thickness, as well as the inlet conditions of both airflows, on the behaviour of desiccant wheels (Ruivo, 2005 and Ruivo et al. 2007[b]). The research done by using such detailed numerical model gives to the manufacturers important guidelines to the optimization of the desiccant dehumidification equipments. Moreover after calibration by comparison with experimental data, the detailed numerical models are also an interesting tool to generate data of global performance of desiccant wheels, namely the outlet state of both airflows or the heat and mass transfer rates for a large set of operating conditions. The achieved global behaviour data can be displayed in a chart or in a table, represented by correlations or be used to test the validity of easy and quick predicting methods. This information is very helpful for a more accurate sizing of the dehumidification and/or cooling installations and to analyse dynamically different solutions, namely to investigate the control strategies that lead to a better energy use.

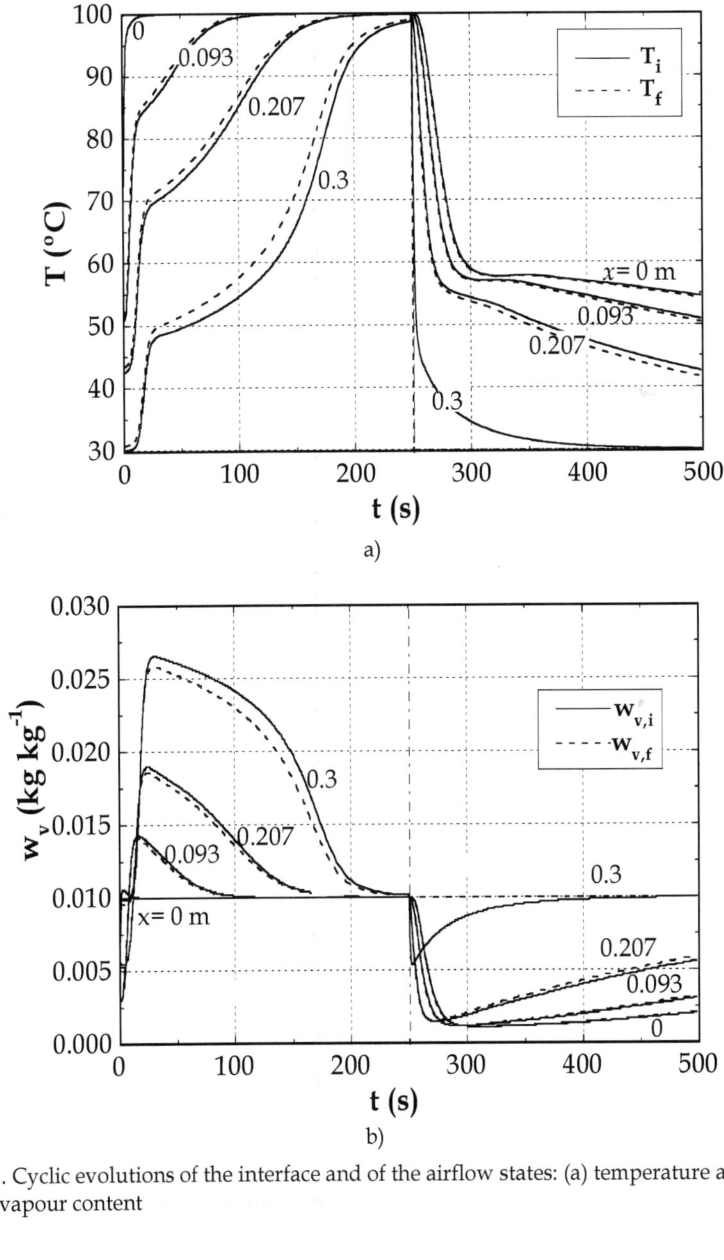

Fig. 11. Cyclic evolutions of the interface and of the airflow states: (a) temperature and (b) water vapour content

The data plotted in Fig. 11 concern to the internal behaviour of a desiccant wheel composed by a compact corrugated matrix with sinusoidal cross section channels. The specific transfer area and the porosity of the matrix are 3198 $m^2 m^{-3}$ and 0.84, respectively. According to Ruivo et al. (2007[b]), the chosen matrix corresponds to cell B3 ($H_{cell} = 1.5$ mm, $P_{cell} = P_{sin} = 3$ mm, $E_p = 0.1$ mm, $H_{sin} = 1.27$ mm), the representative channel of the matrix being modelled with $H_p = 0.05$ mm and $H_c = 0.263$ mm. The channel length is $L_c = 0.3$ m. The rotor speed is 7.2 rotations per hour, that corresponds to $\tau_{cyc} = 500$ s. The desiccant wheel is divided into two equal parts, the adsorption and desorption zones being crossed by counter-current airflows. The desiccant is silica gel. The inlet temperatures of the process and regeneration airflows are 30 °C and 100 °C, respectively. Equal values of the inlet water vapour content (0.01 kg kg^{-1}) and of the mass inlet velocity (1.5 kg s^{-1} m^{-2}) are imposed to both airflows.

The illustrated time evolutions of the states of the interface and of the bulk airflow evidence the abrupt variations in the mode transition. From the temperature and the moisture content evolutions (Figs. 11.a and 11.b), it can be observed that the airflow and the wall are closely in thermodynamic equilibrium in most of the rotor domain, a condition that is sometimes taken as a simplifying hypothesis in the behaviour analysis of ideal desiccant wheels (v., e.g., Van den Bulk, (1985)). It can also be seen that only in the desorption mode the channel wall surface achieves equilibrium with the incoming air, an indication that the regeneration process is completed. This suggests that it is possible to optimise the dehumidification performance of the rotor through the changes of the rotation speed and of the adsorption and desorption zones.

Other cases with different cycle durations were simulated. The registered influence of τ_{cyc} on the global heat and mass transfer rates is shown in Fig. 12. The heat transfer rate exhibits a monotonic decreasing trend with the cycle duration while a maximum value of the mass transfer rate is observed.

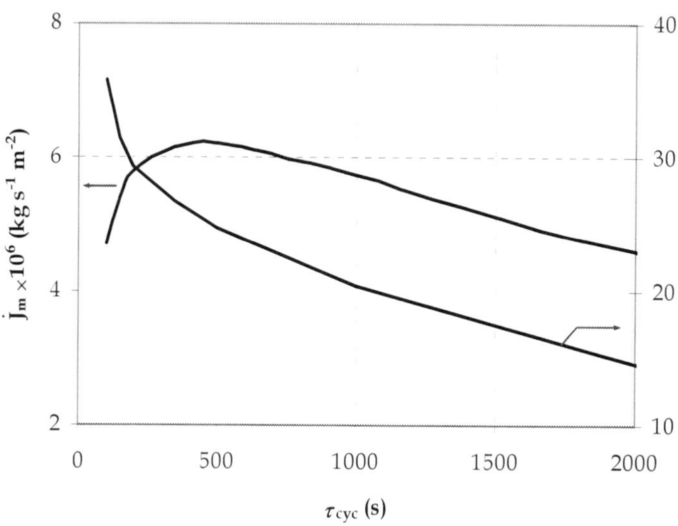

Fig. 12. Heat and mass transfer rates per unit of transfer area of the desiccant wheel

3.2 Test of simplifying assumptions for numerical modelling

Several studies have been carried out to predict the behaviour of desiccant wheels using simplified mathematical models (e.g. Zheng, 1993; Dai et al., 2001; Zhang et al. 2003 and Gao et al. 2005). In most of them, the heat and mass transfers inside the desiccant medium are not described in a detailed way, simplified approaches being adopted instead. The range of validity of such models can be investigated by using experimental techniques and by detailed numerical modelling. The air stream behaviour and its interaction with the desiccant medium are also often treated in a simplified way, namely by assuming a fictitious bulk flow pattern, as well as fictitious heat and mass convection coefficients for the gas side.
When advanced numerical methods are used, solving the complete set of differential transport equations, a number of critical issues still remain, such as the lack of suitable functions to describe the variation of the porous medium properties and the complexity of numerically solving the intrinsically coupled phenomena within the porous desiccant solid and the great time consumption of computational calculations.
In the present section, the numerical detailed model is used to simulate the physical behaviour of the desiccant layer of a wall element of the channel. It is also supposed that the desiccant layer belongs to the channel wall of a compact desiccant matrix, which is crossed by a moist air flow. The hypothesis of one-dimensionality assumed in this study is mainly intended to better identify the effects to be analysed, namely the importance of neglecting the internal heat and mass diffusive resistances. The physical model is schematically represented in Fig. 13.

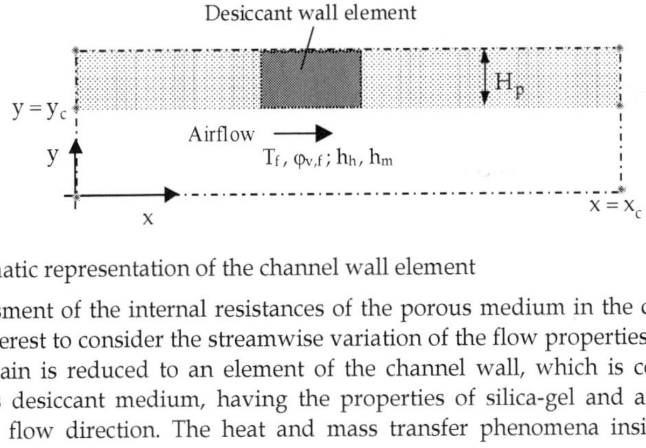

Fig. 13. Schematic representation of the channel wall element

For the assessment of the internal resistances of the porous medium in the cross direction, there is no interest to consider the streamwise variation of the flow properties. Therefore the physical domain is reduced to an element of the channel wall, which is considered as a homogeneous desiccant medium, having the properties of silica-gel and an infinitesimal length in the flow direction. The heat and mass transfer phenomena inside the porous medium are considered only in the y direction.

The air stream in contact with the infinitesimal-length wall element is treated as a well mixed flow (bulk flow), characterised by constant and uniform properties (pressure, temperature and vapour content), thus dispensing the need of solving any conservation equation in the flow domain.
Results of the investigation about two simplifying approaches based on the lumped-capacitance method (Ruivo et al., 2008[b]) are here presented. The first approach corresponds to the theoretical analysis of the system with negligible internal resistances to the heat and mass diffusion, commonly known as the global lumped-capacitance method (approach A- "null resistances"). It is numerically simulated by specifying enough great values of the

diffusion coefficients (the thermal conductivity and the coefficients of Knudsen diffusion and of surface diffusion). The other approach corresponds only to the thermal lumped-capacitance method (approach B-"null thermal resistance"). It is numerically simulated by imposing an enough great value to the thermal conductivity of the desiccant medium.

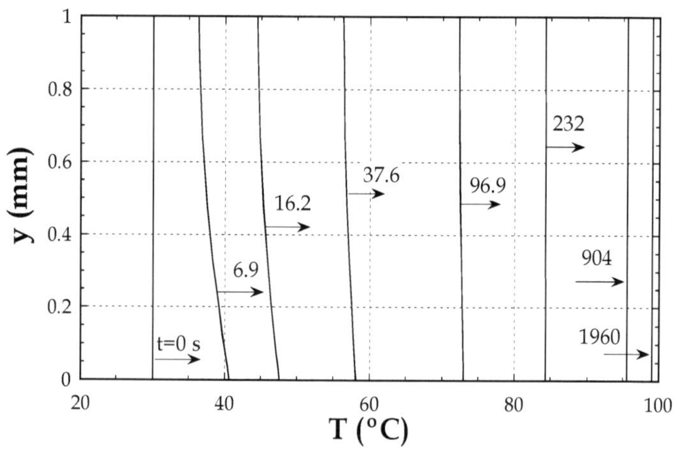

a)

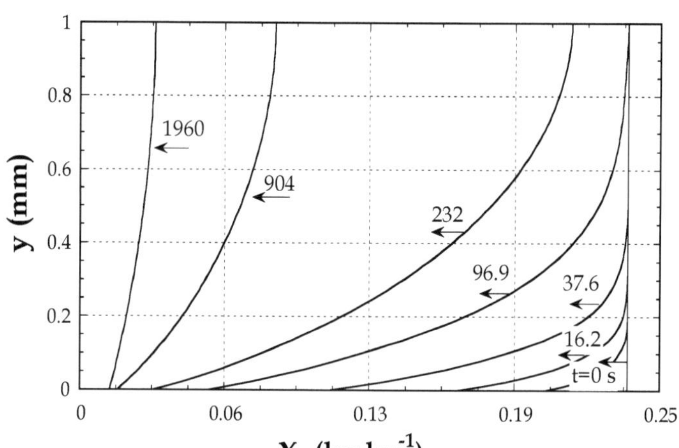

b)

Fig. 14. Time-varying profiles of the dependent variables along the desorption process in a desiccant layer of $H_p = 1$ mm: (a) temperature and (b) adsorbed water content X_ℓ

The numerical tests of both simplifying approaches consists of the analysis of the response of the desiccant wall to a step change of the airflow conditions, starting from a given initial desiccant state and finishing when the desiccant wall achieves the equilibrium with the moist airflow.

The results here presented are for the simulation of the transient process of desorption, where the initial temperature and adsorbed water content in the desiccant were 30°C and 0.2363 kg kg^{-1}, respectively. The airflow conditions were 100°C and 0.01 kg kg^{-1}, respectively for the temperature and the water vapour content. The process pressure was assumed as 101325 Pa.

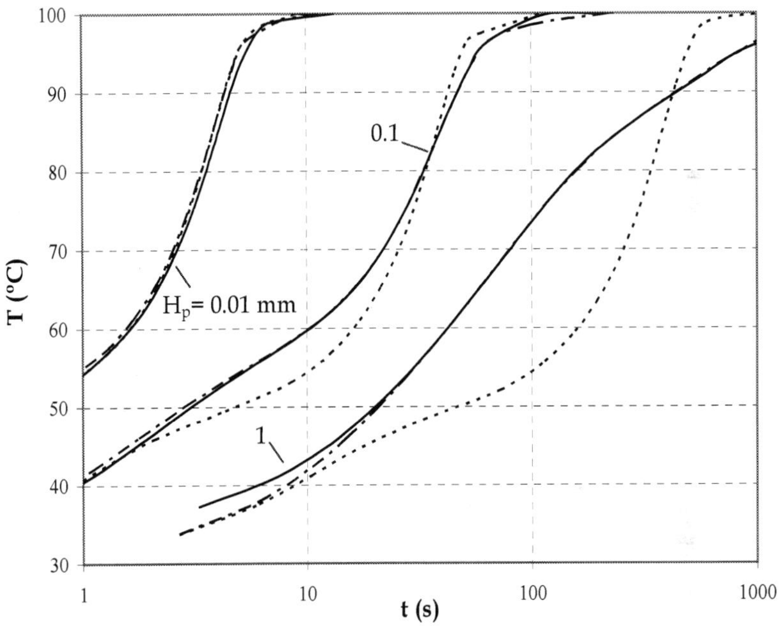

Fig. 15. Predicted temperature at the convective interface considering normal internal resistances (———), null thermal and mass resistances (------ approach A) and null thermal resistance (-·-·-·- approach B)

The value 2.45 was assigned to the Nusselt number corresponding to heat convection between a uniform temperature wall and a fully-developed laminar flow inside a corrugate sinusoidal-type channel of a compact exchanger (Zhang et al., 2003). The channel cross-section area was 4.5 mm^2, with an internal perimeter of 10.6 mm and a hydraulic diameter of 1.69 mm, approximately. The conducted runs covered a wide range of values of the layer thickness in order to study the validity of neglecting internal thermal and mass resistances and only the thermal resistance. The properties of the desiccant medium were referred also to silica gel RD (Pesaran, 1983). A particular case was selected for the analysis of the transient evolutions of the main variables inside a desiccant layer with 1 mm of thickness. Time-evolving profiles are shown in Figs. 14.a and 14.b as calculated with the "normal" internal resistances. It is seen that the temperature field presents small gradients, meaning that the internal resistance to heat diffusion is almost insignificant, contrarily to those restricting the mass diffusion. The gradients of the adsorbed water content are significant during almost all the transient process, mainly near the convective surface.

The diagrams in Figs. 15 and 16 show the time evolutions of the temperature and of the adsorbed water content at the interface determined by the detailed model for the three different scenarios: with normal internal resistances, without internal resistances to the heat and mass and without internal resistance to heat conduction. It can be observed that the results of approach A (dashed lines) show an increasing deviation relatively to those of the detailed model (full lines) for desiccant layers with $H_p \geq 0.1$ mm. Moreover, the results of approach B agree reasonably well with the detailed model for all studied wall thicknesses, unlike those of approach A. This indicates that significant inaccuracies may result when the internal resistance to mass diffusion is neglected, leading to unrealistic estimation of the convection fluxes at the interface.

It can be concluded that the heat and mass lumped capacitance assumption supporting the pseudo gas-side model is acceptable for desiccant layers thinner than about 0.1 mm.

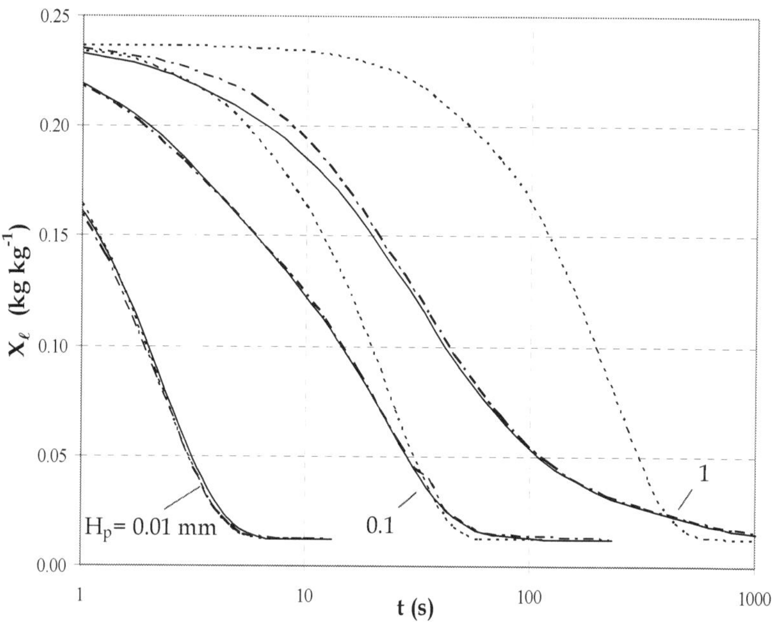

Fig. 16. Predicted adsorbed water content at the convective interface considering normal internal resistances (———), null thermal and mass resistances (------ approach A) and null thermal resistance (-·-·-·- approach B)

3.3 Assessment of effectiveness parameters

Several assessments of the desiccant wheel effectiveness parameters have been recently conducted based on experimental and numerical modelling data. The cases here presented were generated by the detailed numerical model to investigate the influence of the regeneration temperature on the effectiveness parameters. Different inlet states of the regeneration airflow were considered, corresponding to 80, 100 and 120°C of temperature, all with the same value of the water vapour content 0.01 kg kg^{-1}. The value of 1.5 kg s^{-1} m^{-2} is considered for the mass inlet velocity of both airflows. The inlet state of the process airflow is defined by a temperature of 30°C and water vapour content of 0.01 kg kg^{-1}. The desiccant wheel is symmetric. The desiccant is silica gel. The rotor matrix corresponds also to cell B3, i.e, the specific transfer area and porosity of the matrix are 3198 m^{2}m^{-3} and 0.84, respectively. According to Ruivo et al. (2007[b]), the representative channel of the matrix being modelled with H_p =0.05 mm, H_c=0.263 mm. The channel length is L_c=0.3 m.
The rotation speed corresponds to cycle duration of 500 s, close to the optimum value that maximises the dehumidification rate (see Fig. 12). The output results of the detailed numerical model for the different simulated cases are indicated in Table 1. The effectiveness parameters η_h and $\eta_{X\ell}$ presented in Table 2 were calculated, respectively, by the Eqs. 3 and 4.

Case	T_{1in} [°C]	$w_{v,1in}$ [kg kg^{-1}]	T_{2in} [°C]	$w_{v,2in}$ [kg kg^{-1}]	T_{1out} [°C]	$w_{v,1out}$ [kg kg^{-1}]
1	30	0.01	80	0.01	54.89	0.003182
2	30	0.01	100	0.01	60.80	0.002035
3	30	0.01	120	0.01	65.01	0.001478

Table 1. Output results of the detailed numerical model

Case	η_h	$\eta_{X\ell}$
1	0.1519	1.000
2	0.1502	0.999
3	0.1471	0.995

Table 2. Calculated effectiveness parameters η_h and $\eta_{X\ell}$.

The registered inlet regeneration temperature dependence of η_h and $\eta_{X\ell}$ is small, being the constant value corresponding to the average of the values acceptable to use in the in the effectiveness method. It should be remarked that the set of analysed cases does not represent an exhaustive study. Further research should be done based on additional numerical or experimental modelling data.

4. Conclusions

This Chapter gives an overview of the research towards the prediction of the behaviour of desiccant wheels. Ideal and real psychrometric evolutions of process and regeneration airflows were analysed, and the general trend of the influence of the channel length of the matrix and the cycle duration was presented, as well as the procedure to determine the ideal outlet states of both airflows. This procedure takes into account the curves of equilibrium between the moist-air and the desiccant medium as well as the relation between both airflow rates, and enables the identification of the airflow that limits the process of heat and mass transfer (critical airflow). A new pair of independent parameters for the effectiveness of the coupled heat and mass transfer in desiccant wheels is proposed, the values at ideal operating conditions being physically intuitive, i.e. $\eta_h = 0$ and $\eta_{X\ell} = 1$, and independent of the operating conditions.

A numerical model for simulating the cyclic adsorption/desorption process in a representative channel of the desiccant matrix was briefly referred. The model is based on the solution of the differential equations for the conservation of mass and energy. The airflow is treated as a bulk flow, the interaction with the wall being evaluated by using appropriated convective coefficients. The wall domain is treated in detail, considering the internal time-varying fields of variables and properties. To illustrate the potentialities of the model in predicting the internal behaviour of a desiccant wheel, data that are useful for the manufacturer to product optimization, the results of a particular case were presented. Results of a parametric study were also presented, showing the dependence of the global behaviour of the desiccant wheel on the rotation speed. An expected optimum point maximizing the dehumidification rate was determined. An adapted version of the model to simulate the detailed heat and mass transfer during the adsorption process in a wall element of the channel was used to investigate the validity of simplifying assumptions, namely those that neglect the internal heat and mass diffusive resistances. The results of the parametric study show that the internal resistance to mass diffusion is much more important that the internal thermal resistance. This justifies the use of a simplified method assuming no internal thermal gradients in the desiccant wall in the direction normal to the airflow. The use of a lumped capacitance method for the coupled heat and mass transfer is acceptable only for very thin desiccant wall layers.

Another parametric study of the global behaviour of a desiccant wheel at different regeneration temperatures was presented and its results were used to calculate the effectiveness values, a quite negligible dependence of η_h and $\eta_{X\ell}$ relatively to the regeneration temperature being observed. The adoption of the effectiveness method seems to be an interesting tool, at the same time easy and intuitive, for design purposes.

5. References

ASHRAE Handbook. Fundamentals (1989). American Society of Heating, Refrigerating and Air Conditioning Engineers, Inc., ISBN 0-910110-57-3, Atlanta GA

Bird, R.; Stewart, W. & Lighfoot, E. (1960). *Transfer phenomena,* John Wiley & Sons, Inc., New York and London

Cejudo, J; Moreno, R. & Carrillo, A. (2002). Physical and neural network models of a silica-gel desiccant wheel. *Energy and Buildings,* Vol. 34, No. 8, (September 2002) pp. 837-844, ISSN 0378-7788

Çengel, Y. (1998). *Heat Transfer- A practical approach,* McGraw-Hill, ISBN 0-07-115223-7

Dai, Y.; Wang, R. & Zhang, H. (2001). Parameter analysis to improve rotary desiccant dehumidification using a mathematical model. *International Journal of Thermal Sciences,* Vol. 40, No. 4, (April 2001) pp. 400-408, ISSN 1290-0729

Gao, Z.; Mei, V. & Tomlinsom, J. (2005). Theoretical analysis of dehumidification process in a desiccant wheel. *Heat and Mass Transfer,* Vol. 41, No. 11, (July 2005) pp. 1033-1042, ISSN 1033-1042

Kodama, A. (1996). *Experimental study on optimization of a honeycomb rotor continuous adsorber operated with thermal swing,* Ph.D thesis, Faculty of Engineering, Kumamoto University, Japan

Mills, A. (1994). *Heat and mass transfer,* CRC Press, ISBN-0256114439.

Patankar, S. (1980). *Numerical heat transfer and fluid flow,* McGraw-Hill, Washington

Pesaran, A. (1983). *Moisture transport in silica gel particle beds,* Ph.D thesis, University of California, Los Angels

Ruivo, C. (2005). *Modelação numérica dos fenómenos de transferência de calor e de massa em rodas higroscópicas,* Ph.D thesis, University of Coimbra, Coimbra, Portugal, ISBN 972-99204-8-6

Ruivo, C.; Costa, J. & Figueiredo, A. (2006). Analysis of simplifying assumptions for the numerical modeling of the heat and mass transfer in a porous desiccant medium. *Numerical Heat Transfer, Part A Applications,* Vol. 49, No. 9, (October 2006) pp. 851-872, ISSN 1521-0634

Ruivo, C.; Costa, J. & Figueiredo, A. (2007)[a]. On the behaviour of hygroscopic wheels: Part I - channel modelling. *International Journal of Heat and Mass Transfer,* Vol. 50, No. 23-24, (November 2007) pp. 4812-4822, ISSN 0017-9310

Ruivo, C.; Costa, J. & Figueiredo, A. (2007)[b]. On the behaviour of hygroscopic wheels: Part II - rotor performance. *International Journal of Heat and Mass Transfer,* Vol. 50, No. 23-24, (November 2007) pp. 4823-4832, ISSN 0017-9310

Ruivo, C.; Costa, J. & Figueiredo, A. (2008)[a]. Numerical study of the behavior of an elementary desiccant layer of a hygroscopic rotor. *Numerical Heat Transfer, Part A Applications,* Vol. 53, No. 10, (January 2008) pp. 1037-1053, ISSN 1521-0634

Ruivo, C.; Costa, J. & Figueiredo, A. (2008)[b]. On the validity of lumped capacitance approaches for the numerical prediction of heat and mass transfer in desiccant airflow systems. *International Journal of Thermal Sciences,* Vol. 47, No. 3, (March 2008) pp. 282-292, ISSN 1290-0729

Ruivo, C.; Costa, J. & Figueiredo, A. (2009). Validity of pseudo-gas-side-controlled models to predict the behaviour of desiccant matrices. *International Journal of Thermal Sciences,* Vol. 48, No. 11, (November 2009) pp. 2171-2178, ISSN 1290-0729

Van Den Bulk, E.; Mitchell, J. & Klein, S. (1985). Design theory for rotary heat and mass exchangers - I. Wave analysis of rotary heat and mass exchangers with infinite transfer coefficients. *International Journal of Heat and Mass Transfer,* Vol. 28, No. 8, (August 1985) pp. 1575-1586, ISSN 0017-9310

Zhang, X.; Dai, Y. & Wang, R. (2003). A simulation study of heat and mass transfer in a honeycomb rotary desiccant dehumidifier. *Applied Thermal Engineering,* Vol. 23, No. 8, (June 2003) pp. 983-1003, ISSN 1359-4311

Zheng, W. & Worek, W. (1993). Numerical simulation of combined heat and mass transfer processes in a rotary dehumidifier. *Numerical Heat Transfer, Part A Applications,* Vol. 23, No. 2, (March 1993) pp. 211-232, ISSN 1040-7782

23

Oscillatory Regimes of Solutocapillary Marangoni Convection

Konstantin Kostarev[1], Andrew Zuev[1] and Antonio Viviani[2]
[1]Institute of Continuous Media Mechanics,
[2]Seconda Università di Napoli
[1]Russia,
[2]Italy

1. Introduction

It is common knowledge that a fluid motion can be initiated both by the volume (buoyancy) forces, due to density variations in presence of a gravity field, and the surface (capillary) forces, due to variations of surface tension along a fluid/fluid interface. Involving in motion the surface and the near-surface layers the capillary forces gives rise to a volumetric flow generally known as the Marangoni convection. In turn, inhomogeneity of the surface tension may result from its dependence on temperature or concentration of the dissolved surface-tension active (surfactant) component. Such convection, thermocapillary or solutocapillary one, respectively, plays an important role in hydrodynamics and heat/mass transfer of inhomogeneous multi-phase media with the liquid/liquid interface or the free surface between liquid and gas. The Marangoni convection has a marked effect on the intensity of many technological processes encountered in food, chemical, oil, metallurgical and other industries, including those proceeding in microgravity conditions where the gravitational mechanisms of convective motion are weakened or absent. Particular interest on research in this field has quickened in recent years due to new achievements in the development of space technologies and life-support and survival systems for orbital stations. Marangoni phenomena is a problem of great application value for such areas as ecology (purification of water surface from petroleum products), meteorology, biology (motion of bacteria and microbodies), medicine (spreading of pulmonic surfactants during aerosol inhalation in treatment of lung diseases) and so on.

Despite the global abundance of the Marangoni convection the direct experimental investigation of the Marangoni flows in the so-called pure form is a challenging task. The reason is that under normal terrestrial conditions the existence of temperature and concentration gradients in a fluid also leads to development of the gravitational Rayleigh convection with considerably higher intensity than that of the Marangoni convection. Hence the latter proves to be almost suppressed by a more severe convective driving mechanism or veiled by its action. At the same time, under certain conditions, the surface forces may dominate over the buoyancy ones. Such conditions are realized, for example, in shallow layers and fluid films, liquid bridges and zones confined between solid surfaces, and also for small insoluble drops or gas bubbles suspended in liquids. In the past decades, the Marangoni convection in such systems has been the subject of intensive investigation. However, most of the studies are concerned with thermocapillary variant of the Marangoni convection whereas the works on solutocapillary convection are practically absent. The reason is that solutocapillary flows are more complicated than thermocapillary ones and are dependent on a variety of factors impeding their theoretical and numerical analysis as well as experimental examination. Indeed, experimentally the concentration gradients are more difficult to generate and sustain than the temperature gradients. The absence of simple and adequate methods for measuring local surfactant concentration at the free surface also presents serious problems. The other challenge is a non-stationary nature of solutal flows associated with the diffusion processes. The action of some additional side effects related with dissolution of the surfactant in a liquid, its evaporation and transition into a gas phase and adsorption at the interface also add complexity to treatment of this problem.

Despite an obvious similarity of the driving mechanisms, the thermocapillary and solutocapillary Marangoni phenomena were found to be not completely analogous. The key difference is that the characteristic times of the admixture diffusion are hundred times longer than times of heat diffusion and so the values of Marangoni number, characterizing the intensity of solutocapillary convection, increase many times. This changes drastically the ratio of contributions to mass transfer made by competing gravitational and Marangoni mechanisms. As a result, the concentration inhomogeneities in the fluid exist for much longer time, whereas the intensity and duration of the capillary forces at the interface show a manifold increase. The other marked distinction of the solutocapillary convection consists of surfactant adsorption which caused an increase in the local concentration of the surfactant and its redistribution over the surface, equalizing surface tension inhomogeneity. These differences give rise to new phenomena that cannot occur in thermocapillary situation.

Thus, the results of experimental observations show that if an upward-directed temperature gradient is maintained in a liquid, a thermocapillary flow forms around a bubble in the form of an axisymmetric toroidal vortex with a vertical symmetry axis. Under the action of Marangoni forces, the liquid flows along the bubble-free surface to its lower pole and slowly rises at some distance, creating a returning flow. Correspondingly, in any vertical plane drawn through a diameter, two stationary vortices that are symmetric with respect to the bubble are observed. The vertically stratified temperature distribution in the liquid also remains stationary. In contrast to thermal Marangoni convection, the described below our experiments (Birikh *et al.*, 2006, 2008; Bushueva *et al.*, 2008; Kostarev *et al.*, 2004, 2006, 2007, 2008a, 2008b, 2009a, 2009b; Zuev *et al.*, 2006, 2008a, 2008b) revealed the self-oscillatory regimes of solutal convection in an aqueous solution of surfactant with vertically stratified concentration, and the threshold character of excitation of solutocapillary flows near the free bubble surface.

2. Experimental approach and apparatus

Visualization of non-homogeneous distribution of the surfactant concentration in the fluids mixture was made by means of interferometer observations. Investigations were made in a thin liquid layer, filling the gap between two plane-parallel glasses. On the side of the fluid the glasses were covered by a semi-transparent mirror coating to provide repeated passage of the object beam through the fluid in the cavity. Thus, the glass plates served as the walls of the shallow 90×40 mm rectangular cavity (the so-called Hele-Shaw cell) forming the working cell of the Fizeau interferometer, which was adjusted to a single infinitely wide fringe. It is to be noted that this particular structure of the cuvette intended for a simultaneous formation of the reference and a series of the object light beams was first used in (Gustafson *et al.*, 1968, Marsters & Advani, 1973) for studying the mass transfer processes. In such a cuvette, the interference pattern is formed only due to inhomogeneity of the medium enclosed between the mirror coatings of the cuvette walls and is not affected by thermal deformations of the interferometer elements. Therefore the obtained picture reflects the real state of the fluid.

The interferometer allowed visualizing anomalies in the solution concentration as a pattern of isolines of the refraction index, which under isothermal conditions varied proportionally to the concentration. For a liquid layer 2 mm thick, the difference in the values of the refraction index between two adjacent monotonous interference fringes was 0.27×10^{-3}. With account of the non-linear dependence of the refraction index on concentration (Vazquez *et al.*, 1995, 1997) this corresponds to a 0.3–0.7% variation in the solution concentration, depending on the kind of used surfactant. The maximum error of the concentration measurements did not exceed 0.1%. As the thickness of the cavity gap did not exceed 2 mm, which is 10 times smaller than other cavity sizes, the optical inhomogeneity along the propagation path of the object beam are neglected and the structure of the concentration fields and the flow streams is considered to be two-dimensional. The scheme of the experimental setup is shown in Fig. 1. The observations can be made both in reflected and transmitted light. The experimental cell can be oriented either horizontally (position I) or vertically (position II). In tests with horizontally oriented cells the collimator was placed in font of the mirror inclined at an angle of 45°. All experiments were carried out at the ambient temperature (20±1)°C.

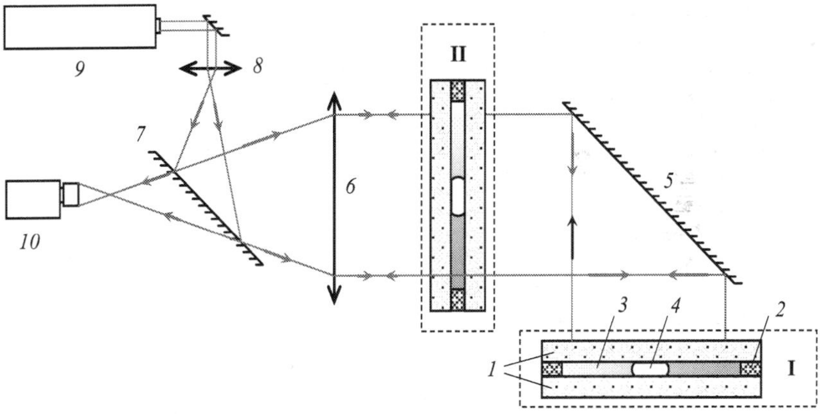

Fig. 1. Experimental setup: *1* - semi-transparent mirror-plated glasses; 2 - sectionalizing plates; 3 - fluid layer; *4* - gas bubble or insoluble droplet; *5* - mirror, *6* - objective; *7* - semitransparent mirror; *8* - diaphragm; *9* - laser; *10* - camera-recorder

3. Two horizontal layers of soluble fluids

In our first experiment (Kostarev *et al.*, 2004) was studied a two-layer system of mutually miscible fluids with different surface tension. For this effect we used water and aqueous solutions of some liquid surfactants such as acetic acid with mass concentration ranging from 0 to 70%, or methyl, isopropyl, and ethyl alcohols with 50–100% concentration. The experimental cells of thickness ≥ 2 mm was first set in a vertical position and filled sequentially by two solutions of various concentration. Due to the small values of the fluids diffusion coefficient these solutions did not mix, only the very narrow diffusion zone between the liquid layers was formed in the cell. After this the cell was placed horizontally. Horizontal positioning of the cell caused a lighter fluid to flow over a heavier one. As a result, we obtained a liquid medium with a vertical surfactant concentration difference. Note that density of the acetic acid is higher and density of the alcohols is lower than that of water. This allowed us to realize situations when a surfactant layer was formed either above or under the water layer, or differently, the situation leading to formation of downward-directed or upward- directed surface tension gradient at the bubble surface.

Before the bubble was injected into the fluid the interference pattern displayed a monochrome image field because in the cell set in a horizontal position the direction of the transmitted optical radiation coincide with the direction of the density gradient of non-disturbed fluid system. Air bubbles, injected into a fluids system by a medical syringe, took the form of a flat disk 5-15 mm in diameter, squeezed between the horizontal cell walls. Due to the vertical concentration difference the solutocapillary Marangoni flow must form along the free lateral surface of the bubble. Under these conditions the surfactant was transported by the solutocapillary forces to the lower (in experiments with the methyl alcohol) or upper (in experiments with acetic acid) areas of the bubble boundary. The transferred lighter or denser liquid fraction has to partly dissolve in the ambient liquid and partly return (ascend or sink, respectively), creating a circulating convective current. However, unlike a thermal convection, which basically had the form of a stationary axially symmetric vertical vortex, the initiated solutal convection was of well-pronounced oscillatory nature.

The evolution of this convection mode is illustrated by a series of interferograms in Fig. 2. At first, after insertion of the air bubble no motion is observed in the surrounding liquid. The concentration field around the bubble does not change either and looks like a uniform field (Fig. 2,*a*). Some time later the bubble turns out to be surrounded by a system of the concentric concentration isolines (Fig. 2,*b*) caused by an abrupt ejection of the excess surfactant accumulated under the action of solutocapillary forces at the lower or upper surface of the bubble. Then, as the buoyancy force restores the vertical density stratification of the solution this disturbance of the concentration field gradually disappears (Fig. 2,*c*) and the interference pattern again changes into a monochrome picture.

In our experiments this process was repeated again and again with a fairly good periodicity and continued until all liquid around the bubble was mixed to a homogeneous medium. The period T of these oscillations (defined as a time interval between two outbursts of the intensive convection) ranged from several seconds to minutes and depended on time, layer thickness, horizontal diameter of the air bubble, the initial concentration gradient and physical properties of the liquids. Fig. 3 shows the variation of the oscillation frequency with time in a two-layer system composed of water and 40% aqueous solution of acetic acid for bubbles with diameters from 2.4 to 15.0 mm. It is seen that as the solution is mixed and the vertical concentration gradient decreases, the period of oscillations initially equaling 10 sec gradually increases (frequency, respectively, decreases), and the oscillations occur more and more seldom, until after approximately 10 min they come to an end. However, the graph does not show any essential effect of the bubble diameter on the oscillation frequency, which only slightly decreases with a growth of the bubble diameter. Moreover, oscillations of smaller bubbles stopped in a shorter time.

a)

b)

c)

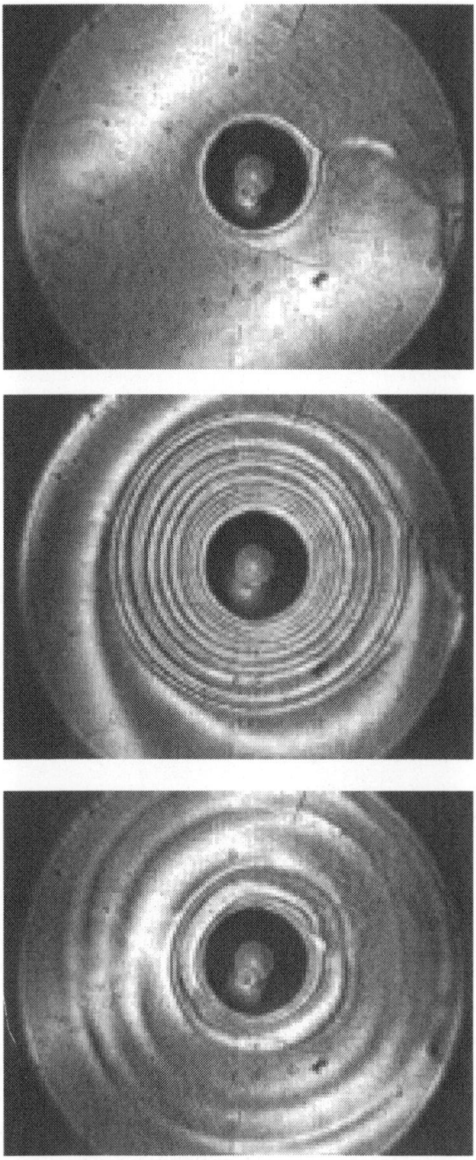

Fig. 2. Evolution of the concentration field disturbance around the air bubble in a horizontal layer of the aqueous isopropanol solution. Layer thickness h = 2.6 mm, concentration difference ΔC = 20%, bubble diameter d = 5.0 mm. t, sec: 0 (*a*), 2.0 (*b*), 10.0 (*c*). View from above

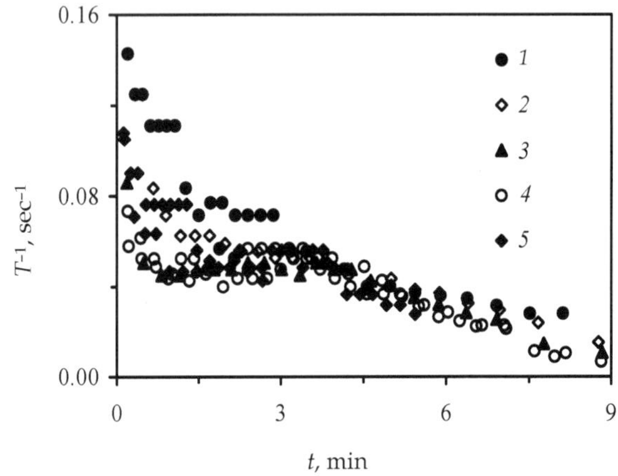

Fig. 3. Time variation of the oscillation frequency in two-layer system composed of water and 40% acetic acid solution for bubbles of various diameters *d*, mm: 15.0 (*1*), 10.7 (*2*), 7.1 (*3*), 5.1 (*4*), 2.4 (*5*)

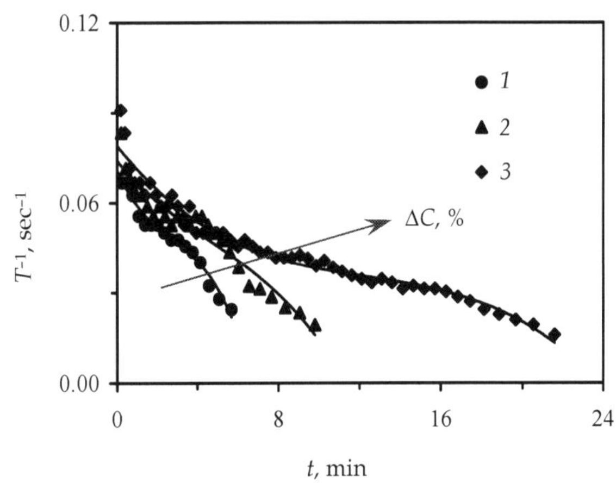

Fig. 4. Time variation of the oscillation frequency in two-layer system composed of water and 40% acetic acid solution. Bubble diameter *d* = 4.8–5.9 mm. Initial concentration difference Δ*C*, %: 20 (*1*), 40 (*2*), 70 (*3*)

Evidently the oscillation frequency is chiefly defined by the layer thickness (i.e., the vertical bubble size) and the concentration gradient. With a decrease of the initial concentration difference (Fig. 4) the oscillation period slightly increases. At the same time, duration (the lifetime) of the oscillations significantly increases with a growth of the initial concentration difference. The specific feature of the observed oscillations is that they decay quite unexpectedly despite the fact that the fluid still preserves some vertical concentration gradient. To a larger extent the discovered phenomenon was found to depend on the choice of the surfactant and its concentrations. Thus, more pronounced oscillations with fairly large periods (hundreds of seconds and more) were observed in the experiments with water and 15-60% solutions of acetic acid. At lower surfactant concentrations the oscillations did not occur at all, whereas at essentially higher concentrations (more than 60%) the oscillations were so frequent that they merged into one continuous oscillation mode in the form of "boiling" of the surfactant-saturated solution around the bubble. On the contrary, in tests with alcohol solutions, the concentration range giving rise to essential oscillations was much narrower and "boiling" was the main mechanism of mass transfer. Evidently, the period and lifetime of the oscillations are defined not by the concentration difference but by the difference in the values of the surface tension, which in most cases depends nonlinearly on the solution concentration.

The experimental findings suggest that solutocapillary flows are initiated in a threshold manner. To ensure their further development it is essential that some parameters specified both by the problem geometry and by physical properties of the examined liquids (mainly surface tension, density and viscosity) should reach critical values. However, such an experimental scheme – a combination of the vertical direction of observations (top view) and the horizontal orientation of the liquid layer – did not allow us to investigate directly the structure of convective flows and the evolution of the vertical distribution of surfactant concentration in the fluid. Therefore, our next study was focused on solutocapillary motion developed around the air bubbles and insoluble drops placed into a thin vertically situated liquid layer filled by aqueous surfactant solution with uniform vertical gradient of concentration.

4. Vertical layer with a stratified solution

The experiments were performed in order to visualize the vertical structure of convective flows and concentration fields (Zuev *et al.*, 2006). The experimental cell was set vertically on its narrow face and filled in turn half by water and half by surfactant solution. The aqueous solutions of isopropyl alcohol with the concentration range from 10 to 50% or of acetic acid with the concentration range from 30 to 70% were used. After some time the diffusion between the fluids led to the formation of an area with a stable vertical stratification of surfactant concentration. The gas bubble or a drop of insoluble fluid was placed in a center of a thin (1.2 mm thick) gap between vertical cell walls with the help of medical syringe. As the initial bubble/drop diameter was larger than the layer thickness, they were squeezed between the cell walls so that took the form of a short cylinders with a free lateral surface. A special wire frame in the form of a flattened thin ring was used to prevent the bubble from buoying up under the action of the Archimedean force and to hold it in a fixed state in the cell center. The frame left the major part of bubble/drop lateral surface free and did not hamper the development of the Marangoni convection. Observations were made on the side of wide cell faces. Due to the very small cell thickness the arising flows and concentration distributions around the bubble/drop interface can be considered to be two-dimensional.

In the absence of the bubble the interference pattern of the concentration field in the stratified solution was presented by a system of horizontal lines. The injection of the bubble essentially disturbed the concentration distribution. Formation of the concentration gradient resulted in the appearance of tangential solutocapillary forces at the bubble lateral surface, which caused the development of the Marangoni convection. Fig. 5 presents the typical interference pattern of the concentration field around the bubble in the heterogeneous fluid mixture, formed at the top by water and at the bottom by the 70% acetic acid solution (downward surfactant concentration gradient). The direction of the surface tension gradient is opposite to the direction of surfactant concentration gradient which implies the origin of solutocapillary capillary forces directed to the upper pole of the bubble. The height of the bubble, which was slightly deformed by the gravitational force, was 3.8 mm, and its width - 6.0 mm. At first, the convective flow near the bubble was quite similar to that observed in thermocapillary experiment (Kostarev *et al.*, 2006). At a distance from the bubble the interference pattern of the concentration field in the solution was presented by a system of horizontal lines. At the bubble surface under the action of tangential solutocapillary forces the solution saturated by the surfactant moves from the lower pole region upwards forming a thin layer round the bubble (Fig. 5,*a*). In contrast to a thermocapillary convection, due to long characteristic times of diffusion the excess of the surfactant transferred along the bubble surface could not fully dissolve in the surrounding fluid. As a result the solution with high surfactant concentration was accumulated near the upper pole in the form of a peculiar "cap", whereas the layer with homogeneous concentration formed around the bubble eliminated the concentration difference between the poles - the source of the Marangoni convection. As soon as the ascending solutocapillary motion along the bubble surface vanished, the dynamic equilibrium of the concentration "cap", which was heavier than the surrounding fluid, was disturbed. Under the action of the Archimedean forces the streams of the acetic acid began to move downward on the right and on the left of the bubble. Due to fluid continuity this motion, in tern, was accompanied by formation of an upward-directed stream under the bubble carrying the solution with a higher surfactant concentration to the lower pole of the bubble (Fig. 5,*b*).

The recovery of the surfactant concentration difference between the bubble poles triggered again the Marangoni forces which abruptly increased the rising flow. As a result an intensive convection motion was initiated near the lateral surface of the cylindrical bubble in the form of two symmetric vortices. During evolution these vortices captured and mixed the larger volume of the solution with high surfactant concentration. As a result the average density of the solution within the vortices increased, they shifted downwards and at a certain moment cut off the surfactant supply from the bubble lower pole (Fig. 5,*c*). Once this happens, the convection motion ceased as fast as it had started. The whole process of evolution and decay of the intensive motion around the bubble lasted not longer than 30 sec. Later under the action of the gravity forces the vertical concentration stratification gradually recovered approaching the initial state. Then, after a time, the whole process was repeated. Such oscillations of the fluid motion occurred with the period of about 1-2 min and could continue as long as the external stratification of the surfactant concentration remained intact – about several hours. Gradually an area of the homogeneous solution formed under the bubble due to a mixing action of the vortex cells and extended in length with time. However as vertical concentration gradient near the bubble still exists, the oscillations are periodically reinitiated, though with less intensity.

In experiments with isopropyl alcohol, in which the gradient of the surface tension had an opposite direction (water is below), the arising structure of the concentration field and flows was bilateral symmetric: the surfactant on the bubble surface was carried to its lower pole and periodically initiated convective vortices rotated in the opposite direction (Fig. 6). In this case, a homogeneous zone was formed over the bubble. The surfactant concentration gradient near the bubble gradually decreased, so did the intensity of the arising vortices.

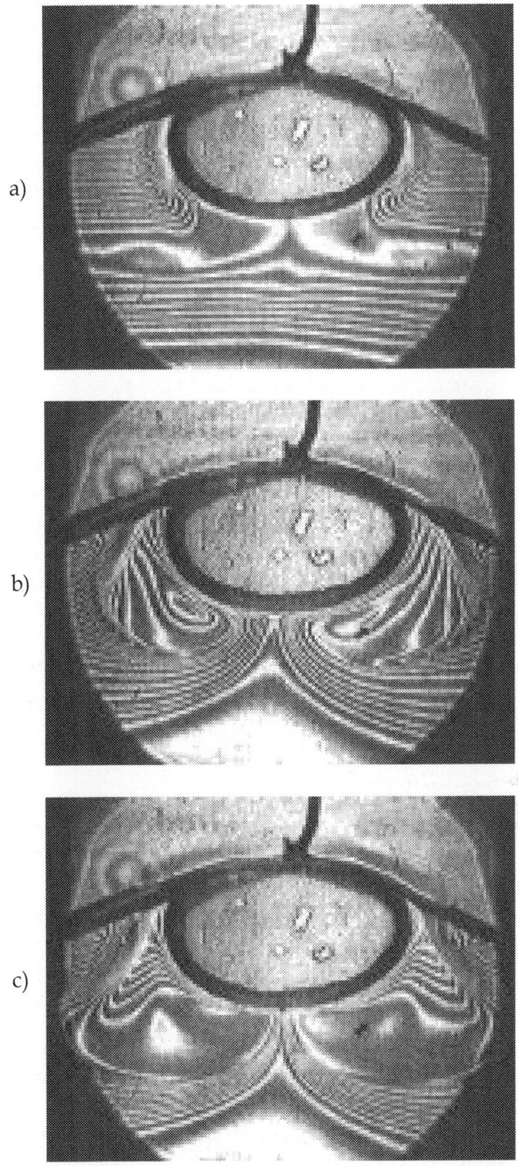

Fig. 5. Interferograms of concentration distribution around the air bubble in vertical layer of stratified acetic acid solution. Concentration difference between the bubble poles ΔC = 10%, vertical diameter of the bubble d = 3.8 mm. Time elapsed from the beginning of a cycle t, sec: 0 (*a*), 4.50 (*b*), 8.25 (*c*). View from the side

a)

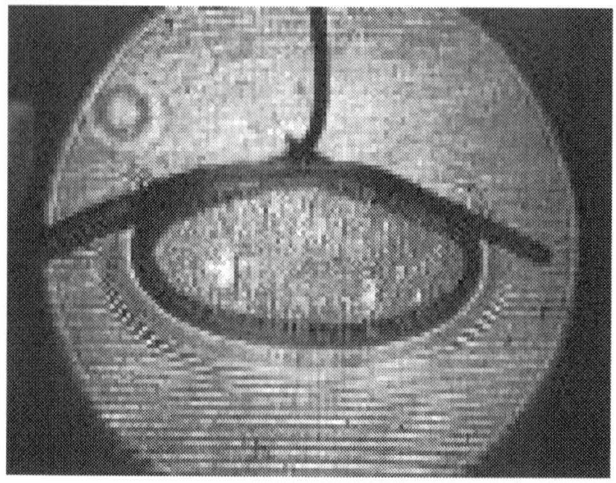

b)

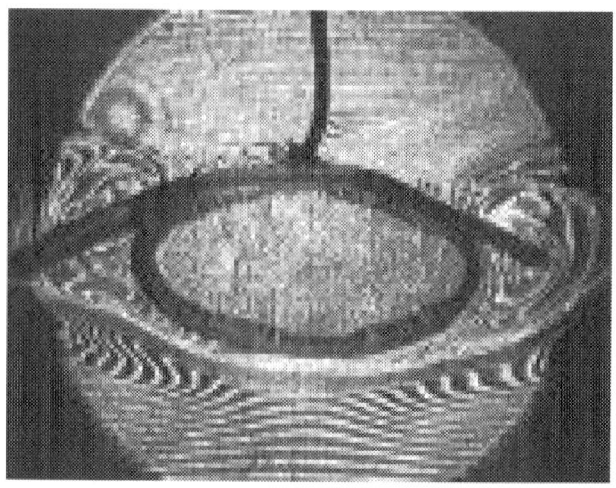

c)

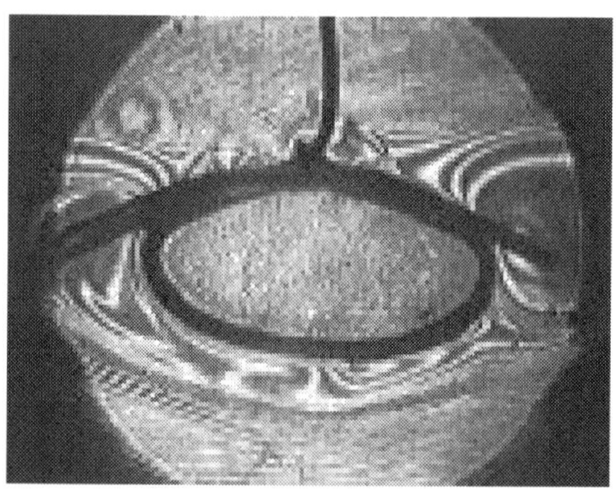

Fig. 6. Interferograms of concentration field around the air bubble in vertical layer of stratified isopropyl alcohol solution. Concentration difference between the bubble poles $\Delta C = 4\%$, vertical diameter of the bubble $d = 3.8$ mm. Time elapsed from the beginning of a cycle t, sec: 0 (*a*), 2.0 (*b*), 10.0 (*c*). View from the side

The videotape recording of interferograms allowed us to trace the evolution of concentration field around the bubble. The measurements of the concentration distribution on the vertical coordinate were made at the time moments preceding the beginning of each cycle of oscillations when the intensity of the convective motion was minimal. In the system "water - 70% acetic acid solution" concentration decreases monotonically with liquid layer height. Then the bubble began to stir actively the surrounding fluid, reducing the concentration difference between the bubble poles and the mean concentration of the solution under the bubble. At the initial time the vertical concentration gradient ∇C was a maximum and equal to 6.5 %/mm. The corresponding values of the diffusion Marangoni *Ma* and Grashof *Gr* numbers were 4.1×10^6 and 1.9×10^3. Then the surfactant concentration gradient diminished rapidly with time, reaching the minimum value 4.4 %/mm ($Ma = 3.0\times10^6$, $Gr = 1.4\times10^3$) approximately 60 min after the beginning of oscillations. After this it again increased approaching some constant asymptotic value of 5.2 %/mm ($Ma = 3.1\times10^6$, $Gr = 2.0\times10^3$).
Analogous studying of the evolution of the concentration fields and the concentration gradient averaged over the bubble height was made for the "water–40% isopropyl alcohol solution" system. In this case in contrast to the solutions of acetic acid a decrease and the following increase of the concentration gradient manifest themselves to a lesser degree. Such a behavior of the gradient is evidently due to the fact that with decrease in the intensity of solutocapillary motion the recover of vertical concentration gradient near the bubble due to gravitation and diffusion occurs much slower in alcohol solutions owing to a greater viscosity and a smaller diffusion coefficient. In experiments with lesser concentrated 20% alcohol solutions such extremum is not observed any longer, and the gradient is merely decreasing in a monotonic manner from 1.0 %/mm ($Ma = 16.8\times10^6$, $Gr = 1.8\times10^3$) to 0.3 %/mm ($Ma = 3.5\times10^6$, $Gr = 0.4\times10^3$).

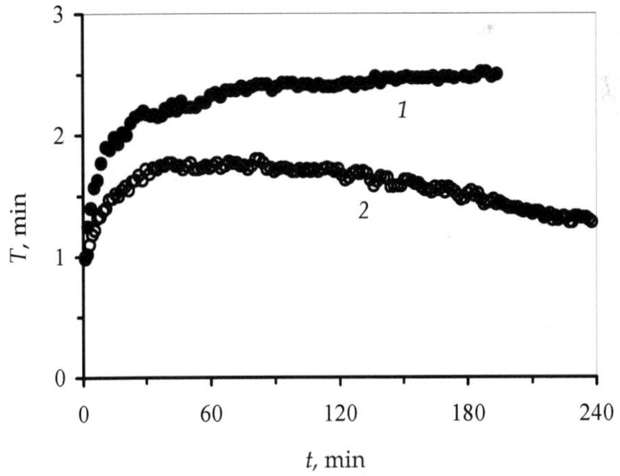

Fig. 7. Time dependence of the oscillation period. *1* - isopropyl alcohol solution, *2* - acetic acid solution

The time dependence of the oscillation period was investigated for different values of the average solution concentration, the concentration gradient and the diffusion Marangoni and

Grashof numbers. It has been found that the period of convective oscillation changed simultaneously with a variation of the concentration gradient. In tests with 20% isopropyl alcohol solution the period increased monotonically with time and reaches the maximum within the first hour, after which it remained unchanged (curve *1* in Fig. 7). In tests with acetic acid solution the oscillation period within the first hour also increased (Fig. 7, curve 2), but then began to decrease, which implies that the time dependence of the period was the inverse of that of the concentration gradient. Thus in both cases there existed proportionality between the frequency (inverse period) of oscillations and concentration gradient of the surrounding fluid. In Fig. 8 the dimensionless frequency, normalized to the concentration Marangoni number, is plotted on the y-coordinate and the dimensionless time is plotted on the x-coordinate. As evident from the graph as soon as the oscillation are set up the ratio $(T / \tau)^{-1} \cdot Ma^{-1}$ turns out to be of the same value for different fluids and remains constant during the whole experiment not depending on the orientation of the concentration gradient (all experimental points for both states are well located on the straight line, which is nearly parallel to the time axis). The obtained result supports the view that just the solutocapillary Marangoni forces are the main initiator of periodical convective motion around the bubble in the heterogeneous solution of fluid surfactant.

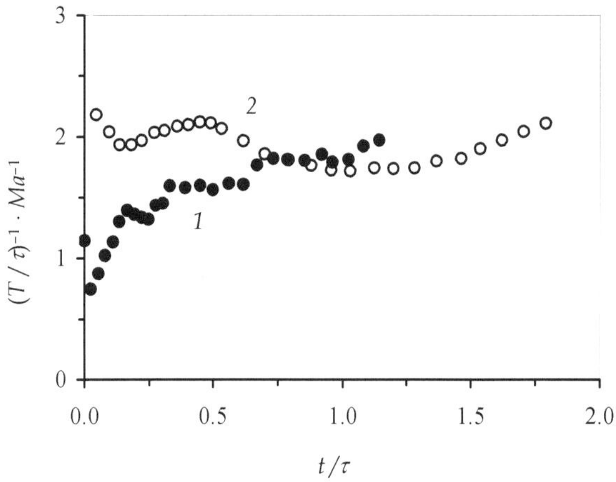

Fig. 8. Dimensionless oscillation frequency. *1* - isopropyl alcohol solution, *2* - acetic acid solution

The described oscillations of the solutal flow were also observed near the interface between two mutually insoluble fluids (Kostarev *et al.*, 2007). For experimental investigation of this case we used the chlorobenzene drop placed in an aqueous solution of isopropanol. Note that isopropanol being surfactant for water and chlorobenzene is readily dissolved in both fluids, whereas the top solubility of pure chlorobenzene in water is 0.05% (at 30° C) only and that of water in chlorobenzene is even less. It means that in the process of mass transfer the drop should absorb only one of the mixture components - a surfactant. Fig. 9 shows a series of interferograms, describing evolution of isopropyl alcohol distribution inside and around the drop of chlorobenzene in water during one of the cycles of the oscillatory mode. The concentration gradient in the stratified isopropyl solution was directed upward. The isopropanol, being soluble in both fluids, diffused easily through the interface in both directions.

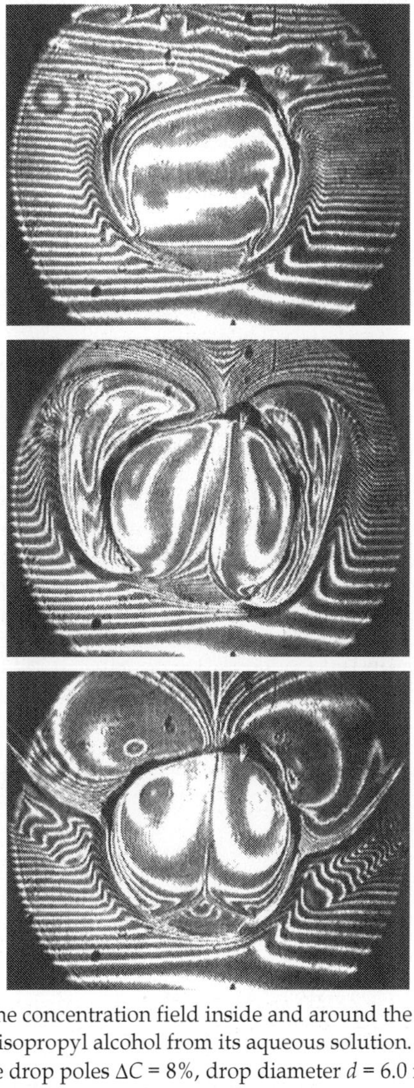

Fig. 9. Interferograms of the concentration field inside and around the drop of chlorobenzene absorbing isopropyl alcohol from its aqueous solution. Difference of alcohol concentration between the drop poles $\Delta C = 8\%$, drop diameter $d = 6.0$ mm. Time elapsed from the beginning of the cycle t, sec: 0 (*a*), 3 (*b*), 18 (*c*)

The ease, with which the surfactant penetrates the surface, imparts some peculiar features to the development of motion in the system of fluids. Thus, a greater part of the drop surface is constantly in slow motion, because, due to absorption of the alcohol by the drop, there is always a gradient of the surfactant concentration at the interface (Fig. 9,*a*). Fig. 9,*b* corresponds to the beginning of the intensive vortex convection motion caused by an influx of the solution with higher alcohol content from the region over the drop. Note that the major mass of the alcohol entrained by the capillary convection is not absorbed by the drop. It remains in the surrounding solution and, being lighter than the adjacent layers, floats up. Separation of the alcohol flux from the drop surface occurs in the region where the action of the capillary forces is balanced by the Archimedean force, which increases as the surfactant moves downward. As in the case with a bubble, the rising solution forms two closed vortex cells. The alcohol penetrating the drop also forms an intensive two-cell flow (Fig. 9,*c*). Capturing an increasingly growing amount of solution with high surfactant concentration, the convective cells outside the drop become more and more light. As a result they move gradually upward until finally they cut off the feeding flux of the alcohol from the drop. This causes the Marangoni flow to slow down abruptly. Simultaneously, inside the drop, the solution of chlorobenzene with rather high alcohol content, held all this time at the lower part of the drop by a large-scale flow, begins to float up. If, by this time, the drop has been already saturated with the surfactant (i.e. within a few cycles after the beginning of the test) the buoying alcohol becomes excessive, and the drop ejects it back into the surrounding solution reaching thereby chemical equilibrium with the environment. Then, as in the case with a bubble, the gravitational force restores the vertical stratification of the surfactant near the drop surface and the cycle is repeated.

Time variation of the period of the flow intensification near the drop is presented in Fig. 10. As measurements show, an increase in the period with time is related to a decrease in the gradient of the surfactant concentration in the surrounding solution due to diffusive and convective mixing. Variation of the surfactant concentration gradient has a two-fold effect: it reduces the intensity of the capillary forces and increases the amount of the surfactant ejected by the drop at the end of the cycle. Fig. 11 presents time dependence of the ratio of the maximum alcohol concentration differences inside and outside the drop. As it is seen, this ratio remains practically constant for the whole period of observation (~3 hours) in spite of a very complicated flow behavior.

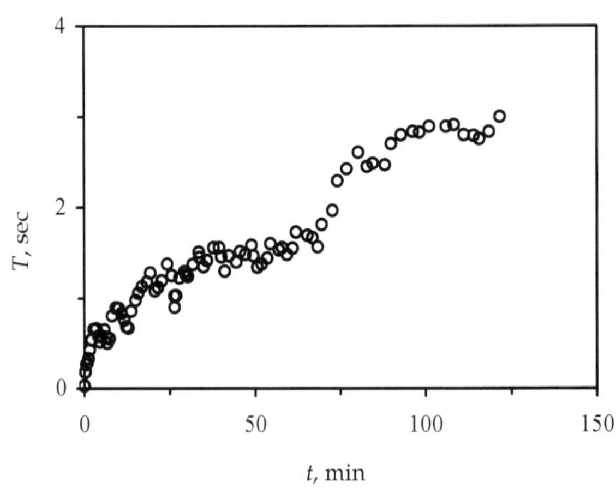

Fig. 10. The period of flow intensification near the drop as a function of time

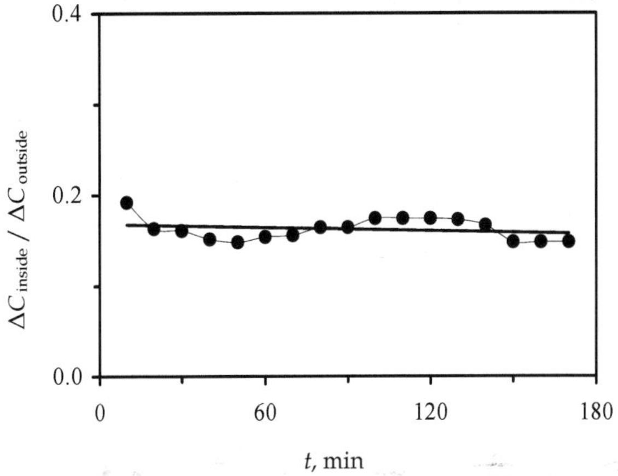

Fig. 11. Relationship between surfactant concentration inside and outside the drop

5. Plane rectangular channel

In our next experiments (Birikh *et al.*, 2006, 2008; Bushueva *et al.*, 2008; Kostarev *et al.*, 2008a, 2009a) the fluid filled a thin gap between two vertically oriented glass plates. Inside the gap the fluid was bounded from above and below by solid plates, which formed a horizontal channel of height $h = 2$ mm and thickness 1.2 mm. The tests were performed for aqueous solutions of methyl, ethyl and isopropyl alcohols. The channel was initially filled with distilled water or uniform alcohol solution with weight percent concentrations ranging from 1 to 10%. An air bubble put into the channel cavity from one of its ends in such a way that it completely bridged over the channel and had one free lateral boundary. Then, from the other end, the cavity was gradually filled with a concentrated (up to 40%) alcohol solution. The fluid density difference thus produced in the channel led to the formation of a rather slow large-scale advective gravitational flow, during which a narrow "tongue" of a more concentrated surfactant solution flowed along the upper channel boundary toward the bubble surface, forming near the surface an area with upward-directed concentration gradient. The value of the concentration gradient was specified by the velocity and concentration of the inflowing surfactant flux, as well as by the initial concentration of the solution in the channel.

The interference patterns of the concentration field near the lateral boundary of the air bubble is shown in Fig. 12 and Fig. 13. In contrast to thermocapillary flow developing in a non-threshold manner at arbitrary small values of the interface temperature gradient, the solutocapillary flow was initiated not at the time when the surfactant "tongue" reached the bubble surface, but with some delay. Thus, in the experiment a time-lag Δt between the arrival of the surfactant flow at the bubble surface and the initiation of the convective vortex was as long as 28 sec. At the same time the difference of the solution concentration ΔC^* between the upper and lower bubble poles caused by a continued motion of the "tongue" reached 2.2%. Then the equilibrium was abruptly disturbed and very rapidly (during about 0.2 sec) a rather intensive Marangoni flow was initiated in the form of a vortex cell, in which the surfactant under the action of solutocapillary forces was carried along the bubble surface toward the lower bubble boundary. Owing to fluid continuity, the arising flow accelerated the flux of the concentrated surfactant solution along the upper boundary of the channel towards the bubble surface, adding thereby intensity to the existing convective vortex. However, the originated vortex cell, entrapping more and more portions of highly concentrated surfactant solution, became increasingly light. Rising up it eventually cut off the arriving jet of surfactant from the top of the bubble. As a result, the vortex flow ceased abruptly and the bubble surface turned out to be surrounded by a thin layer of the surfactant solution having a uniform concentration.

a)
b)
c)
d)
e)
f)
g)

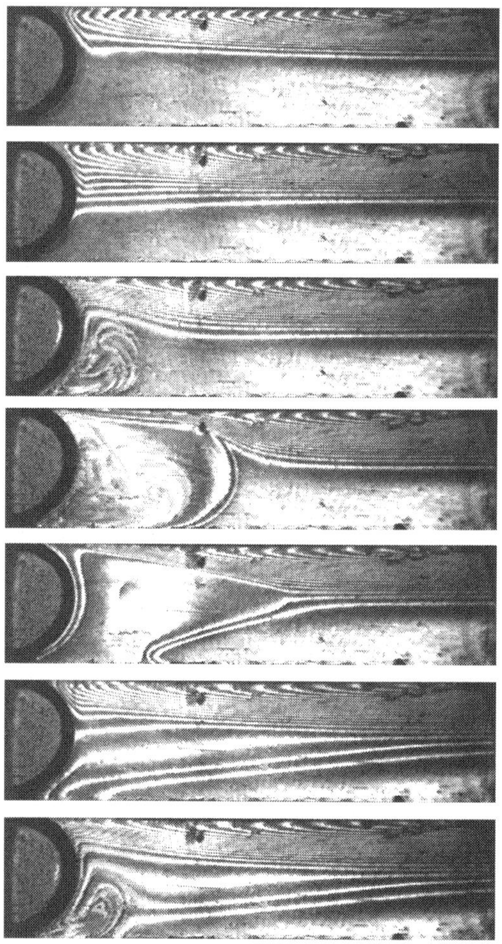

Fig. 12. Interferograms of concentration field evolution around the air bubble in rectangular layer with stratified isopropyl alcohol solution. *t*, sec: 0, 28.0, 28.1, 28.2, 29.2, 34.2, 34.3

a)
b)
c)
d)
e)
f)
g)

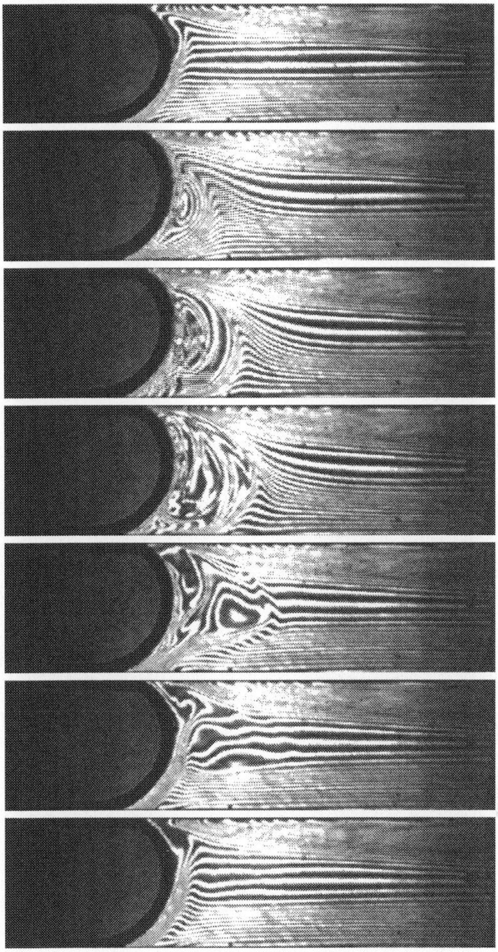

Fig. 13. Interferograms of concentration field evolution around the air bubble in rectangular layer with stratified ethyl alcohol solution. *t*, sec: 0, 0.2, 0.4, 0.8, 5.0, 10.0, 20.0

However, equalizing of the generated horizontal gradient of the surfactant concentration resulted in the development of a slow advective flow. This gravitational motion, by restoring the disturbed vertical stratification of the solution, again draws a concentrated surfactant solution to the upper bubble boundary. As soon as the flux of surfactant touched the bubble surface, the solutocapillary vortex recurred. The cycle repeated iteratively, with the difference that the oscillation period increased with time whereas the intensity of the vortex flow decreased due to a gradual decrease of the vertical concentration gradient. Marangoni convection ceased at the time when the concentration of the solution became almost uniform throughout the whole layer.

A characteristic feature of this process is that during consecutive cycles the capillary flow was initiated at much less concentration differences at the bubble surface (thus, the onset of the second cycle happened at $\Delta C^{*}\sim 0.6\%$). At the same time, the average concentration of the surfactant at the bubble surface gradually increased, which might be one of the explanations for the essential decrease in the concentration difference at the bubble surface at the beginning of each consecutive cycle of the vortex convection motion. In order to verify this suggestion we investigated the dependence of the critical Marangoni numbers *Ma**, defined by the maximum vertical concentration difference between the bubble poles at the moment of formation of the first vortex, on the surfactant concentration in the solution surrounding the bubble. To this purpose, the channel was initially filled not with pure water but with a homogeneous aqueous solutions of alcohols with different initial concentration C_0. It was found that, as C_0 increases (and, respectively, as the surface tension at the bubble surface decreases) the values of *Ma** decrease monotonically. Fig. 14 presents the critical Marangoni numbers at the beginning of different cycles of the vortex flow in solutions of ethanol and isopropanol (points *1* and *2*) as functions of the crispation number *Cr*, characterizing the ratio of viscous and capillary forces. The crispation numbers, in turn, were calculated using the values of the surface tension corresponding to the average surfactant concentration at the air bubble surfaces at the moments of motion intensification. The diagram also presents the results of measurements of *Ma** as a function of *Cr*, obtained from tests for solutions with different initial alcohol concentration (solid line drawn through points *3* for ethanol and *4* for isopropanol). It is seen that the values of *Ma** obtained under various conditions are rather close and all curves are qualitatively coincide.

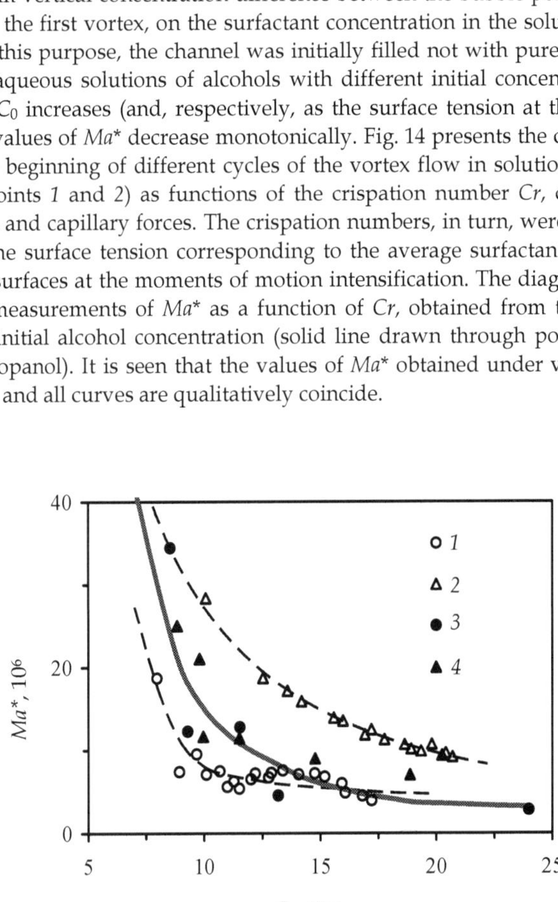

Fig. 14. Critical Marangoni numbers as a function of crispation number

In our experiments we also investigated the time dependence of the oscillation period of the convective flow near the bubble in solutions of ethanol, isopropanol and methanol. In all cases, the oscillation frequency of the flow near the bubble surface was first rather high (accordingly the period *T* was small ~ 5-10 sec). Then, the oscillation period increased monotonically with time, and after some time the oscillations suddenly came to an end. The experimental data obtained were compared with the results of numerical analysis (Birikh *et al.*, 2006). The results of experiments and numerical calculations were found to be in good agreement with respect to the convective flow structure and the oscillation period. The period of the steady-state oscillations decreases with an increase of the Grashof number and depends weakly on the Marangoni number. This dimensionless relationship is represented in Fig. 15. The different points on the plot represent the results of experiments for various alcohol solutions (*1* - ethanol at $C_0 = 40\%$, *2* - ethanol at $C_0 = 20\%$, *3* - isopropanol, *4* - methanol). The solid line corresponds to numerical calculations.

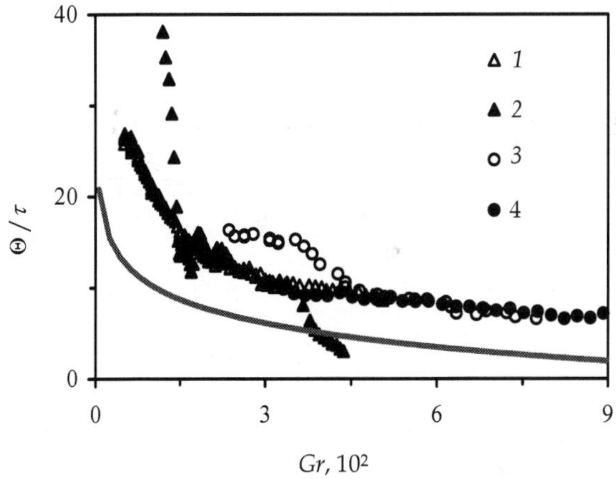

Fig. 15. Dimensionless oscillation period

An analogous situation was observed on the chlorobenzene – aqueous isopropanol solution interface (Kostarev *et al.*, 2008b). A chlorobenzene drop, immiscible in the water, was injected by means of a syringe into a channel cavity from one of its ends in such a way that completely bridged over the channel and formed lateral liquid/liquid boundary. A "tongue" of a concentrated (40%) isopropyl solution flowed along the upper channel boundary toward the drop surface, forming near it an area of nixture with vertical surfactant stratification. In contrast with the convection initiation around the gas bubble, a much more greater surfactant concentration gradient is needed to be created during the time between the moments when the alcohol reaches the droplet surface and before solutocapillary motion begins to develop, as is seen in the interferograms in Fig. 16. The difference of the solution concentration ΔC^* between the upper and lower drop ends equalized to 4.1% against the value of 2.2% in case of gas bubble. The other distinction is that surfactant diffuses through the phase interface into chlorobenzene, thus creating a concentration gradient in the latter also. Initially, this process occurs rather uniformly along the droplet surface, so that a surfactant gradient at the surface itself is absent. Only later, an intense Marangoni convection develops in the solution (at $\Delta t \sim 1$ min after the contact between the surfactant tongue and the droplet surface.

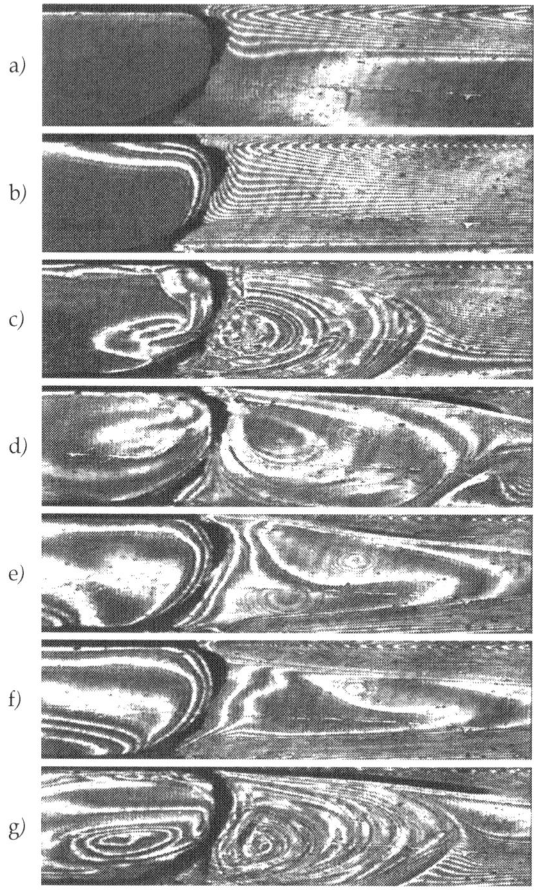

Fig. 16. Interferograms of the concentration field in the experiment with a chlorobenzene droplet in a stratified aqueous isopropanol solution: *t*, sec: 0, 60, 61, 66, 70, 76, 77

6. Conclusion

The performed experiments provided convincing evidence for the origination of intensive solutocapillary Marangoni convection in multiphase systems with inhomogeneous concentration of the soluble surfactant along the phase interface. The specific feature of these fluid flows was that they took place under conditions of weak surfactant diffusion (the values of the Schmidt number in experiments were about 10^3), and that the changes in the spatial concentration of the solution happened mainly due to convective mass transfer. The arising stresses and flows in some ways are similar to thermocapillary ones and are governed by similar relationships, for example, flow intensity is proportional to the surface tension gradient. On the other hand, the solutocapillary phenomena demonstrate some specific features, which are related both to fairly large values of the Marangoni numbers and

to a more complicated character of formation of the surface tension gradient. The mechanism of the transfer (absorption) of a surfactant at the phase interface is distinct from the mechanism maintaining the temperature at the interface. The interface has inertia, and a convective transfer of the surfactant is possible along it, accompanied by surface diffusion. The latter peculiarity is caused by slower (compared to heat transfer) diffusion of the fluid molecules with low surface tension onto the interface and adsorption of the surfactant at the inter-phase boundary. These specific features should be taken into account in constructing theoretical models of the boundary conditions at the interface allowing for the formation of a new, surface phase of the surfactant, controlling a transition of the soluble mixture component from one phase to another. The process of formation of such a surface phase includes two stages. First, the molecules of the surfactant reach the interface of the liquid system via diffusive transport, because the normal component of the convective velocity at the surface is zero due to impermeability of the boundary. This process is followed then by formation of the proper interface in itself, by means of adsorption and desorption.
The experiments demonstrated that the solutocapillary convection displays well-defined non-stationary properties. The interaction between the buoyancy and the Marangoni convective flows is responsible for the onset of auto-oscillation regime of convective motion around the gas bubbles and insoluble drops in a vertically stratified surfactant solutions. The periodical outbursts of solutocapillary flow at the bubbles/drops interface intensify substantially the stirring of the solution and can be used for the control of this procedure. The clarification of the nature of the flows developing close to resting liquid and gaseous inclusions in inhomogeneous surfactant solutions and the discovery of a threshold for the excitation of solutocapillary motion allow explaining many aspects of mass exchange and material structure formation in a number of technological experiments in microgravity, and predicting the behavior of complex systems of liquids and multiphase media in thin channels and layers, and in cavities of complex geometry. The results obtained in the experiments can be used in designing systems of passive homogenization in liquids and in optimizing working regimes of the already operational technology lines in various branches of industry. A special role can be played by the Marangoni effects elaborated here in the design of microsystems for cooling and heat exchange with multi-component mixtures of liquids as a heat agent.

7. Acknowledgements

The work was supported by Russian Foundation of Fundamental Research under project No 09-01-00484, joint project of SB, UB and F-EB of RAS (116/09-C-1-1005) and the program of the Department of Power Engineering, Mechanical Engineering, Mechanics and Control Processes of RAS No 09-T-1-1005.

8. References

Birikh, R.V.; Zuev, A.L.; Kostarev, K.G. & Rudakov, R.N. (2006). Convective self- oscillations near an air-bubble surface in a horizontal rectangular channel. *Fluid Dynamics*, Vol. 41, No. 4, pp. 514–520.

Birikh, R.V.; Rudakov, R.N.; Kostarev, K.G. & Zuev, A.L. (2008). Oscillatory modes of solutocapillary Marangoni convection at a drop-liquid interface. *Proceedings of 6th*

EUROMECH Nonlinear Dynamics Conf., pp. 1-6., Saint Petersburg, Russia, 30 June - 4 July 2008.

Bushueva, K.A.; Denisova, M.O.; Zuev, A.L. & Kostarev, K.G. (2008). Flow development at the surfaces of bubbles and droplets in gradient solutions of liquid surfactants. *Colloid J.*, Vol. 70, No. 4., pp. 416-422.

Gustafson, S.E.; Kjellander R.A.E. & Rolf A.E. (1968). An interferometer for direct recording of refractive index distributions. *Z. Naturforch.*, Vol. 23a, No. 2, pp. 242-246.

Kostarev, K.G.; Zuev, A.L. & Viviani, A. (2004). Oscillatory Marangoni convection around the air bubble in a vertical surfactant stratification. *Comptes Rendus Mecanique*, Vol. 332, No. 1, pp. 1-7.

Kostarev, K.G.; Zuev, A.L. & Viviani, A. (2006). Thermal and concentrational Marangoni convection at liquid/air bubble interface. *J. Applied Mechanics. Transactions ASME*, Vol. 73, No. 1, pp. 66-71.

Kostarev, K.G., Pisarevskaya, N. N., Viviani, A. & Zuev, A. L. (2007). Oscillatory Marangoni convection around bubbles and drops in heterogeneous solutions of surfactants. *Int. J. Microgravity Science & Technology*, Vol. 19, No. 2, pp. 12-17.

Kostarev, K.G.; Zuev, A.L. & Viviani, A. (2008a). Experimental study of convective self-oscillations near the lateral surface of a bubble in a plane rectangular channel. *Acta Astronautica*, Vol. 62, No. 6-7, pp. 431-437.

Kostarev, K.G.; Zuev, A.L. & Viviani, A. (2008b) Experimental studies of concentration convective surfactant mass transfer near a drop-liquid interface. *Proceedings of 19th Int. Symp. on Transport Phenomena*, pp. 1-5, Reykjavik, Iceland, 17-21 August 2008. (In Electronic Conference Proceedings).

Kostarev, K.G.; Zuev, A.L. & Viviani, A. (2009a). Experimental considerations of solutocapillary flow initiation on bubble/drop interface in the presence of a soluble surfactant. *Int. J. Microgravity Science and Technology*, Vol. 21, No. 1-2, pp. 59-65.

Kostarev, K.G.; Zuev A.L. & Viviani A. (2009b). Convective stirring of a stratified surfactant solution by the oscillatory solutocapillary flow. *Proceedings of 4th Int. Conf. on Physics and Control*, pp. 1-8, Catania, Italy, 1-4 September 2009. (In Online Conference Proceedings http://lib.physcon.ru/download /p1930.pdf).

Marsters, G.F. & Advani, A.A. (1973). A tilted plate interferometer for heat transfer studies. *Rev. Sci. Instrum.*, Vol. 44, No. 8, pp. 1015-1018.

Vazquez, G.; Alvarez, E. & Navaza, J.M. (1995). Surface-tension of alcohol plus water from 20-degrees-C to 50-degrees-C. *J. Chem. Eng. Data*, Vol. 40, No. 3, pp. 611-614.

Vazquez, G.; Alvarez, E.; Sanchez-Vilas. M.; Sanjurjo B. & Navaza J.M. (1997). Surface tension of organic acids + water binary mixtures from 20°C to 50°C. *J. Chem. Eng. Data*, Vol. 42, No. 5, pp. 957-960.

Zuev, A.L. & Kostarev, K.G. (2006). Oscillation of a convective flow round the air bubble in a vertically stratified solution of a surfactant. *J. Experimental and Theoretical Physics*, Vol. 130, No. 2, pp. 363-370.

Zuev, A. L.; Kostarev, K.G. & Viviani, A. (2008a). Peculiarities of the solutocapillary convection. *Proceedings of 11th Int. Conf. on Multiphase Flow in Industrial Plants*, pp. 39-46, Palermo, Italy, 7-10 September 2008

Zuev, A.L. & Kostarev, K.G. (2008b). Certain peculiarities of the solutocapillary convection, *Physics-Uspekhi*, Vol. 51, No. 10, pp. 1027-1045.

Aerodynamics of Ceramic Regular Packing for Heat-Massexchenge Processes

Alexandr Pushnov
Department of Engineering, Moscow State University of Environmental Engineering, Staraya Basmannaya street 21 / 4, 105066 Moscow Russia

1. Introduction

The chemical, petroleum and other industries in implementing the removal processes, distillation and purification of gases from the emissions are widely used structured, as well as structural attachment mesh configuration [1-2]. Both of these belong to the regular batch nozzles, forming a three-dimensional spatial multi-channel structure.

Structured packing is usually implemented as a set of individual corrugated sheets assembled in packets (blocks). Themselves with sheets can be made of polymer, ceramic and other materials. Formed in this channel have a complex spatial configuration. The most common structured nozzle of this type in the industry are packing company Sulzer Chemtech (Switzerland).

To clear the air of various pollutants also find application developed at the Vilnius Gediminas Technical University biological plants, the main element of which is a filter with bio-fill [3, 4]. As a sorbent in the biofilter used cheap and available material - pieces of fir bark of various fractions, eg, 35, 25 and 12,5 mm [4, 5]. Performance and prospects of biofilters for air cleaning from harmful impurities doubt. However, the use of apparatus thin layer pieces sorbent indicated linear sizes and fractions makes it difficult to organize an optimal homogeneous structure of the granular layer throughout the cross section of the apparatus described in [4]. At the same time of contact of microorganisms with pollution in different parts of the biofilter, it seems, can vary significantly.

Unlike the bulk of irregular attachments regular structured [1] as the structural attachment [2] have a greater specific surface and at the same time have significantly lower hydraulic resistance. In addition, the structured packing sheet avoid contacting bypass flows due to inherent in bulk irregular layers (eg, rings and saddles Rashig Burley) phenomena wall anisotropy [6, 7, 16, 17].

However, the known sheet structured packing does not have the properties of isotropy. They are due to the peculiarities of its design will organize a system of parallel isolated from each other channels and therefore do not provide a satisfactory cross-mixing of contacting streams. This affects the effectiveness of a column of contact devices of chemical technology, as well as in power (cooling towers) processes.

One possible way to improve a class of structured nozzles is to create three-dimensional isotropic structure on the basis of highly porous cellular materials (HPCM) [8], which in the European Community referred to the term "foam".

Given the emergence of new highly porous cellular materials can offer the following tips for a new classification of heat and mass transfer processes (see Fig. 1).

Recently, a number of publications devoted to studying the possibilities of using HPCM as industrial attachments for the implementation of the processes of heat and mass transfer in the chemical industry.

So in [10] L. Padeste, A. Baiker, J.P. Gabathuler point to prospects of using ceramic foam packaging of cordierite as carriers for catalysts. In their experiments the authors of [10] based on the known method of determining the dwell time distribution of fluid flow in a layer of packing for writing marking substance (label). Moreover, for the greater persuasiveness of their results, the authors [10] conducted experiments for two cases: the sand layer of spheres of diameter 1, 2, 3 and 5 mm, as well as ceramic foam packaging. However, as shown by O. Levenspiel, J.C.R. Turner [11] using the principle of measuring the distribution of time spent using the tags must be complete confidence in the presence of a flat profile of velocity. Contrary to the authors of [10] with reference to the work of [12] for bulk layers of balls, this condition is just not satisfied. On the contrary, as follows from a series of special studies in the timing characteristics of devices with bulk layers of balls and other grains form in a wide range of Reynolds numbers in a layer have the balls characteristic velocity profile with an extreme surge near the walls of the apparatus of the greater looseness of packing of spheres in this area [6, 17]. Incidentally, this is also evidenced by the results of [12]. This circumstance gives rise to a certain degree of doubt and in other results [10] on ceramic foam packaging.

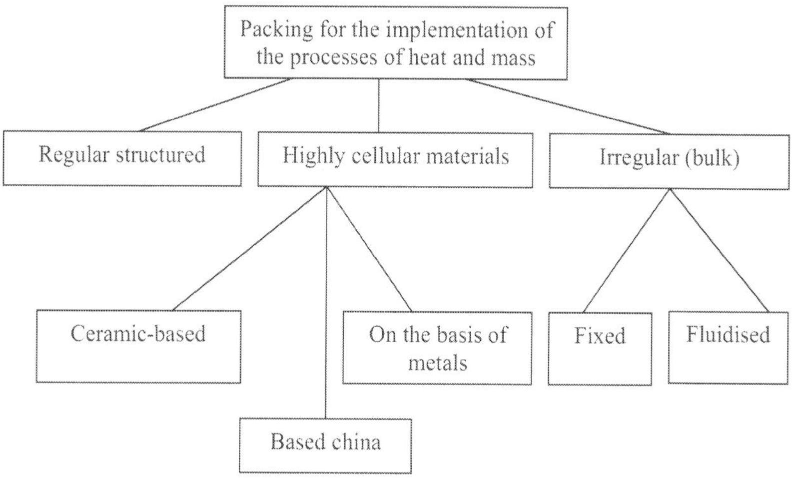

Fig. 1. Classification of packing for the processes of heat and mass transfer

Thus, the problem of studying the aerodynamics of packing on the basis of ceramic HPCM remains relevant.

In another study [13] presented the results of experiments on samples of porous blocks of metallic foam.

However, experimental data on the basic geometric characteristics of porous packing of ceramic materials technology HPCM in the literature is largely absent. Hydrodynamics of ceramic packing HPCM also is not yet sufficiently studied.

This paper presents the results of a study of the aerodynamic and geometric characteristics of the regular porous bits of ceramic materials - namely, pressure loss and the degree of turbulence in a wide range of loads on gas.

2. The structure of highly porous cellular materials

Highly cellular materials represent a new type of porous material. These materials, generally speaking, can be manufactured as a metal and nonmetal on the basis of having the maximum possible porosity and permeability of the reticulated cellular structure of the pore space.

The geometrical structure of these HPCM`s really almost entirely consistent with the definition of isotropy and it could be used to produce a new generation of ceramic packing by powder metallurgy. Since the porous structure of the framework HPCM`s for the manufacture of packing largely determines the appropriate level of physical and chemical properties of the future heads, with its development were considered different options. It was preferred porous polymer materials, namely - widely used in polyurethane foam industry. That this material is completely open, transparent, thin, isotropic, spatial, three-dimensional structure with virtually no local defect closed macropores that may violate isotropy of any fragment of the future attachments.

Another important advantage of polyurethane foam as a basis for the manufacture of a new nozzle is easy to cut the material into separate pieces of any predetermined arbitrary shape (see fig. 2).

It is also important to note a small volume fraction of foam in the total amount of future attachments. According to [8] itself polymeric material occupies no more than 2,5%. Thus, the initial proportion of free volume (separately) a fragment of the foundations of attachment of this polymer could reach 97.5%.

From the viewpoint of the properties of the full isotropic element attachment is particularly important to high initial homogeneity of the structure of polyurethane foam, as well as linear geometric dimensions themselves local micropores, forming the frame of the future attachments. It is important and the lack of closed pore canals.

Samples from the nozzle HPCM produced by slip casting [20]. After heat treatment formed delicate arch-labyrinthine structure (see Fig. 3). Geometric model of the unit cell packing of the space pore HPCM is pentagondodecahedron, whose structure is shown in Fig. 4.

Fig. 2. Photo samples from the packing HPCM

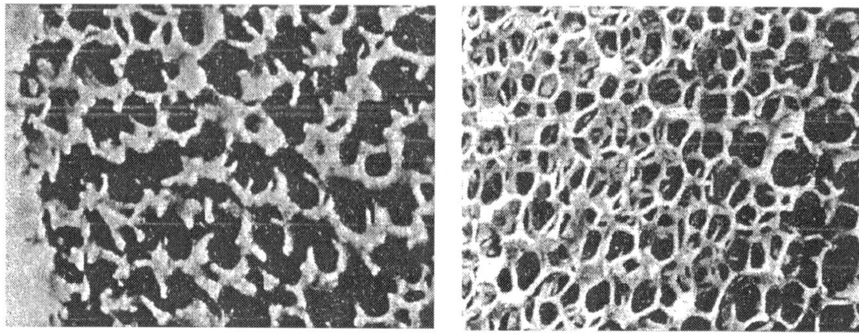

Fig. 3. Arch-labyrinthine structure of highly porous cellular material

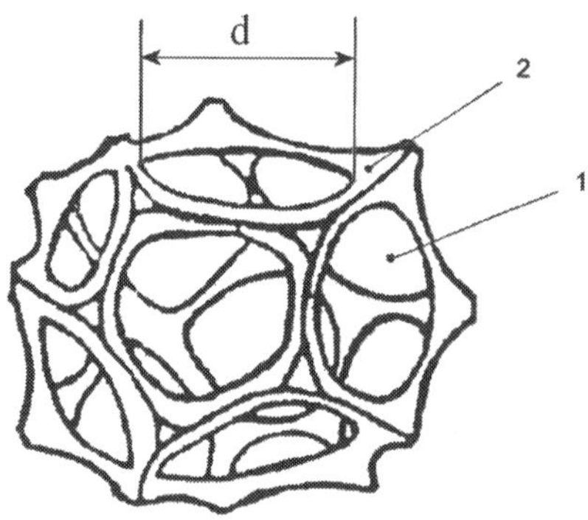

Fig. 4. Structure pentagondodecahedron cell material from the packing HPCM, 1 - pore channel; 2 - wall

3. Geometric characteristics of layer packing

3.1 General

The main geometric characteristics, determining the structure and parameters of heat and mass transfer in a packed column apparatuses include:

- • the proportion of free volume of the layer packing - **ε** (porosity), m^3/m^3;
- • specific surface layer of regular or bulk packing in a unit volume - **a**, m^2/m^3.

The magnitude of the porosity of the layer packing is numerically equal to the value of "live" section of the layer, determines the value of the characteristic velocity of the gas flow in the layer.

Generally speaking, during the gas flow in the layer of bulk (irregular) packing is a "hybrid" hydrodynamic problem. On the one hand, the gas stream flows around the elements of the packing layer. On the other hand, the flow proceeds in the pores between adjacent elements of the packing.

In the case of the packing, made of highly porous cellular material there is reason to consider the flow in a layer of a packing as an internal problem - that is, the flow in the pore channels (see Fig. 4). Gas flows through the layer HPCM through the complicated cross-section, defines the surface of **a** layer of packing per unit volume and the proportion **ε** of free volume (porosity).

3.2 Equivalent diameter packing

In the English literature to refer to the equivalent diameter of the channel often use the term *hydraulic radius*. Here, the term *equivalent diameter of the channel*.
In [22] suggested as the defining geometric size of the channels of complex cross section using the equivalent hydraulic diameter - $\mathbf{d_e}$, equal to:

$$d_e = 4 \cdot F / P \tag{1}$$

where: **F** - cross sectional area of the channel, m^2, **P** - wetted perimeter of the channel, m.
Looking for a gas flow in the packing layer Zhavoronkov and Aerov [23, 24] introduced the concept of equivalent diameter of pore channels is equal to four times the hydraulic radius:

$$d_e = 4 \cdot \varepsilon / a \tag{2}$$

In [22] that the concept of equivalent diameter "is not in general is universal and allows only a few cases to calculate the pressure loss in the channels of different geometry formulas for tubes of circular section. On the other hand, made in [25-27] treatment of the results of experiments to measure the pressure loss in granular layers of various geometric shapes and sizes showed that these results can be generalized single dual-term criterial equation of the form [25]:

$$Eu_m = A + B / \mathrm{Re}_{d_e} \tag{3}$$

In equation (3): **A** = 0,9, **B** = 100; $\mathbf{Eu_m}$ - modified Euler number equal to:

$$Eu_m = \left(\Delta P \cdot \varepsilon^2\right) / \left(H \cdot a \cdot \rho \cdot W_0^2\right) \tag{4}$$

ρ - gas density, kg • s^2/m^4; **H** - height of the layer packing, m; **ΔP** - pressure loss, kg/m^2; $\mathbf{W_0}$ -velocity gas in the expectation of full cross-section of an empty vehicle, m / sec.
Re_{d_e} - Reynolds number, relative to the equivalent diameter of the channel in a layer of packing equal to:

$$\mathrm{Re}_{d_e} = \left(W_0 \cdot d_e\right) / \left(\varepsilon \cdot \nu\right) \tag{5}$$

Here, **ν** - kinematic viscosity of gas, m^2 / s; $\mathbf{W_0}$ - gas flow velocity in the calculation of the total cross section of empty vehicle, m / s.
Thus, the use of equation (3) as the defining geometric parameters of the layer packing equivalent diameter of the channel - d_e with respect to the granular layers and packed columns allows the apparatus with sufficient accuracy to calculate the pressure loss in them. Consequently, we can assume that in packed columns vehicles use the equivalent diameter of the channel is fully justified in carrying out hydraulic and other technological calculations of these devices.

3.3 Methods for determining the basic geometric characteristics of the packing HPCM

With respect to the packing of HPCM most accurate method for determination of the porosity should be considered as a method of weighing fragment packing of known volume. Another widely known method - fill the porous packing water can in this case to make significant errors due to air bubbles remaining in the pore volume packing at filling it with water. For a known specific weight of material value of porosity packing can be determined by the ratio:

$$\varepsilon = 1 - \frac{\gamma_{pack}}{\gamma_m} \tag{6}$$

where - γ_{pack} - weight fragment of the packing, kg/m³;
γ_m - the proportion of the monolith (packing material), kg/m³.
The specific surface of the packing HPCM because of the complexity of the geometry of its forms appropriate to define terms of the known hydraulic resistance element of the packing from the equation Gelperin I.I., Kagan A.M. [14]:

$$a = \left(324 \cdot 10^{-6} \cdot \frac{W_0^2}{\nu^2} + \frac{0{,}04 \cdot \varepsilon^2 \cdot \Delta P}{\nu \cdot H \cdot \rho \cdot W_0}\right)^{0,5} + 18 \cdot 10^{-3} \cdot \frac{W_0}{\nu} \tag{7}$$

where:
W_0 - the rate of gas flow per total cross section of an empty vehicle, m / s;
ν - kinematic viscosity of gas, m² / s;
ρ - density of gas, kg • s²/m⁴;
ΔP - hydraulic resistance of the layer packing, kg / m²;
H - height of a packed layer, m.
Determination **ΔP** for subsequent calculation of specific surface attachment technique [14] should be made as a result of blowing a fragment of the test packing at a rate of gas flow corresponding to the laminar flow regime dominated by viscous forces. This condition is explained by the fact that the measured resistance should be caused only by friction on the surface of the packing.
According to [14] laminar flow in the granular layer correspond to the values of Reynolds numbers:

$$Re_e << 40 \tag{8}$$

The Reynolds number is assigned to an equivalent diameter of grain - **d_e**:

$$Re_e = \frac{W_0 \cdot d_e}{\varepsilon \cdot \nu} \tag{9}$$

$$d_e = \frac{4 \cdot \varepsilon}{a} \tag{10}$$

Geometrical characteristics of some ceramic and metal bits of HPCM presented in Table.

Number	Name	**ε**, m³/m³	**a**, m²/m³	d_e, m
1	Fine-pored	0,85	2700	0,00126
2	Coarse-pored	0,92	1500	0,00245

Table 1. Characteristics of regular ceramic industrial attachments of HPCM according to [21]

Num ber	Material and shape of the packing	The average cell size, d, mm	Specific surface, **a**, m^2/m^3	Fenes tration, ε, m^3/m^3	Bulk density, ρ, kg/m^3	$\mathbf{d_e} \cdot 10^3$, m	Link
1	Nickel, pentagon-dodecahedron	-	570	0,935	512	6,56	[13]
2	Nickel, pentagon-dodecahedron	-	700	0,96	310	5,49	[13]
3	Ceramics, pentagon-dodecahedron	1,5	-	0,8648	540	-	[34]
4	Ceramics, pentagon-dodecahedron	1,5	-	0,875	500	-	[34]
5	Ceramics, pentagon-dodecahedron	1,5	-	0,8306	680	-	[43]
6	Ceramics, pentagon-dodecahedron	1,5	-	0,7929	830	-	[34]
7	Ceramics, pentagon-dodecahedron	3,0	-	0,905	400	-	[34]
8	Ceramics, pentagon-dodecahedron	3,0	-	0,905	390	-	[34]
9	Ceramics	1,0	-	0,97	-	-	[35]
10	Ceramics	-	2373	0,89	605,7	1,5	

Table 2. Geometrical characteristics of packed HPCM

3.4 Dependence of specific surface attachment on the value of equivalent diameter

Below are the results of generalization of experimental data on the basic geometric characteristics of HPCM accessories, as well as various bulk and regular tips for heat and mass transfer processes as they relate to the hydraulic radius (equivalent to the diameter of the channel) packed column apparatus. Our results generalize the results of [1, 21, 28-33] in the form of dependence of specific surface of the packing - **a** the value of equivalent diameter - $\mathbf{d_e}$ represented on the graph (see fig. 5).

Presented on Fig. 5 results shows that an increase equivalent to the diameter of the packing from the $\mathbf{d_e} = 2 \cdot 10^{-3}$ m to $\mathbf{d_e} = 10^{-2}$ m (five times) leads to a decrease in the specific surface area from 1700 m^2/m^3 to 350 m^2/m^3, there are also about 5 times.

Presented in Fig. 5 dependence of the surface - **a** from the equivalent diameter - $\mathbf{d_e}$ for packing of different shapes, bulk materials and regular, from ceramics and metals was universal. Dependence $a = f(d_e)$ for all industrial attachments with a deviation not exceeding ± 10%, described by the equation:

$$d_e = A \cdot (a)^n \tag{11}$$

Here: A = 57319; n = -1,3985.

As seen from the graph shown in Fig. 5, the proposed equation (11) satisfactorily correlates well the experimental data of Kozlov [21] on ceramic head of HPCM.

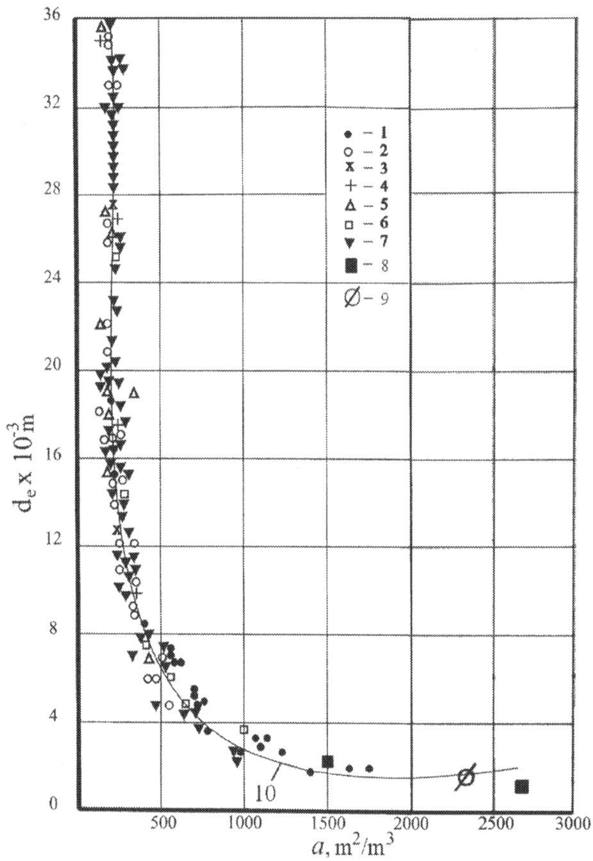

Fig. 5. Dependence of specific surface bulk and regular tips - **a** the value of equivalent diameter - **d_e**; 1 - various bulk packing according to the Polevoy [28]; 2 - bulk packing according Vedernikov et al [29]; 3 - bulk packing according Kolev et al [30]; 4 - bulk packing "Inzhehim - 2000" according to the Laptev and Farahov [31]; 5 - regular packing according to [1, 29]; 6 - bulk packing according to [32]; 7 - bulk packing according to [33]; 8 - ceramic packing of HPCM according to [21]; 9 - ceramic packing of HPCM on the results of our experiments; 10 - calculated according to our equation

4. Experienced stands and methodology for conducting experiments

The study of aerodynamics HPCM samples were carried out on a laboratory setup in the apparatus with a diameter of 100 mm. Height experienced packing layer was 40 mm. The support grid for the packing was carried out in a grid of stainless steel 1 mm thick with a cell size of 7x7 mm. The experimental setup is shown in Fig. 6.

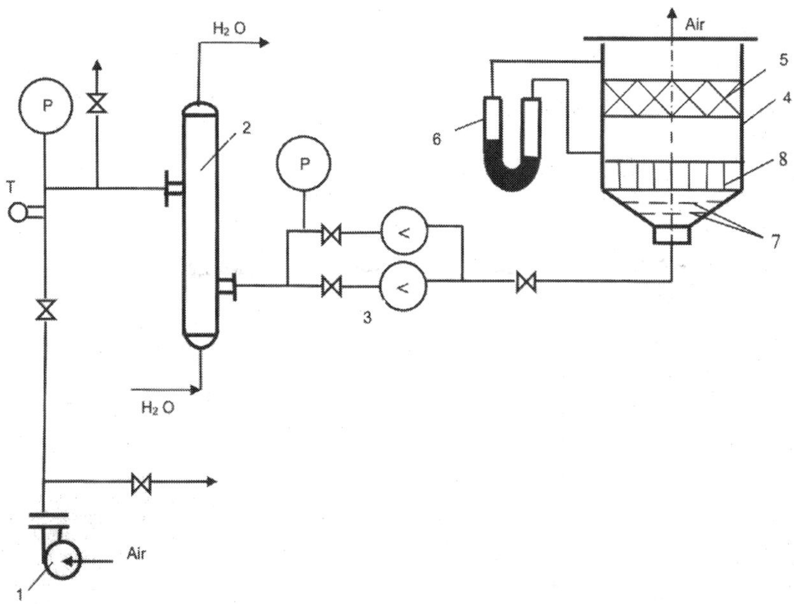

Fig. 6. Scheme of experimental equipment: 1 - air tank; 2 - heater; 3 - diaphragms; 4 - cylinder mechanism ∅ 100 mm; 5 - block of high porosity material sedimentation; 6 – differential pressure gauge, 7 –graded lattice; 8 - honeycomb

During the experiments, controlled the flow of the gas phase and the pressure loss in the layer of the packing. Gas flow in the test apparatus 4 with a packing 5, measured apertures 3, and pressure loss - a standard pressure switch 6 designs TCXA with the price scale division of 1 mm water column.

The experiments were conducted in a range of load variation on the gas, the corresponding average linear velocities in the calculation of the total cross section of empty vehicle from 0,3 to 1,2 m/sec. The set of gas distribution grids honeykomb 7 and 8 in the experimental apparatus 4 in accordance with the recommendations of [7] aligns the velocity field of gas flow at the entrance to the subject block attachments. The velocity profile was monitored during the preliminary experiments in the empty apparatus without attachment with the Pito tube.

Investigation of velocity field and the degree of turbulence on the air outlet of the experimental ceramic samples HPCM performed on a specially designed stand to test the individual elements of the nozzle [15] LEI (r Kaunas, Lithuania) using a precision hot-wire system equipment «DISA 55M».

Use of the equipment and experimental plot of the stand shown in Fig. 7.

Besides the above mentioned hot-wire apparatus used a tape recorder company Lipeks and Fourier analyzer firm Gevlet-Pakkard. Static pressure was measured by sensors and analog device company Gëttingen Baldvin Messtehnik. Pilot plant itself is an open aerodynamic contour. As in the first series of experiments, experimental plot device included a leveling device in the form of gas distribution grids and perforated honeykomb that ensures uniform velocity fields and turbulence intensity at the location of the studied sample packing.
Measurements of the velocity profile and the degree of turbulence performed directly on the output stream from the test element packing at a distance of 10, 30 and 60 mm from the packing exit (see fig. 7).

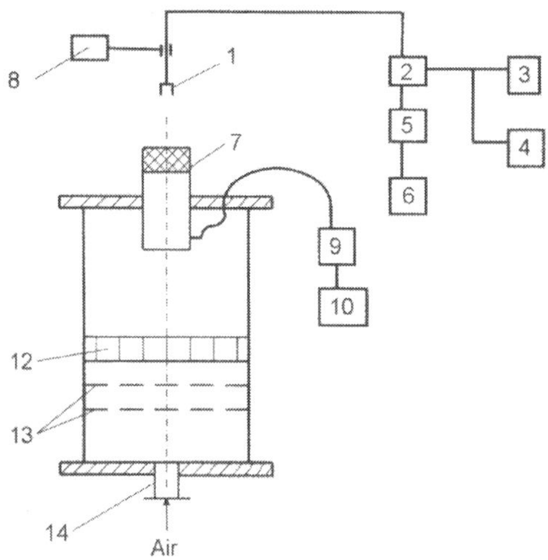

Fig. 7. Experimental section of bench and measuring block-scheme: 1 - velocity gauge; 2 - anemometer; 3 -voltmeter; 4 - quadratic voltmeter; 5 - tape recorder; 6 -analyzer of Furje; 7 - block high porosity material of sedimentation; 8 - mechanism of moving velocity gauge by X-Y axes; 9 - static pressure gauge; 10 - second pressure gauge; 11 - experimental apparatus; 12 - honeycomb; 13 -grid of gas distribution; 14 - entering branch

The diameter of the test sample nozzle was 50 mm, thickness - 20 mm, and pore size in different samples varied in the range from 0,4 to 1,8 mm. In addition to the hydraulic resistance in these experiments were carried out special experiments to measure the dependence of the degree of turbulence of the gas flow from the **Tu** Reynolds number $\mathbf{Re_D}$ samples packing with equivalent pore diameter $\mathbf{d_e}$ = 0.4 mm and 0.9 mm and 1.8 mm.
Here $\mathbf{Re_D}$ - Reynolds number, relative to the diameter of the column **D**:

$$\mathrm{Re}_D = \frac{W_0 \cdot D}{\nu} \tag{12}$$

In this series of experiments, Reynolds numbers were: $\mathbf{Re_D}$ = 6800, 15600, 27700.
The degree of turbulence of the gas flow **Tu** was estimated by the relation:

$$Tu = \sqrt{W_0'^2} \,/\, W_0, \% \tag{13}$$

Here $W_0'^2$ - standard deviation from the mean flow velocity, m / sec.

Fig. 8 shows the basic scheme of the experimental setup RXTU them. Mendeleev [34], designed to measure the hydraulic resistance in the flow of the fluid flow through the samples HPCM.

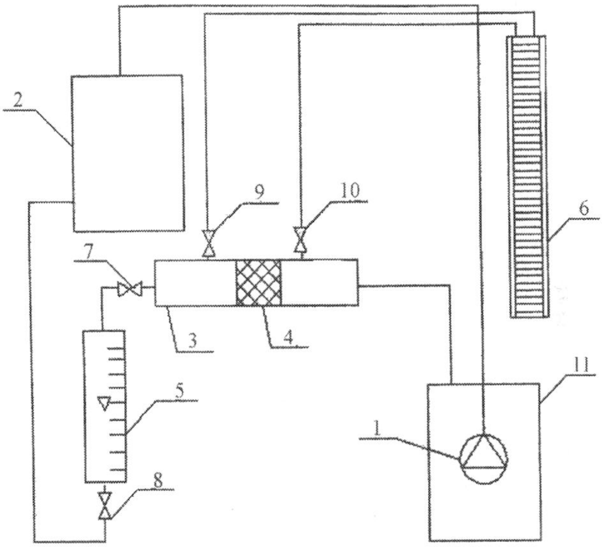

Fig. 8. Experimental setup for determining the hydraulic resistance in the flow of fluid flow through the ceramic samples HPCM produced by slip technology [34]: 1 - drain pump, 2 - pressure tank, 3 - working chamber, 4 - sample of a ceramic carrier based on HPCM, 5 - rotameter, 6 - U-shaped differential manometer, 7 , 8, 9, 10 – valves; 11 - reception tank

5. Hydraulic resistance of the ceramic head HPCM

5.1 Filtration of the gas flow

Fig. 9 presents the results of our experiments to measure the pressure loss **ΔP/H** airflow through the sample HPCM with the following geometrical characteristics: the linear dimensions of time - from 1,0 to 2,0 mm; $\mathbf{d_e}$ = 1,5 mm, weight of 1 m^3 - 605.7 kg / m^3; porosity - 0,89 m^3/m^3. As can be seen, the experimental data fit satisfactorily on a single curve.

Fig. 10 in the semi-logarithmic coordinates shows the dependence of **ΔP/H** on the Reynolds number $\mathbf{Re_D}$ on the results of our experiments [18] with a ceramic packing of HPCM with equivalent pore diameter $\mathbf{d_e}$ = 1,8 mm.

Comparison of hydraulic resistance of the tested samples of ceramic packing HPCM with other types of industrial attachments and dry granular materials is presented in Fig. 11 in logarithmic coordinates in the form of dependence **ΔP/H = f($\mathbf{W_0}$)**. From the Fig. 11 graphs can be seen that the linear velocity of air flow $\mathbf{W_0} \approx$ 0,5 m/s occurs characteristic kink curves, which indicates a change in flow regime of gas flow in a layer of packing. From the Fig. 11 experimental data that the samples from the packing HPCM have an order of magnitude lower hydraulic resistance as compared to granular materials at similar values of $\mathbf{d_e}$.

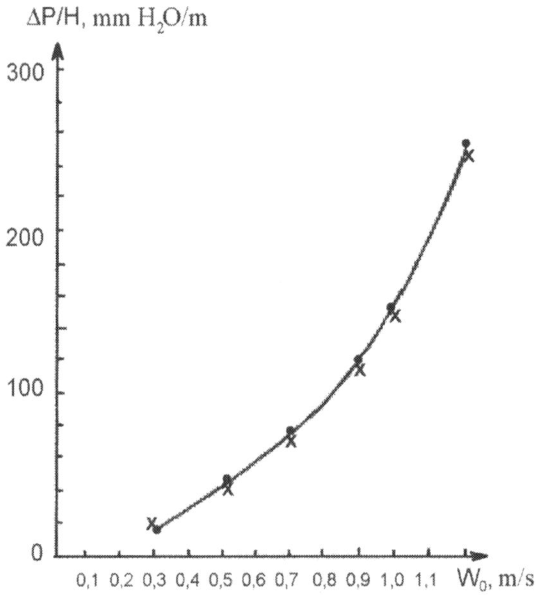

Fig. 9. Dependence of pressure loss **ΔP/H** of the air velocity $\mathbf{W_0}$ dry ceramic head of HPCM; • - first experiments; x - repeated measurements

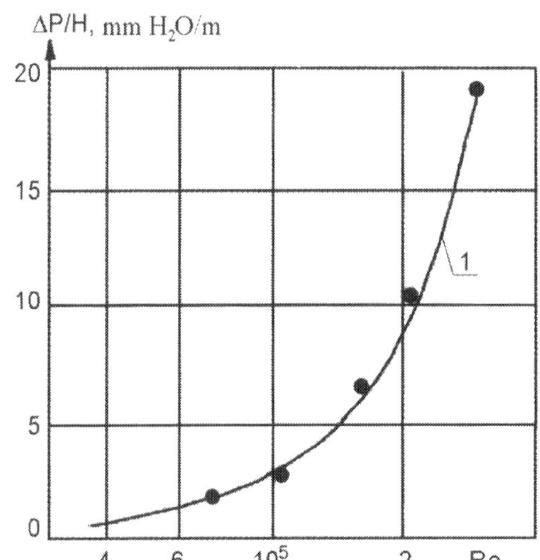

Fig. 10. Dependence $\Delta P / H = f(\mathrm{Re}_D)$ for dry sedimentation from ceramic high porosity material with equivalent diameter of pore $\mathbf{d_e}$= 1.8 mm

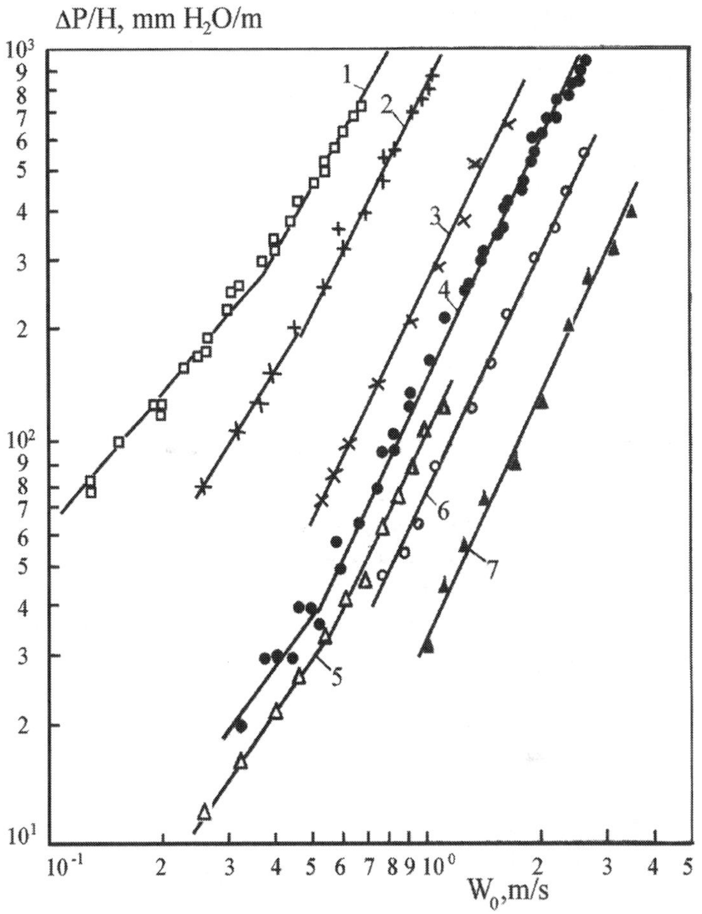

Fig. 11. Dependence $\Delta P / H = f(W_0)$ for various attachments: 1 - coal "SKT-2, cylinders ∅ 1,5 x5 mm; 2 - catalyst tablets ∅ 6x6 mm; 3 - HPCM, $\mathbf{d_e}$ = 0,9 mm; 4 - Block catalyst, $\boldsymbol{\varepsilon}$ = 0,5, $\mathbf{d_{hol}}$. = 4,7 mm; 5 - Ring Rashig, ceramics, 10x10x2 mm; 6 - ring Rashig, metal, 10x10x1 mm; 7 - HPCM, $\mathbf{d_e}$ = 2,75 mm

5.2 Fluid filtration

The study of fluid flow through the samples from HPCM conducted Grunsky and others [34] and Kozlov et al [21]. In this paper Beklemeshev et al [35] in their experiments observed that the values of the pressure drop obtained with the same flow rate of the samples HPCM compiled in module, higher than those for the whole sample, with values exceeding **ΔP** greater, the higher the number blocks in the module. Similar situation was observed in [36] when laying block Honeycomb Catalysts "overlap" when overlapping channels in the blocks reached 10%. It can be assumed that the cause of such phenomena is a violation of the homogeneity of the structure of storage material in the plane of the junction of the samples, since the ends of the jumper of one sample does not coincide with the terminal jumpers another, forming a cross-section with an area smaller than the average cross-sectional sample. We have completed processing of experimental data [35] to study the pressure drop in the samples HPCM in logarithmic coordinates in the form of dependence **ΔP = f (W$_0$)**. This dependence is shown in Fig. 12. As seen from the data shown in this figure, the experimental data [35] describes a system of direct, regularly increase with the velocity - **W$_0$** and height of blocks attachments - **h**. This data **ΔP** for **d$_e$** = 1,26 mm are higher than for **d$_e$** = 2,45 mm.

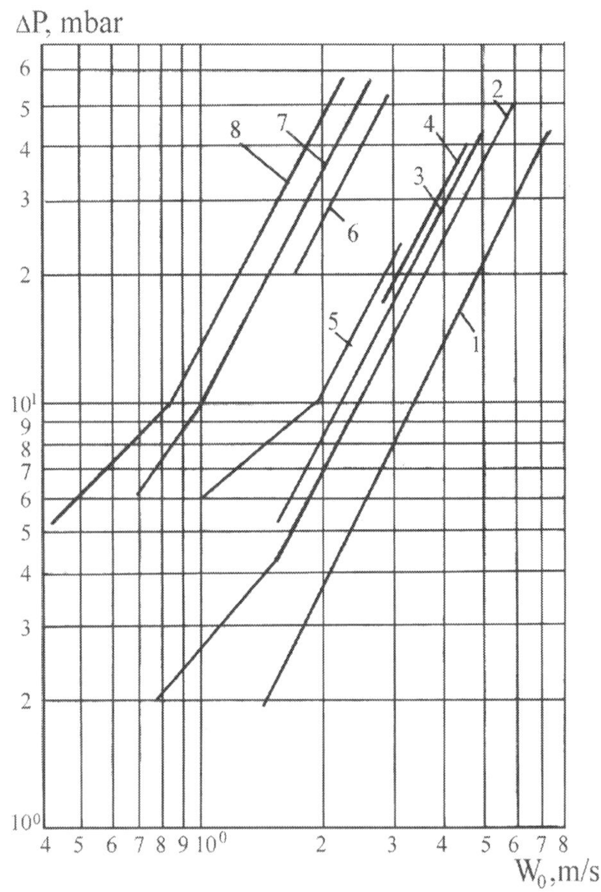

Fig. 12. The dependence of the differential pressure on the rate of filtration of fluid samples of ceramic packing HPCM with different diameter of the cell and at different height blocks attachments: curves 1 ÷ 4 - d$_e$ = 2,45 mm: 1-h = 70mm, 2-h = 140mm, 3-h = 210mm, 4-h = 280mm; curves 5 ÷ 8 - d$_e$ = 1,26 mm: 5-h = 70mm, 6-h = 140mm, 7-h = 210mm, 8-h = 280 mm

In the literature, very little information about the hydraulic characteristics of the packing HPCM filtering liquid at low speeds, which can be explained by the complexity of measuring small pressure drop. Fig. 13 shows the dependence **ΔP = f(W_0)** to filter water through a packing HPCM with $\mathbf{d_e}$ = 1,26 mm at a flow rate 0,01-0,025 m/s, we have constructed according to the Grunsky and others [34] in logarithmic coordinates. The height of the experiments, as is evident (see fig. 13), was different. It may be noted that the data at a block height **h** = 15mm a few fall out of the picture. However, curves 1-3 in this figure show the regular increase **ΔP** with increasing altitude blocks attachments.

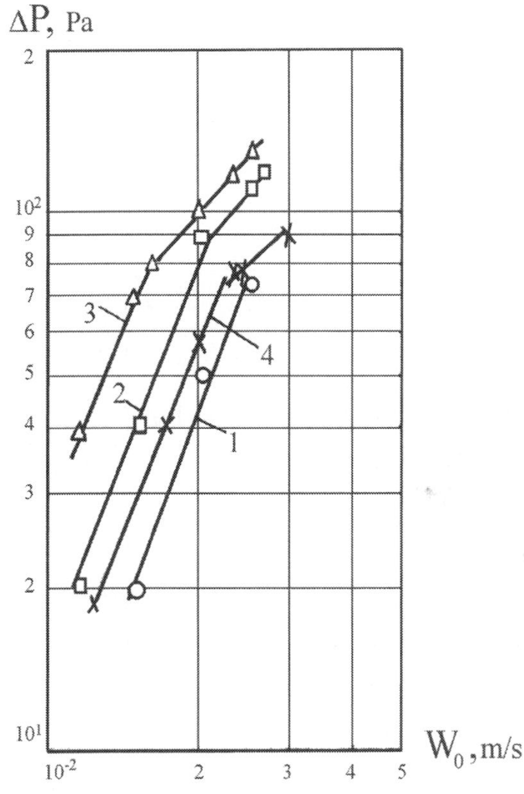

Fig. 13. Dependence of hydraulic resistance **ΔP** of the fluid velocity according to the Grunsky and others [34] for fine-meshed blocks ($\mathbf{d_e}$ = 1,26 mm) packing HPCM varying height **h**: 1-h = 20mm, 2-h = 40mm, 3-h = 60mm, 4-h = 15mm

6. The velocity field and turbulence of the gas flow in the samples from HPCM

Results of preliminary test experiments on the velocity field of airflow in the empty apparatus without attachment shown in Fig. 14.

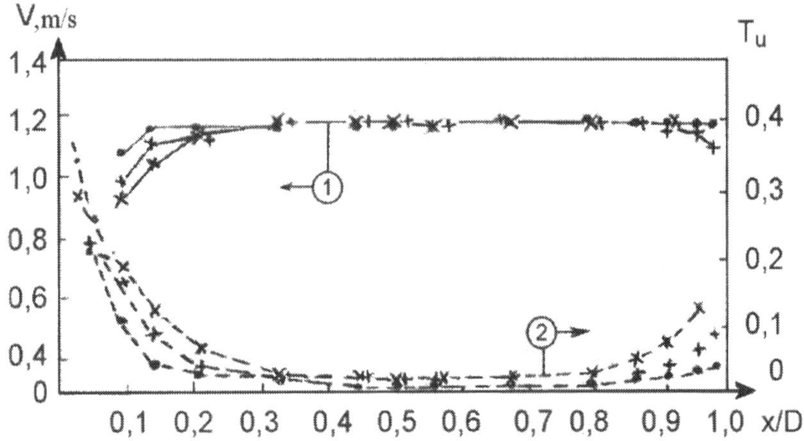

Fig. 14. Profile of creasing longitudinal velocity of stream - **V** and degrees of turbulence - **Tu** in identical sections of empty experimental apparatus without sedimentation at the charge corresponding number $\mathbf{Re_D}$ = 24400 at various distances from a cut tip: • - distance from the packing exit l =10 mm; + - distance from the packing exit l = 30 mm; x — distance from the packing exit l = 60 mm

Fig. 15 shows the profile of the longitudinal air flow - **V** and the degree of turbulence - **Tu** in the wind tunnel layer for a block of ceramic nozzles HPCM Reynolds numbers $\mathbf{Re_D}$ = 6800 and $\mathbf{Re_D}$ = 15600.

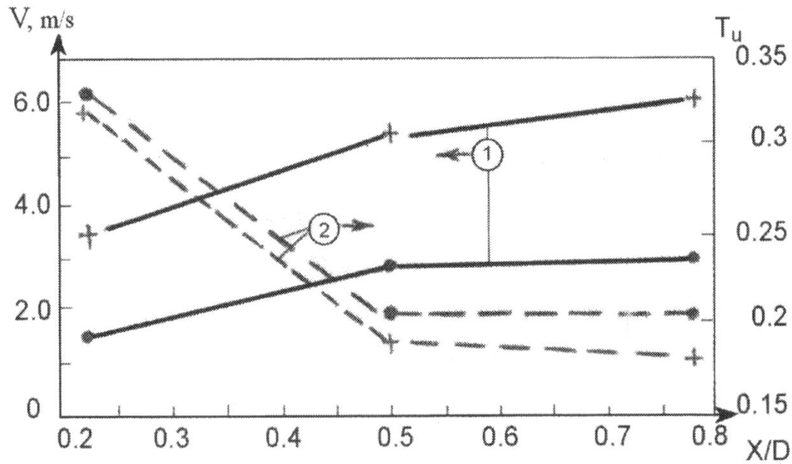

Fig. 15. Longitudinal component of velocity of stream **V** and degrees of turbulence of stream in aerodynamic environment behind the block of sedimentation high porosity material at different numbers $\mathbf{Re_D}$: • - $\mathbf{Re_D}$ = 6 800; + - $\mathbf{Re_D}$ = 15600

As can be seen from the graph presented in Fig. 15, the longitudinal component of velocity - **V** in the aerodynamic wake behind the block of packing HPCM with increasing Reynolds numbers from 15600 to 6800 profile of the longitudinal velocity in the central part of the packing remains essentially flat, as well as the degree of turbulence. Data on **V** and **Tu** at **X/D** <0,25 (at the edges of the packing) are not typical because of defects since at this point the packing (see picture in Fig. 2).

A comparison of data on the value of the degree of turbulence packing HPCM and immobile granular layer. The comparison showed that the magnitude **Tu** in all samples tested packing HPCM in 3 ÷ 10 times lower than the parameter **Tu** of the layer of grains, where he was, according to [6], 30-50%. At the packing HPCM samples tested value **Tu** is in the range from 4,6 to 16% in all-tried a range of Reynolds numbers (see Fig. 16).

In [19] based on the well-known Navie-Stokes equations using the finite element method is considered the hydrodynamic flow structure in the cell, isolated in a layer HPCM. The model is described in detail in [19].

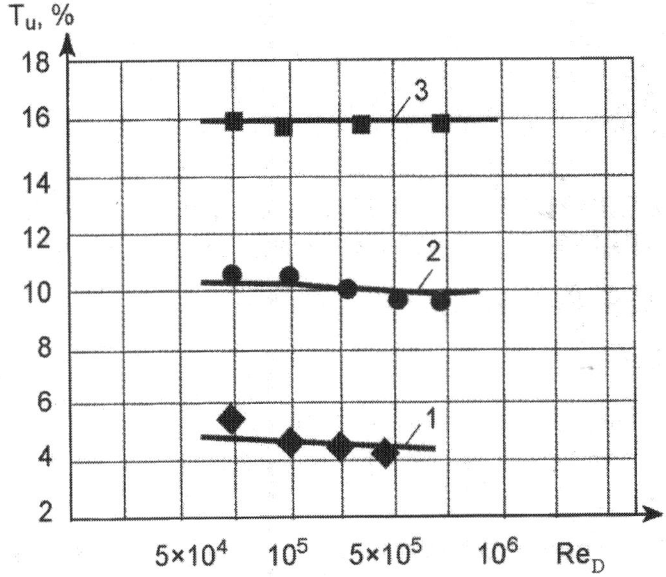

Fig. 16. Dependence of turbulence degree of stream **Tu** on Reynolds number **Re**$_D$ for sedimentation from ceramic high porosity material with different size of equivalent diameter of pours: 1 - d_e = 0,4 mm; 2 - d_e = 0,9 mm; 3 - d_e = 1,8 mm

7. Findings

The results of aerodynamics testing of new ceramic packing of highly porous cellular materials have shown promising ceramic materials for the manufacture of packing HPCM for a wide range of chemical technology processes, including those for hardware design methods for cleaning absorption of harmful gases to protect the ambient air.

It is shown that the profiles of longitudinal velocity and the degree of turbulence of the gas flow in the aerodynamic wake behind a block of ceramic packing HPCM at Reynolds numbers $\mathbf{Re_D}$ = 6800 ÷ 15600 remain virtually flat, which proves the isotropy of the structure of the packing.

Found that the magnitude of hydraulic resistance in samples from the packing HPCM significantly lower compared with a layer of grains of other industrial attachments.

For the first time measured the degree of turbulence in a wind layer for the elements of the ceramic head HPCM. It was found that all samples tested the degree of turbulence is in the range from 4,6% to 16%, which is 3 ÷ 10 times lower compared to conventional granular materials as a layer of cylinders.

8. References

[1] G.B. Dmitrieva, M.G. Berengarten, M.I. Klyushenkova, A.S. Pushnov Efficient structured design of packing for the processes of heat mass transfer / / *Chemical and Petroleum Engineering*, 2005, № 8, pp. 15-17.

[2] T. Komorovich, J. Maghera, D.A. Baranov, M.G Berengarten. High-ring-structural attachments // *Chemical and Petroleum Engineering*, 2001, № 8, pp. 8-19.

[3] Pranas Baltrenas, Aušra Zigmontienė, Rasa Vaiškūnaitė. Oro valumo biotechnologyos. Monografija. Vilnius: *Technika*, 2004, 207p.

[4] Rasa Vaiškūnaitė. RESEARCH ON AND APPLICATION OF THE BIOFILTER WITH ACTIVATED PINE BARK CHARGE FOR AIR PURIFICATION. Summary of Doctoral Dissertation. VGTU. Vilnius: *Technica* 2004.

[5] P. Baltrenas, A. Zagorskis. Modeling of aerodynamic processes in a biofilter // *Chemical and Petroleum Engineering*, 2006, № 5, c. 41-42.

[6] A.S. Pushnov. Gas distribution in the contact and adsorption apparatus with a stationary granular layer: A thesis for the degree of candidate of technical sciences. M.: GIAP, 1987, 353s.

[7] I.E. Idelchik. Aerodynamics of technological devices. (Supply, drainage and flow distribution over the cross section sets) - Moscow: *Machine industry*, 1983, 351s.

[8] S.V. Belov. Porous Metals in Mechanical Engineering. M.: *Machine industry*, 1981., 248 pp.

[9] M.V. Zhelnis, A.S. Pushnov. Filter for purification of molten metal from nonmetallic inclusions. *AS USSR* № 1708491 Al, V22S9/08, BI № 4, 1992.

[10] L. Padeste, A. Baiker, J.P. Gabathuler. Verweilzeitverhalten von Flűssidkeiten in Schaumkeramik - Mischelementen: *Chem. - Ind. - Tech.* 61 (1989) № 6. S. 482-484.

[11] O. Levenspiel, J.C.R. Turner: *Chem. End. Sci.* 25 (1970) S. 1605 - 1609.

[12] E.J. Cairns, J.M. Prausnitz: *Chem. End. Sci.* 51 (1959) S. 1441.

[13] A.C. Polevoy, A.B. Mamaev, A.C. Pushnov, B.B. Uborsky. Properties foam block packing mass-exchange columns high-clean substance. 1991, № 4, pp. 96-106.

[14] I.I. Gelperin, A.M. Kagan. Development of aerodynamic research immobile granular layer // *Chemical Industry*. 1984, № 12, pp. 741-746.

[15] A.S. Pushnov. Rapid method for optimizing the design elements of packing // Abstracts of the All-Union Conference heat and mass transfer equipment - 88. *TSINTIHIMNEFTEMASH*, M.: 1988. c. 137-138.

[16] I.A. Sabanaev, N.I. Bagautdinov. Convective heat exchange in apparatus with the granular layer. *V-th International Scientific Conference Petrochemistry*-99. Abstracts. Nizhnekamsk, 1999, Volume 2, s.179-180.

[17] A.S. Pushnov, P. Baltrenas, A.M. Kagan, A. Zagorskis. Aerodynamics of air pollution control devices with the granular layer. Vilnius: *Technika*, 2010, 289s.

[18] A. Pushnov, A. Sakalauskas. Aerodynamics regular ceramic packing for heat and mass exchange processes in the apparatus protecting the environment // *Journal of Environmental Engineering and Landscape Management*, 2008, 16 (1), p.30-37.

[19] A.S. Shaimardanov, E.M. Koltsova, A.I. Kozlov, A.V. Zhensa, V.A. Kostikov, L.S. Gordeev. Hydrodynamic processes in the high permeability of porous cellular materials // *Theoretical Foundations of Chemical Technology*, 2007, v. 41, № 2, S.148-153.

[20] V.S. Beskov, V.N. Grunsky, A.I. Kozlov, V.I. Vanchurin. Catalysts with a developed external surface // *Chemical Industry Today*, 2009, № 9, pp. 47-51.

[21] A.I. Kozlov. Block honeycomb catalysts and prospects for their use in the Chemistry and Technology of Nitro Compounds // A.I. Kozlov, V.L. Zbarsky, V.N. Grunsky. M.: *Chemistry RAS*. D.I. Mendeleev, 2009, 119 pp.

[22] Hydrodynamics and heat transfer in nuclear power plants (the basis of calculation). V.I. Subbotin, M. Kh. Ibragimov, P.A. Ushakov, V.P. Bobkov, A.V. Zhukov, Y.S. Yur'ev. M.: *Atomizdat*, 1975, 408s.

[23] N.M. Zhavoronkov. Theoretical Foundations of Chemical Technology. M.: *Science*, 2007, 351s.

[24] M.E. Aerov, O.M. Todes, D.A. Narinskaya. Devices with a stationary granular layer. Hydraulic and Thermal basics. L.: *Chemistry*, 1979, 176s.

[25] I.I. Gelperin, A.M. Kagan, G.I. Krinitsina. Hydraulic resistance and the specific surface of granular materials // *Chemical Industry*, 1977, № 2, s.67-68.

[26] A.S. Pushnov, A.M. Kagan, I.I. Gelperin. Hydraulic resistance of the particles in cylindrical and annular in the transitional flow regime // *Powder metallurgy*, 1982, № 3, pp. 41-44.

[27] A.S. Pushnov, A.M. Kagan. Calculation of hydraulic resistance of the fixed cylindrical layers of grains of different size // *Chemical Industry*, 1989, № 9, page 54-57.

[28] A.S. Polevoy. Investigation of high bits in the rectification of dilute solutions based on β-Phenethyl alcohol // *Theoretical Foundations of Chemical Technology*, 1996, Volume 30, № 5, page 473-484

[29] M.I. Vedernikov, L.G. Startsev, Y.L. Yuryev, V.T. Orlov. Examples and problems of mass-transfer processes of chemical technology. *Reference manual. Part IV. Basic physical, chemical and thermal properties of substances*. Ekaterinburg: Urals State. Forestry University, 2009, 154s.

[30] N. Kolev, K. Winkler, R. Drakichev, Z. Brosh. Creating effective tips for the column apparatus on the basis of the theory of mass transfer processes // *Chemical Industry*, 1986, № 8, pp. 489-493.

[31] A.G. Laptev, M.I. Farahov. The separation of heterogeneous systems in packed vehicles. Kazan: Kazan State University, 2006, 342s.

[32] W. Kast, O. Krisher, G. Raynike, K. Vintermantel. Convective heat and mass transfer. Translated from German, Moscow: *Energiya*, 1980, 49s.

[33] I. Ganzinov. Modeling of packed bed scrubber. TUT, 2005, Tampere. Finland. Diplomityö TTKK.

[34] V.N. Gunnskiy, A. Pivkin, A.V. Bespalov, A.I. Kozlov. Hydraulic properties of the block of highly porous media at low speeds of fluid filtration // *Chemical Industry Today*, 2010, № 2, pp. 18-23.

[35] V.N. Antsiferov, A.I. Beklemeshev, V.G. Gilyov, S.E. Porozova, G.P. Shveynik. *Problems of powder materials*. Part II. Highly permeable materials // Yekaterinburg: URO RAS, 2002, 262 pp.

[36] A.V. Bespalov, V.V. Demin, V.S. Beskov. Hydraulic resistance of catalysts of different geometric shapes and sizes // *Theoretical Foundations of Chemical Technology*. 1991, Volume 25, pp. 533-541

25

Topics in Heat and Mass Transfer in Porous Media: Cross-Diffusion, Thermophoresis and Reactive Surfaces

Adrian Postelnicu
Transilvania University of Brasov
Romania

1. Introduction

Coupled heat and mass transfer by natural convection in a fluid-saturated porous medium is a dynamic domain of research, due to many important engineering and geophysical applications, see the books by (Nield & Bejan, 2006), (Ingham & Pop, 1998; 2002; 2005), (Pop & Ingham, 2001), (Ingham et al., 2004) where a comprehensive account of the available information in the field is presented.

In the present context, the following assumptions will be used: a) the fluid and the porous medium are in local thermodynamic equilibrium; b) the flow is laminar, steady and two-dimensional; c) the porous medium is isotropic and homogeneous; d) the properties of the fluid and porous medium are constants; e) the Boussinesq approximation is valid and the boundary-layer approximation is applicable.

Thermal diffusion (thermodiffusion or Soret effect) corresponds to species differentiation developing in an initial homogeneous mixture submitted to a thermal gradient, (Soret, 1880). The energy flux caused by a composition gradient is called Dufour (diffusion-thermo) effect. These effects are considered as second order phenomena, on the basis that they are of smaller order of magnitude than the effects described by Fourier's and Fick's laws, but they may become significant in clear fluids, according to (Eckert & Drake, 1972), but also in porous media in areas such geosciences or hydrology, see (Benano-Melly et al., 2001) and the references quoted there. Consequently, Dufour and Soret effects on convective flows in porous media have been analyzed in a growing body of literature in the last years. However, the majority of the papers issued in the journals, in the last period, do not offer unfortunately a physical basis to calculate Dufour and Soret coefficients.

Restricting our considerations to porous media, let us refer further to the textbook by (Nield & Bejan, 2006) where some basic information are given on pages 42-44. According to the references quoted there

- In most liquid mixtures the Dufour effect is inoperative, but that this may not be the case in gases (Platten & Legros, 1984). This fact was confirmed also by (Mojtabi & Charrier-Mojtabi, 2000), who found that in liquids the Dufour coefficient is an order of magnitude smaller than the Soret effect.
- A not encouraging conclusion was drawn by these later authors, (Mojtabi and Charrier-Mojtabi, 2000), in the sense that for saturated porous media, the phenomenon of cross

diffusion is complicated due to the interaction between the fluid and the porous matrix. This is the reason why accurate values of these coefficients are not available.

(Benano-Melly et al, 2001) analyzed the problem of thermal diffusion in binary fluid mixtures, lying within a porous medium and subjected to a horizontal thermal gradient. The reader is directed to the references quoted in that paper for related information concerning the measurements and the Dufour coefficient.

Another possibility to deal with these effects is to use the theory of thermodynamics of irreversible processes, like in (Li et al., 2006) who considered a strongly endothermic chemical reaction system in a porous medium formed by spherical carbonate pellets in a reactor. However, orders of magnitude are given for these coefficients, in dimensional form, without physical justification and results are reported taking these coefficients as equal. For other references along this direction, the interested reader may consult those presented in the 9th chapter from the book by (Ingham and Pop, 2005).

In a general form the energy and concentration equations are expresed as

$$\frac{(\rho c)_m}{(\rho c)_f}\frac{\partial T}{\partial t} + \mathbf{V}\cdot\nabla T = \nabla\cdot\left(D_T\nabla T + D_{TC}\nabla C\right) \tag{1a}$$

$$\frac{(\rho c)_m}{(\rho c)_f}\frac{\partial T}{\partial t} + \mathbf{V}\cdot\nabla T = \nabla\cdot\left(D_T\nabla T + D_{TC}\nabla C\right) \tag{1b}$$

see (Nield and Bejan, 2006), where ϕ is the porosity of the medium, $D_T = k_m/(\rho c)_f$ is the thermal diffusivity, $D_C = D_m$ is the mass diffusivity, while D_{TC}/D_T and D_{CT}/D_C may be considered as Dufour and Soret coefficients (numbers) of the porous medium. In the last time, basically in all papers dealing with this subject, the previous equations were used in the form

$$u\frac{\partial T}{\partial x} + v\frac{\partial T}{\partial y} = \alpha_m\frac{\partial^2 T}{\partial y^2} + \frac{D_m}{C_s}\frac{k_T}{C_p}\frac{\partial^2 C}{\partial y^2} \tag{2a}$$

$$u\frac{\partial C}{\partial x} + v\frac{\partial C}{\partial y} = D_m\frac{\partial^2 C}{\partial y^2} + \frac{D_m k_T}{T_m}\frac{\partial^2 T}{\partial y^2} \tag{2b}$$

where α_m is the thermal diffusivity, C_p and C_s are the specific heat at constant pressure and concentration susceptibility and k_T is the thermal diffusion ratio. It seems that this form originates from a paper by (Kafoussias & Williams, 1995), where Dufour and Soret effects have been introduced in a free convection boundary-layer past a vertical plate in a viscous fluid

The recent literature offers a couple of papers, where the Dufour and Soret effects are quantified by dimensionless parameters, resulted from a non-dimensionalization procedure, associated usually with the search for similarity solutions of the governing equations of momentum, energy and mass transfer (which are partial differential equations).

1.1 Vertical plate, Darcy model

The Darcy law applies when the Reynolds number is less than 1 (Bear, 1988), otherwise non-Darcy models must be used. Consequently, the basic equations of motion are, in the Darcy formulation

$$\frac{\partial u}{\partial x}+\frac{\partial v}{\partial y}=0 \tag{3}$$

$$u=\frac{gK}{\upsilon}\left[\beta_T\left(T-T_\infty\right)+\beta_C\left(C-C_\infty\right)\right] \tag{4}$$

1.1.1 Prescribed wall temperature and concentration

The boundary conditions of the problem are

$$y=0:\ v=0,\ T=T_w,\ C=C_w \tag{5a}$$

$$y\to\infty:\ u\to 0,\ \ T\to T_\infty,\ C\to C_\infty \tag{5b}$$

(Anghel et al., 2000) looked for the free convection past this common geometry, by using then by (Bejan & Khair, 1985)

$$\psi=\alpha_m Ra_x^{1/2}f(\eta),\ \theta=(T-T_\infty)/(T_w-T_\infty),\ \phi=(C-C_\infty)/(C_w-C_\infty),\ \eta=\frac{y}{x}Ra_x^{1/2} \tag{6}$$

where $Ra_x=gK\beta_T(T_w-T_\infty)x/(\nu\alpha_m)$ is the local Rayleigh number. In order to stay close to (Bejan & Khair, 1985), the stream function ψ is defined as $u=-\partial\psi/\partial y$, $v=\partial\psi/\partial x$. The governing equations become

$$f'=-\theta-N\phi \tag{7}$$

$$\theta''-\frac{1}{2}f\theta'+D_f\phi''=0 \tag{8}$$

$$\frac{1}{Le}\phi''-\frac{1}{2}f\phi'+S_r\theta''=0 \tag{9}$$

where Le, D_f and S_r are Lewis, Dufour and Soret numbers, respectively

$$Le=\frac{\alpha_m}{D_m},\ D_f=\frac{D_m k_T\left(C_w-C_\infty\right)}{C_s C_p\alpha_m\left(T_w-T_\infty\right)},\ S_r=\frac{D_m k_T\left(T_w-T_\infty\right)}{\alpha_m T_m\left(C_w-C_\infty\right)} \tag{10a}$$

while

$$N=\frac{\beta_C\left(C_w-C_\infty\right)}{\beta_T\left(T_w-T_\infty\right)} \tag{10b}$$

is the sustentation parameter, which measures the relative importance of mass and thermal diffusion in the buoyancy-driven flow. We notice that N it is positive for thermally assisting flows, negative for thermally opposing flows and zero for thermal-driven flows. The transformed boundary conditions are

$$f(0)=0,\ \theta(0)=1,\ \phi(0)=1 \tag{11a}$$

$$f' \to 0,\ \theta \to 0,\ \phi \to 0,\ \text{as } \eta \to \infty \tag{11b}$$

We notice that the problem reduces to that formulated by (Bejan and Khair, 1985) when $D_f = 0$ and $S_r = 0$. The parameters of engineering interest are the local Nusselt and the Sherwood numbers, which are given by the expressions

$$Nu_x / Ra_x^{1/2} = -\theta'(0),\ Sh_x / Ra_x^{1/2} = -\phi'(0) \tag{12}$$

Three cases were considered by (Anghel et al., 2000):

- Case I: $Le = 1$, $N = 1$, (D_f, S_r) = ((0.05, 1.2), (0.075, 0.8), (0.03, 2.), (0.037, 1.6), (0.6, 0.1))
- Case II: $Le = 1$, $N = 1$, $D_f = 0.15$, $S_r = 0.4$
- Case III: $Le = 1$, (N, D_f, S_r) = ((0.2, 0.15, 0.4), (0.5, 0.075 0.8), (0.8, 0.03, 2.0)).

Many other subsequent papers used these combinations of Dufour and Soret coefficients. (Postelnicu, 2004a) extended the work by (Anghel et al., 2000), by including the effects of a magnetic field, such as Eq. (7) becomes

$$\left(1 + M^2\right) f' = -\theta - N\phi \tag{13}$$

where M is the magnetic parameter, defined as $M^2 = K\sigma\mu_e^2 H_0^2 / \mu$ The parameters involved in the present problem are M, Le, N, D_f and S_r.

Table 1 presents local Nusselt and Sherwood numbers calculated for each set of parameters. Figs 1 and 2 show the dimensionless velocity, temperature and concentration when $M = 0$ and 1, for the following values of the parameters: $N = 1$, $D_f = 0$, $S_r = 0$, and Le = (1, 2, 4, 6, 8, 10, 100). We remark that as M increases, the thickness of the hydrodynamic/concentration boundary layer increases.

M	Le	N	D_f	S_r	$Nu_x / Ra_x^{1/2}$	$Sh_x / Ra_x^{1/2}$
0	1	0.2	0.15	0.4	0.46331	0.38100
1	1	0.2	0.15	0.4	0.32762	0.26942
2	1	0.2	0.15	0.4	0.20723	0.17044
0	1	0.5	0.075	0.8	0.55508	0.28764
1	1	0.5	0.075	0.8	0.39250	0.20339
2	1	0.5	0.075	0.8	0.24825	0.12866
0	1	0.8	0.03	2.0	0.67028	-0.13736
1	1	0.8	0.03	2.0	0.47936	-0.09712
2	1	0.8	0.03	2.0	0.29976	-0.06142

Table 1. Values of Nusselt and Sherwood numbers in case III

To this end, we notice that, as remarked by Magyari (2010), if one defines a modified Rayleigh number as

$$Ra_{x,M} = gK\beta_T\left(T_w - T_\infty\right)x / \left(\nu\alpha_m\right) / \left(1 + M^2\right) = Ra_x / \left(1 + M^2\right) \tag{14}$$

then the magnetic parameter is eliminated from the analysis and Eqs. (8-9) and (13) reduces to (7-9). One obtains the wall gradients $\theta'(0;M)$ and $\phi'(0;M)$ for non-zero M, which can be calculated from their counterparts $\theta'(0;0)$ and $\phi'(0;0)$ for $M = 0$ by the relationships

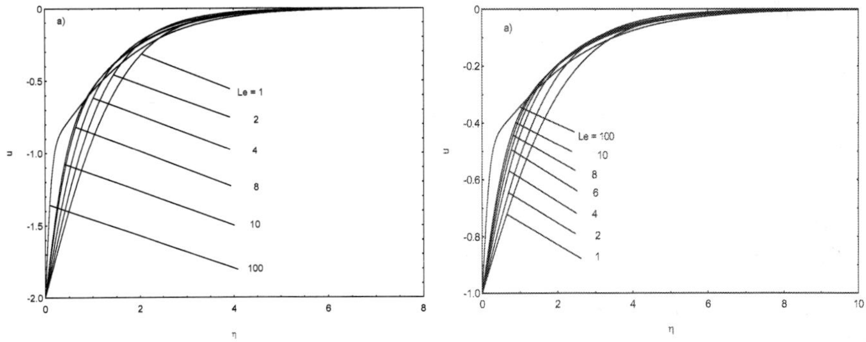

Fig. 1. Variations of velocity across the boundary layer for $N = 1$, $D_f = 0$, $S_r = 0$: $M = 0$ (left), $M = 1$ (right)

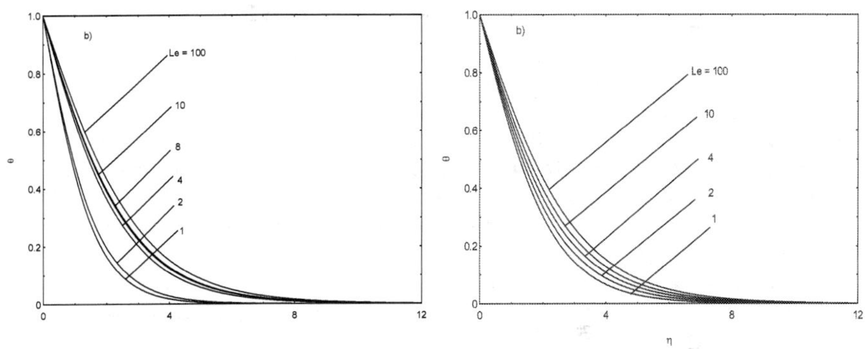

Fig. 2. Variations of temperature across the boundary layer for $N = 1$, $D_f = 0$, $S_r = 0$: $M = 0$ (left), $M = 1$ (right)

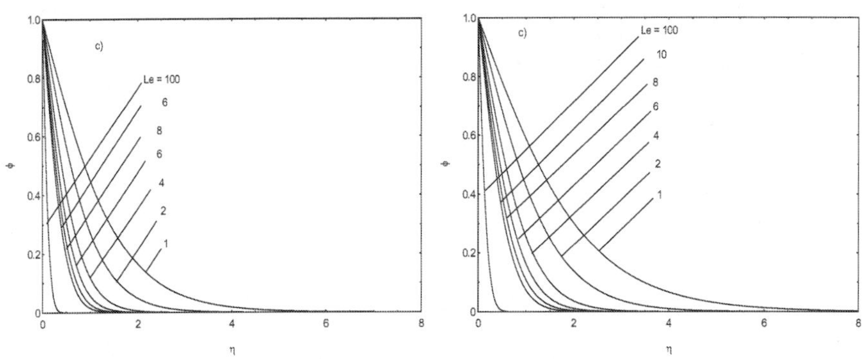

Fig. 3. Variations of the concentration across the boundary layer for $N = 1$, $D_f = 0$, $S_r = 0$: $M = 0$ (left), $M = 1$ (right)

$$\theta'(0;M)=\frac{\theta'(0;0)}{\sqrt{1+M^2}}\ ,\ \phi'(0;M)=\frac{\phi'(0;0)}{\sqrt{1+M^2}} \tag{15}$$

1.1.2 Prescribed wall heat and mass fluxes

At our best knowledge, in a single paper by (Lakshmi Narayana & Murthy, 2007), both Soret and Dufour effects have been considered in a free convection boundary layer along a vertical surface placed in porous medium, subject to wall heat and mass fluxes. In a paper by (Magyari & Postelnicu, 2010a), there is performed a thorough analysis of how wall heat and mass fluxes can be prescribed in order to get similarity solutions.

1.1.3 Other contributions for vertical flows

(Partha et al., 2006) looked for the effect of double dispersion, thermal-diffusion and diffusion-thermo effects in free convection heat and mass transfer in a non-Darcy electrically conducting fluid saturating a porous medium. In a related paper, (Lakshmi Narayana & Murthy, 2006) studied the Soret and Dufour effects in a doubly stratified Darcy porous medium.

More effects, besides thermal-diffusion and diffusion-thermo phenomena, such as thermal dispersion and temperature-dependent viscosity have been introduced by (Afify, 2007a) and (Afify, 2007b) in the analysis of non-Darcy MHD free convection past a vertical isothermal surface embedded in a porous medium. Unfortunately, this kind of approaches seems to be of marginal interest for practical applications, due to the large number of parameters.

(Lakshmi Narayana et al. 2009a; b) analyzed cross-diffusion effects on free convection of Non-Newtonian power-law fluids from a vertical flat plate in saturated porous media.

1.2 Other geometries

- Hiemenz flow with cross-diffusion through a porous medium was analyzed by (Tsai and Huang, 2009), where, by combining also various other effects, such as variable viscosity, heat source, radiation and chemical reaction. Moreover, both wall-temperature and wall-concentration are taken as power-law functions of the coordinates along the plate. The number of problem parameters is well above 10, while their impact on the relevant quantities (skin-friction, Nusselt and Sherwood numbers, velocity, temperature and concentration profiles in the boundary layer) is presented in 7 tables and 8 figures.
- (Cheng, 2009) studied the Soret and Dufour effects on the boundary layer flow due to natural convection heat and mass transfer over a downward-pointing vertical cone in a porous medium saturated with Newtonian fluids. Constant wall temperature and concentration boundary conditions were considered in that paper and similarity equations have been obtained and solved by a cubic spline collocation method.
- Another variation on geometry can be found in the very recent paper by (Rathish Kumar and Krishna Murthy, 2010), on a free convection flow from a corrugated vertical surface in a non-Darcy porous medium, under constant wall temperature and concentration boundary conditions. They used for Forchheimer extended non-Darcy model and performed a similarity transformation followed by a wavy to flat surface transformation. The resulting equations were solved having as parameters Grashof and Lewis numbers, buoyancy ratio, wavy wall amplitude, Soret and Dufour numbers. Comparisons of local and average Nusselt and Sherwood numbers have been presented.

1.3 Mixed convection

Mixed convection in the presence of Soret and Dufour effects was tackled by (Chamkha and Ben-Nakhi, 2008). Additional effects included in that paper were MHD, radiation and permeability of the plate (placed in a porous medium). This time, the set of governing equation is no more reduced to ordinary differential equations, but to partial ones, of parabolic type, through appropriate transformation of variables. (Beg, Bakier and Prasad, 2009) presented a numerical study of magnetohydrodynamic heat and mass transfer from a stretching surface placed in a porous medium with Soret and Dufour effects.

1.4 Stagnation-point flows

Consider the natural convection in a two dimensional stagnation-point flow in a fluid saturated porous medium. The temperature and concentration of the ambient medium are T_∞ and C_∞, respectively, where the wall temperature and concentration are such that $T_w > T_\infty$ and $C_w > C_\infty$. The *x*-coordinate is measured along the body surface and the *y*-coordinate normal to it. The momentum equation takes the form

$$u = \frac{gK}{\nu}\left[\beta_T\left(T - T_\infty\right) + \beta_C\left(C - C_\infty\right)\right]S(x) \tag{16}$$

where S is the shape factor. In general $S(x) = \sin\gamma$, where γ is the angle between the outward normal from the body surface and the downward vertical. For a two dimensional stagnation-point flow, $S(x) = x/l$ where l is an appropriate length scale. (Postelnicu, 2010) considered also the effect of blowing/suction, therefore the boundary conditions of the problem are as in (5), except for $v = v_w$, where v_w is the suction/injection velocity.

In order to get similarity solutions, the stream function is expressed as $\psi = xf(y)$, using now the more common definition $u = \partial\psi/\partial y$, $v = -\partial\psi/\partial x$, the governing equations become after non-dimensionalization

$$f' = \theta + N\phi \tag{17}$$

$$\theta'' + f\theta' + D_f\phi'' = 0 \tag{18}$$

$$\frac{1}{Le}\phi'' + f\phi' + S_r\theta'' = 0 \tag{19}$$

where primes denote differentiation with respect to y. The transformed boundary conditions are

$$f(0) = f_w,\ \theta(0) = 1,\ \phi(0) = 1 \tag{20a}$$

$$\theta \to 0,\ \phi \to 0 \text{ as } y \to \infty \tag{20b}$$

Besides the numerical attack of the problem (17-20), in (Postelnicu, 2010a) it was performed also an asymptotic analysis, for large suction rates.

1.5 Effect of a chemical reaction

Another situation which can arise in practice is a chemical reaction produced in the porous medium. In engineering applications, when chemical reactions occur, they can be either

homogeneous or heterogeneous. Homogeneous chemical reactions occur uniformly in the domain of a given phase, while heterogeneous reactions develop along the boundary of a phase.
Some principles on the modeling of a chemical reaction in a porous medium are discussed by (Nield and Bejan, 2006), pages 38-39. The rate equation for a reagent of concentration C is: $dC_m / dt = -k_m C_m^n$, where $C_m = C/m$ is the concentration in moles per unit volume of the fluid mixture, m denotes the molar mass of the reagent and n is the order of the reaction. Further, the rate coefficient k is in general a function of the temperature, described for instance by the Arrhenius equation. However, in many studies k is considered as a constant. In his book, (Probstein, 1994) stated that most known chemical reactions are of first and second order.
Several recent papers related to this topic have been published. (El-Amin et al. 2008) studied the effects of chemical reaction and double dispersion on the non-Darcy free convection and heat and mass transfer, for constant temperature and concentration at the wall. (Chamkha et al. 2010) carried-out an unsteady analysis of natural convective power-law fluid flow past a vertical plate embedded in a non-Darcian porous medium in the presence of a first order homogeneous reaction.
Let us focus on the chemical reaction in the presence of cross-diffusion effects for natural convection boundary layer flow along a vertical flat plate. Besides Eqs. (2a, 3, 4) which keep their forms, now Eq. (2b) is modified as follows

$$u\frac{\partial C}{\partial x} + v\frac{\partial C}{\partial y} = D_m \frac{\partial^2 C}{\partial y^2} + \frac{D_m k_T}{T_m}\frac{\partial^2 T}{\partial y^2} - K_1 (C - C_\infty)^n \tag{21}$$

The nondimensionalization scheme id the same as in (6), to give same Eqs. (7) and (8), while the concentration equation becomes

$$\frac{1}{Le}\phi'' + f\phi' - \gamma\phi^n + S_r\theta'' = 0 \tag{22}$$

where again primes denote differentiation with respect to η, while Le, D_f, S_r and N are Lewis, Dufour, Soret numbers and sustentation parameter as defined previously. Further, in order to get similarity solutions, the constant dimensionless chemical reaction parameter $\gamma = K_1 x^2 / (\alpha_m Ra_x)$ was introduced in Eq. (22). The transformed boundary conditions are precisely (11). Local Nusselt number and local Sherwood number are given by the same expressions as in (12). The problem has been solved numerically by (Postelnicu, 2007a), using a version of the Keller-box method adapted to solve ordinary differential equations was used.
The parameters involved in the present problem are Le, N, D_f, S_r, γ and n, the last two quantities pertaining to the chemical reaction. As expected, the Sherwood number is much more sensitive than the Nusselt number to the variation of parameters γ and n.
In Figures 3 and 4 Nusselt and Sherwood local numbers vs γ are plotted for $D_f = 0.6$, $S_r = 0.1$, Le = 1, $N = 1$ and $n = (1, 2, 3)$. Then, Fig. 5 depicts the concentration profiles when $D_f = 0.6$, $S_r = 0.1$, Le = 1 and $N = 1$, for various combinations of n and γ. Another case with nonzero Dufour and Soret numbers was considered in Fig. 6: $D_f = 0.15$, $S_r = 0.4$, Le = 1 and $N = 1$, using the same combinations of n and γ, as in Fig. 4. The general behaviour is similar to that depicted in Fig. 4, but now we remark the strong overshoot when $n = 2$ and $\gamma = 1$.

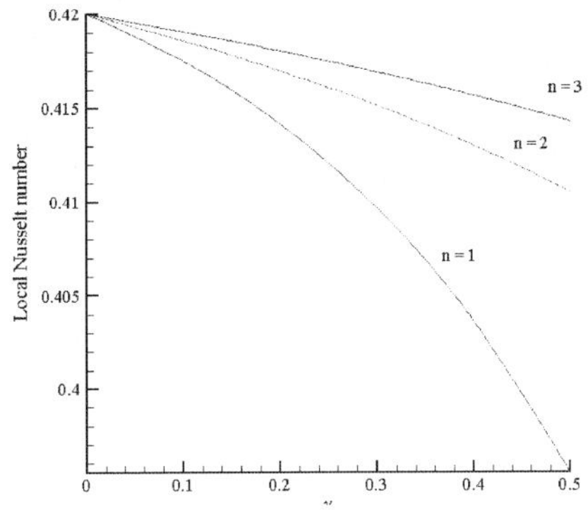

Fig. 3. Nusselt number variation with γ for D_f = 0.6, S_r = 0.1, Le = 1, N = 1 and n = (1, 2, 3)

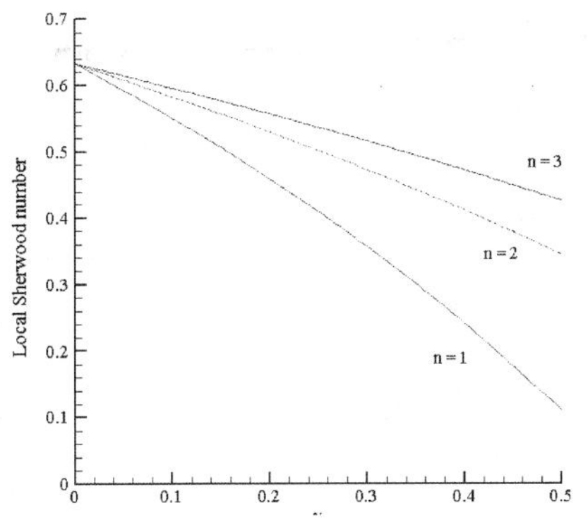

Fig. 4. Sherwood number variation with γ for D_f = 0.6, S_r = 0.1, Le = 1, N = 1 and n = (1, 2, 3)

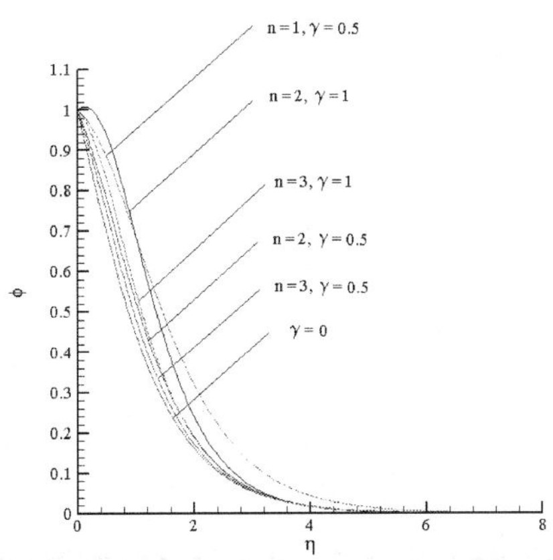

Fig. 5. Concentration profiles for D_f = 0.6, S_r = 0.1, Le = 1 and N = 1

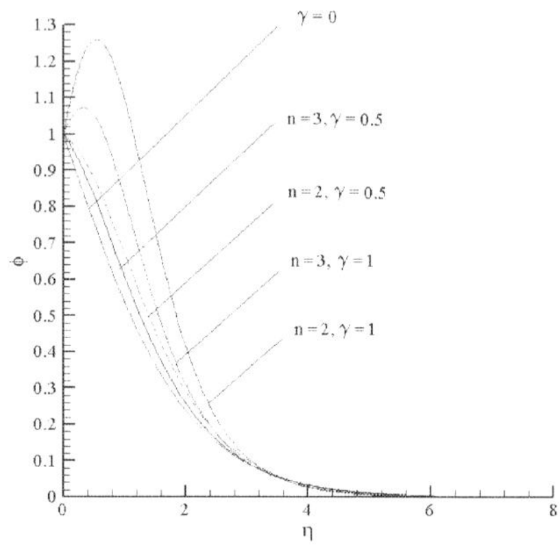

Fig. 6. Concentration profiles for D_f = 0.15, S_r = 0.4, Le = 1 and N = 1

The present problem subject to boundary conditions of prescribed wall heat and mass fluxes is under investigation in a paper by (Magyari and Postelnicu, 2010b).

2. Mass transfer (only) with chemical reaction

In this section we will analyze the convective diffusive-transport with chemical reaction in natural convection in fluid saturated porous medium. A vertical plate is composed of a chemical species maintained at a given concentration and immersed in a fluid-saturated porous medium. Far away from the plate the concentration is constant, at a value C_∞ The species on the plate is first transferred from the plate to the adjacent medium. Following (Postelnicu, 2009a), the basic equations in the boundary layer approximation are

$$\frac{\partial u}{\partial x} + \frac{\partial v}{\partial y} = 0 \tag{23}$$

$$u = \frac{gK}{\upsilon}(C - C_\infty) \tag{24}$$

$$u\frac{\partial C}{\partial x} + v\frac{\partial C}{\partial y} = D_m \frac{\partial^2 C}{\partial y^2} + \dot{C}''' \tag{25}$$

where x and y are the coordinates along and normal to the plate, respectively, the orientation of the x-axis being upwards. The reaction-rate term is modelled by a power-law model

$$\dot{C}''' = -k(C - C_\infty)^n \tag{26}$$

where k is the reaction-rate constant and n is the order of the reaction. The boundary conditions are

$$u = v = 0\,,\ C = C_0(x),\ \text{at}\ y = 0 \tag{27}$$

Similarity solution is sought in the form

$$\psi = \alpha Ra_{x,C}^{1/2} f(\eta)\,,\ \eta = \frac{y}{x} Ra_{x,C}^{1/2}\,,\ \phi = \frac{C - C_\infty}{C_0 - C_\infty} \tag{28}$$

where $Ra_{x,C} = gK\beta(C - C_\infty)x / (\upsilon D_m)$ is the concentration Rayleigh number. Using (28), the problem transforms to

$$f' = \phi \tag{29}$$

$$\frac{1}{Le}\phi'' + f\phi' - \frac{k}{D_m} \cdot \frac{x^2}{Ra_{x,C}}(C - C_\infty)^n \phi^n = 0 \tag{30}$$

The last term in (30) can be rewritten as

$$\varepsilon(x) = \frac{k}{D_m} \cdot \frac{x^2}{Ra_{x,C}}(C - C_\infty)^n \phi^n = \frac{k\upsilon(C - C_\infty)^{n-1} x}{gK_C\beta}\phi^n \tag{31}$$

and, on physical basis, this quantity is small. The boundary conditions are

$$f(0) = 0\,,\ \phi(0) = 1\,,\ \phi(\infty) = 0 \tag{32}$$

Since we are considering a regular perturbation problem, there is no need for matching of layers or for multiple scales. The unknown functions of the problem are expanded as follows

$$f(\eta) = f_0(\eta) + \varepsilon f_1(\eta) + \ldots\,,\ \phi(\eta) = \phi_0(\eta) + \varepsilon\phi_1(\eta) + \ldots \tag{33}$$

and retaining the terms up to the first order, we obtain the following problems]

$$f_0' = \phi_0\,,\ \frac{1}{Le}\phi_0'' + f_0\phi_0' = 0 \tag{34a}$$

$$f_0(0) = 0\,,\ \phi_0(0) = 1\,,\ \phi_0(\infty) = 0 \tag{34b}$$

$$f_1' = \phi_1\,,\ \frac{1}{Le}\phi_1'' + f_0\phi_1' + \phi_0' f_0 - \phi_0^n = 0 \tag{35a}$$

$$f_1(0) = 0\,,\ \phi_1(0) = 0\,,\ \phi_1(\infty) = 0 \tag{35b}$$

Two parameters Le and n are involved in the present problem. A quantity which dictates on the final solution is the small parameter ε.

Le = 0.1 $f_0''(0)$ = -0.19948		Le = 1 $f_0''(0)$ = -0.62756		Le = 10 $f_0''(0)$ = -1.98450		Le = 100 $f_0''(0)$ = -6.27554	
n	$f_1''(0)$	n	$f_1''(0)$	n	$f_1''(0)$	n	$f_1''(0)$
1	-0.218055	1	-0.69979	1	-2.17852	1	-6.53320
2	-0.13900	2	-0.43593	2	-1.32077	2	-3.62634
3	-0.10427	3	-0.32418	3	-0.95866	3	-2.42976
4	-0.08407	4	-0.25950	4	-0.74983	4	-1.75969
5	-0.07066	5	-0.21659	5	-0.61186	5	-1.331435

Table 2. Dimensionless wall shear-stress for n = (1, ..., 5) and Le = 0.1, 1, 10, 100

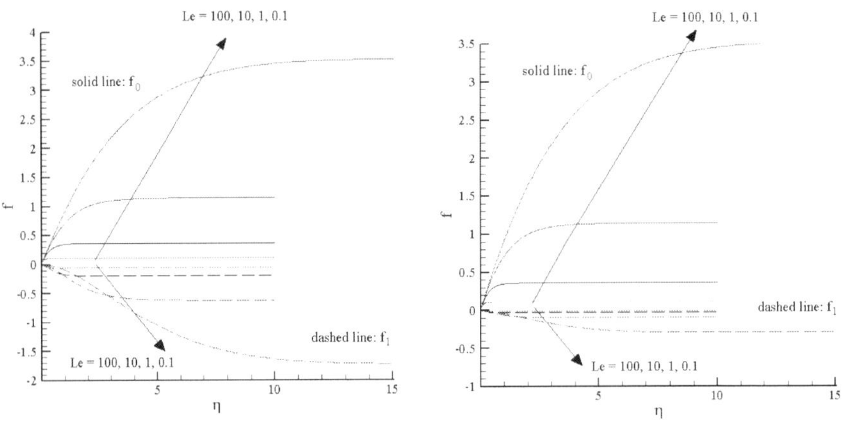

Fig. 7. Dimensionless stream function when n = 1 and n = 3

In Fig. 7 there is shown the variation of the dimensionless stream function across the boundary layer, when n = 1 and 3, for Le = 0.1, 1, 10 and 100. Figs. 8 and 9 depict the velocity profiles across the boundary layer for n = 1 and 3. It is worth to remark that the dimensionless velocity is equal to the dimensionless concentration according to Eq. (7).

It is readily seen from Table 2 and Fig. 7 that as n increases, the impact of the first order solution on the zeroth order solution (which describes the situation without reaction) decreases. In clear fluids, there are regions close to the wall where the flow accelerates, see (Aris, 1965). A similar behaviour is observed for porous media, see for instance Figs. 8 and 9.

A final comment is on the small parameter ε. From our Table 1 and from the figures it is seen that the order of magnitude of the first order solution is roughly the same as that of the zeroth order solution. So, a choice of a maximum 0.01 should be appropriate, in order to get appropriate results. On the other hand, the definition of ε given in (8) does not give us much chances to calculate it on a physical basis. So, it was decided to show graphically the stream function and velocity across the boundary layer for the zeroth and first order solutions.

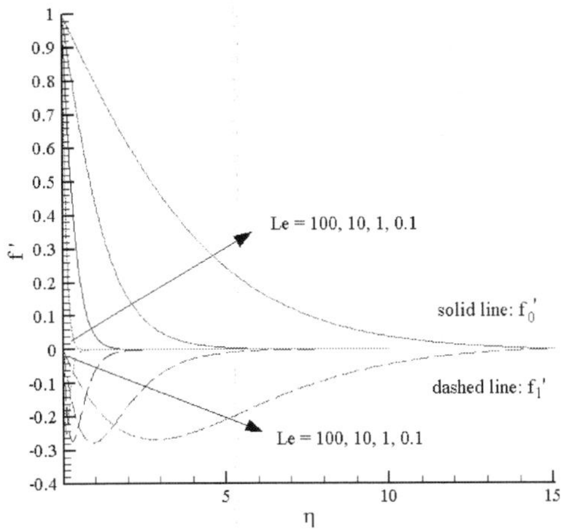

Fig. 8. Velocity profiles, n = 1, zeroth and first order solutions

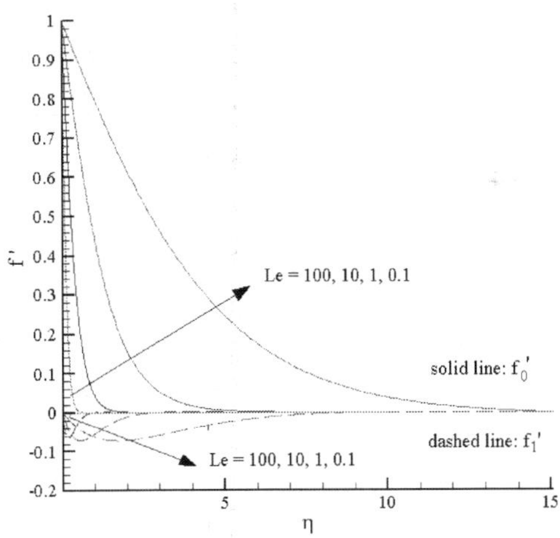

Fig. 9. Velocity profiles, n = 3, zeroth and first order solutions

3. Effects of thermophoresis particle deposition in free convection boundary layers

In the phenomenon of thermophoresis small sized (submicron) particles suspended in an isothermal gas acquire velocities relative to the gas in the direction of decreasing temperature. Thermophoresis is important when the particle sizes are small and the temperature gradients are large. When the wall is cold, the particles tend to deposit on the surface, while when the wall is hot the particles tend to repel from that surface.

3.1 Vertical plate

(Chamka and Pop, 2004) and (Chamka et al, 2006) looked to the effect of thermophoresis particle deposition in free convection boundary layer from a vertical flat plate embedded in a porous medium, without and with heat generation or absorption, respectively.

3.2 Horizontal plate

We are going into details by using the paper by (Postelnicu, 2007b), where it was analyzed the effect of thermophoresis particle deposition in free convection from a horizontal flat plate embedded in a porous medium. The plate is held at constant wall temperature T_w and constant wall concentration C_w. The temperature and concentration of the ambient medium are T_∞ and C_∞, respectively. The x-coordinate is measured along the plate from its leading edge, and the y-coordinate normal to it. The following assumptions are used for the present physical model: a) the fluid and the porous medium are in local thermodynamic equilibrium; b) the flow is laminar, steady-state and two-dimensional; c) the porous medium is isotropic and homogeneous; d) the properties of the fluid and porous medium are constants; e) the Boussinesq approximation is valid and the boundary-layer approximation is applicable.

In-line with these assumptions, the governing equations describing the conservation of mass, momentum, energy and concentration can be written as follows

$$\frac{\partial u}{\partial x}+\frac{\partial v}{\partial y}=0 \tag{36}$$

$$u=-\frac{K}{\mu}\frac{\partial p}{\partial x},\quad v=-\frac{K}{\mu}\left(\frac{\partial p}{\partial y}+\rho g\right) \tag{37}$$

$$u\frac{\partial T}{\partial x}+v\frac{\partial T}{\partial y}=\alpha_m\frac{\partial^2 T}{\partial y^2} \tag{38}$$

$$u\frac{\partial C}{\partial x}+v\frac{\partial C}{\partial y}+\frac{\partial (v_T C)}{\partial y}=D_m\frac{\partial^2 C}{\partial y^2} \tag{39}$$

together with the Boussinesq approximation $\rho=\rho_\infty\left[1-\beta_T(T-T_\infty)-\beta_C(C-C_\infty)\right]$, where the thermophoretic deposition velocity in the y-direction is given by

$$v_T=-k\frac{\nu}{T}\frac{\partial T}{\partial y} \tag{40}$$

where k is the thermophoretic coefficient. We remark that only the velocity component given by (40) is to be considered within the boundary-layer framework. The boundary conditions are

$$y = 0 \; : \; T = T_w \, , \; C = C_w \, , \; v = 0 \tag{41a}$$

$$y \to \infty : \; u \to 0 \, , \;\; T \to T_\infty \, , \;\; C \to C_\infty \tag{41b}$$

Introducing the stream function ψ in the usual way, in order to identically satisfy the continuity equation, and using the dimensionless quantities

$$\psi = \alpha_m Ra_x^{1/3} f(\eta) \, , \; \theta(\eta) = \frac{T - T_\infty}{T_w - T_\infty} \, , \; \phi(\eta) = \frac{C - C_\infty}{C_w - C_\infty} \, , \; \eta = Ra_x^{1/3} \frac{y}{x} \tag{42}$$

equations (36-39) become

$$f'' - \frac{2}{3}\eta\theta' - \frac{2}{3}N\eta\phi' = 0 \tag{43}$$

$$\theta'' + \frac{1}{3} f\theta' = 0 \tag{44}$$

$$\frac{1}{Le}\phi'' + \frac{1}{3} f\phi' + \frac{k\Pr}{N_t + \theta}\left(\theta'\phi' + \theta''\phi - \frac{\phi}{N_t + \theta}\theta'^2\right) = 0 \tag{45}$$

where the sustentation parameter N, the thermophoresis parameter N_t, the local Rayleigh number Ra_x and the Prandtl number Pr are defined as follows

$$N = \frac{\beta_C (C_w - C_\infty)}{\beta_T (T_w - T_\infty)} \, , \; N_t = \frac{T_\infty}{T_w - T_\infty} \, , \; Ra_x = \frac{\rho_\infty g \beta_T K (T_w - T_\infty) x}{\mu \alpha} \, , \; \Pr = \frac{\nu}{\alpha_m} \tag{46}$$

The set of ordinary differential equations (43-45) must be solved along the following boundary conditions

$$f(0) = 0 \, , \; \theta(0) = 1 \, , \; \phi(0) = 1 \tag{47a}$$

$$f'(\infty) = 0 \, , \; \theta(\infty) = 0 \, , \; \phi(\infty) = 0 \tag{47b}$$

Of technical interest is the thermophoretic deposition velocity at the wall, which is given by the expression $V_{tw} = -\frac{k\Pr}{1 + N_t}\theta'(0)$. Some graphs are reproduced below, from the paper by (Postelnicu, 2007b).

Fig. 10 shows the effects of N on concentration profiles for $k = 0.5$, $N_t = 100$, when Le = 10. In comparison with the vertical case, Fig. 2 from (Chamka and Pop, 2004), the behaviour of the concentration profiles shown in our Fig. 2 is quite similar.

The effects of Le and N on thermophoretic deposition velocity V_{tw} can be seen in Fig. 11 when $k = 0.5$ and $N_t = 100$. Once again, it is instructive to compare our results with those obtained by Chamka and Pop (2004), see Fig. 1 from that paper, where the parameters have the same values as ours. The general behaviour is the same, but the values of V_{tw} are larger in present case. In Fig. 12 there is represented the thermophoretic deposition velocity as a function of k and N when Le = 10 and $N_t = 100$. Similar plots may be obtained for other

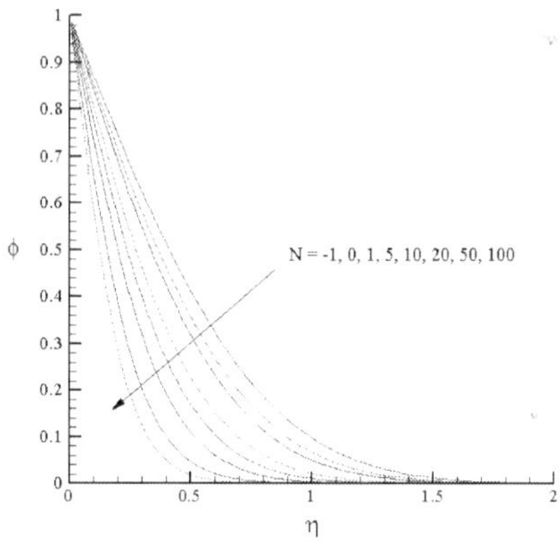

Fig. 10. Effects of N on concentration profiles, Le =10, k = 0.5, N_t = 100

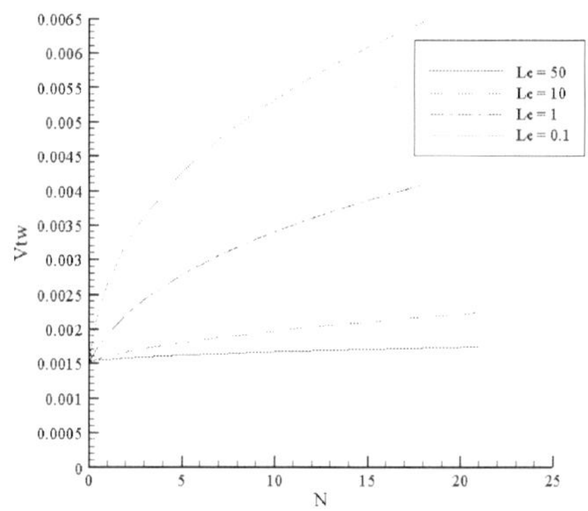

Fig. 11. Effects of Le and N on thermophoretic deposition velocity, k = 0.5, N_t = 100

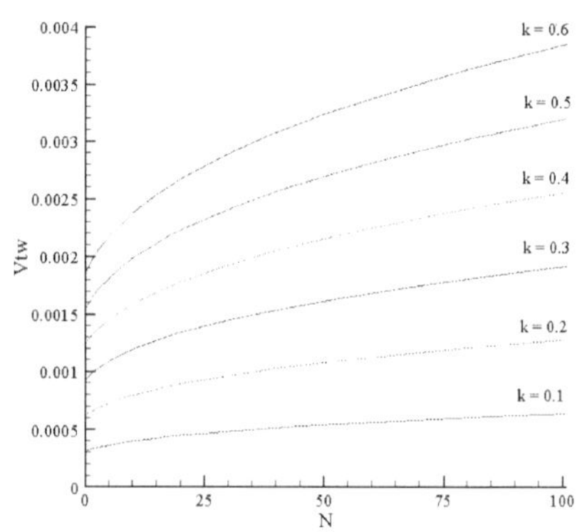

Fig. 12. Effects of k and N on thermophoretic deposition velocity, Le = 10, N_t = 100

values of the Lewis number. The thermophoretic deposition velocity increases as k increases, at a fixed value of N, as in the vertical case.

The problem may be extended on many directions, but the first one seems to be to consider a power law variation of the wall temperature with x: $T = T_w \pm Ax^{\lambda}$, where the "+" and "-" signs are for a heated plate facing upward and for a cooled plate facing downward respectively and A is a positive constant, but the general behaviour portrayed previously remains.

3.3 Other contributions

In a paper by (Chamkha et al., 2004), the steady free convection over an isothermal vertical circular cylinder embedded in a fluid-saturated porous medium in the presence of the thermophoresis particle deposition effect was analyzed.

The effect of suction / injection on thermophoresis particle deposition in a porous medium was studied by Partha (2009). Using again the boundary layer assumptions, but with a non-Darcy formulation, he found that the heat transfer is intensified when second order effects (thermal dispersion and cross-diffusion) are present.

Very recently, (Postelnicu, 2010b) analyzed thermophoresis particle deposition in natural convection over inclined surfaces in porous media. In this case, Eqs. (37) must be replaced with

$$\frac{\partial u}{\partial y} = \frac{g\beta_T K}{\upsilon}\left(\frac{\partial T}{\partial y}\sin\delta - \frac{\partial T}{\partial x}\cos\delta\right) + \frac{g\beta_C K}{\upsilon}\left(\frac{\partial C}{\partial y}\sin\delta - \frac{\partial C}{\partial x}\cos\delta\right) \tag{48}$$

where the angle of inclination of the plate with respect to horizontal is denoted by δ. The problem is no longer amenable to a set of ordinary differential equations, but partial ones, as follows

$$f'' - \frac{2}{3}(\theta' + N\phi') = \xi\left(\theta' + N\phi' - \frac{1}{3}\frac{\partial\theta}{\partial\xi} - \frac{1}{3}N\frac{\partial\phi}{\partial\xi}\right) \tag{49a}$$

$$\theta'' + \frac{1}{3}f\theta' = \frac{1}{3}\xi\left(f'\frac{\partial\theta}{\partial\xi} - \theta'\frac{\partial f}{\partial\xi}\right) \tag{49b}$$

$$\frac{1}{Le}\phi'' + \frac{1}{3}f\phi' + \frac{k\Pr}{N_t + \theta}\left(\theta'\phi' + \theta''\phi - \frac{\phi\theta'^2}{N_t + \theta}\right) = \frac{1}{3}\xi\left(f'\frac{\partial\phi}{\partial\xi} - \phi'\frac{\partial f}{\partial\xi}\right) \tag{49c}$$

subject to the boundary conditions which are essentially (47), at every ξ. The streamwise variable ξ is defined as $\xi = (Ra_x \cos\delta)^{1/3}\tan\delta$,where the local Rayleigh number Ra_x is defined as in (46). This system of partial differential equations is of parabolic type and may be solved by one of the well-known appropriate numerical methods, such as the Keller-box method, Local Nonsimilarity Method, etc. Aiming to throw some insight on the application of the last method to the present problem, we will refer shortly to this aspect.

This method was introduced by (Sparrow et al., 1970), then applied to thermal problems by (Sparrow & Yu, 1971), where a good description of the algorithm may be found. In the so-called 2-equations model, one neglects in a first step the first-order derivatives with respect to ξ in Eqs. (49). In the second step, there is performed the differentiation of (49) with respect to ξ and the second-order derivatives $\partial^2/\partial\xi^2$ are neglected. Proceeding so and introducing the notations

$$F = \frac{\partial f}{\partial \xi},\ \Theta = \frac{\partial \theta}{\partial \xi} \text{ and } \Phi = \frac{\partial \phi}{\partial \xi},$$

we get the system of equations

$$f'' - \frac{2}{3}(\theta' + N\phi') = \xi\left(\theta' + N\phi' - \frac{1}{3}\Theta - \frac{1}{3}N\Phi\right) \quad (50a)$$

$$\theta'' + \frac{1}{3}f\theta' = \frac{1}{3}\xi\left(f'\Theta - \theta' F\right) \quad (50b)$$

$$\frac{1}{Le}\phi'' + \frac{1}{3}f\phi' + \frac{k\Pr}{N_t + \theta}\left(\theta'\phi' + \theta''\phi - \frac{\phi\theta'^2}{N_t + \theta}\right) = \frac{1}{3}\xi\left(f'\Phi - \phi' F\right) \quad (50c)$$

$$f'' - \frac{2}{3}(\theta' + N\phi') = \xi\left(\theta' + N\phi' - \frac{1}{3}\frac{\partial\Theta}{\partial\xi} - \frac{1}{3}N\frac{\partial\Phi}{\partial\xi}\right) \quad (50d)$$

$$\Theta'' + \frac{1}{3}f\Theta' + \frac{1}{3}\theta' F = \frac{1}{3}\xi\left(f'\Theta - \theta' F\right) + \frac{1}{3}\xi\left(F'\Theta - F\Theta'\right) \quad (50e)$$

$$\frac{1}{Le}\Phi'' + \frac{1}{3}\xi\left(F\Phi' - F'\Phi\right) + \frac{1}{3}f\Phi' + \frac{2}{3}\varphi' F - \frac{1}{3}f'\Phi + \frac{k\Pr}{N_t + \theta}\left(\theta'\Phi' + \varphi'\Theta' + \theta''\Phi\right) \quad (50f)$$

$$-\frac{k\Pr}{\left(N_t + \theta\right)^2}\left[\left(\theta'\varphi' + \theta''\varphi\right)\Theta + 2\varphi\theta'\Theta' + \left(\theta'\right)^2\Phi\right] + \frac{2k\Pr}{\left(N_t + \theta\right)^3}\varphi\theta'\Theta = 0$$

that must be solved along the boundary conditions

$$f(\xi,0) = 0,\ \theta(\xi,0) = 1,\ \phi(\xi,0) = 1,\ F(\xi,0) = 0,\ \Theta(\xi,0) = 0,\ \Phi(\xi,0) = 0 \quad (51)$$

$$f'(\xi,\infty) = 0,\ \theta(\xi,\infty) = 0,\ \phi(\xi,\infty) = 0,\ F'(\xi,\infty) = 0,\ \Theta(\xi,\infty) = 0,\ \Phi(\xi,\infty) = 0 \quad (52)$$

Now the problem was reduced to the set of differential ordinary equations (50) that must be solved subject to the boundary conditions (51) and (52) by any standard numerical method.

4. Convective flows on reactive surfaces in porous media

This kind of chemical reactions may undergo throughout the volume of (porous) region, or along interfaces / boundaries of this region. Real-world applications include chemical engineering systems, contaminant transport in groundwater systems, or geothermal processes. The catalytic systems are modeled usually by including the description of the reaction kinetics of the catalytic process and the transport of momentum, heat, and mass coupled to this process. Concerning the transport phenomena, access to the catalyst is determined by the transport of mass and energy in a reactor. In heterogeneous catalysis, the access to the catalyst is maximised through the use of porous structures. Examples of catalytic surface reactions are methane/ammonia and propane oxidation over platinum, see for instance (Song et al., 1991) and (Williams et al, 1991). Our interest in the present section is related to the chemical reactions which take place along interfaces / boundaries of the flow region.

4.1 External flows

It is now recognized that chemical reactions affect buoyancy driven flows at least in two directions: the transition from conduction-reaction regimes to conduction-convection-reaction regimes and the influence of natural convection on the development of the chemical reaction.

Models for convective flows on reactive surfaces in porous media have been proposed for external flows by (Merkin and Mahmood, 1998), (Mahmood and Merkin, 1999), (Minto et al., 1998), (Ingham et al., 1999). In these studies bifurcation diagrams were presented for various combinations of the problem parameters and hysteresis bifurcation curves were identified, whenever they exist.

The study by (Merkin and Mahmood, 1998) was extended by (Postelnicu, 2004b) for porous media saturated with non-Newtonian fluids. We shall follow this later author and we will focus on the free convection near a stagnation point of a cylindrical body in a porous medium saturated with a non-Newtonian fluid. We point-out that many fluids involved in practical applications present a non-Newtonian behaviour. Such practical applications in porous media could be encountered in fields like ceramics production, filtration and oil recovery, certain separation processes, polymer engineering, petroleum production.

The fluid which saturates the porous medium is considered of power-law type. The governing equations of this process are

$$\frac{\partial u^*}{\partial x^*} + \frac{\partial v^*}{\partial y^*} = 0 \tag{53a}$$

$$\left(u^*\right)^n = \frac{gK^*(n)\beta}{\nu^*}(T - T_\infty)\frac{x^*}{l} \tag{53b}$$

$$u^*\frac{\partial T}{\partial x^*} + v^*\frac{\partial T}{\partial y^*} = \alpha_m \frac{\partial^2 T}{\partial y^{*2}} \tag{53c}$$

$$u^*\frac{\partial C}{\partial x^*} + v^*\frac{\partial C}{\partial y^*} = D_m \frac{\partial^2 C}{\partial y^{*2}} \tag{53d}$$

in standard notations, where stars mean dimensional quantities. The x and y-coordinates are taken along the body surface and normal to it, respectively. Moreover, the flow velocity and the pores of the porous medium are assumed to be small so that Darcy's model can be used. The modified permeability $K^*(n)$ is given by

- $K^*(n) = \frac{6}{25}\left(\frac{n\phi}{3n+1}\right)^n \left(\frac{\phi d}{3(1-\phi)}\right)^{n+1}$, according to (Christopher and Middleman, 1965);
- $K^*(n) = \frac{2}{\phi}\left(\frac{d\phi^2}{8(1-\phi)}\right)^{n+1} \frac{6n+1}{10n-3}\left(\frac{16}{75}\right)^{\frac{3(10n-3)}{10n+11}}$, according to (Darmadhikari and Kale, 1985),

where d is the particle diameter and ϕ is the porosity.
Heat is released by the first order reaction

$$A \to B + heat\text{ , rate= } k_0 C \exp\left(-\frac{E}{RT}\right) \tag{54}$$

with a heat of reaction $Q > 0$ which is taken from the body surface into the surrounding fluid-porous medium by conduction. We notice that (54) describes an exothermic catalytic reaction, of Arrhenius type, where the reactant A is converted to the inert product B. Here E is the activation energy, R is the universal gas constant, k_0 is the rate constant, T is the temperature and C is the concentration of reactant A within the convective fluid. This reaction scheme is a realistic one and has been used in the past in modelling of combustion processes, and also for reactive processes in porous media.
The boundary conditions are

$$v^* = 0\,,\; k_m \frac{\partial T}{\partial y^*} = -k_0 Q C \exp\left(-\frac{E}{RT}\right),\; D_m \frac{\partial C}{\partial y^*} = k_0 C \exp\left(-\frac{E}{RT}\right),\text{ on } y^* = 0\,,\; x^* \ge 0 \tag{55a}$$

$$v^* \to 0\,,\;\; T \to T_\infty\,,\; C \to C_\infty \text{ as } y^* \to \infty\,,\; x^* \ge 0 \tag{55b}$$

Using the stream function: $u^* = \partial\psi^* / \partial y^*$, $v^* = -\partial\psi^* / \partial x^*$, we proceed to render the problem in non-dimensional form by introducing the following quantities

$$x = \frac{x^*}{l}\,,\; y = \sqrt{Ra}\frac{y^*}{l}\,,\; \psi = \frac{1}{\sqrt{Ra}}\frac{\psi^*}{\alpha_m}\,,\; \theta = \frac{T - T_\infty}{RT_\infty^2 l}\,,\; \varphi = \frac{C}{C_\infty} \tag{56}$$

where $Ra = \left(\frac{gK^*(n)\beta RT_\infty^2 l^n}{v^* \alpha_m^n}\right)^{1/n}$ is the Rayleigh number and l is a length scale. We obtain

$$\left(\frac{\partial\psi}{\partial y}\right)^n = x\theta \tag{57a}$$

$$\frac{\partial\psi}{\partial y}\frac{\partial\theta}{\partial x} - \frac{\partial\psi}{\partial x}\frac{\partial\theta}{\partial y} = \frac{\partial^2\theta}{\partial y^2} \tag{57b}$$

$$\frac{\partial\psi}{\partial y}\frac{\partial\phi}{\partial x} - \frac{\partial\psi}{\partial x}\frac{\partial\phi}{\partial y} = \frac{1}{Le}\frac{\partial^2\theta}{\partial y^2} \tag{57c}$$

$$\psi = 0\,,\ \frac{\partial \theta}{\partial y} = -\delta \varphi_w \exp\left(\frac{\theta_w}{1+\varepsilon\theta_w}\right),\ \frac{\partial \varphi}{\partial y} = \lambda\delta\varphi_w \exp\left(\frac{\theta_w}{1+\varepsilon\theta_w}\right),\ \text{on } y=0,\ x \ge 0 \tag{58a}$$

$$\frac{\partial \psi}{\partial y} \to 0\,,\ \theta \to 0\,,\ \varphi \to 1\,,\ \text{as } y \to \infty\,,\ x \ge 0 \tag{58b}$$

Looking for similarity solutions, we introduce the following quantities

$$\psi = x f(y)\,,\ \theta = x^{n-1} g(y)\,,\ \varphi = x^{n-1} h(y) \tag{59}$$

that render the problem (57-58) in the form

$$(f')^n = g \tag{60a}$$

$$g'' + f g' - (n-1) f' g = 0 \tag{60b}$$

$$\frac{1}{Le} h'' + f h' - (n-1) f' h = 0 \tag{60c}$$

$$f(0) = 0\,,\ g'(0) = -\lambda h_w \exp\left(\frac{x^{n-1} g_w}{1+\varepsilon x^{n-1} g_w}\right),\ h'(0) = \lambda\delta h_w \exp\left(\frac{x^{n-1} g_w}{1+\varepsilon x^{n-1} g_w}\right) \tag{61a}$$

$$g \to 0\,,\ h \to 1\,,\ \text{as } y \to \infty \tag{61b}$$

Using the transformations $f = {g_w}^{1/2} F(Y)$, $g = {g_w}^n G(Y)$, $h = 1-(1-h_w) H(Y)$, $Y = {g_w}^{1/2} y$, Eqs (60-61) become

$$(F')^n = G \tag{62a}$$

$$G'' + F G' - (n-1) F' G = 0 \tag{62b}$$

$$\frac{1}{Le} H'' + F H' - (n-1)\left[1-(1-h_w)H\right] F' = 0 \tag{62c}$$

$$F(0) = 0\,,\ G(0) = 1\,,\ H(0) = 1 \tag{63a}$$

$$G(\infty) = 0\,,\ H(\infty) = 0 \tag{63b}$$

where now primes denote differentiation with respect to Y. It is worth to remark that the problems in (F, G) and in H are now no more coupled. The last two boundary conditions from (63a) become

$$ {g_w}^{n+1/2} \left(\frac{dG}{dY}\right)_{Y=0} = -\lambda h_w \exp\left(\frac{x^{n-1} g_w}{1+\varepsilon x^{n-1} g_w}\right) \tag{64a}$$

$$-(1-h_w)g_w^{1/2}\left(\frac{dH}{dY}\right)_{Y=0}=\lambda\delta h_w\exp\left(\frac{x^{n-1}g_w}{1+\varepsilon x^{n-1}g_w}\right) \quad (64b)$$

i.e. two equations in the unknowns g_w and h_w. Eliminating h_w between (64) gives

$$\lambda=g_w^{1/2}\exp\left(-\frac{x^{n-1}g_w}{1+\varepsilon x^{n-1}g_w}\right)/\left[\frac{\delta}{C_1}+\frac{(g_w)^{-n}}{C_0}\right] \quad (65)$$

where $C_0=-(dG/dY)_{Y=0}$ and $C_1=(dH/dY)_{Y=0}$. We remark that C_0 depends only on n, while C_1 depends on n, Le, ε, δ and h_w.

a. Case of no reactant consumption

In this case, $\delta=\gamma=0$ and $h_w\equiv 1$ so that Eq. (65) simplifies to

$$\lambda=C_0 g_w^{n+1/2}\exp\left(-\frac{x^{n-1}g_w}{1+\varepsilon x^{n-1}g_w}\right) \quad (66)$$

The critical points on the graphs g_w vs. λ are obtained from the condition: $d\lambda/dg_w=0$ and are given by

$$g_w^{(1,2)}=\frac{1-(2n+1)\varepsilon\pm\sqrt{1-2(2n+1)\varepsilon}}{(2n+1)\varepsilon^2 x^{n-1}} \quad (67)$$

The following conclusions can be obtained from (67)

- For $0<\varepsilon<0.5/(2n+1)$, there are two critical points $\lambda_1=\lambda\left(g_w^{(1)}\right)$ and $\lambda_2=\lambda\left(g_w^{(2)}\right)$.
- At $\varepsilon=0.5/(2n+1)$, there is a hysteresis bifurcation, where the slope becomes vertical.
- For $\varepsilon>0.5/(2n+1)$, g_w increases with λ
- In the case $\varepsilon<<1$, one obtains using (67),

$$g_w^{(1)}\sim\frac{2n+1}{2}\left[1+(2n+1)\varepsilon+...\right]x^{1-n},\ g_w^{(2)}\sim\frac{2-(2n+1)\varepsilon-\frac{1}{2}(2n+1)^2\varepsilon^2+...}{(2n+1)\varepsilon^2}x^{1-n} \quad (68)$$

so that

$$g_w^{(1)}\to\frac{2n+1}{2}x^{1-n}\ \text{and}\ g_w^{(2)}\to\infty\ \text{as}\ \varepsilon\to 0 \quad (69)$$

Some curves g_w vs. λ are represented in Fig. 13, for no reactant consumption, when $\varepsilon=0$ and $\delta=0.5$.

b. General case, reactant consumption

In this case, we have to cope with equation (65). Looking again for the critical points, the condition $d\lambda/dg_w=0$ gives

$$\left(\frac{\delta^2}{C_1}+\frac{g_w{}^{-n}}{C_0}\right)\left[\frac{1}{2}g_w{}^{-1/2}-\frac{x^{n-1}g_w{}^{1/2}}{\left(1+\varepsilon x^{n-1}g_w\right)^2}\right]+\frac{n\delta^2 C_0 C_2}{C_1^3}g_w{}^{-n-1/2}+\frac{n}{C_0}g_w{}^{-n-1/2}=0 \qquad (70)$$

In deriving this equation, it was necessary to compute dC_1 / dg_w, which is obtained after some algebra as

$$\frac{dC_1}{dg_w}=\frac{n\delta C_0 C_2}{C_1}g_w{}^{n-1} \qquad (71a)$$

$$C_2=\frac{dC_1}{dh_w}=\left(\frac{dS}{dY}\right)_{Y=0} \qquad (71b)$$

In order to find the hysteresis bifurcation curve in the (ε, γ) space, with (*n*, *x*, δ) as remainder parameters, it is needed to solve equation (70) together with its derivative with respect to g_w. The following values of the index parameter have been considered: n = 0.25, 0.5, 0.75, 1.5 and 2.0. In each case, parameters ε, δ and Le have been varied as follows: ε = (0, 0.05, 0.1 and 0.15), δ= (0.5 and 1.0), Le = (1, 10 and 100), while x was taken as 0.1, 0.5 and 1.0. We found that the basic shape of the bifurcation diagram does not depend basically on x, once the other parameters are held at fixed values.

Let us proceed firstly with the case n = 0.25. Fig. 14 shows a typical plot g_w as a function of γ for ε= 0, δ= 0.5 and Le = 1. Basically, this types of variation, known also for Newtonian fluids, is found for any combination of (ε, δ, Le) and they do not lead to hystheresis.

The case n = 0.5 presents many interesting features. Figs. 15 and 16 show two cases when the vertical slope becomes infinite for x = 0.1, but not for x = 0.5 and 1.0. We found that this characteristic for any case associated with Le = 1. But for Le = 10 and several cases associated with Le = 100 the vertical slope becomes infinite for any x, see Fig. 17 and 18.

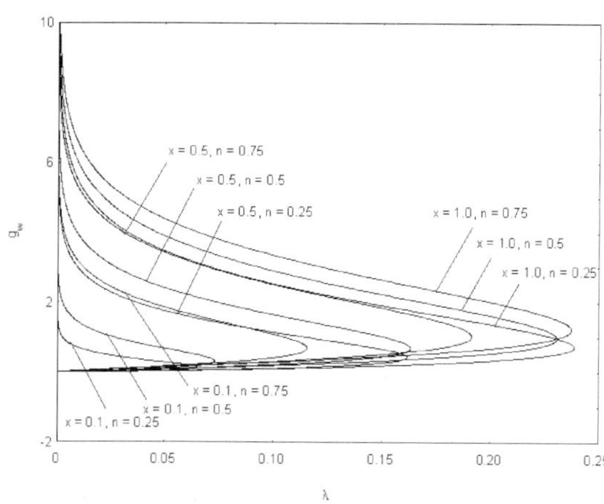

Fig. 13. Curves g_w vs. λ: no reactant consumption, ε= 0, δ = 0.5

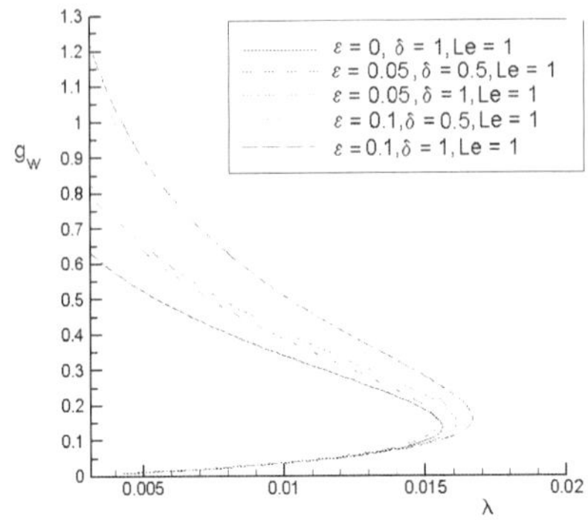

Fig. 14. Bifurcation diagrams for the case of reactant consumption $n = 0.25$, $x = 0.1$

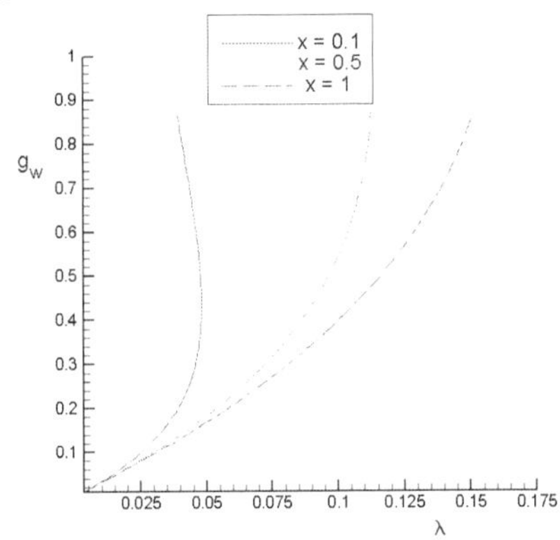

Fig. 15. Bifurcation diagrams for the case of reactant consumption $n = 0.5$, $\varepsilon = 0.1$, $\delta = 1.0$ and $Le = 1$

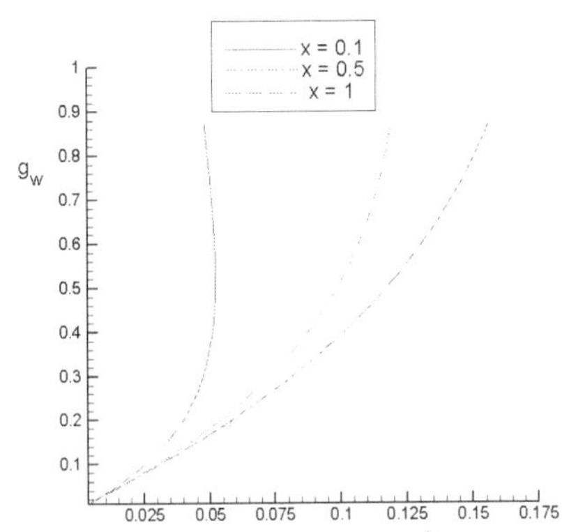

Fig. 16. Bifurcation diagrams for the case of reactant consumption $n = 0.5$, $\varepsilon= 0.15$, $\delta= 1.0$ and $Le = 1$

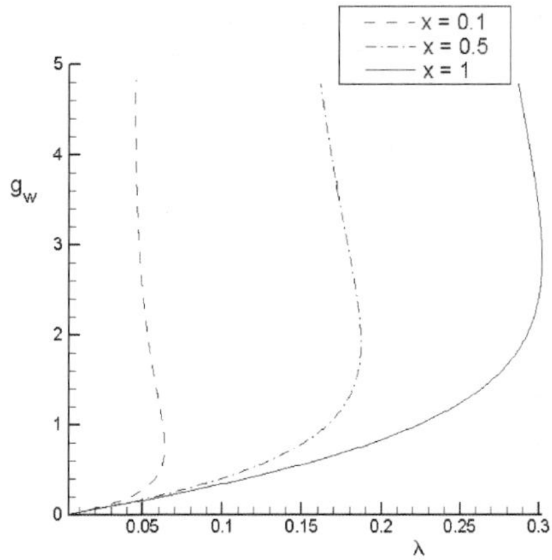

Fig. 17. Bifurcation diagrams for the case of reactant consumption $n = 0.5$, $\varepsilon = 0.15$, $\delta = 1.0$, Le = 10

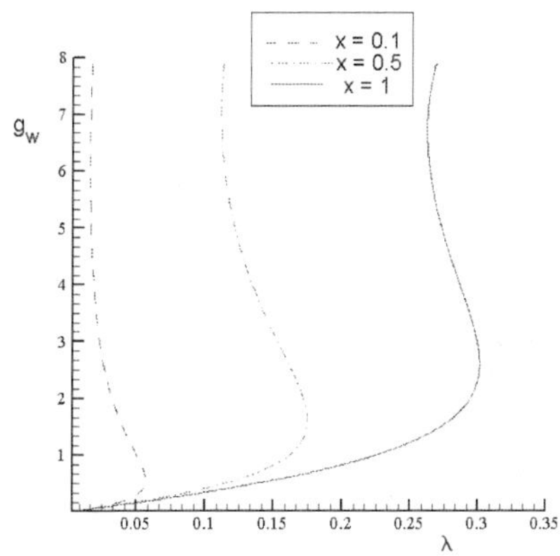

Fig. 18. Bifurcation diagrams for the case of reactant consumption $n = 0.5$, $\varepsilon = 0.1$, $\delta = 1.0$, Le = 100

One kind of behaviour in the case $n = 1.5$ is shown in Fig. 19 specific for Le = 1. A thorough investigation, peformed also for Le = 10 and 100 leads to the conclusion is that there are not chances to obtain hystheresis conditions for $n = 1.5$ irrespective of ε, δ and Le.

Finally the bifurcation diagrams in the case $n = 2.0$ look like in Fig. 20. These shapes are similar to those obtained in the $n = 1.5$ case, for Le = 10 and 100 but this time they are obtained for any Lewis number. We conclude that no hysteresis conditions can be obtained in the case $n = 2$.

In conclusion, we proved that in the case of no reactant consumption it is possible to proceed entirely analytically, and the hysteresis bifurcation was found at a certain value of the activation parameter ε, similarly as in the Newtonian fluids. But in the case of reactant consumption, it is no more possible to use analytical means of analysis. Our numerical runs revealed that the most prolific case in bifurcation diagrams shapes is when the index parameter $n = 0.5$. Cases leading to hysteresis were found to be $n = 0.5$ and $n = 0.75$. Finding hysteresis bifurcation curves is more difficult due to the increased number of parameters as compared to the Newtonian case.

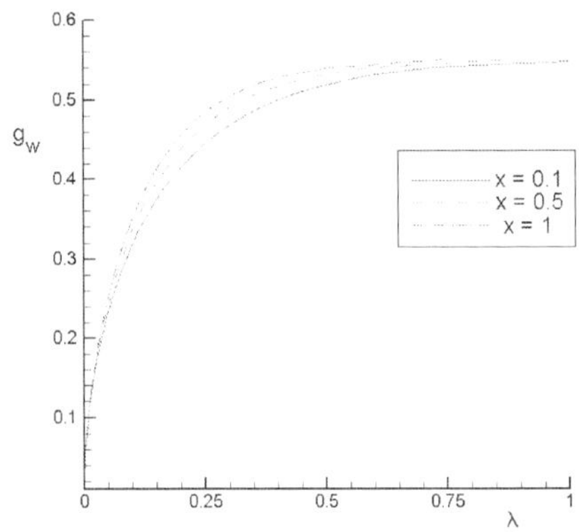

Fig. 19. Bifurcation diagrams for the case of reactant consumption $n = 1.5$, $\varepsilon = 0.05$, $\delta = 1$, Le = 1

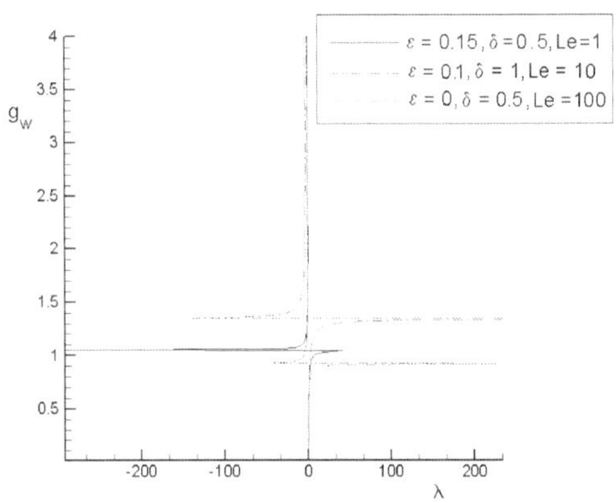

Fig. 20. Bifurcation diagrams for the case of reactant consumption $n = 2.0$, $x = 0.1$

4.2 Onset of convection in a horizontal porous layer driven by catalytic surface reaction on the lower wall

We consider in this subsection the situation when the convective flow in a horizontal porous layer is driven by an exothermic catalytic reaction taking place on the lower surface whereby a reactive species A reacts to form an inert product B. The upper wall is subjected to uniform temperature and concentration, while on the lower wall there is an exothermic surface reaction, whereby reactant P is converted to an inert product B, via the first-order Arrhenius kinetics non-isothermal reaction (54). The surface reaction releases heat, which produces a convective flow close to the surface and, in turn, fresh reactant will replace that used up in the reaction. In this way, an interaction will occur between the convective flow, heat transfer and mass transport of the reactant.

Using usual notations, the governing equations which describe the problem at hand are mass conservation, Darcy's law, equation of energy and that of concentration

$$\frac{\partial \bar{u}}{\partial \bar{x}}+\frac{\partial \bar{v}}{\partial \bar{y}}=0 \tag{72}$$

$$\bar{u}=-\frac{K}{\mu}\frac{\partial \bar{p}}{\partial \bar{x}}, \quad \bar{v}=-\frac{K}{\mu}\frac{\partial \bar{p}}{\partial \bar{y}}+\frac{\rho g \beta K}{\mu}\left(T-T_r\right) \tag{73}$$

$$\frac{\partial T}{\partial \bar{t}}+\bar{u}\frac{\partial T}{\partial \bar{x}}+\bar{v}\frac{\partial T}{\partial \bar{y}}=\alpha_m\left(\frac{\partial^2 T}{\partial \bar{x}^2}+\frac{\partial^2 T}{\partial \bar{y}^2}\right) \tag{74}$$

$$\frac{\partial C}{\partial \bar{t}}+\bar{u}\frac{\partial C}{\partial \bar{x}}+\bar{v}\frac{\partial C}{\partial \bar{y}}=D_m\left(\frac{\partial^2 C}{\partial \bar{x}^2}+\frac{\partial^2 C}{\partial \bar{y}^2}\right) \tag{75}$$

where T_r is a reference temperature and over-bars refer to dimensional quantities. The $\bar{x}$ and $\bar{y}$ axes are taken along the porous layer and normal to it respectively, and the lower wall is located at $\bar{y}=0$. We point out that in writing equations (73), the Boussinesq approximation was invoked and differences in reactant concentration (which may induce buoyancy forces) are assumed to be small. The Darcy model is justifiable when the heat of reaction is small or moderate. Otherwise, when the heat of reaction is large, non-Darcy models must be used.

The thermal boundary conditions on the lower wall are

$$k_T\frac{\partial T}{\partial \bar{y}}=-Qk_0 C\exp\left(-\frac{E}{RT}\right), \quad D\frac{\partial C}{\partial \bar{y}}=k_0 C\exp\left(-\frac{E}{RT}\right) \tag{76}$$

where k_T is the thermal conductivity of the surface, Q is the heat of reaction, which is taken as positive, meaning that heat is taken from the surface into the surrounding fluid-porous medium by conduction. Eliminating the pressure, using the streamfunction ψ and introducing dimensionless quantities, Eqs. (72-75) become

$$\frac{\partial^2 \psi}{\partial x^2}+\frac{\partial^2 \psi}{\partial y^2}=Ra\frac{\partial \theta}{\partial x} \tag{77}$$

$$\frac{\partial \theta}{\partial t}+u\frac{\partial \theta}{\partial x}+v\frac{\partial \theta}{\partial y}=\frac{\partial^2 \theta}{\partial x^2}+\frac{\partial^2 \theta}{\partial y^2} \tag{78}$$

$$\frac{\partial\varphi}{\partial t}+u\frac{\partial\varphi}{\partial x}+v\frac{\partial\varphi}{\partial y}=\frac{1}{Le}\left(\frac{\partial^2\varphi}{\partial x^2}+\frac{\partial^2\varphi}{\partial y^2}\right) \tag{79}$$

see (Postelnicu, 2009b), where $Ra=\frac{\rho g\beta Kh}{\mu\kappa}\cdot\frac{RT_r^2}{E}$ is the Rayleigh number. The boundary conditions in the final model are

$$\psi=0\text{, on both } y=0 \text{ and } y=1 \tag{80a}$$

$$\frac{\partial\theta}{\partial y}=-A\varphi\exp\left(\frac{\theta}{1+\varepsilon\theta}\right),\ \frac{\partial\varphi}{\partial y}=B\varphi\exp\left(\frac{\theta}{1+\varepsilon\theta}\right)\text{, on } y=0 \tag{80b}$$

$$\theta=0\ ,\ \varphi=1\text{, on } y=1 \tag{80c}$$

where

$$A=\left(Qk_0C_0h/k\right)\cdot\left(E/RT_r^2\right),\ B=k_0h/D$$

are dimensionless parameters, while $\varepsilon=RT_r/E$ is the activation energy parameter.
The basic state is motionless $u_b = v_b = \psi_b = 0$ and is characterized by the linear temperature and concentrations profiles

$$\theta_b=ay+b\ ,\ \varphi_b=cy+d\ ,$$

where the constants *a*, *b*, *c*, and *d* are obtained by enforcing the boundary conditions (80). The following transcendental equation is obtained for *a*

$$a+\left(A+Ba\right)\exp\left(-\frac{a}{1-\varepsilon a}\right)=0 \tag{81}$$

where three parameters are involved: *A*, *B* and ε. For *b*, *c* and *d* see (Postelnicu , 2009b).
For the stability analysis, one expresses

$$\psi=\Psi\ ,\ \theta=\theta_b+\Theta=ay+b+\Theta\ ,\ \varphi=\varphi_b+\Phi=cy+d+\Phi \tag{82}$$

where ψ, Θ and Φ are perturbed quantities, $|\Psi|<<1$, $|\Theta|<<1$ and $|\Phi|<<1$, and looking for the solutions in the form

$$\Psi=e^{\lambda t}f(y)\sin kx\ ,\ \Theta=e^{\lambda t}g(y)\cos kx\ ,\ \Phi=e^{\lambda t}h(y)\cos kx \tag{83}$$

the linearised stability problem is governed by the equations

$$f''-k^2f+kRg=0 \tag{84a}$$

$$g''-\left(k^2+\lambda\right)g+akf=0 \tag{84b}$$

$$h''-\left(k^2+\lambda\right)h+ckf=0 \tag{84c}$$

$$f(0)=0\,,\ g'(0)=-Ah(0)\,,\ h'(0)=Bh(0) \tag{85a}$$

$$f(1)=0\,,\ g(1)=0\,,\ h(1)=0 \tag{85b}$$

The eigenvalue problem formulated in (84)-(85) must be solved for the Rayleigh number. It can be shown that the principle of exchange of stability holds, so we can take $\lambda = 0$ in the previous equations. We mention that a problem where the frequency (λ in our case) is real, so that the marginal stability occurs when $\lambda = 0$, is said to obey the principle of exchange of stability. Since there are no analytical solutions of this eigenvalue problem, it will be solved numerically, by minimizing the Rayleigh number over the wave number. The corresponding values of the wavenumber and Rayleigh number are termed critical.
In Fig. 21 there is depicted the variation of the critical wave number with B, for $A = 0.5$ and $\varepsilon = 0.5$. It is seen that, at given Lewis number, the critical wave-number increases with B, while increase of the Lewis number leads to an increase of k_c. Critical Rayleigh number vs B is shown in Fig. 22, when $A = 0.5$ and $\varepsilon = 0.5$. The critical Rayleigh number increases with B, almost linearly for small Lewis number. On the other hand, we remark the usual increase of R_c with Le, a value near 240 being reached by the critical Rayleigh number for $B = 1$, when Le = 100.

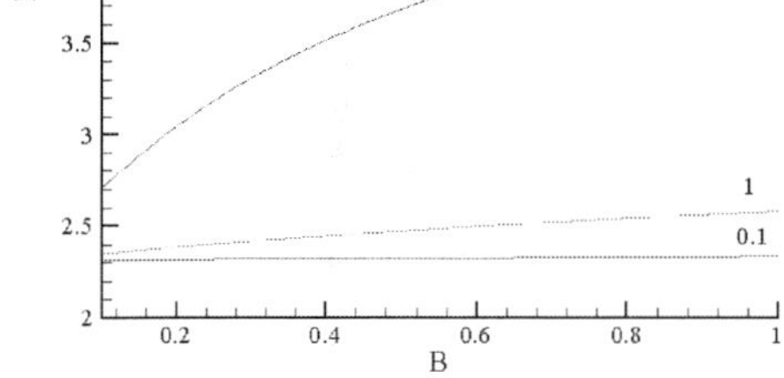

Fig. 21. Variation of the critical wave number with B, for $A = 0.5$ and $\varepsilon = 0.5$

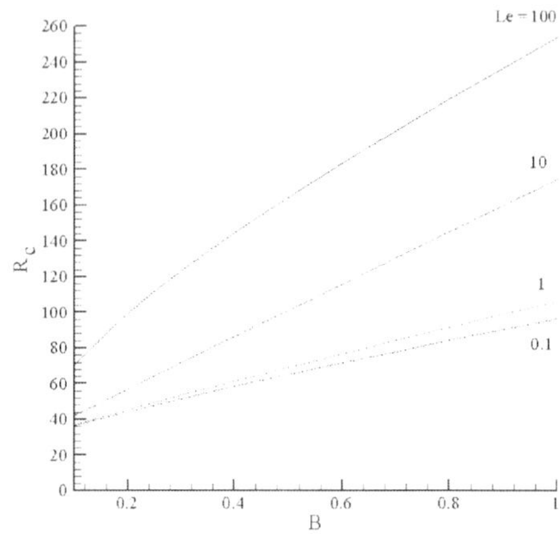

Fig. 22. Critical Rayleigh number vs B, when A = 0.5 and ε = 0.5

5. References

Afify, A. (2007a). Effects of temperature-dependent viscosity with Soret and Dufour numbers on non-Darcy MHD free convective heat and mass transfer past a vertical surface embedded in a porous medium. *Transp. Porous Media* Vol. 66, 391–401

Afify, A. (2007b). Effects of thermal-diffusion and diffusion-thermo on non-Darcy MHD free convective heat and mass transfer past a vertical isothermal surface embedded in a porous medium with thermal dispersion and temperature-dependent viscosity. *Appl Math Modelling* Vol. 31, 1621-1634.

Anghel, M., Takhar, H.S., Pop, I. (2000). Dufour and Soret effects on free convection boundary-layer over a vertical surface embedded in a porous medium. Studia Universitatis Babes-Bolyai, *Mathematica* Vol. XLV, 11-22.

Aris, R. (1965). *Introduction to the analysis of chemical reactors*, Prentice-Hall, Englewood Cliffs, N.J.

Bear, J. (1988). *Dynamics of fluids in porous media*, Dover Publications, New York.

Beg, O.A., Bakier, A.Y., Prasad, V.R. (2009). Numerical study of magnetohydrodynamic heat and mass transfer from a stretching surface to a porous medium with Soret and Dufour effects. *Comput. Mater. Sci.*,Vol. 46, 57-65.

Bejan, A., Dincer, I., Lorente, S., Miguel, A.F., Reis, A.H., *Porous and complex flow structures in modern technologies*. Springer, New York, 2004.

Benano-Melly, L.B., Caltagirone, J.P., Faissat, B., Montel, F., Costeseque, P. (2001). Modelling Soret coefficient measurement experiments in porous media considering thermal and solutal convection, *Int J Heat Mass Transfer* 44: 1285-1297.

Cheng, C-Y (2009). Soret and Dufour effects on natural convection heat and mass transfer from a vertical cone in a porous medium, *Int. Comm. Heat Mass Transfer*, Vol. 36 1020-1024

Chamkha AJ, Ben-Nakhi A (2008). MHD mixed convection-radiation interaction along a permeable surface immersed in a porous medium in the presence of Soret and Dufour's effects. *Heat Mass Transfer*, Vol. 44 845–856.

Chamkha, A. Pop, I. (2004). Effect of thermophoresis particle deposition in free convection boundary layer from a vertical flat plate embedded in a porous medium, *Int. Comm. Heat Mass Transfer*, Vol. 31 421-430.

Chamkha, A., Ali-Mudhaf, A.F., Pop, I. (2006). Effect of heat generation or absorption on thermophoretic free convection boundary layer from a vertical flat plate embedded in a porous medium, *Int. Comm. Heat Mass Transfer* Vol. 32 1096-1102.

Chamkha, A., Jaradat, M., Pop, I. (2004). Thermophoresis free convection from a vertical cylinder embedded in a porous medium, *Int. Jo. Appl. Mech. Eng.*, Vol. 9 471-481.

Chaudhary, M.A., Merkin, J. (1994). Free convection stagnation point boundary layers driven by catalytic surface reactions: I. The steady states,. *J. Eng. Math.* Vol. 28 145-171.

Chaudhary, M.A., Merkin, J. (1996). Free convection stagnation point boundary layers driven by catalytic surface reactions: II. Times to ignition, *J. Eng. Math.*, Vol. 30 403-415.

Chaudhary, M.A. Liffan, A., Merkin, J. (1995). Free convection boundary layers driven by exothermic surface reactions: critical ambient temperature, *Math. Eng. Ind.*, Vol. 5, 129-145.

Cheng, P., Mynkowycz, W.J. (1977). Free convection about a vertical flat plate embedded in a porous medium with application to heat transfer from a dyke, *J. Geophys. Res.*, Vol. 82, 2040-2044.

Christopher, R. V., Middlemann, S. (1965). Power-law flow through a packed tube. *Ind. Engng.Chem Fundls.*, Vol. 4, 424-426.

Dharmadhikari, R.V., Kale, D.D. (1985). Flow of non-Newtonian fluids through porous media. *Chem. Engng..Sci.*, Vol. 40, 527-529.

Eckert, E.R.G., Drake, R.M. (1972). *Analysis of heat and mass transfer*, Mc-Graw-Hill, New York.

Ingham, D.B., Pop, I. (Eds.), (1998). *Transport Phenomena in Porous Media I*, Pergamon, Oxford.

Ingham, D.B., Pop, I. (Eds.), (2002). *Transport Phenomena in Porous Media II*, Pergamon, Oxford.

Ingham, D.B., Pop, I. (Eds.), (2005). *Transport Phenomena in Porous Media III*, Elsevier, Oxford.

Ingham, D.B., Bejan, A., Mamut, E., Pop, I. (Eds.) (2004). *Emerging Technologies and Techniques in Porous Media*, Kluwer, Dordrecht.

Lakshmi Narayana PA, Murthy PVSN (2007) Soret and Dufour Effects in a Doubly Stratified Darcy Porous Medium. *Journal of Porous Media*, Vol. 10 613-624.

Ingham, D.B., Harris, S.D., Pop, I. (1999). Free-convection boundary layers at a three-dimensional stagnation point driven by exothermic surface reaction, *Hybrid Methods in Engineering*, Vol. 1 401-417.

Kafoussias, N.G., Williams, E.W. (1995). Thermal-diffusion and diffusion-thermo effects on mixed free-forced convective and mass transfer boundary layer flow with temperature dependent viscosity, *Int. J. Eng. Sci.*, Vol. 33 1369-1384.

Lakshmi Narayana, P.A., Murthy, P.V.S.N., Postelnicu, A. (2009a) Soret and Dufour Effects on Free Convection of Non-Newtonian Power Law Fluids with Yield Stress from a Vertical Flat Plate in Saturated Porous Media, *Jo of Porous Media*, Vol.12 967-981

Lakshmi Narayana, P.A., Murthy, P.V.S.N., Krishna, P.V.S.S.S.R., Postelnicu, A. (2009b). Free Convective Heat and Mass Transfer in a Doubly Stratified Porous Medium Saturated with a Power-Law Fluid, *Int Jo Fluid Mech Research*, Vol. 36 524-537

Lakshmi Narayana PA, Murthy PVSN (2008). Soret and Dufour effects on free convection heat and mass transfer from a horizontal flat plate in a Darcy porous medium. *Journal of Heat Transfer, Trans of the ASME*, Vol, 130 104504-1–104504-5.

Li, M., Tia, Y., Zhai, Y. (2008). Soret and Dufour effects in strongly endothermic chemical reaction system of porous media, *Trans. Nonferrous Met. Soc. China* , Vol. 16, 1200-1204.

Magyari, E. (2010), private commnication.

Magyari, E., Postelnicu, A. (2010a). Double diffusive natural convection flows with thermosolutal symmetry in porous media in the presence of the Soret-Dufour effects, in press to Transport in Porous Media.

Magyari, E., Postelnicu, A. (2010b). Effect of chemical reaction on a vertical boundary layer flow with heat and mass transfer in porous medium, with prescribed wall heat and mass fluxes (in preparation).

Merkin, J., M.A. Chaudhary, M.A. (1994). Free convection boundary layers on vertical surfaces driven by an exothermic surface reaction, *Quart. J. Appl. Math.*, Vol. 47, 405-428.

Merkin, J., Mahmood, T. (1998). Convective flows on reactive surfaces in porous media, *Transport in Porous Media*, Vol. 33, 279-293.

Mahmood, T., Merkin, J. (1999). The convective boundary-layer flow on a reacting surface in a porous medium, *Transport in Porous Media* Vol. 32 285-298.

Minto, B.J., Ingham, D.B., Pop, I. (1998). Free convection driven by an exothermic on a vertical surface embedded in porous media, *Int. J. Heat Mass Transfer*, Vol. 41, 11-23.

Mojtabi A, Charrier-Mojtabi M C (2000). Double-diffusive convection in porous media. *Handbook of Porous Media* (K. Vafai, ed.), Marcel Dekker, New York, pp. 559–603.

Nield DA, Bejan A (2006). *Convection in Porous Media*, Springer, New York.

Partha, M.K., Murthy P.V.S.N., Raja Sekhar, G.P. (2006) Soret and Dufour effects in a non-Darcy porous medium. *Jo Heat Transfer, Trans of the ASME*, Vol. 128, 605-610.

Partha, M.K. (2009). Suction/injection effects on thermophoresis particle deposition in a non-Darcy porous medium under the influence of Soret and Dufour effects, *Int. J. Heat Mass Transfer*, Vol. 52 1971-1979.

Platten, J.K., Legros, J.C. (1984). *Convection in Liquids*, Springer, New York.

Pop, I., Ingham, D.B. (2001). *Convective Heat Transfer: Mathematical and Computational Modelling of Viscous Fluids and Porous Media.* Pergamon, Oxford.

Postelnicu, A. (2004a). Influence of a magnetic field on heat and mass transfer by natural convection from vertical surfaces in porous media considering Soret and Dufour effects. *Int J Heat Mass Transfer*, Vol.47, 1467-1472.

Postelnicu A. (2004b). Free convection near a stagnation point of a cylindrical body in a porous medium saturated with a non-Newtonian fluid, *Proceedings of International Conference on Applications of Porous Media (ICPAM 2004)*, Evora, Portugal, May 24–27, 2004, Editors: A.H. Reis, A.F. Miguel 231-236.

Postelnicu, A. (2007a). Influence of chemical reaction on heat and mass transfer by natural convection from vertical surfaces in porous media considering Soret and Dufour effects, *Heat Mass Transfer*, Vol. 43, 595-602.

Postelnicu, A. (2007b). Effects of thermophoresis particle deposition in free convection boundary layer from a horizontal flat plate embedded in a porous medium, *Int J Heat Mass Transfer*, Vol. 50, Issues 15-16, pp. 2981–2985.

Postelnicu, A. (2009a). Convective diffusive transport with chemical reaction in natural convection in a porous medium, *Proceedings of the National Conference on Thermodynamics*, Brasov, May 21-22, 2009, Special Issue of the Bulletin of the Transilvania University from Brasov, series B1, Vol. 2(51), 261-264

Postelnicu, A. (2009b). Onset of convection in a horizontal porous layer driven by catalytic surface reaction on the lower wall, *Int. J. Heat Mass Transfer*, Vol. 52, 2466–2470, doi:10.1016/j.ijheatmasstransfer.2009.01.021.

Postelnicu A (2010a). Heat and mass transfer by natural convection at a stagnation point in a porous medium considering Soret and Dufour effects, *Heat Mass Transfer* Vol. 46, 831-840, DOI 10.1007/s00231-010-0633-3.

Postelnicu, A. (2010b). Thermophoresis particle deposition in natural convection over inclined surfaces in porous media, submitted to *Int. J. Heat Mass Transfer.*

Probstein, R.F. (1994). *Physiochemical Hydrodynamics*, John Wiley & Sons New York.

Rathish Kumar, B.V., Krishna Murthy, S.V.S.S.N.V.G. (2010). Soret and Dufour effects on double-diffusive free convection from a corrugated vertical surface in a non-Darcy porous medium. *Transp Porous Media* (online) DOI 10.1007/s11242-010-9549-0

Rees D.A.S. (2000). *The stability of Darcy-Benard convection*, in: K.Vafai (Ed.), Handbook of Porous Media, Marcel Dekker, pp. 521-558.

Song, X., Williams, W.R., Schmidt, L.D., Aris, R. (1991). Steady state and oscillations in homogeneous-heterogeneous reaction systems, *Chem. Eng. Sci.* Vol. 46, 1203-1215.

Sparrow, M., Quack,H., Boerner, C.J. (1970). Non-similarity boundary layer solutions, *AIAA Journal*, Vol. 8 1936-1942.

Sparrow, E.M., Yu, M.S. (1971). Local nonsimilarity thermal boundary-layer solutions, *J. Heat Transfer, Trans. of the ASME*, Vol. 93 328-332.

Soret C (1880) Influence de la temperature sur la distribution des sels dans leurs solutions. *CR Acad Sci* Paris, Vol. 91, 289-291.

Tsai, R. & Huang, J.S. (2009). Heat and mass transfer for Soret and Dufour's effects on Hiemenz flow through porous medium onto a stretching surface, *Int. J. Heat Mass Transfer*, Vol.52 2399-2406.

Vafai, K. Desai, C. Chen, S.C. (1993). An investigation of heat transfer process in a chemically reacting packed bed, *Num. Heat Transfer, Part A*, Vol. 24 127-142.

Vafai, K. (2005). (Ed.*), Handbook of Porous Media*, second ed., Taylor & Francis, New York.

Williams, W.R., Zhao, J., Schmidt, L.D. (1991). Ignition and extinction of surface and homogeneous oxidation of NH_3 and CH_4, *A. I. Che. J.*, Vol. 37 641-649.

26

Mass and Heat Transfer During Thin-Film Evaporation of Liquid Solutions

Janusz Dziak
Wroclaw University of Technology
Poland

1. Introduction

Thin-film evaporation is applied in heat exchangers called thin-film or thin-layer evaporators. Actually thin-layer evaporators may be considered as heat and mass exchangers, since molecules from liquid phase are transferred to gas phase during process of vaporization and the movement of molecules from gas phase to liquid phase is probable during process of condensation. This kind of devices are commonly applied in chemical, pharmaceutical and food industries since 1950. Vertical thin-layer evaporator, example of which is presented in Fig.1, is characterized by small pressure drop and short residence time of the phases in the apparatus, which means also that there is short contact time of the liquid with hot surface of the evaporator wall. These features of the evaporator cause that it is applied for the concentration of heat sensitive liquid solutions especially with high viscosity. Because of low value of pressure drop during gas flow inside the evaporator the boiling temperature of liquid, which is evaporated, depends only on its composition and does not depend on liquid position in the evaporator, which is of great importance in case of evaporation at low pressure.

There are two main types of the thin-layer evaporator, which will be considered below, namely: static type thin-layer evaporator and mechanically agitated thin-layer evaporator. All types are constituted by vertical cylinder heated from the outside, inside which liquid flows off gravitationally on the heated surface. During that flow vaporization of liquid takes place.

If liquid flows off freely, by means of gravity force, without any disturbance, it means, without any mechanical action on flowing liquid inside the evaporator, then such an apparatus is called static type thin-layer evaporator.

If the evaporator is equipped with blades that mix the liquid flowing inside the evaporator, then such an apparatus is called mechanically agitated thin-layer evaporator or mechanical thin-layer evaporator. The blades applied in the evaporator distribute liquid uniformly on the evaporator wall, which prevents the appearance of dry places on the heated surface of the evaporator. The blades could be fixed rigidly to the shaft of the mixer (Fig.1.) or they could be mounted to the rotating shaft in such a way that the blades swing freely and scrape the surface of the evaporator wall. The pressure exerted by the blade on the surface of evaporation depends on revolution speed of the shaft equipped with moving blades. This kind of evaporator is especially suitable for evaporation of liquid solutions that produce crystals during the process of evaporation (Dziak & Skoczylas, 1996).

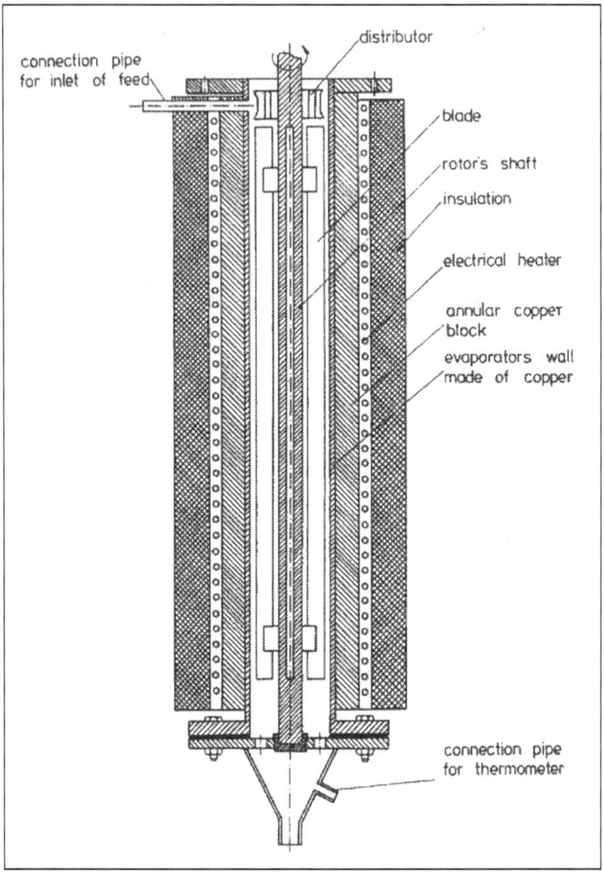

Fig. 1. Vertical section of the mechanical thin-layer evaporator equipped with rotor possessing rigidly fixed blades

One can find in the literature mathematical models of hydrodynamics of thin layer flow of liquid over a vertical wall (El Haj Assad M. & Lampinen M. J., 2002; Chen F.C. & Gao Z., 2004; Ludwig et al., 2008).

There are many works that deal with heat transfer during the process of heating and evaporation of liquids in thin-layer evaporators. Most of them are devoted to individual heat transfer coefficients determination in case of one component liquid treating (Chun K.R. & Seban R.A., 1971; Skoczylas A., 1978; Alhusseini A. et al., 1998). There are mathematical equations that allow determination of individual heat transfer coefficients depending on the conditions of heat transfer. Mechanism of evaporation is quite complicated and depends on many factors such as: liquid physical properties, the roughness of the surface (where vapour bubbles are formed), heat load. Because of that the equations, presented by different authors, for calculation of the individual heat transfer coefficients could give the results that sometimes differ substantially and one should be careful applying them.

Even more complicated is the process of evaporation of liquid solutions i.e. liquid that contains more than one component. At some circumstances during the process of liquid solution evaporation the gradient of concentration in liquid phase appears and it causes some additional difficulties in heat and mass transfer analysis of the process of evaporation. There are not so many works that deal with this problem (Leuthner S. et al., 1999; Krupiczka R. et al., 2004). Some experimental results, ideas and theoretical considerations will be presented below.

2. Theories of mass transfer during the process of thin-layer evaporation

The fundamental theory applied for determination of separation efficiency of the thin-layer evaporator was presented by Billet (Billet R., 1957). Billet presumed that the vapor generated during vaporization of the liquid mixture is removed from the evaporator without any contact with the liquid phase flowing down the tube. Similar point of view was presented by Dieter (Dieter K., 1970), who analyzed the evaporation process both in mechanical thin-layer and molecular evaporators. Molecular evaporator is an apparatus in which the distance between vaporization surface and condensation surface is smaller than mean free path between molecules collision in the gas phase.

Billet's theory is based on the mass balance of more volatile component in differentially small element of *dh*-length in the evaporator (Fig.2), which is constituted by a pipe, heated from the outside, inside which the process of evaporation takes place. Assuming that for this element the liquid and generated vapor are in equilibrium ($y=y^*$) one can write the equation:

$$L \cdot x = (L - dV) \cdot (x - dx) + y^* dV \tag{1}$$

where: L - liquid molar flow at the inlet to the element of dh length, x - more volatile component concentration in liquid at the inlet to the dh element, $(L - dV)$ - the liquid molar flow at the outlet of the dh element, $(x - dx)$ - more volatile component concentration in liquid at the outlet of the dh element, dV- molar flow of vapor produced in the element of dh length of evaporator, y^* - vapor concentration, which is in equilibrium with liquid concentration of more volatile component.

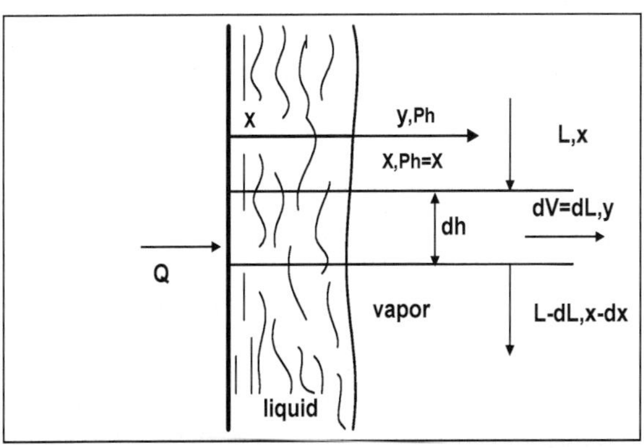

Fig. 2. Scheme of the two-component liquid thin-layer evaporation. Billet's theory

The solution of equation (1) is the integral:

$$\ln\frac{L_F}{L_R}=\int_{x_R}^{x_F}\frac{dx}{y^*-x} \tag{2}$$

Integral (2) can be solved numerically. This equation allows calculation of the residue flowrate from the evaporator L_R, knowing the values of the feed flowrate L_F, its concentration x_F as well as the concentration of residue x_R. It is also necessary to have the vapor-liquid equilibrium data [$y^*= f(x)$].

The average composition of the distillate can be determined from the mass balance equation:

$$D = L_F - L_R \tag{3}$$

$$x_D=\frac{L_F\cdot x_F-L_R\cdot x_R}{D} \tag{4}$$

Mannheim and Passy (Mannheim C.H. & Passy N., 1974) stated that during evaporation process of liquid mixtures, in the vertical thin-layer evaporators, the rectification could appear as an effect of countercurrent flow of liquid and vapor phases inside the evaporator. Similar point of view presented Billet in his later work (Billet, 1983). He also noticed that distillation of liquid mixtures in a thin-layer evaporator could lead to significantly better separation than that obtained in corresponding simple batch distillation. According to him this effect is caused by intimate contact of liquid and gas phases in the evaporator. Anybody of those authors presented experimental verification of their hypothesis.

3. Experimental work on thin-layer evaporation of two-component liquid solutions

An extensive experimental work was proceeded by the author of this chapter to establish the results of thin-layer distillation of two component liquid solutions in mechanical thin-film evaporator (Skoczylas A. & Dziak J., 1990). The investigation was carried out in the mechanical thin-layer evaporator, which some details are presented in Fig.1.

The liquid feed was supplied from the connection pipe to the distributor and then it was distributed uniformly over the surface of the evaporator. Inside the evaporator the liquid was spread over the inner surface of the tube by means of four rotating blades. The clearance between the inner surface of the tube and the edge of the rotor blades might be set continuously in the range 0.5-2.5 mm.

The main geometric dimensions of the evaporator were: inner diameter d=0.04 m, height of the evaporator h=0.57 m, heat transfer area A=0.0716 m^2.

The experimental installation, which was applied in the investigation, is presented in Fig.3. Two component liquid solution from the feed tank (V_1) was pumped to the thin-layer evaporator (T.L.E.) through the rotameter and double pipe heat exchanger (in which the liquid was heated up to the boiling point). The vapor phase flowed up to the condenser (CND) and liquid distillate was collected in the receiver of distillate (V3). The bottom product was collected in the receiver of residue (V2). The evaporator was heated by an electrical heater and the power supplied was regulated by means of changing the input voltage.

The following systems were applied in the experimental work: water-ethylene glycol, methyl alcohol-water. The concentrations of examined liquids were determined by

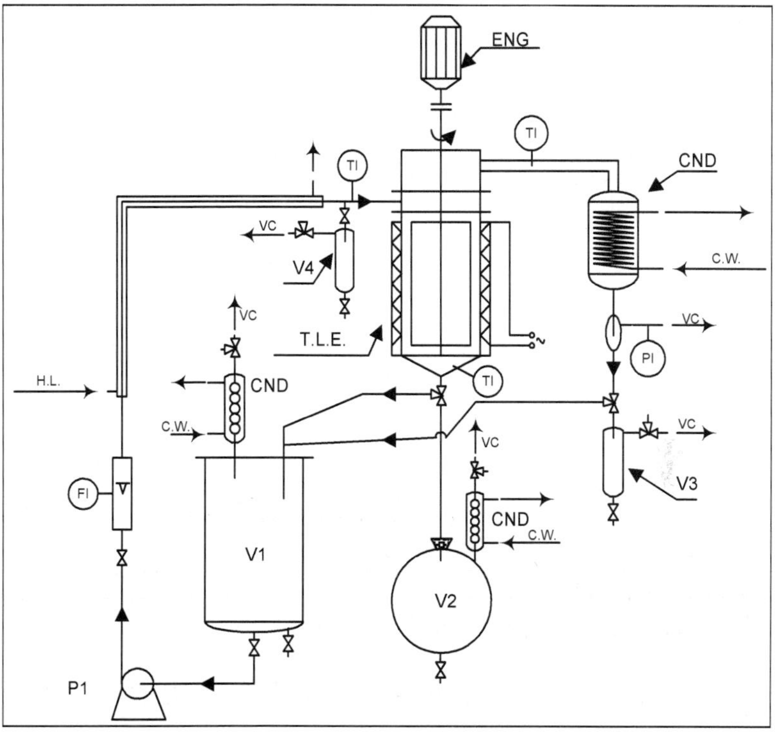

Fig. 3. Scheme of installation for heat and mass transfer examination in thin-layer evaporator. CND-condenser, C.W.-cooling water, H.L.- heating liquid, P1- pump, V1- feed tank, V2- residue tank, V3- distillate tank, V4- feed sample receiver, VC- to vacuum, T.L.E.- thin-layer evaporator, FI- flow indication, PI- pressure indication, TI- temperature indication

measuring density. Altogether 118 measurements of separation efficiency of the mechanical thin-layer evaporator were carried out, which included 77 measurements on a methyl alcohol-water system and 41 measurements on a water-ethylene glycol system. When selecting the systems, the coefficients of relative volatility α_{AB} were considered. The numerical values of these coefficients, at the applied working conditions, amounted to: α_{AB} =5 for methyl alcohol-water system and α_{AB}=120 for the water-ethylene glycol system. The ranges of the experimental conditions are presented in table 1.

The purpose of the work was confrontation Billet's theory results with reality and to establish whether the rectification effect exists during the process of thin-film evaporation. In each experiment the following magnitudes were measured: -moles of feed, distillate and residue per unit time; compositions of feed, distillate and residue; temperatures of liquid at the inlet and outlet of the evaporator; temperature of vapor at the outlet from the evaporator; pressure inside the evaporator. Having these data one could get both experimental and calculated values of separation efficiency.

The vapor-liquid equilibrium data applied in calculations using Billet's theory were taken from the literature (Gmehling J. & Onken U., 1977).

	Methyl alcohol-water system	Water-ethylene glycol system
Moles of feed per unit time (kmol/s)	$1.18 \cdot 10^{-5}$-$3.71 \cdot 10^{-4}$	$6.35 \cdot 10^{-5}$-$8.29 \cdot 10^{-4}$
Mole fraction of the more volatile component in the feed	0.181-0.605	0.048-0.418
Rate of heat flux (kW/m²)	4.25-13.36	3.18-14.97
No. of rotor's revolutions (rev./min.)	200,600,1000	200,600
Pressure (hPa)	1013	33

Table 1. Ranges of the experimental conditions for thin-layer evaporation of two component liquid solutions

The results of experimental investigation are presented in Fig.4 and Fig.5.

Fig.4 shows the dependence between water concentration in distillate x_D and its average concentration in liquid phase x_L, calculated as an arithmetic average of water concentration in the feed and in the residue, obtained in the process of thin-layer evaporation. The dependence was obtained in water-ethylene glycol system. The solid line in this figure describes the theoretical relationship $x_D=f(x_L)$, obtained according to Billet's theory, whereas experimental data are represented by individual points.

If the process of evaporation was accompanied with rectification effect then the experimental results would be above the corresponding theoretical results. Actual concentrations of more volatile component in the distillate would then be higher than corresponding concentrations calculated applying equations 2-4. Figure 3 shows that it is not the case for the process of evaporation in mechanical thin-layer evaporator.

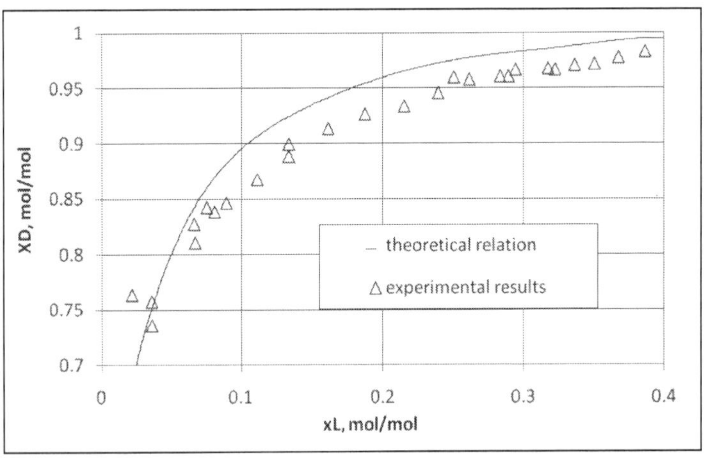

Fig. 4. Concentration of more volatile component in distillate x_D as a function of average concentration of that component in liquid $x_L = (x_F + x_R)/2$. Water-ethylene glycol system

The sets of curves in Fig.5, obtained in methyl alcohol-water system, illustrates the relation between the average mole fraction of methyl alcohol in distillate x_D and the ratio of distillate to feed flowrates D/F for different values of methyl alcohol concentration in the feed. The ratio D/F constitutes the degree of evaporation, obtained in the thin-layer evaporator. Solid lines show theoretical relationships and individual points represent experimental results. It can be seen from Fig.5 that the agreement between theoretical and experimental results is very good. For all measurements, which were performed on methyl alcohol-water system, the mean deviation between x_D data, obtained experimentally, and the theoretical ones amounted to 1.86% and the maximum deviation was 7.94%.
From the comparison of the experimental results and theoretical calculations presented above one can conclude that in the thin-layer evaporators, similar to the design that was used by the author of this work, there is no rectification effect despite countercurrent flow of vapor and liquid phases through the evaporator. Even more, one can notice that in water-ethylene glycol system some phenomenon took place, causing that experimentally obtained values of more volatile component concentration in the distillate are lower than those calculated theoretically.

4. Explanation of the absence of rectification effect during evaporation of liquid solution in the thin-layer evaporator

The absence of the rectification effect during distillation of liquid solutions in the thin-layer evaporator can be explained on the grounds of detailed analysis of hydrodynamics as well as heat and mass transfer. The mechanism of flow of both phases in the thin-layer evaporator is quite different to that in the rectification column. This difference is a consequence of different way of supplying energy to the apparatus. One can compare the way of working of both apparatus: rectification column and thin-layer evaporator taking the advantage of Fig.6.

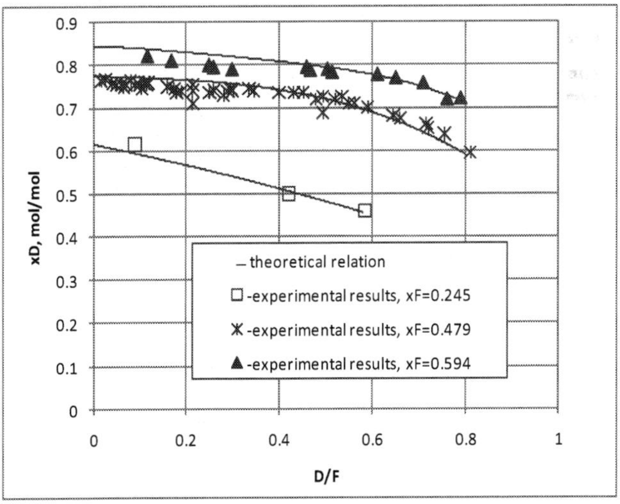

Fig. 5. Relationship between distillate concentration x_D and the ratio of distillate and feed flowrates D/F for different values of the feed concentration x_F. Methyl alcohol-water system

In the case of the rectification column the vapor phase generated in the conventional evaporator is fed to the lower part of the rectification column, whereas the liquid phase generated in the condenser is fed on the top of the column. The cylindrical surface of the rectification column ought to have an adiabatic jacket, whereas in the case of thin-layer evaporator this surface is heated by means of steam or by electrical heater.

Let's examine in detail the mechanism of heat and mass transfer during vaporization of a two component liquid system, both in the rectification column and in the thin-layer evaporator.

In case of rectification column, if the molar heat of vaporization of both components is the same and the column has an adiabatic jacket, then equal-molar flow of the liquid and vapor phases is achieved.

The vapor phase from the lower part of the column, having temperature t_V and concentration of the more volatile component y_V flows up and contacts with the liquid, which flows down. The temperature of liquid t_L is lower than the temperature of the vapor $t_V > t_L$. The concentration of the more volatile component in gas phase y_V is lower than the concentration of more volatile component in gas phase y_V^*, which is in equilibrium with liquid phase (possessing concentration x_L) that contacts with gas. Because of thermodynamic state of the liquid and gas phases, that are in contact, there are conditions for heat and mass transfer. The vapor flowing near by the liquid layer is being cooled and the drop of vapor temperature brings about that part of it is being condensed, mainly the

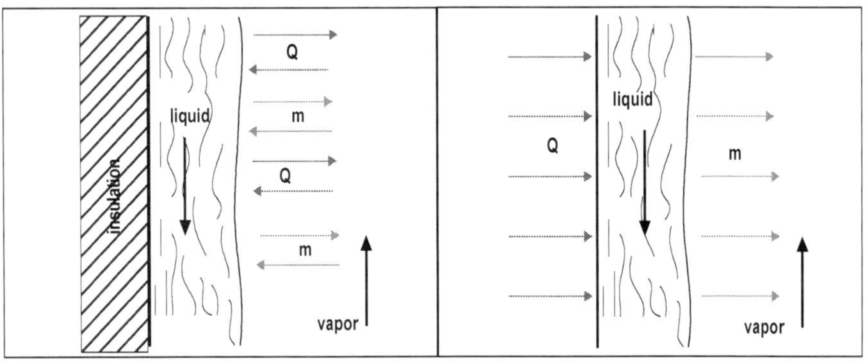

Fig. 6. Comparison of the work of rectification column (a) and thin-layer evaporator (b)

less volatile component. The liberated heat of condensation causes evaporation of an equivalent amount of the liquid phase, mainly the more volatile component. If the molar heat of vaporization of components are the same, and the column has an adiabatic jacket, then during the process of heat and mass transfer no volume changes of liquid and vapor phases occur.

In case of thin-layer evaporator, there is no adiabatic jacket, and its cylindrical wall is just a heating surface of the evaporator. The liquid phase at boiling temperature t_b flows down on the inner surface of the vertical tube, which temperature t_w is higher than the liquid temperature $t_w > t_b$. In the time of liquid flowing down, part of it evaporates, mainly the more volatile component. The volume of vapor created V_V is much greater than the volume of

disappearing liquid V_L and the ratio V_V/V_L is in the range 10^3-10^5. Such a considerable increase of volume accompanying the evaporation process of a liquid on the vertical wall of the tube brings about the increase of pressure in the immediate neighborhood of the liquid layer. As a result, the motion of vapor particles is mainly unidirectional, from the liquid layer to the centre of the tube, and only on the liquid-vapor surface one can expect the existence of equilibrium conditions.
The dynamic pressure exerted by vapor generated during process of evaporation in the thin-layer evaporator could be calculated from equation (5):

$$\Delta p = \frac{w^2 \cdot p_v}{2} \tag{5}$$

and the linear velocity of the vapor phase, perpendicular to the tube's wall can be calculated from equation (6):

$$w = \frac{q}{p_v \cdot r} \tag{6}$$

Where: q is the rate of heat flux, kW/m^2; ρ_V- density of the vapor phase, kg/m^3; r- heat of vaporization, kJ/kg.
The numerical values of w and Δp for applied, by the author, working conditions were in the range:

$$w=3\cdot10^{-3}\div0.25 \text{ m/s}$$

$$\Delta p=6\cdot10^{-6}\div7.7\cdot10^{-4}\text{Pa}$$

The above presented mechanism explains the absence of rectification effect in the thin-layer evaporator, but it does not explain why the concentration of more volatile component in vapor phase is lower than that obtained from theoretical calculations applying Billet's theory, as it is illustrated in Fig.3. This could be explained considering in detail what is going on in liquid phase during two component liquid solution thin-layer evaporation.

5. Gröpp and Schlünder theory of simultaneous heat and mass transfer during thin-layer evaporation of two-component solutions

The process of liquid solution evaporation in thin-layer evaporator is more complex than evaporation of one component liquid. It is connected with the fact of mass transfer resistance arise at some conditions of two-component liquid thin-layer evaporation.
Gröpp and Schlünder (Gröpp U. & Schlünder E.S., 1986) presented theory that allows understanding the process of mass transfer in liquid phase during thin-layer evaporation of liquid solution.
Fig.1 shows schematically the temperature and concentration profiles in the thin-layer of liquid during surface evaporation in the vertical thin-layer evaporator. Two-component liquid (with components A and B) flows along the heated surface and is partially evaporated. Mole fraction of more volatile component in gas phase y_{Aph} and in liquid phase x_{Aph} at interface, as well as temperature at interface t_{ph} are established according to thermodynamic equilibrium conditions. If it is assumed that the vapor generated on liquid surface is directly discharged to the condenser, and there is no mass transfer resistance in the

gas phase, then the distillate concentration equals the gas concentration y_{Aph} at the interface. The mole flowrate of more volatile component in the direction to the interface on liquid side $n_A^{\bullet}$ could be determined from the equation concluded from boundary layer theory:

$$n^{\bullet}{}_A = -\rho_L \cdot \delta_L \cdot \left(\frac{\partial x_A}{\partial s} \right) + n^{\bullet} \cdot x_A \tag{7}$$

Where: ρ_L- liquid density, kmol/m^3; δ_L- liquid diffusivity, m^2/s; x_A- mole fraction of more volatile component in liquid phase; s- boundary layer thickness, m; $n^{\bullet}$- overall mole flowrate, kmol/s

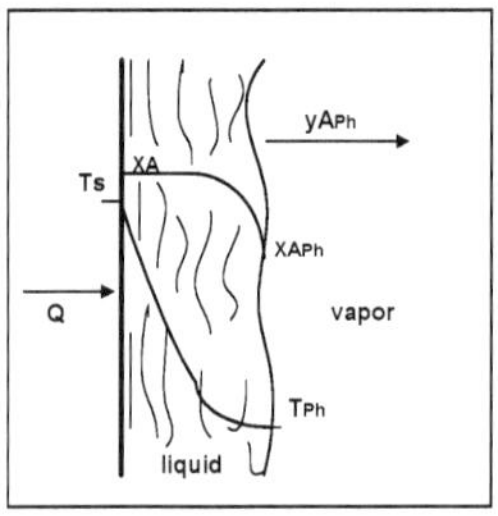

Fig. 7. Thin-layer evaporation accompanied with mass transfer resistance in liquid phase

$$n^{\bullet} = n^{\bullet}{}_A + n^{\bullet}{}_B \tag{8}$$

Mole flowrate of more volatile component could be determined from equation (9):

$$n^{\bullet}{}_A = n^{\bullet} \cdot y_{Aph} \tag{9}$$

Taking advantage of the equation (9) in equation (7), one obtains:

$$n^{\bullet} \cdot y_{Aph} = -\rho_L \cdot \delta_L \cdot \left(\frac{\partial x_A}{\partial s} \right) + n^{\bullet} \cdot x_A \tag{10}$$

And then:

$$-\rho_L \cdot \delta_L \cdot \frac{\partial x_A}{\partial s} = n^{\bullet} \cdot \left(y_{Aph} - x_A \right) \tag{11}$$

Integrating the equation (11) gives:

$$-\frac{\rho_L \cdot \delta_L}{n^{\bullet}} \cdot \int_{x_A}^{x_{Aph}} \frac{dx_A}{y_{Aph} - x_A} = \int_0^s ds$$

$$-\frac{\rho_L \cdot \delta_L}{n^{\bullet}} \cdot [-\ln(y_{Aph} - x_A)]_{x_A}^{x_{Aph}} = s$$

$$-\ln\left(y_{Aph} - x_{Aph} \right) + \ln\left(y_{Aph} - x_A \right) = -\frac{n^{\bullet}}{\rho_L} \cdot \frac{s}{\delta_L} \tag{12}$$

If we define mass transfer coefficient by the equation (13):

$$\beta_L = \frac{\delta_L}{s} \tag{13}$$

and linear velocity in the interface direction by the equation (14):

$$v_L = \frac{n^{\bullet}}{\rho_L} = \frac{q}{\rho_L{}' \cdot r} \quad (14), (15)$$

Where: q- heat load, W/m^2; r- heat of vaporization, J/kg; ρ_L'-liquid density, kg/m^3
Introducing equations (13) and (14) into equation (12) one obtains:

$$\ln \frac{y_{Aph} - x_A}{y_{Aph} - x_{Aph}} = -\frac{v_L}{\beta_L} \tag{16}$$

The equations presented above allow determination of mass transfer coefficient value for specific thin-layer evaporation conditions. Knowing the values of liquid heat of vaporization (r), heat load (q) and liquid density (ρ_L')one can calculate the value of v_L from equation (15). After that, knowing the value of v_L and the values of concentrations in liquid and vapor on the interface boundary y_{Aph}, x_{Aph}, as well as the average concentration in liquid phase x_A, one can determine the value of mass transfer coefficient β_L from relationship (16).

6. Experimental investigation of mass transfer in liquid phase during liquid solution evaporation in thin-layer evaporator

The purpose of the work, carried out by the author, which results are presented below, was checking whether the mass transfer resistances exist during the process of thin-layer evaporation of liquid solutions, and if they do, finding values of mass transfer coefficients for specific conditions of thin-layer distillation. Another objective of the work was determination, if there is a way of mass transfer resistance lowering, if such resistance exists. Isopropanol-water and propylene glycol solutions were chosen for the examination. Both systems possess the components that significantly differ in volatility, which is essential considering the possibility of mass transfer resistance appearance. Additionally the water-propylene glycol solutions possess relatively high viscosity.
The scheme of installation applied in an experimental work is presented in Fig. 2. Two kind of evaporators were used in experiments: one of static type (without blades that mix liquid) and the second equipped with mixer, which blades swing freely and scrape the surface of the evaporator. The main dimensions of both evaporators are presented in table 2.
Ranges of changeability of magnitudes applied in experiments are presented in table 3.

	Static thin-layer evaporator	Thin-layer evaporator equipped with mixer
Heated height h, m	0.56	0.266
Evaporator diameter d, m	0.04	0.06
Heat transfer area A, m^2	0.07	0.05

Table 2. Main dimensions of thin-layer evaporators used in experiments

Magnitude	Isopropanol-water system	water-propylene glycol system
External pressure, Pa	$1{,}013 \cdot 10^5$	3332
Volatility=$P_{more\ volatile}/P_{less\ volatile}$ (t,°C)	2 (82°C)	47 (74°C)
Feed concentration-X_F, mol/mol	0.0678-0.133	0.1788- 0.2475
Sprinkle intensity-m', kg/(ms)	0.0531-0.213	$6.839 \cdot 10^{-3}$- $9.9882 \cdot 10^{-2}$
Heat load-q, W/m^2	8087.9-32809.5	7216.8- 19752
Liquid density-ρ, kg/m^3	878.4-921.4	989.2-1008.4
Liquid viscosity-η, mPa·s	0.3106-0.3449	3.4- 4.8
Mixer revolutions, rev./min.	-	200-800

Table 3. Ranges of changeability of magnitudes applied in experiments

Intensity of heated surface sprinkle by the liquid was calculated from relation (17):

$$m' = m^{\bullet}/O \tag{17}$$

Where: $m^{\bullet}$- average mass flowrate of liquid, kg/s; O- evaporator perimeter, m
Heat load of evaporator heated surface was determined from equation (18):

$$q = Q^{\bullet}/A \tag{18}$$

Where: $Q^{\bullet}$- heat exchanged in the evaporator, kW; A-heat exchange surface area of the evaporator, m^2
The concentrations of examined liquids were determined by measuring density for isopropanol-water system and refractive index for water-propylene glycol system. The vapor-liquid equilibrium data were taken from the literature (Gmehling J.& OnkenU., 1977).

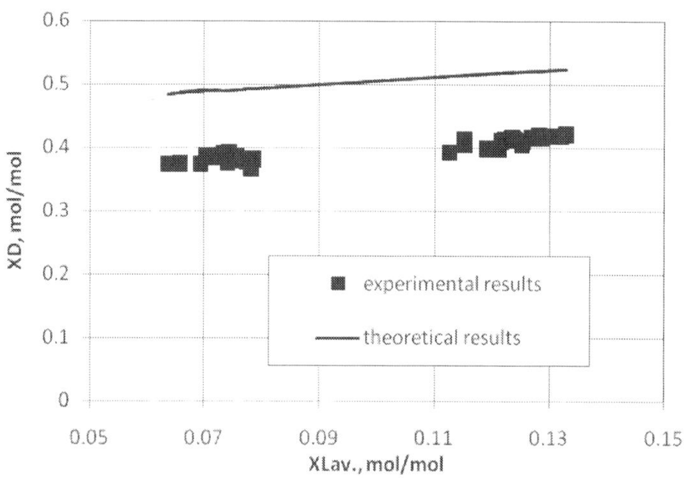

Fig. 8. Results of distillation of isopropanol-water liquid solutions carried out in static type thin-layer evaporator. Dependence between more volatile component concentration in distillate and its average concentration in evaporated liquid

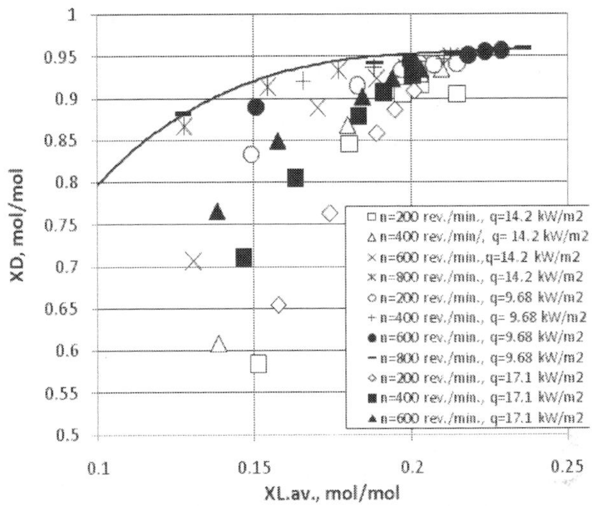

Fig. 9. Results of distillation of water-propylene glycol solutions carried out in wiped-film thin-layer evaporator. Dependence between more volatile component concentration in distillate and its average concentration in evaporated liquid. Solid line represents theoretical calculations

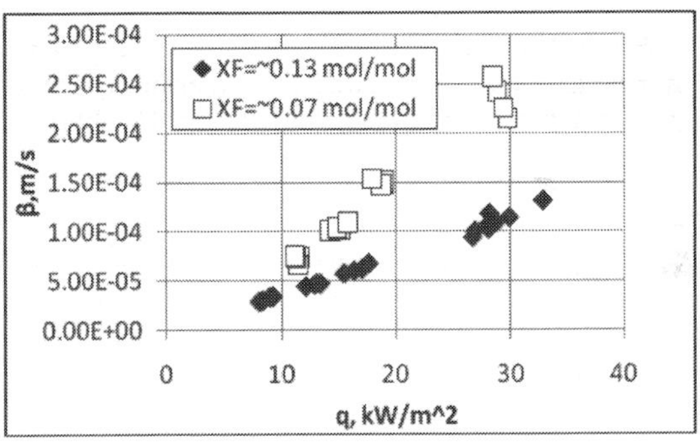

Fig. 10. Individual mass transfer coefficient dependence on process variables. Static type thin-layer evaporator. Isopropanol-water system

In Fig.8 and Fig.9. there is a comparison of the results of distillation in thin-layer evaporator, calculated theoretically (Billet's theory) and determined experimentally. Fig.8 describes results obtained in static type thin-layer evaporator for isopropanol-water system and Fig.9 represents the results obtained applying wiped thin-film evaporator for water-propylene glycol system.

As one can see in Fig.8 and Fig.9 mass transfer resistance during distillation of liquid solutions in the thin-layer evaporator significantly influences the results of distillation. Theoretically calculated compositions of distillate (solid line in Fig. 8 and Fig.9) are not in line with experimental results (individual points). Experimental results are much worse than those anticipated theoretically. It can be also noticed from Fig.9 that mass transfer resistance could be substantially reduced by the action of blades that mix the evaporated liquid film.

One can read in Fig.10 and Fig.11 the values of mass transfer coefficients determined with the use of Gröpp and Schlünder theory.
The dependency of individual mass transfer coefficient on liquid concentration and heat load was noticed in the case of isopropanol-water system.
On the other hand, in case of water-propylene glycol system, clear dependence of individual mass transfer coefficient on liquid flowrate and mixer number revolutions per minute was

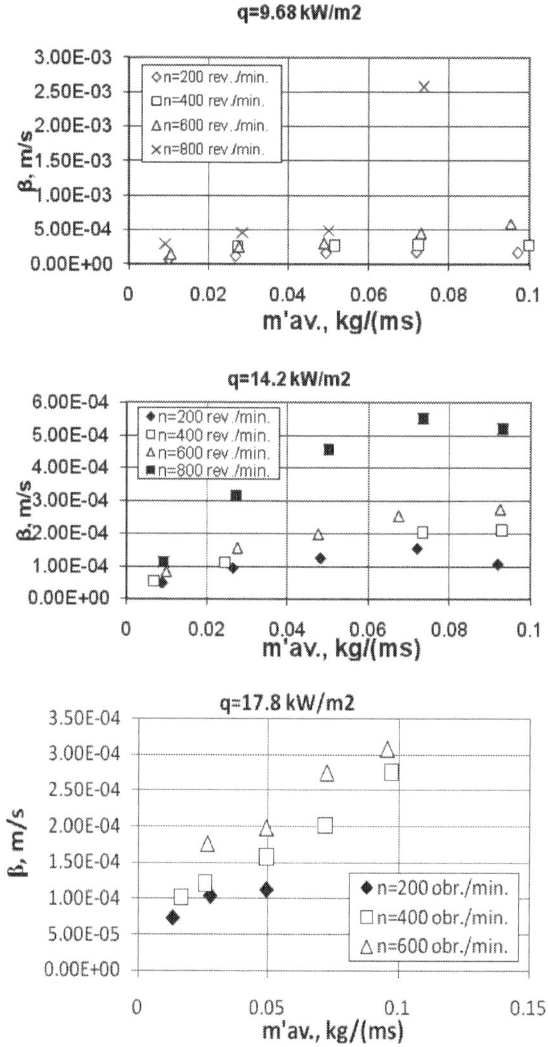

Fig. 11. Individual mass transfer coefficient dependence on process variables. Wiped thin-layer evaporator. Water-propylene glycol system

confirmed. One can notice significant growth of individual mass transfer coefficient with the growth of speed of the mixer rotation. Especially high values of individual mass transfer coefficient were obtained for heat load q=9.68 kW/m^2 and n=800 rev./min.. It could be expected that for low value of heat load and suitable mixer action mass transfer resistance, in liquid phase, could be eliminated.

7. Conclusions

As one can see, from the above presented considerations, the process of liquid solution evaporation, carried out in the thin-layer evaporator, is complicated. The processes of heat and mass transfer, which proceed simultaneously, influence to each other. Simple theory that does not take into account the mass transfer resistance in liquid phase during the process of thin-layer evaporation, becomes unrealistic, especially in the case of evaporation carried out in the evaporator without liquid film mixing. Mass transfer resistances in liquid phase arising during the process of liquid solution thin-layer evaporation could be lowered by applying of liquid-film mixing. Mixing is generally helpful for intensification of mass and heat transfer.

8. References

Alhusseini A. A., Tuzla K. & Chen J.C. (1998). Falling film evaporation of single component liquids, *Int. J.Heat Mass Transfer*, 41, 12, 1623-1632, ISSN: 0017-9310

Billet R. (1957). Trennung von Flüssigkeitsgemischen durch teilweise Destillation. *Chemie Ing. Techn.*, 29, 733

Billet R. (1983). Distillation in thin-film evaporators, *Proc .Int. Symp. Film Techniques Applied to Chem. Engng Operations*, pp. 343-384, Rueil-Malmaison,27-29 April, 1983, France

Chen F.C. & Gao Z. (2004) , An analysis of black liquor falling film evaporation, *Internationa Journal of Heat and Mass Transfer*, 47, 1657-1671, ISSN: 0017-9310

Chun K.R. & Seban R.A. (1971). Heat transfer to evaporating liquid films, *Journal of Heat Transfer*, 93, 391-396, 1971, ISSN: 0022-1481

Dieter K. (1970). Leistungsverhalten von Dünnschichtverdampfern. *Chem. Z.*, 94, 445

Dziak, J. & Skoczylas, A. (1996). The Application of thin-film evaporator for getting solid products from liquid solutions, *Drying'96-Proceedings of the 10th International Drying Symposium (IDS'96)*, vol.A, pp. 447-454, ISBN: 83-86903-07-4, Krakow, Poland, 30 July-2 August 1996, Lodz Technikal University, Lodz

El Haj Assad M. & Lampinen M. J. (2002). Mathematical modeling of falling liquid film evaporation process, *International Journal of Refrigeration*, 25, 985-991, ISSN: 0140-7007

Gmehling J.& OnkenU. (1977). *Vapour -Liquid Equilibrium-Data Collection. Aqueous-Organic Systems.* DECHEMA

Gröpp U. & Schlünder E.S. (1986). The influence of liquid-side mass transfer on heat transfer and selectivity during surface and nucleate boiling of liquid mixtures in a falling film, *Chemical Engineering and Processing*, 20, no.2, 103, ISSN: 0255-2701

Krupiczka R., Rotkegel A. & Ziobrowski Z. (2004). The influence of mass transfer on the heat-transfer coefficients during the film boiling of multicomponent mixtures, *Chemical Engineering and Processing*, 43, 949-954, ISSN: 0255-2701

Leuthner S., Harun Maun A., Fiedler S. & Auracher H. (1999). Heat and mass transfer in wavy falling films of binary mixtures, *International Journal of Thermal Science*, 38, 937-943, ISSN: 1290-0729

Ludwig W., Dziak J., Królikowski L., Kapłon J. & Tuta J. (2008). Modeling of hydrodynamics of thin-layer evaporator with grawitational liquid flow using CFD methods, *Chemical and Process Engineering*,29, 215-220, ISSN: 0208-6425

Mannheim C.H. & Passy N. (1974). *Proc.Int.Symp.Advances in Preconcentration and Dehydration of Foods*, ISBN: 83-204-0182-8, U.K. Applied Science Publishers Ltd.

Skoczylas A. (1978), Heat transfer in film preheating of liquids in Sambay-type heat exchanger, *Chemical and Process Engineering*, 8, 691, ISSN: 0208-6425

Skoczylas A. & Dziak J. (1990). Separation efficiency of thin-layer evaporators. *Computers Chem. Engng*, Vol.14, No.9, pp. 1001-1007, ISSN 0098-1354